SUBMERGED LANDSCAPES OF THE EUROPEAN CONTINENTAL SHELF

Quaternary Paleoenvironments

Cost (European Cooperation in Science and Technology) is a pan-European intergovernmental framework. Its mission is to enable break-through scientific and technological developments leading to new concepts and products and thereby contribute to strengthening Europe's research and innovation capacities.

It allows researchers, engineers and scholars to jointly develop their own ideas and take new initiatives across all fields of science and technology, while promoting multi- and interdisciplinary approaches. COST aims at fostering a better integration of less research intensive countries to the knowledge hubs of the European Research Area. The COST Association, an International not-for-profit Association under Belgian Law, integrates all management, governing and administrative functions necessary for the operation of the framework. The COST Association has currently 36 Member Countries. www.cost.eu.

 COST is supported by
the EU Framework Programme
Horizon 2020

SUBMERGED LANDSCAPES OF THE EUROPEAN CONTINENTAL SHELF

Quaternary Paleoenvironments

Edited by

Nicholas C. Flemming
National Oceanography Centre, Southampton, UK

Jan Harff
University of Szczecin, Poland

Delminda Moura
University of Algarve, Portugal

Anthony Burgess
University of Malta, Malta

Geoffrey N. Bailey
University of York, UK

EUROPEAN COOPERATION
IN SCIENCE AND TECHNOLOGY

WILEY Blackwell

Registered Offices
John Wiley & Sons, Inc., 111 River Street, Hoboken, NJ 07030, USA
John Wiley & Sons Ltd, The Atrium, Southern Gate, Chichester, West Sussex, PO19 8SQ, UK

Editorial Office
9600 Garsington Road, Oxford, OX4 2DQ, UK

For details of our global editorial offices, customer services, and more information about Wiley products visit us at www.wiley.com.

Wiley also publishes its books in a variety of electronic formats and by print-on-demand. Some content that appears in standard print versions of this book may not be available in other formats.

Library of Congress Cataloging-in-Publication Data

Names: Flemming, N. C. (Nicholas Coit)
Title: Submerged landscapes of the European continental shelf : Quaternary paleoenvironments / edited by Nicholas C. Flemming [and four others].
Description: Hoboken, NJ : John Wiley & Sons, Inc., 2017– | Includes bibliographical references and index.
Identifiers: LCCN 2016053352| ISBN 9781118922132 (cloth) | ISBN 9781118927717 (epub)
Subjects: LCSH: Paleoclimatology–Quaternary. | Paleoclimatology–Europe. | Submerged lands–Europe. | Continental shelf–Europe. | Archaeology–Europe.
Classification: LCC QC884.5.E87 S83 2017 | DDC 909/.09633–dc23
LC record available at ttps://lccn.loc.gov/2016053352

Cover images: (Top Image) © Berné, S. & Gorini, C./Elsevier; (Bottom Image) © Courtesy of Ehud Galili/Photo: Itamar Grinberg, Israel
Cover design by Wiley

Set in 11/14pt Adobe Garamond by Aptara Inc., New Delhi, India

Printed and bound by CPI Group (UK) Ltd, Croydon, CR0 4YY

10 9 8 7 6 5 4 3 2 1

Contents

Contributors

Fabrizio Antonioli
ENEA, Roma, Italy
Email: fabrizio.antonioli@enea.it

Marco Anzidei
INGV, Rome, Italy

Geoffrey N. Bailey
Department of Archaeology, University of York, UK
Email: geoff.bailey@york.ac.uk

Richard Bates
Department of Earth Sciences and Scottish Oceans Institute, University of St Andrews, Fife, Scotland, UK
Email: crb@st-andrews.ac.uk

Ole Bennike
Geological Survey of Denmark and Greenland, Copenhagen, Denmark
Email: obe@geus.dk

Yves Billaud
Ministère de la Culture/DRASSM, Marseille, France

CNRS, UMR 5204 Edytem, Université de Savoie, Le Bourget-du-Lac, France
Email: yves.billaud@culture.gouv.fr

Anthony Burgess
University of Malta, Department of Classics and Archaeology, Archaeology Centre, Malta
Email: anthony.burgess.13@um.edu.mt

Isabel Cacho
GRC Geociències Marines, Dept. de Dinàmica de la Terra i de l'Oceà, Universitat de Barcelona, Barcelona, Spain

Miquel Canals
GRC Geociències Marines, Dept. de Dinàmica de la Terra i de l'Oceà, Universitat de Barcelona, Barcelona, Spain
Email: miquelcanals@ub.edu

Lucilla Capotondi
CNR, ISMAR, Bologna, Italy

Glicherie Caraivan
National Research and Development Institute for Marine Geology and Geoecology, Constanta Branch, Romania
Email: glicheriecaraivan@yahoo.com

Laurent Carozza
UMR 5602 Géode Géographie de l'Environnement, Maison de la Recherche de l'Université du Mirail, Toulouse, France

Daniele Casalbore
University La Sapienza, Rome, Italy

José Luis Casamor
GRC Geociències Marines, Dept. de Dinàmica de la Terra i de l'Oceà, Universitat de Barcelona, Barcelona, Spain

Corneliu Cerchia
National Research and Development Institute for Marine Geology and Geoecology, Constanta Branch, Romania

Francesco L. Chiocci
Earth Science Dept. University La Sapienza, Rome, Italy

Kim M. Cohen
Utrecht University, Utrecht, The Netherlands

TNO Geological Survey of the Netherlands, Utrecht, The Netherlands
Email: k.m.cohen@uu.nl

Deltares Research Institute, Utrecht, The Netherlands

Alastair Dawson
Department of Archaeology, School of Geosciences, University of Aberdeen, St Mary's, Scotland, UK
Email: a.g.dawson@dundee.ac.uk

Sue Dawson
Geography, School of Social Sciences, University of Dundee, Scotland, UK
Email: s.dawson@dundee.ac.uk

Robin Edwards
School of Natural Sciences, Trinity College Dublin, Republic of Ireland
Email: Robin.Edwards@tcd.ie

Gilles Erkens
Deltares Research Institute, Utrecht, The Netherlands
Email: gilles.erkens@deltares.nl

Utrecht University, Utrecht, The Netherlands

R. Helen Farr
Southampton Marine and Maritime Institute, Archaeology, University of Southampton, Highfield, UK
Email: R.H.Farr@soton.ac.uk

Nicholas C. Flemming
National Oceanography Centre, Southampton, UK
Email: n.flemming@sheetsheath.co.uk

Ehud Galili
Zinman Institute of Archaeology, University of Haifa, Haifa, Israel

Israel Antiquities Authority, Israel
Email: udi@israntique.org.il

Maria Geraga
Department of Geology & Geoenvironment, University of Patras, Greece

Ana Gomes
Universidade do Algarve, Faculdade de Ciências e Tecnologia, Centro de Investigação Marinha e Ambiental (CIMA), Faro, Portugal
Email: aisgomes@ualg.pt

Andreas Groh
Helmholtz Centre Potsdam, GFZ German Research Centre for Geosciences, c/o DLR Oberpfaffenhofen, Wessling, Germany

Jan Harff
University of Szczecin, Szczecin, Poland
Email: jan.harff@io-warnemuende.de

Marc P. Hijma
Deltares Research Institute, Utrecht, The Netherlands
Email: marc.hijma@deltares.nl

João Horta
Universidade do Algarve, Faculdade de Ciências e Tecnologia, Centro de Investigação Marinha e Ambiental (CIMA), Faro, Portugal

Birgit Hünicke
Helmholtz Centre Geesthacht, Geesthacht, Germany

Pavel Kuprin
Department of Lithology and Marine Geology, M.V. Lomonosov Moscow State University, Moscow, Russia

Galderic Lastras
GRC Geociències Marines, Dept. de Dinàmica de la Terra i de l'Oceà, Universitat de Barcelona, Barcelona, Spain

Gilles Lericolais
IFREMER, DAEI, Issy-les-Moulineaux, France
Email: gilles.lericolais@ifremer.fr

Vasilis Lykousis
Institute of Oceanography, Hellenic Centre for Marine Research, Anavyssos, Greece

Donatella Magri
Plant Biology Dept., University La Sapienza, Rome, Italy

Yossi Mart
Recanati Institute of Maritime Studies, University of Haifa, Haifa, Israel

Nikolay Maslakov
Department of Marine Geology and Mineral Resources of the National Academy of Sciences of Ukraine, Kiev, Ukraine
Email: nikalmas@mail.ru

Martin Meschede
University of Greifswald, Greifswald, Germany

Grażyna Miotk-Szpiganowicz
Polish Geological Institute – National Research Institute, Branch of Marine Geology, Gdańsk, Poland
Email: grazyna.miotk-szpiganowicz@pgi.gov.pl

Garry Momber
Maritime Archaeology Trust, National Oceanography Centre, Southampton, UK
Email: garry.momber@maritimearchaeologytrust.org

Irena Motnenko
Avalon Institute of Applied Science, Winnipeg, Canada
Email: irmot@avalon-institute.org

Delminda Moura
Universidade do Algarve, Faculdade de Ciências e Tecnologia, Centro de Investigação Marinha e Ambiental (CIMA), Faro, Portugal
Email: dmoura@ualg.pt

Yaacov Nir
Rehovot, Israel

Anatoly Pasynkov
Department of Physical Geography and Geomorphology, Geographical faculty, Federal Crimean University named after V.I. Vernadsky, Simferopol, Crimea
Email: anatoly.pasynkov@yandex.ua

Daniela Popescu
Basin Water Administration – Dobrogea Littoral, Constanta, Romania

Alar Rosentau
Department of Geology, University of Tartu, Tartu, Estonia
Email: alar.rosentau@ut.ee

Grigoris Rousakis
Institute of Oceanography, Hellenic Centre for Marine Research, Anavyssos, Greece

Dimitris Sakellariou
Institute of Oceanography, Hellenic Centre for Marine Research, Anavyssos, Greece
Email: sakell@hcmr.gr

Anna Sànchez-Vidal
GRC Geociències Marines, Dept. de Dinàmica de la Terra i de l'Oceà, Universitat de Barcelona, Barcelona, Spain

Julie Satchell
Maritime Archaeology Trust, National Oceanography Centre, Southampton, UK
Email: julie.satchell@maritimearchaeologytrust.org

Evgeny Schnyukov
Department of Marine Geology and Mineral Resources of the National Academy of Sciences of Ukraine, Kiev, Ukraine

Sergio Silenzi
ISPRA, Rome, Italy

Olena Smyntyna
Department of Archaeology and Ethnology, Odessa I.I.
Mechnikov National University, Odessa, Ukraine
Email: smyntyna_olena@onu.edu.ua

Valentin Sorokin
Department of Lithology and Marine Geology, M.V.
Lomonosov Moscow State University, Moscow, Russia
Email: vsorok@rambler.ru

Takvor Soukisian
Institute of Oceanography, Hellenic Centre for Marine
Research, Anavyssos, Greece

Szymon Uścinowicz
Polish Geological Institute - National Research Institute,
Branch of Marine Geology, Gdańsk, Poland
Email: szymon.uscinowicz@pgi.gov.pl

Dina Vachtman
Statoil ASA, Harstad, Norway

Valentina Voinea
National Museum of History and Archaeology of Con-
stanta, Romania
Email: vialia_rahela@yahoo.fr

Henk J.T. Weerts
Cultural Heritage Agency, Amersfoort, The Netherlands
Email: h.weerts@cultureelerfgoed.nl

Kieran Westley
School of Geography and Environmental Sciences, Ulster
University, Coleraine, Northern Ireland, UK
Email: kl.westley@ulster.ac.uk

Caroline Wickham-Jones
Department of Archaeology, School of Geosciences,
University of Aberdeen, St Mary's, Scotland, UK
Email: c.wickham-jones@mesolithic.co.uk

Valentina Yanko-Hombach
Interdisciplinary Scientific and Educational Center of
Geoarchaeology, Marine and Environmental Geology,
Paleonthological Museum, Odessa I.I. Mechnikov
National University, Odessa, Ukraine
Email: valyan@onu.edu.ua

Department of Physical and Marine Geology, Odessa I.I.
Mechnikov National University, Odessa, Ukraine

Avalon Institute of Applied Science, Winnipeg, Canada
Email: valyan@avalon-institute.org

Wenyan Zhang
MARUM - Center for Marine Environmental Sciences,
University of Bremen, Bremen, Germany

Eduardo Zorita
Helmholtz Centre Geesthacht, Geesthacht, Germany

Foreword

As little as ten years ago, the idea of producing a systematic and comprehensive examination of the glacial-maximum geomorphology and terrestrial environment of Europe's continental shelf at a pan-continental scale would have seemed either hopelessly fanciful or impossibly ambitious.

For archaeologists, the existence of a vast and now submerged prehistoric territory exposed at lowered sea level had barely entered the professional and academic consciousness. Traditionally, this underwater realm has been regarded as too inaccessible and difficult to deal with, and too ravaged by destructive processes to preserve more than a vestigial record of archaeological sites or landscape features, a record, moreover, considered unlikely to make any difference to the understanding of world prehistory. Such pioneering studies as exist have been focused mainly at local or national level, and sometimes at a regional level, for example in relation to particular sea basins such as the North Sea or the Baltic.

For Earth scientists, more familiar with large-scale collaboration at a continental or global scale, the continental shelf has been extensively studied in the context of plate tectonics, the extension of continental geology under the sea, national resources, coastal zone management, the exploitation of minerals and hydrocarbons, cable and pipeline route surveys and, to a certain extent, Pleistocene sea-level change. It has also been surveyed and mapped topographically at a resolution sufficient for safe navigation. However, analysis and interpretation of seabed data as evidence of a former terrestrial landscape that has been repeatedly exposed and then submerged by sea-level change has remained somewhat peripheral to their concerns.

The idea that the techniques already developed and the large quantities of data obtained piecemeal for many different applications could be synthesized and interpreted to reconstruct and understand the prehistoric occupation of the continental shelf during the various phases of low sea level was not on anybody's agenda until very recently.

Such is the fate of research questions that fall outside the scope of pre-existing research agendas. Despite boldly expressed aspirations and much protestation of good faith about the virtue of 'interdisciplinarity', successful integration of ideas and methods drawn from many different disciplines remains a formidable challenge; by definition such endeavors lie at the boundaries between more established disciplines and are discounted in consequence, typically falling into the gaps between different conceptual structures, administrative organizations and funding bodies. In turn, the new ideas and agendas required to give momentum to the study of unfamiliar research questions are liable to slow and fitful development. The continental shelf, viewed as a submerged landscape and a former terrestrial environment, is no exception, and has long remained a marginal zone in both a literal and a conceptual sense.

An important recent spur to changing research agendas is the growing importance of understanding sea-level change in a world of climate warming and impending sea-level rise, and the threat this poses to human life and livelihoods on a global scale. The study of Quaternary ice caps, the crustal impact of redistributed masses of ice and water, and the changes of sea level on the continental shelf have been studied in several large global collaborations. But even here, difficulties in the efficient application of standard marine geophysical techniques, compounded by stratigraphic discontinuities and disturbances of seabed sediments, have hampered the study of submerged shoreline features, and more accessible proxy data has often been preferred such as deep-sea sediment cores, shoreline features above modern sea level, or the outputs of Earth-geophysical and climatic modeling. Yet, interpretations based on proxy data and theoretical modeling are only as good as the assumptions that underpin them, and in need of continuous testing and refinement against field data. Since sea level has been lower than present for most of human history on this Earth and in most regions, it follows that most of the relevant data of past shorelines is now likely to be submerged on the seabed. Moreover, sea-level change is only one part of the history of the continental shelf, being inextricably bound up with the geological history of the Earth's crust and the geomorphological transformations at its surface, as the chapters in this volume make clear.

It was precisely this challenge – the need to bring together a multi-national group of individuals with very different scientific and archaeological skills and interests, but a shared interest in the terrestrial environment of the continental shelf and its potential for the preservation of archaeological data relating to human settlements – that brought into being the idea of "Project Deukalion", conceived in 2008 by Nicholas C. Flemming and Dimitris Sakellariou. This rapidly led to the formation of the Deukalion planning group with 16 experts from 8 European countries and the goal of drafting the outlines of a multidisciplinary project.

The Deukalion initiative was subsequently expanded and incorporated into the SPLASHCOS network in 2009. SPLASHCOS (Submerged Prehistoric Archaeology and Landscapes of the Continental Shelf) is the outcome of a proposal to the EU COST (Cooperation in Science and Technology) program, designed to stimulate international and interdisciplinary collaboration. As COST Trans-Domain Action TD0902, SPLASHCOS provided sustained support and funding over a four-year period to bring together in regular meetings archaeologists, geoscientists, geophysicists, paleoclimatologists, oceanographers, sea-level experts and representatives of government organizations and offshore industries from 25 European states and over 100 research institutions and agencies in a concerted effort to develop a new research agenda. Funds also included encouragement and training of early-career researchers, and other dissemination activities (see www.splaschos.org for further details). The momentum created by that initiative continues.

This volume is the first major collaborative publication to result directly from the sustained activities of the SPLASHCOS initiative, and is the product of Working Group 2, Environmental Data and Reconstructions, led by Jan Harff. A second volume based on work initiated by Working Group 1, led by Anders Fischer, on Archaeological Data and Interpretations is in preparation and is intended to examine in detail the evidence of prehistoric archaeology.

In keeping with the nature and aspirations of the SPLASHCOS Action, the authorship of this volume is collaborative, multi-national and multi-disciplinary, and the geographical scope pan-European, dealing with all the major European sea basins, ranging from the Atlantic Ocean and North Sea in the north-west, through the Baltic and the Mediterranean, to the Black Sea in the south-east.

The primary focus is the nature of the continental shelf as it would have existed during periods of lowered sea level as a terrestrial landscape – its variable geographical configuration, topography, sedimentary depositional processes, stratigraphy, climate, ice limits, river drainage, flora and fauna, and its potential and history as a zone of human habitation and a repository of archaeological data. Reconstructing these features is no simple matter, the biggest problem being the multiple ways in which successive cycles of sea-level rise and fall have variously buried, obscured, exposed, eroded or destroyed the material traces of past human activities and their original landscape setting. Hence, a major theme of research and interpretation must be the ways in which sub-sea processes, including commercial activities, have affected the preservation and visibility of formerly terrestrial deposits and their associated archaeological remains.

The problems posed by differential preservation and visibility are not insuperable, nor are they problems unique to the study of submerged land surfaces. Terrestrial archaeologists, too, have increasingly come to realize that the distribution of archaeological sites on land does not reflect in any simple way the distribution of past human activities or past human populations. Rather, such distributions represent a complex interaction between the locations where past peoples left the material traces of their existence, the nature of the activities carried out in these different locations, the manner in which material was discarded, and the various natural and human processes that acted subsequently to transform the land surface.

The term 'landscape taphonomy' is sometimes used to describe this field of research, referring to the variable ways in which the physical features of a past land surface, including land forms, soils and sediments, and the archaeological materials deposited on or in them by successive human generations, have been variously buried, exposed, preserved, mixed, scattered, destroyed or otherwise transformed, whether by processes that are natural in origin or anthropogenic. The problem is especially acute underwater, but no less present on dry land. In both cases, the nature of these transformations remains poorly understood, under-researched, and in need of much greater attention.

The opening four chapters of this volume provide comprehensive overviews at a continental scale of key themes such as geological structure, large-scale tectonic evolution, sea-level change, glacial history, climate, environment and mapping, including the vexed issue of bathymetric maps and databases; these are often assumed to exist already in adequately published or digital form but in reality they are mostly produced for quite different purposes and are too inaccurate or of too coarse a resolution to be of more than limited usefulness for archaeological and paleoenvironmental purposes. The existing accessible and published data provide a broad framework in many regions, but as with so many other examples in the history of science, new research questions will demand the collection of new and more detailed data with methods best adapted to yield the necessary information.

These opening chapters are not confined narrowly to the continental shelf but encompass the continental landmass more widely, and make for illuminating reading, relevant to everyone interested in the natural and human history of the European continent over the past one million years and more.

These themes are followed through in more detail in the subsequent sea-basin chapters, which also include reference to key underwater archaeological finds as examples of preservation mechanisms and the potential for future discoveries. A notable feature is the comprehensive listing of online sources of information. In a digital world, information is increasingly being made available on the internet, either as searchable, accessible and properly maintained databases, or in more ephemeral or inaccessible form, and this is likely to be a growing trend for the future.

Another notable outcome of these chapters is to highlight is the variable nature and extent of the shelf environment in different regions, the variability in geology, oceanographic conditions, geomorphological processes and preservation potential, and the differences of approach best suited to these different conditions. Already the outlines of a more detailed pattern of variability are beginning to emerge more clearly and this will surely provide an important step towards more sharply defined research questions, new field investigations, and improved standards for wide-ranging comparative analysis.

The topics covered in this volume are not only of scientific and intellectual interest, but central to some of the most pressing and practical concerns affecting our collective livelihood, prosperity and sense of common identity in the coming century – understanding of sea-level change and its likely trajectory and human impact, management of a massively increasing volume of digital records, and improved knowledge of how the now-submerged territories of the continental shelf have contributed to the early growth of our civilization.

Much remains to be done, and some of the geophysical knowledge that forms the necessary foundation for new investigation is highly technical. Nevertheless, the interested reader will find this volume an essential starting point for entry into a vast new intellectual, scientific and multidisciplinary territory. It is our hope and expectation that this will create the basis for an expanding field of future research in the coming decades, with improved integration of its multiple sources of information and expertise, an increasing number of participants, a growing investment of resources and funding, a new generation of marine scientists and archaeologists with the proper training to move across the borders of the traditional disciplines, and significant progress in advancing the new discipline of Continental Shelf Prehistoric Research.

Geoffrey N. Bailey
University of York, UK
Chair of SPLASHCOS

Dimitris Sakellariou
Hellenic Centre for Marine Research, Greece
Vice Chair of SPLASHCOS

Preface

This book considers the complex question of how, why, and where prehistoric occupation sites and artifacts on the European continental shelf have survived inundation by the postglacial rise of sea level, and how, why and where in other cases they have been destroyed. It forms an essential prelude to the interpretation of known submerged prehistoric archaeological sites and artifacts on the European continental shelf, which are now known to number in excess of 2600 find spots (see splashcos-viewer.eu).

The seed of the idea for this volume was sown in January 2008, when Dimitris Sakellariou invited Nicholas C. Flemming to present a report on continental shelf prehistoric research to the Hellenic Centre for Marine Research in Athens. In the discussions afterwards the idea emerged of a European-scale project which would be submitted to the European Community Framework-7 grant-awarding system in Brussels, and we hoped to obtain several millions of euros for this work. The sixth World Archaeology Conference (WAC6) was held in Dublin in June 2008, and Dimitris attended that meeting. There was a session on seabed prehistoric research organized by Amanda Evans and Joe Flatman. Less than a month later, in July 2008, the third International Conference on Underwater Archaeology (IKUWA3) was held in London, and Dimitris and Nicholas C. Flemming planned a day's session that started with papers on research already conducted on seabed prehistoric sites, and then devoted half a day to planning an application for research funds to the European Commission. The session was attended by about 80 people, and we created an address contact list, and a core planning group of ten people called the "Deukalion Group", after the hero of the Greek flood myth. At the same conference, Joe Flatman and Amanda Evans started to edit a book illustrating examples of seabed prehistoric sites from all over the world (Evans *et al.* 2014), drawn from the papers presented at WAC6 and IKUWA3.

The Deukalion Group was chaired jointly by Nicholas C. Flemming and Dimitris Sakellariou and successive meetings were held at different European academic research venues during 2008 and 2009. In the event, it proved impractical to plan and manage an integrated research project which could qualify for research funding on the scale that we had originally intended, not least because of a changing emphasis in EU funding towards smaller-scale research projects with a strong policy-relevant theme. Accordingly, Geoffrey N. Bailey proposed that in order to perpetuate and expand the group and provide funds for continued planning meetings and exchange of ideas, an application should be submitted to the COST (Cooperation in Science and Technology) Office in Brussels to fund a four-year Action or research network. Geoffrey N. Bailey chaired special meetings of the Deukalion Group to prepare and submit a proposal, which was approved for funding in 2009, leading to the project known as SPLASHCOS, "Submerged Prehistoric Archaeology and Landscapes of the Continental Shelf", (TD0902), which ran from 2009–2013, with a final conference in Szczecin, Poland in September 2013. The COST funding provided support for meetings, working groups dedicated to advancing thinking on particular themes, websites, publications, communications, data archives, training workshops, and field training schools, but not for the costs of new research projects. In the event, individual research groups working within SPLASHCOS were highly successful in raising financial support for fieldwork during the four years of the project.

Many individual reports and publications have been produced during the SPLASHCOS project. Agencies and research groups from 25 European countries participated, with over 100 individual researchers involved. There were 23 projects funded from various sources, both national and international, amounting to over €20 million in total. Numerous academic articles and reports were published in the refereed literature. The present volume is a concluding publication within the SPLASHCOS schedule.

The editors thank all the authors for their hard work. This book is not a collection of papers where the authors had free rein. They were required to write chapters which addressed a novel range of problems. The energy, commitment, and spirit shown by the authors, and their willingness to respond to the demands of editorial review, has been very encouraging and rewarding, and many of the ideas in this book are the result of the interactions between experts of many disciplines, nationalities and backgrounds at SPLASHCOS meetings. Special thanks are due to Kieran Westley, who acted as coordinator and sub-editor for the group of chapters on the European Northwest Shelf.

In the editing process, Nicholas C. Flemming has led the primary task of reviewing and editing the contributions, with substantial input of intellectual and scientific expertise and revision from Delminda Moura and Jan Harff. Tony Burgess played a critical role in coordinating and formatting the content of the chapters in all stages of preparing the manuscript and Geoffrey N. Bailey critically reviewed the final text to ensure internal consistency, explanation of technical terms and coordination with the archaeological work of the SPLASHCOS project.

We thank the COST Office for their financial and administrative support, and in particular the successive Science Officers who guided our work; Julia Stamm, Geny Piotti, Andreas Obermaier and Luule Mizera, and the COST Rapporteurs Daniela Koleva and Ipek Erzi who attended meetings and gave invaluable advice. We also thank the institutions that hosted the SPLASHCOS plenary meetings: March 2010, University of York, UK; October 2010, Rhodes, Greece, Hellenic Centre for Maritime Research; April 2011, Berlin, Germany, Museum für Asiatische Kunst; October 2011, Zadar, Croatia, University of Zadar; April 2012, Amersfoort, Netherlands,

National Cultural Heritage Agency; September 2012, Rome, University of Rome "La Sapienza"; March 2013, Esbjerg, Denmark, University of Southern Denmark; September 2013, Szczecin, Poland, University of Szczecin. Finally, we thank Cynthianne DeBono Spiteri, who acted as Grant Holder and administrative secretary of SPLASHCOS, and who provided with unfailing patience and good humor the central coordination of the complex network of communications, paperwork and logistics required in the organization of meetings and progress reports.

Reference

Evans, A., Flemming, N. & Flatman, J. (eds.) 2014. *Prehistoric Archaeology of the Continental Shelf: a Global Review*. Springer: New York.

Nicholas C. Flemming
National Oceanography Centre, Southampton, UK

Jan Harff
University of Szczecin, Szczecin, Poland

Delminda Moura
Universidade do Algarve, Faculdade de Ciências e Tecnologia, Centro de Investigação Marinha e Ambiental (CIMA), Faro, Portugal

Anthony Burgess
University of Malta, Department of Classics and Archaeology, Archaeology Centre, Malta

Geoffrey N. Bailey
University of York, Department of Archaeology, York, UK

Acknowledgement

This book is based upon work from COST Action TD0902 SPLASHCOS (Submerged Prehistoric Archaeology and Landscapes of the Continental Shelf), supported by COST. www.splashcos.org.

Chapter 1

Introduction: Prehistoric Remains on the Continental Shelf — Why do Sites and Landscapes Survive Inundation?

Nicholas C. Flemming,[1] Jan Harff,[2] Delminda Moura,[3] Anthony Burgess[4] and Geoffrey N. Bailey[5]

[1] National Oceanography Centre, Southampton, UK
[2] University of Szczecin, Szczecin, Poland
[3] Universidade do Algarve, Faculdade de Ciências e Tecnologia, Centro de Investigação Marinha e Ambiental (CIMA), Faro, Portugal
[4] University of Malta, Department of Classics and Archaeology, Archaeology Centre, Malta
[5] Department of Archaeology, University of York, UK

The Big Question

This book is designed to provide the best partial answer to an apparently clear-cut and uncomplicated question: "Why do some prehistoric sites, settlements, landscapes, and artifacts survive on the sea floor, after inundation by postglacial sea-level rise, when many others are destroyed or scattered by waves and currents?" There are over 2600 known submerged prehistoric sites in European seas (Jöns *et al.* 2016). At the glacial maximum when sea level was at its lowest, at about −130 m, an additional increment of land became available on the European continental shelf estimated at about 40% of the present-day European land area, amounting to an estimated 4 million km^2. The question of what determines the survival or destruction

Submerged Landscapes of the European Continental Shelf: Quaternary Paleoenvironments, First Edition.
Edited by Nicholas C. Flemming, Jan Harff, Delminda Moura, Anthony Burgess and Geoffrey N. Bailey.
© 2017 John Wiley & Sons Ltd. Published 2017 by John Wiley & Sons Ltd.

of archaeological sites and landscape features is therefore a serious one. We have no information on how many sites have already been destroyed, or how many survive and are yet to be discovered, or how much of the original pre-inundation terrestrial landscape has been destroyed beyond recognition.

There is a corollary question: "If we can understand the oceanographic conditions, geological circumstances, changes through time and topographic geometry that most favor the survival of prehistoric settlements during and after marine transgression, can we turn the argument round, and use knowledge of oceanography, ice-cap chronology, and coastal geomorphology to predict where prehistoric sites existed, and where they will be preserved on the sea floor of the continental shelf?"

When people are told that there are thousands of prehistoric sites on the sea floor, human settlements and places of occupation ranging in age from 5000 to more than 100,000 years old, their first reaction is often incredulity or skepticism (Bailey & Flemming 2008). This reaction is usually the case both for expert archaeologists and members of the general public alike. How can it be true that fragile, unconsolidated deposits of cultural remains, charcoal, food debris, scattered stone tools, débitage from flint knapping, wooden hut posts, and bits of bone or fragments of wooden canoes survive first the process of postglacial rising sea level and transit through the surf zone, and then thousands of years submerged under present oceanic and coastal conditions? How and why do they survive, and how can we discover their most probable locations?

In this book we make a beginning on the answers to those questions. The answers are not simple, and it has taken the work of many people from many different academic professions to piece together the whole story so far. There is rapid progress continuing in many of the component sub-disciplines that contribute to this research, and so the present book can only be a snapshot of the present situation. We try to indicate the ways in which changes are taking place.

The aim of the present volume, therefore, is to review the current state of knowledge concerning the environment of the exposed continental shelf during glacial lowering of sea level, the causes of changes of land and sea level, the processes of inundation, the mechanism of attack by waves and currents on anthropogenic deposits in shallow seas, and the circumstances in which such deposits are most likely to be preserved, or to be destroyed. The primary goal is to use an understanding of these processes to illuminate the reasons for the differential survival or destruction of archaeological sites. Additionally, such an investigation also contributes baseline data for reconstructing the environment as it would have existed on the now-submerged shelf, the sorts of resources of food, water supplies, topography, soils, flora and fauna that would have been available for human exploitation on these extensive tracts of new territory, and ultimately their impact on patterns of site location and human settlement, mobility and dispersal.

Throughout the book we refer to identified and studied archaeological sites only as examples to show how they may have been preserved or partially destroyed by different processes. SPLASHCOS (Submerged Prehistoric Archaeology and Landscapes of the Continental Shelf — www.splashcos.org/) has produced a digital database of over 2600 submerged prehistoric sites (Jöns et al. 2016), and a second volume on the archaeological material is currently in preparation. By using practical examples we hope to base the argument on facts in the field, rather than over-reliance on theoretical hypotheses.

The aim of this introductory chapter is to highlight some general issues affecting the survival or destruction of archaeological features, summarize the recent history of collaboration associated with the SPLASHCOS project and its predecessor, the Deukalion Planning Group project, which gave rise to the work underpinning this volume, explain the rationale for the organization of the volume, outline its contents, and set out the conventions and standards used in presenting information.

General Issues

In the early days of Deukalion and SPLASHCOS in 2008–9, the meetings were dominated by archaeologists and prehistorians, and they frequently wanted to know if it was possible to predict the survival and location of seabed prehistoric sites accurately, so that they could do their work more efficiently and at lower cost. This question could not be answered quickly, and always dropped down the agenda. The oceanographers, sedimentologists, climate experts and technologists felt frustrated by the intractability of the problem, and were not only unable to answer it, but even unable to explain why they could not answer it.

Recently published books on individual sites, or conference volumes (e.g. Flemming 2004; Benjamin *et al.* 2011; Evans *et al.* 2014), confine themselves to the study of prehistoric sites which, by definition, have survived. They do not consider the probable or possible distribution of sites which have not survived or have not been found. A few paragraphs at most may be devoted to considering why particular artifacts have survived at the site being studied. And yet, as scholars from many disciplines start to consider the prehistoric continental shelf as an integrated whole, whether European or global, the same questions recur again and again: where should we search for anthropogenic signals from the periods of glacial-maximum low sea levels? Where would searching be pointless? Where were people living on the continental shelf, and why? What conditions favor survival of anthropogenic deposits and signals? Where will archaeological deposits be buried under tens of meters of modern sediments, and hence difficult to find or excavate? Where will deposits have been eroded away? And where will the overburden of protecting sediments have been eroded away by submarine channels to a sufficient extent to expose material without destroying it? Are there predictable patterns of survival of sites? Where are fragile sites now exposed to erosion so that immediate study or preservation are needed? How can we maximize the efficiency of a search strategy? What instrumental techniques will provide useful environmental data? What geophysical and sedimentary data already exist to help us define favorable conditions for survival of deposits?

The subsidiary questions continue: do we know enough accurately about the positions of the sea level and the land surface at different dates? Can we define the positions of ancient shorelines, river valleys, and shallow coastal lagoons and marshes? Do we know where there are submerged caves or rock shelters that people might have lived in? Can we reconstruct the fine gradations of landscapes, vegetation and fauna extending away from the edge of the ice sheet through the areas of tundra and temperate forests to the Mediterranean or to tropical climates and vegetation zones on the continental shelf? When we have identified drowned prehistoric deposits in context, can we reconstruct the immediate landscape, fauna and flora in the adjacent foraging and hunting area?

The questions listed above, which are by no means exhaustive, illustrate the fundamental and potentially exciting aspects of continental shelf prehistoric archaeology. The answers are not themselves directly archaeological in nature, but rather paleoenvironmental. Yet archaeologists need the answers to these questions, and those answers in their turn depend on extensive interdisciplinary collaboration.

The survival and discovery of submerged prehistoric sites on the continental shelf implies two distinct defining issues: firstly the nature of the archaeological sites themselves, their age, function, cultural associations and technology, and the choice by ancient peoples of favored locations for settlements, camp sites or more ephemeral locations where materials were discarded; and secondly the circumstances of abandonment of the material, its burial or exposure while on land, followed by the processes of inundation and possible destruction or survival. The process of burial and preservation of sites, deposits, or single artifacts depends on complex interactions between environmental forces, coastal geodynamics, coastal configuration, geochemical processes, and biological interactions as shown throughout the chapters of this book for a wide range of conditions and circumstances. The process of inundation by rising sea level and the traverse of the surf zone across an archaeological site are likely to be destructive, or lead, at the very least, to a local scattering of artifacts and other materials. This is not always so, since in some cases sea-level rise may be accompanied by extreme wave attenuation caused by local topography and accumulation of sediments that can protect and partially or totally bury the archaeological material. Nevertheless the likelihood of destruction must be assumed in the analysis, and we should expect that survival of intact sites is likely to be the minority occurrence in most regions and most geodynamic situations.

As the argument progresses it becomes clear that the conditions for survival depend not only on regional climatic and oceanographic conditions, but critically upon topographic land forms and seabed morphology on a scale of meters to a few kilometers in the immediate vicinity of the stratified deposit during inundation and in the following few millennia (see, for example, Flemming 1963; Gagliano *et al* 1982; Belknap & Kraft 1981; 1985; Belknap 1983; Flemming 1983; Waters 1992; Flemming 2004; Harff *et al.* 2007; Benjamin 2010; TRC Environmental Corporation 2012; Evans *et al.* 2014). In addition, the exposure to marine forces, waves, currents, erosion, and burial has to be considered. Thus analysis of each situation depends upon quantitative hydrodynamic laws and mechanisms, but is locally site-specific, and success depends on having very accurate topographic data on the paleocoasts and the sea floor throughout the area of the modern continental shelf. This puts a huge demand upon availability of accurate high-resolution seabed data; although much bathymetric data is already available, it is often not adequate for the kind of analysis required. These data problems are reviewed generically in a chapter on the availability of digital data (Chapter 4), and each regional-sea analysis summarizes the available data sources, with a special emphasis on electronically available maps, core data, and displays.

Recent History of Collaboration

The authors and editors of this book have worked for many years on the task of finding and studying submerged prehistoric sites on the continental shelf of Europe (e.g. Flemming 1968; Galili & Weinstein-Evron 1985; Long *et al.* 1986; Galili *et al.* 1993; Antonioli & Ferranti 1994; Fischer 1995; Momber 2000; Harff & Lüth 2007). However, only in the last decade have collaboration and the exchange of information between different research groups resulted in a European-scale collaboration. During the 1980s and 1990s, various regional groups were already combining disciplines and sharing data to integrate knowledge of climate change, sea-level change, sediment movements and the discovery and interpretation of submerged prehistoric settlements.

Particularly strong groups developed to study the Danish archipelago and the straits between Denmark and southern Sweden and Norway, Kattegat and Skagerrak (Andersen 1985; Fischer 1995; Pedersen *et al.* 1997), and the German islands of the southern Baltic (Harff & Lüth 2007; 2011). In the southern North Sea the archaeologists in the Netherlands have a long tradition of working with the fishermen who trawl up Pleistocene megafauna bones and occasional human artifacts (Louwe Kooijmans 1970-71; van Kolfschoten & Vervoort-Kerkhof 1985). In 2002, a conference was held in London bringing together prehistorians from both sides of the North Sea, resulting in a volume on the known finds on the seabed, and outlining national and agency policies (Flemming 2004). This conference in turn led to a joint Anglo–Dutch initiative to promote collaboration between the national cultural heritage agencies in the two countries (Peeters *et al.* 2009), and to further discussion of collaborative initiatives at the IKUWA 3 (International Congress on Underwater Archaeology) held in London in 2008 (www .nauticalarchaeologysociety.org/shop/ikuwa-3-beyond-boundaries-3rd-international-congress-underwater-archaeology) and the subsequent establishment of the Deukalion Planning Group and the SPLASHCOS Action to further consolidate plans for international collaboration.

At the meeting of the SPLASHCOS Working Group 2 in York in April 2010 it was suggested that a review was needed of the environmental conditions in each European sea basin, which would provide prehistorians and other scientists collaborating with them with the background information they required to understand the preservation and destruction of submerged prehistoric sites in their region. From this proposal grew the present book.

Outline of this Book

We start with a series of three thematic chapters analyzing different environmental marine and coastal processes as they may affect the original location and then the survival of submerged prehistoric sites. These conclude with a brief summary of data sources and types.

Europe, the Mediterranean, and the Black Sea constitute a geographical area that is subject to many different types of vertical earth movement and relative sea-level change on time and space scales that overlap with prehistoric archaeological events. Chapter 2 therefore starts by addressing geological and tectonic processes that are controlled by plate tectonic movements on timescales of tens of millions of years. Notwithstanding the relatively slow rates of movement, the vertical changes over tens to hundreds of thousands of years have a profound effect on coastlines, topography, and sedimentary basins or depocenters like the North Sea, or the configuration of the Aegean Sea basin. In the Mediterranean region the convergence of Africa and Eurasia leads to both active mountain building and regions of rapid subsidence on archaeological timescales. The overriding climatic events of the last two million years have been the multiple and recurrent glacial ice caps on the northern continents, associated with growth and decay of more local ice accumulations on mountain ranges such as the Pyrenees, Alps, Apennines and Carpathians. Each phase of increasing ice volume on the land created an equivalent drop of global sea level, and exposure of large areas of what is now the continental shelf. For this reason Chapter 2 provides a thorough overview of the regional tectonics of Europe, the mechanisms of ice cap formation, the driving forces that determine the growth and decay of the ice sheets, the calculation of sea-level change, and the response of the Earth's crust to the redistribution of ice and water.

Chapter 3 starts with a definition of the continental shelf, the shallow flooded margin of the continent, beyond which the sea floor plunges into true oceanic depths of thousands of meters. We then consider how the multiple causes of fluctuating sea level affect the evolution of the continental shelf, and its sediment cover. Special attention is given to the process of sea-level rise transgressing across a prehistoric occupied area. Since the majority of known submerged prehistoric sites have been found in sea water shallower than 5 m, and some have survived in this situation for many thousands of years, it is immediately apparent that the destruction of sites is not inevitable, even when they are potentially exposed to wave action. This empirical fact is used to examine the coastal geodynamic circumstances which have protected sites. Wave and current actions have the potential to destroy some sites while preserving others separated by a few kilometers along the coast. The effects of sediment burial and chemical changes in the submerged environment provide the final stages of preservation.

Chapter 4 outlines the present sources of data needed to describe the seabed and its composition and sub-surface sedimentary stratigraphy. The data demands for reconstructing the paleoenvironment are extreme, even by modern industrial standards. Prehistoric archaeological sites are constrained by significant topographic features which need to be defined to a resolution of a few meters or less, while the hunting and foraging strategy and seasonal migration patterns of a Paleolithic family group extended over a range of tens or hundreds of kilometers. Thus we need very high-resolution seabed data over a very large geographical range. This is not available for all parts of the European continental shelf. Where data are available at sufficiently high resolution the data volumes are so great that electronic media are essential, and modern data management and data-merging techniques are needed.

Chapters 2–4 provide background data on processes that recur to differing extents at a regional and local scale in later regional chapters, where the authors go into more detail.

In order to investigate oceanographic, paleoclimatic and environmental processes in a consistent way, throughout the varying climatic zones of Europe and the Mediterranean, we break the area down into discrete sea basins each of which is examined in a separate chapter. Within each basin the conditions of tectonics, geomorphology, tide, wind strength, water temperature, occurrence of sea ice, wave height and so on can be analyzed in a holistic way from coast to coast. If the same factors were analyzed, for example, by national zones of jurisdiction, each sea basin would be broken into arbitrary sections by lines through the center. Oceanographic conditions do not respect national boundaries. Our approach also serves to integrate the prehistoric archaeology within coherent environmental conditions across exposed areas of the sea floor that are now flooded.

The core of the book follows with 14 chapters describing the regional sea basins of Europe in terms of their past and present geomorphological and oceanographic processes on the continental shelf, with selected examples of known submerged prehistoric sites. We start with a review of the Baltic Sea and its potential to preserve prehistoric sites (Chapter 5), and then work southwards in a counter-clockwise direction round the Atlantic margins of Europe (Chapters 6–11) through to the Mediterranean (Chapters 12–15) and the Black Sea (Chapters 16–18).

The authors for each regional chapter are usually teams combining earth scientists, marine geoscientists and prehistoric archaeologists. Even with this pooling of skills it is obviously not possible to provide a complete or exhaustive coverage or analysis of the complex forces, local conditions, particular regional changes of climate, etc., which would be needed ideally for each sea basin. The complex of processes, regional expression of glacio-eustatic sea-level change and local isostatic response, the regional expression of wind–wave energy, ocean basin currents and wind-driven currents, wave climate and previous wave climates under different environmental conditions, sediment transport, changing river patterns and so on, create a requirement for a large book for each sea basin, not just a chapter.

We have had to compromise. The solution has been to summarize the processes in a high-level aggregated manner, citing a full reference list of the major published integrating studies, and to combine this with access to electronic sources of archived data. The latter point is fundamental. We are trying to help the reader to understand multiple interacting spatial processes at scales from less than one meter up to hundreds of kilometers. Printing maps on a book page would not help, or is not sufficient. Thus the decision was taken early on in preparing this book to cite electronic sources wherever possible. These URL citations are grouped at the end of each section within a chapter, and refer the reader to maps, archives, sediment core data, ice-cap models, and other large datasets. Most of the citations are to major institutions and governmental or international agencies which are not going to disappear, but some of the more exotic projects and data sources, which are still important for this book, may not be active more than a few years

after publication, although all are current as of December 2016. The intention in these chapters is to provide the reader with a rapid guide to the key literature on the major sea basins, and an authoritative summary of up-to-date information, while accepting that it is impossible to go into detail, or consider all nuances.

Conventions and Standards

Each chapter is designed so that it can be accessed electronically as an independent publication. Thus all acronyms are explained within each chapter, and cited references are included at the end of each chapter.

One of the main aims of the chapters on regional sea areas in this volume is to provide the reader with a guide to the sources of data in many different disciplines. Where specimen maps or screen dumps are shown, it is appropriate that they should be shown in their original format. Thus there is considerable variation in the style of graphics.

It may surprise the reader to discover how uncertain some marine geoscience data are regarding different phases of the Late Pleistocene landscape and environment. In attempting to recreate the coastal and continental shelf environment at the human scale we need to consider many different sources of data, geological, glaciological, sedimentary, etc., and combine them. Thus the uncertainty or errors at the end of the research in one discipline become errors in the input to the next stage of integration — they propagate — and the uncertainty may be magnified. Many different consequences may arise if two adjacent ice sheets did or did not actually meet and merge; a large ice lake may or may not have persisted for thousands of years if a particular ridge was eroded or not; a particular river may or may not have been crossable at a particular date, depending on the rate of melting of an upstream ice sheet, and the raft-building technology of the time. There is uncertainty as to the ability of early humans or hominins to adapt to life in a cold climate, and the earliest dates of seafaring and exploitation of marine and coastal resources.

Multiple sources of data can be integrated with some rigor by intensive research at a relatively restricted regional level, as in the SINCOS projects ("Sinking

Coasts: Geosphere, Ecosphere and Anthroposphere of the Holocene Southern Baltic Sea", Harff & Lüth 2007; 2011), but such a combination of professional skills from prehistoric archaeology to mathematical oceanographic modeling is still rare.

So far as it is possible, we have ensured that the authors of different chapters use similar terminology and conventions regarding symbols, assumed sea-level history, gross European climate changes and sources of European-scale data. The described sea areas have slight overlaps, and there should be no major contradictions at the borders. Authors of the chapters on regional sea areas were provided with an agenda or proforma of topics to be covered, but these are definitely not identical in importance from the icy coasts of the Gulf of Bothnia to the palm-fringed shores of the Middle East. Each team of authors has tended to develop or emphasize additional themes which seemed to them important, or for which they had special skills. This gives an added individuality or color to each chapter, which we have not tried to avoid.

The Last Glacial Maximum is referred to frequently as the LGM, often followed by a bracket of dates such as 21 ka BP to 18 ka BP (Before Present). These dates are not always the same, since the same global cause has different responses depending on the latitude or other regional variables and we have not requested authors to standardize unless precisely the same event is referred to more than once in the same chapter. The so-called LGM can be defined variously as the maximum volume of global ice on land, the time or period of maximum reduction in global sea level, the time of the lowest temperature reached in a region, or the maximum extent of continental and shelf ice in a region. These events may have considerable duration, and local terminology describing the Pleistocene may have adopted a conventional date, or bracket of dates, within the maximum range of about 27 ka BP to 17 ka BP. In general, the initiation of final melting and the start of global sea-level rise is usually quoted in the range 21 ka BP to 18 ka BP.

As explained above, many of the specialized fields summarized in this book have large uncertainties in them, sufficient to cause further problems when used as the input for another level of integration or analysis.

Thus different authors may have used different ice-limit models, and maps showing these differences occur in different chapters. It is not possible at present to decide which model is correct, or the most nearly correct, and some models will be most accurate in certain characteristics, and less accurate in others. This is not a school textbook in which we present the student with a smooth unruffled picture of the best approximation to modern knowledge. Sources are cited which may give different interpretations of the data, and the reader must draw conclusions as to which is most useful. This applies equally to different interpretations of regional relative sea-level curves and to best estimates of shorelines at different dates, such as in the Black Sea, and to models of ice volumes and directions of ice flow. Models of these parameters are evolving continuously and becoming more accurate.

The terms 'large scale' and 'small scale' should not need explaining, but in recent decades they have become reversed from their original meaning in common parlance in some communities. In this book we use the original meaning as follows. A small scale map is one in which objects and features look small. Thus, a map or chart at a scale of 1:5 million represents 1 km on the ground as 0.2 mm on the map, i.e. very small. It is a small scale map. A local plan at a scale of 1:1,000 represents 1 km on the ground as 1 m on paper, so that features such as houses and roads appear as large recognizable objects with shape and width. It is a large scale map. Notwithstanding this convention, numerical modeling communities describe the process of using the output boundary equations and variables from a small-scale model of a large oceanic area to drive a more local coastal large-scale model with higher resolution as "downscaling" or "downsizing", when in fact the scale at the local level is being enlarged.

Geological and archaeological terminology used in some regions contains localized and not universally used terms, but these are still the correct form and will be found in the regional literature cited for that area. Such terms are defined in context within each chapter where used. In addition, there may be small differences in spelling and word order between names in the text and embedded in images, an inevitable occurrence when examining such a wide and diverse geographical range. Each chapter on regional sea areas also contains an extensive bibliography of local literature, sometimes in languages other than English.

Statements regarding dates appear in several formats. The statement that something occurred so many thousand years ago should be interpreted as an estimate in calendar years before the date of publication and can be expressed using 'ka' (or 'Ma' for millions of years). The terms 'kyr/Myr' on the other hand refer to lengths of time measured in thousands/millions of years. There is no implication as to method of dating or calibration errors. Dates throughout the book are frequently cited as having been determined by radiocarbon ^{14}C measurements. The use of the acronym BP (Before Present) necessarily implies that the method of dating has been radiocarbon, and that the date is measured as time before 1950. Authors have been requested to cite whether radiocarbon dates have been calibrated (signified by the use of 'cal BP') or not, and where possible to include the laboratory reference and standard deviation. This additional information is provided in some cases, but not in most cases. Where there is no information on calibration, it must be assumed that the assessment is uncalibrated, and possibly that there is a margin of uncertainty. The fact that a cited date is uncalibrated is sometimes followed by the statement that this correlates roughly with so many years ago, even when accurate calibration data are not cited. Issues of precision and accuracy are often more critical in archaeological studies than in the more general geomorphological subjects discussed in the present volume, since interpretation can be significantly affected by margins of error in the correlation of separate events or the measurement of their duration. Formal agreements on standards of presentation of ^{14}C data are summarized by Stuiver and Pollach (1977) Reimer et al. (2013) and Millard (2014).

Physical oceanographers are concerned about the exact concentration of salinity in sea water since, combined with the temperature, it largely determines the density and thus the stratification of layers of water, as well as what species can easily live in it. Until the late 1970s, the standard expression of measured salinity was parts per thousand (per mill), expressed by the symbol ‰. During that decade oceanographers devised and later adopted the new Practical Salinity Scale, and dispensed with the symbol, on the basis that the salinity was expressed as a pure number in terms of the ratio of weight of salt to the weight of water. The new international standard TEOS-10 (Thermodynamic Equation of Seawater 2010) uses absolute salinity values in g/kg (Millero et al. 2008). Although the use of the old salinity symbol ‰ (per mill) has been discouraged since 1978, it has been customary in a large part of the oceanographic and popular literature to use this expression, and readers unaware of the complex theoretical changes in the professional world of marine physics, expect to see such a symbol. The focus on the precise thermodynamic properties of water of different salinities has led to further definitions of different scales based on functions of salinity (Wright et al. 2011). For the present work the implications of the different scales and units are not significant, and we have not asked authors to confirm whether they have used one scale convention or another. Where the symbol ‰ appears it is because the authors of that chapter used it, or cited literature that used it. Where salinity is cited purely as a number with no symbol or unit, the same applies.

Conclusion

We hope that the first question, why some sites survive, is answered successfully to a considerable extent by this book. The reverse question, can we predict where the sites are and find them, is much more difficult. The readers of this book should be in a better position to address this question by the time they finish, and future generations of researchers may refine the arguments still further, and achieve more reliable answers. Improvements in technology, and the growth of large accessible archives of digital data, as well as the development of highly sophisticated models, will make it easier to reconstruct accurate paleotopographies, showing drowned river valleys, soil types, vegetation, and coastlines that are at present under the sea. The use of the knowledge in this book enables the prediction of locations of high probability for prehistoric sites, but still usually leaves the researcher with a further stage of detailed field survey and analysis before anthropogenic indicators can be located (e.g. Wessex Archaeology 2011; Weerts et al. 2012; Moree & Sier 2014; 2015). This work can be expensive:

hence the incentive to improve the models and their predictive accuracy and reliability.

References

Andersen, S. H. 1985. Tybrind Vig. A preliminary report on a submerged Ertebølle settlement on the west coast of Fyn. *Journal of Danish Archaeology* 4:52-69.

Antonioli, F. & Ferranti, L. 1994. La Grotta sommersa di Cala dell'Alabastro (S. Felice Circeo). *Memorie della Società Speleologica Italiana* 6:137-142.

Bailey, G. N. & Flemming, N. C. 2008. Archaeology of the continental shelf: Marine resources, submerged landscapes and underwater archaeology. *Quaternary Science Reviews* 27:2153-2165.

Belknap, D. F. 1983. Potentials of discovery of human occupation sites on the continental shelf and nearshore coastal zone. In *Proceedings of the Fourth Annual Gulf of Mexico Information Transfer Meeting.* 15th – 17th November 1983, New Orleans, pp. 382-387.

Belknap, D. F. & Kraft, J. C. 1981. Preservation potential of transgressive coastal lithosomes on the U.S. Atlantic Shelf. *Marine Geology* 42:429-442.

Belknap, D. F. & Kraft, J. C. 1985. Influence of antecedent geology on stratigraphic preservation potential and evolution of Delaware's barrier systems. In Oertel, G. F. & Leatherman, S. P. (eds.) *Marine Geology (Special Issue — Barrier Islands)* 63:235-262.

Benjamin, J. 2010. Submerged prehistoric landscapes and underwater site discovery: Reevaluating the 'Danish Model' for international practice. *Journal of Island & Coastal Archaeology* 5:253–270.

Benjamin, J., Bonsall, C., Pickard, C. & Fischer, A. (eds.) 2011. *Submerged Prehistory.* Oxbow Books: Oxford.

Evans, A., Flatman, J. & Flemming, N.C. (eds.) 2014. *Prehistoric Archaeology on the Continental Shelf: a Global Review.* Springer: New York.

Fischer, A. (ed.) 1995. *Man and Sea in the Mesolithic. Coastal Settlement Above And Below Present Sea Level.* Oxbow Books: Oxford.

Flemming, N. C. 1963. Underwater caves of Gibraltar. In Eaton, B. (ed.) *Undersea Challenge: a Report of the Proceedings of the Second World Congress of Underwater Activities.* pp. 96-106. British Sub-Aqua Club: London.

Flemming, N. C. 1968. Holocene earth movements and eustatic sea-level change in the Peloponnese. *Nature* 217:1031-1032.

Flemming, N. C. 1983. Survival of submerged lithic and Bronze Age artefact sites: a review of case histories. In Masters, P. & Flemming, N. C. (eds.) *Quaternary Coastlines and Marine Archaeology.* pp. 135-174. Academic Press: London and New York.

Flemming, N. C. (ed.) 2004. *Submarine Prehistoric Archaeology of the North Sea.* CBA Research Report 141. Council of British Archaeology: York.

Gagliano, S. M., Pearson, C. E., Weinstein, R. A., Wiseman, D. E. & McClendon, C. M. 1982. *Sedimentary Studies Of Prehistoric Archaeological Sites: Criteria For The Identification Of Submerged Archaeological Sites Of The Northern Gulf of Mexico Continental Shelf.* Coastal Environments Inc: Baton Rouge.

Galili, E. & Weinstein-Evron, M. 1985. Prehistory and paleoenvironments of submerged sites along the Carmel coast of Israel. *Paléorient* 11:37-52.

Galili, E., Weinstein-Evron, M., Hershkovitz, I. *et al.* 1993. Atlit-Yam: A prehistoric site on the sea floor off the Israeli coast. *Journal of Field Archaeology* 20:133-157.

Harff, J. & Lüth, F. (eds.) 2007. *SINCOS — Sinking Coasts: Geosphere, Ecosphere and Anthroposphere of the Holocene Southern Baltic Sea.* Bericht der Römisch-Germanische Kommission: Special Issue 88.

Harff, J. & Lüth, F. (eds.) 2011. *SINCOS II — Sinking Coasts: Geosphere, Ecosphere and Anthroposphere of the Holocene Southern Baltic Sea II.* Bericht der Römisch-Germanische Kommission: Special Issue 92.

Harff, J., Lemke, W., Lampe, R. *et al.* 2007. The Baltic Sea coast — A model of interrelations among geosphere, climate, and anthrosphere. *Geological Society of America Special Papers* 426:133-142.

Jöns, H., Mennenga, M. & Schaap, D. 2016. *The SPLASHCOS-Viewer. A European Information system about submerged prehistoric sites on the continental shelf. A product of the COST Action TD0902 SPLASHCOS (Submerged Prehistoric Archaeology and Landscapes of the Continental Shelf).* Accessible at splashcos-viewer.eu/

Long, D., Wickham-Jones, C. R. & Ruckley, N. A. 1986. A flint artefact from the northern North Sea. In

Roe, D. A. (ed.) BAR International Series (No. 296) *Studies in the Upper Palaeolithic of Britain and North West Europe*. pp. 55-62. British Archaeological Reports: Oxford.

Louwe Kooijmans, L. P. 1970-71. Mesolithic bone and antler implements from the North Sea and from the Netherlands. *Berichten van de Rijksdienst voor het Oudheidkundig Bodemonderzoek* 20-21:27-73.

Millard, A. R. 2014. Conventions for reporting radiocarbon determinations. *Radiocarbon* 56:555-559.

Millero, F. J., Feistel, R., Wright, D. G. & McDougall, T. J. 2008. The composition of Standard Seawater and the definition of the Reference-Composition Salinity Scale. *Deep-Sea Research Part I: Oceanographic Research Papers* 55:50-72.

Momber, G. 2000. Drowned and deserted: a submerged prehistoric landscape in the Solent, England. *International Journal of Nautical Archaeology* 29:86-99.

Moree, J. M. & Sier, M. M. (eds.) 2014. *Twintig meter diep! Mesolithicum in de Yangtzehaven — Maasvlakte te Rotterdam. Landschapsontwikkeling en bewoning in het Vroeg Holoceen. BOORrapporten 523.* Bureau Oudheidkundig Onderzoek Rotterdam (BOOR): Rotterdam.

Moree, J. M. & Sier, M. M. (eds.) 2015. Twenty metres deep! The Mesolithic Period at the Yangtze Harbour site — Rotterdam Maasvlakte, the Netherlands. Early Holocene landscape development and habitation. In Moree, J. M. & Sier, M. M. (eds.) *Interdisciplinary Archaeological Research Programme Maasvlakte 2, Rotterdam. BOORrapporten 566.* pp. 7-350. Gemeente Rotterdam / Rijksdienst voor het Cultureel Erfgoed & Port of Rotterdam: Rotterdam.

Pedersen, L. D., Fischer, A. & Aaby, B. (eds.) 1997. *The Danish Storebaelt since the Ice Age: Man, Sea and Forest.* Storebaeltsforbindelsen: Copenhagen.

Peeters, H., Murphy, P. & Flemming, N.C. (eds.) 2009. *North Sea Prehistory Research and Management Framework (NSPRMF) 2009.* English Heritage and RCEM: Amersfoort.

Reimer. P. J., Bard, E., Bayliss, A. *et al.* 2013. INTCAL13 and MARINE13 Radiocarbon age calibration curves 0-50,000 Cal BP. *Radiocarbon* 55:1869-1887.

Stuiver, M. & Polach, H. A. 1977. Discussion: Reporting of ^{14}C data. *Radiocarbon* 19:355-363.

TRC Environmental Corporation 2012. *Inventory and Analysis Of Archaeological Site Occurrence On The Atlantic Outer Continental Shelf.* US Dept. of the Interior: New Orleans.

van Kolfschoten, T. & Vervoort-Kerkhof, Y. 1985. Nijlpaarden van Nederlandse bodem en uit Noordzee. *Cranium* 2:36-44.

Waters, M. R. 1992. *Principles of Geoarchaeology: A North American perspective.* University of Arizona Press: Tuscon.

Weerts, H., Otte, A., Smit, B. *et al.* 2012. Finding the needle in the haystack by using knowledge of Mesolithic human adaptation in a drowning delta. In Bebermeier, W., Hebenstreit, R., Kaiser, E. & Krause, J. (eds.) *Landscape Archaeology — Proceedings of the International Conference (Special Volume 3).* 6th – 8th June, Berlin, pp. 17-24.

Wessex Archaeology 2011. *Seabed Prehistory: Site Evaluation Techniques (Area 240) Synthesis.* Wessex Archaeology: Salisbury.

Wright, D. G., Pawlowicz, R., McDougall, T. J., Feistel, R. & Marion, G. M. 2011. Absolute Salinity, "Density Salinity" and the Reference-Composition Salinity Scale: present and future use in the seawater standard TEOS-10. *Ocean Science* 7:1-26.

Chapter 2
Sea Level and Climate

Jan Harff,[1] Nicholas C. Flemming,[2] Andreas Groh,[3] Birgit Hünicke,[4] Gilles Lericolais,[5] Martin Meschede,[6] Alar Rosentau,[7] Dimitris Sakellariou,[8] Szymon Uścinowicz,[9] Wenyan Zhang[10] and Eduardo Zorita[4]

[1] University of Szczecin, Szczecin, Poland
[2] National Oceanography Centre, Southampton, UK
[3] Helmholtz Centre Potsdam, GFZ German Research Centre for Geosciences, c/o DLR Oberpfaffenhofen, Wessling, Germany
[4] Helmholtz Centre Geesthacht, Geesthacht, Germany
[5] IFREMER, DAEI, Issy-les-Moulineaux, France
[6] University of Greifswald, Greifswald, Germany
[7] Department of Geology, University of Tartu, Tartu, Estonia
[8] Institute of Oceanography, Hellenic Centre for Marine Research, Anavyssos, Greece
[9] Polish Geological Institute — National Research Institute, Branch of Marine Geology, Gdańsk, Poland
[10] MARUM — Center for Marine Environmental Sciences, University of Bremen, Bremen, Germany

Introduction

After two decades of research on the archaeology, climate and environment of the drowned landscapes of local and subregional parts of the European continental shelf, one of the first tasks of the SPLASHCOS Action was to attempt a comprehensive and integrated view of the shelf and its marginal seas, including the human component. The way in which the changing conditions on the continental shelf determined where humans lived, and the survival or destruction of their archaeological remains, will be discussed in later chapters. However, everything that follows is determined and influenced by the changing climate and changing sea level, and this will be analyzed first.

The European continental shelf was subject to dramatic environmental changes after the last glaciation. Continental ice sheets disappeared and melt water and new drainage systems shaped the periglacial landscape. Dammed meltwater lakes filled morphological depressions. Sea level rose and the transgression flooded large parts of the continental shelf, turning former freshwater lakes into brackish-marine ones. In this chapter we will give an overview of the environmental changes to the European shelf and its marginal seas during the Late Pleistocene to the Middle Holocene. The driving forces of these environmental changes are complex and interdependent. The most important variables are climatically-controlled eustatic sea-level change and vertical dislocation of the Earth's crust. Eustatic change is to be regarded mainly as a function of the change of ice/water volume during the glacial cycles and of steric effects (thermal expansion or contraction of sea water due to changing temperature) during the Holocene. Vertical crustal movement expresses variable rates of subsidence and uplift induced by lithospheric dynamics, as well as more localized tectonic movements. Superimposed on these effects of internal geodynamic forces (expressed within the tectonic regionalization of Europe) are external forces: loading and unloading of the crust by ice and water. These changes in load have to be considered as an effect of global climate. Climatic fluctuations in the Earth's history cause accumulation of inland ice shields during cold (glacial) periods and a draining of melt water to marine basins during warm (interglacial) periods. In the context of relative sea-level change, depending on the tectonic setting, but also on the geographic position, three subregions can be distinguished for Europe (Fig. 2.1): the Baltic basin, the North Sea and Atlantic shelf, and the Mediterranean basin together with the Black Sea.

The first section of this chapter is devoted to the regional tectonics of Europe. The age of the consolidation of the basement together with the plate tectonic setting will serve here as the main parameters determining coastal formation.

The next section reviews the fluctuation of glacial and interglacial stadia as an effect of orbital parameters

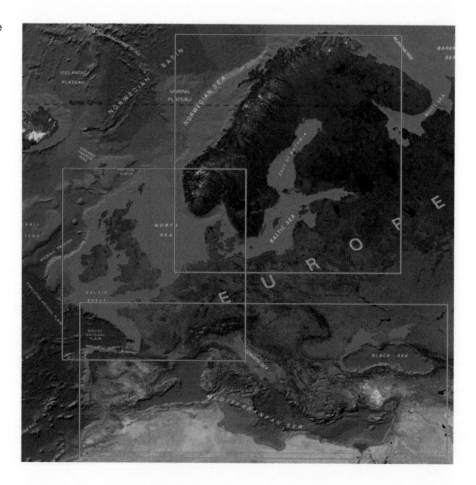

Fig. 2.1 Geographical overview of Europe including shelf areas, and marginal sea sub-regions to be treated separately in this chapter. Map credit GEBCO.

of the Earth around the Sun. These processes are interpreted as driving forces of ice-sheet dynamics and eustatic fluctuations. We will examine, in particular, eustatic change during the Last Glacial Cycle (LGC). As recent human populations have been mainly affected by postglacial sea-level rise and isostatic effects of the decaying ice sheet, the last deglaciation will be highlighted in particular. We present a regional model, which classifies European coastal regions according to whether they are regions of relative sea-level rise (sea transgression) or sea-level fall (regression of the sea). In isostatically and tectonically stable or subsiding coastal areas the postglacial paleogeographic history is mainly a result of sea transgression. Only if the rate of uplift of the crust exceeds the eustatic sea-level rise does a regression of the sea take place, resulting in the emergence of land formerly covered by the sea. This interplay of loading and its effect on relative sea-level change will be explained by a numerical model, before addressing regional paleogeographic reconstructions.

The next sections describe the development of the Baltic Sea, the North Sea and the Atlantic shelf, based on numerical sea-level scenarios constrained by observational data. In semi-enclosed and partly enclosed basins such as the Baltic Sea and the Black Sea basins the models cannot be linked to global sea-level scenarios. Here the water level is controlled by the drainage of the surrounding land including the inflow of melt water from decaying ice sheets. Therefore, the subregional paleogeographic reconstructions given for these regions are derived from the interpretation of sedimentological proxy data and mapping of paleoshorelines.

In areas of inundation the migrating shoreline reworks the surface of the paleolandscape. For times of relatively rapid sea-level rise — as during the Late Pleistocene and Early Holocene — we find that shorelines tend to preserve their former shape better than in periods of slow sea-level rise (Late Holocene) when hydrographic forcing reworked intensely the substrate of the coastal zone and eventually destroyed many archaeological sites.

For integrated modeling approaches of these processes aimed at predicting where human settlements are likely to have been preserved, information about the atmospheric forcing of coastal dynamics is needed. The data can

be derived from long-term climate modeling. In the last section of this chapter, we explain the basics of the ECHO-G global climate model, together with an overview of Late Holocene changes in European climate with special attention to wind forces. In the second part of this section, we give an overview of long-term coastal morphogenesis, using the southern Baltic Sea sandy barrier coast as an example.

Tectonic and Geological Setting of Europe with Special Reference to the Shelf and Marginal Seas

The continental crust of Europe can be subdivided into four different structural units which differ in age and tectonic history (Fig. 2.2). The oldest complex is the East European Platform representing an old cratonic structure mainly built in Proterozoic times. This complex formed the continent 'Baltica', which is taken as an independent tectonostratigraphic unit in most paleogeographic reconstructions of the Late Proterozoic and Early Paleozoic (e.g. Lawver *et al.* 2009). Various Phanerozoic orogenic events added crustal provinces to this continental plate and it became part of the Eurasian plate. The crustal provinces get younger in sequence from north to south.

Proto-Europe

The oldest crustal units of Europe belong to the East European Platform which comprises the Baltic Shield, the Russian Platform and the Ukrainian Shield (Fig. 2.2). These units mainly consist of highly metamorphosed and igneous Precambrian rocks. The southern boundary of Baltica is clearly confined by the Tornquist-Teisseyre Zone running in a southeast–northwest direction from the Carpathians into the area of the North Sea. In the northwestern part, this zone is split into a northern branch, named the Sorgenfrei-Tornquist Zone, and a southern branch. This zone is one of the most important suture zones in Europe.

Fig. 2.2 Tectonostratigraphic units of Europe. Red colors = Precambrian basement; green colors = Caledonian orogeny; blue colors = Variscan orogeny; yellow colors = Alpine orogeny: MKZ = Mid-German Crystalline Zone. From Meschede (2015), modified and supplemented after Kossmat (1927), Franke (1989), Martínez Catalán et al. (2007) and others.

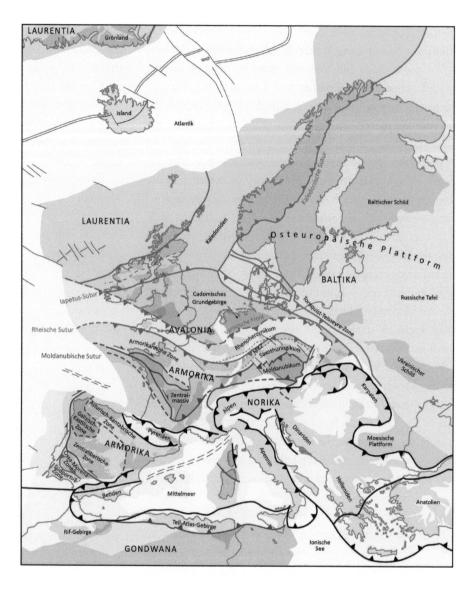

Paleo-Europe

The northwestern boundary of the East European Platform is marked by the mountain range of the Caledonides (*Caledonia* [lat.] — Scotland) which expands from Norway to Scotland and Ireland (Fig. 2.2). The other side of this mountain range abuts against the North America-Greenland Platform which is part of the old continent Laurentia (named after the Saint Lawrence River in North America). A small remnant of Laurentia is still located in the north-westernmost corner of Scotland. However, because of Cenozoic rifting and opening of the Atlantic Ocean, the predominant part of Laurentia is today situated on the American side of the Atlantic Ocean.

Baltica and Laurentia collided in Ordovician and Silurian times forming the Caledonian mountain range. Between the two continents the Iapetus Ocean opened in the Late Precambrian and was completely subducted by the end of the Ordovician. Today a deep crustal level of the collisional orogen is exposed because of intense erosion and isostatic rebound since the Caledonian orogeny. A small continental block named Avalonia (after the peninsula of Avalon on the Canadian island, Newfoundland) was accreted as a terrane to Baltica during the Ordovician before the collision of Laurentia and

Baltica. Between Baltica and Avalonia, the Tornquist Ocean, a small lateral branch of the Iapetus Ocean, was closed. Both Baltica and Avalonia collided in the Ordovician as a unit forming with Laurentia the new and bigger continent Laurussia. Avalonia had originally been part of Gondwana. It was separated from this large continental mass in the south during the Early Paleozoic and subsequently migrated towards the north as a terrane. As the result of collision with a continent, it accreted to the continental margin. Avalonia comprises Newfoundland located today on the American side of the Atlantic Ocean, the southern parts of Great Britain and Ireland, the basement of northern Germany today covered by younger sediments, and the Variscan units of the Rhenohercynian (Rhine–Harz mountains zone). Avalonia is connected to Baltica along the Tornquist-Teisseyre Zone.

The Baltic Sea is not an ocean in the sense of plate-tectonic definitions. It is completely underlain by continental crust which mostly belongs to the East European Platform. The maximum depth of the Baltic Sea is about 460 m (Landsort Deep) compared to more than 5000 m in the oceanic crust realms of the Mediterranean Sea and more than 2000 m in the Black Sea. The area south of the Tornquist-Teisseyre Zone belongs to the continental crust added to Europe during the Caledonian orogeny.

The most recent tectonic movements in the Baltic Sea region were induced by the glaciation of the Northern Hemisphere in the Quaternary. The repeated emergence and removal of a kilometer-thick ice cover led to isostatic rebound, which triggered earthquakes mainly occurring at normal faults. However, the strike-slip characteristics of recent earthquakes shown by fault-plane-solutions (e.g. the 2008 earthquake of Malmö, Sweden, magnitude 4.7) may indicate a reactivation of faults in the Teisseyre-Tornquist Zone.

Meso-Europe

The Variscan (or Hercynian) mountain range comprises the southern part of Avalonia and large parts of Armorica, an extensively-stretched terrane composed of several tectonic units and different rocks of various ages. These units are Iberia, the Armorican Massif, the French Central Massif, most parts of the Vosges and the Black Forest, and the Bohemian Massif as well as the Saxothuringian unit in the sense of Kossmat (1927).

The Variscides are a multiply subdivided mountain range across Europe which was mainly formed in Devonian and Carboniferous times. As with the Caledonides, only a part of this mountain range is located today in Europe while the other, bigger part continues as the Appalachian range on the western side of the Atlantic Ocean. In Middle Europe the Variscides are subdivided into three units, from north to south the Rhenohercynian, the Saxothuringian, and the Moldanubian (Kossmat 1927). The northernmost part belongs to Avalonia, but the main portion of the Variscides is part of Armorica which was also once connected to Gondwana. Armorica rifted away from Gondwana during the Silurian and migrated independently towards the north. During the Variscan orogeny, Armorica was welded to Avalonia, which at that time was already part of Laurussia. The Rheic Ocean between Avalonia and Armorica was closed during the Variscan orogeny. Sediments of the Rhenohercynian were deposited at the northern margin of this ocean where an area with thinned continental crust was formed at the passive continental margin of Laurussia (former Avalonia). The Rheic Ocean is subducted beneath Armorica. The volcanic arc at the northern boundary of Armorica belongs to the Saxothuringian. It is represented today in the Mid-German Crystalline Zone MGCZ (marked MKZ on Fig. 2.2). Sediments of the Saxothuringian are made up of similar sedimentary facies as the Rhenohercynian. The Moldanubian Zone is characterized by highly metamorphosed basement rocks and a long-lasting Cambrian to Devonian carbonate platform (named the Teplá–Barrandian). Both the Saxothuringinan and Moldanubian are part of Armorica, which is bordered in the south by the Moldanubian suture zone. In the south and south-east, some units exist which probably did not disconnect completely from Gondwana during the Variscan orogeny. A part of these terranes with strong affinity to Gondwana is the Alpine terrane Norica. The formation of the Variscan mountain range marks the combination of all the large continental lithospheric plates into the supercontinent Pangea.

Neo-Europe

The youngest affiliation of crustal provinces to the Eurasian continent was the Alpine mountain range with its southern-associated microplates. However, the interpretation, number and shape of the microplates involved in the Alpine orogeny is still under debate. In this overview, the Adriatic–Apulian plate and the Anatolian plate are subdivided. Besides the Alps, the mountain range formed during the Alpine orogeny comprises also the Carpathians, the Dinarides, the Hellenides, the Apennines, the Tell-Atlas mountains, the Betic Cordillera and the Pyrenees (Fig. 2.2). The Alpine orogeny started at the end of the Cretaceous and is still not completed.

The Mediterranean is thus a remnant ocean in its eastern part, where the oldest oceanic crust which is still in place can be observed (Müller *et al.* 2008). The oceanic crust was formed in the Tethys Ocean during the Triassic. The western part of the Mediterranean consists of young oceanic crust which formed when Corsica and Sardinia rifted away from Europe during the Miocene. The Black Sea also contains part of the trapped oceanic crust from the Tethys Ocean. With an age of Early Cretaceous it is slightly younger than in the eastern Mediterranean.

Quaternary Climate and Sea-level Change

The global evolution of climate during the Quaternary has been analyzed and described by numerous authors, among them Brooks (1926), Milankovitch (1930), Crowley and North (1991) and Ruddiman (2008). In this section we review mainly the results published by Rahmstorf (2002), Plant *et al.* (2003), Carlson (2011), Wanner *et al.* (2011) and Grant *et al.* (2012).

The climate history of the Quaternary is considered an alternation of glacial and interglacial periods. Milankovitch (1930) described mathematically the periodical variation of the Earth's orbital parameter around the Sun as the cause of changes in insolation and consequently external energy supply to the Earth.

According to this theory (Berger & Loutre 1991) it is assumed that cyclic changes in boreal summer insolation are responsible for the cyclicity in the build-up of continental ice shields during the glacial periods and their interglacial decay. This process leads to a cyclic redistribution of water from the ocean to the continents where it is stored during the glacial periods as ice, and from where it returns by melting during the interglacials to the oceanic (and terrestrial) basins. This interdependence between boreal insolation, continental ice volume and global sea level may — according to Carlson (2011) — be further amplified, since ocean circulation and atmospheric CO_2 concentration may also be influenced by the Milankovitch Cycles (compare Fig. 2.3). Atmospheric temperature reconstructions are mainly based on oxygen isotope proxies from Antarctic

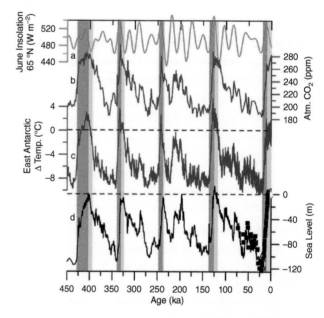

Fig. 2.3 Climate and sea level of the last 450 kyr; (a) June insolation at 65°N (Berger & Loutre 1991); (b) Atmospheric CO_2 concentrations from Antarctic ice cores (Siegenthaler et al. 2005); (c) East Antarctic change in temperature (dashed line = present-day temperature) after Jouzel et al. (2007); (d) Sea-level records from the Red Sea (black line), black squares = individual sea-level estimates, dashed line = present-day sea level (Clark et al. 2009; Rohling et al. 2009), gray bars = deglaciations; yellow bars = interglaciations. Carlson (2011). Reproduced with permission from Nature Education.

and Greenland ice cores, whereas sea-level changes can be traced back by radiometrically-dated corals (Lambeck & Chappell 2001; Waelbroeck *et al.* 2002) or changes in the isotopic signature of planktonic foraminifera shells of the semi-enclosed basins (Red Sea, Mediterranean Sea) reflecting changes in paleo-salinities because of periodic isolation from ocean circulation (Siddall *et al.* 2004; Rohling *et al.* 2009; Grant *et al.* 2012).

In particular, studies of oxygen isotopes of Red Sea and Mediterranean Sea foraminifera have led to a detailed picture of global sea-level change during the LGC. Grant *et al.* (2012) have developed a probabilistic model of the LGC sea-level change based on a statistical analysis of empirical data including an estimation of errors from different sources. The authors show that the timing of ice-volume fluctuations (within a centennial response time) correlates well with the Antarctic and Greenland climate. In Fig. 2.4 the relative sea-level probabilistic model (RSL$_{Pmax}$), RSL data and the first derivative (as an expression of sea-level change) are depicted. Time spans of rapid sea-level changes (rise) are marked by red arrows. It is remarkable that rates of sea-level rise during

ice-volume reduction phases have reached more than 1.2 m per century.

The sea-level curve in Fig. 2.4 reveals a cyclic pattern of sea-level dynamics during the last glaciation which can be roughly correlated with the boreal summer insolation (Berger & Loutre 1991) supporting the theory of a cause-and-effect relation, but climate variations in higher resolution need a more complex interpretation. In Fig. 2.5, sea surface temperature (SST) as reconstructed from sediment proxies for the southern Atlantic Ocean and oxygen isotope signatures for the Greenland ice for Marine Isotope Stage 3 (MIS 3) are depicted. There is obviously a cyclic change between warmer and colder periods, whereby the driving forces have to be sought in the nonlinear processes of ocean circulation (Rahmstorf 2002). The appearance of warm events within the Dansgaard-Oeschger (D-O) fluctuations during the LGC follows according to Alley *et al.* (2001) a periodicity of ca. 1500 years (and its multiples 3000 and 4500 years) which is mainly explained by changes in the North Atlantic overturning. Heinrich events play the role of the cold counterparts of D-O events with an irregular

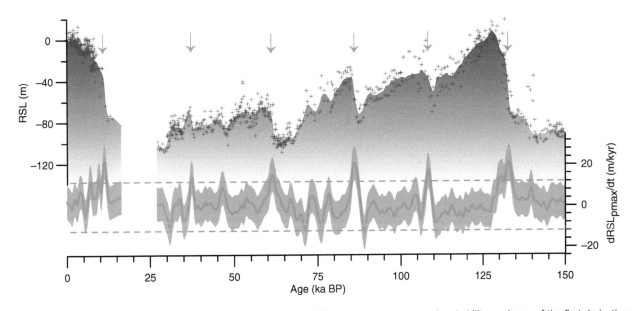

Fig. 2.4 Stochastic relative sea-level model (RSL$_{Pmax}$, gray shading), RSL data (blue crosses) and probability maximum of the first derivative of RSL (red) with 95% confidence interval (pink shading). Rates of sea-level change of +12 m/kyr and −12 m/kyr are indicated (dashed lines). Red arrows mark peaks in sea-level rises of more than 12 m/kyr. Grant, K. M., *et al.* (2012). Reproduced with permission of McMillan Publishers Ltd.

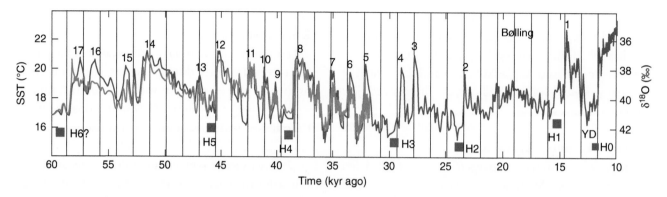

Fig. 2.5 Temperature reconstructions from ocean sediments and Greenland ice. Proxy data from the subtropical Atlantic (green) and from the Greenland ice core GISP2 (blue) show several Dansgaard-Oeschger (D-O) warm events (numbered). The timing of Heinrich events is marked in red. Gray lines at intervals of 1,470 years illustrate the tendency of D-O events to occur with this spacing, or multiples thereof. Rahmstorf (2002). Reproduced with permission from MacMillan Publishers Ltd.

return period of about 10,000 years. Ice rafted debris (IRD) in the marine North Atlantic sediments points to lines of evidence for massive iceberg discharges from the Laurentian ice sheet during these cold periods. The occurrence is mainly explained by a shift in the circulation pattern of the northern Atlantic Ocean so that the North Atlantic Deep Water (NADW) formation is interrupted and Antarctic water advances into the entire deep North Atlantic basins.

During the time span between 26,500 and 19,000–20,000 years ago, the last glaciation reached its maximum in continental ice extension. This time span — the Last Glacial Maximum (LGM) — is linked to an extreme impact of ice sheets on the Earth's climate and surface morphogenesis, resulting in extreme droughts, desertification, and exposure of most of today's continental shelf area due to a dramatic fall in sea level.

After the LGM, climatic change brought about the last deglaciation, extending from 20 ka BP to 6 ka BP whereby the retreat of all Earth's ice sheets caused a sea-level rise of 125 m to 130 m (Fig. 2.6). Figure 2.6f depicts the changes in sea level with extreme values involving many meters of sea-level rise over time spans of centuries. The first increase in the rate of sea-level rise at 20 ka BP to 19 ka BP to ~1.4 cm/year followed shortly after the initial rise in boreal summer insolation (Fig. 2.3a) (Yokoyama *et al.* 2000; Clark *et al.* 2009). The increase in the rate of sea-level rise — Meltwater Pulse (MWP) 1A

(Fairbanks 1989) — correlates roughly with warming of the North Atlantic region during the Bølling warm period ~14.6 ka BP (Fig. 2.6b). During MWP-1A, with values of about 2.8 cm/year, the rate of sea-level rise reached its maximum of the last deglaciation. During the Younger Dryas stadial, cold climatic conditions and droughts returned for a relatively brief period, also referred to as the 'Big Freeze', between 12.9 ka BP and 11.7 ka BP, influenced obviously by the collapse of the North American ice sheets. The Younger Dryas cooling stadial is reflected by a deceleration of sea-level rise before another acceleration starting around 11.7 ka BP together with the Holocene warming. During the Holocene, the climate fluctuated between positive and negative temperature and precipitation anomalies. Based on IRD investigation of North Atlantic sediments, Bond *et al.* (1997; 2001) claim the appearance of a quasi-periodic cycle of cold events during the Holocene with a period of 1470± 500 years, similar to the period of D-O events. These Bond-Cycles, numbered from 0 to 9, are believed to be the result of insolation being driven southward, and eastward advection of cold, ice-bearing surface waters from the Nordic and Labrador seas, but discussions about a theory for the cause of Bond-Cycles are still ongoing. In order to gain a deeper understanding of these climate fluctuations, Wanner *et al.* (2011) analyzed a global set of Holocene time series from paleoclimate onshore proxy data. The authors identified six specific cold events (8200, 6300, 4700, 2700, 1550

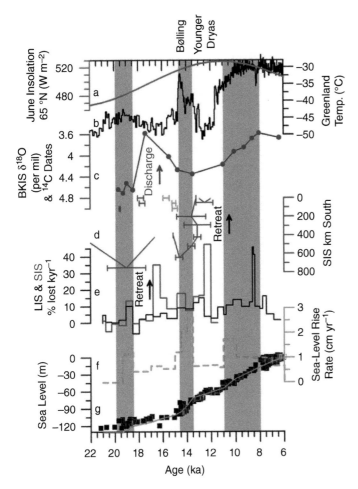

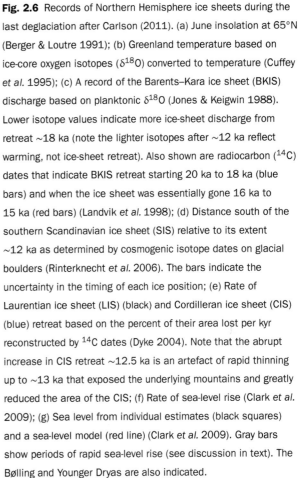

Fig. 2.6 Records of Northern Hemisphere ice sheets during the last deglaciation after Carlson (2011). (a) June insolation at 65°N (Berger & Loutre 1991); (b) Greenland temperature based on ice-core oxygen isotopes ($\delta^{18}O$) converted to temperature (Cuffey *et al.* 1995); (c) A record of the Barents–Kara ice sheet (BKIS) discharge based on planktonic $\delta^{18}O$ (Jones & Keigwin 1988). Lower isotope values indicate more ice-sheet discharge from retreat ~18 ka (note the lighter isotopes after ~12 ka reflect warming, not ice-sheet retreat). Also shown are radiocarbon (^{14}C) dates that indicate BKIS retreat starting 20 ka to 18 ka (blue bars) and when the ice sheet was essentially gone 16 ka to 15 ka (red bars) (Landvik *et al.* 1998); (d) Distance south of the southern Scandinavian ice sheet (SIS) relative to its extent ~12 ka as determined by cosmogenic isotope dates on glacial boulders (Rinterknecht *et al.* 2006). The bars indicate the uncertainty in the timing of each ice position; (e) Rate of Laurentian ice sheet (LIS) (black) and Cordilleran ice sheet (CIS) (blue) retreat based on the percent of their area lost per kyr reconstructed by ^{14}C dates (Dyke 2004). Note that the abrupt increase in CIS retreat ~12.5 ka is an artefact of rapid thinning up to ~13 ka that exposed the underlying mountains and greatly reduced the area of the CIS; (f) Rate of sea-level rise (Clark *et al.* 2009); (g) Sea level from individual estimates (black squares) and a sea-level model (red line) (Clark *et al.* 2009). Gray bars show periods of rapid sea-level rise (see discussion in text). The Bølling and Younger Dryas are also indicated.

and 550 BP). These events are plotted in Fig. 2.7 together with Bond events 0 to 6. Looking for the reasons of the cooling events, Wanner *et al.* (2011) mention a complex system of driving forces including meltwater flux into the North Atlantic, low solar activity, explosive volcanic eruptions, fluctuations of the thermohaline circulation, and internal dynamics in the North Atlantic and Pacific area.

The influence of climatic fluctuation on eustatic sea-level during the Holocene is a matter of debate and depends on the resolution of sea-level reconstructions. It is generally accepted that Holocene warming caused a

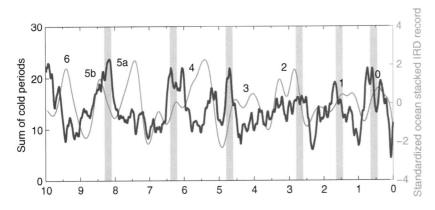

Fig. 2.7 Weighted and smoothed curve representing the sum of cold periods (x-axis: time in ka BP). Red curves with the numbers in each figure show the standardized ocean-stacked ice-rafted debris (IRD) record representing the Bond events no. 0–6 (Bond *et al.* 2001). The six vertical blue bars denote the time of the cold events after Wanner *et al.* (2011: fig. 8).

continuous slow sea-level rise. The final stages of deglacia-tion during the Early Holocene caused an increase in the mass of water in the sea, while the slower sea-level rise after ca. 6000 BP is predominantly the result of thermal expansion of the marine water mass. On the regional to local scale, the sea level has to be regarded as the relative change of the water level as a result of the interaction of eustatic changes and glacio-isostatic adjustment (GIA). A special phenomenon is the formation of dammed meltwater reservoirs in front of the retreating ice sheets. The water level of these basins may rise rapidly, but may also fall dramatically because of sudden drainages. This phenomenon together with the effects of glacio-isostatic and hydro-isostatic effects will be discussed in the following sections.

The Quaternary and European Sedimentary Environments

The landscape of northern Europe was generally shaped by the Quaternary glaciations. Glaciers carved the Earth's surface, and enormous masses of glacial, glacio-fluvial, eolian, and also organic sediments were transported and deposited during the Quaternary. Most of the recent drainage system of Europe can be traced back to glacial or interglacial periods. Figure 2.8a shows the extent of the four European ice sheets (Weichselian Glaciation, Warthe phase of the Saalian Glaciation, Saalian Glaciation, and Elsterian Glaciation).

The change between glacial and interglacial periods led to a change of load on the continents caused by the build-up and decay of ice masses, resulting in local depression and rebound of the Earth's crust, whilst the global sea level dropped and rose according to changes in water volume. After the LGM, these processes caused 'drowned' fjord coastlines and uplifted beaches along the northern European coasts (the degree of drowning or uplift related to GIA effects), and the formation of the Irish Sea, the present North Sea, the English Channel, and the tilting of the Baltic Sea basin. During the ice advances from the north-east, the European drainage system — directed originally to the

North Sea basin — was reorganized. Ice, wind and fluvial transport caused a general southward-directed redistribution of fluvial, glacial and eolian sediments. This process of re-organization of sediment distribution changed cyclically with the alternations from glacial to interglacial environment and vice-versa.

Four broad environmental domains, characterized by their depositional histories, can be distinguished in Europe according to Plant et al. (2003) in Fig. 2.8b. The source region for glacial sediments — the Fennoscan-dian region — underwent a net loss of material by glacial scouring. The environment is determined here by crystalline bedrocks, glacial psammites and tills, U-shaped valleys, fjords and lakes. Much of northwest and central Europe, south to 50°N, is to be regarded as a glacial sink of sediments consisting of glacial tills, glacio-fluvial and eolian sediments. South of 50°N, fluvial sediments and loess rest on an older pre-Quaternary basement. Regionally, the Alpide orogenic belt (Alps and Pyrenees), hosted glacial ice sheets and interrupted the general European pattern of the Quaternary sedimentary environment. Fluvial and mass movement sediments prevail in the Mediterranean area.

Glacio-isostatic Adjustment and Relative Sea-level Modeling

The solid Earth deforms under the influence of changing loads on its surface. These deformations occur on different timescales and originate from different regions of the Earth's interior. The timescales vary from instantaneous to response times of several thousand years, depending on the period for which the load was acting on the Earth. Instantaneous, elastic reactions are conducted by the lithosphere as well as the Earth's mantle. Time-dependent viscous reactions originate solely from the mantle and correspond to the redistribution of viscous mantle material beneath the load. Hence, as a good approximation, the viscoelastic solid Earth can be modeled as a Maxwell body, which is the combination in series of an elastic (Hooke) model and a viscous (Newton) model. The former can be represented

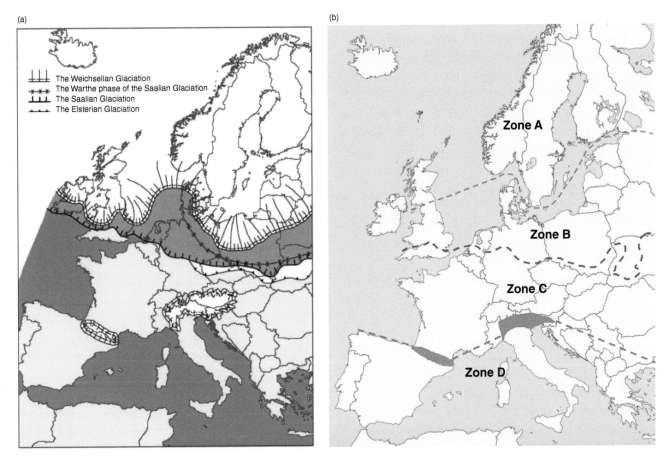

Fig. 2.8 Scandinavian ice-sheet dynamics with special respect to the Weichselian Glaciation Regional Quaternary depositional environments of Europe: (a) Maximum extent of the European ice sheets: the Weichselian Glaciation, the Warthe phase of the Saalian Glaciation, the Saalian Glaciation, and the Elsterian Glaciation (from Anderson & Borns 1997); (b) Depositional environments: Zone A – Pre-Quaternary bedrock, rock-, sand- and gravel-based tills, hummocky topography, U-shaped valleys, fjords and lakes; Zone B – complex of interbedded loessic eolian sands, soils, tills, and fluvial glacial sediments; Zone C – periglacial zone with fluvial glacial outwash sediments, loess and periglacial weathering; Zone D – fluvial and mass-movement deposits; gray shaded areas = upland glaciations. Modified from Plant *et al.* (2003: fig. 2).

by a spring while the latter can be illustrated by a dashpot.

Surface loads vary on a wide range of timescales. For example, tidally-induced redistributions of ocean water or changing atmospheric loads vary with frequencies of hours to several days and will only induce elastic deformations in the earth. In contrast, long-term changes of surface loads, such as during the waxing or waning of ice sheets, will induce both elastic and viscous deformation. These deformations are isostatic and will continue until the Earth has returned to a state of equilibrium. If the Earth's response is induced by glacial

load changes it is referred to as glacio-isostatic adjustment (GIA).

The deglaciation of an ice sheet invokes a more complex response of the Earth than a purely viscoelastic deformation, however. In addition, on a global scale, average sea level will rise due to the meltwater influx (eustatic sea-level change). But the mean sea surface is an equipotential surface of the Earth's gravity field. Hence, changes in the Earth's gravity potential due to the decreasing ice mass and the redistributed mantle material will cause a highly non-uniform change in sea level. Moreover, the relocated water mass is another surface load which induces additional crustal deformation and an additional deformation potential. A gravitationally self-consistent description of these interactions is given by the sea-level equation (SLE). The basic idea of the SLE dates back to the work of Woodward (1888), while the SLE was formulated by Farrell and Clark (1976). According to Peltier (1998) (see also Groh *et al.* (2011)) the SLE reads as:

$$\delta S(\theta, \lambda, t) = C(\theta, \lambda, t) \cdot \{\delta G(\theta, \lambda, t) - \delta R(\theta, \lambda, t)\}$$

$$= C(\theta, \lambda, t) \cdot \left\{ \int \left[\int \int (g^{-1}\Phi(\psi, t - t') \right. \right.$$
$$\left. \left. - \Gamma(\psi, t - t')) \cdot L(\theta', \lambda', t')d\sigma \right] dt' \right.$$
$$\left. + g^{-1}\Delta\Phi(t) \right\} \tag{1}$$

At a given position (co-latitude: θ, longitude: λ) and time (t) the change in RSL (δS), which is the sea level relative to the Earth's crust, is the difference between the change of the geoid (δG) and the vertical crustal deformation (δR). The geoid is a surface of constant gravitational potential and corresponds to the mean sea level. $C(\theta,\lambda,t)$ is the so-called ocean function which is unity over the ocean and zero over land. In general, C is time-dependent as well since land areas can subside below sea level or ocean regions can be uplifted above sea level while an ice sheet melts or aggregates. In more detail, the SLE reveals that δG and δR result from the temporal and spatial convolution of the Green functions for the gravitational potential Φ and for the vertical deformation Γ with the load L. The conservation of the total mass of ice and water is assured by the purely time-dependent term $g^{-1} \delta\Phi(t)$ where g is the gravitational acceleration. The Green functions depend on the spherical distance ψ between the position under consideration (θ,λ) and the load element (θ',λ') as well as on the viscoelastic properties of the Earth. The load distribution L incorporates load changes of the ice and of the changing sea level δS itself. Hence, this linear integral equation has to be solved iteratively. In this basic form the SLE does not account for effects induced by changes in the rotational feedback of the Earth.

Thus, modeling of GIA-induced phenomena, such as present-day RSL changes or the temporal evolution of RSL, by solving the SLE, requires several different input data sets. First, an ice-load history describing the temporal and spatial evolution of the ice load is needed. Second, a viscoelastic Earth model is required to solve for the Green functions. Ice-load histories are available on global and regional scales and are reconstructed from geological, geomorphological and archaeological evidence of the former ice extent and of the past sea level. A few models incorporate dynamic ice models, too. Usually the Earth model is jointly inferred with the ice-load history (Lambeck *et al.* 1998a). Instrumental records of present-day RSL changes from tide gauges or of present-day crustal deformations from global positioning systems (GPS) are often used to constrain or validate the inferred ice-load histories and Earth models (e.g. Lambeck *et al.* 1998b; Milne *et al.* 2001). One widely-used global ice-load history is ICE-5G (Peltier 2004). The corresponding viscoelastic Earth model VM2 was chosen in a way that best fits the utilized observational data set in a global sense. Another global model was developed by Kurt Lambeck. The Fennoscandian component of this model is described by Lambeck *et al.* (1998a). In contrast to VM2 the corresponding Earth model is particularly constrained to the Fennoscandian dataset. Hence, this Earth model cannot be used within global studies but is most appropriate for studies in Fennoscandia. Both models provide an ice-load history from the LGM period until the present. The results are well constrained by the rich observational data set for this period. A global model of sea level and ice volumes based on observations distant from the ice sheets has been published by Lambeck *et al.* (2014).

Figure 2.9 shows the present-day GIA-induced vertical crustal deformations over Europe. The underlying calculations are based on the ice-load history ICE-5G and the corresponding Earth model VM2. The utilized software (Spada & Stocchi 2007) solves the SLE for a radially stratified, incompressible, non-rotating Earth in the spectral domain and does not account for a time-dependent ocean function. A generalized version of the VM2 model, consisting of an elastic lithosphere (120 km thickness) and three viscous mantle layers (viscosities:

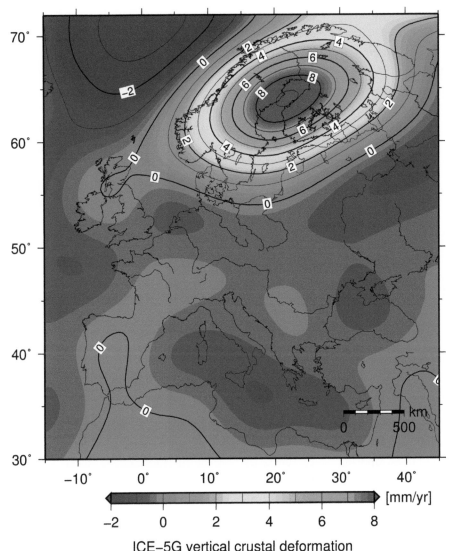

Fig. 2.9 Present-day vertical crustal deformation rates caused by changing ice and ocean water loads since LGM according to ICE-5G (VM2). A. Groh with data from Peltier (2004). Reproduced with permission of Annual Reviews.

ICE–5G vertical crustal deformation

upper mantle: $0.5*10^{21}$ Pa∗s; transition zone: $0.5*10^{21}$ Pa∗s; lower mantle: $2.0*10^{21}$ Pa∗s) was used. The ice-load history is given in 1 kyr steps and predicts a global eustatic sea-level rise of about 127 m since the LGM at 21 ka BP. The deformation pattern reveals a large uplift dome at the former center of glaciation around the Gulf of Bothnia with a maximum rate of uplift of more than 9 mm/year. This uplifting region is surrounded by a zone of slight subsidence. This zone corresponds to the collapsing peripheral bulge from where the mantle material is still being redistributed back to central Fennoscandia. As in all former glaciated regions, the crustal deformations are the dominating signal and GIA-induced changes of the

geoid are relatively small. In Fennoscandia they do not exceed the 0.5 mm/year level. Hence, the present-day fall in RSL due to GIA reaches a maximum of about 8 mm/year. Over ocean regions further away from the former ice sheet the dominating part of the present-day crustal deformation originates from the additional water load due to meltwater influx. Thus, regions like the Black Sea and the Mediterranean Sea experience subsidence. This is clearly revealed by Fig. 2.9.

From the previous discussion it is apparent that different parts of the European coastline and shallow marginal seas have experienced different degrees of vertical change of relative sea level and different rates of change as a result of combinations of tectonics, isolation of lake basins, climate change, atmospheric circulation, and glacio-isostatic adjustment. In the following sections we consider different regions briefly, and these processes will be discussed in more detail in later chapters.

The Baltic Glacio-isostatic Adjustments

The expansion and retreat of the Fennoscandian ice sheet is relatively well recorded by geological proxies as are also the effects of relative sea-level changes. Therefore, models described in the preceding sections have been applied to the Baltic area for several years in order to describe the effect of deglaciation to the region (Harff *et al.* 2007; 2011; Lambeck *et al.* 2010). But, for earlier periods of the glacial cycle the observational evidence can be poor. Nevertheless, Lambeck *et al.* (2010) have used their inversion method (Lambeck *et al.* 1998a) and the rich data set for LGM and post-LGM periods to constrain the behavior of the ice sheet (Fig. 2.10). These constraints, the sparse observational evidence and assumptions on the prevailing basal conditions are further used to extrapolate back to earlier times, resulting in an ice-load history for the preceding MIS 3 (~60–30 ka BP).

After the Weichselian ice sheet had covered the whole Baltic basin during the LGM, the ice retreated and melt water filled the basin at a time when the North Atlantic was still covered by sea ice. This first phase in the postglacial history of the Baltic basin named the Baltic Ice Lake (BIL) lasted from 16 ka BP to 11.7 ka BP. The retreating ice enabled a connection between the Baltic basin and the Skagerrak area along the Central Swedish Depression for a short phase (the Yoldia Sea 11.7–10.7 ka BP) before the GIA uplift disconnected the Baltic basin again from the Global Ocean and another freshwater phase (Ancylus Lake) began at 10.7 ka BP lasting until 8 ka BP. The environment changed completely at about 8 ka BP when, due to the collapsing lithospheric bulge in the southern Baltic area and continuous eustatic sea-level rise, marine water masses transgressed through the Kattegatt into the Baltic basin

(Littorina transgression) converting the former freshwater reservoir to a permanent brackish-marine environment (Andrén 2011; Harff *et al.* 2011). During the Atlantic (8–5 ka BP) and the Subatlantic (2.5–0 ka BP) periods, coastline development in the Baltic basin was the result of a complex interaction between eustatic sea-level rise and GIA. In the south, subsiding land and eustatic sea-level rise caused a permanently retreating coast. In the north, eustatic rise was overcompensated by GIA uplift, resulting in continuous regression of the sea (Fig. 2.11).

The water (sea) level history reconstructed for the Narva-Luga basin in the eastern Gulf of Finland impressively demonstrates the interplay between isostatic uplift and eustatic water-level change (Rosentau *et al.* 2013; Fig. 2.12). After the Baltic Ice Lake stage and a 25 m drop in water level by drainage of the ice-dammed lake, marine waters entered into the basin via the Central Swedish Depression for the duration of the Yoldia Sea period.

With the onset of the Ancylus Lake period, connection to the Global Ocean was closed off, and the water level rose permanently (Ancylus transgression) because of the inflow of melt water from the Fennoscandian ice sheet. By 9.8 ka BP the Ancylus Lake began to drain and the water level dropped by more than 10 m (Ancylus regression). After 8.5 ka BP the curve describes a eustatically controlled sea-level rise because of an open connection of the Baltic basin to the marine realm of the North Sea area. The end of the transgression is marked by a well-developed raised beach around 7.3 ka BP. The following sea-level drop in the curve of Fig. 2.11 is caused by the GIA-uplift of the Fennoscandian Shield which started to outpace eustatic sea-level rise in the Narva-Luga area.

North Sea and Atlantic Shelf from the Last Glacial Maximum to the Atlantic Period

Model results have been produced based on the inversion of sea-level data from the British Isles and northern

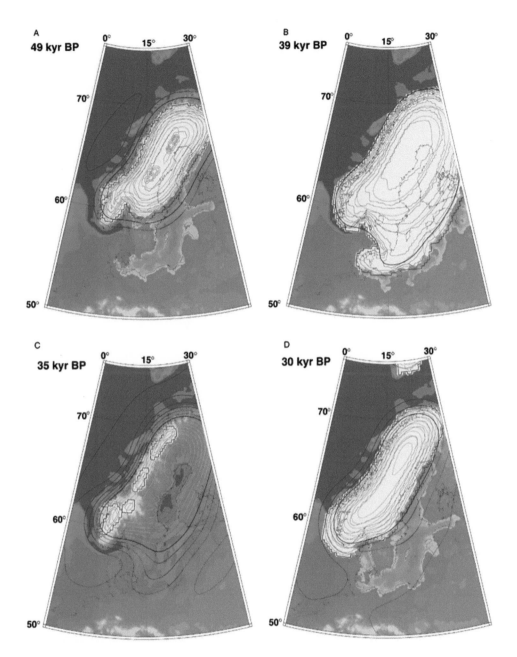

Fig. 2.10 Paleogeographical reconstructions for three principle interstadials and stadials during MIS 3. Ice-thickness contours: 200 m. Positive relative sea-level contours (orange), negative ones (red). Red stars mark thresholds of ice-marginal lakes. Lambeck *et al.* (2010). Reproduced with permission of John Wiley & Sons Ltd.

Europe, complemented with inversions of data from North America and from areas far from the former ice sheets (Peltier *et al.* 2002; Milne *et al.* 2006; Shennan *et al.* 2006; Hanebuth *et al.* 2011). The shoreline reconstructions are thus predictions based on these solutions without taking into account the effect of local tectonics,

erosion and sedimentation. Normally, these results are compared with independent observational evidence of paleoshorelines not used in the solutions; if there are major discrepancies, this additional information is used to improve upon the solution. These model results have not yet been compared with such additional data and are

Fig. 2.11 Areas of transgression (blue) and regression (red) of the Baltic Sea since the Littorina transgression ca. 8 ka BP, modified from Harff *et al.* (2007). Narva-Luga area (Fig. 2.12) and Darss-Zingst Peninsula (Fig. 2.20) are marked.

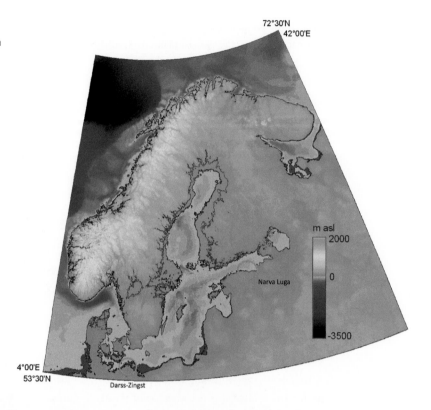

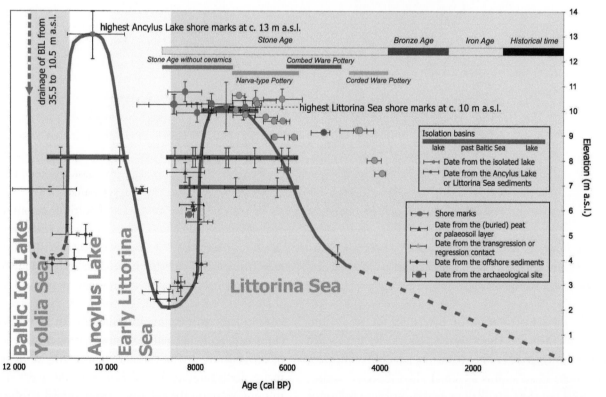

Fig. 2.12 Relative water-level change curve for Narva-Luga area (location marked in Fig. 2.11) in the eastern Gulf of Finland. Modified from Rosentau *et al.* (2013).

therefore a first iteration. Accordingly, we will consider the probable magnitude of the unquantified factors, and then consider the predicted results at different time intervals.

The North Sea and extended Northwest Shelf of Europe consist of a series of shallow semi-enclosed basins and marginal seas typically in the depth range of 50 m to 100 m. They are profoundly influenced by Atlantic storms from the west, and tidal amplitudes are frequently of the order of 2 m to 5 m, and over 10 m in some limited areas, with associated strong currents (Neill *et al.* 2009; 2010). The oceanographic conditions result in vigorous erosion and redistribution of coastal sediments. These include actively mobile sedimentary features such as sand waves, sand banks, and sand ribbons on the sea floor, as well as massive accumulations of sediments on the outer shelf of the western approaches in the Celtic Sea (Lericolais *et al.* 2003; Toucanne *et al.* 2009) brought down by the Channel River during glacial melting and deposited during sea-level rise. Gibbard (1988) provides a longer timescale for the analysis of the European rivers on the continental shelf. The calculated GIA movements and the present position of ancient shorelines and areas of drowned terrestrial landscape previously exposed at different dates could therefore be influenced by, or concealed by, local and regional tectonics, erosion, and extensive redeposition of postglacial marine sediments. In particular, they will not necessarily be detectable by bathymetric surveying alone, and will in many cases only be detectable by sub-bottom profiling or 3D seismic surveying (Bakker *et al.* 2012).

Tectonic activity on land and on the seabed in the British area of the Northwest Shelf, which is the largest proportion, has been a minor factor during the Quaternary (Barton & Woods 1984; Thorne & Watts 1989), although active graben formation and crustal extension and thinning of the central North Sea occurred during the Jurassic-Cretaceous. Crustal extension, rifting, and graben subsidence were at least partially counterbalanced by continuous infilling of the depressions by sediments. The Rhine graben does not continue into the North Sea. In contrast, the English Channel area has been subject to compression due to the Alpine orogeny (Lagarde *et al.* 2003). For the last 730,000 years the average subsidence

of the central North Sea has been 0.4 m/kyr (Cameron *et al.* 1992), amounting to 8 m in 20,000 years, while the uplift of the Channel region has been 0.1 m/kyr (Lagarde *et al.* 2003), amounting to 2 m in 20,000 years. In both cases the tectonic factor is an order of magnitude smaller than the glacio-isostatic changes in the same period, and, at least in the case of the North Sea, subsidence as an effect on bathymetry may be offset by sediment accumulation.

Consistent with this observation, the recorded and current seismicity of the British Isles is very low, and furthermore a proportion of the recorded epicenters correlates with crustal adjustment arising from postglacial rebound (Musson 2007). The earthquake records in the UK have been assessed by Musson (2007) and there have only been 35 events over Local Magnitude (LM) 4 since 1850, and none above LM 5.4 (Musson 2007, fig. 6).

Tsunamis are another factor which can change the bathymetric parameters and alter or conceal paleoshorelines. The Storegga event (Weninger *et al.* 2008) created waves which struck the coast of Scotland and the northern slopes of the Dogger Bank approximately 8.2 ka BP. The major effect on seabed bathymetry was the additional mass of sediment in deep water off the Norwegian continental slope, and thus outside the interest of this study. No accurate calculation has been made of the quantitative effect on the submerged landscape in coastal waters, but Weninger *et al.* (2008, figs. 4 & 5) show the areas most strongly impacted by the Storegga tsunami in the contemporaneous North Sea coastal waters, and on the northern flanks of the Dogger Bank.

When the relative velocities of vertical tectonics and glacio-isostatic earth movements compared with the vertical rate of rise of sea level produce a local relative stillstand, wave energy and currents produce cliffs, erosion, beach ridges, terraces, and active sediment transport. There should thus be identifiable submerged shorelines against which predicted GIA sea-level models can be verified or corrected. However, this is not a simple exercise since, in an area of extreme postglacial isostatic adjustment, the shore features occur at different depths and different times against the tilting land surface. This problem is addressed for the central North Sea by Ward *et al.* (2006).

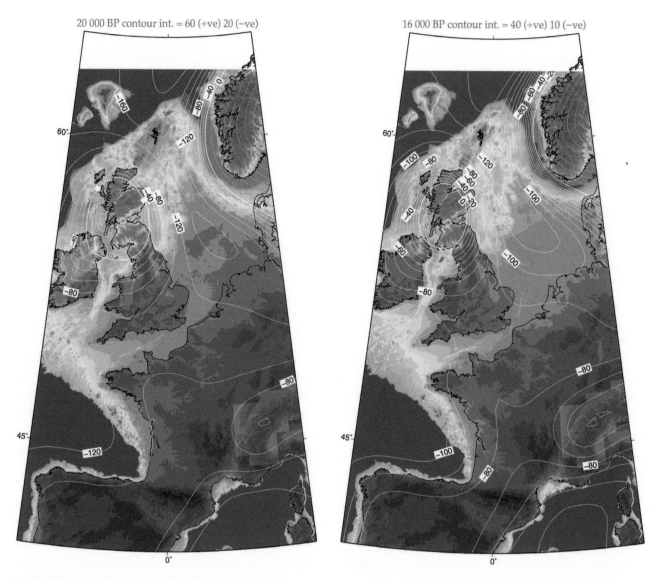

20 000 BP contour int. = 60 (+ve) 20 (−ve)   16 000 BP contour int. = 40 (+ve) 10 (−ve)

Fig. 2.13 Paleogeographic models of the North Sea and parts of the Northeast Atlantic shelf 20, 16, 12, 10, 8, and 6 ka BP. Models by K. Lambeck based on Lambeck (1995), Lambeck and Purcell (2001), and Lambeck *et al.* (2010).

The complex task of identifying continuous shorelines in a terrain of very low relief during postglacial isostatic tilting is also facilitated by the detailed sub-surface maps produced by Gaffney *et al.* (2007) and Bakker *et al.* (2012).

The models shown below have not been corrected for the factors mentioned above, and a complete comparison with paleoshorelines detected offshore has not been carried out. The first iteration results are therefore illustrated below. The impact of sea-level changes on human population movements into and out of the British

Isles on a longer timescale is discussed extensively by Pettit and White (2012).

The results included and shown in Fig. 2.13 are for the following epochs:

1 T = 20,000 BP[†] representing the late LGM interval. This is shortly after the maximum ice extent across the North Sea when retreat from the North Sea back to

[†] In this chapter, in regards to model dates, a timescale of calendar years BP (= 1950) is used, as model time is unrelated to any ^{14}C calibration

12 000 BP contour int. = 30 (+ve) 10 (−ve)

10 000 BP contour int. = 10 (+ve) 5 (−ve)

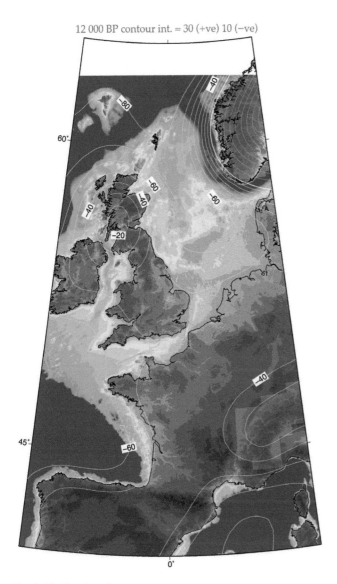 

Fig. 2.13 (Continued)

Norway and Scotland has already started. Northward retreat across England and the Irish Sea has also started before this time.

Ireland and England are tenuously connected via a very low land bridge across the southern Irish Sea across which flows the spillover from the ice-dammed lake to the north.

Much of the North Sea is exposed, including the locations east of Shetland where Late Glacial artifacts have been found on the present sea floor (Long *et al.* 1986).

2 T = 16,000 BP representing the Late Glacial Period. The sea is beginning to advance across the North Sea floor but quite slowly at first because the rebound is approximately keeping up with the rising water.

3 T = 12,000 BP. The Dogger Bank is still connected to a substantial exposed area of the southern North Sea, but the Irish Sea is now widely flooded.

4 T = 10,000 BP, representing the separation of Dogger Bank and the early phase of the opening of the

8 000 BP contour int. = 5 (+ve) 2 (−ve)

6 000 BP contour int. = 2 (+ve) 1 (−ve)

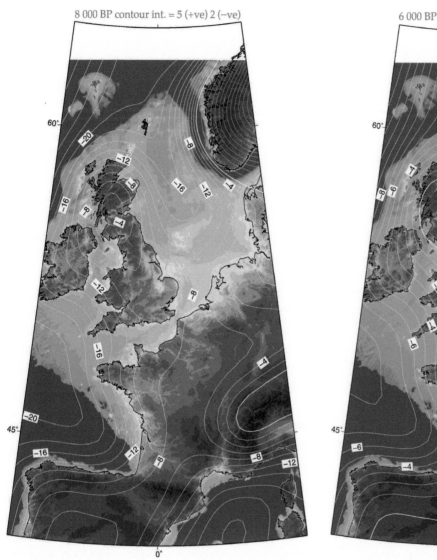
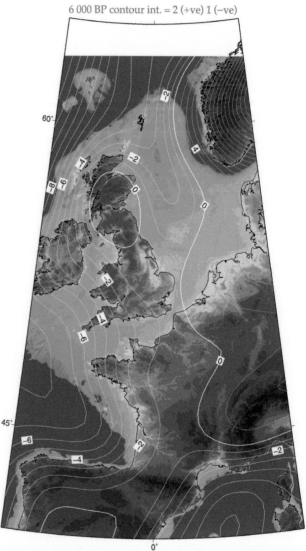

Fig. 2.13 *(Continued)*

English Channel which is completed just before 9000 BP.

5 T = 8000 BP. By 8000 years ago the shorelines have approximated their present locations but in the low-lying areas of the Netherlands, Belgium, Denmark and Germany, sea level continues to rise and there is extensive inundation after about 7500 BP.

6 T = 6000 BP. Present shorelines widely achieved, and uplift continues inland in the areas of maximum glacial loading at the time of the LGM.

Tectonic Controls: The Mediterranean Sea

Paleogeographic models of the Mediterranean and Black Sea area have been presented by Lambeck and Purcell (2005; 2007) and Lambeck *et al.* (2011). But, for most of the areas along the Mediterranean coastline, sea-level models deviate from field observations and geological data. The reason for this discrepancy between models

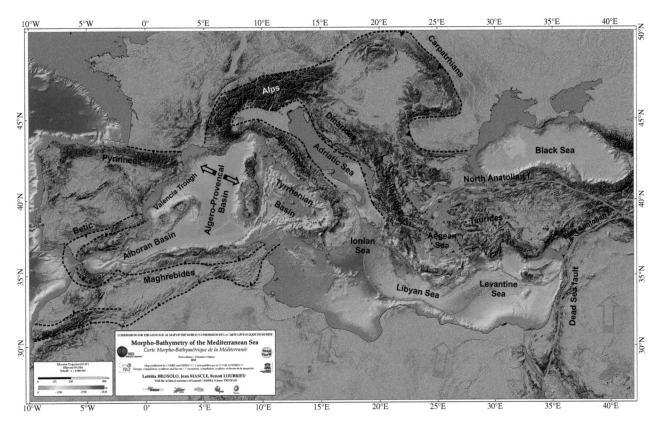

Fig. 2.14 Main geotectonic boundaries and features drawn on the Morpho-Bathymetry Map of the Mediterranean Sea (CGMW – Commission for the Geological Map of the World/UNESCO) Brossolo *et al.* (2012). Adapted with permission.

and observations is the ongoing geodynamic and tectonic processes and crustal deformation over most of the Mediterranean region. Active processes create a dynamic environment with long- and short-term, vertical and horizontal, crustal movements which are superimposed on and modify the effect of sea-level fluctuations. Comparison of observations and field data, including different geomorphological and archaeological sea-level markers, with the predicted sea-level curves provides estimates of the vertical tectonic contribution to relative sea-level change. Long-term vertical tectonic rates of movement can be inferred from the elevation of the 124 ka last interglacial highstand (MIS 5.5) marker (Ferranti *et al.* 2006; Antonioli *et al.* 2009a,b). This marker, where developed and preserved, offers an excellent tool to estimate patterns of differential displacement within adjacent tectonic blocks (Lambeck *et al.* 2011).

Active tectonics in the Mediterranean and adjacent areas is predominantly driven by the convergence between the African and the Eurasian plates. Figure 2.14 shows the complex geotectonic structure developed due to the collision of these continental plates (compare also with the more general map of Fig. 2.2). Subduction of oceanic crust and/or collision of continental blocks goes on along the Betic-Rif orogenic belt, the Calabrian Arc, the Hellenic Arc and the Cyprus Arc. Active movements along strike-slip fault boundaries of adjacent plates or microplates add more complexity to the overall geodynamic structure and kinematics of the Mediterranean and create a puzzle of crustal blocks independently moving, either uplifting or subsiding.

The eastern Mediterranean basins, namely the Ionian, Libyan and Levantine basins, are underlain by the last remnants of the Mesozoic Tethys Ocean, which is currently being consumed beneath the Calabrian Arc in the western Ionian Sea, the Hellenic Arc in the eastern Ionian, Libyan and western Levantine seas and the Cyprus Arc in the east Levantine Sea.

Evolution of the north–south trending Levantine coast of the Mediterranean's eastern edge is dominated by the Dead Sea Strike-Slip Fault (DS) and its secondary branches. This runs through the Middle East from south to north and accommodates the northward movement of the Arabian plate caused by the opening of the Red Sea. Large, destructive, magnitude >6 historical and recent earthquakes all along the DS are evidence of enhanced tectonic activity along the Dead Sea Fault and subsequent movements which have shaped the morphology of the Levantine coastline throughout the Quaternary. The oceanic floor of the Levantine basin is being consumed beneath Cyprus, along the Cyprus Arc.

The Hellenic Arc and the extending Aegean region behind it are tectonically and kinematically the most active areas in Europe. The main ongoing processes which dominate the geodynamic evolution of the Aegean and eastern Ionian seas are: the westward extrusion of the Anatolian continental block along the dextral strike-slip North Anatolian Fault boundary; the NNE subduction of the eastern Mediterranean lithosphere beneath the south-westward migrating Hellenic Arc at a rate of 3 cm/year to 4 cm/year; the resulting SSW–NNE extension of the Aegean back-arc region; the collision of northwestern Greece with the Apulian block in the northern Ionian Sea, particularly to the north of the Kephallinia Fault; and the incipient collision with the Libyan promontory south of Crete (e.g. McKenzie 1970; 1978; Dewey & Şengör 1979; Le Pichon & Angelier 1979; Le Pichon et al. 1982; Meulenkamp et al. 1988; Mascle & Martin 1990; Meijer & Wortel 1997; Jolivet 2001; Armijo et al. 2004; Kreemer & Chamot-Rooke 2004).

The building up of stresses along the boundaries and in the interior of the crustal blocks leads to extensive deformation in the upper crust, expressed by very high seismicity and vertical and horizontal movements accommodated by normal, thrust and strike-slip faults. Continuous, long-term, tectonic uplift and/or subsidence by 1 m/kyr or more is evident at many places along the Hellenic Arc and within the Aegean region and

has been documented with mapping and dating of uplifted Pleistocene marine terraces or submerged pro-delta prograding sequences (e.g. Armijo et al. 1996; Lykousis et al. 2007; Lykousis 2009). Short-term, incremental, vertical movements in the Late Holocene modify the modeled post-LGM sea-level rise (Lambeck & Purcell 2007), as postulated by uplifted or submerged paleoshorelines observed in numerous places along the Aegean coastline as in Crete, Rhodes, the Gulf of Corinth, Evia Island, the Aegean Islands, etc. (e.g. Pirazzoli et al. 1989; Kontogianni et al. 2002; Evelpidou et al. 2012a,b). The largest, abrupt, vertical tectonic dislocation recorded in the Mediterranean Sea occurred along the uplifting Hellenic Arc and resulted from the 365 AD earthquake with magnitude >8 which uplifted and tilted the western half of Crete Island by up to 8 m (Shaw et al. 2008).

Quaternary tectonics in the Aegean, particularly in central Greece, has led to the formation of WNW–ESE trending neotectonic grabens which presently form elongate gulfs cutting across the Alpine structure of the Hellenides mountain chain: the 900 m deep Gulf of Corinth, the 450 m deep North Evia Gulf, the 400 m deep West Saronikos Gulf and the shallower Amvrakikos, South Evia and Pagasitikos gulfs have one characteristic in common: they are connected to the open sea through narrow and shallow straits which were exposed above the sea level during the LGM and possibly during earlier periods of low sea level. These presently marine water bodies were isolated lakes during the LGM with water-level considerably higher than the contemporaneous sea level (Perissoratis et al. 1993; Lykousis & Anagnostou 1993; Richter et al. 1993; Kapsimalis et al. 2005; Lykousis et al. 2007; Sakellariou et al. 2007a,b). Thus, the Holocene water/sea-level rise curve of these areas deviates significantly from the curves in adjacent areas.

The Calabrian Arc in the western Ionian Sea and central Mediterranean Sea marks the last phase in the subduction of the Ionian oceanic basin beneath the eastwards migrating Apennines-Maghrebides orogenic belt (Carminati & Doglioni 2005). It is characterized by outward (east–southeastward) migration and frontal compression, arc-parallel extension, relative rotation of crustal fragments and fast uplift of onshore and shelf areas since the Mid Pleistocene (Westaway 1993; Sartori 2003; Viti *et al.* 2011).

The western Mediterranean comprises four, young (less than 30 Ma old), extensional basins (Biju-Duval & Montadert 1977): the Alboran basin, developed behind the westward-migrating Betic-Rif Arc; the Valencia Trough; the Algero-Provençal basin; and the Tyrrhenian basin, which opened progressively behind the eastward-moving Apennines-Maghrebides Arc after the final collision of the Iberian and Eurasian plates along the Pyrenees (Carminati & Doglioni 2005). The Tyrrhenian Sea is the result of rifting, back-arc extension and crustal thinning of the Alpine-Apennine orogenic belt above the westward-subducting Ionian oceanic lithosphere below the Calabrian Arc. Eastward migration of the latter led to the initiation of spreading and formation successively of the Vavilov and Marsili oceanic basins (Kastens *et al.* 1988; Kastens & Mascle 1990; Sartori 1990; Jolivet 1993). Subduction-related volcanism migrated from west to southeast, namely from Sardinia to the presently active Aeolian Island Arc (Serri 1997).

Deformation in the west Mediterranean occurs mostly in the Alboran basin and along the North African Margin (Stich *et al.* 2005; 2007). In the south Iberian Margin, crustal deformation is mainly driven by the NW–SE convergence (4–5 mm/year) of the African and Eurasian plates (e.g. Argus *et al.* 1989). Convergence is accommodated over a wide deformation zone distributed among a number of active tectonic structures which are characterized by low to moderate seismicity (e.g. Buforn *et al.* 1995; 2004; Stich *et al.* 2005; 2007; 2010). Active faulting activity in the southeast Iberian Margin is dominated by the eastern Betic Shear Zone strike-slip system, a 450 km-long fault zone, the splays of which, the Bajo Segura Fault to the north and the Carboneras Fault to the south, with slip-rates of about 1.3 mm/year (Moreno 2011), extend into the sea. The Yussuf Fault, a 250 km-long, composite, dextral strike-slip fault, marks the boundary between the Alboran Sea and the Algerian Margin, and is capable of generating up to magnitude 7.4 earthquakes. The Al Idrissi Fault, running NNW–SSE in the Alboran Sea, is a left-lateral strike-slip structure (Martínez-García *et al.* 2011; Bartolomé *et al.* 2012) characterized by enhanced seismic activity with up to

magnitude 6 earthquakes or larger. Both aforementioned faults in the Alboran Sea, along with other minor ones, have contributed significantly to the morphological configuration of the coastal areas.

A Hydrologic System: The Black Sea since MIS 2

The level of the Black Sea, to a certain extent, was controlled more by the regional climate than by global eustatic changes. From these interpreted results, Lericolais *et al.* (2009) have proposed a curve representing the water-level fluctuation in the Black Sea since the LGM. Today, in the light of the new results on ^{14}C calibration obtained by Soulet *et al.* (2011a) and leading to a revised calendar age for the last reconnection of the Black Sea to the Global Ocean at 9000 cal BP (Soulet *et al.* 2011b), it is possible to provide a calendar age sea-level curve for the Black Sea since the LGM (Fig. 2.15, see also Chapter 17, pages 484–485).

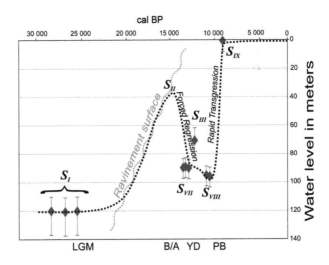

Fig. 2.15 Water-level fluctuation in the Black Sea since the LGM, deduced from the observations recovered on the northwestern Black Sea shelf. LGM = Last Glacial Maximum; B/A=Bølling-Allerød; MWP-1a = Meltwater Pulse 1a; YD = Younger Dryas; PB = PreBoreal. S$_I$ to S$_{IX}$ are the sequences interpreted and dated from the Romanian Black Sea shelf. Modified from Lericolais *et al.* (2009), revised version Lericolais *et al.* (2011).

The transition of the Black Sea system from a lacustrine to a marine environment is perhaps one of the best records of climate change on the European continent. Back at the Last Glacial Maximum, ca. 20,000 years ago, the Black Sea was probably a giant freshwater lake as proposed by Arkhangel'sky and Strakhov (1938) and confirmed by Soulet *et al.* (2010), and its water level stood around 120 m below today's level. The studies carried out on the Danube deep-sea fan (Popescu *et al.* 2001; Lericolais *et al.* 2013) confirm that the last channel–levée system developed during MIS 2. Sediments supplied by the Danube River were transported to the deep basin through the Viteaz canyon (Popescu *et al.* 2004). The water brought to the Black Sea after the MWP-1A at approximately 14,600 cal BP (Fairbanks 1989: 18; Bard *et al.* 1990) is supposed to have been sufficiently important that the water level rose to between –40 m and –20 m, where the Dreissena layers were deposited. This last value for the transgression upper limit would have brought the level of the Black Sea even higher to the level of the Bosporus sill, and possible inflow of marine species like Mediterranean dinoflagellate populations can be envisaged (Popescu *et al.* 2003). The rise in the Black Sea water level, which stayed between fresh to brackish conditions, stopped deep-sea fan sedimentation.

Palynological studies conducted on BlaSON cores (Popescu *et al.* 2003) show that during the Younger Dryas a cool and drier climate prevailed. Northeastern rivers converged on the North Sea and the Ancylus Lake (Baltic Sea) (Jensen *et al.* 1999) giving reduced river input to the Black Sea resulting in a receding shoreline there. These assumptions are consistent with some evaporative drawdown of the Black Sea and correlated with the evidence of an authigenic aragonite layer present in all the cores studied (Giunta *et al.* 2007). This lowered sea level in the Black Sea persisted afterwards, evidenced by (1) continuously dry climatic conditions in the region which had started around 13,000 to 10,000 cal BP, and (2) dune formation between 10,500 and 9000 cal BP on the desiccated north-western Black Sea shelf at –100 m. The Younger Dryas climatic event had lowered the Black Sea water level and the presence of the coastal sand dunes and wave-cut terrace confirm this lowstand. Preservation of these sand dunes and buried small incised

valleys can be linked with a rapid transgression where the ravinement processes related to the water-level rise have had no time to erode sufficiently the sea bottom (Ahmed Benan & Kocurek 2000; Lericolais *et al.* 2004). Around 9000 cal BP, the surface waters of the Black Sea suddenly attained present-day conditions owing to an abrupt flooding of the Black Sea by Mediterranean waters, as shown by dinoflagellate cyst records (Popescu 2004) and as supposed by Ryan *et al.* (1997; 2003) who proposed 7500 cal BP — subsequently modified because no age reservoir correction was available until the work by Soulet *et al.* (2011a). Furthermore, Soulet *et al.* (2011b) and Nicholas *et al.* (2011) demonstrate that the Black Sea 'Lake' reconnection occurred in two steps, as follows: (1) Initial Marine Inflow (IMI) dated at 9000 cal BP followed by (2) a period of 900 to 1000 years, of increasing basin salinity that led to the disappearance of lacustrine species (DLS). This last event can also be correlated with the beginning of the sapropel deposit which is widespread and synchronous across slope and basin floor. The Black Sea basin would have been flooded in ~1000 years, equalizing water levels in the Black Sea and Sea of Marmara. Such a sudden flood would have preserved lowstand marks on the Black Sea's northwestern shelf. Furthermore, the model developed by Siddall *et al.* (2004) suggests that about 60,000 m^3 of water per second must have flowed into the Black Sea basin after the sill broke and it would have taken of the order of 33 years to equalize water levels in the Black Sea and the Sea of Marmara.

Holocene Climate and Coastal Morphodynamics

Climate

Climate models are one of the most powerful tools to simulate future climates. They have evolved in recent decades to become complicated software codes that try to represent as realistically as possible the different subsystems of the Earth's climate — the atmosphere, the ocean, the cryosphere, the terrestrial biosphere, the global carbon cycle, and others. Usually they are described as Atmosphere–Ocean General Circulation Models (AOGCM) or, more generally, Earth System Models (ESM) when they include carbon cycle and dynamic vegetation models. The models are driven by estimates of external forcing factors that comprise the greenhouse gases, volcanic activity, solar irradiance, land use and the configuration of the Earth's orbit. In the pre-industrial Holocene, the most important drivers were solar, volcanic and orbital forcing (Schmidt *et al.* 2011).

In spite of the complexity of present climate models, they still have clear limitations. One is their spatial resolution; at about 200 km, it is not capable of capturing important aspects of regional climates such as the effect of coastlines and mountain ranges. Also, important dynamic processes like atmospheric turbulence, clouds and precipitation are represented in a simplified form. Most of the uncertainties in future climate projections stem from the different representation of the effects of clouds in the radiative and moisture balance in the climate models (Bindoff *et al.* 2007; Medeiros *et al.* 2008; Lauer & Hamilton 2013).

The second limitation pertinent to the simulation of sea-level changes is the lack of a proper land-ice model in virtually all current climate models. This implies that the estimation of the eustatic contribution to sea-level changes requires additional inputs, for instance specific models of the polar ice caps and mountain glaciers, which in turn are partially driven by the simulated changes in air temperature. However, this approach cannot completely capture important processes for the mass balance of ice caps and glaciers. The resolution of the simulated temperature field is not adequate for the requirements of a glacier model. Glaciers are also partially driven by precipitation, which is not well represented in global climate models. Also, air temperature directly drives the surface melt over ice caps (van den Broeke *et al.* 2009) but its influence on the ice rheology is much more subtle and not totally understood (Joughin *et al.* 2012).

As a result, as reflected in the series of reports of the Intergovernmental Panel on Climate Change (IPCC) (Church *et al.* 2013), global climate models are more capable of simulating the thermal expansion of the water column of the global oceans. The other contributions to

mean sea level due to climate changes mentioned above have to be estimated by other means. This limitation pertains not only to the globally-averaged sea level, but also to the large-scale spatial variations of sea-level change due to the self-gravitational effect (Mitrovica *et al.* 2001). Due to this effect, ice melting from the Antarctic ice sheet is most strongly felt in the sea level of the Northern Hemisphere. It is smaller in the Southern Hemisphere and sea level may even fall near Antarctica when land-ice melts there. Likewise, the melting of Greenland ice is most strongly felt in the sea level in the Southern Hemisphere, whereas its direct effect in western Europe is much smaller, about 10% to 20% of the global sea-level rise caused by Greenland ice melting (Mitrovica *et al.* 2001). More complicated spatial patterns result from the melting of mountain glaciers, which are concentrated in a few regions scattered around the globe — the Himalayas, the Andes and Alaska. For the focus of this book on western European coasts, it is thus mainly the melting of the Antarctic ice sheet, in particular from the West Antarctic ice sheet, that is of relevance. However, it is not easy to estimate the different contributions of polar ice caps and glaciers to global ice melting, given a global level of warming or cooling.

In seas like the Baltic and the Mediterranean, mean sea-level changes at monthly to decadal timescales can be modulated by prevailing wind stress and air pressure gradients (barometric effect). Also, coastal sediment transport is strongly modulated by prevailing wind forcing. These are generally considered to be much more realistically simulated by climate models.

Therefore, when considering the sea level simulated by present climate models it has to be borne in mind that the different processes that are involved in sea-level changes may be simulated with different levels of uncertainty, ranging from the most uncertain, that related to mass loss from polar ice caps, to the relatively most certain, that related to the thermal expansion of the water column.

Most of the paleoclimate simulations with comprehensive atmosphere–ocean coupled models are time-slice simulations. The models run with external drivers frozen-in to the values attained at a certain time, for instance 6000 BP. The aim of these simulations is thus

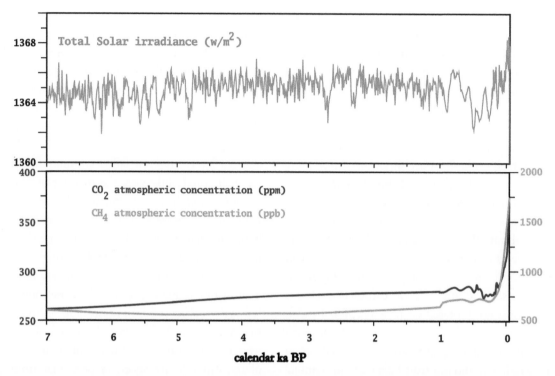

Fig. 2.16 Time series of model ECHO-G forcings (orbital forcing, solar variations and greenhouse gases). These data are derived from the analysis of cosmogenic isotopes in ice cores and tree rings (solar irradiance, Solanki *et al.* 2004), and of the air bubbles trapped in polar ice (Indermühle *et al.* 1999). There is still some uncertainty regarding the amplitude of past solar variations and thus this series has to be considered as one possibility among others. Schmidt *et al.* (2011).

not to simulate climate evolution through time, but to generate a certain number of years of stationary climate at that particular point in time. For instance, the Climate Model Intercomparison Project Phase 5 (CMIP5), whose simulations are being used in the Fifth Assessment Report of the IPCC, contains a set of simulations denoted Mid Holocene that are time-slice simulations.

On the other hand, in transient simulations the external forcing is continuously changed to represent the real changes of the external forcing that occurred in the past. There are only very few transient simulations with AOGCMs covering the Holocene or even the period between the Holocene Thermal Maximum — about 6000 years ago — and the present. These simulations have been conducted with AOGCMs with a horizontal resolution coarser than the resolution used for future climate projections, due to the demanding computing resources for a simulation several thousand model-years long. Nevertheless, these simulations are indeed useful to

estimate the range of magnitude of the relevant processes that may have influenced sea level in the past. In the following we present some select aspects of one of these simulations that may be relevant for sea-level and coastal thermodynamics in West European coastal seas.

The model ECHO-G, composed of the atmospheric model ECHAM4 and the ocean model HOPE was used to simulate the last 7000 years, driven by orbital forcing, solar variations and greenhouse gases (Fig. 2.16, see also Hünicke *et al.* (2011)).

Figure 2.17 shows AOGCM ECHO-G simulation results of Northern Hemisphere mean temperature in winter and summer, with a horizontal resolution of 3.75 degrees. The model was driven by estimations of part orbital forcing, solar variations and greenhouse gases. In this simulation no volcanic forcing was included due to the lack of reliable reconstructions over the past millennia. However, it is generally believed that volcanic forcing may affect temperatures at decadal or maybe multi-decadal

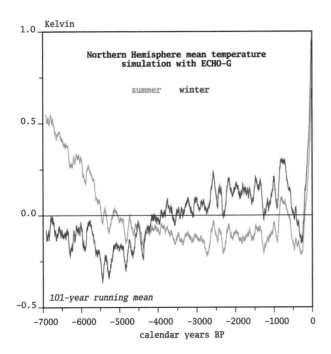

Fig. 2.17 Northern Hemisphere mean temperature in winter and summer simulated by the simulation with the atmosphere-ocean general circulation model ECHO-G driven by orbital, solar and greenhouse gas forcings since 7 ka BP.

timescales, but not produce multi-centennial or longer temperature trends.

The variations in the near-surface temperature are of the order of 1° K at most, which would translate into an expansion of the water column of the order of a few tenths of a centimeter, depending on how deep into the ocean the temperature variations would penetrate. This thermal expansion is an order of magnitude smaller than the estimated sea-level variations over the last 7000 years in the Baltic Sea region due to melting land ice and isostatic glacial rebound, and thus it likely played a minor role.

The evolution of the summer temperature in the Northern Hemisphere, with warm summers around the Holocene Thermal Maximum and decreasing temperatures thereafter, is consistent with orbital forcing, which caused higher summer insolation in the Northern Hemisphere at high latitudes — the perihelion occurred in the summer season, compared to January today. It is also consistent with pollen-based reconstructions in Europe (Davis *et al.* 2003). On the other hand, the evolution of winter temperatures, though similarly

consistent with orbital forcing, is not totally in agreement with the evidence provided by proxy records, which indicate warmer winters in the Holocene Thermal Maximum in Europe. Most climate models from the CMIP5 model suite are not capable of producing these reconstructed warm mean winter temperatures in the Mid Holocene, this being still an open question. Probably, other mechanisms additional to pure orbital forcing may have played a role in winter.

More important for sea-level variations in regional seas and for nearshore sediment transport are changes in the prevailing wind strength and direction. The wind regime in the North Atlantic-European sector can be described by the dominating patterns of variations of the sea-level-pressure field, of which the best known is the North Atlantic Oscillation (NAO). The NAO index statistically describes the meridional pressure gradient between the subtropical anticyclone and the subpolar low pressure cell, and thus also describes the strength of the westerly winds in western Europe. The same simulation with the AOGCM ECHO-G indicates that, in the winter season, the state of the NAO may have been turning progressively more negative (weakening pressure gradient and weakening zonal winds) over the last 7000 years, although the millennial-scale trend is weak (Fig. 2.18). This is in qualitative agreement with the findings of Olsen *et al.* (2012). On the other hand, the summer NAO has undergone a weakening trend from the beginning of the simulation to as recently as 4000 BP, strengthening thereafter. During the whole simulation, more importantly, the summer NAO index has been clearly negative with respect to present values, meaning that the meridional sea-level-pressure gradient would have been weaker than present, and thus the zonal summer winds over western Europe would have been accordingly weaker as well.

It is these weakened zonal winds in summer that most strongly affect the annual mean winds. Figure 2.19 displays the annual mean 10-m winds in the Mid Holocene (6500–5500 BP) as deviations relative to the twentieth-century mean. The wind deviations over northern Europe are easterly, meaning the mean westerly winds would have been weaker than present. On the other hand, southern Europe and the Mediterranean region

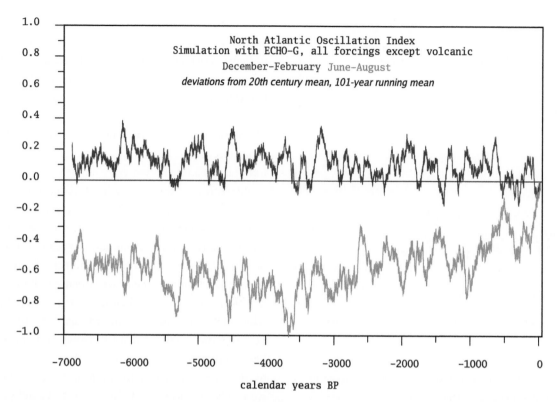

Fig. 2.18 Index of the North Atlantic Oscillation since 7 ka BP simulated by the model ECHO-G in winter (December–February) and summer (June–August).

Fig. 2.19 Changes in the annual mean 10-m winds in the period 6.5–5.5 ka BP relative to present simulated by the model ECHO-G.

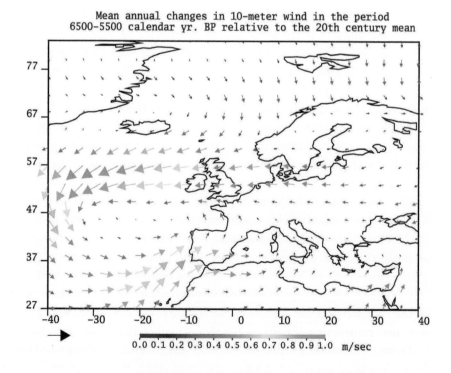

experienced stronger westerlies, with a stronger influence of the air masses of North Atlantic.

Coastal morphodynamics

A globally stabilized mean sea level in the Mid-to-Late Holocene restricts sea-surface wave actions to a small-range coastal area and a limited vertical range (i.e. from low-lying coastal lands with altitude less than ~10 m above mean sea level to the storm wave base at mid-shelf that is ~100 m below the mean sea level). Coasts built up by soft sediments such as Quaternary deposits are constantly shaped by winds, tides and waves and, on a longer timescale, can shift landward or seaward due to oscillations of the sea level and variations in the sediment supply, destroying also existing archaeological sites. Integrated high-resolution morphodynamic modeling approaches can help to reconstruct the development of paleocoastal landscapes, climate and glacio-isostasy, and can demonstrate favorable conditions for the preservation of paleolandscapes.

Existing coastal morphodynamic models can be classified into three types: (1) process-based; (2) behavior-oriented and (3) a hybrid of the former two types. The advantages as well as shortcomings of the first two model types have been widely revealed and discussed (e.g. de Vriend 2001; Fagherazzi & Overeem 2007). Although high-resolution process-based models are reliable in capturing short-term (from hourly to daily scale) coastal morphodynamics, their application to the longer-term (from annual up to millennial scale) is severely hindered. The reason for this originates not only from the numerical errors induced by the solutions for the partial differential equations, but also from insufficient knowledge of the complexity of the natural system. Great efforts have been made by researchers in recent decades to improve the reliability of high-resolution coastal morphodynamic models, e.g. 'reduction' concepts are introduced by de Vriend et al. (1993a,b) to reduce the errors of process-based models in long-term simulation; techniques of morphological update acceleration have been proposed by Roelvink (2006) to increase the long-term computational efficiency. For complex coastal systems, hybrid modeling, which combines advantages both of process-based and behavior-oriented models, has shown promising results for long-term modeling (e.g. Jiménez & Sánchez-Arcilla 2004; Karunarathna et al. 2008).

Recently Zhang et al. (2010; 2012) have developed a modeling methodology for the simulation of long-term morphological evolution of wave-dominated coasts and applied it successfully to hindcast the morphological evolution of two barrier island systems in the southern inundated Baltic Sea (the Darss-Zingst Peninsula and the Swina Gate barrier) on a centennial (Zhang et al. 2011) and a millennium scale (Zhang et al. 2014), respectively. The methodology consists of three major components: (1) an analysis of the key boundary conditions driving the morphological evolution of the study area based on statistical analysis of meteorological and sedimentological data; (2) a multi-scale high-resolution hybrid model in which 'reduction' concepts, techniques for morphological update acceleration and approaches for maintaining computational stability are implemented; and (3) a sensitivity study in which a large number of simulation iterations are carried out to derive an optimum parameter setting for the model.

A case study on the Darss-Zingst Peninsula is briefly introduced here to explain the use of high-resolution models for investigation of medium-to-long-term coastal morphodynamics. For details of the work the reader is referred to Zhang et al. (2014). In order to initiate a historical hindcast of coastal morphological evolution, Zhang et al. use a paleo-Digital Elevation Model (DEM) serving as the initial condition and reconstructed from a compilation of recent digital elevation data sets, a eustatic sea-level curve, an isostatic map and dated sediment cores. Representative wind series have been generated based on a statistical analysis of paleo-wind data from a simulation with the coupled atmosphere–ocean general circulation model ECHO-G over the last 7000 years. These wind data were calibrated by proxies from lithostratigraphic studies of sediment cores from the central Baltic Sea, and used as climate driving conditions for the morphodynamic model. Based on the reconstructed paleo-DEM and the representative climatic driving conditions, the model is applied to reconstruct the Holocene morphogenesis of the Darss–Zingst Peninsula since 6000 BP. Simulation results (Fig. 2.20) indicate that

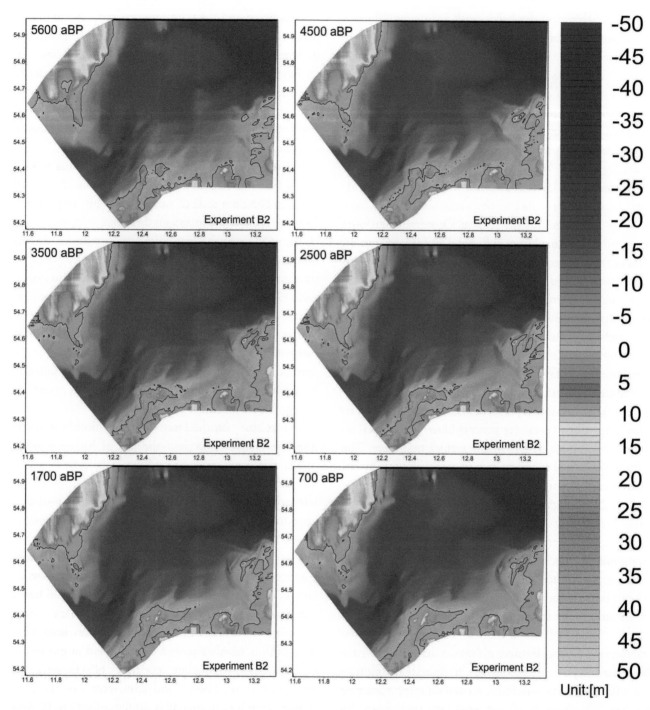

Fig. 2.20 Simulated morphological evolution (color scale: m.a.s.l.) of the Darss-Zingst Peninsula (location marked in Fig. 2.11) since 6 ka BP. Modified from Zhang *et al.* (2014).

the development of the barrier system is a combination of long-term effects of climate change, isostatic crustal movement, wave dynamics and eolian transport with short-term effects of extreme wind events, i.e. storms. The information derived from the modeling study may provide valuable prerequisites for planning and executing archaeological surveys on the continental shelf.

Conclusion

The environment of the European continental shelf changed drastically after the Last Glacial Maximum: inland ice melted and the resulting drainage system re-shaped the landscape. Ice lakes formed in front of the decaying glaciers, the rising sea level caused a submergence of large parts of the continental shelf, and freshwater lakes converted to brackish-marine water reservoirs. The most important processes for the change of coastlines comprise the interrelation of climatically controlled eustatic sea-level rise and vertical crustal displacement of the Earth's crust. The latter is dependent on the position of a site under investigation related to (1) the decaying ice shield, and (2) the tectono-stratigraphic units of Europe. Coastlines on Proto-, Paleo- and Meso-Europe are mainly influenced by glacio- and hydro-isostatic adjustment interacting with eustatic change. In Neo-Europe — that is in particular along the coasts of the Mediterranean Sea — regional and local tectonics due to the collision of Africa and Eurasia determine the pattern of uplift and subsidence of crustal units of different spatial dimensions. So, individual relative sea-level curves reflect the puzzle of tectonic units of the Mediterranean Sea. Generally on the European shelf, external forces causing glacio-(hydro)-isostatic adjustments, and eustatic changes are superimposed on the internally induced tectonic dynamics. Additionally, on a shorter timescale, processes of sediment dynamics driven by atmospheric and hydrographic forces have to be considered.

According to the factors affecting changes of relative sea level, the different driving forces that have shaped the European shorelines and the variable exposed areas of the continental shelf can be separated both temporally as well as regionally.

Glacio-isostatic adjustment caused by unloading (melting inland ice) and loading (seawater volume increase) was the dominant factor affecting coastline change from the Late Glacial Period to the Atlantic Period. Water-level change of dammed isolated freshwater lakes was a major factor influencing paleogeographic patterns in the Baltic and Black seas from the Late Pleistocene to the Boreal.

Climatically controlled eustatic sea-level rise was the major factor determining the inundation of the shelf in the North Sea and Atlantic during deglaciation, extending up to the Subatlantic Period.

Along Mediterranean coasts, tectonic deformation of the Earth's crust was superimposed on the postglacial sea-level rise continuously since the glacial maximum.

Coasts built-up of soft materials such as Pleistocene deposits are reshaped by the dynamics of nearshore processes driven by atmospheric circulation. These forces have had their main effect on coastline development mainly since the sea-level rise slowed during the Subatlantic Period.

Integrated modeling approaches of climate and glacio-isostasy can demonstrate favorable conditions for the preservation of submerged paleolandscapes. These approaches may provide valuable prerequisites for planning and executing archaeological surveys on the continental shelf. These models describe changes in Earth's climate during the Late Pleistocene and the corresponding melting of continental ice shields marking the end of the last glaciation. The parameterization of the load due to the redistribution of water by the melting of continental ice sheets is accomplished by the interpretation of geological proxy data. Such models compute the changing rates of relative sea level along the continental margins and display the transgression or regression of the coastline. For times of relatively rapid sea-level rise — as during the Late Pleistocene and Early Holocene — paleolandscapes are more likely to preserve their former shape than during periods of slow sea-level rise. From the Mid-to-Late Holocene — a period of relatively slow sea-level rise — migrating highly dynamic shorelines rework intensively the substrate of the coastal zone, eventually destroying many archaeological sites and burying others. The application of long-term morphodynamic sediment models can help to identify paleocoasts with higher potential for the conservation of archaeological sites and paleolandscapes.

References

Ahmed Benan, C. A. & Kocurek, G. 2000. Catastrophic flooding of an aeolian dune field: Jurassic Entrada and Todilto Formations, Ghost Ranch, New Mexico, USA. *Sedimentology* 47:1069-1080.

Alley, R. B., Anandakrishnan, S. & Jung, P. 2001. Stochastic resonance in the North Atlantic. *Paleoceanography* 16:190-198.

Andersen, B. G. & Borns Jr., H. W. 1997. *The Ice Age World*. Scandinavian University Press: Oslo.

Andrén, T., Björck, S., Andrén, E., Conley, D., Zillén, L. & Anjar J. 2011. The Development of the Baltic Sea Basin During the Last 130 ka. In Harff, J., Björck, S. & Hoth, P. (eds.) *The Baltic Sea Basin*. pp. 75-97. Springer: New York.

Antonioli, F., Ferranti, L., Fontana, A. *et al.* 2009a. Holocene relative sea-level changes and vertical movements along the Italian coastline. *Quaternary International* 206:102-133.

Antonioli, F., Amorosi, A., Correggiari, A. *et al.* 2009b. Relative sea-level rise and asymmetric subsidence in the northern Adriatic. *Rendiconti online della Società Geologica Italiana* 9:5-7.

Argus, D. F., Gordon, R. G., DeMets, C. & Stein, S. 1989. Closure of the Africa-Eurasia-North America plate motion circuit and tectonics of the Gloria Fault. *Journal of Geophysical Research* 94(B5):5585-5602.

Arkhangel'sky, A. D. & Strakhov, N. M. 1938. *Geological Structure and Evolution of the Black Sea*. Geological Institute of the Russian Academy of Sciences of the USSR: Moscow-Leningrad (in Russian).

Armijo, R., Meyer, B., King, G. C. P., Rigo, A. & Papanastassiou, D. 1996. Quaternary evolution of the Corinth rift and its implications for the Late Cenozoic evolution of the Aegean. *Geophysical Journal International* 126:11-53.

Armijo, R., Flerit, F., King, G., & Meyer, B. 2004. Linear elastic fracture mechanics explains the past and present evolution of the Aegean. *Earth and Planetary Science Letters* 217:85-95.

Bakker, M., Meekes, S., van Heteren, S. *et al.* 2012. Towards a three-dimensional geological model of the North Sea subsurface. In Kranenburg, W. M., Horstman, E. M. & Wijnberg, K. M. (eds.) *NCK-days 2012: Crossing borders in Coastal Research — Jubilee Conference Proceedings (University of Twente)*. 13th – 16th March 2012, Enschede (the Netherlands), pp. 81-84.

Bard, E., Hamelin, B., Fairbanks, R. G. & Zinder, A. 1990. A calibration of the ^{14}C timescale over the past 30,000 years using mass spectrometric U-Th ages from Barbados corals. *Nature* 345:405-410.

Bartolomé, R., Gràcia, E., Stich, D. *et al.* 2012. Evidence for active strike-slip faulting along the Eurasia-Africa convergence zone: Implications for seismic hazard in the southwest Iberian Margin. *Geology* 40: 495-498.

Barton, P. & Woods, R. 1984. Tectonic evolution of the North Sea basin: crustal stretching and subsidence. *Geophysical Journal of the Royal Astronomical Society* 79:987-1022.

Berger, A. & Loutre, M. F. 1991. Insolation values for the climate of the last 10 million years. *Quaternary Science Reviews* 10:297-317.

Biju-Duval, B. & Montadert, L. (eds.) 1977. *Structural History of the Mediterranean Basins*. Editions Technip: Paris.

Bindoff, N. L., Willebrand, J., Artale, V. *et al.* 2007. Observations: Oceanic climate change and sea level. In Solomon, S., Qin, D., Manning, M. *et al.* (eds.) *Climate Change 2007: The Physical Science Basis. Contribution of Working Group I to the Fourth Assessment Report of the Intergovernmental Panel on Climate Change*. Cambridge University Press: Cambridge/New York.

Bond, G., Showers, W., Cheseby, M. *et al.* 1997. A pervasive millennial-scale cycle in North Atlantic Holocene and glacial climates. *Science* 278: 1257-1266.

Bond, G., Kromer, B., Beer, J. *et al.* 2001. Persistent solar influence on North Atlantic climate during the Holocene. *Science* 294:2130-2136.

Brooks, C. E. P. 1926. *Climate Through The Ages: A Study of the Climatic Factors and Their Variations.* Ernest Benn Ltd: London.

Brossolo, L., Mascle, J. & Loubtrieu, B. 2012. *Morpho-Bathymetric map of the Mediterranean Sea (scale 1:4,000,000), 1st Edition.* CGMW/UNESCO.

Buforn, E., Sanz de Galdeano, C. & Udías, A. 1995. Seismotectonics of the Ibero-Maghrebian region. *Tectonophysics* 248:247-261.

Buforn, E., Bezzeghoud, M., Udías, A. & Pro, C. 2004. Seismic sources on the Iberia-African Plate boundary and their tectonic implications. *Pure & Applied Geophysics* 161:623-646.

Cameron, T. D., Crosby, A., Balson, P. S. *et al.* 1992. *UK Offshore Regional Report: The Geology of the Southern North Sea.* HMSO for the British Geological Survey: London.

Carlson, A. E. 2011. Ice sheets and sea level in Earth's past. *Nature Education Knowledge* 3(10):3.

Carminati, E. & Doglioni, C. 2005. Europe — Mediterranean tectonics. In Selley, R. C. Cocks, L. R. M. & Plimer, I. R. (eds.) *Encyclopedia of Geology.* pp. 135-146. Elsevier: Oxford.

Church, J. A., Clark, P. U., Cazenave, A. *et al.* 2013. Sea Level Change. In Stocker, T. F., Qin, D., Plattner, G.-K. *et al.* (eds.) *Climate Change 2013: The Physical Science Basis. Contribution of Working Group I to the Fifth Assessment Report of the Intergovernmental Panel on Climate Change.* Cambridge University Press: Cambridge/New York.

Clark, P. U., Dyke, A. S., Shakun, J. D. *et al.* 2009. The last glacial maximum. *Science* 325:710-714.

Crowley, T. J. & North, G. R. 1991. *Paleoclimatology.* Oxford University Press: New York.

Cuffey, K. M., Clow, G. D., Alley, R. B., Stuiver, M., Waddington, E. D. & Saltus, R. W. 1995. Large Arctic temperature change at the Wisconsin-Holocene glacial transition. *Science* 270:455-458.

Davis, B. A. S., Brewer, S., Stevenson, A. C. & Guiot, J. 2003. The temperature of Europe during the Holocene reconstructed from pollen data. *Quaternary Science Reviews* 22:1701-1716.

de Vriend, H. J. 2001. Long-term morphological prediction. In Seminara, G. & Blondeaux, P. (eds.) *River,*

Coastal and Estuarine Morpho-dynamics. pp. 163-190. Springer: Berlin Heidelberg.

de Vriend, H. J., Copabianco, M., Chesher, T., De Swart, H. E., Latteux, B. & Stive, M. J. F. 1993a. Approaches to long-term modelling of coastal morphology. *Coastal Engineering* 21:225-269.

de Vriend, H. J., Zyserman, J., Nicholson, J., Roelvink, J. A., Pechon, P. & Southgate, H. N. 1993b. Medium-term 2DH coastal-area modelling. *Coastal Engineering* 21:193-224.

Dewey, J. F. & Şengör, A. M. C. 1979. Aegean and surrounding regions: Complex multiplate and continuum tectonics in a convergent zone. *Geological Society of America Bulletin* 90:84-92.

Dyke, A. S. 2004. An outline of North American deglaciation with emphasis on central and Northern Canada. In Ehlers, J. & Gibbard, P. L. (eds.) *Quaternary Glaciations: Extent and Chronology Part II: North America.* p. 373-424. Elsevier: Amsterdam.

Evelpidou, N., Vassilopoulos, A. & Pirazzoli, P. A. 2012a. Holocene emergence in Euboea island (Greece). *Marine Geology* 295-298:14-19.

Evelpidou, N., Vassilopoulos, A. & Pirazzoli, P. A. 2012b. Submerged notches on the coast of Skyros Island (Greece) as evidence for Holocene subsidence. *Geomorphology* 141-142:81-87.

Fagherazzi, S. & Overeem, I. 2007. Models of deltaic and inner continental shelf landform evolution. *Annual Review of Earth and Planetary Science* 35: 685-715.

Fairbanks, R. G. 1989. A 17,000-year glacio-eustatic sea level record: influence of glacial melting rates on the Younger Dryas event and deep-ocean circulation. *Nature* 342:637-642.

Farrell, W. E. & Clark, J. A. 1976. On postglacial sea level. *Geophysical Journal of the Royal Astronomical Society* 46:647-667.

Ferranti, L., Antonioli, F., Mauz, B. *et al.* 2006. Markers of the last interglacial sea-level high stand along the coast of Italy: Tectonic implications. *Quaternary International* 145-146:30-54.

Franke, W. 1989. Tectonostratigraphic units in the Variscan belt of central Europe. *Geological Society of America Special Papers* 230:67-90.

Gaffney V., Thomson, K. & Fitch, S. (eds.) 2007. *Mapping Doggerland: The Mesolithic Landscapes of the Southern North Sea.* Archaeopress: Oxford.

Gibbard, P. L. 1988. The history of the great northwest European rivers during the past three million years. *Philosophical Transactions of the Royal Society, London* B318:559-602.

Giunta, S., Morigi, C., Negri, A., Guichard, F. & Lericolais, G. 2007. Holocene biostratigraphy and paleoenvironmental changes in the Black Sea based on calcareous nannoplankton. *Marine Micropaleontology* 63:91-110.

Grant, K. M., Rohling, E. J., Bar-Matthews, M. *et al.* 2012. Rapid coupling between ice volume and polar temperature over the past 150,000 years. *Nature* 491:744-747.

Groh, A., Dietrich, R. & Richter, A. 2011. Geodetic evidence and modelling of sea-level changes and load-induced crustal deformations in the southern Baltic Sea. In Harff, J. & Lüth, F. (eds.) *Sinking Coasts — Geosphere, Ecosphere and Anthroposphere of the Holocene Southern Baltic Sea II.* pp. 17-40. Bericht der Römisch-Germanische Kommission (vol. 92).

Hanebuth, T. J. J., Voris, H. K., Yokoyama, Y., Saito, Y. & Okuno, J. 2011. Formation and fate of sedimentary depocentres on Southeast Asia's Sunda Shelf over the past sea-level cycle and biogeographic implications. *Earth-Science Reviews* 104:92-110.

Harff, J., Lemke, W., Lampe, R. *et al.* 2007. The Baltic Sea coast — A model of interrelations among geosphere, climate, and anthroposphere. *Geological Society of America Special Papers* 426:133-142.

Harff, J., Endler, R., Emelyanov, E. *et al.* 2011. Late Quaternary climate variations reflected in Baltic Sea sediments. In Harff, J., Björck, S. & Hoth, P. (eds.) *The Baltic Sea Basin.* pp. 99-132. Springer: Berlin Heidelberg.

Hünicke, B., Zorita, E. & Haeseler, S. 2011. Holocene climate simulations for the Baltic Sea Region — application for sea level and verification of proxy data. In Harff, J & Lüth, F. (eds.) *Sinking Coasts — Geosphere, Ecosphere and Anthroposphere of the Holocene Southern Baltic Sea II.* pp. 211-249. Bericht der Römisch-Germanische Kommission (vol. 92).

Indermühle, A., Stocker, T. F., Joos, F. *et al.* 1999. Holocene carbon-cycle dynamics based on CO_2 trapped in ice at Taylor Dome, Antarctica. *Nature* 398:121-126.

Jensen, J. B., Bennike, O., Witkowski, A., Lemke, W. & Kuijpers, A. 1999. Early Holocene history of the southwestern Baltic Sea: The Ancylus Lake stage. *Boreas* 28:437-453.

Jiménez, J. A. & Sánchez-Arcilla, A. 2004. A long-term (decadal scale) evolution model for microtidal barrier systems. *Coastal Engineering* 51:749-764.

Jolivet, L. 1993. Extension of thickened continental crust, from brittle to ductile deformation: examples from Alpine Corsica and Aegean Sea. *Annali di Geofisica* 36:139-153.

Jolivet, L. 2001. A comparison of geodetic and finite strain pattern in the Aegean, geodynamic implications. *Earth and Planetary Science Letters* 187:95-104.

Jones, G. A. & Keigwin, L. D. 1988. Evidence from Fram Strait (78°N) for early deglaciation. *Nature* 336:56-59.

Joughin, I., Alley, R. B. & Holland, D. M. 2012. Ice-sheet response to oceanic forcing. *Science* 338:1172-1176.

Jouzel, J., Masson-Delmotte, V., Cattani. O. *et al.* 2007. Orbital and millennial Antarctic climate variability over the past 800,000 years. *Science* 317:793-796.

Kapsimalis, V., Pavlakis, P., Poulos, S. E. *et al.* 2005. Internal structure and evolution of the Late Quaternary sequence in a shallow embayment: The Amvrakikos Gulf, NW Greece. *Marine Geology* 222-223: 399-418.

Karunarathna, H., Reeve, D. & Spivack, M. 2008. Long-term morphodynamic evolution of estuaries: An inverse problem. *Estuarine, Coastal and Shelf Science* 77:385-395.

Kastens, K. A. & Mascle, J. 1990. The geological evolution of the Tyrrhenian Sea: An introduction to the scientific results of ODP Leg 107. In Kastens, K. A., Mascle, J., Aurroux, C. *et al. Proceedings of the Ocean Drilling Program, Scientific Results, 107: College Station, TX (Ocean Drilling Program).* pp. 3-26.

Kastens, K. A., Mascle, J., Auroux, C. *et al.* 1988. ODP Leg 107 in the Tyrrhenian Sea: insights into passive margin and back-arc basin evolution. *Geological Society of America Bulletin* 100:1140-1156.

Kontogianni, V. A., Tsoulos, N. & Stiros, S. C. 2002. Coastal uplift, earthquakes and active faulting of Rhodes Island (Aegean Arc): Modeling based on geodetic inversion. *Marine Geology* 186:299-317.

Kossmat, F. 1927. Gliederung des varistischen Gebirgsbaues. *Abhandlungen des Sächsischen Geologischen Landesamts* 1:1-39.

Kreemer, C. & Chamot-Rooke, N. 2004, Contemporary kinematics of the southern Aegean and the Mediterranean Ridge. *Geophysical Journal International* 157:1377-1392.

Lagarde, J. L., Amorese, D., Font, M., Laville, E. & Dugué, O. 2003. The structural evolution of the English Channel area. *Journal of Quaternary Science* 18:201-213.

Lambeck, K. 1995. Late Devensian and Holocene shorelines of the British Isles and the North Sea from models of glacio-hydro-isostatic rebound. *Journal of the Geological Society* 152:437-448.

Lambeck, K. & Chappell, J. 2001. Sea-level change through the last glacial cycle. *Science* 292:679-686.

Lambeck, K. & Purcell, A. P. 2001. Sea-level change in the Irish Sea since the Last Glacial Maximum: constraints from isostatic modelling. *Journal of Quaternary Science* 16:497-506.

Lambeck, K. & Purcell, A. 2005. Sea-level change in the Mediterranean Sea since the LGM: model predictions for tectonically stable areas. *Quaternary Science Reviews* 24:1969-1988.

Lambeck, K. & Purcell, A. 2007. Palaeogeographic reconstructions of the Aegean for the past 20,000 years: Was Atlantis on Athens' doorstep? In Papamarinopoulos, St. P. (ed.) *The Atlantis Hypothesis: Searching for a Lost Land.* pp. 241-257. Heliotopos Publications: Santorini.

Lambeck, K., Smither, C. & Johnston, P. 1998a. Sea-level change, glacial rebound and mantle viscosity for northern Europe. *Geophysical Journal International* 134:102-144.

Lambeck, K., Smither, C. & Ekman, M. 1998b. Tests of glacial rebound models for Fennoscandinavia based on instrumented sea- and lake-level records. *Geophysical Journal International* 135:375-387.

Lambeck, K., Purcell, A., Zhao, J. & Svensson, N.-O. 2010. The Scandinavian Ice Sheet: from MIS 4 to the end of the Last Glacial Maximum. *Boreas* 39:410-435.

Lambeck, K., Antonioli, F., Anzidei, M. *et al.* 2011. Sea level change along the Italian coast during the Holocene and projections for the future. *Quaternary International* 232:250-257.

Lambeck, K., Rouby, H., Purcell, A., Sun, Y. & Sambridge, M. 2014. Sea level and global ice volumes from the Last Glacial Maximum to the Holocene. *Proceedings of the National Academy of Sciences* 111:15296-15303.

Landvik, J. Y., Bondevik, S., Elverhoei, A. *et al.* 1998. The last glacial maximum of Svalbard and the Barents Sea area: Ice sheet extent and configuration. *Quaternary Science Reviews* 17:43-75.

Lauer, A. & Hamilton, K. 2013. Simulating clouds with global climate models: A comparison of CMIP5 results with CMIP3 and satellite data. *Journal of Climate* 26:3823-3845.

Lawver, L. A., Dalziel, I. W. D., Norton, I. O. & Gahagan, L. M. 2009. PLATES 2009 Atlas of Plate Reconstructions (750 Ma to present day). *PLATES Progress Report No. 325-0509.* University of Texas Technical Report No. 196.

Le Pichon, X. & Angelier, J. 1979. The Hellenic arc and trench system: a key to the neotectonic evolution of the eastern Mediterranean area. *Tectonophysics* 60: 1-42.

Le Pichon, X., Angelier, J. & Sibuet, J.-C. 1982. Plate boundaries and extensional tectonics. *Tectonophysics* 81:239-256.

Lericolais, G., Auffret, J-P. & Bourillet, J. F. 2003. The Quaternary Channel River: Seismic stratigraphy of its palaeo-valleys and deeps. *Journal of Quaternary Science* 18:245-260.

Lericolais, G., Chivas, A. R., Chiocci, F. L. *et al.* 2004. Rapid transgressions into semi-enclosed basins since the Last Glacial Maximum. In IGC-IUGS-UNESCO (ed.) *32nd International Geological Congress (Abstracts).* 20th – 27th August 2004, Florence, p. 1124.

Lericolais, G., Bulois, C., Gillet, H. & Guichard, F. 2009. High frequency sea level fluctuations recorded in the Black Sea since the LGM. *Global and Planetary Change* 66:65-75.

Lericolais, G., Guichard, F., Morigi, C. *et al.* 2011. Assessment of Black Sea water-level fluctuations since the Last Glacial Maximum. In Buynevich, I., Yanko-Hombach, V., Gilbert, A. & Martin, R. (eds.) *Geology and Geoarchaeology of the Black Sea Region: Beyond the Flood Hypothesis.* GSA Special Paper 473:33-50.

Lericolais, G., Bourget, J., Popescu, I. *et al.* 2013. Late Quaternary deep-sea sedimentation in the western Black Sea: New insights from recent coring and seismic data in the deep basin. *Global and Planetary Change* 103:232-247.

Long, D., Wickham-Jones, C. R. & Ruckley, N. A. 1986. A flint artefact from the northern North Sea. In Roe, D. A. (ed.) BAR International Series (No. 296) *Studies in the Upper Palaeolithic of Britain and North West Europe.* pp. 55-62. British Archaeological Reports: Oxford.

Lykousis, V. 2009. Sea-level changes and shelf break pro-grading sequences during the last 400 ka in the Aegean margins: Subsidence rates and palaeogeographic implications. *Continental Shelf Research* 29:2037-2044.

Lykousis, V. & Anagnostou, C. 1993. Sedimentological and paleogeographic evolution of the Saronic Gulf during the Late Quaternary. *Bulletin of the Geological Society of Greece* 28(1):501-510 (in Greek).

Lykousis, V., Sakellariou, D., Moretti, I. & Kaberi, H. 2007. Late Quaternary basin evolution of the Gulf of Corinth: Sequence stratigraphy, sedimentation, fault-slip and subsidence rates. *Tectonophysics* 440: 29-51.

Martínez Catalán, J. R., Arenas, R., Díaz García, F. *et al.* 2007. Space and time in the tectonic evolution of the northwestern Iberian Massif: Implications for the Variscan belt. *Geological Society of America Memoirs* 200:403-423.

Martínez-García, P., Soto, J. I. & Comas, M. C. 2011. Recent structures in the Alboran Ridge and Yusuf fault zones based on swath bathymetry and sub-bottom profiling: evidence of active tectonics. *Geo-Marine Letters* 31:19-36.

Mascle, J. & Martin, L. 1990. Shallow structure and recent evolution of the Aegean Sea: A synthesis based on continuous reflection profiles. *Marine Geology* 94:271-299.

McKenzie, D. P. 1970. Plate tectonics of the Mediter-ranean region. *Nature* 226:239-243.

McKenzie, D. 1978. Active tectonics of the Alpine-Himalayan belt: the Aegean Sea and surrounding regions. *Geophysical Journal International* 55:217-254.

Medeiros, B., Stevens, B., Held, I. M. *et al.* 2008. Aqua-planets, climate sensitivity, and low clouds. *Journal of Climate* 21:4974-4991.

Meijer, P. T. & Wortel, M. J. R. 1997. Present-day dynamics of the Aegean region: A model analysis of the horizontal pattern of stress and deformation. *Tectonics* 16:879-895.

Meschede, M. 2015. *Geologie Deutschlands — Ein-prozessorientierter Ansatz.* Springer Spektrum: Berlin Heidelberg.

Meulenkamp, J. E., Wortel, M. J. R., van Wamel, W. A., Spakman, W. & Hoogerduyn Strating, E. 1988. On the Hellenic subduction zone and the geodynamical evolution of Crete since the late Middle Miocene. *Tectonophysics* 146:203-215.

Milankovitch, M. 1930. *Mathematische Klimalehre und Astronomische Theorie der Klimaschwankungen (Hand-buch Klimatologie I A).* Gebrüder Borntraeger: Berlin.

Milne, G. A., Davis, J. L., Mitrovica, J. X. *et al.* 2001. Space-Geodetic Constraints on Glacial Isostatic Adjustment in Fennoscandia. *Science* 291:2381-2385.

Milne, G. A., Shennan, I., Youngs, B. A. R. *et al.* 2006. Modelling the glacial isostatic adjustment of the UK region. *Philosophical Transactions of the Royal Society (Series A)* 364:931-948.

Mitrovica, J. X., Tamisiea, M. E., Davis, J. L. & Milne, G. A. 2001. Recent mass balance of polar ice sheets inferred from patterns of global sea-level change. *Nature* 409:1026-1029.

Moreno, X. 2011. *Neotectonic and Paleoseismic Onshore-Offshore Integrated Study of the Carboneras Fault (East-ern Betics, SE Iberia).* Ph.D Thesis. Universitat de Barcelona, Spain.

Müller, R. D., Sdrolias, M. Gaina, C. & Roest, W. R. 2008. Age, spreading rates, and spreading asymmetry of the world's ocean crust. *Geochemistry, Geophysics, Geosystems* 9:Q04006.

Musson, R. M. W. 2007. British Earthquakes. *Proceedings of the Geologists' Association* 118:305-337.

Neill, S. P., Scourse, J. D., Bigg, G. R. & Uehara, K. 2009. Changes in wave climate over the northwest European shelf seas during the last 12,000 years. *Journal of Geophysical Research* 114:C06015.

Neill, S. P., Scourse, J. D. & Uehara, K. 2010. Evolution of bed shear stress distribution over the northwest European shelf during the last 12,000 years. *Ocean Dynamics* 60:1139-1156.

Nicholas, W. A., Chivas, A. R., Murray-Wallace, C. V. & Fink, D. 2011. Prompt transgression and gradual salinisation of the Black Sea during the early Holocene constrained by amino acid racemization and radiocarbon dating. *Quaternary Science Reviews* 30:3769-3790.

Olsen, J., Anderson, J. N. & Knudsen, M. F. 2012. Variability of the North Atlantic Oscillation over the past 5200 years. *Nature Geoscience* 5:808-812.

Peltier, W. R. 1998. Postglacial variations in the level of the sea: Implications for climate dynamics and solid-Earth geophysics. *Reviews of Geophysics* 36:603-689.

Peltier, W. R. 2004. Global glacial isostasy and the surface of the ice-age Earth: The ICE-5G (VM2) model and GRACE. *Annual Review of Earth and Planetary Sciences* 32:111-149.

Peltier, W. R., Shennan, I., Drummond, R. & Horton B. 2002. On the postglacial isostatic adjustment of the British Isles and the shallow viscoelastic structure of the Earth. *Geophysical Journal International* 148:443-475.

Perissoratis, C., Piper, D. J. W. & Lykousis, V. 1993. Late Quaternary sedimentation in the Gulf of Corinth: the effects of marine-lake fluctuations driven by eustatic sea level changes. pp. 693-744. *Special Publications of the National Technical University of Athens (dedicated to Prof. A. Panagos)*.

Pettit, P. & White, M. 2012. *The British Palaeolithic: Human Societies at the Edge of the Pleistocene World*. Routledge: Abingdon/New York.

Pirazzoli, P. A., Montaggioni, L. F., Saliege, J. F., Segonzac, G., Thommeret, Y. & Vergnaud-Grazzini, C. 1989. Crustal block movements from Holocene shorelines: Rhodes Island (Greece). *Tectonophysics* 170:89-114.

Plant, J. A., Reeder, S., Salminen, R. *et al.* 2003. The distribution of uranium over Europe: geological and environmental significance. *Applied Earth Science (Trans. Inst. Min. Metal. B)* (vol. 112(3)).

Popescu, I., Lericolais, G., Panin, N., Wong, H. K. & Droz, L. 2001. Late Quaternary channel avulsions on the Danube deep-sea fan, Black Sea. *Marine Geology* 179:25-37.

Popescu, I., Lericolais, G., Panin, N., Normand, A., Dinu, C. & Le Drezen, E. 2004. The Danube submarine canyon (Black Sea): morphology and sedimentary processes. *Marine Geology* 206:249-265.

Popescu, S. M. 2004. Sea-level changes in the Black Sea region since 14 ka BP. In IGC-IUGS-UNESCO (ed.) *32nd International Geological Congress* (vol. 2, part 2) *Abstracts*. 20th–28th August 2004, Florence, p. 1426.

Popescu, S. M., Head, M. V. & Lericolais, G. 2003. Holocene Black Sea environments according to palynology. *Geological Society of America Abstracts with Programs* (vol. 35 (6)). 2nd – 5th November 2003, Seattle, p. 462.

Rahmstorf, R. 2002. Ocean circulation and climate during the past 120,000 years. *Nature* 419:207-214.

Richter, D. K., Anagnostou, Ch. & Lykousis, V. 1993. Aragonitishe Whiting-Ablagerungen in Plio-/Pleistozaenen Mergelsequenzen bei Korinth (Griechenland). *Zentralblatt für Geologie und Paläontologie Teil 1*. 6:675-688.

Rinterknecht, V. R., Clark, P. U., Raisbeck, G. M. *et al.* 2006. The last deglaciation of the southeastern sector of the Scandinavian ice sheet. *Science* 311:1449-1452.

Roelvink, J. A. 2006. Coastal morphodynamic evolution techniques. *Coastal Engineering* 53:277-287.

Rohling, E. J., Grant, K., Bolshaw, M. *et al.* 2009. Antarctic temperature and global sea level closely coupled over the past five glacial cycles. *Nature Geoscience* 2:500-504.

Rosentau, A., Muru, M., Kriiska, A. *et al.* 2013. Stone Age settlement and Holocene shore displacement in the Narva-Luga Klint Bay area, eastern Gulf of Finland. *Boreas* 42:912-931.

Ruddiman, W. F. 2008. *Earth's Climate: Past and Future* (2nd Ed). W. H. Freeman & Co.: New York.

Ryan, W. B. F., Pitman III, W. C., Major, C. O. *et al.* 1997. An abrupt drowning of the Black Sea shelf. *Marine Geology* 138:119-126.

Ryan, W. B. F., Major, C. O., Lericolais, G. & Goldstein, S. L. 2003. Catastrophic flooding of the Black Sea. *Annual Review of Earth and Planetary Sciences* 31:525-554.

Sakellariou, D., Lykousis, V., Alexandri, S. *et al.* 2007a. Faulting, seismic-stratigraphic architecture and Late Quaternary evolution of the Gulf of Alkyonides Basin — East Gulf of Corinth, Central Greece. *Basin Research* 19:273-295.

Sakellariou D., Rousakis, G., Kaberi, H. *et al.* 2007b. Tectono-sedimentary structure and Late Quaternary evolution of the north Evia gulf basin, central Greece: Preliminary results. *Bulletin of the Geological Society of Greece* 40(1):451-462.

Sartori, R. 1990. The main results of ODP Leg 107 in the frame of Neogene to recent geology of peri-Tyrrhenian areas. In Kastens, K.A., Mascle, J., Aurroux, C. *et al. Proceedings of the Ocean Drilling Program, Scientific Results, 107: College Station, TX (Ocean Drilling Program).* pp. 715-730.

Sartori, R. 2003. The Tyrrhenian back-arc basin and subduction of the Ionian lithosphere. *Episodes* 26:217-221.

Schmidt, G. A., Jungclaus, J. H., Ammann, C. M. *et al.* 2011. Climate forcing reconstructions for use in PMIP simulations of the last millennium (v1.0). *Geoscientific Model Development* 4:33-45.

Serri, G. 1997. Neogene-Quaternary magmatic activity and its geodynamic implications in the Central Mediterranean region. *Annali di Geofisica* 40:681-703.

Shaw, B., Ambraseys, N. N., England, P. C. *et al.* 2008. Eastern Mediterranean tectonics and tsunami hazard inferred from the AD 365 earthquake. *Nature Geoscience* 1:268-276.

Shennan, I., Bradley, S., Milne, G., Brooks, A., Bassett, S. & Hamilton, S. 2006. Relative sea-level changes, glacial isostatic modelling and ice-sheet reconstructions from the British Isles since the Last Glacial Maximum. *Journal of Quaternary Science* 21:585-599.

Siddall, M., Pratt, L. J., Helfrich, K. R. & Giosan, L. 2004. Testing the physical oceanographic implications of the suggested sudden Black Sea infill 8400 years ago. *Paleoceanography* 19:PA1024.

Siegenthaler, U., Stocker, T. F., Monnin, E. *et al.* 2005. Stable carbon cycle–climate relationship during the Late Pleistocene. *Science* 310:1313-1317.

Solanki, S. K., Usoskin, I. G., Kromer, B., Schüssler, M. & Beer, J. 2004. Unusual activity of the Sun during recent decades compared to the previous 11,000 years. *Nature* 431:1084-1087.

Soulet, G., Delaygue, G., Vallet-Coulomb, C. *et al.* 2010. Glacial hydrologic conditions in the Black Sea reconstructed using geochemical pore water profiles. *Earth and Planetary Science Letters* 296:57-66.

Soulet, G., Ménot, G., Garreta, V. *et al.* 2011a. Black Sea "Lake" reservoir age evolution since the Last Glacial — Hydrologic and climatic implications. *Earth and Planetary Science Letters* 308:245-258.

Soulet, G., Ménot, G., Lericolais, G. & Bard, E. 2011b. A revised calendar age for the last reconnection of the Black Sea to the global ocean. *Quaternary Science Reviews* 30:1019-1026.

Spada, G. & Stocchi, P. 2007. SELEN: A Fortran 90 program for solving the "sea-level equation". *Computers & Geosciences* 33:538-562.

Stich, D., Mancilla, F. & Morales, J. 2005. Crust mantle coupling in the Gulf of Cadiz (SW Iberia). *Geophysical Research Letters* 32:L13306.

Stich, D., Mancilla, F., Pondrelli, S. & Morales, J. 2007. Source analysis of the February 12th 2007, M_W 6.0 Horseshoe earthquake: Implications for the 1755 Lisbon earthquake. *Geophysical Research Letters* 34:L12308.

Stich, D., Martín, R. & Morales, J. 2010. Moment tensor inversion for Iberia–Maghreb earthquakes 2005–2008. *Tectonophysics* 483:390-398.

Thorne, J. A. & Watts, A. B. 1989. Quantitative analysis of North Sea subsidence. *The American Association of Petroleum Geologists Bulletin* 73:88-116.

Toucanne, S., Zaragosi, S., Bourillet, J. F. *et al.* 2009. Timing of massive 'Fleuve Manche' discharges over the last 350 kyr: insights into the European ice-sheet oscillations and the European drainage network from MIS 10 to 2. *Quaternary Science Reviews* 28:1238-1256.

van den Broeke, M., Bamber, J., Ettema, J. *et al.* 2009. Partitioning recent Greenland mass loss. *Science* 326:984-986.

Viti, M., Mantovani, E., Babbucci, D. & Tamburelli, C. 2011. Plate kinematics and geodynamics in the Central Mediterranean. *Journal of Geodynamics* 51:190-204.

Waelbroeck, C., Labeyrie, L., Michel, E. *et al.* 2002. Sea-level and deep water temperature changes derived from benthic foraminifera isotopic records. *Quaternary Science Reviews* 21:295-305.

Wanner, H., Solomina, O., Grosjean, M., Ritz, S. P. & Jetel, M. 2011. Structure and origin of Holocene cold events. *Quaternary Science Reviews* 30:3109-3123.

Ward, I., Larcombe, P. & Lillie, M. 2006. The dating of Doggerland — post-glacial geochronology of the southern North Sea. *Environmental archaeology* 11:207-218.

Weninger, B., Schulting, R., Bradtmöller, M. *et al.* 2008. The catastrophic final flooding of Doggerland by the Storegga Slide tsunami. *Documentia Praehistorica* 35:1-24.

Westaway, R. 1993. Quaternary uplift of southern Italy. *Journal of Geophysical Research: Solid Earth* 98(B12):21741-21772.

Woodward, R. S. 1888. On the form and position of the sea level. *United States Geological Survey Bulletin* 48:87-170.

Yokoyama, Y., Lambeck, K., De Deckker, P., Johnston, P. & Fifield, L. K. 2000. Timing of the Last Glacial Maximum from observed sea-level minima. *Nature* 406:713-716.

Zhang, W. Y., Harff, J., Schneider, R. & Wu, C. 2010. Development of a modeling methodology for simulation of long-term morphological evolution of the southern Baltic coast. *Ocean Dynamics* 60:1085-1114.

Zhang, W. Y., Harff, J. & Schneider, R. 2011. Analysis of 50-year wind data of the southern Baltic Sea for modelling coastal morphological evolution — a case study from the Darss-Zingst Peninsula. *Oceanologia* 53:489-518.

Zhang, W. Y., Schneider, R. & Harff, J. 2012. A multi-scale hybrid long-term morphodynamic model for wave-dominated coasts. *Geomorphology* 149-150: 49-61.

Zhang, W. Y., Harff, J., Schneider, R., Meyer, M., Zorita, E. & Hünicke, B. 2014. Holocene morphogenesis at the southern Baltic Sea: Simulation of multi-scale processes and their interactions for the Darss–Zingst peninsula. *Journal of Marine Systems* 129: 4-18.

Chapter 3

Non-Cultural Processes of Site Formation, Preservation and Destruction

Nicholas C. Flemming,[1] Jan Harff[2] and Delminda Moura[3]

[1] National Oceanography Centre, Southampton, UK
[2] University of Szczecin, Szczecin, Poland
[3] Universidade do Algarve, Faculdade de Ciências e Tecnologia, Centro de Investigação Marinha e Ambiental (CIMA), Faro, Portugal

Introduction

The previous chapter described changes of sea level and climate at the global scale with an emphasis on the timescale of the last 20,000 to 120,000 years (Harff *et al.* this volume). Here we look more closely at how a relatively rising, constant, or falling sea level threatens to destroy a prehistoric deposit, and under what immediately local conditions it is likely to survive in a way that preserves information, preferably in the context of its terrestrial landscape, which can be interpreted by archaeologists. Later chapters will describe the coastal and shelf conditions favorable to prehistoric occupation, and

the local oceanographic conditions, in different parts of European seas, both during the last marine transgression, and at present, as they determine the real and potential survival and discovery of prehistoric sites.

This chapter addresses the generic problem of how some prehistoric sites and remains on the sea floor are preserved through the vicissitudes of inundation by transgressing surf and subsequent submergence, and why others are destroyed. How do any sites survive in such a hostile environment? Can we predict where they will occur and survive? If we can predict their probable survival, can we find them by scientific or deterministic planned searching? Submerged prehistoric sites have been

Submerged Landscapes of the European Continental Shelf: Quaternary Paleoenvironments, First Edition.
Edited by Nicholas C. Flemming, Jan Harff, Delminda Moura, Anthony Burgess and Geoffrey N. Bailey.

found on European coasts of all types in terms of geology and oceanographic conditions, so can we derive general rules for survival? This is gradually becoming possible, and we make a contribution, based on currently known sites, toward that objective. Predictive modeling has been tried with varying degrees of success and the specific examples will be mentioned in the regional chapters that follow.

Prehistoric archaeological sites and artifacts have been found and identified in the intertidal zone on coastal foreshores, and offshore in deeper water since at least 100 years ago (Reid 1913; Blanc 1940; Steers 1948: 339; Flemming 1968; 1983; Fischer 1995; Benjamin *et al.* 2011; Evans *et al.* 2014; Flemming *et al.* 2014). Since the beginning of proactive searches for submerged prehistoric sites in the 1960s, the question of the favorable conditions for their survival and detection has been a recurrent theme (e.g. Ruppé 1978; 1980; Gagliano *et al.* 1982; Flemming 2002).

Thousands of submerged prehistoric sites are known in the European seas scattered from the beach zone to the outer edges of the continental shelf (Jöns *et al.* 2016). Offshore prehistoric sites are found off all the coasts of Europe, though in very different spatial concentrations. In this book we will examine a small sample of sites in the chapters on regional sea areas, in order to study their geomorphological and hydrodynamic situation, but not to interpret the archaeology. Understanding how and why sites are destroyed or survive on the continental shelf, and their relationship to the surrounding landscape is an essential precursor and accompaniment to archaeological interpretation and will also assist in future discovery, research, conservation, and management.

The cultural significance of this substantial archaeological resource has been considered by Bailey (2011), Benjamin *et al.* (2011), and Evans *et al.* (2014) and the literature cited in those volumes, and the importance of the cultural research on this material will not be considered in any detail in the present work. Suffice it to say that the exposed continental shelf at glacial maxima added 40% to the land area of Europe, and that the human and pre-human occupation of this large land area has a profound impact on our understanding of how the European continent was occupied and exploited by successive hominins throughout the last million years and the origins of early maritime skills and exploitation of marine resources.

The earliest discoveries of offshore prehistoric sites were completely random and by chance (e.g. Burkitt 1932) and it was not until the late 1960s that scholars started trying to understand how and why prehistoric archaeological material survived on the seabed, and how further sites might be discovered by proactive prediction and search (e.g. Emery & Edwards 1966; Harding *et al.* 1969; Louwe Kooijmans 1970-71; Iversen 1973; Wreschner 1977; Fladmark 1975; Bowdler 1977; Ruppé 1980; Skaarup 1980; Masters & Flemming 1983; Pearson *et al.* 1986; Fischer 1995; Pedersen *et al.* 1997).

With the improved understanding of physical oceanography during and after the Second World War, especially in computing the statistics of wave climate and extremes, improvements in seabed acoustic mapping techniques (e.g. Stride 1963; Belderson *et al.* 1972; Fleming 1976; Kenyon *et al.* 1981; Stoker & Bradwell 2009) and a greater understanding of the chronology of Pleistocene glacially-driven sea-level changes (Shackleton & Opdyke 1973), it became possible to envisage purposeful searching for submerged continental-shelf prehistoric sites, and the construction of an integrated interpretation of prehistoric occupation of the shelf. It was optimistically expected that, given an understanding of the kinds of sites that occur in a region on the present land, knowledge of the local oceanographic conditions and wind fetch, a history of past climate change and sea level, and a map of the local continental shelf topography, it would be possible to create computer numerical models that would predict site occurrence, which could then be confirmed by acoustic survey and seabed sampling or diving. This would reduce the costs of search, and reduce the delay in waiting for chance finds. It would also reduce uncertainties, and hence costs, in licensing offshore concessions and the obligation to protect cultural heritage.

This goal has remained as the ideal, and indeed is part of the motivation of the present book. But the achievement of the ideal has turned out to be more difficult than expected in the 1970s and 80s. Predictive models have revealed areas of high probability and low probability for the discovery of sites (e.g. Pedersen *et al.* 1997; Benjamin 2010; Flemming *et al.* 2014: 63-65), but the application of them in a wide variety of coastal environments has not led to a deterministic ability to locate sites accurately. The conditions of the shallow seas in the Danish archipelago have proved uniquely productive of submerged prehistoric sites (e.g. Jessen 1938; Salomonsson 1971; Welinder 1971; Iversen 1973; Skaarup 1980; Andersen 2013) and it has been possible in this case to develop predictive models that save time when divers are searching the seabed. On the Mediterranean coast of Israel, reduced sand supply and periodic storms result in fresh exposures of prehistoric remains which are found by routine repeated surveys. In other parts of the European seas the discovery of exposed artifacts on the seabed or buried landscapes and artifacts in morpho-sedimentary structures has been more variable and problematic, and remains strongly influenced by chance industrial activity and intensive local knowledge of the seabed combined with repeated visual surveys.

The present chapter, indeed this whole book, approaches the problem from a Baconian rather than a Cartesian perspective. If we cannot predict the occurrence and survival of sites from a few logical first principles, can we assemble a mass of empirical data that provides the evidence for a set of useful pragmatic rules? Can we find patterns in large volumes of field data? Such rules or generalizations would show areas of parallel or similar conditions where sites are most likely to survive, and other areas where they are unlikely to survive. In due course, such experience may be used to strengthen rigorous predictive models, which can then be tested against the data and calibrated.

We have the evidence of thousands of sites that do survive (Fig. 3.1) and that have been discovered. While the precise geomorphological conditions have not always been reported, it is possible to see in a general way what conditions favor site survival and discovery. We can learn from this what factors may be important in future improved predictive models.

We are concerned with the empirical facts that describe, or can be deduced from, the processes of burial and survival in the sedimentary column of known archaeological deposits and terrestrial landscapes which have then been inundated by rising sea level. It is tempting to regard this as an extension of the study of taphonomy, which would be a convenient one-word

definition of the topic: literally "the science of burial", but the strict definition and usage of taphonomy refers to the conversion of living matter to the fossilized lithic state (Efremov 1940; Lyman 2010), and is regarded as a component of paleontology. It would be unjustifiable to misappropriate this term wholesale, notwithstanding that it is used fairly widely by archaeologists in the general sense of the process of burial and subsequent preservation or deformation of all types of materials including inorganic artifacts. The phrases "site formation" "site formation processes" and "palimpsests" have considerable associated intellectual literature (e.g. Wood & Johnson 1978; Binford 1980; Schiffer 1987; Stein 2001; Bailey 2007; Holdaway & Wandsnider 2008), and it is clear that this concept is close to the concerns of this book. The term "site formation processes" includes the various cultural and behavioral attributes that lead to the deposition and accumulation of material remains in a certain way, or in certain patterns and frequencies, and the subsequent human activities which may disturb the initial deposits through site re-use and the creation of palimpsest effects. The changes after that point are termed "non-cultural site formation processes", and it is only the non-cultural processes that concern us here, although it is obvious that modern coastal and offshore industries can and do disturb archaeological sites. In practical terms, since we know that wave and current action have destroyed many, and probably the majority, of prehistoric sites which once existed below present sea level, the phrase "non-cultural site formation, survival, destruction and discovery" is most appropriate. The word "taphonomy" will be used from time to time as shorthand for this expression provided that no misunderstanding can occur. Where the word "taphonomy" has been used by the authors of other chapters, it has not been altered. At this stage we are concerned only with the mechanistic and sedimentary biogeochemical analysis of how sites survive, or do not, and we are not extending the interpretation to consider the way in which this biases the subsequent state of knowledge about the culture and behavior of the people who created the sites. That bias manifestly occurs and must be considered by others. The large number of seabed sites that have been lost is probably not a random and unbiased sample of all the sites that existed below present sea level. Similarly, in this book,

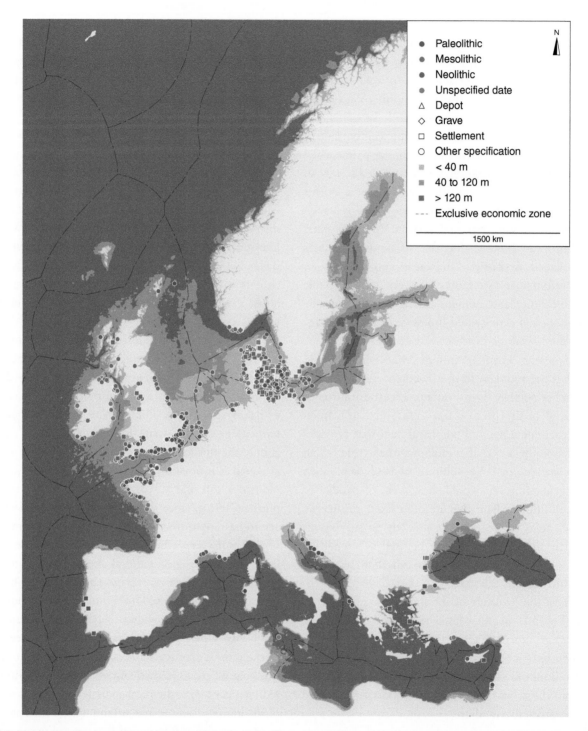

Fig. 3.1 Distribution of submerged prehistoric settlements and artifacts over 5000 years old on the coasts and in the coastal seas of Europe, reported to the SPLASHCOS database. Map courtesy of Hauke Jöns.

we are not attempting to understand or predict the cultural reasons why prehistoric peoples would choose one site rather than another for a settlement, kill site, camp, flint quarry or other activity. We are primarily concerned with understanding how or if a site, once formed, will survive marine inundation, or be destroyed. In addition, by obtaining or measuring the data that allows reconstruction of the terrestrial paleoenvironment, we hope to provide the means to a better understanding of patterns of living, foraging, hunting, mobility, migration, dispersal and adaptation to change.

The regional chapters that follow in this book analyze the geological, geomorphological, and oceanographic processes that are relevant to this problem in each area. Particular prehistoric sites will be referred to briefly, and readers wishing to obtain further details on these sites should consult the SPLASHCOS viewer (Jöns *et al.* 2016).

The sequence within the present chapter will be to examine the following themes:

1 the geology of the continental shelf, its origin, and the present shelf morphology and geomorphic forcing processes;
2 the effect of the sea-level changes described in Chapter 2 on the profile and sediments of the continental shelf;
3 how and why sites survive or are destroyed in the long, intermediate, or short term;
4 site formation and coastal landscapes and processes;
5 diagenesis of deposits in sub-aerial and coastal conditions;
6 case reviews of how sites survive in different example environments;
7 threats to known submerged prehistoric sites.

The Continental Shelf Profile, Landscape, and Factors Determining Site Survival on Different Timescales

The continental shelf is defined as the sea floor and underlying rocks from the coast seawards, having an outer edge and break of slope at a depth of 150 m below present sea level. The continental shelf is a global feature although, in places of rapid tectonic uplift and active or convergent margins, it may be extremely narrow or vestigial. Viewed simply, it is the cumulative effect over many millions of years of the marginal sinking blocks of continental crust and listric faulting on passive or divergent margins and thick sediment layers stacked on top of the sinking blocks by the sediment outflow from eroding and down-cutting rivers, glaciers, ice caps, and coastal erosion. Shepard (1963: 105–174) attempted to explain the origin and form of continental shelves before the development of plate tectonics theory, and provided an excellent descriptive summary, but the mechanism was much clearer once plate tectonic models showed the distinction between divergent margins which are spreading apart or diverging with an expanding oceanic crust area between them, and convergent margins where continental rocks are overriding the subducted oceanic crust or residual marine basins are being eliminated. The processes are summarized briefly by Seibold and Berger (1996: 45–57). A global review of the present state of knowledge of continental shelves is provided by Chiocci and Chivas (2014).

In the European-Mediterranean area both divergent and convergent margins exist and the distinction helps us to understand the extent of land which would be exposed at low Pleistocene sea levels. The Atlantic Margin of Europe from southern Spain to Norway is a divergent margin, spreading away from the Americas, with creation of new oceanic crust at the Mid-Atlantic Ridge. Thus there have been extensional tectonics, the formation of shallow marginal sea basins, and thick accumulation of sediments from the continental rivers, from glacial outwash, and coastal erosion. The result is a broad continental shelf off most coasts and hence a large area that was exposed at the times of glacial maxima and low sea level. In contrast, the Mediterranean sea basin as a whole is convergent and the distance between North Africa and southern Europe is reducing by about 5 mm/year (Sakellariou *et al.* this volume). The crustal response to the convergence and the decreasing distance between two continents is complex at the scale of 100 km to 1000 km, with active mountain building on many

Fig. 3.2 Map of Europe and the Mediterranean showing the continental shelf in gray, defined in this case as having a depth of <200 m. Reproduced with permission from GEBCO.

coasts, increasing depth of central basins, subduction of the relict oceanic crust under island arcs, lateral displacement of microplates, and active faulting and volcanics. The result is a narrow continental shelf on most Mediterranean coasts, typically less than 10 km, with exceptions in the Golfe du Lion, northern Adriatic, north and northwest Black Sea, and the Gulf of Sirte, North Africa (Fig. 3.2).

The topography of the wider divergent-margin European continental shelves is itself complex, although the undulating relief seldom exceeds 10 m to 30 m in amplitude. The average gradient is of the order of 1:1000 (0.1°) over many tens of kilometers, although rock outcrops, fossil beaches and shorelines, sand dunes, glacial moraines, river valleys, and other features can cause local highs and depressions, and even vertical cliffs. Figure 3.3

shows the bathymetry of a typical marginal sea, the North Sea, which is within the northwest European continental shelf. The details of the submerged or buried terrestrial features will be discussed in later chapters.

By contrast, Fig. 3.4 shows part of the continental shelf of eastern Spain (Mediterranean shelf), where a narrow shelf typically 10 km to 20 km wide is dissected by deep-water canyons approaching up to the modern shoreline.

Effect of sea-level variation on the shelf profile

During the last 4 million years the global sea level has been falling from an average position about 25 m to

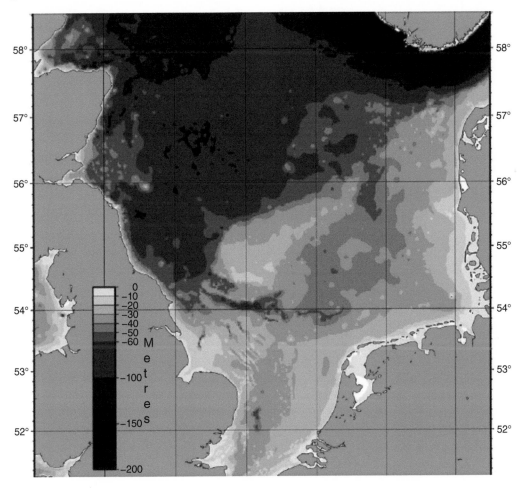

Fig. 3.3 Bathymetry of the North Sea (courtesy Geotek). Note that in the southern part of the sea the depth seldom exceeds 50 m, and that only in narrow depressions. This boundary of amplitude of relief extends over distances of 200 km to 400 km.

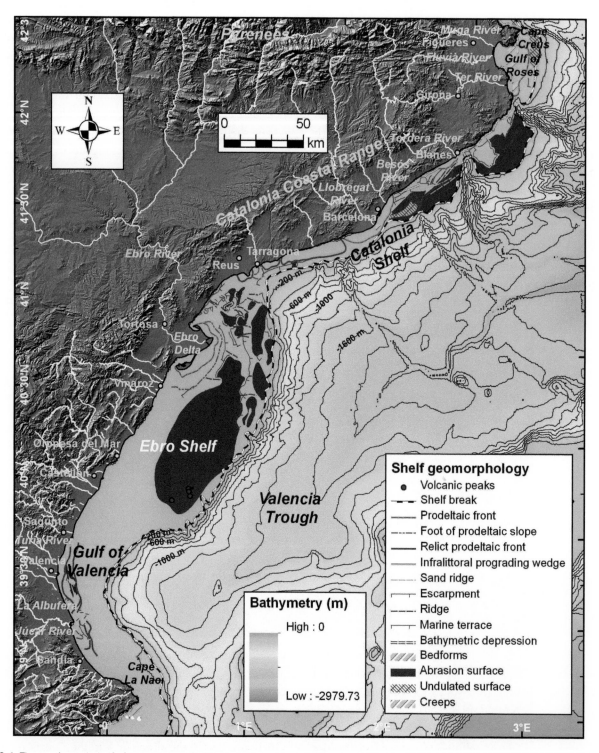

Fig. 3.4 The continental shelf of northeastern Spain, showing a narrow shelf and rapid descent into a deep trough. Lobo *et al.* (2014). Reproduced with permission from the Geological Society of London.

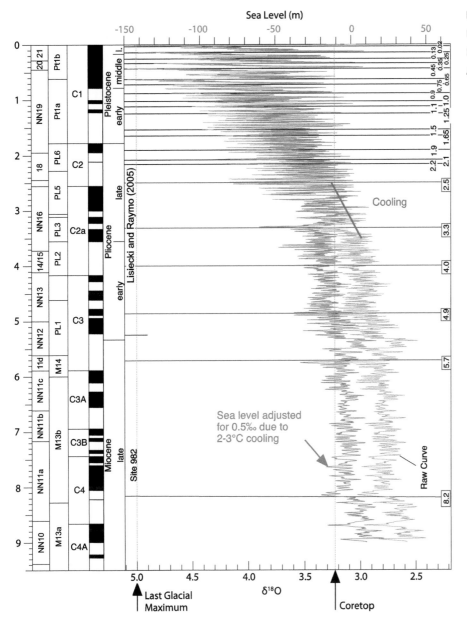

Fig. 3.5 Global sea-level trend for the last 9 million years. Miller *et al.* (2005). Reproduced with permission from the AAAS.

50 m above present sea level to 50 m below present sea level (Miller *et al.* 2005) with an oscillating vertical amplitude of the order of 50 m to 100 m (Fig. 3.5). The early higher levels created terraces and raised beaches and cliffs even in areas which have not experienced tectonic uplift. The larger glacial fluctuations of the last 1 million years, and especially the last 20,000 years, have been analyzed in the previous chapter.

Since the shoreward parts of the shelf and its hinterland above present sea level are generally erosional in origin, with visible cliff retreat in some places, it is logical to expect that almost all fragile features on the currently submerged inner shelf would be destroyed or eroded through time, while those further seawards would be buried by deposits of sediment transported seawards. On timescales of the order of 1 million years or more this is probably true, though fossil animal remains and, theoretically, early hominin traces, could be preserved in drowned valleys and depocenters which have been infilled with many tens of meters of sediments. Pliocene fossils of terrestrial fauna have been recovered by trawlers in the North Sea (Post 2010). Low-lying alluvial coastal flood-plains and large deltas also provide environments within which Paleolithic deposits can be preserved below present sea level, either inland, or a short distance offshore.

To an approximation, the total time that the sea level has been between any two vertical intervals can be represented by a Gaussian curve (Flemming 1968) distributed symmetrically about present sea level, which is close to the average for the last few million years (Fig. 3.5) (Miller *et al.* 2005). The more time that the sea has cumulatively occupied a certain level, the further we would expect the coastal process to have eroded landwards relative to the mean gradient envelope of the land surface, taking into account the tectonic nature of the margin. Also, the more sediment would be expected to be transported seawards below the given sea level. This is a close approximation to what we see today, where a low-gradient shelf is incised into the continental gradient with a maximum landward extent at close to present sea level, and an increased gradient above that point.

In order to survive multiple glacial cycles, an archaeological deposit would benefit from originating in an area of gradual subsidence and becoming buried either by rapid or slow sediment accumulation in a depocenter. If the burial is deep, the anthropogenic signal is only likely to be discovered by marine industrial operations. This is supported by the discovery of anthropogenic materials dating from 340 ka (Wessex Archaeology 2011; Tizzard *et al.* 2014) in the sediments of the southern North Sea. Gaffney *et al.* (2007) have shown how re-interpretation of commercial geophysical seismic records can be used to identify sub-bottom buried terrestrial landscape topography, including rivers, creeks, shorelines, marshes, and near-coast islands (Gaffney *et al.* 2007). Sites with Mesolithic and Neanderthal skull fragments have been discovered in the southern North Sea (e.g. Hublin *et al.* 2009). Nevertheless, even this broad generalization has exceptions, and the hand axes found by Bruno Werz in Table Bay, South Africa, have probably survived several glacial cycles while buried in several meters of sand and earth on an open exposed low-gradient shore (Werz & Flemming 2001), and an assemblage of flints off Cap Lévi, near Cherbourg has a date range of

100 ka to 50 ka, suggesting several periods of inundation and re-exposure (Cliquet *et al.* 2011). Another example of long-term survival of hominin remains on the continental shelf outside Europe is the recovery by fishermen of an archaic hominin jawbone from a depth of 60 m to 120 m in the Taiwan Strait, provisionally dated to the timespan 450 ka to 10 ka (Chang *et al.* 2015). Because so few anthropogenic remains have been found offshore older than 100 ka, it has been necessary to include non-European examples to illustrate the conditions. In several of the following chapters on regional sea areas the authors have identified cumulative depocenters offshore where analogous archaeological materials could survive from before the Last Glacial Maximum (LGM).

Assessments of How and Why Sites Survive or are Destroyed in the Long, Intermediate, or Short Term

In general terms it is convenient to discuss the rates of potential erosion and destruction of the coastal zone on three timescales. These are (1) the multiple glacial cycles and stadials of the whole Quaternary with periods of 20,000 years to 1–2 million years; (2) the processes that occur within the fluctuations of sea level and climate in a single glacial stadial and Dansgaard-Oeschger (D-O) events, that is 100 years to 20,000 years; and (3) the processes that occur during periods of approximately constant sea level, from weeks to 100 years, or a few thousand years. The boundaries between these time periods are very flexible, but it is necessary to break the complexity down into manageable time segments, and we recognize that the rates of processes will vary significantly between regions and on differing geological substrates, with different rates of erosion, weathering, and geochemical processes.

The broadest timescale of multiple glacial cycles has been discussed in the previous section, and the number of known sites offshore that have survived for such durations, more than one complete glacial cycle or half-cycle, is still very small. An important systematic methodology for the analysis and interpretation of stratigraphic sequences has been developed by Catuneanu *et al.* (2011) which is useful in understanding the varied structure of the sedimentary accumulations of the continental shelf, but the scale and time resolution of the analysis tends to be in units of many millions of years, making it difficult to appreciate the finer analogous processes that have occurred within the Pleistocene. The section on coastal and shallow water siliciclastic settings (Catuneanu *et al.* 2011: 198 *et seq.*) is the most relevant to geoarchaeology. This approach can be refined, and there is every hope that more sites and drowned landscapes earlier than the LGM will be found, especially during commercial operations that extract sediments from the sea floor, but, for the time being, such finds are sparse.

Effects of coastal and shallow water processes on the timescale of 100 years to 20,000 years

We consider here the timescale which would embrace a stadial oscillation within a glacial cycle, a D-O event, or the deglaciation since the LGM, in terms of how coastal and other marine processes might preserve or destroy anthropogenic material and terrestrial landscapes. We assume that a prehistoric archaeological deposit consists initially of unconsolidated strata of terrestrial deposits, within which hominin/human indicators exist, or a single artifact such as a flint tool or carved antler. Within caves or underground karstic river channels, it is possible for hominin indicators to be cemented to the walls or floor of the cave by speleothem accumulations, or carved into the rock, but this will be treated later as a special case in Chapter 12 and associated Annex (Canals *et al.* — this volume). In most situations the anthropogenic material, before it is inundated, is in clastic deposits of soil, sand, gravel, or cohesive sediments such as mud, clay, or peat. In these cases direct destructive wave attack by waves of the order of 1 m to 2 m significant wave height (H_s) can, in general, partially erode the site with each single breaking wave on an exposed shore. Since H_s can be as much as 10 m in the Mediterranean, and 20 m in the Atlantic, it

will be important to analyze how some prehistoric sites or terrestrial soils survive the first storm that strikes their location. This discussion will be conducted later in the section on the shortest time period.

As will be shown in the consideration of short-term processes in the next section, the preservation of sites in the very short term depends greatly on the local topography upon which the sea impinges, the response and behavior of the lateral sediment transports on the coastline, and equilibrium gradient of the shore. Masters and Flemming (1983) showed by reviewing the preserved seabed prehistoric sites known at that date that topographic features such as lagoons, local rock outcrops, small offshore islands, sand spits and barriers, sheltered estuaries, and stable dissipative low-gradient beaches, etc., could provide protection for prehistoric sites in shallow water, or on the beach for many centuries. Thus, in the present section, we need to consider to what extent these defensive barriers and protective shields can be destroyed on a longer timescale.

It is difficult to specify general rates of change and rates of destruction or preservation on this timescale, and complex models would be needed in each specific location to test the propositions put forward here. For example, the rates of change, and the three-dimensional response of a low-gradient sediment-rich coast to rising sea level, are discussed in the previous chapter (Zhang *et al.* 2010; 2011; 2012; Harff *et al.* this volume). But the topographic feature that protects a site could be a massive granite rock ridge, as in the case of La Mondrée, Fermanville, at Cap Lévi near Cherbourg (Scuvée & Verague 1988; Cliquet *et al.* 2011), a low-gradient oceanic sandy beach as in the case of Table Bay, Cape Town (Werz & Flemming 2001) and on the Atlantic coast of Florida (Cockerell & Murphy 1978), a large rocky near-coastal island such as the Isle of Wight protecting the Solent Channel, or a network of postglacial sandy islands, as in the case of the southern Baltic (Lübke 2001; 2002a,b; 2003; Harff & Lüth 2007; 2011; Harff *et al.* 2007).

If the protecting screen is hard rock or a rocky coastal island, the sheltered site(s) could be preserved for more than one glacial cycle. Cohesive mud and clay has proved to be a good preserving medium (Aldhouse-Green *et al.* 1992; Bell & Neumann 1997). If the screen is provided by a sediment gradient, peat, or a sand spit screening a lagoon, ria valley or estuary, the site may be buried and preserved, or a change of sea level could radically alter the sediment supply to unconsolidated landforms, and destroy the archaeology at the same time.

There is a great deal of work and research modeling to be done here. It is worth considering that the processes operating on this timescale are distinct from the effects over many glacial cycles, and from those at the very short term. The latter will be discussed in the next section. On this intermediate timescale, also, prehistoric archaeological deposits need to survive centuries or millennia of exposure to terrestrial weather, precipitation, and geomorphological processes, but these are the same in generic form to those applying to sites found on the present landmass, and will be discussed later only in the context of the coastal environment immediately prior to inundation.

Processes in the short term: weeks to 100–1000 years

It is not possible here to expand in detail on the mechanisms of waves in shallow water, or the different ways in which shoaling waves of different wavelengths impact on coasts of different offshore gradients. These factors are outlined in relatively simple terms in standard textbooks such as Davis and Fitzgerald (2004), and more academically by Holthuijsen (2007). Coastal processes and geomorphology are described in more detail by works such as Carter (1988) and Davidson-Arnott (2010). The point which needs emphasizing and elucidating here is that, in order to survive marine inundation as a primary deposit, an unconsolidated prehistoric archaeological deposit must be protected by natural circumstances to such an extent that the wave forces striking it directly in an erosive manner are very weak indeed, or almost zero. Put bluntly: how does an unconsolidated prehistoric site withstand the impact of the first storm waves that strike the shore at the same level in its neighborhood as the sea level rises? Even if the level is rising at several meters per century, the site will be potentially exposed to the dynamic oscillations of breaking waves for the order of 100 years. A protective situation can be created

either by the three-dimensional topographic protection of the coastline preventing large waves reaching the site, or by the offshore gradient interacting with the waves so that their internal water movements are dissipative at the seabed and are not erosive, and the beach+foreshore is in equilibrium, or by longshore drift and beach progradation providing an overlying screen, or bar and lagoon. These processes are often combined. As the shore adjusts geomorphologically to the impact of the rising sea, some sites are preserved by the local response, while another site a few kilometers along the shore can be destroyed by the same regional wind-wave-current regime.

A further way in which this protection can be enhanced is when the archaeological stratum is already buried in several meters of soil, sand, peat or rock-fall before the sea rises over it, and the cover is sufficiently thick that the wave action does not erode down to the archaeological deposit. This certainly can occur, but then it is unlikely that the archaeological material will be discovered until either the overburden is eroded away, or excavation for industrial purposes brings the archaeology to light. Events of this kind include the discovery of the so-called Viking Flint (Long et al. 1986); the A240 Paleolithic site (Tizzard et al. 2014); the dredging of Early Neolithic material at Port-Leucate harbor, Roussillon, France (Geddes et al. 1983); and the dredging of the Yangtze Maasvlakte 2 harbor (Moree & Sier 2014). This possible sequence of events should be borne in mind when offshore industrial activities are being planned.

In the more general case, the deposit is not adequately protected by overburden, and is in direct contact with the surface of the sea or shallow sea water. Even in the most unlikely circumstances, for example Atlantic waves breaking on a rocky foreshore, fragments can be trapped between massive boulders for centuries, as in the case of jewelry from Armada wrecks on the coast of Ireland. At a constant sea level a clastic beach or coastline can reach an equilibrium form which may change slightly with the seasons, or before and after storms, but which remains almost constant for decades or centuries. On a longer timescale, as described above, cliffs may erode, sand spits and deltas build forwards, but archaeological sites in areas which do not suffer massive and constant erosion will be safe, even if they become buried. Masters and Flemming (1983) brought together archaeologists and oceanographers to consider this problem in 1981, and a chapter in that book by Flemming (1983: 135–173) reviewed the 31 submerged prehistoric sites, or suspected sites, known at that time from the published literature. It was apparent from that review (Flemming 1983: 161–163) that a major factor in defining the characteristics of every site was the protective topography of the coastline within a few hundred meters horizontally around the site. All sites were in shallow water, less than 10 m, except for two sites at about 20 m in Florida sinkholes (Murphy 1978; Clausen et al. 1979). The favorable sites were classified into six categories: (1) Ria, lagoon, and estuarine; (2) Sheltered alluvial coasts; (3) Exposed equilibrium or accumulating beaches; (4) Submerged sea caves; (5) Karstic caves and sinkholes; and (6) Islands and archipelagos. The common factor in all cases except case (3) is the protection provided by limited wind fetch, with the fetch restricted to a few kilometers or less by the local topography.

There are thus two propositions which seem self-evidently probable on brief consideration, but which are probably unsound, and which need careful checking to see whether they are universally or even sometimes applicable. The first is that given a constant sea level, the majority of, if not all, prehistoric sites will be destroyed quite quickly by a combination of wave action and currents; the second proposition, which is not exactly a correlate of it, but is consistent, is that when a site is transgressed by rising or falling sea level it is most likely to survive if the rate of change of vertical level is rapid.

A striking conclusion from the 1983 review, especially given the debate in recent years about the importance of maximum rate of change of sea level in order to preserve sites, is that all the known open-sea coastal sites known at that date are in very shallow water, in locations where the relative sea level has been constant to within 1 m or 2 m for the last 5000 years. That is, thousands of years of storms at a near-constant sea level had not destroyed them yet, although we cannot know how many sites have already been lost. It is possible that we are seeing some sites in a narrow window between their exposure and subsequent destruction. Repeated observation of some sites has revealed varying rates of continued slow

destruction, even in locations sheltered from waves. For example, tidal currents can erode cohesive deposits even when there is very effective protection from large waves, as in the case of Bouldnor Cliff (Momber et al. 2012: 261). Seabed vegetation such as eelgrass stabilizes the sediments and reduces erosion, but if pollution damages the vegetation, then erosion increases (Fischer 2011). We can add to the early evidence cited above the results of the SPLASHCOS survey, which is a register of submerged prehistoric sites, using data that had been accumulated for decades previously by local experts (Jöns et al. 2016). A large number of prehistoric sites in the SPLASHCOS Database are at present in water shallower than 5 m or are in the intertidal zone, with the waves and tidal currents washing over them every day (Fig. 3.6). As noted above, we may be seeing some sites for a short period between exposure and erosion, but many others appear to be stable in spite of their closeness to present sea level. Since storm waves can cause oscillatory water motion at the seabed down to depths of many meters this analysis includes all sites down to 5 m. The SPLASHCOS Database in March 2014 contained records of over 2400 prehistoric sites, and of these 331 were submitted with the depth information accurately quantified. Of these 331 the number reported as occurring in water shallower than 5 m is 238, that is 71%. Although the database reports a few sites which occur in depths of 20 m to 100 m or more, it is thus apparent that the great majority are in shallow water. Table 3.1 shows the distribution of the 238 sites by country, and they are well distributed in many different climates, geological zones, and environments. Furthermore, the Danish data which describe many sites that are known to be very shallow had not yet been entered in the database.

It follows that an inundated prehistoric site, usually of unconsolidated — although possibly cohesive — terrestrial soils and gravels intermixed with anthropogenic materials, can, in the right circumstances, survive at or close to the sea level, and can survive in this position when the sea level remains constant for several thousand years. If the sea level is rising and coastal and shallow marine sediments are redistributed, this period of initial survival can be followed either by burial or by inundation at a depth and location such that the residual currents and

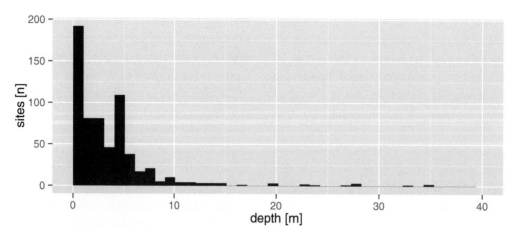

Fig. 3.6 Histogram of the submerged prehistoric sites from the SPLASHCOS Action showing number of sites for which the depth is known in bands of 1 m depth. The number of submerged prehistoric sites for which the depth is known is N = 632. The most common depth band is at less than 1 m, or intertidal, and most sites are shallower than 5 m. The peak in the band 4.01 m to 5.0 m is probably due to the common tendency to pick round numbers. The number of sites at depth 5 m or shallower is 509, that is 80% of the total. Data and graphic from Mennenga Moritz and Hauke Jöns. Reproduced with permission.

wave action are sufficiently attenuated by the topography to ensure survival. While a relatively rapid rise of sea level may not protect sites by sufficiently reducing the time exposed to breaking waves, a rapid rate of rise after transgression may result in rapid burial, which promotes survival in the longer term. The conditions for complete and undisturbed survival are determined by chances of local topography, but the same can be said of the survival of stratified prehistoric sites on land. Some survive, but many are lost. The study of the stability or fragility of known submerged prehistoric sites is now a matter of great interest.

We must assume that many sites are indeed destroyed by rising sea level, or during a constant sea-level stillstand, but the easy assumption that they are all destroyed, and that no site can survive close to a constant sea level, must be rejected. Something other than a rapid rate of change of sea level is providing the additional protection, and this is the complex three-dimensional response of the coast to the forces of waves and tides, as discussed above.

The second common assertion (e.g. Belknap 1983; Belknap & Kraft 1985; Waters 1992; Ballard 2008: 184-185; TRC Environmental 2012) is that, since it is assumed to be obvious that sites at or close to sea level are liable to be destroyed by waves, then it follows that a site is more likely to survive the process of inundation if the sea level is changing quickly, say 2.5 m per century in a peak surge of glacial melting, a meltwater pulse (Stanford *et al.* 2011), rather than the long-term deglaciation average of about 0.6 m to 1.0 m per century since the LGM. The first part of this assumption is often true, though far from always, as shown in the previous paragraphs. The question is, given conditions which do expose a site to erosion during marine transgression, will a faster rate of rise save it from destruction? There is one set of circumstances where this can be correct, and that is where prehistoric deposits are already buried by many meters of overlying terrestrial or later marine sediments, so that the incoming marine forces have to erode a large distance inland and then downwards before damaging the archaeological layer. While the specific quantification of this process will vary from site to site, it is apparent that a deeply buried site could survive for longer, and that a rapid rise of sea level would help survival. However, if a site is that deeply buried, it may not be discovered later when it is also deep under the sea.

In the more general case where an unconsolidated prehistoric deposit is located on the land surface, and the stratigraphic layers are a few meters thick, a single storm wave of a height of 2 m to 5 m striking the deposit would,

TABLE 3.1 NUMBER OF SUBMERGED PREHISTORIC SITES REPORTED WITH A KNOWN DEPTH BELOW MEAN SEA LEVEL (OR BELOW MEAN HIGH TIDE IN TIDAL SEAS), AND DISTRIBUTED BY COUNTRY, AND SHALLOWER THAN 5 M DEEP IN THE SEA.

Country where depth data is available	Sites depth <=5.0m	Total sites (with depth data)
Belgium	3	3
Bulgaria	8	19
Croatia	9	11
Cyprus	0	2
Denmark	83	113
France	74	84
Germany	54	68
Greece	2	2
Israel	17	20
Italy	5	6
Netherlands	0	3
Northern Ireland (UK)	25	26
Norway	17	17
Portugal	4	4
Rep. of Ireland	16	17
Sweden	26	42
Turkey	11	20
UK – England	138	156
UK – Scotland	12	14
UK – Wales	5	5

other things being equal, begin to erode the deposit, and an extensive area could be eroded or scattered within a few years at most. Since the waves can exert erosive force to a depth of several meters, it would be irrelevant whether the site was exposed to the wave action for 100 years of rapid sea-level change, or 500 years of slow sea-level change. The site would be destroyed in both cases. A rapid rate of sea-level rise is therefore not, in most cases, a sufficient factor promoting survival of prehistoric deposits. Cohesive sediments such as compacted mud, clay, or peat may resist erosion for a short while, but on a timescale of hundreds of years do not provide protection.

Rapid vertical rise of sea level cannot of itself significantly increase the chance of a site surviving inundation. However, if it is combined with a low gradient, rapid

horizontal transgression of the surf zone, protective topography, and is followed by a prolonged period of shallow-water coastal processes, it does enhance survival. Ballard (2008: 184–185) proposes that a combination of thick overburden and relatively rapid rise of transgressing sea level could achieve this effect. This is a possible scenario, as noted above at the intermediate timescale. Indeed, the proposition in the previous section is that the structural protection provided by landscape coastal forms will survive for longest if it is not attacked repeatedly at slightly different levels over tens of thousands of years. This is another way of saying that the survival is more likely if the sea level does not keep re-occupying the same vertical interval, creating an integrated and cumulative level of destruction, and passes it by with a rapid rate of change relative to the intermediate timescale.

The apparently simple proposition that sites are most likely to survive during the phases of the post-LGM rise of sea level that were most rapid has not yet been put to the test, and there are difficulties in establishing such a correlation empirically. The two most cited periods when rapid rise might lead to well-preserved prehistoric sites are Meltwater Pulses 1A and 1B at approximately 90 m depth 15,000 years ago, and 50 m depth 11,000 years ago respectively (ignoring local glacio-isostatic adjustment and local tectonics). As demonstrated above, very few sites have been found deeper than 50 m, and the population density in the Paleolithic was probably much lower than in later periods. Thus the chances of acquiring the statistical sample to demonstrate this assertion are poor. The implications of rate of rise of sea level are discussed by Belknap and Kraft (1981; 1985), and by TRC Environmental Corporation (2012: 20 *et seq.*), but the significant effect of local topography is also recognized.

It follows from this preliminary discussion that for most prehistoric sites in most locations, the apparent coincidence of the sea level and the site deposit for a given time, longer or shorter (on this timescale), is not the defining factor in survival or destruction. Other combinations of factors create the protective screen that ensures that a surprising number of sites do indeed survive for many thousands of years close to a constant sea level. As we have shown, those factors are the local coastal topography, refraction and diffraction of wave energy, gradient of land or gradient of the beach, the offshore seabed gradient, the fetch over which the wind can blow straight onto the site, and a variety of other site-specific characteristics which can, almost counter-intuitively, allow one site to be protected for millennia, while destroying another nearby site in a year or less.

One of the most common formulae for assessing the quantitative effect of rising sea level is the application of the Bruun Rule (Bruun 1962). This suggests that, as the sea level rises, the sediments at or just above the sea level are eroded and deposited down-slope, according to a simple formula, and a new beach gradient established. If this were universally true, no prehistoric deposits would survive transgression, unless very deeply buried. In contrast to the over-simplified Bruun Rule, quantification of coastal geomorphological changes must be based on three-dimensional source-to-sink transport models. The two-dimensional Bruun Rule neglects longshore transport and should not be recommended for the purposes of this book. An alternative to the Bruun concept was given by Deng *et al.* (2014) with the so-called Dynamic Equilibrium Shore Model (DESM). The basic concept of the model is a dynamic equilibrium of the coastal cross-shore profiles adapting to sediment mass balancing in a semi-enclosed coastal area, in which the unknown parameters of the cross-shore profile shapes are calculated by numerical iterations.

The discussion above has already cast doubt on the automatic assumptions implied by the Bruun Rule. At this point it is worth quoting fully (with the authors' permission) the abstract from a paper by Cooper and Pilkey (2004). It is difficult to state the case more succinctly or effectively:

> In the face of a global rise in sea level, understanding the response of the shoreline to changes in sea level is a critical scientific goal to inform policy makers and managers. A body of scientific information exists that illustrates both the complexity of the linkages between sea-level rise and shoreline response, and the comparative lack of understanding of these linkages. In spite of the lack of understanding, many appraisals have been undertaken that employ a concept known as the "Bruun Rule". This is a simple two-dimensional model of shoreline response to rising sea level. The model has seen near-global application since its original formulation in 1954. The concept provided an advance in understanding of the coastal system at the time of its first publication. It has, however, been superseded by numerous subsequent findings and is now invalid.
>
> Several assumptions behind the Bruun Rule are known to be false and nowhere has the Bruun Rule been adequately proven; on the contrary several studies disprove it in the field. No universally applicable model of shoreline retreat under sea-level rise has yet been developed. Despite this, the Bruun Rule is in widespread contemporary use at a global scale both as a management tool and as a scientific concept. The persistence of this concept beyond its original assumption base is attributed to the following factors:
>
> 1 appeal of a simple, easy to use analytical model that is in widespread use;
> 2 difficulty of determining the relative validity of 'proofs' and 'disproofs';
> 3 ease of application;
> 4 positive advocacy by some scientists;
> 5 application by other scientists without critical appraisal;
> 6 the simple numerical expression of the model;
> 7 lack of easy alternatives.
>
> The Bruun Rule has no power for predicting shoreline behavior under rising sea level and should be abandoned. It is a concept whose time has passed. The belief by policy makers that it offers a prediction of future shoreline position may well have stifled much-needed research into the coastal response to sea-level rise. (Cooper & Pilkey 2004:157).

The data and discussion of the previous sections of this chapter are compatible with the position taken by Cooper and Pilkey (2004). Furthermore, the statement by Cooper and Pilkey is consistent with the approach to numerical modeling of coastal geomorpho-dynamics

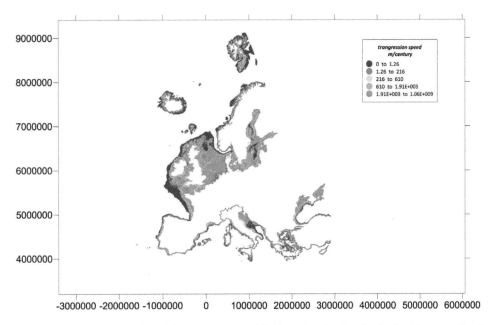

Fig. 3.7 The rate of horizontal transgression of the rising sea level post-LGM in meters/century is obtained by computing the gradient of the region and the rate of rise of sea level. The gradient also determines wave characteristics in shallow water and the rate at which sediments can be transported down-slope. Graphic by F. Chiocci is fig. 2.10 from Flemming *et al.* (2014).

summarized in Chapter 2 (Zhang *et al.* 2010; 2011; 2012; Harff *et al.* this volume).

The significance of seafloor gradient and rate of transgression of the rising sea level is shown in Fig. 3.7. The plot is obtained by combining the generalized curve for relative rate of rise of sea level with the local average gradient to produce the rate of horizontal transgression in meters per century. The variation in local gradients from near-vertical to near-horizontal has a dramatic effect on the rate of horizontal transgression. After marine transgression, during sea-level rise over a low gradient, a prehistoric site will be located in a shallow-water environment, and potentially sheltered by local undulations of the topography. During sea-level fall, those conditions will apply when the water is shallow before the site is exposed to the terrestrial conditions, and after exposure the site will be in a low-gradient landscape, and hence probably low rates of terrestrial erosion. The majority of sites reported to the SPLASHCOS Database (as shown in Fig. 3.1) are concentrated in the areas of low gradient and sediment accumulation. Relatively fewer sites have been found so far on the coasts of steeper gradient, although this is partly due to the necessary

correlate that the total area of submerged shelf is small. The potential for discovery of sites on a steep rocky shelf is further reduced by the fact that deeper water necessarily means an older site, extending back into the Paleolithic, when population densities were probably lower, and hence fewer sites are likely to have existed before inundation. The strip within which Mesolithic and Neolithic sites could have existed is generally less than half the shelf width on a steep gradient.

Site Formation and Coastal Landscapes and Processes

Geomorphological processes on the shore and close to the coast combine with wave and current action to generate changes and forces tending to destroy or bury prehistoric sites. Processes will be described very briefly. We consider in sequence:

high latitude rock weathering
glacial erosion and transport
high latitude, eolian erosion and transport

mid latitude rock weathering
fluvial erosion and transport
early diagenesis
substrate and coastal physiography

High latitude (>50°N) rock weathering

In high latitudes the physical weathering of rock largely dominates in comparison to chemical weathering. Frost weathering is a common mechanism where temperature frequently falls below zero °C. Discontinuities like pores, cracks, joints and bedding planes make it easy for water to penetrate, and ice wedgings go deep into the rocks. Hence, a large amount of sediment is produced due to the rock breakdown. This mechanism is more efficient where successive freezing and thawing occurs due to seasonal or diurnal temperature contrasts. The amount of sediment produced by frost weathering that is then available to be eroded and transported depends on the geological substrate. Physical weathering is more effective in coarse textures (e.g. granite, gneiss) than in fine textures like basalts (Konishchev & Rogov 1993; Schaetzl & Anderson 2005). For instance, the breakdown of the very old and deformed crystalline rocks of the Baltic Shield produced a remarkable volume of detrital sediment throughout the Quaternary. Those glaciogenic sediments are currently exposed in the cliffs of the southern region of the Baltic basin and cover the Baltic Sea floor (where they also occur as eskers) and on the eastern margin of Scotland.

In contrast, the weathering products of chalk are very different from the granular disintegration of the crystalline rocks. Chalk texture is also vulnerable to frost weathering and furthermore, the landscape morphological features tend to be preserved as a fossilized surface. This is because secondary porosity is developed into the chalk due to frost action and the water infiltration is then higher than surface runoff so that the valleys become dry with no fluvial erosion. This is observable in the chalks along the English Channel (French 2007).

Thermokarst is a process of thawing of permafrost in high latitude regions and was developed on the continental shelves of periglacial areas during the LGM marine regression. It then influenced coastline evolution during the Holocene transgression. Thermokarst geoforms such as lakes and depressions formed ca. 11 ka to 8.5 ka and evolved into coastal lagoons leading to the development of an indented coastline physiography during the Laptev Sea shelf flooding (Romanovskii *et al.* 2000).

Glacial erosion and transport

The Quaternary glaciations were mainly a Northern Hemisphere phenomenon and ice sheets extended far southward in Europe during the LGM, covering the Norwegian continental shelf and part of the North Sea basin (Williams *et al.* 1993). Glaciers have larger transport capacity than rivers due to the greater viscosity of ice relative to water in liquid phase. As a consequence, glaciers transport large amounts of heterometric material from boulders to mud. However, the distance of the glacier transport, up to tens of kilometers, is less than that produced by rivers. Glaciers drain regions where snow accumulates and since they attain either altitudes or latitudes where temperature lies below the freezing point, the coarser sediments deposit in a chaotic way (moraines). Glaciers may reach continental margins where the deposits may potentially accumulate, either as submerged glacial features, or so as to construct paleo-shorelines. Submerged moraines, drumlins and eskers are morphological features related to glacial activity and their occurrence, dimension and orientation on northern shelves are good proxies for paleocoastlines. Drumlins and esker genesis relate to changes in the net ice flow and drainage under the ice respectively, with the former being parallel to the ice flow. Thus, when occurring underwater as are the eskers reported in the Baltic Sea, they mean that the region was exposed during previous glacial times and this can be used to facilitate reconstruction of the coastal landscape. The evolution is determined by the successive glaciers as they retreat and spread. However, moraines are successively eroded, transported and deposited, making it often difficult to establish clear relative chronologies. Large amounts of coarse fluvio-glaciogenic sediments transported onto the exposed shelf during the LGM formed a barrier complex during the postglacial sea-level rise and may be preserved to survive wave erosion, as happened in the northeastern English Channel (Mellett *et al.* 2012).

Eolian erosion and transport

Wind was an important geomorphic agent during the Pleistocene cold events, acting over unvegetated surfaces on the exposed shelves in periglacial areas during the LGM low sea level (Lowe & Walker 1984). Chalk outcrops along the English Channel are veneered by loess accumulated mainly during the last glaciation (French 2007). Silt blown by the wind in periglacial arid regions is entrained from glaciogenic deposits like eskers, braided channels, glacial outwash plains and lake shores (van Huissteden *et al.* 2001; Dietrich & Seelos 2010). Sand dunes or sand sheets in cold dry environments as in the southeastern Netherlands were deposited during the Mid and Late Pleistocene between ca. 14 ka and 12 ka during the permafrost degradation and increased aridity (Kasse 1997). Submerged beach deposits off the coast of Normandy and Brittany in depths of 50 m to 90 m and as far out as the Hurd Deep consist of conglomerates cemented by loess (Antoine *et al.* 2003; Murton & Lautridou 2003; Lefort *et al.* 2011; 2013). The majority of the more widespread distribution of loess which blew south from the English coast onto the floor of the English Channel has been eroded by the rising sea level since the LGM.

Mid latitude (50°N–30°N) rock weathering

Rock weathering in mid latitudes includes all the weathering mechanisms both chemical and physical, and the extent and rate are highly dependent on the substrate properties and climatic conditions. These show high variability influenced by several variables, among them the distance from the sea, and the relationship between the dominant wave direction and the elevation, orientation and topography of coastal mountains. Karstification in carbonate rocks is the paradigmatic example of the relationship between lithology and weathering process. Dissolution acting in limestone and dolomite leads to the development of dissolution landscapes. Karstic landscapes may have great social impact due to the

lack of surface drainage and patches of soil, which are formed by the insoluble residue (sand, silt and clay) of carbonate rocks confined inside solution cavities. Karstic caves provide attractive refuges and were successively over time occupied by human and animals. Nevertheless, detrital load transported by streams traversing limestone and dolomite lithologies does not represent an important sedimentary contribution to continental shelves since the major component of these rocks (Ca^{2+}) is highly soluble. On the contrary, karstic landscapes act as a trap for detrital sediment in karstic depressions and underground conduits. During high discharge events those sediments may be transported by subterraneous streams and reach the coast. Coastal karst is successively fossilized and exhumed due to mean sea-level variations and, during erosional phases, large amounts of detrital sediment contribute to feed the adjacent beaches and shelf. Rocks of phaneritic texture like granite and gneiss vulnerable to hydrolysis provide large amounts of detrital sediments to be transported by rivers.

Fluvial erosion, transport and coastline evolution

Fluvial landscapes are the most widespread in mid latitude areas, and continental margins are the main reception areas of the materials eroded from the mainland. The thickness of the sedimentary coverage of the continental shelves is a balance between the sedimentary input, the marine climate and the space for accommodation (Catuneanu *et al.* 2011), the latter governed mainly by tectonics and sediment compaction. Some detrital sediments over the shelves are inherited from past lowstands of sea level in areas where the contemporaneous fluvial sedimentary load is insufficient to cover them. Moreover, the shelf morphology together with the sedimentary input to the coast determines the coastline position. As a result, for the same uniform global changes of sea level, the coastline behavior and evolution are highly dependent on several variables other than the sea level.

The thickness of detrital sediments transported onto the shelf has a dual role in the preservation of human remains and landscape structures. When the deposition rate is high, both structures and traces of human activity are quickly buried, thus becoming better protected from erosive processes such as wave and current action later. However, the thick sedimentary coverage hinders both their detection and recovery. This aspect is well illustrated in the Danube Delta where, due to the combined effect of the delta subsidence and the high sedimentation rate, the Late Prehistory layer should be now buried at depths greater than 4 m to 5 m (Dimitriu 2012). The thickness of sediments accumulated during the Pliocene and the Pleistocene epochs increases eastwards from 100 m in the Danube Delta proximal area to 1 km in its deep-sea cone (Dinu *et al.* 2005). Thus, the compaction of such thick sedimentary sequences may cause the subsidence of the delta, which in turn influences the regional relative mean sea level. Detrital sedimentary input may be a first order factor in determining coastline evolution of confined sea basins. This is well illustrated in the Black Sea, a marginal sea that receives the massive sedimentary load of rivers draining the Alpine orogenic chains, among them the Danube, the Dnieper and the Don (e.g. Winguth *et al.* 1997; Dinu *et al.* 2005; Dimitriu 2012; Rosyan *et al.* 2012). Similarly, the semi-enclosed Mediterranean basin exhibits several environmental indicators favorable to the preservation of morpho-sedimentary features underwater, such as its marine climate, sub-aerial climate and sedimentary input from the drainage basin catchment area.

The perisutural basin of the Po River Plain that received the drainage from the Alps and Apennines is filled by up to 800 m of sediments accumulated during the Pliocene and Quaternary, from which up to 40 m thickness accumulated during the last ca. 19 kyr (Dinelli *et al.* 2012). Large amounts of sediments from the Po Plain on the Adriatic sea shelf led to the formation of barrier-lagoon systems, which migrated landward during the postglacial sea-level rise, and two of those barrier-lagoon systems are now submerged and preserved due to a rapid sea-level rise (up to 60 mm/year) and a marine climate different from the present (Storms *et al.* 2008). The coastline was ca. 20 km to 30 km landward from the present one at ca. 7 ka during the maximum Holocene transgression. After that maximum transgression, the sea-level rise decelerated and the large amount of sediments transported by the Po River led to the seaward migration of the coastline due to delta progradation (Amorosi *et al.* 2003). This trend still continues at least since 1600 AD with different migration rates and some episodes of landward migration (Marabini 1997). Thus, while the global trend of the coastline is landward retreat (until reaching a geomorphic limit) in response to the mean sea-level rise, in areas with strong sedimentary input to the coastline the trend can be in the opposite direction. Morphosedimentary bodies like sand bars, barriers and dunes formed on the continental shelves due to strong sedimentary supply may be preserved depending on marine conditions, rate of sea-level rise and early diagenesis.

Early diagenesis

The underwater genesis and preservation of morpho-sedimentary bodies has several requisites: (1) sediment available to be transported by waves, currents and wind to form spits, barriers and dunes; (2) rapid sea-level rise and; (3) early diagenesis of the sand. The combination of these environmental variables is identified as responsible for the preservation of submerged shorelines even on a steep and high-energy shelf (Salzmann *et al.* 2013). Although early diagenesis is often reported as restricted to subtropical conditions (Salzmann *et al.* 2013), it may occur in temperate climate at mid latitudes showing marked seasonality depending on the carbonate content of sands e.g. Cornwall, UK (Howie 2009), the Bay of Biscay (Arrieta *et al.* 2011), and the Algarve in southern Portugal (Moura *et al.* 2007; 2011).

Sandstones of high content of carbonate (up to 80%) both from shells and from erosion of the adjacent carbonate rocky cliffs occur submerged in the southernmost continental shelf of Portugal forming a series of paleo-spit bars and barriers roughly parallel to the current coastline (Infantini *et al.* 2012). During glacial sea-level falls and continental shelf exposure to aerial conditions, sandy bodies (barrier, beachrocks and eolianites) were rapidly cemented due to the high content of calcium carbonate and the well-marked seasonality. The high porosity of the sand allows an easy circulation of fresh water during the wet season and the consequent dissolution of the carbonate fraction. During the hot and dry season the water evaporates rapidly and therefore carbonate precipitates as chemical cement leading to the formation of carbonate sandstones (Arrieta *et al.* 2011). Cementation of beach sands may be as rapid as within 30 years, potentiated by biological activity, such as that reported in Corsica on an artificial shore produced by debris discharge of asbestos (Bernier *et al.* 1997). However, beachrock should be used carefully as a proxy for reconstruction of coastlines because of the rate of sea-level variations in macrotidal sites where tidal range can reach several meters. Eolianites may also be used to reconstruct paleocoastline evolution based on the variation of the sand recruitment area to form dunes (Moura *et al.* 2007).

Eolian dunes accumulated extensively over the Northern Hemisphere continental shelves during the cold and dry climate anomalies of the Younger Dryas and the 8.2 ka Event (Alley & Ágústsdóttir 2005 and the cited literature). Like beachrock and sandy barriers, eolianites may also be preserved underwater and the potential for preservation is directly correlated to the rapidity of diagenesis and sea-level rise.

Substrate and coastal physiography

Rocks are metastable phases that undergo weathering when exposed to sub-aerial conditions. Landscapes evolve more or less rapidly depending on the climatic conditions and substrate. Carbonate rocks are extremely vulnerable to chemical attack producing typical karstic landscapes where caves are frequent. Wet conditions, joints, faults, crevices and thin bedding layers favor the genesis of karstic caves. Caves are propitious morphological features to preserve human remains on land, often close to the coast, and many archaeological sites have been reported in karstic caves on land (e.g. Joris 2002; Ontañón 2003; Karkanas *et al.* 2007; Carrión *et al.* 2008; Kuhn *et al.* 2009; Nakazawa *et al.* 2009). However, the preservation and discovery of prehistoric material in submerged karstic environments has been problematic (see, for example, the discussion of this subject by Billaud in Canals *et al.* this volume, Annex, and Flemming and Antonioli (2017)). The sub-aerial evolution of karst includes the collapse of dolines and caves to produce karstic breccias, and

several archaeological remains have been found within breccias resulting from cave collapses (e.g. Yeshurun *et al.* 2007). Besides karstic landscapes being the most favorable geomorphic context to produce caves, several other lithologies other than carbonate ones can produce caves or at least small cavities if affected by fractures or eroded dikes and sills.

Karstic aquifer behavior (porous and fractured) permits the influx of large amounts of fresh water to the coast conducted through galleries or fractures, producing freshwater springs in the marine environment. For instance, in arid or semi-arid regions such as the Mediterranean basin and Atlantic south Iberian coast, karstic aquifers represent the main groundwater drainage network (PNUE 2004). This process should play an important role in providing freshwater resources on the exposed continental shelf during lowstands. Whereas freshwater tables fell on the continental areas as a consequence of relative mean sea-level (RMSL) fall and climatic aridity, the availability of surface fresh water increased on the emerged shelf (Faure *et al.* 2002).

Differential erosion producing different types of topographic relief is very frequent in carbonate landscapes due to the occurrence of conspicuous outcropping harder dolomites relative to mechanically and chemically less resistant limestones and marls. Coastal promontories are often sculpted around dolomites more resistant to wave attack whereas bays are cut into softer limestones. This leads to a highly crenulated coastline physiography as on the southern coast of Portugal, where waves and littoral current propagation show a strong morphological control (Bezerra *et al.* 2011; Moura *et al.* 2007; 2011; this volume; Horta *et al.* 2013; Rocha *et al.* 2013).

Crenulated coasts provide a geomorphic context which favors the genesis of spit bars anchored in headlands and, in warm Mediterranean climates, beachrock in the more sheltered bays. Headlands are natural obstacles to longshore drift and sediments are deposited downdrift due to current deceleration (Jackson *et al.* 2005; Backstrom *et al.* 2009; Jackson & Cooper 2010; Silva *et al.* 2010). Beach behavior within a headland-beach system is of fundamental interest for coastal evolution since it represents a buffer to wave attack on the backing cliff (Benavente *et al.* 2002). In these various ways the nature and structures of the substrate have a major influence on coastal processes.

Selected Case Examples

It is not practical to provide an exhaustive set of geomorphological examples for how sites have survived in different locations for different periods of time. More examples and types of prospective location will be described in subsequent chapters. The following selection provides brief geomorphological data on sites in increasing age.

Golfe du Morbihan, Brittany, France, 5000 BP to 4500 BP

Standing stone circles occur on the shore of the island of Er Lannic in a massive flooded ria bay. The lowest standing menhir is situated about 6 m below mean high water spring (MHWS), with very restricted wind fetch, but high tidal range (Giot 1968; Giot et al. 1979; Prigent et al. 1983: 307; Cassen et al. 2011).

Pavlopetri, southern Greece, 5000 BP

At Pavlopetri (Harding et al. 1969; Henderson et al. 2011), there are 8 hectares of well-preserved Early Bronze Age ruins dated 5000 BP to 3000 BP that have survived on the sea floor sloping down from the beach to a depth of over 3 m, and probably associated with a shoreline at −5 m. The location of the town is at the head of the large Bay of Vatika, which is screened in all directions from maximum storm-fetch waves by headlands and islands, so that the maximum wave height reaching the location is less than 2 m, based both on local observation and numerical wind-wave refraction+shoaling models. A submerged ridge of fossil eolianite dune curves out from the shore, breaking the surface at several locations, and terminating in the island of Pavlopetri, which is about 80 m × 50 m in size. The submerged town is in the triangular sea area between the ridge and the present shoreline. Wave action within this sheltered zone is very limited, seldom exceeding

0.5 m in amplitude and with short wavelength. Moving bands of seabed sand banks have covered large parts of the town at different dates in the last 45 years, and the exposed areas seem to be increasingly damaged by boring species and corrosion. The topography has protected the ruins to an extraordinary degree, but biological action and casual souvenir collecting by tourists suggests that the site may be unrecognizable in another hundred years.

Southwest Baltic, Danish and German coasts

Many hundreds of submerged prehistoric sites have been found with the earliest settlements about 8000 years old (Fischer 1995; 2002; 2011; Lübke 2002a,b; 2003; 2004; Harff & Lüth 2011; Jöns 2011; Lübke et al. 2011). The Baltic as a whole is sheltered from Atlantic oceanic storm waves, and has minimal tides. The southwest Baltic archipelago further restricts wind fetch to a few kilometers, and sites studied on the sea floor were usually protected by islands, sand banks or peninsulas, now submerged, before site inundation. Re-working of the coastal sediments during sea-level rise further protected some sites, while eroding others. Freshwater conditions have preserved wooden objects such as tool shafts, canoes, paddles, fish weirs, and hut posts.

Western UK and Severn Estuary, 8000 BP to 6000 BP

The intertidal mudflats of western England have revealed a wide range of prehistoric materials and imprints of Mesolithic to Bronze Age origin, 8000 BP to 6000 BP (Aldhouse-Green et al. 1992; Bell & Neumann 1997; Bell 2007). The finds include human footprints of children and adults, and the hoof-prints of cattle, sheep and pigs as well as wooden remains, trackways, posts, and fish traps. The substrate is laminated silt that retains fossilized seasonal variations, overlain by mobile layers of soft mud and sediments that are partially eroding. The gradients are extremely low, and the archaeological signals survive in spite of the high spring tidal range of 14 m and exposure to storm waves from the west.

Atlit-Yam, Israel, Pre-Pottery Neolithic, 8000 BP

The stretch of coast between Dor and Haifa is bordered by calcified eolianite dune ridges known as *kurkar* which are closely parallel to the shore, both on land and offshore (Galili *et al.* 1993; 2004). The offshore kurkar ridges have protected the settlements from wave action on the site during inundation, and the remains were subsequently buried by the longshore drift of sand supplied by the Nile River. Mid-twentieth-century damming of the Nile has reduced the sand supply, so that Early Neolithic villages are now being exposed in water depths of the order of 10 m to 15 m. Local fishermen, and professional and amateur diving archaeologists inspect the shallow seabed frequently to detect remains revealed after storms. Finds include hearths, charcoal, human burials, freshwater wells, hut foundations, food remains, and other organic materials.

Bouldnor Cliff, Isle of Wight, UK, 8000 BP

The submerged Mesolithic site is at a depth of 11 m on the north shore of the Isle of Wight, protected from the storm waves of the English Channel and the wind-wave fetch from the Atlantic to the south-west by a narrow channel. The Solent is 4.5 km across in the region of the site so that locally generated wind waves are limited. There are strong tidal currents which continue to erode the archaeological strata which are embedded in peat and clay. The site has been monitored for 20 years so that the erosion is documented (Momber 2006; Momber *et al.* 2012). The archaeological remains include timbers, cut wood, worked lithics, string, a hearth, and seeds of grain.

Rotterdam Port, the Netherlands, North Sea, 30,000 BP to 10,000 BP

During the extension of Rotterdam port known as Yangtze Maasvlakte 2, layers of alluvial sediments 10 m to 20 m thick were dredged out of the coastal waters of the North Sea, and Mesolithic and Paleolithic remains up to 30,000 years old were identified and partially recovered (Weerts *et al.* 2012; Moree & Sier 2014). Site distribution was reconstructed on the buried dune ridges. This extensive project confirms both the concentration of prehistoric Mesolithic communities on the estuarine and deltaic dunes, adjacent to marine and freshwater resources, and the potential for preservation of such sites as they were deeply buried in deltaic sediments during sea-level rise.

Cap Lévi, Anse de La Mondrée, France, 100 ka

Paleolithic flint artifacts were recovered in the late 1970s by divers on the east side of the granite ridge and rocky islet known as Biéroc, projecting from Cap Lévi near the village of Fermanville, east of Cherbourg (Scuvée & Verague 1988). Subsequent analysis of the lithics and dating of sediments has shown that the site at a depth of 18 m to 20 m was occupied sporadically between approximately 100 ka to 50 ka (Cliquet *et al.* 2011). Collections of flint cores, worked tools, and débitage show that flints were knapped on the site. The tidal currents reach speeds of 5 knots, but the high granite ridge prevents scour at the seabed, and equally protects the deposits from the wind waves and Atlantic storms driven in from the west.

A240 concession, East Anglia, UK, North Sea, 300 ka to 250 ka

Dredging for aggregate gravels in 2009 revealed Acheulean handaxes and Levallois flakes from a concession area defined as A240, 11 km off the coast of East Anglia in the North Sea, at a depth of 16 m to 35 m. The area was protected from further commercial extraction, and archaeological tests and examination were conducted by Wessex Archaeology (Tizzard *et al.* 2014). Core samples revealed stratified river gravels, estuarine, coastal and terrestrial sediment layers, showing marine transgressions and regression, and lithic tools were recovered from a layer that could be dated to approximately 250 ka. While some lithics may be derived from older layers, the majority owe their survival to burial in a steadily accumulating marine basin, protected from the open Atlantic Ocean, and with regional low gradients making erosion minimal at the site. At the date of writing (2016), this site constitutes the oldest seabed archaeological site in European seas.

Value of Understanding the Submerged Landscape: Sites in Context

Although the first indication that a submerged prehistoric site exists is often the chance retrieval of artifacts, such deposits cannot be understood without considering the context in terms of the overall stratigraphy of the anthropogenic components, and their relationship to the topography and ecology of the immediate surroundings. This is a commonplace of archaeological investigation, but in the offshore situation it needs to be stressed because it usually entails the survey and sampling of the seabed for hundreds of meters, or several kilometers, around the site, requiring acoustic survey systems and seabed sampling equipment. Analysis of sites in environmental and landscape context reveals patterns that facilitate further site prediction, and provides understanding of foraging and hunting strategies, as well as prehistoric access to raw materials and fresh water. The regional chapters which follow Chapter 4 provide sources of data facilitating the reconstruction of drowned landscapes.

Environmental and Industrial Threats to Known Sites and Preserved Submerged Landscapes

Known submerged prehistoric sites are at risk from coastal erosion in shallow locations, from seabed erosion in open water, and from industrial excavation or tourist and professional looting. In addition, we can assume that there are thousands of offshore prehistoric deposits, either exposed on the sea floor, or buried in the sedimentary column, that are unknown as yet, and are similarly at risk. Thousands more have already been destroyed. The preceding discussion has shown how unexpectedly robust prehistoric sites are in the marine environment, but that discussion was structured to counter the supposition that nothing at all will survive inundation. In practice, we must recognize the precarious balance of environmental forces, such that some sites do survive and can be studied, while others will certainly be destroyed, and it is probably not possible to save them.

In shallow coastal waters out to a depth of about 10 m, many sites are found by fishermen, sports divers, and amateur archaeologists, all of whom report finds to the authorities (Flemming *et al.* 2014). Research agencies and academic bodies can then usually respond and conduct investigations. In deeper water, finds to a depth of 20 m to 30 m can still be examined by divers using compressed air, but deeper than 40 m requires specialists in mixed gas diving. At depths of 30 m to 40 m or more, and tens of kilometers offshore, work by research or conservation agencies become expensive, and surveys to find prehistoric sites are rare. However, bottom trawl nets recover paleontological materials (see for example Mol *et al.* 2008), while aggregate dredging and pipelaying extract massive samples from the sea floor, within which prehistoric traces may be found. Thus, in spite of the risk of damage, and provided that there is routine communication with archaeological experts, the offshore industries help to locate, and ultimately protect, prehistoric traces on the continental shelf that would never otherwise be found. A reciprocal relationship is needed whereby offshore operators are required to assess the potential for prehistoric materials and to monitor discoveries in their concession areas. The Regional Seas chapters of this book document coastal change processes, whether accumulation or erosion, and hence the potential threat to known sites.

Conclusion

Given suitable local topography and oceanographic exposure, field evidence shows that prehistoric remains can survive for hundreds, or even a few thousand, years in the shallow surf zone at constant sea level. The protecting factors are topographic, morphodynamic, and progradational, and examples have been provided. This observation and deduction contradicts the common assumption that sites cannot survive close to a steady sea level, and can only survive when they are deeply buried and the sea level is changing rapidly.

For longer periods, prehistoric archaeological and paleontological remains have the greatest potential to survive on the seabed, and through multiple glacial cycles of sea-level change, if they have been deposited in a sedimentary depocenter with a steady rate of sediment accumulation. The prime example of this on the scale of the whole Quaternary is the central to southern North Sea. On the timescale of the last glacial cycle, the southern Baltic has provided long-term accumulation and protection for the survival of thousands of sites.

The preservation of underwater landscapes, particularly paleoshorelines in constructive environments, is favored by: (1) high sedimentary input to the coast; (2) calm marine climate with low wave and current energy; (3) early diagenesis and; (4) sea level not repeatedly re-occupying the same or a similar level. Crenulated coasts where headlands act as natural obstacles to longshore drift are favorable environments for the genesis of sandy barrier complexes, and hence burial or screening of prehistoric sites.

At the intermediate timescale of hundreds of years to tens of thousands of years, the probability that a site will survive is determined by the ability of its protective topographic screen to resist multiple transgressions and regressions, with wave attacks delivered at different levels and possibly from different directions and in different climates through multiple cycles of sea-level change. The cumulative time that the sea surface stays at, or re-occupies, the same vertical zone within ±10 m of the site increases the probability that it will be destroyed.

Analysis of the effect of rising sea level on a coastal landscape must be conducted with fully three-dimensional modeling, replicating the lateral processes of coastal erosion and transport of sediments.

The selected examples outlined above provide an illustration of the wide range of environmental and geo-dynamic situations in which submerged prehistoric sites have survived.

The drowned sites, landscapes, and ecosystems discussed in following chapters are all from the European and Mediterranean seas, and hence in mid latitude to near-polar environments. All the circumstances discussed in this chapter, and indeed throughout this book, will not be applicable directly to tropical or fully polar environments, where different processes occur.

References

Aldhouse-Green, S., Whittle, A., Allen, J. R. L. *et al.* 1992. Prehistoric human footprints from the Severn Estuary at Uskmouth and Magor Pill, Gwent, Wales. *Archaeologia Cambrensis* 141:14-55.

Alley, R. B. & Ágústsdóttir, A. M. 2005. The 8k event: cause and consequences of a major Holocene abrupt climate change. *Quaternary Science Reviews* 24:1123-1149.

Amorosi, A., Centineo, M. C., Colalongo, M. l., Pasini, G., Sarti, G. & Vaiani, S.C. 2003. Facies architecture and Latest Pleistocene–Holocene depositional history of the Po Delta (Comacchio area), Italy. *Journal of Geology* 111:39-56.

Andersen, S. H. 2013. *Tybrind Vig: submerged Mesolithic settlements in Denmark.* (Jutland Archaeological Society Publications). Aarhus University Press: Højbjerg.

Antoine, P., Catt, J., Lautridou, J. -P. & Sommé, J. 2003. The loess and coversands of northern France and southern England. *Journal of Quaternary Sciences* 18:309-318.

Arrieta, N., Goienaga, N., Martínez-Arkarazo, I. *et al.* 2011. Beachrock formation in temperate coastlines: Examples in sand-gravel beaches adjacent to the Nerbioi-Ibaizabal Estuary (Bilbao, Bay of Biscay, North of Spain). *Spectrochimica Acta Part A, Molecular and Biomolecular Spectroscopy* 80:55-65.

Backstrom, J. T., Jackson, D. W. T. & Cooper, J. A. G. 2009. Shoreface morphodynamics of a high-energy, steep and geologically constrained shoreline segment in Northern Ireland. *Marine Geology* 257:94-106.

Bailey, G. N. 2007. Time perspectives, palimpsests and the archaeology of time. *Journal of Anthropological Archaeology* 26:198-223.

Bailey, G. N. 2011. Continental shelf archaeology: where next? In Benjamin, J., Bonsall, C., Pickard, C. & Fischer, A. (eds.) *Submerged Prehistory*. pp. 311-331. Oxbow Books: Oxford.

Ballard, R. (ed.) 2008. *Archaeological Oceanography*. Princeton University Press: Princeton, New Jersey.

Belderson R. H., Kenyon, N. H., Stride, A. H. & Stubbs, A. R. 1972. *Sonographs of the Sea Floor: A Picture Atlas*. Elsevier Publishing Co.: Amsterdam.

Belknap, D. F. 1983. Potentials of discovery of human occupation sites on the continental shelf and nearshore coastal zone. In *Proceedings of the Fourth Annual Gulf of Mexico Information Transfer Meeting*. 15th – 17th November 1983, New Orleans, pp. 382-387.

Belknap, D. F. & Kraft, J. C. 1981. Preservation potential of transgressive coastal lithosomes on the U.S. Atlantic Shelf. *Marine Geology* 42:429-442.

Belknap, D. F. & Kraft, J. C. 1985. Influence of antecedent geology on stratigraphic preservation potential and evolution of Delaware's barrier systems. In Oertel, G. F. & Leatherman, S. P. (eds.) *Marine Geology (Special Issue — Barrier Islands)* 63:235-262.

Bell, M. 2007. *Prehistoric Coastal Communities. The Mesolithic in western Britain*. CBA Research Report 149. Council for British Archaeology: York.

Bell, M. & Neumann, H. 1997. Prehistoric intertidal archaeology and environment in the Severn Estuary, Wales. *World Archaeology* 29:95-113.

Benavente, J., Del Río, J., Anfuso, G., Gracia, F. J. & Reyes, J. L. 2002. Utility of morphodynamic characterisation in the prediction of beach damage by storms. *Journal of Coastal Research (Special Issue 36, International Coastal Symposium Proceedings Northern Ireland)*. pp. 56-64.

Benjamin, J. 2010. Submerged prehistoric landscapes and underwater site discovery: Reevaluating the 'Danish Model' for international practice. *Journal of Island & Coastal Archaeology* 5:253–270.

Benjamin, J., Bonsall, C., Pickard, C. & Fischer, A. (eds.) 2011. *Submerged Prehistory*. Oxbow Books: Oxford.

Bernier, P., Guidi, J. -B. Böttcher, M. E. 1997. Coastal progradation and very early diagenesis of ultramafic sands as a result of rubble discharge from asbestos excavations (northern Corsica, western Mediterranean). *Marine Geology* 144:163-175.

Bezerra, M. M., Moura, D., Ferreira, Ó. & Taborda, R. 2011. The influence of wave action and lithology on sea cliff mass movements in central Algarve coast, Portugal. *Journal of Coastal Research* 27:162-171.

Binford, L. R. 1980. Willow smoke and dog's tails: Hunter gatherer settlement systems and archaeological site formation. *American Antiquity* 45:4-20.

Blanc, A. C. 1940. Industrie musteriane e paleolithiche superiore nelle dune fossile e nelle grotte litorannee del Capo Palinuro. *R.C. Reale Accademia d'Italia* 10(7):1.

Bowdler, S. 1977. The coastal colonisation of Australia. In Allen, J., Golson, J. & Jones, R. (eds.) *Sunda and Sahul: Prehistoric Studies in Southeast Asia, Melanesia and Australia*. pp. 205-246. Academic Press: London.

Bruun, P. 1962. Sea-level rise as a cause of shore erosion. *Journal of the Waterways, Harbors and Coastal Engineering Division* 88:117-130.

Burkitt, M. C. 1932. A Maglemose Harpoon dredged up recently from the North Sea. *MAN* 32:118.

Carrión, J. S., Finlayson, C., Fernández, S. *et al.* 2008. A coastal reservoir of biodiversity for Upper Pleistocene human populations: palaeoecological investigations in Gorham's Cave (Gibraltar) in the context of the Iberian Peninsula. *Quaternary Science Reviews* 27:2118-2135.

Carter, R. W. G. 1988. *Coastal Environments: An Introduction to the Physical, Ecological and Cultural Systems of Coastlines*. Academic Press: London.

Cassen, S., Baltzer, A., Lorin, A., Fournier, J. and Sellier, D. 2011. Submarine Neolithic Stone Rows near Carnac (Morbihan), France: Preliminary results from acoustic and underwater survey. In Benjamin, J., Bonsall, C., Pickard, C. & Fischer, A. (eds.) *Submerged Prehistory*. pp. 99-110. Oxbow Books: Oxford.

Catuneanu, O., Galloway, W. E., Kendall, C. G. St C. *et al.* 2011. Sequence stratigraphy: Methodology and Nomenclature. *Newsletters on Stratigraphy* 44:173-245.

Chang, C. -H., Kaifu, Y., Takai, M. *et al.* 2015. The first archaic *Homo* from Taiwan. *Nature Communications* 6:Article number 6037.

Chiocci, F. L. & Chivas, A. R. (eds) 2014. *Continental Shelves of the World: Their Evolution During the Last Glacio-Eustatic Cycle.* Memoirs (41). Geological Society: London.

Clausen, C. J., Cohen, A. D., Emiliani, C., Holman, J. A. & Stipp, J. J. 1979. Little Salt Springs, Florida: a unique underwater site. *Science* 203:609-614.

Cliquet, D., Coutard, S., Clet, M. *et al.* 2011. The Middle Palaeolithic underwater site of La Mondrée, Normandy, France. In Benjamin, J., Bonsall, C., Pickard, C. & Fischer, A. (eds.) *Submerged Prehistory.* pp. 111-128. Oxbow Books: Oxford.

Cockerell, W. A. & Murphy, L. 1978. 8 SL17: Methodological approaches to a dual component marine site on the Florida Atlantic coast. In Barto-Arnold, J. (ed.) *Beneath the Waters of Time — Proceedings of the 9th Conference on Underwater Archaeology.* pp. 175-182. Texas Antiquities Committee Publication No. 6.

Cooper, J. A. G. & Pilkey, O. H. 2004. Sea-level rise and shoreline retreat: time to abandon the Bruun Rule. *Global and Planetary Change* 43:157-171.

Davidson-Arnott, R. 2010. *Introduction to Coastal Processes and Geomorphology.* Cambridge University Press: Cambridge.

Davis, R. A., & Fitzgerald, D. M. 2004. *Beaches and Coasts.* Blackwell Publishing: Oxford.

Deng, J., Zhang, W., Harff, J. *et al.* 2014. A numerical approach for approximating the historical morphology of wave-dominated coasts — A case study of the Pomeranian Bight, southern Baltic Sea. *Geomorphology* 204:425-443.

Dietrich, S. & Seelos, K. 2010. The reconstruction of easterly wind directions for the Eifel region (Central Europe) during the period 40.3-12.9 ka BP. *Climate of the Past* 6:145-154.

Dimitriu, R. G. 2012. Geodynamic and hydro-geological constraints regarding the extension of the prospective archaeo-cultural area within the northern Romanian coastal zone. *Quaternary International* 261:32-42.

Dinelli, E., Ghosh, A., Rossi, V. & Vaiani, S.C. 2012. Multiproxy reconstruction of Late Pleistocene-Holocene environmental changes in coastal successions: microfossil and geochemical evidences from the Po Plain (Northern Italy). *Stratigraphy* 9:153-167.

Dinu, C., Wong, H.K., Tambrea, D. & Matenco, L. 2005. Stratigraphic and structural characteristics of the Romanian Black Sea shelf. *Tectonophysics* 410:417-435.

Efremov, J. A. 1940. Taphonomy: a new branch of Paleontology. *The Pan-American Geologist* 74:81-93.

Emery, K. O. & Edwards, R. L. 1966. Archaeological potential of the Atlantic continental shelf. *American Antiquity* 31:733-737.

Evans, A., Flatman, J. & Flemming, N.C. (eds.) 2014. *Prehistoric Archaeology on the Continental Shelf: a Global Review.* Springer: New York.

Faure, H., Walter, R. C. & Grant, D. R. 2002. The coastal oasis: ice age springs on emerged continental shelves. *Global and Planetary Change* 33:47-56.

Fischer, A. (ed.) 1995. *Man and Sea in the Mesolithic. Coastal settlement above and below present sea level.* Oxbow Books: Oxford.

Fischer, A. 2002. Food for feasting? An evaluation of explanations of the neolithisation of Denmark and southern Sweden. In Fischer, A. & Kristiansen, K. (eds.) *The Neolithisation of Denmark — 150 Years of Debate.* Sheffield Archaeological Monographs 12. pp. 343-393. J.R. Collis Publications: Sheffield.

Fischer, A. 2011. Stone Age on the Continental Shelf: an eroding resource. In Benjamin, J., Bonsall, C., Pickard, C. & Fischer, A. (eds.) *Submerged Prehistory.* pp. 298-310. Oxbow Books: Oxford.

Fladmark, K. R. 1975. *A Palaeoecological Model for Northwest Coast Prehistory.* National Museums of Canada: Ottawa.

Fleming, B. W. 1976. Side-scan sonar: a practical guide. *International Hydrographic Review* 53:65-92.

Flemming, N. C. 1968. Derivation of Pleistocene marine chronology from morphometry of erosion profiles. *Journal of Geology* 76:280-296.

Flemming, N. C. 1983. Survival of submerged lithic and Bronze Age artefact sites: a review of case histories. In Masters, P. & Flemming, N. C. (eds.) *Quaternary Coastlines and Marine Archaeology.* pp. 135-174. Academic Press: London and New York.

Flemming, N. C. 2002. *The scope of Strategic Environmental Assessment of North Sea areas SEA3 and SEA2 in regard to prehistoric archaeological remains.* UK Department of Trade and Industry: London.

Flemming, N. C. & Antonioli, F. 2017. Prehistoric archaeology, palaeontology, and climate change indicators from caves submerged by change of sea level. In Campbell, P. (ed.) *The Archaeology of Underwater Caves.* pp. 22-38. Highfield Press: Southampton.

Flemming N. C., Çağatay M. N., Chiocci F. L. *et al.* 2014. *Land Beneath the Waves: Submerged Landscapes and Sea Level Change. A Joint Geoscience-Humanities Research Strategy for European Continental Shelf Prehistoric Research.* Chu N.-C. & McDonough N. (eds.) Position paper 21 of the European Marine Board: Ostend, Belgium.

French, H. M. 2007. *The Periglacial Environment* (3rd Ed). John Wiley & Sons: Chichester.

Gaffney V., Thomson, K. & Fitch, S. (eds.) 2007. *Mapping Doggerland: The Mesolithic Landscapes of the Southern North Sea.* Archaeopress: Oxford.

Gagliano, S. M., Pearson, C. E., Weinstein, R. A., Wiseman, D. E. & McClendon, C. M. 1982. *Sedimentary Studies of Prehistoric Archaeological sites: Criteria for the Identification of Submerged Archaeological Sites of the Northern Gulf of Mexico Continental Shelf.* National Park Service: Baton Rouge, LA.

Galili, E., Weinstein-Evron, M., Hershkovitz, I. *et al.* 1993. Atlit-Yam: A prehistoric site on the sea floor off the Israeli coast. *Journal of Field Archaeology* 20:133-157.

Galili, E., Lernau, O. & Zohar, I. 2004. Fishing and coastal adaptations at Atlit-Yam — A submerged PPNC Fishing village off the Carmel coast, Israel. *'Atiqot* 48:1-34.

Geddes, D. S., Guilaine, J. & Monaco, A. 1983. Early Neolithic occupation of the submerged continental plateau of Roussillon (France). In Masters, P. M. & Flemming, N. C. (eds.) *Quaternary Coastlines and Marine Archaeology.* pp. 175-188. Academic Press: London.

Giot, P.-R. 1968. La Bretagne au péril des mers holocènes. In Bordes, F. & de Sonneville-Bordes, D. (eds.) *La Préhistoire, problèmes et tendances.* pp. 203-208. CNRS: Paris.

Giot, P. -R., L'Helgouac'h, J. & Monnier, J. -L. 1979. *Préhistoire de la Bretagne.* Ouest-France: Rennes.

Harding, A. F., Cadogan, G. & Howell, R. 1969. Pavlopetri: an underwater Bronze Age town in Laconia. *The Annual of the British School at Athens* 64:113-142.

Harff, J. & Lüth, F. (eds.) 2007. *SINCOS — Sinking Coasts: Geosphere, Ecosphere and Anthroposphere of the Holocene Southern Baltic Sea.* Bericht der Römisch-Germanische Kommission: Special Issue 88.

Harff, J. & Lüth, F. (eds.) 2011. *SINCOS II — Sinking Coasts: Geosphere Ecosphere and Anthroposphere of the Holocene Southern Baltic Sea II. Bericht der Römisch-Germanische Kommission:* Special Issue 92.

Harff, J., Lemke, W., Lampe, R. *et al.* 2007. The Baltic Sea coast — A model of interrelations among geosphere, climate, and anthroposphere. *Geological Society of America Special Papers* 426:133-142.

Henderson, J. C., Gallou, C., Flemming, N. C. & Spondylis, E. 2011. The Pavlopetri Underwater Archaeology Project: investigating an ancient submerged town. In Benjamin, J., Bonsall, C., Pickard, C., & Fischer, A. (eds.) *Submerged Prehistory.* pp. 207-218. Oxbow Books: Oxford.

Holdaway, S. & Wandsnider, L. (eds.) 2008. *Time in Archaeology: Time Perspectivism Revisited.* pp. 13-30. University of Utah Press: Utah.

Holthuijsen, L. H. 2007. *Waves in Oceanic and Coastal waters.* Cambridge University Press: Cambridge.

Horta, J., Moura, D., Gabriel, S. & Ferreira, O. 2013. Measurement of pocket beach morphology using geographic information technology: the MAPBeach toolbox. *Journal of Coastal Research* (Special Issue 65):1397-1402.

Howie, F. M. P. 2009. Beachrock development along the north coast of Cornwall. *Geoscience in South-West England* 12:85-94.

Hublin, J. -J., Weston, D., Gunz, P. *et al.* 2009. Out of the North Sea: the Zeeland Ridges Neandertal. *Journal of Human Evolution* 57:777-785.

Infantini, L., Moura, D. & Bicho, N. 2012. Utilização de ferramentas SIG para o estudo da morfologia submersa da Baía de Armação de Pêra. In Almeida, A. C., Bettencourt, A. M., Moura, D., Monteiro-Rodrigues, S. & Alves, M. I. C. (eds.) *Environmental Changes and Human Interaction along the Western Atlantic Edge.* pp. 227-241. APEQ: Coimbra.

Iversen, J, 1973. *The Development of Denmark's Nature Since the Last Glacial.* Geological Survey of Denmark V. series No. 7-c.

Jackson, D. W. T. & Cooper, J. A. G. 2010. Application of the equilibrium planform concept to natural beaches in Northern Ireland. *Coastal Engineering* 57:112-123.

Jackson, D. W. T., Cooper, J. A. G. & del Rio, L. 2005. Geological control of beach morphodynamic state. *Marine Geology* 216:297-314.

Jessen, K. 1938. Some west Baltic pollen diagrams. *Quartär* 1:124-139.

Jöns, H. 2011. Settlement development in the shadow of coastal changes — case studies from the Baltic Rim In Harff, J., Björck, S. & Hoth, P. (eds.) *The Baltic Sea Basin.* pp. 301-336. Springer-Verlag: Berlin-Heidelberg.

Jöns, H., Mennenga, M. & Schaap, D. 2016. *The SPLASHCOS-Viewer. A European Information system about submerged prehistoric sites on the continental shelf. A product of the COST Action TD0902 SPLASHCOS (Submerged Prehistoric Archaeology and Landscapes of the Continental Shelf).* Accessible at splashcos-viewer.eu/

Joris, C. 2002. The Magdalenian assemblages from Ardèche (France) in the Mediterranean Basin context. *L'Anthropologie* 106:99-134.

Karkanas, P., Shahack-Gross, R., Ayalon, A. *et al.* 2007. Evidence for habitual use of fire at the end of the Lower Paleolithic: Site-formation processes at Qesem Cave, Israel. *Journal of Human Evolution* 53:197-212.

Kasse, C. 1997. Cold-climate Aeolian sand-sheet formation in North-Western Europe (c. 14–12 ka); a response to permafrost degradation and increased aridity. *Permafrost and Periglacial Processes* 8:295-311.

Kenyon, N. H., Belderson, R. H., Stride, A. H. & Johnson, M. A. 1981. Offshore tidal sandbanks as indicators of net sand transport and as potential deposits. In Nio, S. D., Schuttenhelm, R. T. E. & van Weering, T. C. E. (eds.) *Holocene Marine Sedimentation in the North Sea Basin.* pp. 257-268. Blackwell Publishing Ltd: Oxford.

Konishchev, V. N. & Rogov, V. V. 1993. Investigation of cryogenic weathering in Europe and Northern Asia. *Permafrost and Periglacial Processes* 4:49-64.

Kuhn, S. L., Stiner, M. C., Guleç, E. *et al.* 2009. The early Upper Paleolithic occupations at Üçağizli Cave (Hatay, Turkey). *Journal of Human Evolution* 56:87-113.

Lefort, J. -P., Danukalova, G. A. & Monnier, J.-L. 2011. Origin and emplacement of the loess deposited in northern Brittany and under the English Channel. *Quaternary International* 240:117-127.

Lefort, J. -P., Danukalova, G. & Monnier, J.-L. 2013. Why the submerged sealed beaches, last remnants of the low stands of the Upper Pleistocene regression, are better expressed in the western than in the eastern English Channel? *Geo-Eco-Marine* 19:5-16.

Lobo, F. J., Ercilla, G., Fernández-Salas, L. M. & Gámez, D. 2014. The Iberian Mediterranean shelves. In Chiocci, F. L & Chivas, A. R (eds.) *Continental Shelves of the World: Their Evolution During the Last Glacio-Eustatic Cycle (Geological Society Memoir No. 41).* pp. 147-170. Geological Society: London.

Long, D., Wickham-Jones, C. R. & Ruckley, N. A. 1986. A flint artefact from the northern North Sea. In Roe, D. A. (ed.) *BAR International Series (No. 296) Studies in the Upper Palaeolithic of Britain and North West Europe.* pp. 55-62. British Archaeological Reports: Oxford.

Louwe Kooijmans, L. P. 1970-71. Mesolithic bone and antler implements from the North Sea and from the Netherlands. *Berichten van de Rijksdienst voor het Oudheidkundig Bodemonderzoek* 20-21:27-73.

Lowe, J. J. & Walker, M. J. C. 1984. *Reconstructing Quaternary Environments.* Longman: London.

Lübke, H. 2001. Eine hohlendretuschierte Klinge mit erhaltener Schäftung vom endmesolithischen Fundplatz Timmendorf-Nordmole, Wismarbucht, Mecklenburg-Vorpommern. *Nachrichtenblatt Arbeitskreis Unterwasserarchäologie* 8:46-51.

Lübke, H. 2002a. Steinzeit in der Wismarbucht: ein Überblick. *Nachrichtenblatt Arbeitskreis Unterwasserarchäologie* 9:75-87.

Lübke, H. 2002b. Submarine Stone Age settlements as indicators of sea-level changes and the coastal evolution of the Wismar Bay area. *Greifswalder Geographische Arbeiten* 27: 202-210.

Lübke, H. 2003. New investigations on submarine Stone Age settlements in the Wismar Bay area. In Kindgren, H., Knutsson, K., Larsson, L., Loeffler, D. & Åkerlund,

A. (eds.) *Mesolithic on the Move — Proceedings of the Sixth International Conference on the Mesolithic in Europe*, 4th – 8th September 2000, Stockholm, pp. 69-78.

Lübke, H. 2004. Spät- und endmesolithische Küstensiedlungsplätze in der Wismarbucht — Neue Grabungsergebnisse zur Chronologie und Siedlungsweise. *Bodendenkmalpflege in Mecklenburg-Vorpommern (Jahrbuch 2004)* 52:83-110.

Lübke, H., Schmölcke, U. & Tauber, F. 2011. Mesolithic Hunter-Fishers in a Changing World: a case study of submerged sites on the Jäckelberg, Wismar Bay, northeastern Germany. in Benjamin, J., Bonsall, C., Pickard, C. & Fischer, A. (eds.) *Submerged Prehistory*. pp. 21-37. Oxbow Books: Oxford.

Lyman, R. E. 2010. What taphonomy is, what it isn't, and why taphonomists should care about the difference. *Journal of Taphonomy* 8:1-16.

Marabini, F. 1997. The Po River Delta evolution. *Geo-Eco-Marina* 2:47-55.

Masters, P. M. & Flemming, N. C. (eds.) 1983. *Quaternary Coastlines and Marine Archaeology*. Academic Press: London.

Mellett, C. L., Hodgson, D. M., Lang, A., Mauz, B., Selby, I. & Plater, A. J. 2012. Preservation of a drowned gravel barrier complex: A landscape evolution study from the north-eastern English Channel. *Marine Geology* 315-318:115-131.

Miller, K. G., Kominz., M. A., Browning, J. V. *et al.* 2005. The Phanerozoic record of global sea-level change. *Science* 310:1293-1298.

Mol, D., Vos, J. de, Bakker, R. *et al.* 2008. *Mammoeten, neushoorns en andere dieren van de Noordzeebodem.* Wetenschappelijke Biblioteek: Diemen.

Momber, G. 2006. Mesolithic occupation: 11m below the waves. In Hafner, A., Niffler, U. & Ruoff, U. (eds.) *The New View: Underwater Archaeology and the Historic Picture*. pp. 56-63. Archaeologie Schweiz: Basel.

Momber, G., Bailey, G. & Moran, L. 2012. Identifying the archaeological potential of submerged landscapes. In Henderson, J. (ed.) *Beyond Boundaries — Proceedings of the 3rd International Congress on Underwater Archaeology*. 9th – 12th July 2012, London, pp. 257-268.

Moree, J. M. & Sier, M. M. (eds.) 2014. *Twintig meter diep! Mesolithicum in de Yangtzehaven — Maasvlakte te Rotterdam. Landschapsontwikkeling en bewoning in het Vroeg Holoceen. BOORrapporten 523.* Bureau Oudheidkundig Onderzoek Rotterdam (BOOR): Rotterdam.

Moura, D., Veiga-Pires, C., Albardeiro, l., Boski, T., Rodrigues, A. L. & Tareco, H. 2007. Holocene sea level fluctuations and coastal evolution in the central Algarve (southern Portugal). *Marine Geology* 237:127-142.

Moura, D., Gabriel, S. & Jacob, J. 2011. Coastal morphology along the Central Algarve rocky coast: Driver mechanisms. *Journal of Coastal Research* (Special Issue 64):790-794.

Murphy, L. 1978. 8 SLO 19: Specialised methodological, technological and physiological approaches to deep water excavation of a prehistoric site at Warm Mineral Springs, Florida. In Barto-Arnold, J. (ed.) *Beneath the Waters of Time — Proceedings of the 9th Conference on Underwater Archaeology*. pp. 123-128. Texas Antiquities Committee Publication No. 6.

Murton, J. B. & Lautridou, J-P. 2003. Recent advances in the understanding of Quaternary periglacial features of the English Channel coastlands. *Journal of Quaternary Science* 18:301-307.

Nakazawa, Y., Straus, L. G., González-Morales, M. R., Solana, D. C. & Saiz, J. C. 2009. On stone-bolling technology in the Upper Paleolithic: behavioral implications from an Early Magdalenian hearth in El Mirón Cave, Cantabria, Spain. *Journal of Archaeological Science* 36:684-693.

Ontañón, R. 2003. Sols et structures d'habitat du Paléolithique supérieur, nouvelles données depuis les Cantabres: la Galerie Inférieure de La Garma (Cantabrie, Espagne). *L'Anthropologie* 107:333-363.

Pearson, C. E., Kelley, D. B., Weinstein, R. A. & Gagliano, S. M. 1986. *Archaeological Investigations on the Outer Continental Shelf: A Study Within the Sabine River Valley, Offshore Louisiana and Texas*. Coastal Environments, Inc: Baton Rouge.

Pedersen, L. D., Fischer, A. & Aaby, B. 1997 (eds.). *The Danish Storebaelt since the Ice Age: Man, Sea and Forest*. Storebaeltsforbindelsen: Copenhagen.

PNUE (Programme des Nations Unies pour l'Environment) 2004. *L'eau des Méditerranéens: Situation et perspectives.* MAP Technical Report Series No. 158. PNUE/PAM: Athens.

Post, K & Kompanje, E. J. O 2010. A new dolphin (Cetacea, Delphinidae) from the Plio-Pleistocene of the North Sea. *Deinsea* 14:1-14.

Prigent, D., Vissent, L., Morzadec-Kerfourn, M. T. & Lautridou, J-P. 1983. Human occupation of the submerged coast of the Massif Armoricain and postglacial sea level changes. In Masters, P. M. & Flemming, N. C. (eds.) *Quaternary Coastlines and Marine Archaeology.* pp. 303-324. Academic Press: London and New York.

Reid, C. 1913. *Submerged Forests.* Cambridge University Press: Cambridge.

Rocha, M., Silva, P., Michallet, H., Abreu, T., Moura, D. & Fortes, J. 2013. Parameterizations of wave nonlinearity from local wave parameters: a comparison with field data. *Journal of Coastal Research* (Special Issue 65):374-379.

Romanovskii, N. N., Hubberten, H. -W., Gavrilov, V. *et al.* 2000. Thermokarst and land-ocean interactions, Laptev sea region, Russia. *Permafrost and Periglacial Processes* 11:137-152.

Rosyan, R. D., Goryachkin, Yu. N., Krylenko, V. V., Dolotov, M. V., Krylenko, M. V. & Godin, E. A. 2012. Crimea and Caucasus accumulative coasts dynamics estimation using satellite pictures. *Turkish Journal of Fisheries and Aquatic Sciences* 12:385-390.

Ruppé, R. J. 1978. Underwater site detection by use of a coring instrument. In Barto-Arnold, J. (ed.) *Beneath the Waters of Time — Proceedings of the 9th Conference on Underwater Archaeology.* pp. 119-121. Texas Antiquities Committee Publication No. 6.

Ruppé, R. J. 1980. The archaeology of drowned terrestrial sites: a preliminary report. *In Bureau of Historical Sites and Properties Bulletin 6.* pp. 35-45. Division of Archives, History and Records Management: Tallahassee.

Salomonsson, B. 1971. Malmotraktens förhistoria. In Bjurling, O. (ed.) *Malmö Stads Historia I.* pp. 15-170. Allhems Forlag: Malmö.

Salzmann, L., Green, A. & Cooper, J. A. G. 2013. Submerged barrier shoreline sequences on a high energy,

steep and narrow shelf. *Marine Geology* 346:366-374.

Schaetzl, R. J. & Anderson, S. 2005. *Soils: Genesis and Geomorphology.* Cambridge University Press: Cambridge.

Schiffer, M. B. 1987. *Formation Processes of the Archaeological Record.* University of New Mexico Press: Albuquerque, NM.

Scuvée, F. & Verague, J. 1988. *Le gisement sous-marin du Paléolithique moyen de l'anse de La Mondrée à Fermanville, Manche.* CEHP-Littus: Cherbourg.

Seibold, E. & Berger, W. H. 1996. *The Sea Floor.* Springer: Berlin and New York.

Shackleton, N. J. & Opdyke, N. D. 1973. Oxygen isotope and palaeomagnetic stratigraphy of Equatorial Pacific core V28-238: oxygen isotope temperature and ice volumes on a 10^5 and 10^6 year scale. *Quaternary Research* 3:39-55.

Shepard, F. P. 1963. *Submarine Geology* (2nd Ed). Harper and Row: New York.

Silva, R., Baquerizo, A., Losada, M. Á. & Mendoza, E. 2010. Hydrodynamics of a headland-bay beach — nearshore current circulation. *Coastal Engineering* 57:160-175.

Skaarup, J. 1980. Undersoisk stenhalder. *Tiddskriftet* 1:3-8.

Stanford, J. D., Hemingway, R., Rohling, E. J., Challenor, P. G., Medina-Elizalde, M. & Lester, A. J. 2011. Sea-level probability for the last deglaciation: A statistical analysis of far-field records. *Global and Planetary Change* 79:193-203.

Steers, J. A. 1948. *The Coastline of England and Wales.* Cambridge University Press: Cambridge.

Stein, J. K. 2001. A review of site formation processes and their relevance to geoarchaeology. In Goldberg, P., Holliday, V. T. & Ferring, C. F (eds.) *Earth Sciences and Archaeology.* pp. 37-51. Springer: New York.

Stoker, M. & Bradwell, T. 2009. Neotectonic deformation in a Scottish fjord, Loch Broom, NW Scotland. *Scottish Journal of Geology* 45:107-116.

Storms, J. E. A., Weltje, G. J., Terra, G. J., Cattaneo, A. & Trincardi, F. 2008. Coastal dynamics under conditions of rapid sea-level rise: Late Pleistocene to Early Holocene evolution of barrier-lagoon systems on the northern Adriatic shelf (Italy). *Quaternary Science Reviews* 27:1107-1123.

Stride A. H. 1963. Current-swept sea floors near the southern half of Great Britain. *Quaternary Journal of the Geological Society* 119:175-197.

Tizzard, L., Bicket, A. R., Benjamin, J. & De Loecker, D. 2014. A Middle Palaeolithic site in the southern North Sea: investigating the archaeology and palaeogeography of Area 240. *Journal of Quaternary Science* 29:698-710.

TRC Environmental Corporation 2012. *Inventory and analysis of archaeological site occurrence on the Atlantic outer continental shelf.* U.S. Dept. of the Interior: New Orleans.

van Huissteden, Ko (J.), Schwan, J. C. G. & Bateman, M. D. 2001. Environmental conditions and paleowind directions at the end of the Weichselian Late Pleniglacial recorded in aeolian sediments and geomorphology (Twente, Eastern Netherlands). *Netherlands Journal of Geosciences* 80:1-8.

Waters, M. R. 1992. *Principles of Geoarchaeology: A North American perspective.* University of Arizona Press: Tuscon.

Weerts, H., Otte, A., Smit, B. *et al.* 2012. Finding the needle in the haystack by using knowledge of Mesolithic human adaptation in a drowning delta. In Bebermeier, W., Hebenstreit, R., Kaiser, E., & Krause, J. (eds.) *Landscape Archaeology — Proceedings of the International Conference (Special Volume 3).* 6th – 8th June, Berlin, pp. 17-24.

Welinder, S. 1971. Överåda — A pitted ware culture site in eastern Sweden. *Meddelande från Lunds historiska museum* 1969-70:5-98.

Werz, B. E. J. S. & Flemming, N. C. 2001. Discovery in Table Bay of the oldest handaxes yet found demonstrates preservation of hominid artefacts on the continental shelf. *South African Journal of Science* 97(May/June):183-185.

Wessex Archaeology 2011. *Seabed Prehistory: Site Evaluation Techniques (Area 240) Synthesis.* Wessex Archaeology: Salisbury.

Williams, M. A. J., Dunkerley, D. L., De Deckker, P., Kershaw, A. P. & Stokes, T. J. 1993. *Quaternary Environments.* Edward Arnold: London.

Winguth, C., Wong, H. K., Panini, N., Dinu, C., Georgescu, P. & Ungureanu, G. 1997. Upper Quaternary sea level changes in the northwestern Black Sea: Preliminary results. *Geo-Eco-Marina* 2:103-113.

Wood, W. R. & Johnson, D. L. 1978. A survey of disturbance processes in archaeological site formation. *Advances in Archaeological Method and Theory* 1:315-381.

Wreschner, E. E. 1977. Neve-Yam. A submerged Late Neolithic settlement near Mount Carmel. *Eretz Israel* 13:260-271.

Yeshurun, R., Bar-Oz, G. & Weinstein-Evron, M. 2007. Modern hunting behavior in the early Middle Paleolithic: Faunal remains from Misliya Cave, Mount Carmel, Israel. *Journal of Human Evolution* 53:656-677.

Zhang, W. Y., Harff, J., Schneider, R. & Wu, C. 2010. Development of a modeling methodology for simulation of long-term morphological evolution of the southern Baltic coast. *Ocean Dynamics* 60:1085-1114.

Zhang, W. Y., Harff, J. & Schneider, R. 2011. Analysis of 50-year wind data of the southern Baltic Sea for modelling coastal morphological evolution — a case study from the Darss-Zingst Peninsula. *Oceanologia* 53:489-518.

Zhang, W. Y., Schneider, R. & Harff, J. 2012. A multi-scale hybrid long-term morphodynamic model for wave-dominated coasts. *Geomorphology* 149-150:49-61.

Chapter 4

Standard Core Variables for Continental Shelf Prehistoric Research and Their Availability

Nicholas C. Flemming
National Oceanography Centre, Southampton, UK

Introduction: The Concept of Core Variables

The collection of seabed data and its archival and analysis to document or recreate submerged terrestrial landscapes suffers from a paradox: the experts who have been collecting seabed data for many decades have seldom thought of identifying terrestrial features or storing the data in a way that preserves the submerged terrestrial signal; while experts in Pleistocene geomorphology on land are not usually familiar with marine geophysical data sources. This chapter considers the routine sources of seabed data and how best to extract paleolandscape information from them.

For seabed prehistoric research we need to transform the usual methods of describing the characteristics of the seabed and sub-seabed structures into descriptors of submerged Pleistocene terrestrial landscapes in a systematic way. We are analyzing in most detail the last 20,000 years, the period during which the sea level rose from about −135 m to the present level about 5000 years ago, but earlier terrestrial features need to be included in some projects. Ideally, we would like to reconstruct the post-Last Glacial Maximum (LGM) landscape in detail at intervals of 2000 years, showing the paleo-sea level, coastline, and drainage patterns etc., with a local spatial resolution of 50 m to 100 m. Aspects of this type of work are carried out frequently at a local scale in academic research papers (e.g. Lambeck *et al.* 2004;

Submerged Landscapes of the European Continental Shelf: Quaternary Paleoenvironments, First Edition.
Edited by Nicholas C. Flemming, Jan Harff, Delminda Moura, Anthony Burgess and Geoffrey N. Bailey.
© 2017 John Wiley & Sons Ltd. Published 2017 by John Wiley & Sons Ltd.

Hubbard *et al.* 2009; Peeters *et al.* 2009; Stoker & Bradwell 2009), but there is no standard terminology, data set, or classification, that permits a search for "Submerged paleolandscapes", and components thereof. For example, even if a bathymetric data set when plotted out showed linear depressions that are probably drowned river valleys, this would have to be done by eye, and it would not be possible to search digitally for submerged rivers. If we consider large varieties of scale, and possibly a third dimension of valleys buried under sediment, the advantages of being able to search automatically would be substantial.

Each sea basin of the European area has characteristics which are unique, and together they span a wide range of conditions. The Baltic, for example, is a low salinity, estuarine, periglacial environment with a low tidal range; the Atlantic coasts are high-energy environments with a high tidal range; and the eastern Mediterranean has relatively high salinity, low precipitation, limited tidal range and low sediment accretion. To describe the seabed conditions thoroughly in each locale will require detailed and high-resolution knowledge of the sediments, paleoshorelines, tectonics, core stratigraphy, palynology, foraminifera, ice-cap thickness, ice flow lines, ice edge, and so on. The subsequent chapters in this volume on each sea basin will consider variables at this specificity.

Nevertheless, there are some variables which are fundamental to the study and understanding of all areas, and which are almost always available, albeit at relatively poor resolution. Such low or modest resolution presentations also provide a broad overview of the landscape covering wide areas.

The concept of Core Variables defines the highest possible resolution of each of the selected variables which can be obtained relatively cheaply and with free, or cheap, access, and which is available uniformly in all European sea areas. Some allowances will be made for availability of different scales or resolutions in some sea areas. In each case, the highest resolution data may only be available by contacting the data originator.

The publication of such a core list can also be used to put pressure on data holders to reduce restrictions on use, and increase the resolution of freely available data sets. Many of the data types required for prehistoric

research are required in a form which is not commercially valuable, and therefore it should be possible to increase the availability by working closely with data holders and data managers.

The steps for defining and selecting a Core Variable are to:

1 review the existing data types and integrated data sets which exist in a standard form already at a relevant resolution, and covering most of the European shelf;
2 identify the data or product holders for (1) and access restrictions if any;
3 establish whether the existing integrated data sets have sufficient resolution to reveal submerged terrestrial landscape features;
4 document future plans and projects for refining resolution and/or increasing freedom of access to the data;
5 interact with data holders and management groups to promote improvements which will support future continental shelf prehistoric research.

Suggested Core Variables for All European Sea Areas

In the following discussions regarding access to European-scale data sets and metadata through the EMODnet system, I am grateful to Alan Stevenson who has provided some of the text and graphics based on his contribution to the European Marine Board Publication Position Paper 21 "Land beneath the Waves" (Flemming *et al.* 2014:88-92, Chapter 5). Since 2008, data have been aggregated and compiled for the European Regional Seas under the European Commission's EMODnet program. The network consists of a consortium of organizations within Europe that assembles marine data, data products and metadata from diverse sources in a uniform way. The main purpose of EMODnet is to unlock fragmented and hidden marine data resources and to make these available to individuals and organizations (public and private), and to facilitate investment in sustainable coastal and offshore activities through improved access to quality-assured, standardized and harmonized marine data. EMODnet is an initiative from the European Commission Directorate-General for Maritime Affairs and Fisheries (DG MARE) as part of its Marine Knowledge 2020 strategy (ec.europa.eu/maritimeaffairs/policy/marine_knowledge_2020).

There are presently eight sub-portals in operation that provide access to marine data and metadata under the following themes: Bathymetry, Geology, Physics, Chemistry, Biology, Coastal Mapping, Seabed Habitats and Human Activities, accessible from the EMODnet website (www.emodnet.eu). The sub-portals providing access to integrated data of most value to continental shelf Pleistocene landscapes are: Bathymetry, Geology, and to a certain extent Seabed Habitats and Human Activities. The Coastal Mapping sub-portal also has potential in this regard, but is currently in its early stages and thus difficult to evaluate. Bids and proposals to expand the Geology sub-portal to include reconstructions of the submerged landscapes of the European continental shelf at various time-frames have been formally submitted and are currently being evaluated (2017).

The following variables are fundamental to submerged landscape analysis, and are also supported at the European level by programs such as SeaDataNet, EMODnet, and Geo-Seas. Within this context, we can envisage the following seven Core Variables as key supports for submerged Quaternary landscape research:

1 bathymetry, to support landscape and topography reconstruction;
2 coastlines, accurate digital coastal data since many prehistoric sites are in shallow water;
3 solid geology, bedrock outcrops, surface mapping;
4 sediments, seabed sediment characteristics, sediment thickness;
5 seabed geomorphology, features and terrestrial landscape;
6 sub-bottom acoustic data, availability and access;
7 sub-bottom core metadata and grab metadata.

Although currently not supported directly, it should be possible in the near future to search for and identify data in terms of seabed features, as discussed previously.

Integrated data sets and metadata across basins will help to avoid the requirement for experts in each country to search for high-quality data on a piecemeal basis.

This is particularly important near median lines and boundaries. It is not yet feasible to include paleo-sea-level data as a Core Variable since, as shown in Chapter 2, the science of interpreting field data and generating realistic glacio-isostatic adjustment (GIA) models, combined with local tectonics and other earth movements, means that a centralized and agreed archive of spatially high-resolution site-specific local relative sea-level curves is still not achievable. Large volumes of data are available in widely scattered sources and programs, but qualifying factors are often needed in their interpretation.

Characteristics and Sources for Core Variables

Bathymetry

This is the most fundamental and, at first sight, the simplest variable. Reliable navigational charts have been published for over 100 years, and numerous industries have carried out detailed modern seabed surveys for specialized purposes. One might expect that a topographic map of the seabed, comparable to a topographic contoured map on land at a scale of 1:50,000, should be readily available at a modest cost, showing valleys, outcrops, gradients, sand banks, at least in coastal waters out to a depth of 100 m to 200 m.

This is not the case, either on paper or digitally. Published navigational charts are not scientific representations of or approximations to the seabed, with the implied scientific margins of error and probability. Nor do they usually resolve the shape of features such as valleys or rock outcrops on the seabed. A navigational chart published by a national hydrographic office is a legal instrument governed by national and international regulations, which, if used properly, ensures that a ship will not strike the seabed by mistake, and that it can proceed safely from Port 'A' to Port 'B'. In order to achieve their legal purpose, the compilers of charts make the following assumptions and adjustments:

i) the benchmark from which depths are measured is the lowest level that the tide drops to in that area, independently of meteorological effects (Lowest

Astronomical Tide, LAT). In tidal waters this is a different reference level for each chart sheet. The LAT benchmark is cross-referred relative to a fixed terrestrial geodetic benchmark or ordnance datum. The tidal amplitude is also shown on the paper chart;

ii) the assumed sound velocity for calibration of echo-sounding surveys in shelf seas is set at a standard speed which tends to make the seabed seem slightly shallower than it really is. The correction for the actual regional and climatic variation of sound speed in sea water is only made for water depths greater than 200 m;

iii) even on the larger-scale paper charts, say 1:50,000, actual depth soundings are only displayed at intervals of about 2 cm on the chart sheet, or 1 km on the seabed. This does not reveal small irregularities of the seabed, and gives a smooth undulating impression of the topography;

iv) for gridding and contouring digital terrain model (DTM) computations, the depth used for each cell is the shallowest recorded depth in that cell, not the average depth. This creates a safe product for navigation, but results in a further 'shoal bias' (i.e. an emphasis on shallow areas) for the entire smoothed topography. The Master of a vessel does not wish to be told that the estimated depth beneath the keel is 'X' m with a standard deviation of ±5 m. The form of the seabed represented by a hydrographic chart is in effect a 'drape' lying across the shallowest points, and this is logical for safe navigation;

v) the further effect of these conventions is that no abrupt changes of depth, and no steep gradients, appear on a navigational chart. Charts never show a vertical cliff underwater, but they often exist (Flemming 1972; Collina-Girard 2002);

vi) there is no need for a navigational chart to indicate unusual depths, but there is every need to show unusual shallows. Thus narrow depressions, small valleys, gullies, and features such as karstic sink-holes seldom show on charts. Only the broadest depressions and major drowned valleys or shelf-edge canyons appear when the printed soundings are contoured;

vii) in the continental shelf waters of many European countries there are large parts of the seabed which have not been sounded in the last 50 years or more (Fig. 4.1). Even the most well-surveyed areas have usually only been surveyed with single-beam echo sounders with tracks hundreds of meters apart, and only small areas have been surveyed with multibeam. Some European countries now have active national programs to increase the application of multibeam surveying (e.g. UK program MAREMAP).

For the reasons listed, the accuracy of delineation of the topography of the seabed using published charts is limited. The raw data holdings in the national hydrographic offices would permit more accurate products, but these are not needed for safety of navigation, and therefore a case has to be made for the necessary expenditure. Commercial organizations can thus obtain more accurate and complete data sets on request, and for a fee, and compile special seabed maps for their particular applications.

The requirements of the research community in continental shelf prehistoric archaeology are very demanding, even by commercial standards. We need to be able to look at the topography over wide coverage areas, of the order of 10 km to 100 km, so as to appreciate the general trend of features such as ridges, banks, terraces, paleocoastlines and valleys, and then, ideally, to zoom in to features at a scale of 10 m to 50 m, so as to resolve fossil beaches, cliffs, cliff-terrace junctions, gullies, minor stream beds, moraines, coral terraces, or shell middens. This ideal is not at present achievable, although it is technically feasible.

This section therefore addresses the availability of data sets which provide some uniformity over wide areas, where the data quality control is documented (metadata), and freedom of access is generous, if not always free.

Assuming that the bathymetry itself is available in a form which describes the topography of the seabed, there are further complications in detecting and describing the drowned terrestrial landscape. These are caused by (a) the overlying postglacial marine sediments deposited on the LGM terrestrial surface, and (b) coastal wave action

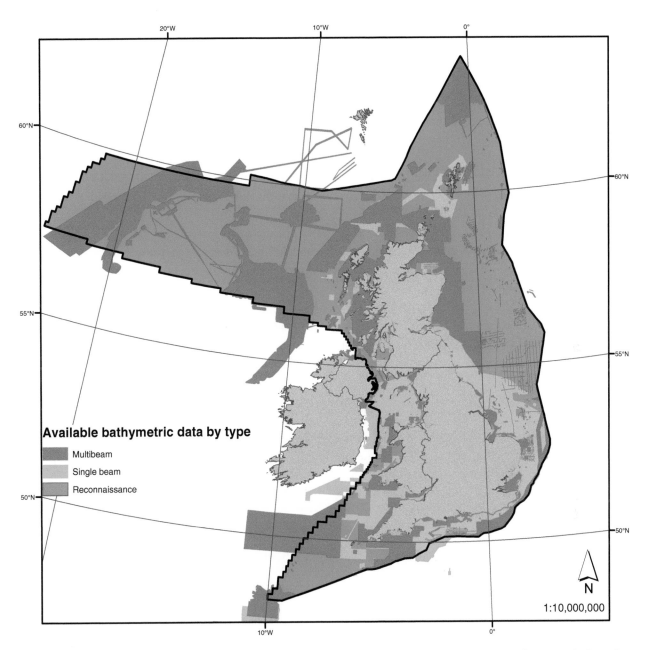

Fig. 4.1 Map showing typical progressive improvement in resolution and accuracy of seabed bathymetric surveys. Gray-green tint is modern multibeam that would resolve abrupt drowned landscape features; yellow is the area mapped by single-beam sonar; the red area has been mapped using reconnaissance soundings. This map only shows data available in the public domain, and does not show all commercial surveys. The map contains public sector information, licensed under the Open Government Licence v2.0, from UKHO, BGS. Map compiled by Rhys Cooper, BGS.

during marine transgression and modern bottom currents and wave action, which may have eroded the original LGM terrestrial surface away (see for example, Bridgland 2002: 30). Thus the topographic surface defined by the bathymetry comprises three types of terrain: (i) the original unaltered LGM terrestrial land surface; (ii) modern marine sediments of unknown thickness covering the original land surface (which may also have been eroded); and (iii) exposed areas where the original LGM land surface has been eroded away.

Notwithstanding these uncertainties, the topography of the seabed as it is now must be the starting point of submerged landscape analysis. The URL for access to EMODnet generalized European bathymetry is: www.emodnet-hydrography.eu/content/content.asp?menu=0030005_000000.

The overall objective of the EMODnet Bathymetry Project is to fill in the gaps of the EU's low-resolution bathymetry map and to assemble a complete inventory of high-resolution seabed mapping data held by public and private bodies. Working together with research institutes, monitoring authorities and hydrographic offices, the project will collect hydrographic data sets and compile DTMs at a resolution of 0.125 arcminutes by 0.125 arcminutes for each geographical region (Fig. 4.2). The DTMs are then loaded and integrated into a spatial database with a powerful, high-end bathymetric

data-products viewing and downloading service that is complemented by Web Map Services (WMS) to serve users and to provide map layers to, for example, the other EMODnet portals, the prototype European Atlas of the Seas, and the broad-scale European Marine Habitats map. The portal includes a metadata discovery and access service that gives clear information about the hydrographic survey data used for the DTM, their access restrictions and distributors. The system also includes a mechanism for requesting access to basic measurements data.

The resolution provided by EMODnet digital terrain modeled bathymetry is at present 0.25 arcminutes, or approximately 500 m. This is much better than global contoured charts such as General Bathymetric Chart of the Oceans (GEBCO) or the 2-minute resolution of the Naval Research Laboratory Digital Bathymetry Data Base (DBDB2 — see also www7320.nrlssc.navy.mil/DBDB2_WWW/), but is still rather crude for the

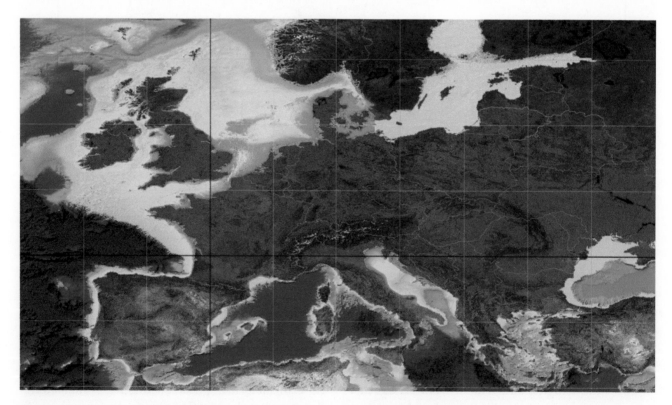

Fig. 4.2 Bathymetry from EMODnet bathymetry portal, 2015. This is a searchable data set that can be zoomed in to a resolution of 0.25 x 0.25 arcminutes. It also provides information on sources of higher resolution digital data.

purposes of accurate identification of small river valleys, moraines, or paleoshoreline features such as a lagoon or a dune ridge. Discussions are continuing to find ways of releasing higher resolution bathymetry in a form which would have no commercial value. At present, the EMODnet Bathymetry portal provides procedures for requesting higher resolution data through SeaDataNet, and the data provider will probably require the user to sign a licensing contract which restricts the use and publication of the data. The following URL gives maps at 1:250,000 scale for UK adjacent seas: www.bgs.ac.uk/products/digbath250/sample.html

Other international data resources include the International Hydrographic Organization (IHO), which coordinates the activities of national hydrographic offices.

The Baltic Sea Bathymetry Database (BSBD) is an effort to gather data in one place and distribute it for the areas of all Baltic Sea countries. This website offers complete, homogeneous and up-to-date Baltic Sea bathymetry data from all Baltic Sea national hydrographic offices under the umbrella of the Baltic Sea Hydrographic Commission. The BSBD project is co-financed by the European Commission Trans-European Transport Network (TEN-T). Land heights and water depths have been calculated for two regular spherical grids from available data. The data are available at the following URL: www.io-warnemuende.de/topography-of-the-baltic-sea.html

For the Mediterranean there are many different examples of published bathymetric datasets but they tend to concentrate on the deepwater gross topography, and not on the smaller features of the continental shelf. Much of the international collaboration is through long-established scientific bodies concerned with the geophysics of the deep ocean floor, and tends to leave the continental shelf to the navigational authorities, with the problems already listed above. One of the files in the above compilation is the IHO-IOC-ICBM 1000 m and 500 m resolution multibeam of the deep-sea Mediterranean, but this does not provide data on the shelf.

The bathymetry of the whole Mediterranean at a scale of 1: 1,000,000 is available in 10 sheets, downloadable from the following website, and reproduced in condensed format as Fig. 4.3: www.ngdc.noaa.gov/mgg/ibcm/images/93001.jpg

The chart set is described by the US National Centers for Environmental Information (NCEI — formerly the National Geophysical Data Center (NGDC)) as follows:

This bathymetric chart is published by the Charts Division of the Head Department of Navigation and Oceanography in Russia under the authority of the Intergovernmental Oceanographic Commission (IOC) of UNESCO. Originally compiled from ninety 1:250,000 scale British Admiralty Mercator plotting sheets for oceanic soundings, it presents the bathymetry and land topography as 200 m contours with supplemental 20 m, 50 m, and 100 m contours at sea. Several versions are available; a 10-sheet set (each 74 x 90 cm) at 1:1,000,000 scale at 38 degrees N, and a photo-reduction to 1:5,000,000 scale. This chart, with an inset of the entire Black Sea at half scale, is also the basemap for a Geological/Geophysical series with Bouguer gravity anomalies (IBCM-G), seismicity (IBCM-S), thickness of Plio-Quaternary sediments (IBCM-PQ), unconsolidated bottom surface sediments (IBCM-SED), and magnetic anomalies (IBCM-M). Digital variants and explanatory brochures for these series are in preparation. The digitized marine contours are available on CD-ROM as part of the British Oceanographic Data Centre's GEBCO Digital Atlas (International Bathymetric Chart of the Mediterranean: A Regional Ocean Mapping Project of the Intergovernmental Oceanographic Commission. www.ngdc.noaa.gov/mgg/ibcm/ibcm. html. Maintained by NOAA/NGDC WDC for MGG, Boulder).

Relevance of electronic charts: ECDIS

International Hydrographic Organization regulations require that ships carry either an up-to-date set of hard-copy paper charts, or an approved Electronic Chart and Display System (ECDIS). Performance standards for electronic charts were adopted in 1995, by resolution

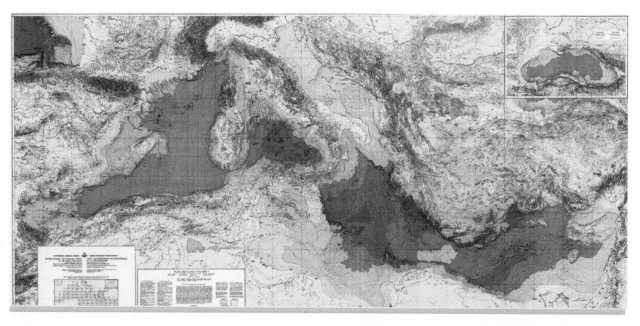

Fig. 4.3 Small-scale reproduction of the IOC-IBCM bathymetry of the Mediterranean. Credit with permission: IBCM Project, Intergovernmental Oceanographic Commission, International Hydrographic Organization, and Head Department of Navigation and Oceanography, Russian Federation.

A.817 (19), which was amended in 1996 by resolution MSC.64 (67) to reflect back-up arrangements in case of ECDIS failure. Additional amendments were made in 1998 by resolution MSC.86 (70) to permit operation of ECDIS in Raster Chart Display System (RCDS) mode, that is, substituting a direct copy of the raster paper chart for a fully interactive electronic system. The amendments to Resolution A.817 (19) state that some ECDIS equipment may operate in RCDS mode when the relevant chart information is not available in vector mode.

Since ECDIS and RCDS displays are intended for safe navigation, the reservations outlined previously in terms of absolute correct depths relative to a stable datum or benchmark and mean sea level all apply. ECDIS/RCDS displays can be a convenient way to scan and search bathymetric data, but they suffer from shoal bias, and may not reveal details of regionally increased depth, or sudden changes of gradient.

Available multibeam surveys, areas covered

EMODnet will develop a metadata map showing the areas covered by all available multibeam surveys on the European shelf. This will combine data on governmental and commercial multibeam surveys. Access to the data is not yet negotiated or under discussion. Figure 4.1 shows a typical national map illustrating the proportion of the national area of jurisdiction which has been mapped systematically with multibeam. The proportion mapped by multibeam is small, and, since the swath width is narrower for shallow water, the rate of coverage for the shallow parts of the European shelf will take many years.

Coastlines and coastal processes

The majority of submerged prehistoric sites found so far are in shallow water, often less than 10 m deep, as shown in Chapter 3. Where the coast itself is complex, with marshes, lagoons, extensive mudflats, or scattered rocky shoals, it is often difficult to obtain a large-scale coastal map which is sufficiently accurate to locate the site in a realistic context in relation to its surroundings. This is especially true for identifying the boundaries of the upper and lower limits of the tidal zone or wave action. Many prehistoric sites have been found intertidally, indicating that there has been a relative change of land–sea level.

Digital representation of the coastline cartographically is complex because the coast itself is complex, and probably contains self-similar patterns at all scales like a fractal. Where the coast is steep and the shore narrow, it is a relatively simple matter to define the coastline by a convention such as the mean sea level, or high-water mark of the Highest Astronomical Tide. The representation digitally is then a matter only of grid resolution. In practice, especially where gradients are low, or there are many coastal rock outcrops, reefs, or marshy islands, for example in the Waddenzee or the Danube Delta, the definition of a map with lines representing the boundary between land and sea becomes extremely complex.

The European Commission DG MARE has invited tenders to develop a prototype digital map of Europe's coastal zone (the land/sea boundary) for inclusion in EMODnet. The contract will have a particular focus on use of high-resolution topographic/bathymetric data and the development of European standards for relevant mapping datums, including a defined coastline boundary at Highest Astronomical Tide and Lowest Astronomical Tide levels. As a prototype, the contract will deliver maps for selected parts of Europe to demonstrate the use and integration of available data and how this can be interfaced with data for terrestrial and deeper water mapping.

Digitized coastlines which are globally available (GEBCO, Central Intelligence Agency World Data Bank II, Flanders Marine Institute/Vlaams Instituut voor de Zee (VLIZ), Global Self-consistent, Hierarchical, High-resolution Geography Database (GSHHG), etc.) are crude when magnified to the local level, often with resolution in steps of 5 km to 10 km. The present EMODnet bathymetry and digitized coastline is at 500 m, but even this is very coarse when studying a local site. Because of the growing political pressure for sophisticated multi-agency coastal management, some countries have programs dedicated to improving the cartographic representation of the coastal zone, linking terrestrial mapping, coastal aerial photography and space remote-sensed (RS) images of the intertidal zone, to charts, and bathymetry, with consistent benchmark datum and geoid (e.g. Guariglia *et al.* 2006). These programs are not complete.

A comparison of a Google Earth RS image and the 1:25,000 map of the same coastal section illustrates the problem. Figure 4.4 shows Cap Lévi, near Cherbourg in northern France, where there is a submerged Paleolithic site offshore (Cliquet *et al.* 2011) just east of Biéroc. The aerial photograph taken at an intermediate tidal level shows the rocky foreshore and some offshore rocks breaking the surface. Figure 4.5 shows the same area on the 1:25,000 scale map.

The grid of blue lines in Fig. 4.5 shows 1-km spacing. The map represents the coast quite well, with outlines of the rocks at a lower tidal level than in the photograph. The navigational chart for the area shows less coastal detail than the Institut national de l'information géographique et forestière (IGN) map. Digital representation of the coastline at 500 m would remove all the significant detail, and even at 100 m much would be concealed. Thus a European digitized coastline which was useful for coastal submerged and intertidal prehistoric research would have to contain nested scales from the low-resolution overall data set to regional and subregional high resolution.

Coastal processes: coastal behavior and erosion

The EMODnet-Geology Project aims to classify all coastal types in each country at 1:250,000 scale including information on rates of sedimentation and erosion. The central parameter in the final description of coastal behavior is the rate of shore-normal coastline migration. The starting point for information compilation is the EUROSION (a European initiative for sustainable coastal erosion management) database supplemented by data held by the EMODnet partners and sources such as *Coastal Erosion and Protection in Europe* (Pranzini & Williams 2013) which includes information on Europe's coast on a country-by-country basis.

EUROSION was a project commissioned by the European Commission, which started in January 2002 and ended in May 2004. The project integrated natural and human-induced causes of erosion (storms, seismic movements, reduced sediment supply from rivers due to dams, coastal defenses etc.); different uses of both

Fig. 4.4 Cap Lévi, northern France, near Cherbourg. There is an underwater Paleolithic archaeological site in this area (Cliquet *et al.* 2011), but accurate definition of the coastline is problematic, due to the large tidal range, and the complex topography. Image from Google Earth, IGN France, Tele Atlas. Reproduced with permission.

the terrestrial and marine environments of the coast, ranging from biodiversity and landscape conservation to tourism, industry and transport; and the different levels of management from the local level up to European and regional sea management. Through supporting the Integrated Coastal Zone Management (ICZM) Practitioners Network and facilitating access to relevant data and information, EUROSION offers a follow-up to the EU demonstration program on ICZM.

Solid substrate geology and Quaternary geology

The EMODnet-Geology portal is constructed by a group of national geological survey organizations to provide

access to data and metadata held by each organization. The data and map products include information on the seabed substrate (see Fig. 4.6 for example) including rate of accumulation of recent sediments; the seafloor geology (bedrock and Quaternary geology) and all boundaries and faults that can be represented at a compilation scale of 1:250,000 wherever possible. There is an index map showing the scale at which each country has compiled geological maps.

The portal also includes information on the lithology and age of each geological unit at the seabed, for minerals, and for geological events and their probability of occurrence. All the interpreted information is based on primary information owned by the project partners, supplemented with other information available in the public domain.

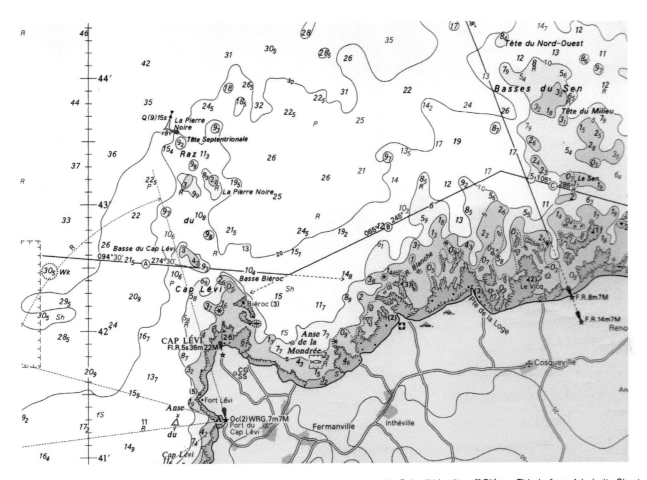

Fig. 4.5 Navigational chart representation of Cap Lévi and the bathymetry close to the Paleolithic site off Biéroc. This is from Admiralty Chart 1106 at a scale of 1:50,000. The scale on the left is in arcminutes, where 1 minute = 1850 m. Printed soundings are typically separated by several hundred meters.

A map of available offshore Quaternary geology information is being compiled using data held by the geological survey organizations of Europe. The proposed map and supporting information will include the age, lithology, genesis, geomorphology, and information about the Quaternary geology at the time of the LGM, where possible. The work will link to the IQUAME 2500 project.

The European national geological services are collaborating to produce offshore data in a project called Geo-Seas. Additionally, at a fully international level, agencies are collaborating through a portal network called OneGeology, available at the following URLs: www.geo-seas.eu/content/content.asp?menu= 0030011_000000 and portal.onegeology.org/

Sediments

Recent sediments on the continental shelf are important factors in locating prehistoric sites, either on the exposed seabed, or buried in the sedimentary column, see for example Fitch *et al.* (2005), Gaffney *et al.* (2007) and van Heteren *et al.* (2014). The following variables are of particular relevance:

1 sediment grain size, type, and sorting on the seabed;
2 thickness to the bottom of modern marine sediments, LGM;
3 thickness to the base of the Pleistocene.

The British Geological Survey (BGS) collaborated with its opposite numbers in the Netherlands and Norway

Fig. 4.6 EMODnet solid seabed geology portal. Example of bedrock stratigraphy map for northern Europe (EMODnet-Geology Project). From www.emodnet-geology.eu/geonetwork/srv/dut/catalog.map. Reproduced with permission.

during the 1980s to produce a series of seabed sediment maps of the North Sea at a scale of 1:250,000. These maps, and the associated cores, are a useful tool for assessing the archaeological potential and sensitivity of areas of the sea floor, providing classification of surface sediments by grain size, thickness of active marine sediments, thickness of Holocene deposits, standard cross-sections, information on tidal currents, sand waves and sand ripples, carbonate percentage, and other items of information which vary from sheet to sheet. Some sheets, but not all, include copious technical notes, sections, core profiles, and analyses of sources, references, and comments on the various facies. All sheets show positions of platforms and pipelines at date of publication. The data sets are available electronically under license, with information at the following site: www.bgs.ac.uk/discoverymetadata/13605549.html

In general, seabed sediment data at low resolution can be obtained from EMODnet, and high-resolution data by contacting national geological agencies. An example of the EMODnet data access display is shown in Fig 4.7.

European-scale data on Quaternary marine and coastal features

This section is concerned with European-scale data sets and mapping programs which include Quaternary landscape features. A later section will review the interpretation of individual submerged features. The International Union for Quaternary Science (INQUA) has five commissions (Coastal & Marine Processes; Humans & the Biosphere; Palaeoclimates; Stratigraphy & Chronology; Terrestrial Processes, Deposits & History) which provide leadership in different spheres of research, and are responsible for ensuring that INQUA scientists remain at the forefront of their fields. INQUA currently funds projects of relevance to marine archaeology under each theme. These include:

– MEDFLOOD: MEDiterranean sea-level change and projection for future FLOODing (www.medflood.org/);
– PALSEA2: PALeo-constraints on SEA-level rise 2 (people.oregonstate.edu/~carlsand/PALSEA2/Home.html);
– Humans and biosphere: Modeling human settlement, fauna and flora dynamics in Europe during the Mid-Pleistocene Revolution (1.2–0.4 Ma);
– Cultural and paleoenvironmental changes in Late Glacial to Middle Holocene Europe: gradual or sudden?
– DIG – 1st Workshop on DInaric Glaciation: Early/Middle Pleistocene glaciations of the northeast Mediterranean;

Fig. 4.7 EMODnet seabed sediments. From www.emodnet-geology.eu. Reproduced with permission.

– SEQS (Section on European Quaternary Stratigraphy) Framing European Quaternary Stratigraphy;
– UNESCO-IUGS-IGCP521-INQUA501 WG12. Black Sea-Mediterranean corridor during the last 30 ky: sea level change and human adaptation.

The IQUAME 2500 Project is a joint initiative of the Commission for the Geological Map of the World (CGMW)/INQUA/BGR — *Bundesanstalt für Geowissenschaften und Rohstoffe* [German Federal Institute for Geosciences and Natural Resources], which is reviewing the International Quaternary Map of Europe (Asch & Müller (2008). Through common partners such as the BGS, links are being made between the IQUAME 2500 Project such that the EMODnet-Geology Project group will provide the marine input to the map thereby developing a baseline Quaternary geology map of both the land and marine areas of Europe. The projects will compile information on the age, lithology and genesis of Quaternary deposits as well information on glacial maxima.

Sub-bottom acoustic data

Each chapter on regional sea areas in this volume will provide examples of the use and interpretation of sub-bottom acoustic data in the search for buried Quaternary landscapes and shorelines. In the EMODnet system sub-bottom profiler data and metadata are accessed through the Seabed Habitats sub-portal. Enormous volumes of sub-bottom data are held by commercial organizations in support of offshore engineering, foundations, port excavation, pipelaying, and the construction of wind farms. Examples of the interpretation of sub-bottom acoustic profiles and 3D reconstructions for research into prehistoric landscapes are provided by Fitch *et al.* (2005), Gaffney *et al.* (2007; 2009; Figs. 4.8 & 4.9), and van Heteren *et al.* (2014) for the North Sea. Examples from the Mediterranean include Lykousis (2009), and Lobo *et al.* (2014).

Seabed core data

The EuroCore project holds data on cores taken from sea depths greater than 200 m, and therefore these are not relevant to Pleistocene low sea-level landscapes. Cores taken on the continental shelf are physically held in stores operated by national geological services, and the metadata are also held in comprehensive searchable data systems at the national level. The example for the BGS catalogue of marine sample sand cores is at this address: www.bgs.ac.uk/discoverymetadata/13603048.html and French core data can be identified in this database: www.ifremer.fr/sismer/UK/catal/base/edmed_an.htql? CBASE=GMCAR

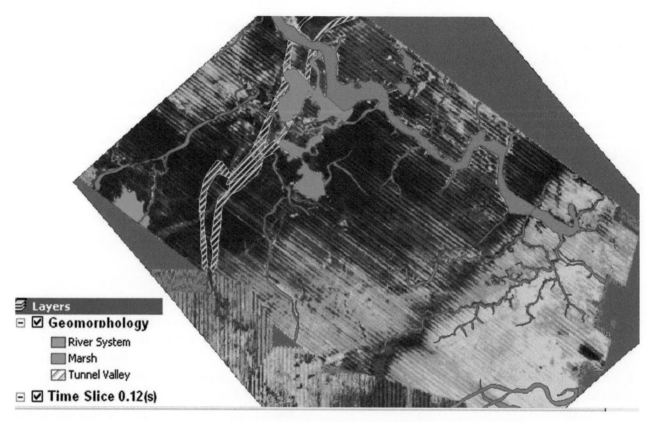

Fig. 4.8 Three types of paleo-geomorphology revealed by reprocessing commercial seismic data from the central North Sea. A tunnel valley scoured by melt water under a previous ice sheet has been infilled by sediment and is then overlain by a low gradient river and marshes. Image provided by V. Gaffney.

Fig. 4.9 Buried river-drainage system revealed with the sediment column of the central North Sea. Gaffney *et al.* (2007). Reproduced with permission.

TABLE 4.1 EXAMPLES OF SEABED FEATURES OF INTEREST RELATING TO FORMERLY TERRESTRIAL LANDSCAPES.

Submerged terrestrial landscape features which may be identified	Dynamic modern seabed marine sediment morphology, bedforms
River valleys	Sand banks
Glacial valleys	Sand waves
Braided and sinuous patterns of river drainage	Sand or gravel ripples
Moraines, eskers, glacial deposit features	Mobile sand dunes or barchan features
Ice cap margin thrust features	Density flows of sediments, and sediment fans
Ice tunnels	Turbidity flows
River deltas	Sand ribbons, transport streams
Cliffs, either sea coast eroded or vertical fault scarps	Shell gravels
Cliff-terrace junctions	Iceberg scour marks
Periglacial features, patterned ground, roches moutonnées, etc.	Methane pock-marks
Coastal dune ridges, often calcified or indurated eolianite	Low gradient silt or mud
Lagoons and marshes, coastal islands and coastal sand bars	Lag deposits (possibly derived from terrestrial mixed sediments)
Freshwater lake basins, ice–dammed lake basins	Tsunami deposits
Peat bog	Bedrock outcrops, scarps, bedding, etc.
Coral terraces, or vermettid terrace	
Caves, karstic sinkholes, sea caves in cliffs	
Submerged freshwater springs	
Drowned forests	
Beach terraces	
Beachrock	

Examples of Seabed Features Relating to Terrestrial Landscapes

Geomorphological or dynamic features of the seabed may themselves be related either to modern marine sediments or to previous terrestrial landscapes. Thus the identification and classification of features is a necessary stage in constructing a paleolandscape. As explained in the section on bathymetry, navigational charts and low-resolution digital terrain mapping can give an excellent impression of the gross topography of large sea areas, but they are not so effective at revealing individual geomorphological features. Objects of interest are shown in Table 4.1.

The types of landforms or marine bedforms that can potentially be identified from a seabed topographic map depend on the procedures used to compile the map, and its resolution. Lead-line soundings from over 100 years ago produce widely-spaced point soundings, and therefore could only reveal topographic highs and lows at the coarsest level. If point soundings are 1 km apart, the smallest hill or valley which can be defined is a curve

2 km to 3 km across. Analogue paper-recording echo sounding, common from the 1930s to 1980s, provided a continuous visual record of depth and gradient, and thus, if required, could be used to identify the sharp changes in gradient which often indicate structural surface features. Digital recording single-beam echo sounding does not show changes of gradient unless a deliberate effort is made to view profiles and plots of the raw data. Even then, the recorded data points may be separated by 100 m or more. Most plotting, gridding, contouring and presentation algorithms remove abrupt changes of gradient, and result

in smooth curves, but special programs to detect rapid changes of gradient can be applied (see pages 99–100). Side-scan sonar, used widely from the 1960s to the present day, also reveals the edges of topographic structures, but is not often used to support topographic mapping. Swath bathymetric or multibeam surveys do reveal high-resolution topography with the edges of features defined by changes of gradient, (Figs. 4.8–4.11) but only if the data are not aggregated into cells and gridded with smoothing or coarse algorithms. The raw data must be examined to find the tell-tale edges of features.

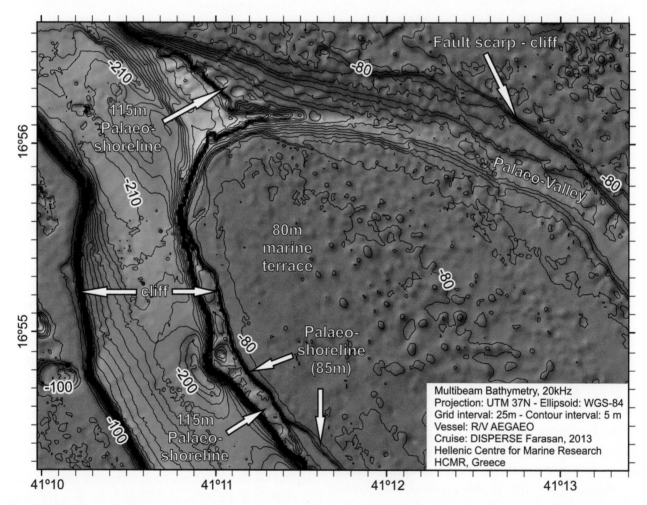

Fig. 4.10 Multibeam image of rugged rocky topography with steep cliffs, close to the Farasan Islands, Red Sea. This image was compiled during a survey of prehistoric submerged landscapes during the DISPERSE (Dynamic Landscapes, Coastal Environments and Human Dispersals) project. Note the ability of the technology to reveal vertical and near-vertical features, and to delineate abrupt changes of gradient and flat terraces. The landscape exhibits faulting due to underlying salt karst, coastal erosion terraces from past low sea levels, and drainage courses indicating past river flow. Image HCMR, courtesy of Sakellariou et al. (2015). From the Proceedings of the 11th Panhellenic Symposium of Oceanography and Fisheries. Reproduced with permission.

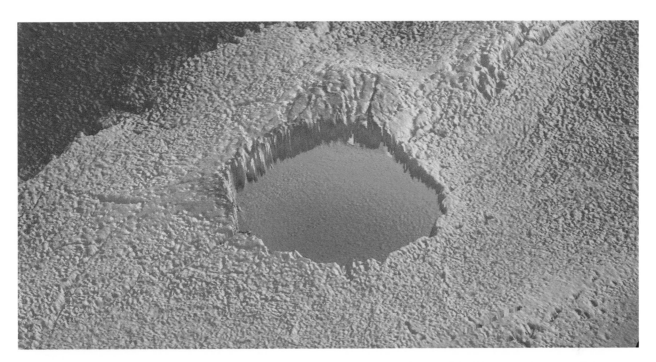

Fig. 4.11 A recent survey by the University of Malta and the UK National Oceanography Centre (NOC) of limestone ridges near Malta unexpectedly revealed the clear form of a collapsed sinkhole. Tim Le Bas (NOC), from Micallef *et al.* (2013). Reproduced with permission from John Wiley & Sons Ltd.

Many geomorphological features which we identify instantly on land are defined by a line-boundary which is the intersection of two surfaces of different gradients. Whether it is a scarp-fault of sheer rock, a river bank, the junction between a talus scree and the underlying terrace, or the flat alluvial-filled basin of a dead lake, the feature is defined by line boundaries, although they may not be straight lines, or confined to a single plain. High-resolution multibeam is providing the data that resolves these indicators, but it will take many years to map the whole European shelf by this method. An example of a feature which is revealed with sharp edges and vertical surfaces is one of the submerged karstic sinkholes recently discovered off the coast of Malta (Fig. 4.11).

The discrepancy in information produced by widely spaced soundings regarding geomorphological bedforms, whether marine or terrestrial in origin, can further be illustrated by Figs. 4.9 to 4.13. Each of these images has been chosen to show the true detail available to analyze bedform, and the extremely false impression that would be obtained by having a representation based on point soundings spaced by 1 km to 2 km.

Automatic Feature Recognition

There is a substantial literature on the association between terrestrial landscape and occupied site distribution for a wide range of prehistoric periods, although precise individual site location cannot be predicted from underwater landscape. It follows that the ability to recognize geomorphological features underwater automatically would be a useful and efficient tool if it could be used reliably with digital data sets. This is not yet possible, but rapid progress is being made in the study of automatic geomorphometrics. Goodwyn *et al.* (2010) use a rule-based system and predictive modeling to classify areas of the UK seabed as potentially rich or poor in containing prehistoric sites, and similar studies have been conducted for Dutch and German waters to indicate and classify areas that need extra protection. For image interpretation on land, research has been conducted to extract features automatically such as rivers, stream beds, watershed, roads, buildings, etc. (for example, see Quackenbush (2004)). Automatic recognition of coastlines is also considered by Guariglia *et al.* (2006).

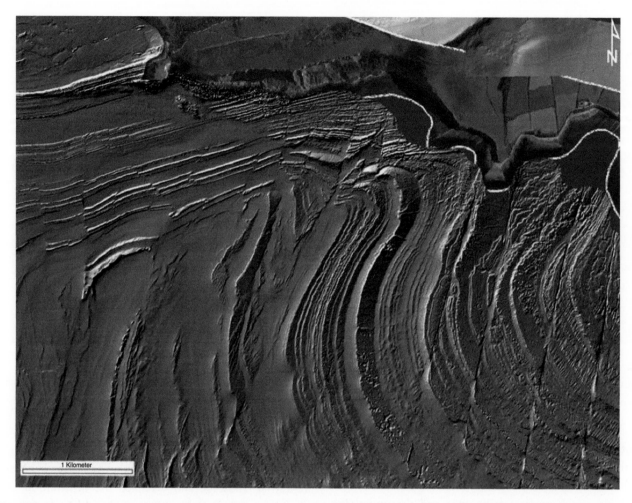

Fig. 4.12 The plunging anticline in Kimmeridge Bay, Dorset coast, English Channel. Data binned in 1-m intervals, and then 'hill-shaded'. The structure of bedrock swept almost clean of marine sediments is apparent. Image courtesy of J.Dix, University of Southampton.

Given high-resolution digital terrain data on land with a resolution of 1.0 m to 0.25 m, geomorphometric analysis can extract the underlying lithological geology and fracture patterns (Coblentz *et al.* 2014). Micallef *et al.* (2007; 2013) refer extensively to similar literature in their analyses of the digital terrain data around Malta. This note can only provide a brief mention of this subject, and the bibliographies of the papers cited provide access to the wider literature.

Recommendations

Digital archives of core environmental seabed variables on a European scale provide the potential support for a wide range of research topics in continental shelf prehistoric research projects, both at the basin scale linking large regional groups of coastal states, and more locally where two or three states may wish to collaborate on prehistoric sites at high resolution. Additionally, a shared system with common access would make such research projects easily accessible to third parties. Through EMODnet, SeaDataNet, and other European programs and projects, the data management work necessary to achieve this objective has started. The process of merging national and commercial data sets, with the associated necessity for checking accuracy and metadata conventions, will continue for several years. Bathymetry is most nearly available in the required form; geological variables,

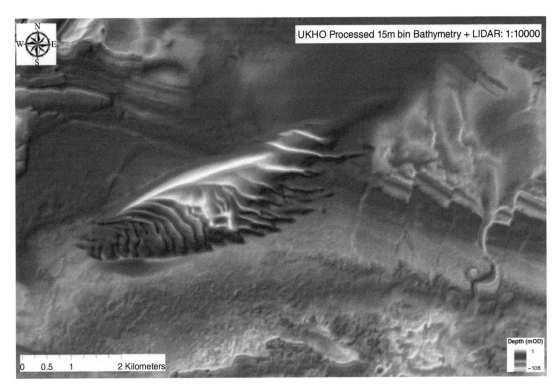

Fig. 4.13 This color-coded bathymetric image of the seabed near Portland Bill, Dorset coast, shows the dune-rippled surface of the Brambles Bank, underlying bedrock with dipping strata and fault offsets, and a sinuous eroded river valley. Image courtesy of J. Dix, University of Southampton, from UKHO processed data.

both bedrock and sedimentary, are in preparation and well advanced. Coastline mapping is receiving research funding and investment, but seems to be still a problem at the detail required. At the subregional and local levels where very large scales and high resolution are required, national and institutional data sets are necessary and often available in hard copy, but not always digitized.

References

Asch, K. & Müller, A. 2008. *International Quaternary Map of Europe*. Bundesanstalt für Geowissenschaften und Rohstoffe, in cooperation with the INQUA Commission for the International Quaternary Map of Europe.

Bridgland, D. R. 2002. Fluvial deposition on periodically emergent shelves in the Quaternary: example records from the shelf around Britain. *Quaternary International* 92:25-34.

Cliquet, D., Coutard, S., Clet, M. *et al.* 2011. The Middle Palaeolithic underwater site of La Mondrée, Normandy, France. In Benjamin, J., Bonsall, C., Pickard, C. & Fischer, A. (eds.) *Submerged Prehistory*. pp. 111-128. Oxbow Books: Oxford.

Coblentz, D., Pabian, F. & Prasad, L. 2014. Quantitative Geomorphometrics for Terrain Characterization. *International Journal of Geosciences* 5:247-266.

Collina-Girard, J. 2002. Underwater mapping of Late Quaternary submerged shorelines in the Western Mediterranean Sea and the Caribbean Sea. *Quaternary International* 92:63-72.

Fitch, S., Thomson, K., & Gaffney, V. L. 2005. Late Pleistocene and Holocene depositional systems and the palaeogeography of the Dogger Bank, North Sea. *Quaternary Research* 64:185-196.

Flemming, N. C. 1972. Relative chronology of submerged Pleistocene marine erosion features in the

Western Mediterranean. *The Journal of Geology* 80:633-662.

Flemming N. C., Çağatay M. N., Chiocci F. L. *et al.* 2014. *Land Beneath the Waves: Submerged Landscapes and Sea Level change. A Joint Geoscience-Humanities Research Strategy for European Continental Shelf Prehistoric Research.* Chu N.-C. & McDonough N. (eds.) Position paper 21 of the European Marine Board: Ostend, Belgium.

Gaffney V., Thomson, K. & Fitch, S. (eds.) 2007. *Mapping Doggerland: The Mesolithic Landscapes of the Southern North Sea.* Archaeopress: Oxford.

Gaffney V., Fitch, S. & Smith, D. 2009. *Europe's Lost World: the Rediscovery of Doggerland.* CBA Research Report 160. Council for British Archaeology: York.

Goodwyn, N., Brooks, A. J. & Tillin, H. 2010. *Waterlands: Developing Management Indicators for Submerged Palaeoenvironmental Landscapes.* Report prepared under the Marine Aggregates Levy Sustainability Fund (Ref No. MEPF 09/P109). ABP Marine Environmental Research.

Guariglia, A., Buonamassa, A., Losurdo, A. *et al.* 2006. A multisource approach for coastline mapping and identification of shoreline changes. *Annals of Geophysics* 49:295-304.

Hubbard, A, Bradwell, T, Golledge, N. *et al.* 2009. Dynamic cycles, ice streams and their impact on the extent, chronology and deglaciation of the British–Irish ice sheet. *Quaternary Science Reviews* 28:758-776.

Lambeck, K., Anzidei, M., Antonioli, F., Benini, A. & Esposito, A. 2004. Sea level in Roman time in the Central Mediterranean and implications for recent change. *Earth and Planetary Science Letters* 224:563-575.

Lobo, F. J., Ercilla, G., Fernández-Salas, L. M. & Gámez, D. 2014. The Iberian Mediterranean shelves. In Chiocci, F. L & Chivas, A. R (eds.) *Continental Shelves of the World: Their Evolution During the Last Glacio-Eustatic Cycle (Geological Society Memoir No. 41).* pp. 147-170. Geological Society: London.

Lykousis, V. 2009. Sea-level changes and shelf break prograding sequences during the last 400 ka in the Aegean margins: Subsidence rates and palaeogeographic implications. *Continental Shelf Research* 29: 2037-2044.

Micallef, A., Berndt, C., Masson, D. G. & Stow, D. A. V. 2007. A technique for the morphological characterization of submarine landscapes as exemplified by debris flows of the Storegga Slide. *Journal of Geophysical Research F (Earth Surface)* 112: F02001, doi:10.1029/2006JF000505.

Micallef, A., Foglini, F., Le Bas, T. *et al.* 2013. The submerged paleolandscape of the Maltese Islands: Morphology, evolution and relation to Quaternary environmental change. *Marine Geology* 335:129-147.

Peeters, H., Murphy, P. & Flemming, N.C. (eds.) 2009. *North Sea Prehistory Research and Management Framework (NSPRMF) 2009.* English Heritage and RCEM: Amersfoort.

Pranzini, E. & Williams, A. (eds.) 2013. *Coastal Erosion and Protection in Europe.* Routledge: Abingdon.

Quackenbush, L. J. 2004. A review of techniques for extracting linear features from imagery. *Photogrammetric Engineering & Remote Sensing* 70:1383-1392.

Sakellariou, D., Bailey, G., Rousakis, G. *et al.* 2015. Quaternary geology, tectonics and submerged landscapes of the Farasan continental shelf, Saudi Arabia, South Red Sea: Preliminary results. *Proceedings of the 11th Panhellenic Symposium of Oceanography and Fisheries,* 13th – 17th May 2015, Mytilene, Lesvos Island, Greece, pp. 985-988.

Stoker, M. & Bradwell, T. 2009. Neotectonic deformation in a Scottish fjord, Loch Broom, NW Scotland. *Scottish Journal of Geology* 45:107-116.

van Heteren, S., Meekes, J. A. C., Bakker, M. A. J. *et al.* 2014. Reconstructing North Sea palaeolandscapes from 3D and high-density 2D seismic data: An overview. *Netherlands Journal of Geosciences* 93:31-42.

Chapter 5
The Baltic Sea Basin

Alar Rosentau,[1] Ole Bennike,[2] Szymon Uścinowicz[3] and Grażyna Miotk-Szpiganowicz[3]

[1] *Department of Geology, University of Tartu, Tartu, Estonia*
[2] *Geological Survey of Denmark and Greenland, Copenhagen, Denmark*
[3] *Polish Geological Institute — National Research Institute, Branch of Marine Geology, Gdańsk, Poland*

Introduction

This chapter gives a brief introduction to the Baltic Sea, which is a special sea in several respects. It is one of the largest brackish-water inland seas in the world and it has undergone a complex and unique development after the last deglaciation. It is located between central and northern Europe, from 53°N to 66°N and from 10°E to 30°E and is connected to the Global Ocean by shallow and narrow straits, and receives a large volume of fresh water from the drainage catchment. The Baltic basin is fairly shallow with a mean depth of 54 m and with a number of smaller and larger sub-basins. The salinity varies widely, from about 10‰ at the surface and 33‰ at the bottom near the Danish Straits to practically fresh water at the easternmost Gulf of Finland and northernmost Bay of Bothnia.

The whole basin was glaciated during the Last Glacial Maximum, and the main phases after the last deglaciation are the Baltic Ice Lake, the Yoldia Sea, the Ancylus Lake and the Littorina Sea, which are described in this chapter. The geomorphological and hydrodynamic conditions of the southern Baltic have been favorable for the survival and discovery of submerged prehistoric archaeological deposits and single artifacts, and it is important to analyze the reasons for this survival, and to evaluate lessons that could be applied elsewhere.

The Baltic Sea has been studied for over three centuries by researchers from the circum-Baltic countries. For example, the first records of water level began in 1703 in St. Petersburg and the world's longest time series of water level at Stockholm began in 1774. Indeed, Andreas Celsius sought to understand and evidence the gradual change in sea level in the Baltic in the 1740s (Celsius 1743). The first studies of the vegetation history of the region were conducted in the 1830s, and studies of the bedrock geology were also underway at this time (Lyell 1835). Thousands of papers have been published, and it is beyond the scope of this short paper to review all this literature. However, we try to summarize some of the most important data.

General description

The Baltic Sea is a semi-enclosed intracontinental sea surrounded by Scandinavia and by the lowlands of central and eastern Europe. The total area of the Baltic Sea (without the Kattegat (22,287 km^2)), is 392,978 km^2 (Leppäranta & Myrbeg 2009) with a catchment area about four times larger than the area of the sea itself (Fig. 5.1). The sea is shallow (average ca. −54 m) with shallow and narrow connections with the North Sea and Atlantic Ocean by the Danish Straits. The Baltic Sea is one of the largest brackish inland seas in the world.

The climate is strongly diversified. The western and southern parts of the Baltic Sea are strongly influenced by the Atlantic Ocean and therefore maritime and temperate climate types prevail. The eastern and northern parts are characterized by a more continental type of climate with colder winters. This is reflected in the distribution of sea ice: during a typical winter the northeastern parts are ice-covered. A characteristic feature of the Baltic Sea waters is a marked thermohaline stratification. A strongly developed halocline at a depth of around 30 m to 40 m in the Arkona basin to around 80m to 90 m in the Eastern Gotland basin separates surface waters of low salinity from waters of higher salinity found at deeper levels. The temperature of the deeper water masses is fairly constant, whereas the temperature of the surface waters varies considerably over the year. There is abundant freshwater runoff from the land and limited inflow from the open ocean, which gives rise to a strong and almost permanent horizontal salinity gradient. The salinity varies from close to zero at river mouths in the north to 15‰ to 20‰ in the deepest basins to 25‰ to 30‰ in the Kattegat. The tidal range is about 10 cm close to the Danish Straits, only 2 cm to 5 cm in most of the Baltic Sea, up to 10 cm in the northern Gulf of Finland (Leppäranta & Myrberg 2009) and even 17 cm to 19 cm in the easternmost part of the Gulf of Finland (Medvedev et al. 2013). The tidal currents may reach 7 cm/sec (Lilover et al. 2014), which

are of clear importance in deeper areas but negligible in the wave-dominated nearshore.

The northern part of the Baltic Sea is situated within the Precambrian Baltic Shield, whereas the southern part lies on the East European Platform and on the West European Platform (Fig. 5.2). During the Pleistocene, the Scandinavian ice sheet advanced several times into the area of the present Baltic Sea and eroded the bedrock and deepened the Baltic basin. During the Eemian interglacial (130–115 ka), the hydrography of the Baltic Sea was significantly different from the Holocene. A seaway existed between the Baltic basin and the White Sea to the north through Karelia during the first ca. 2.5 kyr of the interglacial. The beginning of the present-day Baltic Sea is related to the retreat of the last Scandinavian ice sheet. In the southwestern part of the present Baltic Sea, the ice sheet melted and the first embryo of the Baltic Ice Lake formed approximately 16 thousand years ago (Andrén et al. 2011).

Coastlines

Coasts of the Baltic Sea are strongly diversified. The southern and eastern coasts of the Baltic Sea are dominated by sandy barriers (spits) and coastal cliffs. Cliffs are built mainly of Quaternary glacial and glacio-fluvial deposits. Cretaceous, Paleogene and Neogene cliffs occur only locally. Miocene sandy and silty deposits outcrop on the western coast in Gdańsk Bay, whereas Eocene and Oligocene sandy and clayey deposits outcrop in the cliffs on the Sambian Peninsula (eastern coast of Gdańsk Bay). Outcrops of Cretaceous deposits occur in coastal cliffs to the west, on the islands of Rügen and Møn. In Stevns Klint in Denmark, the boundary between Cretaceous soft chalk and overlying Paleogene hard limestone is exposed. Cliffs formed in rocks of Devonian age occur on the coast of Latvia, and in Estonia cliffs with Cambrian–Ordovician–Silurian–Devonian rocks are found. The skären-type coastal zone in Sweden and Finland is dominated by Precambrian crystalline rocks, but beaches of Holocene sand and gravel are also seen. The coastline in this part is highly irregular, with numerous small islands. Coastal erosion dominates many parts of the southern and eastern coast

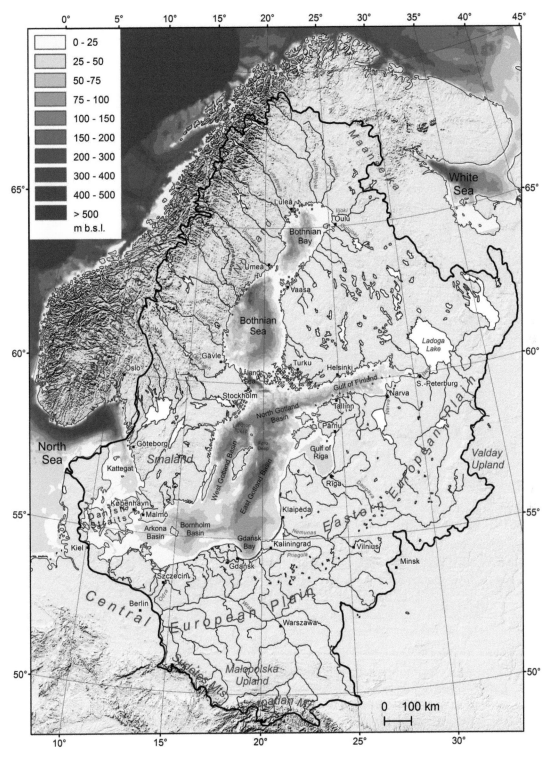

Fig. 5.1 Baltic Sea catchment area and the division of the Baltic Sea into major hydrographical regions and basins according to Leppäranta & Myrbeg (2009). Land area: shaded terrain model after USGS GTOPO30. Bathymetry of the Baltic Sea after iowtopo2. Modified from Uścinowicz (2014).

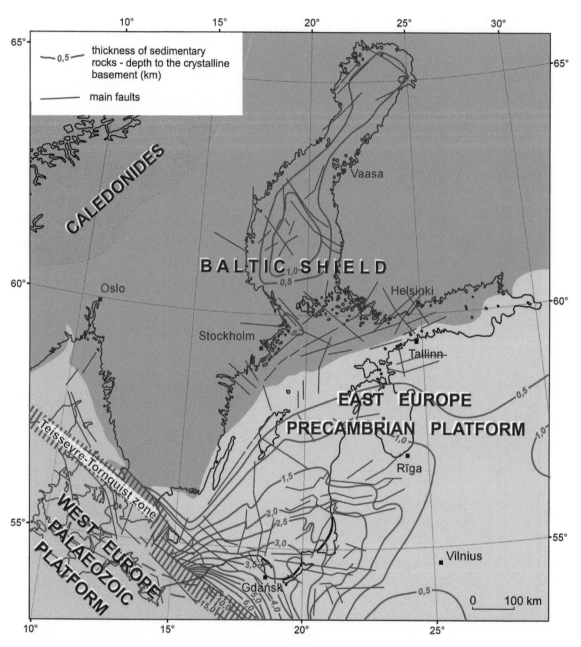

Fig. 5.2 Main tectonic features of the Baltic Sea region. Uścinowicz (2014).

of the Baltic Sea, while on the uplifting rocky coasts of Sweden and Finland erosion is not a problem.

Bathymetry

An extensive overview of the hydrographical conditions in different parts of the Baltic Sea is given in the book *Physical Oceanography of the Baltic Sea* (Leppäranta & Myrberg 2009). A characteristic feature of the Baltic Sea

is its shallowness. It has a mean depth of 54 m and the greatest depth of only 459 m is in Landsort Deep between Stockholm and Gotland (Leppäranta & Myrberg 2009). A number of basins divided by seabed elevations and escarpments are distinguished in the Baltic Sea (Fig. 5.1). The Danish Straits (including Little Belt, Great Belt, and the Öresund) are the gates to the Baltic Sea with water depths mostly less than 20 m. The transition zone between the Baltic Sea and the North Sea and between brackish and oceanic watermasses is the Kattegat which is sometimes included as part of the Baltic Sea (Leppäranta & Myrberg 2009). In the southwestern Baltic Sea, east of the Danish Straits, two basins can be distinguished: the Arkona basin with a maximum depth of 53 m, and the Bornholm basin with a maximum depth of 105 m. The Bornholm basin is connected with Gdańsk Bay by the Słupsk furrow or sill with a maximum depth of approximately 95 m (Figs. 5.1 & 5.7). Gdańsk Bay (maximum depth 114 m) together with the Eastern (maximum depth 249 m), Western (maximum depth 459 m) and Northern Gotland basins (maximum depth 150 m) form the Gotland Sea. The Gotland Sea together with the Arkona Basin, and the Bornholm Basin form the Baltic Sea Proper. Eastern and northern gulfs of the Baltic Sea consist of the Gulf of Riga (maximum depth 51 m), the Gulf of Finland (maximum depth 123 m) and the Gulf of Bothnia (including the Åland Sea, the Archipelago Sea, the Sea of Bothnia, and Bothnian Bay; maximum depth 293 m) (Leppäranta & Myrberg 2009).

Data sources

There are four publically available bathymetric data sets that provide the best cover of the entire Baltic Sea. The General Bathymetric Chart of the Oceans (GEBCO) global dataset is at a relatively low resolution, but it combines both regional-scale bathymetry and terrestrial topography. More recently, the European Marine Observation and Data Network (EMODnet) has made available bathymetric data for the European shelf at a resolution of 0.25 arcminutes (i.e. 15 arcseconds). EMODnet also provides metadata on the underlying higher resolution local data sets used to create the down-sampled continental-scale coverage. The dataset 'iow-topo2' provides a digitized topography of the Baltic Sea together with surrounding terrestrial topography (from the Geotopo30 dataset). The resolution is 2 minutes with respect to longitude, and 1 minute to latitude. This is approximately 1 nautical mile, or 2 km in both cases. The Baltic Sea Bathymetry Database (BSBD) has the objective of gathering in one place the water-depth data for the sea areas of all Baltic Sea countries and of distributing data products. It offers complete, homogeneous and up-to-date Baltic Sea bathymetry data from 'official' sources:

that is, all Baltic Sea national hydrographic offices that come under the umbrella of the Baltic Sea Hydrographic Commission. Details of these can be found at the following URLs:

EMODnet Hydrography Portal: www.emodnet-hydro graphy.eu/

GEBCO bathymetry: www.gebco.net/

Digital Topography of the Baltic Sea: www.io-warnemuende.de/topography-of-the-baltic-sea.html

Baltic Sea Bathymetry Database: data.bshc.pro

Earth Sciences

Pre-Quaternary geology

The Baltic Sea covers an area with geological structures similar to the land that surrounds it. The northern part of the Baltic Sea is situated within the Precambrian Baltic Shield, whereas the southern part lies on the Precambrian East European Platform. A small southwestern part of the Baltic Sea lies on the Paleozoic West European Platform that is separated from the East European Platform by the Teisseyre-Tornquist Zone (Fig. 5.2).

In those parts of the Baltic Sea that are situated on the Baltic Shield, Precambrian (mainly Proterozoic) crystalline rocks outcrop in coastal areas and on bottom highs (Fig. 5.3). Precambrian rocks are found below Quaternary sediments in Bothnian Bay, the Bothnian Sea and off the northern coasts of the Gulf of Finland in the area of the Archipelago Sea and off the eastern coasts of Sweden, as well as off the eastern shores of the Kattegat. Jotnian sandstones of Proterozoic age occur in Bothnian Bay and in the Bothnian Sea, in depressions of the Baltic Shield surface and are also locally exposed on the coasts, overlying Proterozoic magmatic rocks (Lundqvist & Bygghammar 1994). Locally, in the northern part of Bothnian Bay and western part of the Bothnian Sea, Cambrian sandstones and Ordovician limestones overlie Jotnian sandstones (Fig. 5.3).

The surface of the Precambrian basement rocks lowers towards the south and the south-east and forms the crystalline foundation of the East European Platform (Figs. 5.2 & 5.3). In the northern part of Estonia, Precambrian basement rocks are present at a depth of

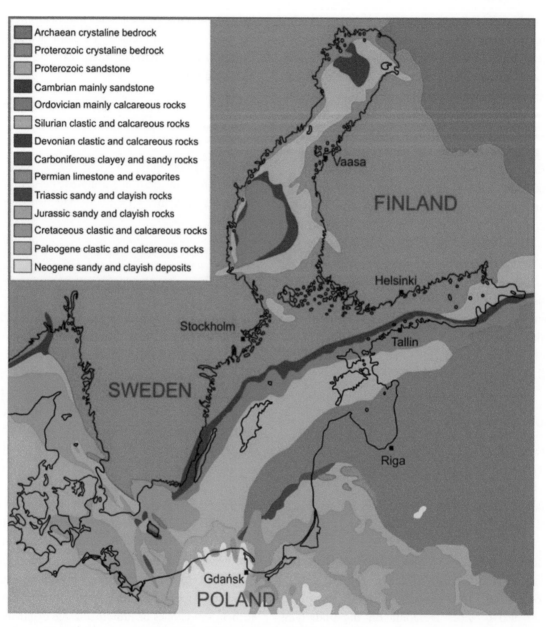

Fig. 5.3 Pre-Quaternary geology of the Baltic Sea region. Modified from Uścinowicz (2014).

approximately 0.2 km to 0.4 km, whereas in Gdańsk Bay they occur at a depth of 3.0 km to 3.5 km. Along the southern coast of the Baltic Sea, at the northeastern side of the Teisseyre-Tornquist Zone, the surface of the Precambrian rocks lies at a depth of around 6.0 km to 6.5 km (Fig. 5.2; Pokorski & Modliński 2007).

In the area of the East European Platform, the thickness of the Phanerozoic sedimentary cover increases towards the south-east and the south and at the same time the surface of the Precambrian basement rocks lowers. In the north and north-west, Cambrian sedimentary rocks underlie Quaternary sediments, whereas in the south Neogene sediments are seen below Quaternary sediments. In the northern part of the Baltic Proper, Paleozoic sedimentary rocks lie directly under the Quaternary cover, whereas in the eastern part the Paleozoic rocks are covered by Mesozoic and Cenozoic sedimentary rocks and sediments.

At the western side of the Teisseyre-Tornquist Zone, in the area of the West European Paleozoic Platform, the surface of the Precambrian basement rocks lies at a depth of approximately 10 km to 15 km (Fig. 5.2). Above the basement rocks lie folded sedimentary rocks of Early Paleozoic (Cambrian, Ordovician and Silurian) age. The Paleozoic rocks are covered by Paleozoic, Mesozoic and locally Paleogene strata. These strata form two platform complexes: a Carboniferous–Devonian complex and a Permo–Mesozoic complex, which at the turn of the Cretaceous and Paleogene periods were divided into smaller units. On the bottom of the Baltic Sea, to the west of the Teisseyre-Tornquist Zone, Mesozoic and Paleogene sediments are present under the Quaternary sediments.

Pleistocene and Holocene sediment thickness

The Quaternary deposits in the Baltic Sea consist mainly of Pleistocene glacial, glacio-fluvial and lacustrine sediments, and of Holocene lacustrine and marine sediments. The thickness of the Pleistocene sediments in the Baltic Sea varies widely. The thinnest accumulations of Pleistocene deposits, with a thickness of less than 10 m, occur in areas that were dominated by glacial erosion. Thick accumulations of Pleistocene deposits occur locally, for example in deep tunnel valleys incised in the pre-Quaternary bedrock. Pleistocene glacial sediments over large areas of the Baltic Sea, both in deep basins and along their shallower peripheries, are represented by a single layer of till from the last glaciation. Locally on the Baltic Sea bed, both in its southern and northern parts, glacio-fluvial deltas and eskers occur.

The postglacial (Late Pleistocene and Holocene) sediment cover and the stratigraphic units are generally the same over the whole area of the Baltic Sea. Postglacial sediments of silt and clay in the bottoms of the sedimentary basins form three major lithostratigraphic units: brown Baltic clay of glacio-lacustrine origin (Baltic Ice Lake), gray Baltic clay of brackish-water and lacustrine origin (Yoldia Sea and Ancylus Sea), and olive-gray Baltic mud of marine and brackish-water origin (Littorina and Post-Littorina Sea) (Fig. 5.4).

The thickness of brown Baltic clay (varved, laminated and homogeneous sediments) lying concordantly on the irregular surface of till is highest in the southern parts of

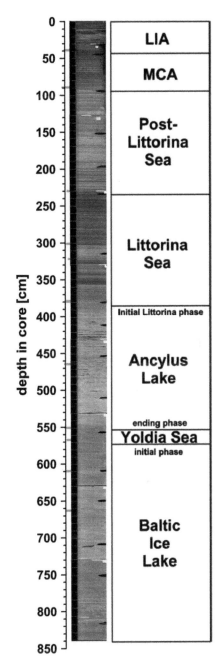

Fig. 5.4 A 'composite' sediment core from the eastern Gotland basin (water depth ca. 240 m) showing typical Late Glacial and Holocene deep-water sediments. The core has been compiled from three individual cores. Note that within the Post-Littorina Sea stage, climatic stages Medieval Climate Anomaly (MCA) and Little Ice Age (LIA) are also shown. Modified from Harff *et al.* (2011) with permission from Springer Science and Business Media.

the Baltic Sea where the unit reaches a thickness of around 10 m. Gray Baltic clay lying concordantly on brown clay reaches a thickness of up to 5 m to 8 m and olive-gray Baltic mud lying discordantly on gray clay reaches a thickness of 4 m to 5 m or locally more than 6 m.

Sand and gravel of the same ages as the fine-grained basin sediments occur on the slopes of sedimentary basins and in nearshore areas. In particular, marine and littoral sediments formed in the Middle and Late Holocene are widespread. In major areas of the shallow-water zone, the thickness of marine sand and gravel is usually less than 2 m, and on large areas it does not reach even 1 m. Only locally does the thickness of marine sand exceed 3 m. Thicker layers, occurring along the southern and eastern Baltic Sea coasts, are related to spits (barriers) that have developed during the Atlantic (Littorina) transgression. In addition, other types of Late Pleistocene and Holocene sediments, such as gyttja, lake marl, peat and delta deposits are found locally in the Baltic basin.

The current rate of accumulation of mud varies from 0.5 to 2 mm/year (Winterhalter et al. 1981; Pempkowiak 1991; Walkusz et al. 1992; Cato & Kjellin 1994; Szczepańska & Uścinowicz 1994; Hille et al. 2006; Mattila et al. 2006). In the central parts of the sub-basins the accumulation rate is higher than on their peripheries, but also regional differences occur. Sandy and sandy-gravelly sediments are deposited in the littoral zone and in shallow-water areas that are impacted by storm waves. Information about the thickness of the dynamic layer that is mobilized during storms is provided by the vertical distribution of caesium-137 in sediment cores. In Gdańsk Bay, at depths of 10 m to 30 m, caesium-137 is present in sandy layers down to 0.2 m to 1.3 m, which indicates that this interval is exposed to transport processes during storms. This depth interval is consistent with the thickness of the dynamic layer of sand as determined from the height of sand waves.

Data sources

The EMODnet-Geology project brings together different geological data about sea-floors including information about:

seabed substrate;

sediment accumulation rate;

seafloor geology;

bedrock lithology;

bedrock stratigraphy;

coastline migration;

mineral resources (oil and gas, aggregates, metallic minerals);

geological events (earthquakes, submarine landslides, volcanic centers).

The website for the EMODnet-Geology project can be found at: www.emodnet.eu/geology

Basic substrate maps of the Baltic Sea basin were produced during a number of habitat mapping projects in connection with research projects such as BALANCE (Kotilainen & Kaskela 2011).

Climate and ice-sheet history

The Baltic Sea basin experienced several glaciation events related to the major climatic shifts during the Last Glacial Period (Marine Isotope Stage (MIS) 4 to 2; Fig. 5.5). During advances and retreats, iceberg calving and freshwater input as well as shifting sea-ice conditions occurred, and the ice sheets recurrently impacted the North Atlantic thermohaline circulation and thereby also the climate of northwest Europe, as indicated by the paleoclimatic records from the North Atlantic Margin and the Greenland ice sheet (Andrén et al. 2011).

Based on detailed correlations and dating of the south-western Baltic glacial stratigraphy, Houmark-Nielsen and Kjær (2003) and Houmark-Nielsen (2010) concluded that the southwest Baltic basin may have experienced two major ice advances during MIS 3, at ca. 50 ka and 30 ka (Figs. 5.5 & 5.6). On Kriegers Flak in the southwestern Baltic Sea, a brackish-water deposit is found which shows that glacio-marine waters inundated the Baltic Sea basin shortly after the 50 ka advance, or perhaps in the Late Saalian (Bennike & Wagner 2010; Anjar et al. 2012). According to glacial rebound modeling the Baltic Sea basin was partially glaciated around 40 ka and became ice-free during the Ålesund Interstadial at around 36 ka (Lambeck et al. 2010). The

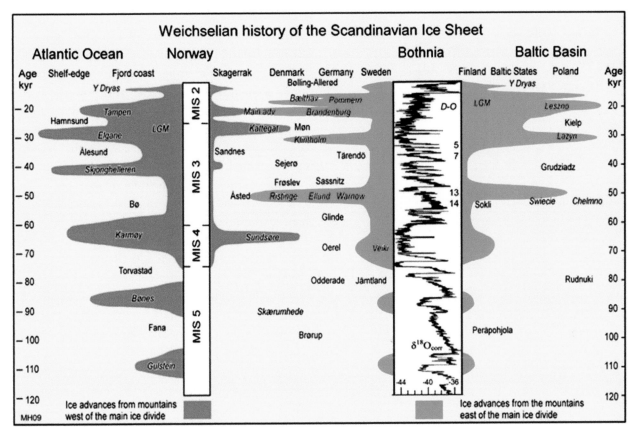

Fig. 5.5 Time–distance diagram of the Scandinavian ice sheet during the Last Glacial Period. Stadials with ice growth and interstadials with ice retreat are indicated from the two main source areas, that is, west (blue) and east (red) of the main ice divide, and are matched against the NGRIP ice-core isotope record with selected Dansgaard-Oeschger (D-O) events. Houmark-Nielsen (2010). Reproduced with permission from John Wiley & Sons.

Last Glacial Maximum (LGM) during MIS 2, at around 20 ka (Fig. 5.7) represented the coldest phase of the last glacial–interglacial cycle with the largest ice volume, which was estimated to be equivalent to a sea-level drop of ca. 120 m (Fig. 5.7; Lambeck *et al.* 2010). The center of ice was located in the Bothnian Sea area (Fig. 5.8) where the highest observed evidence of shore erosion in Fennoscandia has been noted (Berglund 2004).

The beginning of the present-day Baltic Sea is related to the retreat of the last Scandinavian ice sheet (Fig. 5.9). The retreat of the Scandinavian ice sheet from the LGM limit was irregular and was influenced by Late Glacial climatic oscillations of the Bølling/Allerød Interstadial and the Younger Dryas cold period. In the southwestern part of the present Baltic Sea, the ice sheet melted and the first embryo of the Baltic Ice Lake formed approximately

16 ka BP (Andrén *et al.* 2011), whereas the deglaciation of the northern Baltic Sea took place much later, around 10.5 ka BP (Lindén *et al.* 2006).

The Late Glacial Period can be divided into five chronozones: three periods with cold, arctic climate (Oldest, Older and Younger Dryas stadials) and two periods of warmer climate (Bølling and Allerød interstadials). The boundaries between them, which are connected with climatic changes, were defined by radiocarbon ages by Mangerud *et al.* (1974). Currently, the chronology is usually defined from ice-core chronologies (e.g. Björck *et al.* 1998; Rasmussen *et al.* 2006). Information about the time span prior to 14.7 ka cal BP corresponding to the Oldest Dryas, is limited, but the presence of steppe-tundra vegetation with a high proportion of heliophytes (Litt *et al.* 2001) indicates low temperatures, and Coope

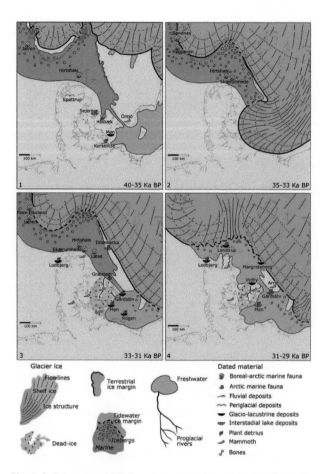

Fig. 5.6 Paleogeographical evolution of the southwestern Baltic Sea basin during MIS 3. Houmark-Nielsen (2010). Reproduced with permission from John Wiley & Sons.

Fig. 5.7 Extent of ice sheets (Svendsen *et al.* 2004) and 120 m lower sea level in northern Europe during the LGM. Note that the maximum glacial limit was reached at different times in different sectors. Topographic data from NASA. Adapted with permission.

et al. (1998) suggested that the temperature during the warmest month of the year was 9°C to 11°C in northern Germany. During the Bølling interstadial (14.7–14.05 ka cal BP) there was a strong temperature gradient from central Poland and central Germany towards the ice-sheet margin, and the mean July temperature dropped from about 19°C to about 10°C according to fossil beetle data (Coope *et al.* 1998). In Denmark the vegetation was dominated by pioneer species, and included shrubs and dwarf-shrubs, whereas no trees are recorded (Bennike *et al.* 2004a; Mortensen *et al.* 2011). The Older Dryas stadial was a short episode that lasted only about 150 years (14.05–13.9 ka cal BP) and which was again characterized by a cold and continental climate (Berglund *et al.* 1994).

During the Allerød interstadial (13.9–12.7 ka cal BP) the mean July temperature in the southeastern part of the Baltic region was around 12°C to 15°C (Heiri *et al.* 2014) and about 9°C in central Sweden (Coope *et al.* 1998). *Betula pubescens* migrated into Denmark and southern Scandinavia and open birch forests formed during the late Allerød (Mortensen *et al.* 2014). Later on open pine forests developed south of the Baltic basin.

During the Younger Dryas cooling (12.7–11.7 ka cal BP) woodland became more open and dispersed and even disappeared over large areas, where it was replaced by pioneer vegetation with dwarf-shrubs and herbs. Soils became unstable and mineral-rich sediment was deposited in many lakes. Younger Dryas temperatures in the region were interpreted from beetle remains (Coope *et al.* 1998) suggesting that the mean July temperature was 8°C to 10°C. Using the head capsules of midge larvae, a mean July temperature of 9°C to 12°C was inferred for the eastern part of the Baltic region (Heiri *et al.* 2014). Multi-proxy investigations of lacustrine sediments from central Poland indicated that the early part of Younger Dryas was drier and the later part more humid (Goslar *et al.* 1993), whereas in Germany the later part of the Younger Dryas may have been drier than the earlier part (Zolitschka *et al.* 1992).

There was a rapid rise in temperature at the Younger Dryas/Holocene boundary, around 11.7 ka cal BP and the temperature may have increased by 4°C over 100 years. This temperature rise led to temperatures comparable to those of the present day in the Early Holocene

Fig. 5.8 Extent of ice sheets (limit marked by white line) and modeled ice thicknesses in northern Europe at the LGM around 20,000 years ago. Note the center of ice over Scandinavia and the Baltic where the ice was about 2700 m thick. Svendsen *et al.* (2004). Reproduced with permission from Elsevier.

(Walker 1995). In northwest Europe, the steppe-tundra communities of the Younger Dryas were replaced within less than 500 years by woodlands with *Betula* and *Pinus* (Berglund *et al.* 1994). An abrupt temperature rise is also registered in Coleopteran data, according to which the mean July temperature quickly reached 14°C to 16°C in southern Sweden, and around 17°C in Poland (Coope *et al.* 1998).

The Holocene can be divided into three periods, Early, Middle and Late Holocene (Walker *et al.* 2012).

Early Holocene (ca. 11.7–8.2 ka cal BP)
During the Early Holocene, pollen records show steadily increasing temperatures. Summer insolation was higher than at present (Berger & Loutre 1991) and the continentality in the region was also more marked (Ralska-Jasiewiczowa 2004). Also the seasonality was marked by shorter transition periods between summer and winter

(Birks 1986). Rising temperatures led to stabilized forest communities, mainly with tree birch (*Betula* sect. *Albae*) and pine (*Pinus sylvestris*). The forests were open with plenty of heliophytes. Gradually, the role of tree birch decreased and tree birch was replaced by more warmth-demanding species, first by hazel (*Corylus avellana*) and later by elm (*Ulmus*), oak (*Quercus*), ash (*Fraxinus*) and lime (*Tilia*). The immigration of these taxa, in combination with rising temperatures and soil stabilization, led to the development of mesophilous mixed deciduous forests.

Middle Holocene (ca. 8.2–4.2 ka cal BP)
Based on pollen data Seppä *et al.* (2009) suggested that the annual mean temperature in northern Europe was about 2.0°C to 2.5°C higher than at present during the earlier part of the Holocene Thermal Maximum at 8 ka cal BP to 6 ka cal BP. This is the same temperature deviation that was inferred from plant macrofossil

Fig. 5.9 Retreat chronology of the Scandinavian ice sheet from the Baltic Sea basin area (from Uścinowicz 2014). LGM = Last Glacial Maximum; F = Frankfurt Phase; PZ = Poznan Phase; C = Chodzież Phase; PO = Pomeranian Phase; B = Baltija Phase; R = Rosenthal Phase; SL = South Lithuanian Phase; H-WS = Halland–West Skane Phase; G = Gardno Phase; ML = Middle Lithuanian Phase; SB = Słupsk Bank Phase; NL = North Lithuanian Phase; H-L = Haanja–Luga Phase; SMB = Southern Middle Bank Phase; O = Otepää Phase; P-N = Pandivere–Neva Phase; P = Palivere Phase; S I–III = Salpausselkä Phases.

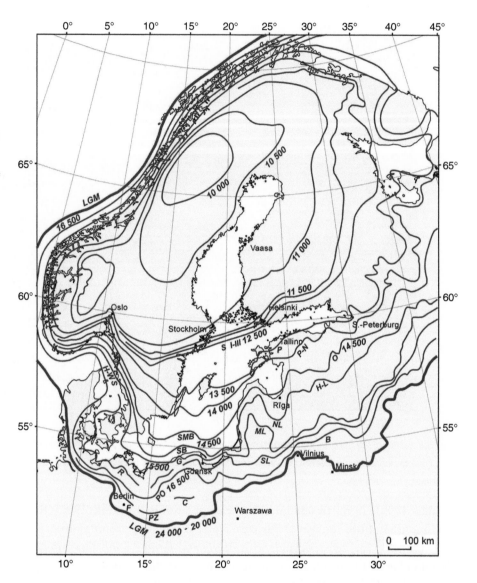

studies by Andersson (1902). During the same time period, the mean July temperature was about 1.5°C higher than at present (Seppä *et al.* 2009). During the Holocene Thermal Maximum, many species extended their geographical ranges farther north than at present. For example, hazel (*Corylus avellana*), the water plants *Trapa natans* and *Najas minor* as well as the European pond turtle (*Emys orbicularis*) spread north in the Baltic region (Andersson 1902; Donner 1995; Bennike *et al.* 2001; Sommer *et al.* 2007; Seppä *et al.* 2015). The arctic tree line also expanded north, and the alpine tree line in the region was higher than at present. High summer temperatures are also indicated by the former more widespread occurrence in the forests of mistletoe (*Viscum album*) and ivy (*Hedera helix*), which extended farther north than today (Huntley & Birks 1983). The presence of mistletoe indicates warm summers and autumns in Poland, with a mean temperature of the warmest month above 16°C (Ralska-Jasiewiczowa 2004). The presence of ivy indicates that the mean temperature of the coldest month was above −2°C (Iversen 1944).

Late Holocene (ca. 4.2–0 ka cal BP)

Based on pollen records and compared with a chironomid-based record, two $\delta^{18}O$ curves from lake sediments and a macrofossil record, Seppä *et al.* (2009) concluded that the Late Holocene was marked by a long-term cooling trend. However, three periods with higher temperatures dating to 5 ka cal BP to 4 ka cal BP, 3 ka cal BP to 1 ka cal BP and to the last 150 years, and two periods with lower temperatures dated to 3.8 ka cal BP to 3 ka cal BP and to 500 cal BP to 100 cal BP were found. The latter corresponds to the Little Ice Age which was the coldest period during the Late Holocene. During the Little Ice Age, glaciers in Sweden and Norway expanded, and winter ice cover on the Baltic Sea and on lakes and rivers in the regions lasted much longer than at present. The Late Holocene climate changes led to a growth of peat bogs and to changes in forest composition. However, in many regions, anthropogenic effects on the natural vegetation were much more pronounced than the climatic effects.

Data sources

The best starting point for paleoclimate and ice-sheet history data is the published literature, such as the above or chapters in the book *Baltic Sea Basin* (Harff *et al.* 2011).

In connection with the *Integrated Ocean Drilling Program (IODP) expedition 347: Baltic Sea Paleoenvironments* (2013) new data about climate, sea level and ice-sheet change will be available during the next few years, data which will be uploaded to the ODP expedition 347 webpage (www.ecord.org/expedition347/).

Development of the Baltic Sea: alternating lake and marine stages

After the last deglaciation the Baltic Sea experienced two lake stages (Baltic Ice Lake and Ancylus Lake) alternating with marine stages (Yoldia Sea and Littorina Sea; Fig. 5.10).

Baltic Ice Lake (16–11.7 ka BP)
During the deglaciation of the Baltic Sea basin, a first embryo of the Baltic Ice Lake formed in the southwestern part of the basin approximately 16,000 years ago (Fig. 5.11). During the initial stage of the Baltic Ice Lake,

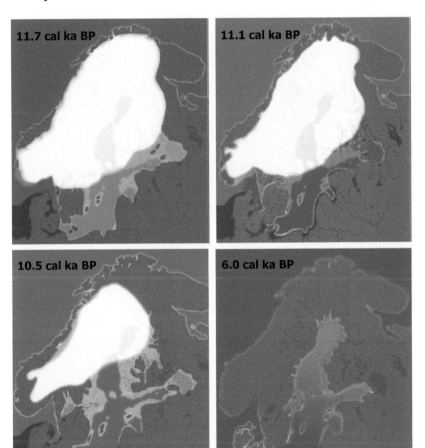

Fig. 5.10 Postglacial development of the Baltic Sea basin with alternating lake and marine stages. Andren *et al.* (2011). Reproduced with permission from Springer Science and Business Media.

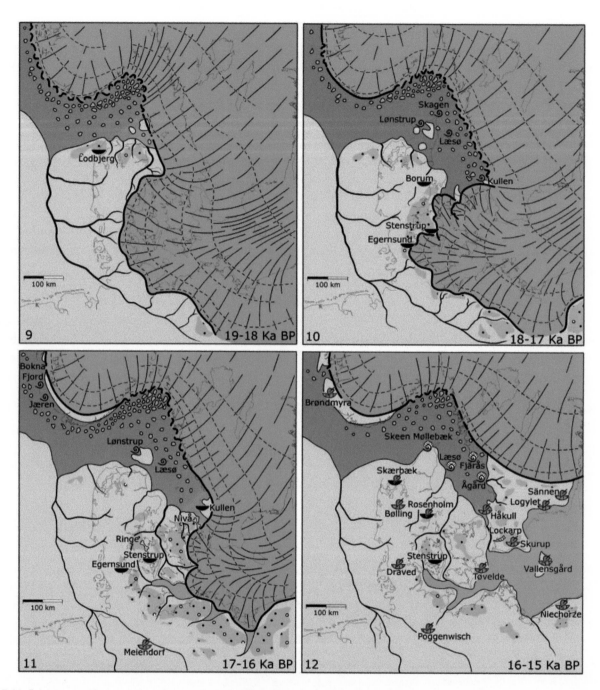

Fig. 5.11 Paleogeographical evolution of the southwestern Baltic Sea basin during the last deglaciation. Houmark-Nielsen & Kjær (2003). Reproduced with permission from John Wiley & Sons. For legend please see Fig. 5.6.

it was most likely at the same level as the ocean. However, as the isostatic rebound of the outlet in the Öresund threshold area between Copenhagen and Malmö — made up of glacial deposits on top of limestone bedrock — was greater than the sea-level rise, the Öresund outlet river eroded its bed in step, at the same pace, with the emerging land. When the fluvial down-cutting reached the flint-rich limestone bedrock, the erosion ceased more or less completely. This is possibly an important turning point in the Baltic Ice Lake development: the uplift of the

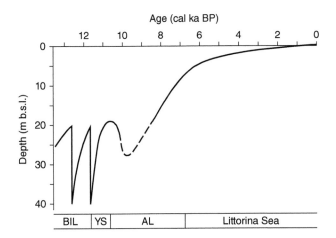

Fig. 5.12 Tentative curve showing relative shore-level changes in the Arkona basin during the Late Glacial and the Holocene. BIL = Baltic Ice Lake stage; YS = Yoldia Sea stage; AL = Ancylus Lake stage. Bennike & Jensen (2013).

threshold lifted the level of the Baltic Ice Lake above sea level and the updamming of this large glacial lake started (Björck 1995). As a result of the ice recession in the area immediately north of Mt. Billingen, the first drainage of the Baltic Ice Lake took place at ca. 13 ka BP (Fig. 5.12; Bennike & Jensen 2013). In connection with the Younger Dryas cooling, the ice-sheet margin advanced and the area north of Mt. Billingen was blocked again, causing a second updamming period. At the end of the Younger Dryas, the ice-sheet margin retreated from the threshold at Mt. Billingen, which caused a new sudden lowering of the Baltic Ice Lake level of ca. 25 m down to sea level. This drainage must have had a huge impact on the whole circum-Baltic environment, with large coastal areas suddenly sub-aerially exposed, large changes in fluvial systems, considerable reworking of previously laid-down sediments, as well as the establishment of a large land bridge between Denmark and Sweden (Fig 5.10).

Yoldia Sea (11.7–10.7 ka BP)
During the Yoldia Sea stage, saline water entered from the Kattegat through the fairly narrow and shallow straits in the south-central Swedish lowland into the Baltic. The duration of this brackish-water phase was rather short and lasted probably not more than 350 years. Due to the high uplift rate in south-central Sweden, the strait

area rapidly became shallower, which together with the large outflow of fresh water prevented saline water from entering the Baltic. The Yoldia Sea was transformed into a huge freshwater lake (Fig 5.10).

Ancylus Lake (10.7–9.8 ka BP)
As the straits in the south-central Swedish lowland were closed, the next updammed stage started in the history of the Baltic basin. In the southern part of the Baltic the Ancylus Lake transgression is recorded from submerged pine trees and peat deposits dated between 11.0 ka BP and 10.5 ka BP, whereas areas to the north experienced a continued regression, but at a lower rate than before.

The transgression and flooding in the south as a consequence of a 'tipping bathtub effect' inevitably resulted in a new outlet in the south. Since Öresund had been uplifted more than other potential outlet/sill areas farther south, these southern areas were now lower than Öresund. What might now have followed is described by Björck et al. (2008). Available data indicate that the Darss sill area, between Germany and Denmark, eventually became the new outlet from the Baltic. However, the transition from the Ancylus Lake to the Littorina Sea is not well understood. For example, there is little evidence for erosion in the German or Danish straits (Lemke et al. 1999; 2001).

Initial Littorina Sea (9.8–8.5 ka BP) and Littorina Sea (8.5 ka BP–present)
Remarkable changes in the Baltic Sea and its ecosystem took place when the freshwater Ancylus Lake was transformed first to a brackish-water stage and later to a marine stage. Early evidence of saline water ingression into the Baltic basin after the Ancylus Lake stage is documented from the Blekinge archipelago (Berglund et al. 2005) and from the Bornholm basin in the southern Baltic around 9.8 ka cal BP (Andrén et al. 2000; 2011). In the Danish area the first sign of saline ingression is recorded around 10 ka cal BP (in the northern Öresund; Bennike et al. 2012) and around 8.8 ka cal BP (in the Storebælt & Lillebælt: Bennike et al. 2004b; Bennike & Jensen 2011). Around 8.5 ka cal BP the Baltic Sea became a brackish-water basin with significantly increased primary production (Andrén et al. 2011) establishing favorable

conditions for human exploitation of marine resources and development of coastal settlements (Jöns 2011).

In some studies, the Post-Littorina Sea or Limnea Sea stage (since ca. 4.5 ka BP) has also been identified according to a decrease in salinity and changes in sediment composition (see Fig. 5.4).

Postglacial rebound, relative shore-level changes and evolution of coastlines

Melting of the Scandinavian ice sheet changed the mass load on the continent and in nearby oceans and seas, including the Baltic Sea, and resulted in isostatic adjustments throughout the region. Records of relative sea-level change differ from place to place because of the interaction between eustatic sea-level rise and postglacial rebound (Fig. 5.13). Relative sea-level changes have been studied in the Baltic Sea region based on geological records, as summarized by Lambeck et al. (1998) and Harff et al. (2005; 2011). For the last few hundred years, relative sea-level changes have been recorded by tide gauge measurements complemented by repeated leveling (Ekman 1996; 2009; Kakkuri 1997; Douglas & Peltier 2002). Recently, continuous point positioning (time series of the coordinates) from the BIFROST (Baseline Inferences from Fennoscandian Rebound Observations, Sea-level and Tectonics) permanent GPS network has also become available for determination of crustal movements with respect to the Earth center of mass (Fig. 5.14).

The differential glacio-isostatic uplift in the southern and the northern parts of the Baltic Sea has influenced both sedimentation processes and evolution of coasts. In the northern part of the Baltic Sea, north of a line from Stockholm to Turku, i.e. in the areas of the Bothnian Sea and Bothnian Bay, the isostatic uplift has exceeded the eustatic rise of sea level throughout the whole period after deglaciation. Therefore, in this part of the Baltic, permanent regression occurred (e.g. Lindén et al. 2006) and regression still continues at present (Fig. 5.13). Areas of seabed originally located underwater gradually emerged and new areas still continue to emerge from the sea. As a result of the uplift, landforms of glacial and glacio-fluvial relief created under the ice sheet and

in front of the ice sheet, and sediments deposited in the early phases of the Baltic Sea development, first came into the range of the erosive impact of waves, and then they emerged above the sea surface.

The highest raised coastlines in the region, on the northwestern coasts of the Bothnian Sea, have been elevated to an altitude of about 270 m above the present sea level since the time of deglaciation while the highest Littorina Sea coastlines are at about 150 m above present sea level (Fig. 5.13; Berglund 2004).

Areas of the southern part of the Baltic Proper adjacent to the coasts of southern Denmark, Germany, Poland and Lithuania, experienced transgression in the Holocene (Fig. 5.13), except during the period of rapid shore-level variations near the end of the Pleistocene. The water flooded previous land areas, eroding relief forms, glacial and glacio-fluvial sediments, and also sediments deposited in terrestrial environments after the initial ice-sheet retreat (e.g. Uścinowicz 2003; 2006). In certain places on the bottom of this part of the Baltic Sea, gyttja, lake marl, peat and in situ tree trunks occur below marine sediments or exposed on the sea floor.

The coast of the central part of the Baltic Proper, i.e. the coasts of Sweden from Scania to Stockholm and the coasts of Latvia, Estonia and southern parts of Finland, has a more complex history (Fig. 5.13). In the Atlantic Period, coastal transgression took place when the rate of eustatic sea-level rise exceeded the local rate of isostatic uplift. When the rate of eustatic sea-level rise slowed down towards the end of the Atlantic Period, the uplift processes started to prevail and coastal marine regression began (e.g. Miettinen 2004; Veski et al. 2005; Yu et al. 2007; Rosentau et al. 2013).

Paleogeographic reconstructions

Studies of the Baltic Sea paleocoastlines and sea-level change provide insight into glacio-isostatic adjustment processes for the development of paleogeographical reconstructions and for a better understanding of paleoenvironments of early human settlements in the Baltic Sea area. Such studies can help to define the contemporaneous sea levels for prehistorical settlement sites (Harff et al. 2005; Schmölcke et al. 2006; Rosentau

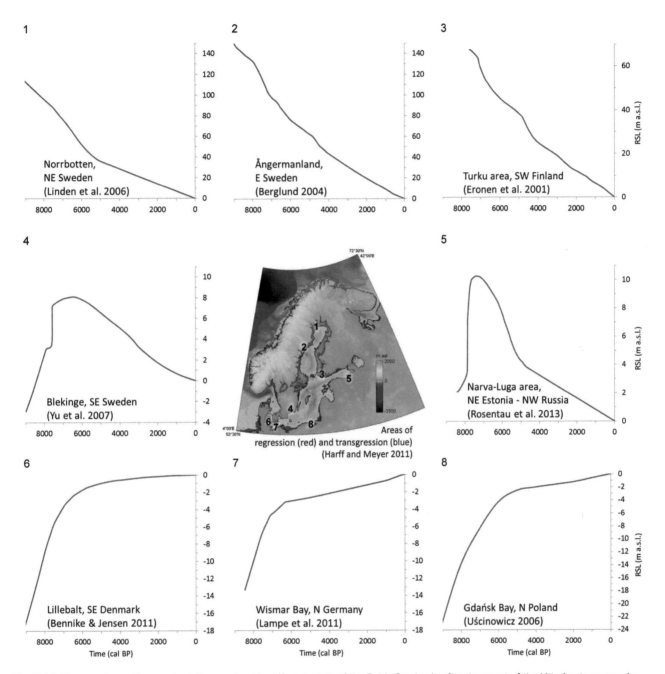

Fig. 5.13 Change of coastlines and relative sea level in different parts of the Baltic Sea basin after the onset of the Littorina transgression about 8000 years ago. Red colors = areas of regression, blue colors = areas of transgression. Compiled after data by Eronen *et al.* (2001); Berglund (2004); Lindén *et al.* (2006); Uścinowicz (2006); Yu *et al.* (2007); Bennike & Jensen (2011); Harff & Meyer (2011); Lampe *et al.* (2011); Rosentau *et al.* (2013). Note that the relative sea-level curves have different vertical scales. From Rosentau *et al.* (in press).

et al. 2011; 2013) and to predict new settlement site locations (Hörnberg *et al.* 2005). Conversely, well-dated coastal settlements can also provide important information about sea-level change (Jöns 2011).

The use of digital terrain models (DTM) and GIS-based (Geographic Information System) spatial calculations has substantially raised the effectiveness and reconstruction quality of the paleowater bodies in

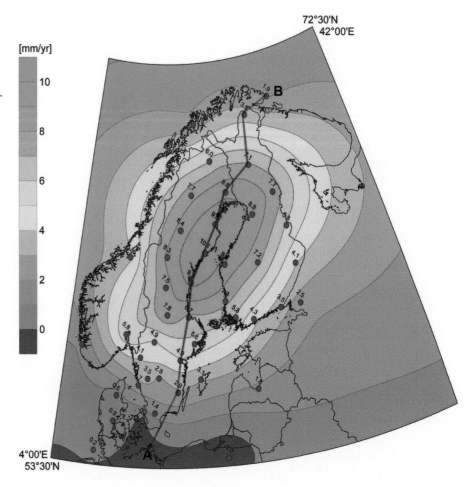

Fig. 5.14 Interpolated surface of present-day crustal uplift rates (in mm/year) according to BIFROST permanent GPS network data by Lidberg *et al.* (2007) with indication of proxy data. Rosentau *et al.* (2012).

formerly glaciated areas with possibilities to process large data sets and make analyses of higher accuracy. Such paleo-reconstructions are based on spatial calculations in which glacio-isostatically deformed water-level surfaces, derived from the shore displacement curve data and/or paleoshoreline elevations, are subtracted from the DTM. The regional reconstructions have provided new information about drainage history and volumetric changes of the past Baltic Sea (Jakobsson *et al.* 2007) while the local scale reconstructions are relevant for a better understanding of environmental conditions around Stone Age settlements (Rosentau *et al.* 2011).

For regional or basin-scale paleo-reconstructions, it is important to consider the effect of the uneven glacio-isostatic adjustment, which can be solved by using glacio-isostatic adjustment (GIA) models (Lambeck *et al.* 1998; 2010) or by using available empirical data on paleoshorelines or relative sea-level data sets combined

with geostatistical modeling (Harff *et al.* 2005; Jakobsson *et al.* 2007). GIA models can also produce paleo-reconstructions using fragmentary observational data and produce quantitative models with predictive capabilities. However, even in relatively small areas the glacial rebound can produce significant tilt of paleoshorelines, which is important to consider in paleogeographic reconstructions. In the peripheral area of the glacial rebound region in Pärnu Bay (Gulf of Riga) the 9000-year-old paleoshoreline is tilting with a gradient ca. 0.2 m/km where it is located at present about 5 m below present sea level in the southeastern part of the study area, and has been uplifted to an elevation of ca. 5 m above present sea level in the north-west (Fig. 5.15).

High-resolution elevation data derived from side-scan sonar or multibeam echo sounders to map seabed features, or airborne LiDAR (Light Detection and Ranging) data to map shallow waters and terrestrial areas,

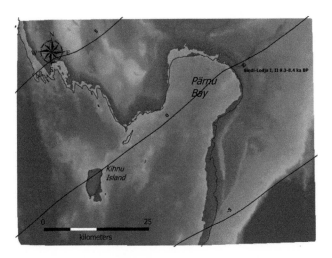

Fig. 5.15 Paleogeographic reconstruction of the Initial Littorina Sea shoreline during the Mesolithic Sindi-Lodja settlements at about 9 ka cal BP with indication of water-level isobases (m above sea level). From Rosentau et al. (2011). Note that the paleoshoreline at 9 ka cal BP was located below the present sea level in southern part of the study area, but has been uplifted above the present sea level in the north.

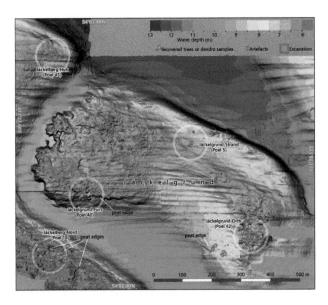

Fig. 5.16 Relief map of the Jäckelgrund in Wismar Bay, southern Baltic, based on multibeam echosound data. The white circles show areas where submerged Stone Age sites have been located. Lübke et al. (2011). Reproduced from Oxbow Books.

are important especially for local-scale reconstructions to detect special features like submerged river valleys, paleoshorelines, and so on. In Wismar Bay, the paleogeographic situation of the 8400–8000-year-old submerged Jäckelberg-Huk and younger Mesolithic sites have been well visualized by using multibeam echosound data (Fig. 5.16). By combining orthophoto and multibeam echo sounder imaging along the Hanö Bay coast, it has been possible to reconstruct the submerged part of the ancient Verkean River (Fig. 5.17).

Evidence for Submerged Terrestrial Landforms

Geological and archaeological studies in the southern Baltic Sea have revealed more than 1200 submerged sites of Stone Age artifacts (Jöns et al. 2016) in Danish (Fischer 2011), German (Lübke et al. 2011), Swedish waters (Hansson et al. in press) and Polish waters. Stone Age settlements and artifacts are found on the bottom of the present-day Baltic Sea as a result of the Holocene

sea-level rise (Bennike & Jensen 2011). In addition to archaeological sites, there are many examples of Holocene submerged terrestrial coastal landscapes, river valleys and drowned forests on the Baltic Sea bed, which reflect past sea-level changes and indicate potential places to look for new prehistoric settlement sites.

Many sites of submerged terrestrial deposits (paleolakes and peatlands) of Holocene age located in the

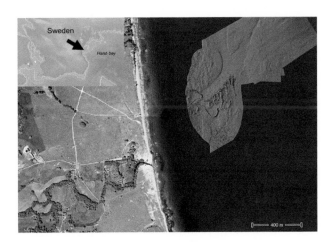

Fig. 5.17 A submerged river valley in Hanö Bay off southern Sweden. Modified from Hansson et al. (in press)

Fig. 5.18 Submerged Mid Holocene forest with tree trunks in Gdańsk Bay in the southern Baltic Sea. Arrows in (B) mark the locations of the tree trunks and the white box the location of the underwater photography in (C). Uścinowicz *et al.* (2011).

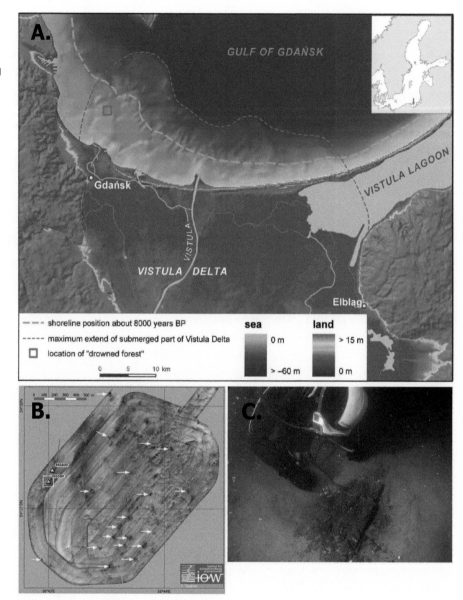

Polish southern Baltic at depths shallower than 30 m are described by Uścinowicz (2003). Submerged tree stumps rooted in sediments of an ice-marginal lake currently lie 23 m below sea level in the area located around 15 km north of Ustka (central part of the Polish coast). Radiocarbon dates of one pine trunk gave ages between 10.9 ka cal BP and 10.7 ka cal BP, corresponding to the time just before the Ancylus Lake transgression (Uścinowicz 2014). A drowned forest with preserved *in situ* rooted tree stumps (Fig. 5.18) was also discovered in Gdańsk Bay ca. 6 km to 7 km offshore in a water depth of 16 m to 17 m (Uścinowicz *et al.* 2011). The

thickness of marine sand at the site is from a few to a dozen centimeters. Below the sand, gyttja with peat intercalations and wood fragments occurs. Sixteen fragments of alder trunks and one of an oak trunk have been collected there. Radiocarbon dating of five selected samples gave ages around 8.8 ka cal BP, corresponding to the time of the Initial Littorina Sea stage. Younger alder tree stumps, with ages between 5.5 ka cal BP and 3.5 ka cal BP, are also known from the Vistula lagoon at depths of around 2 m (Łęczyński *et. al.* 2007).

Submerged tree stumps have also been located at three sites off the Lithuanian coast; two of them were found in

Fig. 5.19 Submerged Early Holocene tree trunk off the Lithuanian coast. Zulkus & Girininkas (2012). Reproduced with permission.

the neighborhood of Klaipėda and the third one south of Klaipėda (Damusyte *et al.* 2004; Zulkus & Girininkas 2012). They are separated by 5 km and 22 km (from north to south). The southernmost site is called RF-I (Relict Forest-I); it is located close to Juodkrante, on a sandy bottom at a depth of 26m to 29 m. Three stumps with roots in till with an overlying sandy layer about 15 cm thick have been discovered (Fig. 5.19). Stumps rise to a height of 0.5 m to 1.5 m above the sea floor and they remain in good condition, and measure from 0.4 m to almost 1 m in diameter. The spacing between the stumps varies from 6 m to 8 m. Two dated pine stumps gave ages of 10.95 ka cal BP and 10.09 ka cal BP. On a stony bottom at the other site (RF-II), in a depth of 14.5 m, a wooden stump, rising 30 cm above the seabed, was found. There is a sandy layer of 5 cm with clay loam slush around it. Two huge branches are situated deep in the clay loam. The stump yielded an age of 7.83 ka cal BP. RF-III appears to be on a sandy bottom, in a water depth of 11 m. A single stump of a tree 33 cm to 35 cm in diameter was discovered and it rises up to 50 cm above the seabed (Zulkus & Girininkas 2012).

Submerged tree stumps are fairly common in Swedish, German and Danish waters. In Sweden, one of the best documented sites is Hanö Bay, from where pine stumps from depths of 13 m to 14 m have been dated to the Early Holocene (Gaillard & Lemdahl 1994). Off the coast of northern Germany, tree stumps are found at a number of sites. A large number of finds from Wismar Bay have been dated (Harff & Lüth 2007; Lampe *et al.* 2011). In Denmark, tree stumps are known from many areas by local fishermen, but only a few have been sampled and dated (e.g. Fischer 1995).

In the transitional zone between transgressive and regressive shore displacement (Fig. 5.13), terrestrial landscapes may also have been preserved on the seabed from the periods of low relative shore level during the Yoldia and Initial Littorina Sea stages (Andrén *et al.* 2011). Based on sediment core data, a low sea level at about 18 m BSL during the Yoldia Sea stage in Järnavik Bay in the Blekinge archipelago at about 11.1 ka cal BP was reported (Andren *et al.* 2011). This low sea-level episode is also supported by radiocarbon dates of submarine peat at Hävang at 15 m BSL and pine stumps at Langörgen (3 m BSL) and Lövdalen (13–14 m BSL) situated east of Haväng in eastern Skåne, and dated to ca. 11 ka cal BP to 10.8 ka cal BP (Gaillard & Lemdahl 1994). From the Hävang area there is also evidence of an Initial Littorina Sea lagoonal environment surrounded by a pine-dominated forest, which was inhabited by Mesolithic humans around 9 ka cal BP (Hansson *et al.* in press). Based on peat layers, Markov (1931) and Yakovlev (1925) supposed that during the Initial Littorina Sea stage, the water level in the eastern Gulf of Finland was lower than at present (2.4 m BSL and 6 m BSL). In the coastal zone of the Gulf of Riga buried peat layers have been documented in the bottom of the Pärnu River valley at a depth of 1m to 5 m BSL and dated to between 9.6 ka cal BP and 8 ka cal BP (Rosentau *et al.* 2011).

Potential for Prehistoric Archaeological Site Survival

Preservation of underwater terrestrial landscapes and prehistoric archaeology in the Baltic Sea basin area seems to be controlled mainly by Quaternary glaciations, past relative sea- and lake-level changes and by high- or low-energy marine conditions.

During the Pleistocene, ice advanced several times into the area of the present Baltic Sea, eroding the bedrock and sediments and deepening the Baltic Sea basin. Most of the bedrock erosion and shaping of major over-deepened troughs was probably accomplished during the first glaciations of the Quaternary. Younger glaciations mainly removed sediments deposited by previous glacial cycles, reducing the thickness of the Quaternary succession and locally incising the bedrock surface. Over the last glacial cycle, ca. 20 m to 90 m of sediments (and locally more) was removed in the zones of most active erosion (Amantov et al. 2011). However, in a few places pre-LGM terrestrial sediments have also been preserved on the bottom of the Baltic. In the process of planning for the Kriegers Flak wind farm, west of the Arkona basin, terrestrial organic sediments, sandwiched between glacial diamicts, have been discovered (Anjar et al. 2012). Paleoenvironmental reconstructions show that between 42 ka cal BP and 36 ka cal BP (MIS 3), sedimentation occurred in shallow lakes and wetlands on Kriegers Flak. The vegetation was tundra-like, or forest tundra-like, possibly with birch and pine in sheltered locations, and a mean July temperature of ca. 10°C was suggested (Anjar et al. 2012). From the same interstadial period several finds of bones and teeth of the woolly mammoth are also known from the Baltic seabed and coastal zone (Fig. 5.6). Preservation of pre-LGM prehistoric archaeological sites in the Baltic Sea basin area is therefore possible, but due to extensive glacial erosion and redeposition the probability of finding such sites is very low.

Past relative sea- and lake-level changes determine the depth range and also broadly the areas in the Baltic where it would be logical to search for submerged prehistoric sites. Major Late Glacial water-level fluctuations are related to the first and the final drainages of the Baltic Ice Lake when the lake level dropped in the southern Baltic down to 40 m to 50 m BSL, which gives a lower limit for the search for anthropogenic sites (Fig. 5.12; Uścinowicz 2006; Bennike & Jensen 2013). In the northern part of the Baltic Sea, north of a line from Stockholm to Turku, i.e. in the areas of the Bothnian Sea and Bothnian Bay, the isostatic uplift caused permanent regression since the deglaciation and therefore all prehistoric sites are uplifted above present sea level. In the slowly uplifting areas in southern Sweden, Latvia, Estonia and northwest Russia, underwater prehistoric settlement sites could be preserved and therefore can be searched for from the periods of low water level before the Ancylus Lake and Littorina Sea transgressions around 11 ka BP and 9 ka BP, respectively.

The great majority of the known submerged prehistoric settlement sites in the Baltic are associated with the beginning of the Littorina Sea stage, when exploitation of marine resources began. This period also coincides with a time of rapid relative sea-level rise (Fig. 5.13) with better preservation potential of prehistoric sites. For periods of relatively rapid sea-level rise — as during the beginning of the Littorina transgression — paleolandscapes can be mapped by topographic surveys, preserving their former shape better than in periods of slow sea-level rise (Late Holocene), when locally more stable and prolonged hydrographic forces intensely reworked the substrate of the coastal zone, eventually destroying many archaeological sites (Harff & Hünicke 2011).

The high-energy conditions on open coasts may easily destroy archaeological sites in the coastal and nearshore zones down to depths somewhat greater than the closure depth even though the beach profile may not appear to change. The most important weather variables that influence the dynamics of the water masses of the microtidal Baltic Sea are atmospheric pressure and wind. Wind leads to the formation of wind waves and currents and, together with atmospheric pressure variations, storm surges and wave set-up, is the major driver of mixing of the upper layers of the sea. In the coastal zone, wind induces coastal currents, and upwelling and downwelling. In the Baltic Sea the wind is characterized by high variability, both in terms of velocity and direction. Over the Baltic Proper, winds from the south-west and west dominate all year round and constitute 35% to 40% of winds, depending on the season (Mietus 1998). Although the majority of storm winds (>15 m/sec) come from the south-west and the west, strong NNW winds occur (at least in the northern Baltic Proper) with an appreciable frequency (Soomere & Keevallik 2001). Winds from the north-west and the south are also relatively common over the Baltic Proper and constitute respectively 7% to 15% and 7% to 14% of all winds. In the north, over the Bothnian Sea and Bothnian Bay, winds from the south and south-west are most frequent (30–55%). Winds from the north and north-east are also common (20–30%; Defant 1972).

Currents and storm waves are the most important agents for erosion of the coast and seabed as well as transport of sediments, and near-bottom currents with velocities exceeding 1 m/sec can occur. During strong storms, waves with significant heights of up to 9.5 m (single waves exceptionally over 15 m) and average periods exceeding 10 seconds may occur in the Baltic Sea (Schmager *et al.* 2008; Tuomi *et al.* 2011). The largest significant wave heights observed in different parts of the Baltic Sea range from 4.46 m (Soomere *et al.* 2012) at a relatively sheltered location at the Darss sill, up to 7.4 m (Pettersson *et al.* 2010) in the southern Baltic Proper and 8.2 m (Tuomi *et al.* 2011) in the northern Baltic Proper. The periods can be 10 seconds to 13 seconds in length (Tuomi *et al.* 2011; Schmager *et al.* 2008). Strong winds and low pressure may lead to high water levels, which can — together with large waves — cause large

and even catastrophic damage to the shore. Maximum sea levels in the southwestern part of the Baltic Proper have exceeded the zero reference level by 2.83 m in Warnemünde and 2.11 m in Kołobrzeg (Richter *et al.* 2011).

Numerical simulations of wave conditions for both idealized ice-free situations (1970–2007: Soomere & Räämet 2011) and realistic seasonally ice-covered situations (2001–2007: Tuomi *et al.* 2011) have shown that the distribution of the mean and maximum values of the significant wave heights has clearly evident east–west asymmetry that reflects a similar asymmetry of the predominant directions of strong winds (Figs. 5.20 & 5.21). The highest simulated extreme waves occur in the areas with the highest overall wave intensity (Fig. 5.21):

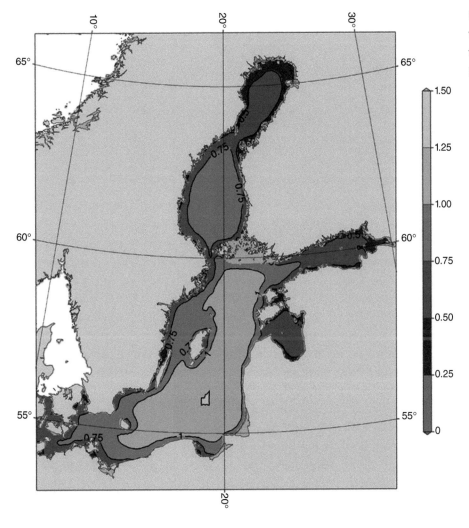

Fig. 5.20 Numerically simulated mean values of significant wave height (m) in the Baltic Sea. Tuomi *et al.* (2011). Reproduced with permission.

Fig. 5.21 Numerically simulated maximum values of significant wave height (m) showing areas with highest extreme waves. Tuomi *et al.* (2011). Reproduced with permission.

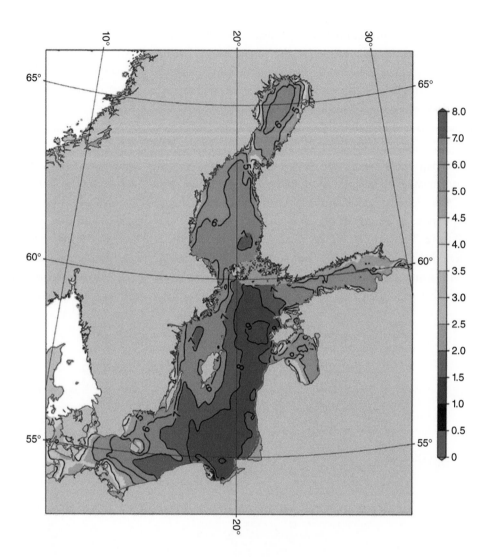

to the south of Gotland and in the northeastern and southeastern Baltic Proper. The Danish waters and the westernmost part of the southern Baltic coast in the German and Pomeranian bights and western part of Gdańsk Bay have clearly less wave impact in terms of both mean and maximum wave heights (Figs. 5.20 & 5.21). These high wave loads exerted over millennia to the southeastern and eastern Baltic Sea coasts have led to extensive changes in the coastline, including substantial straightening of several sections, erosion and re-deposition of remarkable masses of sediment and formation of massive spits (such as the Curonian or Hel Spit), barriers and coastal lakes.

The magnitude of the impact of waves on nearshore and coastal sediments depends on a multitude of factors.

This impact extends from the dune toe (during extreme water levels) down to so-called closure depth (Kraus 1992; Schwarzer *et al.* 2003). This depth depends on both wave height and period, and increases rapidly if the wave periods increase. While the dependence on the wave height of the depth down to which waves regularly and substantially impact the seabed is roughly linear, this dependence on wave periods is highly nonlinear. For example, the maximum near-bottom orbital velocity in 1-m-high waves at a depth of 3 m decreases from about 0.7 m/sec down to about 0.15 m/sec when the wave period decreases from 4 to 2 seconds. For this reason the long-period wakes of contemporary high-power vessels may create dangers for otherwise sheltered underwater archaeological sites (e.g. Soomere 2005).

In limited fetch conditions, the wave height is roughly proportional to the wind velocity, the square root of the distance over which the wind can blow creating waves on the sea (the fetch), and the duration of time during which the wind blows (e.g. Holthuijsen 2007). For any given fetch and wind velocity there is a maximum wave height and associated period which is achieved after a certain time. In the inner Danish waters the fetch from shore to shore is often less than 10 km to 15 km, and often in the range of only 1 km to 5 km. The coast is irregular, with headlands, spits, bays, inlets, and small islands, so that the fetch is sometimes even less than 1 km. This means that the waves cannot be large or long, which provides maximum protection against erosion of archaeological sites. In addition, the seabed gradients are often low and the basins between the islands shallow, so that wave refraction and bottom friction reduce the wave energy that reaches the shallow water close to the coast.

The coastline of Denmark is over 7000 km long, of which ca. 2000 km is on the North Sea side, and the rest on the Baltic side. For almost all that 5000 km length, the normal fetch to the shore is in the range of 10 km to 20 km, and often as low as 1 km to 5 km. If the wind is blowing through a channel, such as the Öresund, the fetch may be 100 km to 150 km, but only if the wind direction is exactly aligned along the length of the strait. Even in this case it is common for a large part of the wave energy to be gradually redirected towards the coasts via depth-induced refraction and internal directional spreading in the wave field, and the effective fetch is much shorter for most of the coast. A wind of 50 knots blowing across a fetch of 10 km would take about 1 hour to generate a wave with 2 m significant wave height, after which the waves would not get any bigger. The predominant wave period in such conditions would only be about 4 seconds, that is about one-third of a long-lasting storm of the same strength in the open part of the Baltic Sea. The favorable combination of fetch-limited wave heights and particularly wave periods is the main physical reason why the impact of waves upon the nearshore seabed within the Danish archipelago is so low, and the erosion rate of archaeological sites located at even quite shallow depths is very low.

Conclusion and Recommendations

The Baltic Sea, one of the largest brackish-water inland seas in the world, has undergone a complex and unique development after the last deglaciation and serves as a natural laboratory to study interactions between climate, sea-level change and prehistoric human adaptions. During the Pleistocene, ice advanced several times into the area of the present Baltic Sea, eroding the bedrock and sediments and deepening the Baltic Sea basin. The preservation of pre-LGM prehistoric archaeological sites in the Baltic Sea basin area is therefore possible, but due to the extensive glacial erosion and re-deposition, the probability of finding such sites is very low. From the perspective of submerged landscape taphonomy, the key characteristics of the study area are deglaciation and shore displacement history, past climate and the modern physical conditions.

During deglaciation the recession of the Scandinavian ice sheet changed the mass load on the continent and in nearby oceans, including the Baltic Sea, and resulted in isostatic adjustments throughout the region. In the northern Baltic Sea, in the areas of the Bothnian Sea and Bothnian Bay, uplift has exceeded the eustatic water-level rise throughout the period since deglaciation. Marine/lake regression has therefore occurred, leading to the emergence of the sea floor and its transformation into a terrestrial landscape, a process that still continues today. Areas of southern Denmark, and the coasts of Germany, Poland and Lithuania, were affected by marine/lake transgression during the full Holocene. The central Baltic Sea coast, from Scania to Stockholm in Sweden and the coasts of Latvia, Estonia and southern coasts of Finland, had a more complex Holocene shore-displacement history. Coastal transgressions took place in connection with the Ancylus Lake updamming, when rapid eustatic sea-level rise exceeded the local rate of uplift at the beginning of the Littorina Sea stage. Coincident with the slowing of eustatic rise towards the end of the Atlantic Period, uplift processes started to prevail, leading to coastal regression. Due to the isostatic and eustatic processes and the periodic updamming and drainage of the Baltic basin, the shore displacement development is rather complicated and requires high-resolution relative shore-level curves to be reconstructed. Recent discoveries of submerged tree stumps off the Swedish, Lithuanian, and Polish coasts provide the opportunity for more precise relative shore-level and paleogeographic reconstructions.

Large changes in the Baltic Sea ecosystem took place in the Mid Holocene when the freshwater Ancylus Lake drained into the ocean allowing saline water to enter the Baltic. Around 8.5 ka cal BP, the Baltic Sea became a brackish-water basin with significantly increased primary production, establishing favorable conditions for human exploitation of marine resources and the development of coastal settlements. Transformation of the subsistence strategy of prehistoric humans is reflected in the system of settlements which were located close to the Littorina shorelines and can be used as sea-level index points. Data from submerged archaeological sites from the Baltic Sea and from the coastal zone, and data on relative sea-level changes that provide detailed information about paleoshoreline positions, can be used to model the development of the submerged landscapes and to predict settlement site locations. Predictive modeling will play an increasing role in both the research and management of the offshore cultural heritage. In the case of the Baltic Sea it requires high-resolution reconstructions of the ice sheet, the ice margin, the periglacial climate, together with accurate sea-level curves, GIA models, and paleovegetation data. The modern coast configuration, especially within the Danish archipelago, exhibits areas where the wind fetch is limited to a few tens of kilometers or even less, so that the significant wave height is limited and erosion reduced. These conditions would also have applied in the past during marine transgressions, reducing erosion during inundation. Somewhat similar conditions may have existed in the central North Sea during the main phases of marine transgression.

Acknowledgments

Editor Dr Nicholas C. Flemming (University of Southampton) provided many useful recommendations and improvements of the entire chapter. Prof. Tarmo Soomere (Estonian Academy of Sciences) provided useful information and suggestions about wave dynamics in the Baltic Sea. Merle Muru (University of Tartu) helped with the preparation of some of the figures. The authors express their thanks to these people for their help in improving this chapter. We thank also our colleagues who provided their figures for the current chapter.

References

Amantov, A., Fjeldskaar, W. & Cathles, W. 2011. Glacial Erosion/Sedimentation of the Baltic Region and the Effect on the Postglacial Uplift. In Harff, J., Björck, S. & Hoth, P. (eds.) *The Baltic Sea Basin.* pp. 53-71. Springer-Verlag: Berlin-Heidelberg.

Andersson, G. 1902. Hasseln i Sverige fordom och nu. *Sveriges Geologiska Undersökning Ser.* C (3):1-168.

Andrén, E., Andrén, T. & Sohlenius, G. 2000. The Holocene history of the southwestern Baltic Sea as reflected in a sediment core from the Bornholm Basin. *Boreas* 29:233-250.

Andrén, T., Björck, S., Andrén, E., Conley, D., Zillén, L. & Anjar, J. 2011. The Development of the Baltic Sea Basin During the Last 130 ka. In Harff, J., Björck, S. & Hoth, P. (eds.) *The Baltic Sea Basin.* pp. 75-97. Springer-Verlag: Berlin-Heidelberg.

Anjar, J., Adrielsson, L., Bennike, O. *et al.* 2012. Palaeoenvironments in the southern Baltic Sea during Marine Isotope Stage 3: a multiproxy reconstruction. *Quaternary Science Reviews* 34:81-92.

Bennike, O. & Jensen, J. B. 2011. Postglacial relative shore level changes in Lillebælt, Denmark. *Geological Survey of Denmark and Greenland Bulletin* 23:37-40.

Bennike, O. & Jensen, J. B. 2013. A Baltic Ice Lake lowstand of latest Allerød age in the Arkona Basin, southern Baltic Sea. *Geological Survey of Denmark and Greenland Bulletin* 28:17-20.

Bennike, O. & Wagner, B. 2010. Amino acid analysis of pre-Holocene foraminifera from Kriegers Flak in the Baltic Sea. *Geological Survey of Denmark and Greenland Bulletin* 20:35-38.

Bennike, O., Jensen, J. B. & Lemke, W. 2001: Late Quaternary records of *Najas* spp. (Najadaceae) from the southwestern Baltic region. *Review of Palaeobotany and Palynology* 114:259-267.

Bennike, O., Sarmaja-Korjonen, K. & Seppänen, A. 2004a. Reinvestigation of the classic late-glacial Bølling Sø sequence, Denmark: chronology, macrofossils, Cladocera and chydorid ephippia. *Journal of Quaternary Science* 19:465-478.

Bennike, O., Jensen, J. B., Lemke, W., Kuijpers, A. & Lomholt, S. 2004b. Late- and postglacial history of the Great Belt, Denmark. *Boreas* 33:18-33.

Bennike, O., Andreasen, M. S., Jensen, J. B., Moros, M. & Noe-Nyegaard, N. 2012. Early Holocene sea-level changes in Øresund, southern Scandinavia. *Geological Survey of Denmark and Greenland Bulletin* 26:29-32.

Berger, A. & Loutre, M. F. 1991. Insolation values for the climate of the last 10 million years. *Quaternary Science Reviews* 10:297-317.

Berglund, M. 2004. Holocene shore displacement and chronology in Ångermanland, eastern Sweden, the Scandinavian glacio-isostatic uplift centre. *Boreas* 33:48-60.

Berglund, B. E., Björck, S., Lemdahl, G., Bergsten, H., Nordberg K. & Kolstrup, E. 1994. Late Weichselian environmental change in southern Sweden and Denmark. *Journal of Quaternary Science* 9:127-132.

Berglund, B. E., Sandgren, P., Barnekow, L. *et al.* 2005. Early Holocene history of the Baltic Sea, as reflected in coastal sediments in Blekinge, southeastern Sweden. *Quaternary International* 130:111-139.

Birks H. J. B. 1986. Late Quaternary biotic changes in terrestrial and lacustrine environments, with particular reference to north-west Europe. In Berglund, B. E. (ed.) *Handbook of Holocene Palaeoecology and Palaeohydrology.* pp. 3-65. Wiley & Sons Ltd: Chichester & New York.

Björck, S. 1995. A review of the history of the Baltic Sea, 13.0-8.0 ka BP. *Quaternary International* 27:19-40.

Björck, S., Walker, M. J. C., Cwynar, L. C. *et al.* 1998. An event stratigraphy for the Last Termination in the North Atlantic region based on the Greenland ice-core record: a proposal by the INTIMATE group. *Journal of Quaternary Research* 13:283-292.

Björck, S., Andrén, T. & Jensen, J. B. 2008. An attempt to resolve the partly conflicting data and ideas on the Ancylus-Littorina transition. *Polish Geological Institute Special Papers* 23:21-26.

Cato, I. & Kjellin, B. 1994. Quaternary deposits on the sea floor. In Fredén, C. (ed.) *National Atlas of Sweden — Geology.* pp. 150-153. Sveriges Nationalatlas Förlag: Stockholm.

Celsius, A. 1743. Anmärkning om vatnets förminskande så i Östersiön som Vesterhafvet. Kongl. *Swenska Wetenskaps Academiens Handlingar* 4:33-50.

Coope, G. R., Lemdahl, G., Lowe, J. J. & Walkling, A. 1998. Temperature gradients in northern Europe during the Last Glacial-Holocene transition (14-9 ^{14}C kyr BP) interpreted from coleopteran assemblages. *Journal of Quaternary Science* 13:419-433.

Damusyte, A., Bitinas, A., Kiseliene, D., Mapeika, J., Petrodius, R. & Pulkus, V. 2004. The tree stumps in the south eastern Baltic as indicators of Holocene water level fluctuations. In IGC-IUGS-UNESCO (ed.) *32nd International Geological Congress (Abstracts II).* 20th – 27th August 2004, Florence, p. 1167.

Defant, F. 1972. Klima und Wetter der Ostsee. *Kieler Meeresforschungen* 28:1-30.

Donner, J. J. 1995. *The Quaternary History of Scandinavia.* Cambridge University Press: Cambridge.

Douglas, B. C. & Peltier, W. R. 2002. The puzzle of global sea-level rise. *Physics Today* 55:35-40.

Ekman, M. 1996. A consistent map of the postglacial uplift of Fennoscandia. *Terra Nova* 8:158-165.

Ekman, M. 2009. *The Changing Sea Level of the Baltic Sea During 300 Years: A Clue to Understanding the Earth.* Summer Institute for Historical Geophysics: Åland Islands (Sweden).

Eronen, M., Glückert, G., Hatakka, L., van de Plassche, O., van der Plicht, J. & Rantala, P. 2001. Rates of Holocene isostatic uplift and relative sea-level lowering of the Baltic in SW Finland based on studies of isolation contacts. *Boreas* 30:17-30.

Fischer, A. 1995. An entrance to the Mesolithic world below the ocean. Status of ten years' work on the Danish sea floor. In Fischer, A. (ed.) *Man and Sea in the Mesolithic.* pp. 371-384. Oxford Books: Oxford.

Fischer, A. 2011. Stone Age on the Continental Shelf: an eroding resource. In Benjamin, J., Bonsall, C., Pickard,

C. & Fischer, A. (eds.) *Submerged Prehistory.* pp. 298-310. Oxbow Books: Oxford.

Gaillard, M-J & Lemdahl, G. 1994. Early-Holocene coastal environments and climate in southeast Sweden: a reconstruction based on macrofossils from submarine deposits. *The Holocene* 4:53-68.

Goslar, T., Kuc, T., Ralska-Jasiewiczowa M. *et al.* 1993. High-resolution lacustrine record of the Late Glacial/Holocene transition in central Europe. *Quaternary Science Reviews* 12:287-294.

Hansson, A., Nilsson, B., Sjöström, A. *et al.* (in press). A submerged Mesolithic lagoonal landscape in the Baltic Sea, south-eastern Sweden — Early Holocene environmental reconstruction and shore-level displacement based on a multiproxy approach. *Quaternary International.* dx.doi.org/10.1016/j.quaint.2016.07.059

Harff, J. & Hünicke, B. 2011. Global sea level rise and changing erosion. In *UNESCO Scientific Colloquium on factors impacting the Underwater Cultural Heritage.* 13th December 2011, Leuven. Royal Library of Belgium: Brussels.

Harff, J. & Lüth, F. (eds.) 2007. *SINCOS — Sinking Coasts: Geosphere, Ecosphere and Anthroposphere of the Holocene Southern Baltic Sea.* Bericht der Römisch-Germanische Kommission: Special Issue 88.

Harff, J. & Meyer, M. 2011. Coastlines of the Baltic Sea — zones of competition between geological process and a changing climate: examples from the southern Baltic. In Harff, J., Björck, S. & Hoth, P. (eds.) *The Baltic Sea Basin.* pp. 149-164. Springer-Verlag: Berlin-Heidelberg.

Harff, J., Lampe, R., Lemke, W. *et al.* 2005. The Baltic Sea — A model ocean to study interrelations of geosphere, ecosphere, and anthroposphere in the coastal zone. *Journal of Coastal Research* 21:441-446.

Harff, J., Björck, S. & Hoth, P. (eds.) 2011. *The Baltic Sea Basin.* Springer-Verlag: Berlin-Heidelberg.

Heiri, O., Brooks, S. J., Renssen, H. *et al.* 2014: Validation of climate model-inferred regional temperature change for Late-Glacial Europe. *Nature Communications* 5:4914.

Hille, S., Leipe, T. & Seifert, T. 2006. Spatial variability of recent sedimentation rates in the Eastern Gotland Basin (Baltic Sea). *Oceanologia* 48:297-317.

Holthuijsen, L. H. 2007. *Waves in Oceanic and Coastal waters.* Cambridge University Press: Cambridge.

Hörnberg, G., Bohlin, E., Hellberg, E. *et al.* 2005. Effects of Mesolithic hunter-gatherers on local vegetation in a non-uniform glacio-isostatic land uplift area, northern Sweden. *Vegetation History and Archaeobotany* 15:13-26.

Houmark-Nielsen, M. 2010. Extent, age and dynamics of Marine Isotope Stage 3 glaciations in the southwestern Baltic Basin. *Boreas* 39:343-359.

Houmark-Nielsen, M. & Kjær, K. H. 2003. Southwest Scandinavia, 40–15 kyr BP: palaeogeography and environmental change. *Journal of Quaternary Science* 18:769-786.

Huntley, B. & Birks, H. J. B. 1983. *An Atlas of Past and Present Pollen Maps for Europe: 0-13,000 years ago.* Cambridge University Press: Cambridge.

Iversen, J. 1944. Viscum, Hedera and Ilex as climate indicators. A contribution to the study of the Post-Glacial temperate climate. *Geologiska Föreningens i Stockholm Förhandlingar* 66:463-483.

Jakobsson, M., Björck, S, Alm, G., Andrén, T., Lindeberg, G. & Svensson, N-O. 2007. Reconstructing the Younger Dryas ice dammed lake in the Baltic Basin: Bathymetry, area and volume. *Global and Planetary Change* 57:355-370.

Jöns, H. 2011. Settlement Development in the Shadow of Coastal Changes — Case Studies from the Baltic Rim. In Harff, J., Björck, S. & Hoth, P. (eds.) 2011. *The Baltic Sea Basin.* pp. 301-336. Springer-Verlag: Berlin-Heidelberg.

Jöns, H., Mennenga, M. & Schaap, D. 2016. The SPLASHCOS-Viewer. *A European Information system about submerged prehistoric sites on the continental shelf. A product of the COST Action TD0902 SPLASHCOS (Submerged Prehistoric Archaeology and Landscapes of the Continental Shelf).* Accessible at splashcos-viewer.eu/

Kakkuri, J. 1997. Postglacial deformation of the Fennoscandian crust. *Geophysica* 33:99-109.

Kotilainen, A. T. & Kaskela, A. M. 2011. Geological modelling of the Baltic Sea and marine landscapes. *Geological Survey of Finland (Special Paper 49)* pp. 293-303.

Kraus N. C. 1992. Engineering approaches to cross-shore sediment processes. In Lamberti, A. (ed.) *Proceedings* of Short Course on Design and Reliability of Coastal Structures (23rd International Conference on Coastal Engineering). 4th – 9th October 1992, Venice, pp. 175-209.

Lambeck, K., Smither, C. & Ekman, M. 1998. Tests of glacial rebound models for Fennoscandinavia based on instrumented sea- and lake-level records. *Geophysical Journal International* 135:375-387.

Lambeck, K., Purcell, A., Zhao, J. & Svensson, N.-O. 2010. The Scandinavian Ice Sheet: from MIS 4 to the end of the Last Glacial Maximum. *Boreas* 39:410-435.

Lampe, R., Naumann, M., Meyer, H., Janke, W. & Ziekur, R. 2011. Holocene Evolution of the Southern Baltic Sea Coast and Interplay of Sea-Level Variation, Isostasy, Accommodation and Sediment Supply. In Harff, J., Björck, S. & Hoth, P. (eds.) *The Baltic Sea Basin*. pp. 233-251. Springer-Verlag: Berlin-Heidelberg.

Łęczyński, L., Miotk-Szpiganowicz, G., Zachowicz, J., Uścinowicz, S. & Krąpiec, M. 2007. Tree stumps from the bottom of the Vistula Lagoon as indicators of water level changes in the Southern Baltic during the Late Holocene. *Oceanologia* 49:245-257.

Lemke, W., Jensen, J. B., Bennike, O., Witkowski, A. & Kuijpers, A. 1999. No indication of a deeply incised Dana River between Arkona Basin and Mecklenburg Bay. *Baltica* 12:66-70.

Lemke, W., Jensen, J. B., Bennike, O. Endler, R., Witkowski, A. & Kuijpers, A. 2001. Hydrographic thresholds in the western Baltic Sea: Late Quaternary geology and the Dana River concept. *Marine Geology* 176:191-201.

Leppäranta, M. & Myrberg, K. 2009. *Physical Oceanography of the Baltic Sea*. Springer: Berlin.

Lidberg, M., Johansson, J. M., Scherneck, H.-G. & Davis, J. L. 2007. An improved and extended GPS-derived 3D velocity field of the glacial isostatic adjustment (GIA) in Fennoscandia. *Journal of Geodesy* 81:213-230.

Lilover, M.-J., Pavelson, J. & Kõuts, T. 2014. On the nature of low-frequency currents over a shallow area of the southern coast of the Gulf of Finland. *Journal of Marine Systems* 129:66-75.

Lindén, M., Möller, P., Björck, S. & Sandgren, P. 2006. Holocene shore displacement and deglaciation chronology in Norrbotten, Sweden. *Boreas* 35:1-22.

Litt, T., Brauer, A., Goslar, T. *et al.* 2001. Correlation and synchronisation of Lateglacial continental sequences in northern central Europe based on annually laminated lacustrine sediments. *Quaternary Science Reviews* 20:1233-1249.

Lübke, H., Schmölcke, U. & Tauber, F. 2011. Mesolithic Hunter-Fishers in a Changing World: a case study of submerged sites on the Jäckelberg, Wismar Bay, northeastern Germany. in Benjamin, J., Bonsall, C., Pickard, C. & Fischer, A. (eds.) *Submerged Prehistory*. pp. 21-37. Oxbow Books: Oxford.

Lundqvist, T. & Bygghammar, B. 1994. The Swedish Precambrian. In Fredén, C. (ed.) *National Atlas of Sweden — Geology*. pp. 16-21. Geological Survey of Sweden: Uppsala.

Lyell, C. 1835. *Principles of Geology*. John Murray: London.

Mangerud, J., Andersen, S. T., Berglund, B. E. & Donner, J. J. 1974. Quaternary stratigraphy of Norden, a proposal for terminology and classification. *Boreas* 3:109-126.

Markov, K. K. 1931. *Razvitiye reliefa severo-zapada Leningradskoy oblasti* [Development of the relief in the northwestern part of the Leningrad District]. Moscow/Leningrad. (in Russian)

Mattila, J., Kankaanpää, H. & Ilus, E. 2006. Estimation of recent sediment accumulation rates in the Baltic Sea using artificial radionuclides ^{137}Cs and ^{239,240}Pu as time markers. *Boreal Environment Research* 11:95-107.

Medvedev, I. P., Rabinovich, A. B. & Kulikov, E. A. 2013. Tidal oscillations in the Baltic Sea. *Oceanology* 53:526-538.

Miettinen, A. 2004. Holocene sea-level changes and glacio-isostasy in the Gulf of Finland, Baltic Sea. *Quaternary International* 120:91-104.

Mietus, M. (co-ordinator) 1998. *The Climate of the Baltic Sea Basin (marine meteorology and related oceanographic activities)* Report No. 41. World Meteorological Organisation: Geneva.

Mortensen, M. F., Birks, H. H., Christensen, C. *et al.* 2011. Lateglacial vegetation development in

Denmark — New evidence based on macrofossils and pollen from Slotseng, a small-scale site in southern Jutland. *Quaternary Sciences Reviews* 30:2534-2550.

Mortensen, M. F., Henriksen, P. S. & Bennike, O. 2014. Living on the good soil: relationships between soils, vegetation and human settlement during the late Allerød period in Denmark. *Vegetation History and Archaeobotany* 23:195-205.

Pempkowiak, J. 1991. Enrichment factors of heavy metals in the Southern Baltic surface sediments dated with ^{210}Pb and ^{137}Cs. *Environment International* 17: 421-428.

Pettersson, H., Lindow, H. & Schrader, D. 2010. *Wave climate in the Baltic Sea 2009*. HELCOM Baltic Sea Environment Fact Sheets 2010. Available at: www.helcom.fi

Pokorski, J. & Modliński, Z. (eds.) 2007. *Geological Map of the Western and Central Part of the Baltic Depression Without Permian and Younger Formations* (scale 1:750000). Państwowy Instytut Geologiczny: Warsaw.

Ralska-Jasiewiczowa, M. 2004. Early Holocene — 10,000-7500 ^{14}C yr BP (ca. 11,500-8300 cal yr BP). In Ralska-Jasiewiczowa, M., Latalowa, M., Wasylikowa, K. *et al.* (eds.) 2004. *Late Glacial and Holocene History of vegetation in Poland Based on Isopollen Maps*. pp. 394-399. Polish Academy of Sciences: Krakow.

Rasmussen, S. O., Andersen, K. K., Svensson, A. M. *et al.* 2006. A new Greenland ice core chronology for the last glacial termination. *Journal of Geophysical Research - Atmospheres* 111:D06102.

Richter, A., Groh, A. & Dietrich, R. 2011. Geodetic observation of sea-level change and crustal deformation in the Baltic Sea region. *Physics and Chemistry of the Earth, Parts A/B/C* 53-54:43-53.

Rosentau, A., Veski, S., Kriiska, A. *et al.* 2011. Palaeogeographic model for the SW Estonian coastal zone of the Baltic Sea. In Harff, J., Björck, S. & Hoth, P. (eds.) *The Baltic Sea Basin*. pp. 165-188. Springer-Verlag: Berlin-Heidelberg.

Rosentau, A., Harff, J., Oja, T. & Meyer, M. 2012. Postglacial rebound and relative sea level changes in the Baltic Sea since the Litorina transgression. *Baltica* 25:113-120.

Rosentau, A., Muru, M., Kriiska, A. *et al.* 2013. Stone Age settlement and Holocene shore displacement in the Narva-Luga Klint Bay area, eastern Gulf of Finland. *Boreas* 42:912-931.

Rosentau, A., Muru, M., Gauk, M. *et al.* (in press). Sea-level change and flood risks at Estonian coastal zone. In Harff, J., Furmanczyk, K., von Storch, H. (eds.). *Coastline Changes of the Baltic Sea from South to East — Past and Future Projection. Coastal Research Library 19.* Springers International Publishing AG.

Schmager, G., Fröhle, P., Schrader, D., Weisse, R. & Müller-Navarra, S. 2008. Sea state and tides. In Feistel, R., Nausch, G. & Wasmund, N. (eds.) *State and Evolution of the Baltic Sea, 1952-2005: A Detailed 50-Year Survey of Meteorology and Climate, Physics, Chemistry, Biology, and Marine Environment.* pp. 143-198. John Wiley & Sons: New Jersey.

Schmölcke, U., Endtmann, E., Klooss, S. *et al.* 2006. Changes of sea level, landscape and culture: A review of the south-western Baltic area between 8800 and 4000 BC. *Palaeogeography, Palaeoclimatology, Palaeoecology* 240:423-438.

Schwarzer, K., Diesing, M., Larson, M., Niedermeyer R.-O, Schumacher, W. & Furmanczyk, K. 2003. Coastline evolution at different time scales — examples from the Pomeranian Bight, southern Baltic Sea. *Marine Geology* 194:79-101.

Seppä, H., Bjune, A. E., Telford, R. J., Birks, H. J. B. & Veski, S. 2009. Last nine thousand years of temperature variability in Northern Europe. *Climate of the Past* 5:523-535.

Seppä, H., Schurgers, G., Miller, P. A. *et al.* 2015. Tree tracking a warmer climate: the Holocene range shift of hazel (*Corylus avellana*) in northern Europe. *The Holocene* 25:53-63.

Sommer, R. S., Persson, A., Wieseke, N. & Fritz, U. 2007. Holocene recolonization and extinction of the pond turtle, (*Emys orbicularis* (L., 1758), in Europe. *Quaternary Science Reviews* 26:3099-3107.

Soomere, T. 2005. Fast ferry traffic as a qualitatively new forcing factor of environmental processes in non-tidal sea areas: A case study in Tallinn Bay, Baltic Sea. *Environmental Fluid Mechanics* 5:293-323.

Soomere, T. & Keevallik, S. 2001. Anisotropy of moderate and strong winds in the Baltic Proper. *Proceedings of the Estonian Academy of Sciences, Engineering* 7:35-49.

Soomere, T. & Räämet, A. 2011. Long-term spatial variations in the Baltic Sea wave fields. *Ocean Science* 7:141-150.

Soomere, T., Weisse, R. & Behrens, A. 2012. Wave climate in the Arkona Basin, the Baltic Sea. *Ocean Science* 8:287-300.

Svendsen, J. I., Alexanderson, H., Valery, I. A. *et al.* 2004. Late Quaternary ice sheet history of northern Eurasia. *Quaternary Science Reviews* 231229-1271.

Szczepańska T. & Uścinowicz, S. 1994. *Geochemical Atlas of the Southern Baltic.* Państwowy Instytut Geologiczny: Warsaw.

Tuomi, L., Kahma, K. K. & Pettersson, H. 2011. Wave hindcast statistics in the seasonally ice-covered Baltic Sea. *Boreal Environment Research* 16:451-472.

Uścinowicz, S. 2003. Relative sea level changes, glacio-isostatic rebound and shoreline displacement in the southern Baltic. *Polish Geological Institute Special Papers* (vol. 10).

Uścinowicz, S. 2006. A relative sea-level curve for the Polish southern Baltic Sea. *Quaternary International* 145-146:86-105.

Uścinowicz, S. 2014. The Baltic Sea continental shelf. In Chiocci, F. L & Chivas, A. R (eds.) *Continental Shelves of the World: Their Evolution During the Last Glacio-Eustatic Cycle (Geological Society Memoir No. 41).* pp. 69-89. Geological Society: London.

Uścinowicz, S., Miotk-Szpiganowicz, G., Krąpiec, M. *et al.* 2011. Drowned forests in the Gulf of Gdańsk (southern Baltic) as an indicator of the Holocene shoreline changes. In Harff, J., Björck, S. & Hoth, P. (eds.) *The Baltic Sea Basin.* pp. 219-231. Springer-Verlag: Berlin-Heidelberg.

Veski, S., Heinsalu, A., Klassen, V. *et al.* 2005. Early Holocene coastal settlements and palaeoenvironment on the shore of the Baltic Sea at Pärnu, southwestern Estonia. *Quaternary International* 130:75-85.

Walker, M. J. C. 1995. Climatic changes in Europe during the last glacial/interglacial transition. *Quaternary International* 28:63-76.

Walker M. J. C., Berkelhammer, M., Björck, S. *et al.* 2012. Formal subdivision of the Holocene Series/Epoch: a discussion paper by a Working Group of INTIMATE (Integration of ice-core, marine and terrestrial records) and the Subcommission on Quaternary Stratigraphy (International Commission on Stratigraphy). *Journal of Quaternary Science* 27:649-659.

Walkusz, J., Roman, S. & Pempkowiak, J. 1992. Contamination of the southern Baltic surface sediments with heavy metals. *Bulletin of the Sea Fisheries Institute* 125:33-37.

Winterhalter, B., Flodén, T., Ignatius, H., Axberg, S. & Niemistö, L. 1981. Geology of the Baltic Sea. In Voipio, A. (ed.) *The Baltic Sea.* pp. 1-121. Elsevier Scientific: Amsterdam.

Yakovlev, S.A. 1925. *Nanosy I relief goroda Leningrada I ego okrestnostey* [Sediments and relief of the city of Leningrad and its environs]. Nauchno-meliorativnyi Institut: Leningrad (in Russian).

Yu, S. -Y., Berglund, B. E., Sandgren, P. & Lambeck, K. 2007. Evidence for a rapid sea-level rise 7600 yr ago. *Geology* 35:891-894.

Zolitschka B., Haverkamp B. & Negendank J. F. W. 1992. Younger Dryas oscillation — varve dated microstratigraphic, palynological and palaeomagnetic records from Lake Holzmaar, Germany. In Bard, E. & Broecker, W. S. (eds.) *The Last Deglaciation: Absolute and Radiocarbon Chronologies* (vol. 2). pp. 81-101. Springer Verlag: Berlin.

Zulkus, V. & Girininkas, A. (eds.) 2012. *The Coasts of the Baltic Sea 10,000 Years Ago.* Klaipeda University: Lithuania (in Lithuanian with English summary).

Chapter 6
The Northwest Shelf

Kieran Westley

School of Geography and Environmental Sciences, Ulster University, Coleraine, Northern Ireland, UK

Introduction

The Northwest Shelf forms an extended plateau extending off continental Europe and surrounding the British Isles. Politically it encompasses the territorial waters of Norway, Denmark, Germany, the Netherlands, Belgium, France, the United Kingdom (covering the devolved administrations of England, Scotland, Wales and Northern Ireland) and the Republic of Ireland. Its geographic configuration means that it can be divided into several semi-enclosed sea basins: the North Sea, the northern North Sea and Atlantic Northwest Approaches, the Irish Sea and Atlantic Margin, and the English Channel/La Manche (Fig. 6.1). Each basin will be discussed individually in the four following chapters. This chapter provides an introductory overview covering the key forces influencing taphonomic processes on a shelf-wide scale.

Taphonomic Processes

Bathymetry is generally shallow, with the majority of the shelf less than 100 m in depth (Fig. 6.1). The largest expanse of shallow shelf is located in the southern North Sea where depths of <50 m are common. Smaller shallow (<–50 m) plateaux are also located in the eastern Irish Sea and on the flanks of the eastern English Channel. Islands are not common and tend to form isolated examples with the exception of archipelagos off western and northern Scotland and barrier chains in the eastern North Sea. The largest bathymetric deep is the Norwegian Trench which reaches a maximum depth of 700 m and separates the main body of the shelf from the Scandinavian Peninsula. Smaller localized deeps (down to ca. –100 m to –150 m) also run down the center of the English Channel and Irish Sea. Bathymetric highs are formed by banks and ridges, the largest of which is the Dogger Bank in the North Sea.

Submerged Landscapes of the European Continental Shelf: Quaternary Paleoenvironments, First Edition.
Edited by Nicholas C. Flemming, Jan Harff, Delminda Moura, Anthony Burgess and Geoffrey N. Bailey.
© 2017 John Wiley & Sons Ltd. Published 2017 by John Wiley & Sons Ltd.

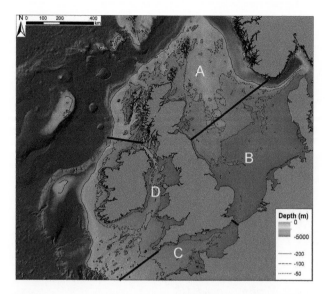

Fig. 6.1 Overview map of the Northwest Shelf showing bathymetry and subdivision into marginal seas. (A) northern North Sea and Atlantic Northwest Approaches; (B) North Sea; (C) English Channel; (D) Irish Sea and Atlantic Margin. Bathymetry has been derived from the EMODnet Hydrography portal (www.emodnet-hydrography.eu). Reproduced with permission.

There are also numerous smaller banks, for instance in the southwest Celtic Sea.

Seabed substrate varies across the Northwest Shelf with variations driven by a number of factors, not least modern oceanographic and coastal processes, and antecedent geology. Large zones of the North Sea consist of sand with isolated mud and gravel, the latter particularly prevalent off eastern England. By contrast, coarse and mixed sediment dominates the English Channel while the Irish Sea has a highly variable substrate including zones of sand, mixed sediment and mud. The Northwest Approaches are typified by sand, coarse sediment and exposed bedrock (broad-scale assessment based on EMODnet-Geology mapping — see 'Data Sources', page 141, for link).

The Northwest Shelf lies in the path of Atlantic low-pressure systems which drive winds mainly from the west and south. The strongest winds (and hence highest wave energy) tend to be experienced on the open Atlantic coasts with a large fetch, such as northwest Scotland and western Ireland. By contrast, lower wave energy is experienced within the Irish Sea and western North

Sea due to the sheltering effect of the British and Irish landmasses (Fig. 6.2a). Wind and wave energy is also seasonally variable and tends to be stronger in winter than summer (Neill *et al.* 2009).

The semi-enclosed basins, variable bathymetry and complex coastal configurations (e.g. the presence of many large bays and inlets) within the study area combined with the rotation of the oceanic tidal wave around amphidromic points within the basins results in variable tidal regimes (Fig. 6.2b). This is most noticeable when comparing opposing sides of the semi-enclosed basins. For instance, the western Irish Sea has average tides of <4 m, while on its eastern boundary, tidal amplification in the inlets of Liverpool Bay and the Bristol Channel, allows ranges up to 8 m and 14 m respectively. Similarly, in the English Channel the French coast has the megatidal (up to 14 m range) Gulf of St. Malo in contrast to the mainly mesotidal (<6 m) south coast of England. Tidal range within the North Sea tends to be smaller (generally not more than 4 m but with local exceptions) and with the smallest range on its eastern side (Davis & Fitzgerald 2004; Neill *et al.* 2010). The constricting effect of straits, variable bathymetry and semi-enclosed basins also affects the tidal current with generally strong currents in the Irish Sea and English Channel but reduced velocities in the more open North and Celtic Seas (Neill *et al.* 2010).

The primary influence on antecedent geology on the Northwest Shelf has been Quaternary climate change, specifically multiple glaciations and cycles of exposure and submergence created by accompanying sea-level change. Northwest Europe has been subject to multiple (potentially up to seven) episodes of glaciation over the past million years, though only the final ones — the Elsterian (MIS 12), Saalian (MIS 6) and Weichselian (MIS 2) — have been identified with confidence on the continental shelf (Böse *et al.* 2012)[1]. Glacial stages were characterized by ice sheets over Scandinavia, Britain and Ireland which, at their maximum size, extended

---

[1] For the purposes of this chapter, climatic intervals will be identified with their generally used northwest European stage names and/or their equivalent Marine Isotope Stages (MIS). For example: the Weichsalian rather than Devensian or Midlandian in the British and Irish terminology respectively

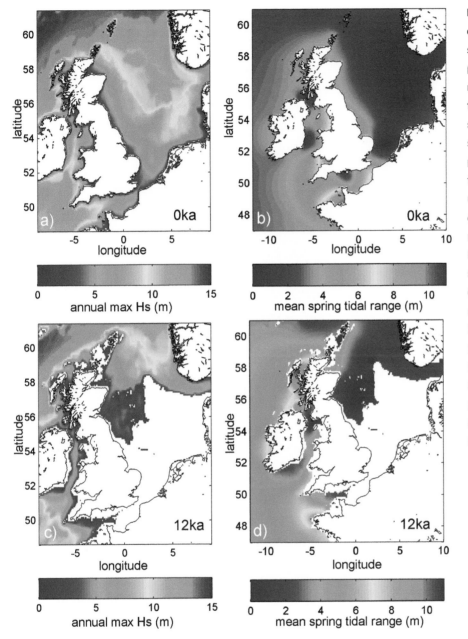

Fig. 6.2 Modeled oceanographic conditions on the Northwest Shelf showing the difference between present-day conditions and 12 ka when much of the North Sea and coastal fringes of the British Isles were subaerially exposed. (A) Maximum annual significant wave height (m) for the present day (from Neill *et al.* 2009: fig. 7); (B) Mean spring tidal range (m) under present-day conditions (from Neill *et al.* 2010: fig. 4. Reproduced with permission of Springer Science and Business Media.); (C) Maximum annual significant wave height (m) for 12 ka (from Neill *et al.* 2009: fig. 8); (D) Mean spring tidal range (m) for 12 ka (from Neill *et al.* 2010: fig. 4. Reproduced with permission of Springer Science and Business Media.). All images courtesy of Simon Neill; (A) and (C) reproduced with permission of John Wiley & Sons; (B) and (D) reproduced with permission of Springer.

onto the continental shelf. The accompanying low sea level exposed large expanses of shelf, though given the climatic conditions, these formed polar deserts or tundra environments which may not have been amenable to hominin occupation. However, long periods between glacial maxima and minima with lower-than-present sea levels, cool, but not totally inhospitable conditions and frequent short warm interstadials, meant that habitable landscapes were often present. With warming came

ice retreat, and the development of more habitable landscapes for flora, fauna and humans. Eventually, the sea-level rise accompanying climate warming would have drowned most of the exposed shelf environments, creating a similar geography to the present. Note that this is only a broad generalized overview. In all cases, paleolandscape developments were controlled on a local to regional level by timing of ice expansion and retreat, isostasy, relative sea-level change, oceanographic and

Fig. 6.3 Scenarios of ice-sheet growth/decay across the Northwest Shelf from the LGM onwards. Clark *et al.* (2012): fig. 17), with permission from Elsevier; image courtesy of Chris Clark.

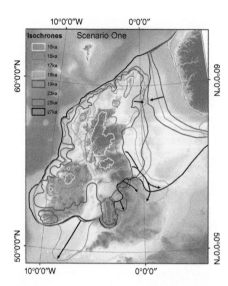 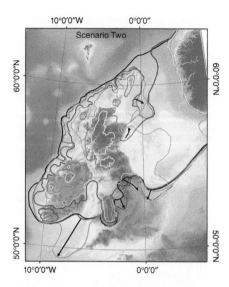

climatic conditions and sediment supply (see for example Cohen *et al.* 2012; 2014).

Paleoenvironmental Change

Reconstructing the precise timing and extent of glacial, sea-level, paleoenvironmental and paleogeographic changes for the Northwest Shelf remains a work-in-progress. Sufficient data now exist to allow time-stepped reconstructions of these changes since the end of the Last Glacial Maximum (LGM: ca. 24–21 ka)[2] and numerous examples exist in the published literature which illustrate these on a shelf scale (e.g. Peltier 1994; Lambeck 1995; Shennan *et al.* 2002; Carr *et al.* 2006; Brooks *et al.* 2008; 2011; Clark *et al.* 2012). The most recent published studies depict the British, Irish and Scandinavian ice sheets as confluent across the North Sea and Irish Sea by ca. 27 ka with an additional localized advance into the Celtic Sea between ca. 23 ka and 20 ka (Fig. 6.3). Under this scenario, the only parts of the Northwest Shelf escaping glaciation were the English Channel and southern North Sea (though this may have been covered by an ice-dammed lake) (Clark *et al.* 2012). Ice retreat began in earnest from ca. 19 ka with exposure of the Celtic Sea followed by the southern Irish Sea and central

North Sea. Retreat was asynchronous across the shelf and subject to localized re-advances such as a potential extension down the coast of eastern England around ca. 17 ka. Only from ca. 16 ka onward was the present shelf almost entirely ice-free, with glaciers limited to presently terrestrial highlands (Clark *et al.* 2012).

In terms of sea-level change, the LGM witnessed major shelf exposure only in the southern and central North Sea, English Channel and eastern Irish Sea (Peltier 1994; Lambeck 1995; Brooks *et al.* 2011). Much of the northern North Sea and Northwest Approaches were isostatically depressed, while exposure on the western Atlantic Margin and western Irish Sea was limited to a narrow fringe around Ireland (Brooks *et al.* 2011; Fig. 6.4). This overall picture remained broadly steady until ca. 15 ka when shelf flooding became increasingly rapid. The first major incursions occurred in the eastern Irish Sea and western English Channel, with major flooding of the North Sea and eastern English Channel between 13 ka and 10 ka. In the northern areas experiencing isostatic adjustment, relative sea level either rose or fell depending on the local history of ice loading. By ca. 8 ka to 6 ka, the modern geography of the Northwest Shelf was largely attained (Brooks *et al.* 2011; Sturt *et al.* 2013). Recent models also suggest that these changes were not only significant in geographical terms, but also exerted a strong influence on the development of the

---

[2] All dates presented are in calendar years before present.

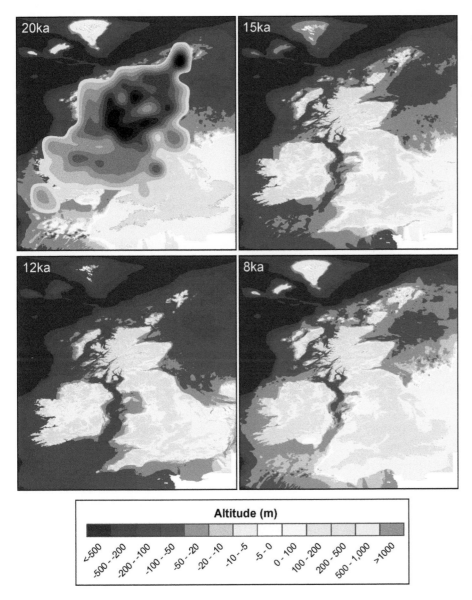

Fig. 6.4 GIA-modeled paleogeographic change at selected intervals since the LGM. Images courtesy of Tony Brooks; see Bradley *et al.* (2011); Brooks *et al.* (2011), for model description.

modern wave and tidal regime (e.g. Uehara *et al.* 2006; Neill *et al.* 2009; 2010; see also Fig. 6.2c,d).

Whilst the broad-scale patterns of change are reasonably well established, the details of ice-sheet chronology and local sea-level histories are still subject to a degree of uncertainty. Holocene patterns of relative sea-level change are generally well constrained by field data and are unlikely to undergo radical revision in the future. In contrast, relative sea level during deglaciation and the earliest Holocene is more poorly resolved due to limitations in the accuracy and availability of datable

evidence (e.g. Shennan & Horton 2002), much of which is currently submerged and/or buried under significant thicknesses of more recent sediment. Consequently, there is scope for future refinement of sea-level curves and associated paleogeographies during these earlier time intervals when changes were large and rapid.

Similarly, as the collection of additional field data further refines our view of ice-sheet growth and decay, some revision of existing ice-sheet models used to drive sea-level simulations will become necessary. As a consequence of the iterative nature of this process, a single, consensus view of ice-sheet history has yet to emerge and the existing patterns will continue to be modified in the coming years (e.g. compare scenarios in Fig. 6.3 and Clark *et al.* 2012). Fortunately, the long-wavelength response of the Earth to loading/unloading means that the ice-sheet models employed in sea-level simulation are of coarser resolution than the more detailed ice-sheet reconstructions derived from geomorphological/stratigraphic evidence (e.g. Clark *et al.* 2012). In practice, this means that fine-scale alterations, such as short-lived advance/retreat of thin ice, have minimal impacts on the resulting sea-level simulations, although they may be important for the interpretation of resulting paleogeographic reconstructions. Of much greater significance will be improvements in the accurate determination of ice-sheet thickness throughout the Last Glacial since this remains poorly constrained in many areas (Chiverrell & Thomas 2010). Since ice-sheet thickness is the important loading term in sea-level models, the ice sheets presented in modeling papers tend to show terrain-corrected ice-sheet thickness (e.g. Brooks *et al.* 2008; 2011). In contrast, geomorphologically-based reconstructions tend to present ice sheets in terms of their lateral extent and surface elevation (e.g. Ballantyne *et al.* 2011; Clark *et al.* 2012). These differences in emphasis and spatial resolution can give the false impression that large discrepancies exist between what are, in a practical sense at least, relatively similar ice-sheet histories.

Prior to the last glacial cycle, the level of uncertainty in interpretation increases due to a reduction in the quantity and resolution of data. This results from various factors including the erosive effect of the last glacial cycle, the deep burial or submergence of the relevant sediments and limitations in existing dating methods. Moreover, over these longer timescales, additional influences on sea-level and paleogeographic change came into play, notably the slow subsidence of the North Sea basin during the Middle Pleistocoene (Funnell 1995) and the breaching of the Weald-Artois Ridge spanning the Dover Strait by glacial melt water — currently dated to sometime between MIS 12 and MIS 6 (Gibbard 1995; Bridgland 2002; Toucanne *et al.* 2009) — both of which were instrumental in the transformation of the British Isles from a European peninsula to a group of islands. Thus, reconstructions for these earlier periods tend to be limited to snapshots, rather than time-stepped sequences, for example showing highstand/lowstand-only paleogeographies (e.g. Stringer 2006; Hijma *et al.* 2012) or maximum ice extents (e.g. Huuse & Lykke-Anderson 2000; Toucanne *et al.* 2009).

Evidence Base

Key factors in the archaeological importance of the Northwest Shelf are its size and shallowness. The existence of broad shallow areas means that vast tracts of land were exposed during sea-level lowstands, creating enormous potential for development of landscapes suitable for human occupation. Moreover, some of the earliest European hominin sites are found in eastern England (Pakefield: ca. 700 ka; Happisburgh: ca. 800–900 ka (Parfitt *et al.* 2005; 2010)) and therefore indicate that this landscape must have been colonized from a very early date. It is unlikely that it was occupied continuously for more than a few thousand years at a time because the aforementioned climate changes probably resulted in multiple episodes of abandonment and re-colonization, as attested to by the cyclical nature of British Paleolithic occupation (White & Schreve 2000; Stringer 2006; Pettit & White 2012). This in turn means that archaeological evidence from the continental shelf could potentially make a major contribution in addressing issues of human mobility, migration and dispersal, and responses or adaptations to climate change (Peeters *et al.* 2009; Bell *et al.* 2013; Westley *et al.* 2013). Importantly, from an archaeological and paleoenvironmental perspective, there is clear evidence of paleolandscape preservation in the form of submerged river channels, associated terrace deposits, peats, terrestrial/freshwater sediments and faunal remains from across the Northwest Shelf (e.g. Mol *et al.* 2006; Gaffney *et al.* 2007; Dix & Sturt 2011). Verifiable human occupation is also apparent in the form of lithic or organic artifacts recovered from the shelf, including fragments of a Neanderthal skull (e.g. Hublin *et al.* 2009; Momber *et al.* 2011; Peeters 2011; Moree & Sier 2014; Peeters & Momber 2014; Tizzard *et al.* 2011; 2014). The paleolandscape, paleoenvironmental and archaeological evidence covers periods from the Early–Mid Pleistocene to the Holocene, thus demonstrating that even pre-Last Glacial material can survive both glaciation and more than one episode of submergence.

The number and time span of the glacial and sea-level changes which have taken place mean that the shelf sedimentary and stratigraphic sequence is complex and fragmentary. Effective scientific research therefore requires a firm knowledge of seabed and sub-seabed conditions which must be obtained from large-scale marine geophysical and geotechnical surveys. Fortunately, the Northwest Shelf is one of the best studied in Europe owing to its intensive industrial use in the modern era; for instance, offshore oil and gas, industrial-level trawling, aggregates extraction, massive harbor expansion and most recently offshore renewable energy developments, are all activities which have taken place here. With these sources, and over a century of academic interest (e.g. Reid 1913), there is now a considerable body of data relevant to taphonomic, geological, paleoenvironmental and archaeological studies which the following chapters will synthesize.

Conclusion

The above description represents a broad overview of the Northwest Shelf and taphonomic processes, and masks a great deal of regional variability. Its size and geographical configuration mean that the impact of modern oceanographic processes (e.g. wind, waves, tides) varies spatially. Past taphonomic processes also varied across space and time, most noticeably in terms of glaciation and relative sea-level change. The upshot is that submerged landscape preservation will vary considerably across the Northwest Shelf on a regional to local level, a fact that will become clear through the following chapters dealing with the individual sea basins. Indeed, at present, extant paleolandscape evidence is concentrated in a few hotspots, such as the southern North Sea (see for example the recent overview in Peeters & Cohen 2014). Although this is at least partly reflective of the history of scientific investigation, it is highly likely that there is a strong taphonomic influence as well.

Before proceeding to the shelf-wide weblinks/data sources and individual sea basin chapters, the reader should be reminded that this is a field in which rapid developments are being made, for instance in ice-sheet reconstruction, sea-level modeling and paleoenvironmental and paleogeographic reconstruction (e.g. Bradley *et al.* 2011; Clark *et al.* 2012, Cohen *et al.* 2012; Hijma *et al.* 2012; Sturt *et al.* 2013). In some instances, these developments have proceeded at different speeds either geographically or by discipline. Given the wide areas that will be covered in the next chapters and the fact that a multidisciplinary subject like submerged landscape archaeology relies on different disciplines building on the work of one another, it is inevitable that inconsistencies will exist between interpretations since different researchers will be influenced by, or have access to, different levels or types of data. An attempt at consensus would be proven wrong before it was completed, and the reader should therefore bear this in mind through the rest of the chapters, if confronted with apparent contradictions in interpretation or reconstruction.

Data Sources

The following sources and links represent data which cross the boundaries of the individual sea basins. Basin-specific sources can be found within the relevant chapters.

Bathymetry

Two main publically available data sets cover the entire Northwest Shelf. The General Bathymetric Chart of the Oceans (GEBCO) global data set is relatively low resolution (30 arcseconds) but combines both regional-scale bathymetry and terrestrial topography. More recently, the European Marine and Data Observation Network (EMODnet) has made available bathymetric data for the European shelf at a resolution of 0.25 arcminutes (i.e. 15 arcseconds). EMODnet also provides metadata on the underlying higher-resolution local data sets used to create the downsampled continental-scale coverage. Commercially available higher-resolution products also cover the Northwest Shelf, notably the SeaZone Solutions Ltd (based on UK Hydrographic Office data) and Olex (based on accumulated soundings from Olex users) data sets.

EMODnet Hydrography Portal: www.emodnet-hydrography.eu/
GEBCO bathymetry: www.gebco.net/
Seazone Solutions Ltd: www.seazone.com
Olex: www.olex.no/index_e.html

Substrate/geology/geomorphology

At a European level, the EMODnet program has begun to amalgamate national data sets on the physical properties of the seabed. This has been heavily driven by biological habitat mapping and includes seabed geology and substrate data which are accessible via the EMODnet-Geology or OneGeology (Europe) data portals. Relevant datasets include a harmonized 1:1,000,000 surficial and bedrock geological map (for terrestrial areas), a 1:1,000,000 seabed substrate map, and a 1:5,000,000 seabed lithology map. An alternative project, again on a Europe-wide level, is the Geo-Seas project. This publishes and maintains data catalogues submitted by national providers (e.g. national geological surveys) and is therefore a good starting point for data on seabed geology. Locations of actual seabed sample sites (e.g. grabs, cores) are available via the EU-SEASED Portal. Finally, low-resolution shapefiles of coastal geology, geomorphology and patterns of coastal change are available from the EUROSION project, a Europe-level exercise of coastal erosion mapping.

OneGeology Europe: onegeology-europe.brgm.fr/geoportal/viewer.jsp
EMODnet-Geology portal: www.emodnet-geology.eu/
Geo-Seas portal: www.geo-seas.eu/
EU-SEASED portal: www.eu-seased.net/
EUROSION: www.eurosion.org/

Oceanographic processes

Information on oceanographic processes comes from two main sources. Firstly, there are actual observations of waves and tide at specific stations. An amalgamated network of European stations is identifiable via the EMODnet Physics portal. Secondly, there are computer-modeled wave and tidal regimes. These have the advantage that they can create shelf-scale outputs which extend beyond and between the accurate but spatially restricted station observations. The models themselves are sophisticated and generally restricted to specialist oceanographic, meteorological or commercial organizations which use them for a variety of projects including ocean forecasting or hindcasting (e.g. Saulter & Leonard-Williams 2011), marine habitat mapping (e.g. Cameron & Askew 2011) and offshore renewable energy research (ABPMer 2008). Examples of such organizations include the National Oceanography Centre - Liverpool (UK) and the UK Met Office.

EMOD Physics portal: www.emodnet-physics.eu/
UK Met Office: www.metoffice.gov.uk/
National Oceanography Centre, Liverpool: www.pol.ac.uk/

Quaternary paleoenvironments

The majority of information dedicated to Quaternary environmental change can be found within the published literature. Several portals provide access to the underlying data and tend to be organized by specialist theme. Paleoecological examples include the European Pollen Database (containing pollen records from across Europe), NEOTOMA (a global database which includes pollen, plant macros, mammals and mollusks) and the Bugs Coleopteran Ecology Package (BUGCEP), a downloadable software package and database for analysis of fossil beetle remains. Paleoceanographic data, principally from deep-sea cores, is accessible via the PANGAEA portal, while the NOAA Paleoclimatology portal has records from across the world, including northwest Europe, which cover a range of environmental proxies.

European Pollen database: www.europeanpollendatabase.net/
NEOTOMA: www.neotomadb.org/
Bugs Coleopteran Ecology Package: www.bugscep.com/
PANGAEA: www.pangaea.de/
NOAA Paleoclimatology: www.ncdc.noaa.gov/paleo/paleo.html

Acknowledgments

Dr. Robin Edwards (Trinity College, Dublin) is thanked for his useful observations of ice-sheet reconstruction and GIA modeling. Editor Nicholas C. Flemming is thanked for his guidance, support and patience throughout the manuscript production.

References

ABPMer 2008. *Atlas of UK Marine Renewable Energy Resources*. Technical Report prepared for the UK Department of Business, Enterprise and Regulatory Reform (BERR). Available at www.renewables-atlas.info/.

Ballantyne, C. K., McCarroll, D. & Stone, J. O. 2011. Periglacial trimlines and the extent of the Kerry-Cork Ice Cap, SW Ireland. *Quaternary Science Reviews* 30:3834-3845.

Bell, M., Warren, G., Cobb, H. *et al.* 2013. The Mesolithic. In Ransley, J. Sturt, F., Dix, J. Adams, J. & Blue, L. (eds.) *People and the Sea: a Maritime Archaeological Research Agenda for England*. CBA Research Report 171. pp. 30-49. Council for British Archaeology: York.

Böse, M., Luthgens, C., Lee, J. R. & Rose, J. 2012. Quaternary glaciations of northern Europe. *Quaternary Science Reviews* 44:1-25.

Bradley, S. L., Milne, G. A., Shennan, I. & Edwards, R. 2011. An improved glacial isostatic adjustment model for the British Isles. *Journal of Quaternary Science* 26:541-552.

Bridgland, D. R. 2002. Fluvial deposition on periodically emergent shelves in the Quaternary: example records from the shelf around Britain. *Quaternary International* 92:25-34.

Brooks, A. J., Bradley, S. L., Edwards, R. J., Milne, G. A., Horton, B. & Shennan, I. 2008. Postglacial relative sea-level observations from Ireland and their role in glacial rebound modelling. *Journal of Quaternary Science* 23:175-192.

Brooks, A. J., Bradley, S. L., Edwards, R. J. & Goodwyn, N. 2011. The palaeogeography of Northwest Europe during the last 20,000 years. *Journal of Maps* 7:573-587.

Cameron, A. & Askew, N. (eds.) 2011. *EUSeaMap — Preparatory Action for Development and Assessment of a European Broad-Scale Habitat Map Final Report*. Available at jncc.gov.uk/euseamap.

Carr, S. J., Holmes, R., van der Meer, J. M. & Rose, J. 2006. The Last Glacial Maximum in the North Sea Basin: micromorphological evidence of extensive glaciation. *Journal of Quaternary Science* 21:131-153.

Chivrrell, R. C. & Thomas, G. S. P. 2010. Extent and timing of the Last Glacial Maximim (LGM) in Britain and Ireland: a review. *Journal of Quaternary Science* 25:535-549.

Clark, C. D., Hughes, A. L. C., Greenwood, S. L., Jordan, C. & Sejrup, H. P. 2012. Pattern and timing of retreat of the last British-Irish Ice Sheet. *Quaternary Science Reviews* 44:112-146.

Cohen, K. M., MacDonald, K., Joordens, J. C. A, Roebroeks, W. & Gibbard, P. L. 2012. The earliest occupation of north-west Europe: a coastal perspective. *Quaternary International* 271:70-83

Cohen, K. M., Gibbard, P. L. & Weerts, H. J. T. 2014. North Sea palaeogeographical reconstructions for the last 1 Ma. *Netherlands Journal of Geosciences* 93:7-29.

Davis, R. A., & Fitzgerald, D. M. 2004. *Beaches and Coasts*. Blackwell Publishing: Oxford.

Dix, J. K. & Sturt, F. C. 2011. *The Relic Palaeo-landscapes of the Thames Estuary*. Report prepared for the Marine Aggregate Levy Sustainability Fund.

Funnell, B. M. 1995. Global sea-level and the (pen-) insularity of late Cenozoic Britain. In Preece, R. C. (ed.) *Island Britain: a Quaternary Perspective*. pp. 3-13. Geological Society: London.

Gaffney V., Thomson, K. & Fitch, S. (eds.) 2007. *Mapping Doggerland: The Mesolithic Landscapes of the Southern North Sea*. Archaeopress: Oxford.

Gibbard, P. L. 1995. The formation of the Strait of Dover. In Preece, R. C. (ed.) *Island Britain: a Quaternary Perspective*. pp. 15-26. Geological Society: London.

Hijma, M. P., Cohen, K. M., Roebroeks, W., Westerhoff, W. E. & Busschers, F. S. 2012. Pleistocene Rhine-Thames landscapes: geological background for hominin occupation of the southern North Sea region. *Journal of Quaternary Science* 27:17-39.

Hublin, J.-J., Weston, D., Gunz, P. *et al.* 2009. Out of the North Sea: the Zeeland Ridges Neandertal. *Journal of Human Evolution* 57:777-785.

Huuse, M. & Lykke-Andersen, H. 2000. Overdeepened Quaternary valleys in the eastern Danish North Sea: morphology and origin. *Quaternary Science Reviews* 19:1233-1253.

Lambeck, K. 1995. Late Devensian and Holocene shore-lines of the British Isles and the North Sea from models of glacio-hydro-isostatic rebound. *Journal of the Geological Society* 152:437-448.

Mol, D., Post, K., Reumer, J. W. F. *et al.* 2006. The Eurogeul — first report of the palaeontological, palynological and archaeological investigations of this part of the North Sea. *Quaternary International* 142-143:178-185.

Momber, G., Tomalin, D., Scaife, R., Satchell, J. & Gillespie, J. (eds.) 2011. *Mesolithic occupation at Bouldnor Cliff and the Submerged Prehistoric Landscapes of the Solent.* CBA Research Report 164. Council for British Archaeology: York.

Moree, J. M. & Sier, M. M. (eds.) 2014. *Twintig meter diep! Mesolithicum in de Yangtzehaven — Maasvlakte te Rotterdam. Landschapsontwikkeling en bewoning in het Vroeg Holoceen. BOORrapporten 523.* Bureau Oudheidkundig Onderzoek Rotterdam (BOOR): Rotterdam.

Neill, S. P., Scourse, J. D., Bigg, G. R. & Uehara, K. 2009. Changes in wave climate over the northwest European shelf seas during the last 12,000 years. *Journal of Geophysical Research* 114:C06015.

Neill, S. P., Scourse, J. D. & Uehara, K. 2010. Evolution of bed shear stress distribution over the northwest European shelf during the last 12,000 years. *Ocean Dynamics* 60:1139-1156.

Parfitt, S. A., Barendregt, R. W., Breda, M., *et al.* 2005. The earliest record of human activity in northern Europe. *Nature* 438:1008-1012.

Parfitt, S. A., Ashton, N. M., Lewis, S. G., *et al.* 2010. Early Pleistocene human occupation at the edge of the boreal zone in northwest Europe. *Nature* 466:229-233.

Peeters, H. 2011. How wet can it get? — approaches to submerged prehistoric sites on the Dutch continental shelf. In Benjamin, J., Bonsall, C., Pickard, C. & Fischer, A. (eds.) *Submerged Prehistory.* pp. 55-65. Oxbow: Oxford.

Peeters, H., Murphy, P. & Flemming, N. C. (eds.) 2009. *North Sea Prehistory Research and Management Framework (NSPRMF) 2009.* English Heritage and RCEM: Amersfoort.

Peeters, J. H. M. & Cohen, K. M. 2014. Introduction to North Sea submerged landscapes and prehistory. *Netherlands Journal of Geosciences* 93:3-5.

Peeters, J. H. M. & Momber, G. 2014. The southern North Sea and the human occupation of northwest Europe after the Last Glacial Maximum. *Netherlands Journal of Geosciences* 93:55-70.

Peltier, W. R. 1994. Ice Age paleotopography. *Science* 265:195-201.

Pettit, P. & White, M. 2012. *The British Palaeolithic: Human Societies at the Edge of the Pleistocene World.* Routledge: Abingdon/New York.

Reid, C. 1913. *Submerged Forests.* Cambridge University Press: Cambridge.

Saulter, A. & Leonard-Williams, A. 2011. *Assessment of Significant Wave Height in UK Coastal Waters — 2011 Update.* Report prepared for the Maritime and Coastguard Agency by the Met Office.

Shennan, I. & Horton, B. 2002. Holocene land- and sea-level changes in Great Britain. *Journal of Quaternary Science* 17:511-526.

Shennan, I., Peltier, W. R., Drummond, R. & Horton, B. 2002. Global to local scale parameters determining relative sea-level changes and the post-glacial iso-static adjustment of Great Britain. *Quaternary Science Reviews* 21:397-408.

Stringer, C. 2006. *Homo Britannicus: the incredible story of human evolution.* Penguin Books: London.

Sturt, F., Garrow, D. & Bradley, S. 2013. New models of North West European Holocene palaeogeography and inundation. *Journal of Archaeological Science* 40:3963-3976.

Tizzard, L., Baggaley, P. A. & Firth, A. J. 2011. Seabed Prehistory: investigating palaeolandsurfaces with

Palaeolithic remains from the southern North Sea. In Benjamin, J., Bonsall, C., Pickard, C. & Fischer, A. (eds) *Submerged Prehistory*. pp. 65-74. Oxbow: Oxford.

Tizzard, L., Bicket, A. R., Benjamin, J. & De Loecker, D. 2014. A Middle Palaeolithic site in the southern North Sea: investigating the archaeology and palaeogeography of Area 240. *Journal of Quaternary Science* 29:698-710.

Toucanne, S., Zaragosi, S., Bourillet, J. F., *et al.* 2009. Timing of massive 'Fleuve Manche' discharges over the last 350 kyr: insights into the European ice-sheet oscillations and the European drainage network from MIS 10 to 2. *Quaternary Science Reviews* 28:1238-1256.

Uehara, K, Scourse, J. D., Horsburgh, K. J., Lambeck, K. & Purcell, A. P. 2006. Tidal evolution of the northwest European shelf seas from the Last Glacial Maximum to the present. *Journal of Geophysical Research* 111:C09025.

Westley, K., Bailey, G., Davies, W. *et al.* 2013. The Palaeolithic. In Ransley, J., Sturt, F., Dix, J., Adams, J. & Blue, L. (eds) *People & the Sea: a Maritime Archaeological Agenda for England*. CBA Research Report 171. pp. 10-29. Council for British Archaeology: York.

White, M. J. & Schreve. D. C. 2000. Island Britain — Peninsular Britain: palaeogeography, colonisation and the Lower Palaeolithic settlement of the British Isles. *Proceedings of the Prehistoric Society* 66:1-28.

Chapter 7
The North Sea

Kim M. Cohen,[1,2,4] Kieran Westley,[3] Gilles Erkens,[1,4] Marc P. Hijma[4] and Henk J.T. Weerts[5]

[1] Utrecht University, Utrecht, The Netherlands
[2] TNO Geological Survey of the Netherlands, Utrecht, The Netherlands
[3] School of Geography and Environmental Sciences, Ulster University, Coleraine, Northern Ireland, UK
[4] Deltares Research Institute, Utrecht, The Netherlands
[5] Cultural Heritage Agency, Amersfoort, The Netherlands

Introduction

The southern and central North Sea basins (Fig. 7.1) represent some of the most important areas globally both for the discipline of submerged prehistory and the wider study of the Paleolithic and Mesolithic (Peeters *et al.* 2009). From their margins come the earliest (as yet) evidence of the hominin occupation of northern Europe in the form of the East Anglian sites of Pakefield and Happisburgh. Both sites are part of the Cromerian Complex stage of the early Middle Pleistocene, and indicate hominin occupation potentially as early as 900 ka to 800 ka, in a coastal plain setting (Parfitt *et al.* 2005; 2010). From this stage onwards, evidence from other British and European sites (including coastal plain sites e.g. Clacton) also shows occupation on a semi-continuous basis with an increase in intensity from Marine Isotope Stage (MIS) 13 (ca. 500 ka) onwards (Pettitt & White 2012). Even so, throughout the Pleistocene the pattern remains one of repeated pulses of (re)colonization and abandonment and with population shifts involving multiple hominin species including *Homo sapiens*, *Homo neanderthalensis*, *Homo heidelbergensis* and the as-yet unidentified occupants of Pakefield and Happisburgh, possibly *H. antecessor* (Stringer 2006; Ashton *et al.* 2014; Roebroeks 2014).

Strongly implicated in this are the former coastal plains of the North Sea, serving not only as a migration pathway into and out of the British Isles at various stages in the Pleistocene (Paleolithic; Parfitt *et al.* 2010; Cohen *et al.* 2012; Roebroeks 2014), but as a vast now-submerged lowland offering its own attractions, and, in the Early Holocene, potentially supporting its own major populations (Mesolithic Doggerland; Coles 1998; Peeters

Submerged Landscapes of the European Continental Shelf: Quaternary Paleoenvironments, First Edition.
Edited by Nicholas C. Flemming, Jan Harff, Delminda Moura, Anthony Burgess and Geoffrey N. Bailey.
© 2017 John Wiley & Sons Ltd. Published 2017 by John Wiley & Sons Ltd.

Fig. 7.1 Map of the North Sea, showing general bathymetry and modern political geography. Key seabed toponyms are indicated by letters, key archaeological localities by numbers. The bathymetry has been derived from the EMODnet Hydrography portal: www.emodnet-hydrography.eu. Reproduced with permission.

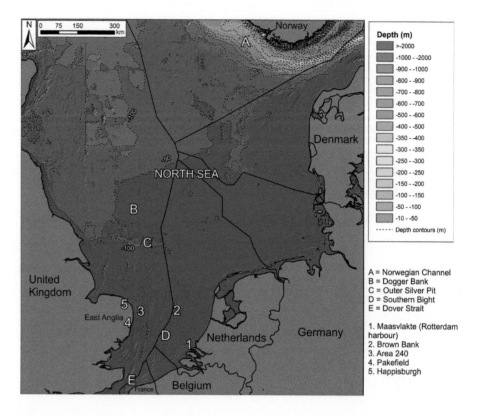

et al. 2009; 2014). Crucially, there is strong evidence that the present seabed and sub-seabed within the North Sea contain a rich and well-preserved archaeological record as well as extensive remains of the former associated paleolandscapes and paleoenvironments. Abundant lithic and organic implements, faunal and hominin remains, and peat deposits have been trawled or dredged from the southern North Sea over the past decades. Among the finds are tonnes of Pleistocene and Holocene bones found offshore of the Netherlands, mainly faunal but also including hominin remains (Mol *et al.* 2006; Hublin *et al.* 2009), assemblages of Middle Paleolithic handaxes from off East Anglia (Tizzard *et al.* 2011; 2014), and an *in situ* Mesolithic assemblage from ca. 20 m water depth in Rotterdam harbor in an area that was just offshore before recent harbor expansion (Weerts *et al.* 2012; Borst *et al.* 2014; Moree & Sier 2014; 2015; Vos *et al.* 2015).

The North Sea is one of the best-studied shelves in the world from a geological perspective. Research triggered by the presence of natural resources such as hydrocarbons and aggregates, and, more recently, by the development of offshore wind farms has generated a vast amount of geoscience data. It is the subject of numerous

publications that are relevant to the study of submerged prehistoric landscapes, their Quaternary evolution and their taphonomy. The last decade in particular has seen an increased number of publications dealing directly with paleolandscapes. The work of Gaffney *et al.* (2007; 2009) demonstrated the use of reprocessed merged industrial 3D-seismics in surveying large swaths of the central North Sea, focusing on the Late Glacial and Early Holocene. The multinational North Sea Prehistory Research and Management Framework (Peeters *et al.* 2009) inspired various collaborative efforts, including a geological background paper for hominin occupation of the southern North Sea (Hijma *et al.* 2012). In 2014, a special issue of the *Netherlands Journal of Geoscience* was devoted to the subject (Peeters & Cohen 2014). New and groundbreaking studies are also underway. In January 2015, the annual discussion meeting of the Quaternary Research Association focused exclusively on the Quaternary of the North Sea, treating a mixed academic and industry audience to many geological and archaeological highlights. In particular, research funded by the Forewind consortium in preparation for a 7 GW wind farm on Dogger Bank will bring our geological and archaeological understanding of the central North Sea to a new level (Forewind Consortium: www.forewind.co.uk/).

Harmonized research into the North Sea as a contiguous sea basin is complicated by its partition into sectors governed by different political entities. Figure 7.1 shows not only the generalized bathymetry, but also the division of the North Sea shelf among seven European countries (France, Belgium, the Netherlands, Germany, Denmark, Norway and the UK). The German part of the continental shelf within the 12 nautical mile zone does not fall under the federal government but under the individual states, adding three more authorities: Lower Saxony, Hamburg and Schleswig-Holstein. Earth science data (high-resolution bathymetry, core logs, 2D and 3D seismics) are at present available in vast numbers (Figs. 7.2 & 7.3) but are highly diverse, collected using multiple techniques and stored in different databases, and owned and maintained by different institutions in each of the countries. In addition, the various data sets have been gathered by workers from many different research traditions. Finally, data is stored not only in national and institutional databases but in less accessible databases

held by the commercial users of the North Sea (e.g. oil, renewables and aggregates industries) and their specialist contractors. At the time of writing, various national and international initiatives include efforts to collate and harmonize data and data products from the North Sea, for example EMODnet (European Marine Observation and Data Network 2014). Many international initiatives focus on key national-level data sources (e.g. those managed by national geological surveys). At national level, initiatives try to unlock larger sets of legacy data, including that held by commercial parties (offshore engineering, aggregate mining) for future public use, for example, the UK Crown Estate Marine Data Exchange for offshore renewables (www.marinedataexchange.co.uk/).

Distilling taphonomic information from such a diverse set of databases is a challenge. This chapter offers a general summary of the Quaternary history and geoarchaeological taphonomy of the southern and central North Sea continental shelf, as a starting point to understanding the taphonomy of submerged archaeology from paleolandscapes preserved below the North Sea floor. For further information, the interested reader is

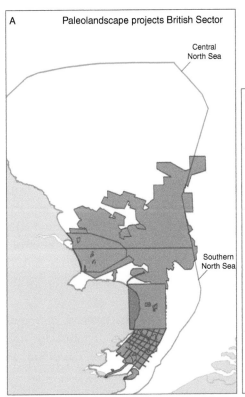
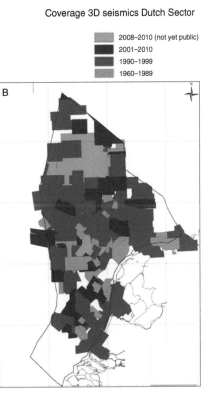

Fig. 7.2 (A) Coverage of paleolandscape projects (gray polygons) in the UK sector of the North Sea (from Bicket 2013: fig. 8). Map includes the '3D seismic central North Sea Mega Merge' area of Gaffney *et al.* (2007); (B) Coverage of 3D seismics in the Dutch sector of the North Sea (www.nlog.nl) (from Erkens *et al.* 2014). Note that many of the seismic cubes, especially in the south, are not suitable for the identification of shallow preserved landscapes, due to the shallow water depths. Colors indicate the year of collection. In the first 5 to 7 years after collection, the collected data are often confidential (red blocks date to 2008).

Fig. 7.3 (A) Example of density of borehole data, and; (B) 2D seismic line data in the Dutch sector of the North Sea as available from the databases of the Geological Survey of the Netherlands. Erkens *et al.* (2014).

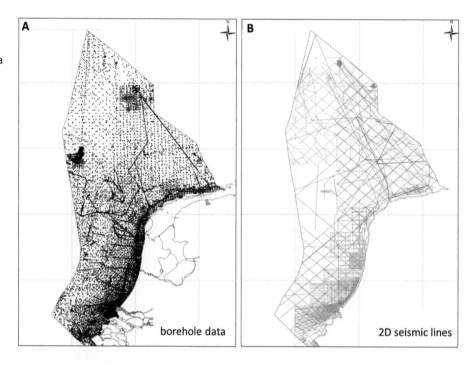

borehole data

2D seismic lines

directed to the references herein, which provide the underlying information in more detail than is possible here. One entry point is the *Audit of Current State of Knowledge of Submerged Palaeolandscapes and Sites*, carried out by Wessex Archaeology (Bicket 2013) on behalf of English Heritage, for the UK sector. The aforementioned integrative articles provide further entry points (Gaffney *et al.* 2007; Hijma *et al.* 2012; papers in Peeters & Cohen 2014) while earlier review papers or volumes include Flemming (2002; 2004). For nearshore and estuarine processes around the North Sea, an entry point is Martinius and van den Berg (2011). For sea-level research, and related matters such as mapping and dating the drowning of landscapes due to sea-level rise, a handbook edited by Shennan *et al.* (2015) is a good entry point.

Physical Geography and Modern Environment

The present-day North Sea is a semi-enclosed epicontinental shelf sea occupying an area of ca. 750,000 km². Water depth is generally shallow and deepens as one moves north and north-east (Fig. 7.1).

The shallowest area — the Southern Bight, located between southeast England and the Low Countries — is less than ca. 40 m deep. North of this, the central North Sea covering the zone between England and Denmark is characterized by depths of ca. 40 m to 100 m. These areas are shallow seas owing to sustained delivery of sediment over millions of years, outpacing subsidence of the North Sea as a tectonic basin. This filled the basin gradually from the south-east to the north-west, with glaciations modifying the surface in the last million years (see sections below, pages 152–161).

The deepest areas of the shelf are located between Scotland and Norway; water depths here are between ca. 100 m to 200 m and there is a major bathymetric depression in the form of the Norwegian Channel which reaches maximum depths of ca. 700 m (e.g. Huthnance 1991; Paramor *et al.* 2009). This trench is a continuation of the Oslo fjord and was the path of a major ice stream in the Last Glacial and at least four earlier glaciations. Its depth is inherited from ice-age subglacial erosion (Sejrup *et al.* 2003).

Generalized (low resolution) bathymetric data indicates a relatively featureless seabed, typified by large expanses of low gradient terrain. The exceptions are upstanding features which range in size and scale from the numerous long and narrow sand banks of the Southern Bight to the plateau-like expanse of the Dogger Bank in the central North Sea. Some bathymetric depressions are also evident on the shelf, most obviously the Outer Silver Pit, a west–east oriented valley incised to depths of ca. 80 m. Smaller depressions, such as the Silver Pit, Sole Pit and Well Hole, are also present (Briggs *et al.* 2007).

The North Sea is more sheltered than other parts of the European shelf owing to the surrounding landmasses. In general, the seabed substrate varies depending on the hydrodynamic conditions described above in conjunction with bathymetry, inherited geomorphology and substrate type. Based on broad-scale mapping, most of the North Sea floor is characterized by sand. Widespread occurrence of coarser sediments is limited to the eastern English coast. Patchy areas with abundant coarse material also occur off Denmark, Germany and the Netherlands where glacial sediment has been winnowed by marine processes. Patchy gravelly areas also occur in the very south off Belgium (Flanders) and southeast England (Kent) in the entrance of the Dover Strait, due to marine winnowing of fluvial deposits. Fine-grained muds or muddy sands occur over other parts of the basin. Past mapping of the seabed substrate and collated maps based on these (e.g. EMODnet) suffer from the crude resolution of the data which was available at the time. Present-day surveys typically allow major improvements in resolution, and can identify finer-grained patches within coarser sand seas and sandy patches within muddy areas, thus helping to resolve inherited landforms (e.g. Tizzard *et al.* 2014).

Tides are the dominant control in shelf hydrodynamics and proceed anticlockwise rotating around amphidromic points in the Southern Bight, west of Denmark and at the southern tip of Norway (Otto *et al.* 1990). Consequently, the timing of tidal cycles varies across the basin, as does the tidal range, which is larger on the British coast compared to continental Europe (Otto *et al.* 1990; Huthnance 1991). Tidal currents vary correspondingly and are therefore greater along the western side of the sea basin (Huthnance 1991; Neill *et al.* 2010). Despite the dominance of tidal forces, storms and wind-driven currents also play a part (van der Molen 2002). Dominant wind directions are from the west and south with wind speeds highest in the northern part of

the North Sea and decreasing south and east (Neill *et al.* 2010). When combined with bathymetry, these forces set up bed shear stresses which are greatest in the shallow nearshore regions of the southern North Sea (i.e. off East Anglia, Belgium, the Netherlands, Germany and southern Denmark) (Neill *et al.* 2010). This is of relevance to paleolandscapes and taphonomy because it controls the height and trough depth of bedform fields ('sand waves') which have become established over the last 6000 years (van der Molen 2002). The reworking depth of these sand-wave fields affects preservation of Late Glacial and Early Holocene land surfaces especially where they are not capped by compacted muds or peat.

Spatial variability in inherited topography, geology, past sea-level change, glaciation and modern oceanographic conditions results in significant natural coastal variation which has, more recently, been amplified by human influence. The bounding northern shorelines (Scotland and Norway) are steep, often rocky, with fjords and large inlets (Barne *et al.* 1997). The Norwegian coast is particularly rugged and complex, indented with numerous bays, inlets and fjords, and dotted with many small islands. Further south, the coastlines become lower gradient and softer and are no longer rocky. The availability of sediment is the result of prolonged supply from the large northwest European rivers and the modification of these rivers by glaciations.

Proceeding in an anticlockwise direction, the coastline of northeast England and the northern half of East Anglia comprises relatively linear unconsolidated cliffs (bluffs) of glacial sediment which, at present, are eroding (Barne *et al.* 1995a,b). It is in fact this cliff retreat which allowed discovery of the Pakefield and Happisburgh sites (Stringer 2006). The southern half of East Anglia and southeast England are characterized by a more indented shoreline dominated by large estuaries (e.g. Thames, Medway) and their associated salt marshes (Barne *et al.* 1998; Allen 2000). Many of these estuaries represent drowned valleys flooded by rising sea level after the end of the Last Glacial.

The opposing side of the Southern Bight is characterized by extensive linear sand beaches, commonly backed by dunes. Major drowned valleys existed in this area too (e.g. the Rhine-Meuse valley of Busschers *et al.* 2007; Hijma & Cohen 2011; the Central Netherlands

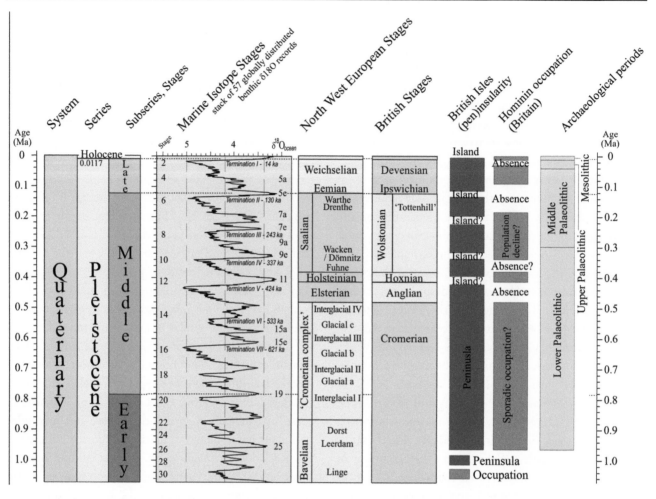

Fig. 7.4 Correlation chart for the North Sea basin for the last 1 million years, showing chronostratigraphic stages and terms. Modified from Gibbard and Cohen (2008), with additional archaeological information from Pettitt and White (2012).

Rhine valley of Peeters *et al.* 2015a), beside newly-formed transgressive tidal inlet systems (Rieu *et al.* 2005; Hijma *et al.* 2010). However, the matured Holocene barrier system means that from ca. 4500 years ago (Beets & van der Spek 2000), these no longer show up as interruptions in the coastline. Major breaks in this shoreline, which encompasses Belgium and the Netherlands, occur in the southwest Netherlands as a series of estuaries of the Scheldt, Meuse and Rhine (e.g. van den Berg *et al.* 1996; Martinius & van den Berg 2011). The large number of estuaries here originates from Medieval storm surge incursions that repeatedly caused loss of cultivated coastal peatland (e.g. Vos & van Heeringen 1997, Vos 2015).

Moving north and east along the Netherlands coast through north Germany and into southern Denmark, the coast is marked by barrier islands (the Frisian Islands)

which protect the muddy back-barrier lagoon of the Waddenzee. The islands are associated with tidal deltas and tidal inlets, and the Wadden environment with salt marshes, intertidal sandy shoals and tidal mudflats (e.g. Oost 1995; van der Spek 1996; Allen 2000; Behre 2004; Streif 2004; Chang *et al.* 2006; Vos 2015). A break in the island chain occurs where the Weser and Elbe exit into the North Sea; these too occupy drowned Pleistocene valleys (e.g. Streif 2004; Alappat *et al.* 2010). Continuing around the coast, western Jutland is typified by mainland sand beaches, lagoons and barriers, with the latter becoming increasingly attached to the mainland as one moves north (Pedersen *et al.* 2009; Fruergaard *et al.* 2015), and farthest north by cliff sections exposing landforms from the Last Glacial with a sandy seabed in front (Anthony & Leth 2002).

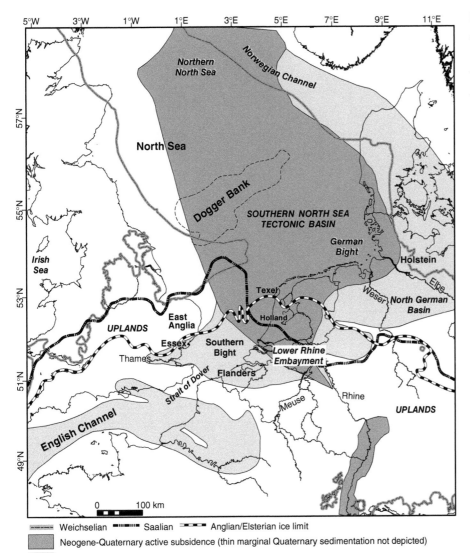

Fig. 7.5 Location of tectonically subsiding areas (depocenters) in the North Sea region, and maximum limits for the Weichselian, Saalian and Anglian/Elsterian glaciations (see Fig. 7.4). Hijma et al. (2012).

‐‐‐‐‐ Weichselian ▪▪▪▪▪ Saalian ▰▰▰▰ Anglian/Elsterian ice limit

Neogene-Quaternary active subsidence (thin marginal Quaternary sedimentation not depicted)

Palaeogene former basins during the Middle Eocene

Quaternary Background and Paleogeographic Framework

Early Pleistocene

The present configuration of the southern North Sea basin is a relatively recent phenomenon. This section outlines the broad landscape and paleogeographical changes taking place in the southern North Sea region over the past ca. 1 million years, coincident with the first hominin occupation, presently confirmed as dating to the Early-Middle Pleistocene transition (Fig. 7.4). Throughout this period, the fluctuating Quaternary climate, sea level and extent of glaciation together with regional tectonics reshaped the landscape on multiple occasions and progressively over multiple glacial-interglacial cycles. From ca. 800 ka onwards the length and intensity of glacial periods increased with attendant effects on ice-sheet extent and glacio-eustatic sea level (Zachos et al. 2001). While preceding ocean volume changes are estimated to have been less than 70 m, from the Middle Pleistocene onward, glacial maxima saw glacio-eustatic lowstands as deep as 120 m to 150 m (Rohling et al. 2009). These figures broadly apply to the North Sea too, although the gravitational effects of the nearby ice masses meant that the regional lowstands were some 10 m to 20 m less deep than the global average.

Fig. 7.6 Drainage basin and delta of the Miocene–Pliocene–Early Pleistocene Eridanos river system. Background map depicts former terrestrial lowland environments in Middle Pleistocene glacially-excavated areas that are nowadays occupied by seas. Reproduced from Cohen et al. (2014).

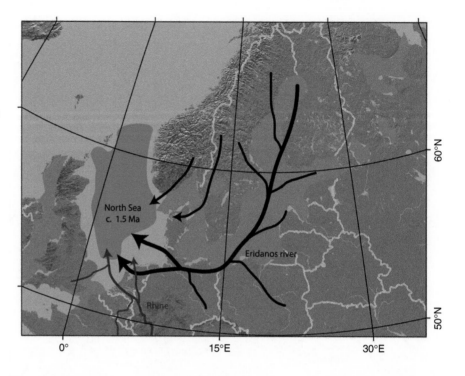

The climatic changes since the onset of the Middle Pleistocene have controlled not only the paleogeography of the study area, but also the preservation of the sedimentary deposits (and attendant archaeological evidence contained within, i.e. taphonomy) that provide evidence of past landscapes. The remainder of this section will outline the links that exist between geological controls and taphonomy. This prepares for the Submerged Landscape and Preservation/Taphonomy dedicated sections in the second half of the chapter (pages 160–175).

A first major control on long-term geological preservation and paleogeographic change is the tectonic setting. Much of the North Sea area is part of a subsiding basin, initiated by reactivation of a Triassic-Jurassic rift system early in the Miocene (25–20 Ma) (see overview in Cloetingh et al. 2005). The basin is aligned SE–NW running from onshore (the Netherlands, northwest Germany, southern Denmark) to offshore, reaching as far north as southern Norway (Fig. 7.5). No major changes in the tectonic regime have taken place during the Quaternary, with The Netherlands' onshore depocenters subsiding at rates averaging up to 0.3 m/kyr (Kooi et al. 1998), increasing to 0.6 m/kyr in the central North Sea offshore.

Subsidence allowed the accumulation of Quaternary sediments up to several hundred meters thick within the depocenter (Cameron et al. 1992; Gatliff et al. 1994). By contrast, mild uplift predominated around the 'shoulders' of the basin: East Anglia, Flanders and the parts of the North Sea between them, as it also did for Mesozoic uplands of the Weald-Artois anticline (e.g. van Vliet-Lanoë et al. 2000; 2002; Cloetingh et al. 2005). This affected paleogeographical evolution and affects paleolandscape taphonomy in the southern North Sea (Cohen et al. 2012; Hijma et al. 2012). The sediment stored in the North Sea depocenter has mainly been deposited by the major northwest European rivers (e.g. Rhine, Thames) and the former Eridanos River (Overeem et al. 2001). This latter ancient watercourse used to drain a catchment spanning the Baltic and Fennoscandia (Fig. 7.6). In doing so, it supplied immense quantities of sediment to the North Sea enabling considerable delta plain progradation. Some 80% of the total accumulation is from this river system (Cameron et al. 1992). The end result was the formation of a wide fluvio-deltaic coastal plain that, at highstand, extended some 100 km northward from the Weald-Artois anticline, and up to 300 km at lowstand. As such, these lowlands were a

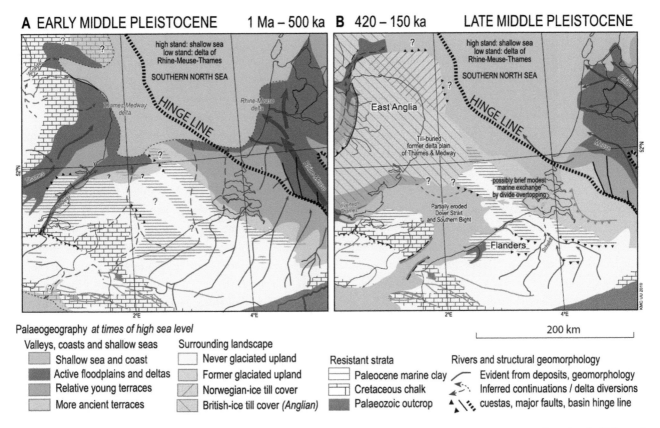

Palaeogeography *at times of high sea level*

Valleys, coasts and shallow seas
- Shallow sea and coast
- Active floodplains and deltas
- Relative young terraces
- More ancient terraces

Surrounding landscape
- Never glaciated upland
- Former glaciated upland
- Norwegian-ice till cover
- British-ice till cover *(Anglian)*

Resistant strata
- Paleocene marine clay
- Cretaceous chalk
- Palaeozoic outcrop

Rivers and structural geomorphology
- Evident from deposits, geomorphology
- Inferred continuations / delta diversions
- cuestas, major faults, basin hinge line

200 km

Fig. 7.7 Scenario maps for paleogeographies of the southern North Sea in the Middle Pleistocene. (A) Interglacial configuration for highstands between 1 Ma and 0.5 Ma (i.e. Cromerian Complex, Fig. 7.4); (B) Interglacial configuration for highstands between ca. 0.42 Ma and 0.17 Ma (i.e. MIS 11, MIS 9, MIS 7; Fig. 7.4). Hijma *et al.* (2012).

considerable part of the broad land bridge that connected Britain and continental Europe even during highstands (Funnell 1995; Hijma *et al.* 2012).

From about 1 Ma onwards, enlarged Fennoscandian ice masses began destroying the Eridanos system and, in tandem, important highstand coastline reconfiguration commenced in the North Sea area (Gibbard 1995; Cohen *et al.* 2014). These events mark the Early-Middle Pleistocene transition in the study area, and provide a second key control on the lowland environments around the North Sea.

Middle Pleistocene

By 1 million years ago, the chalk hills across the modern Dover Strait still formed a distinct southern boundary to the study area (Figs. 7.6 and 7.7a). This was the time when hominins first reached the North Sea, during the interglacial highstands of the Cromerian Complex (MIS 21–13; Fig. 7.4). Loss of the Eridanos sediment supply at that time meant that the fluvio-deltaic plain was moribund and its surface subsiding (Cameron *et al.* 1992). The trend of coastal progradation had ceased and highstand shorelines were shifting ever closer towards their present position (Zagwijn 1979; Funnell 1996). Farther south, between the chalk hills and the North Sea highstand shoreline, a considerable stretch of terrestrial lowland habitat was present, linking southern Britain and the European mainland (Funnell 1995; Gibbard 1995). Importantly, marine incursions via the English Channel to the south did not breach the Dover Strait area at this time (Fig. 7.7a).

The highstand shoreline along the southern rim of the North Sea, running from the Netherlands to East Anglia, can be seen as a continuous deltaic coastal plain supported by the extensions of the Rhine, Meuse, Scheldt, Medway and Thames rivers of the time (Cohen *et al.* 2012). The early sites of Pakefield and Happisburgh were located on the western edge of this coastal plain. Paleoecological data indicate that the environment during (part of) the earliest occupation (ca. 900–800 ka; Happisburgh) was dominated by boreal forest (e.g. pine, spruce) and modestly cooler than these localities are today. Late phases of occupation within the Cromerian Complex stage (Pakefield), took place during periods of interglacial climax, in areas characterized by marsh, oak woodland and open grassland, modestly warmer than these locations are today (Parfitt *et al.* 2005; 2010). As the Cromerian Complex was the time of earliest demonstrable hominin occupation of the region (Lower Paleolithic; Fig. 7.4), and as sites from that period are encountered in lowland deposits and paleoenvironments (Parfitt *et al.* 2010), it is paleoanthropologically relevant to generate information on past distribution and present preservation of such deposits in the nearshore and offshore in the modern North Sea (Peeters *et al.* 2009).

The state of preservation/taphonomy of deposits in coastal plain environments, is controlled by various processes, both syn-sedimentary (during the interglacial concerned) and post-sedimentary (in later glacials and interglacials). With the increased amplitude of glaciations during the Cromerian Complex, the proximity of the North Sea to centers of glaciation increased glacio-hydro-isostatic adjustment (GIA) effects. This had a sustaining effect on relative sea-level rise into highstands, causing transgressions in the North Sea to commence relatively late in the interglacial.

While post-glacial transgressions were initially fast, they subsequently slowed down and continued for a long period. This meant that each interglacial following a large glaciation saw its highstand coastal rims characterized by extended periods of sediment aggradation. This, in turn, produced coastal plain architectures with favorable properties for enclosing archaeological sites (Cohen *et al.* 2012). As the furthest inland rims of the coastal plain are the last to erode, even from late in the interglacial, sites located there would still have had a good chance of burial by continued coastal aggradation. This enhances the long-term preservation potential for sites from the second half of the interglacial in the North Sea region

(whereas in the tropics and subtropics, where GIA differs, this effect unfortunately does not exist, Cohen *et al.* 2012). In the southern North Sea, GIA effects thus caused late transgressions that gave hominins dwelling in the coastal plain enough time to reach this far north (see also Late Glacial and Holocene, pages 161–164). Furthermore, GIA effects helped to ensure that sites from the second half of the interglacial would still be preserved and located relatively far inland.

The depths of preservation of former coastal plain landscapes in the long run are controlled by tectonic subsidence regimes. In regions lacking tectonic subsidence, highstand coastal plain deposits from past interglacials underwent coastal erosive attack in younger interglacials each time a new highstand North Sea was established. Therefore, around the North Sea the problem of preserving coastal plain archaeology does not lie in its being preserved during the ca. 10 kyr of an interglacial, but its longer-term preservation thereafter (Cohen *et al.* 2012). In that respect, the role of the Anglian glaciation in preserving Cromerian Complex coastal plain deposits appears to be essential. It is the water-laid nature of these glacial deposits that explains why the tills buried rather than incorporated the older deposits in this particular region, outside the preservation-favourable tectonic area (Fig. 7.5). This relatively favorable preservation control applies to areas offshore East Anglia, once covered by Anglian till, and is a taphonomic reason to pay special attention to the proglacial situation of the Anglian glaciation. A further reason (Gibbard 1988) is that besides controlling the preservation of Cromerian Complex deposits, the Anglian glaciation affected the drainage networks of many of the rivers around the southern North Sea (see next pages).

Cold stages are apparent for the pre-MIS 12 Cromerian Complex, with glaciation likely encroaching on the northern margin of the North Sea basin (e.g. MIS 16) (Rose 2009; Böse *et al.* 2012). However, the first major verifiable expansion of lowland ice across the study area occurred during MIS 12 (Fig. 7.4; Pawley *et al.* 2008; Lee *et al.* 2012). On at least three occasions in the last 500,000 years, ice sheets originating in Britain and Scandinavia extended across the sub-aerially exposed landscape of the North Sea basin. These are the well-documented Anglian/Elsterian, Saalian (at maximum glaciation called the Drenthe substage) and Weichselian limits across the North Sea (Laban 1995; Praeg 2003; Carr *et al.* 2006; Busschers *et al.* 2008; Sejrup *et al.* 2009; Lee *et al.* 2012; Moreau *et al.* 2012). Glaciation was also substantial in MIS 10 (Elsterian or earliest Saalian) and MIS 8 (early Saalian), with British ice covering northern England and Scandinavian ice covering Denmark and reaching into the central North Sea and the German Bight, but leaving the southern North Sea ice-free.

The impact of glaciation on the landscape was considerable, not only in ice-covered areas but also in the foreland that received abundant glacial outwash, and was marked by severe periglacial frost conditions mobilizing the surficial sediments on slopes and in rivers. Prior to the large glaciations, many river systems, like the Thames, Rhine, Elbe, Meuse, Scheldt and (now extinct) northern English rivers Ancaster, Bytham and proto-Trent, had drained northwards into the North Sea, reaching the Eridanos Delta lowland (Gibbard 1988; Rose 1994; Funnell 1996; Rose *et al.* 2002; Cohen *et al.* 2014). Each large glaciation forced many of the river courses around the North Sea into new, more southerly positions, temporarily or permanent. Valley floors of abandoned courses became used by underfit smaller streams (e.g. Gibbard *et al.* 2013; Peeters *et al.* 2015a), and many of them were eventually buried by periglacial colluvium, glacial diamictons (tills) or marine deposits. By governing abandonment and burial of valleys and contemporaneous creation of new pathways, shifting of rivers played a major double role in recording and preserving the archaeology from the various Paleolithic periods (Fig. 7.4), especially in the southern North Sea (Hijma *et al.* 2012).

Areas directly under the ice were remodeled on each occasion and imprinted with a variety of glacial features including both depositional elements, such as till sheets, and erosional landforms, such as tunnel valleys (see 'Evidence of submerged landscapes on the shelf', pages 166–169). Former ice margins are marked by features such as moraines, ice-push ridges and outwash fans and indicate that, at maxima, ice coalesced across the North Sea. This happened for the first time in MIS 12 (Toucanne *et al.* 2009), during the Anglian glaciation,

when British and Scandinavian ice sheets reached a maximum extent and covered all but the lower part of the Southern Bight. In comparison, Saalian ice cover, 250,000 years later, was less extensive over England but more extensive in the Netherlands. Over the southern North Sea region however, the Anglian and Saalian glaciations were comparable in coverage (Figs. 7.5 & 7.8).

During the Last Glacial (Weichselian), glaciation along the northeast English coast reached relatively far south. Over the central North Sea, however, Scandinavian ice reached only as far as the Dogger Bank, and areas such as the German Bight remained ice-free (Houmark-Nielsen & Kjær 2003). Even though the Scandinavian and British ice sheets may have converged over the North Sea, large tracts of its central and southern sections in the last 140,000 years have remained unglaciated (Fig. 7.5; Ehlers & Gibbard 2004; Clark *et al.* 2012). The Dover Strait region has never been glaciated.

During periods of maximum glaciation, large proglacial lakes are envisaged in the southern North Sea (Fig. 7.8), blocked by the ice-front to the north and upland topography to the south (Gibbard 1995; Toucanne *et al.* 2009; Hijma *et al.* 2012; Murton & Murton 2012; Cohen *et al.* 2014). The lakes are inferred/reconstructed both in the older Anglian and Saalian glaciations (parts of MIS 12 and MIS 6 respectively) and the younger Weichselian glaciation (Last Glacial, parts of MIS 2). Partial sedimentary evidence for these lakes exists. For the Anglian, water-laid ice marginal deposits are part of Norfolk's Anglian sequences (Lunkka 1994) and the Elsterian of the northern Netherlands contains subaqueous outwash fans of subglacially-entrained micaceous sands (Peelo Formation, Laban & van der Meer 2011; Lee *et al.* 2012). That this was part of one larger lake held up by a single ice-front (Fig. 7.8a) is inference, supported by the maximum ice-limit reconstructions for this period (Fig. 7.5).

For the Saalian (Fig. 7.8b), water-laid proglacial and ice-marginal deposits are known from the Cleaver Bank region (Laban 1995; Moreau *et al.* 2012) and the maximum-limit ice-marginal Rhine at initial stages grades to a high base level, inferred to be from the same lake (Busschers *et al.* 2008; Peeters *et al.* 2015a). Note that in these ice-limit zones, push-moraines and subglacial

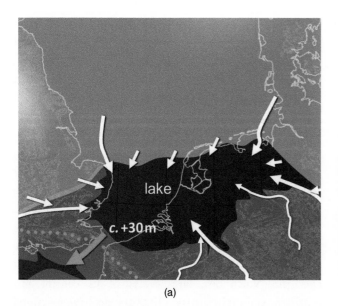 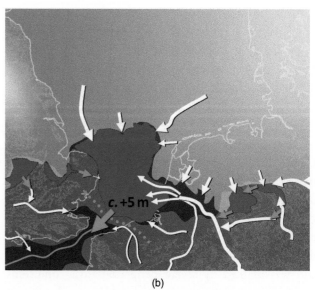

(a)   (b)

Fig. 7.8 Presumed proglacial lake extents in the southern North Sea, (A) within MIS 12, and; (B) within MIS 6. From Gibbard and Cohen (2015). Based on Gibbard (1995; 2007), Busschers *et al.* (2008) and Moreau *et al.* (2012). See also Murton and Murton (2012).

features were also created and the interrelations between these landforms, water-laid proglacial features, and water-laid deposits from deglaciation phases have long been difficult to separate in regional mappings (e.g. Laban 1995). Seismic interpretation appears to be changing this recently (e.g. Moreau *et al.* 2012). From the central North Sea, indications for water-laid proglacial and ice marginal deposition are reported in Laban (1995) for the central North Sea in the Dogger Bank region. Ephemeral formation and drainage of the Weichselian lakes to the Norwegian Sea is evident from $\delta^{18}O$ signals in deep marine sediments (Lekens *et al.* 2006). We can infer that between sudden drainage events towards the north-west, these central North Sea lakes overspilled towards the south, and that their southern shores must be sought north of the Brown Bank follows from mapping of the Southern Bight region (Hijma *et al.* 2012).

Figure 7.7b shows an interglacial highstand situation with a half-eroded land bridge for the time period between MIS 12 and MIS 6. It highlights that a terrestrial connection continued to exist between East Anglia and the Netherlands, even after initial erosive cutting in the Dover Strait had begun (Funnell 1995; Hijma *et al.* 2012). This terrestrial connection was narrow and the saddle could potentially be overtopped by shallow seas during true interglacial highstands (i.e. as high as at present, if

not a few meters higher), since for most of the MIS 12 to MIS 6 period, sea level was below such levels (White & Schreve 2000).

The position of the southern lake limits and the elevations of lake levels in Fig. 7.8 are tentative estimates. In part, the positions are supported by, for example, the distribution of estuarine sediments from the interglacial highstands that Fig. 7.7b summarizes as a single scenario (with further evidence from biostratigraphy: Meijer & Preece 1995; Roe & Preece 2011). The (initial) lake level elevation for the Anglian is supported by a major channel fill deposit at Wissant (Roep *et al.* 1975), interpreted to be a spillway fragment — but estimates of tectonic uplift and GIA at the time of and since the Anglian are also factored in. Similarly, the initial level at the time of the Saalian maximum is estimated, factoring in GIA and tectonic subsidence (Busschers *et al.* 2008).

That the ice sheets produced vast amounts of melt water at their maximum stage and that these gathered in ice-marginal rivers and were routed towards the North Sea is evident from onshore glacio-fluvial features, to which northward-flowing rivers from England and mainland Europe added further discharge. That major flows of water were routed south through the Dover Strait at lowstands, is evident from seafloor geomorphology of the English Channel (Gupta *et al.* 2007). The greater importance of glaciations during MIS 12 and MIS 6 in eroding the Dover Strait land bridge, compared to that of other glacials, is supported by sediment delivery rates as measured off the English Channel floor in records collected in deep marine settings (Toucanne *et al.* 2009).

The waters collected in the lakes shown in Fig. 7.8 were routed southward as lake overspill. This is used to explain the formation of the Dover Strait as the carving of a proglacial-lake spillway valley (Roep *et al.* 1975; Smith 1985). In turn, this would explain the signals of biogeographical insularity seen in paleoenvironmental records from Britain over a series of interglacials and give an indication of the first glaciation that was extensive enough to induce this change (Gibbard 1995).

The degree of importance attributed to proglacial spillage in opening the Dover Strait is a matter of interpretation. The explanations entertained here (based on Gibbard 2007; Gupta *et al.* 2007; Busschers *et al.* 2008; Hijma *et al.* 2012; Gibbard & Cohen 2015 — see Murton & Murton 2012 for an independent review) attribute maximum glaciation lake spillage a relatively large role in a short period (the Anglian and Saalian glacial maximum together span perhaps 10,000 to 20,000 years). Erosion by normal rivers and tributaries, estuarine processes and coastal cliff retreat are each attributed a smaller role despite being in operation over much longer time spans in the last 450,000 years. This is because both vertical erosion and the sudden carving of a new north-to-south valley have to be explained. Spillage would be the process to create a new path (as a headward cutting tributary could also do, but would take much more time), while river and estuarine erosion would be processes that deepen existing paths. Cutting of the spillway may have been catastrophic (Smith 1985; Gupta *et al.* 2007; i.e. an event within a glacial) or may have been gradual (i.e. going on for several thousand years during a glacial). At the time that the land bridge (sill) was >100 km wide (Figs. 7.7a & 7.8a), gradual retrogradation is more probable (i.e. in the Anglian), and by the time that the sill had shortened and lowered (Figs. 7.7b & 7.8b), a final catastrophic event is probable (i.e. in the Saalian).

In the aftermath of the Saalian deglaciation (the situation following Fig. 7.8b), the proglacial valley in the Southern Bight remained in use by the River Meuse. Discharge from the Rhine system had contributed to erosion of the Dover Strait land bridge at the times when proglacial lakes existed, but stopped doing so during the Saalian deglaciation (Busschers *et al.* 2008; Peeters *et al.* 2015a). As a lowstand river, however, it would only start to use the Axial Channel escape route out of the North Sea basin later in the Last Glacial (Fig. 7.9).

Details in this theory explaining the landscape evolution of the southern North Sea in the last 1 Ma have bearings on the taphonomy of the North Sea floor. They determine how major downcutting and/or reworking episodes operated in a given period. Hijma *et al.* (2012), for example, explore scenarios of valley downcutting at the end of the Saalian and during the Weichselian, that included trade-offs between proglacial spillway and periglacial river modes and intensities of erosion. Similarly, the depiction of the coastline configurations for the highstands between the Anglian and the Saalian (Fig. 7.7) would require scenario exploration of such trade-offs for these two glaciations (and possible further events in MIS 10 and MIS 8).

Late Pleistocene up to the Last Glacial Maximum

For the extent of the North Sea in the Last Interglacial and the situation of the North Sea floor in the Last Glacial (e.g. Fig. 7.9), abundant evidence is available that is independent of scenarios and interpretations regarding the erosional history of the Dover Strait in the Middle Pleistocene.

Figure 7.9a shows the situation of the southern North Sea during the Last Interglacial (MIS 5; Fig. 7.5). As a result of the cumulative effects of tectonic setting and progressive glaciations (see previous section, pages 154–159, and Figs. 7.5–7.8), the Last Interglacial was the first interglacial when Britain was an island (Bridgland & D'Olier 1995; Funnell 1995; Gibbard 1995; Meijer & Preece 1995). It was also the first interglacial when the North Sea had an extent and coastal configuration similar to today's (Streif 2004). At full highstand, the North Sea was a substantial barrier to biota, as expressed in terrestrial biostratigraphical differences (including hominin absence) between Britain and the Continent in the Last Interglacial (Funnell 1995). The Last Interglacial proper (regionally defined as the Eemian/Ipswichian;

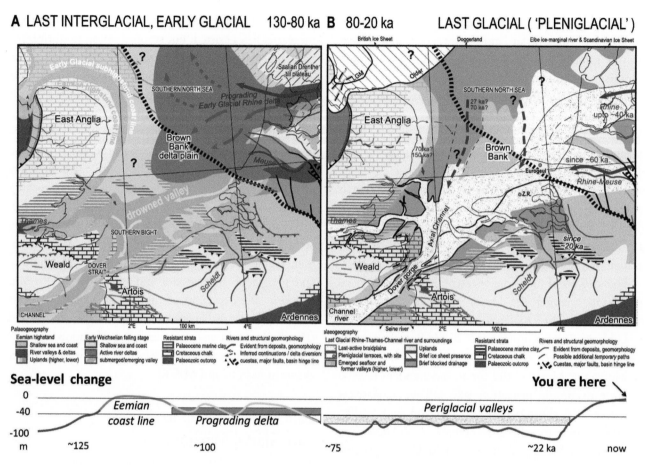

Fig. 7.9 Reconstructed paleogeography of the southern North Sea in the Late Pleistocene, with a tentative sea-level curve for this timeframe and study area. (A) Last Interglacial highstand and onset of sea-level fall in MIS 5 (ca. 130–80 ka); (B) Last Glacial lowstand of MIS 4 to MIS 2 (ca. 80–20 ka). Hijma *et al.* (2012).

Fig. 7.4) was relatively short (just ca. 12 kyr, of which ca. 6 kyr was true highstand; Zagwijn 1983; Sier *et al.* 2015; Peeters *et al.* 2016). It was followed by slow sea-level fall with the Rhine Delta building out into the North Sea (Fig. 7.9a). In the later parts of MIS 5, this helped to create a lowland land bridge in the shallow southern North Sea allowing hominins to (re)occupy England from northwest Europe. A Neanderthal skull fragment found offshore of the southwest Netherlands, on grounds of morphological resemblance to such skulls in northwest France (Hublin *et al.* 2009) and the setting and inferred taphonomy of the offshore find location (Hijma *et al.* 2012), would be from this time.

As global climate cooled further into the Last Glacial and global sea level fell from ca. −25 m (MIS 5a, 85 ka) to ca. −90 m (MIS 4, 70 ka), an increasing expanse

of the southern and central North Sea floor became exposed (Fig. 7.9b). Once again, river valleys traversed the exposed shelf and once again, the north of the basin became glaciated. It is not fully certain if the British and Scandinavian ice sheets coalesced in the northern North Sea, but it is suspected that they did for a brief period (Carr 2004; Carr *et al.* 2006; Lekens *et al.* 2006; Sejrup *et al.* 2009; Clark *et al.* 2012). The northern half of Denmark and the area of the Dogger Bank are home to large complexes of ice-pushed moraines that mark the limits of the Scandinavian ice sheet in the Last Glacial. British ice reached down along Lincolnshire to the north Norfolk coast, leaving tills and subglacially-carved lows such as the Outer Silver Pit (Fig. 7.1).

Parts of the Danish and German sectors may have hosted proglacial lakes for shorter intervals, with similar formative mechanisms to those of the Anglian and Saalian in the southern North Sea (see 'Middle Pleistocene' pages 154–159; Fig. 7.8). In this region, the Elbe ice-marginal river contributed water, as did outwash channels sourced from Scandinavia. At times of coalescent ice sheets and proglacial ponding, overflow of lake water must have been routed southward towards the river systems in the area of the present-day Southern Bight, Dover Strait and English Channel (Clark *et al.* 2012). Figure 7.9b shows possible spillage pathways joining fluvial drainage in the Southern Bight and the gorge of the Dover Strait. For the preservation of Early Glacial records in the southern North Sea, the lake spillage from the north may be relevant. The mapping of the pathways implies that large parts of the original regressive Rhine Delta Plain (Brown Bank Formation) have been eroded by waters from the north (compare Figs. 7.9a and 7.9b) and not just as a consequence of normal erosion by the Rhine. Recognizing northerly lakes and southerly spillage pathways could be important for taphonomic assessment of the contents of erosive lags (e.g. Glimmerveen *et al.* 2006) in fluvial deposits such as appear on the eastern side of Brown Bank (also see 'Evidence of submerged landscapes on the shelf' pages 166–169.

From the final stages of the Last Glacial (the last 18 kyr, if not the last 27 kyr), the Elbe ice-marginal system shows up as a valley system on the North Sea floor (Figge 1980; Houmark-Nielsen & Kjær 2003; Alappat *et al.* 2010), directed to the Norwegian Channel and graded to a base level of –70 m or deeper (Fig. 7.1). On the grounds of position, depth and grading to relatively low base levels matching contemporaneous sea level, this indicates that the ice stream from the Oslo fjord had withdrawn and no longer blocked the Elbe's northward drainage. Any Last Glacial major proglacial lake stage in the central North Sea must therefore date from before the functioning of that bathymetrically expressed last valley.

Late Glacial and Holocene

Postglacial inundation of the North Sea in the last 20,000 years, at a shelf scale, is typically visualized using output of GIA models (e.g. Lambeck 1995; Shennan *et al.* 2000; Vink *et al.* 2007; Sturt *et al.* 2013; Fig. 7.10).

Besides numerical descriptions of the Earth's geophysical response to changing loads, GIA models use present-day bathymetry/topography, reconstructions of ice-sheet advance/retreat and suites of sea-level index points (SLIPs) as input data (see Milne 2015). The models then iteratively solve the values of properties describing the response of the Earth's crust and mantle in response to shifted loads of ice and water over the planet.

For ice-mass centers such as Scandinavia and Scotland the modeled GIA comprises subsidence during ice build-up and rebound during deglaciation, tailing out in postglacial times. For ice-sheet peripheral regions such as the central and southern North Sea (in this context dubbed 'Near Field Regions'), it means upwarping during ice build-up and accelerated subsidence during deglaciation (e.g. Lambeck 1995; Shennan *et al.* 2000; Vink *et al.* 2007). Combined with the inland position of the southern North Sea on the continental shelf, this means that in each interglacial, this area is transgressed relatively late, but when it does, rates of sea-level rise are relatively high ('drowns last, sinks fast'). This allows sufficient time for a boreal-to-temperate vegetation cover to develop and substantial postglacial soil formation to take place on the fresh surfaces of depositional landscapes inherited from the preceding glacial, before such surfaces are transgressed and drowned. This is a geographical property of the North Sea basin that affects the taphonomy of terrestrial surfaces buried and preserved by transgressive units. It is most evident for Late Glacial and Holocene drowned landscapes, for which radiocarbon dating provides an accurate independent age control on the timing of interglacial warming and marine transgression (e.g. Törnqvist *et al.* 2015), but applies equally to, for example, the basal surfaces of coastal plain units of the Cromerian Complex stage (e.g. page 156) and the Eemian stage (e.g. page 160).

GIA modeling results tend also to be expressed as a time-stepped series of maps that show the intersection of the sea surface with the warped topography/bathymetry (e.g. Fig. 7.10). In recent decades, GIA models have been improved and several generations of GIA-modeled North Sea drowning histories exist (e.g. Sturt *et al.* 2013 is an upgrade of Lambeck 1995). Both regionally and globally resolved versions of GIA models exist. Peltier (2004),

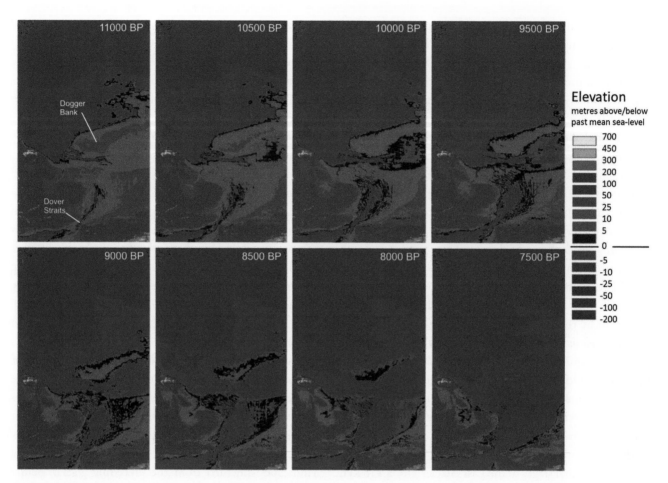

Fig. 7.10 GIA-modeled transgression of the North Sea during the Early Holocene. The Dogger Bank transforms into an island and rapid marine incursion occurs from both the north and the south. Figure created using supplementary images from Sturt *et al.* (2013), with permission from Elsevier.

for example, gives a global solution, whereas Lambeck (1995), Vink *et al.* (2007), Brooks *et al.* (2011) and Sturt *et al.* (2013) give regional solutions (see also Peltier *et al.* 2002; Steffen & Wu 2011).

Of the regional GIA-model solutions entertained for the North Sea, some optimize for the SLIP dataset around Britain (e.g. Lambeck 1995; Shennan *et al.* 2000; Peltier *et al.* 2002; Ward *et al.* 2006; Sturt *et al.* 2013), while others see the region as the periphery of Scandinavia and optimize using different SLIP datasets (e.g. Lambeck *et al.* 2006; Vink *et al.* 2007). Overlap in the SLIP datasets used also occurs, especially for offshore data points from the central North Sea. Furthermore, unlike the uplifted coastal areas of Scotland and south Scandinavia where the SLIP datasets cover both the Late Glacial (when rates of GIA were relatively fast) and Holocene (when GIA began

to tail out), in peripheral areas to the south, SLIP datasets are biased to the Middle and Late Holocene. Efforts to synchronize the data sets in use by the various sea-level research, GIA modeling and offshore archaeology groups of the different countries around the North Sea are underway and include database protocol activities (Hijma *et al.* 2015).

GIA-modeled paleocoastline maps for the North Sea are sensitive to the choice of model setup and completeness of SLIP datasets that are used. In particular, for the Early Holocene (11–8 ka), there is considerable variation in the timing of drowning of the parts of the central North Sea now 20 m to 50 m deep, between different GIA-model studies — especially between areas that are spatially peripheral to one study but central in another. Dates that are collected for local geological and archaeological reasons should be used to verify the GIA model-predicted shorelines and, when related to timing of transgression, should feed into SLIP databases to improve future GIA-paleogeographical modeling.

Despite differences in approach, all GIA-modeled map series of recent decades show an initial period of relative stability in shoreline position around the northern and central North Sea between ca. 21 ka to 14 ka, when isostatic warping kept pace with rising glacio-eustatic sea level. The southern North Sea at this stage remained dry land and hosted river valleys that continued towards the English Channel. From ca. 14 ka onwards, marine transgression was rapid throughout, flooding the exposed land from the north and from the south via the Dover Strait. The most recent GIA modeling study calculates that the terrestrial connection between Britain and Europe began to be overtopped at ca. 9.5 ka and the final large island in the central North Sea (the present Dogger Bank) was submerged by ca. 8 ka to 7.5 ka (Sturt et al. 2013; Fig. 7.10).

Independent of the SLIP datasets used in the GIA modeling, marine sedimentary indicators show the establishment of mixed southerly and northerly sources of water in the Skagerrak by ca. 8.5 ka (Streif 2004; Gyllencreutz 2005). This indicates when water depths in a corridor between the southern and central North Sea had increased enough to allow establishment of current systems similar to today (see also van der Molen 2002) and thus when marine erosion and reworking processes that were the final taphonomic control on preserved drowned landscapes and archaeological sites began.

From 7.5 ka, Fig. 7.10 shows the shape of the North Sea as similar to that today. In part, this is an artefact of the bathymetric/topographic data used as input to generate the maps, which show recent features that simply did not exist during the modeled time steps. For instance, these data include the morphology of beach-barrier systems that formed in the last 7.5 kyr. The coastal zones of the southern North Sea, in particular the sandy shores of Flanders, Zeeland, Holland, the Waddenzee and the German Bight show backstepping of the coastal system until 6.5 ka (southwest Netherlands: Beets & van der Spek 2000; Hijma et al. 2010; Hijma & Cohen 2011) or ca. 5 ka (Waddenzee, German Bight: Oost 1995; van der Spek 1996; Streif 2004; Vos 2015) and maturing barrier systems thereafter. Major parts of the presently

barrier-protected Waddenzee and Dutch coastal plain were, at 7.5 ka, exposed to the open North Sea and the true coastline was positioned inland of that depicted in Fig. 7.10.

These effects are also relevant to the western side of the southern North Sea. For instance, although the model does account for Late Holocene infilling of the low-lying East Anglian fens (by using borehole-derived isopach maps instead of modern topography: Sturt et al. 2013), it does not incorporate erosion and retreat of the unconsolidated cliffs along the outer coast of East Anglia (e.g. Dong & Guzzetti 2005). For the coastal zones of the North Sea, geological mapping and paleogeographical reconstruction documenting their evolution is well developed, but incorporation of such mapping in GIA-modeled shorelines to cover not only the offshore, but the full North Sea, is not (Cohen et al. 2014). This most affects taphonomic assessments for the period between 8 ka and 3 ka, especially in nearshore, tidal inlet and estuarine areas. For taphonomy farther offshore, the GIA-modeled visuals for the North Sea are of more direct use.

Because of the gentle gradient of the shallow shelf, a stepwise acceleration between 8.45 ka and 8.25 ka (as initiated by the meltwater pulse triggering the 8.2 ka Event) may have been of particular importance to the drowning of the central North Sea at the turn of the Middle to Late Mesolithic. The onset of this sea-level rise acceleration is radiocarbon dated in transgressed terrestrial sediments in the southern North Sea off and below Rotterdam (Hijma & Cohen 2010). The structure of this sea-level jump, which occurred on top of a background rate of about a meter of rise per century, reveals that at two moments within these two centuries, sea level jumped by a meter (because of events on the other side of the Atlantic Ocean, where Canadian lakes Agassiz and Ojibway drained as Hudson Bay and the Tyrrell Sea became ice-free; Teller et al. 2002). Along the shores of the contemporaneous North Sea (e.g. Vos et al. 2015: Maasvlakte area, off Rotterdam harbor), a meter of sea-level rise would have occurred within the lifespan of a human generation (say 20–25 years), and another such meter-scale jump about 150 years later. In discussions of coastal Mesolithic archaeology, transgression as a push factor to inland migration and contact between coastal and terrestrial

Mesolithic societies is often mentioned (e.g. Smith *et al.* 2011). Superimposed events such as the 8.45 ka to 8.25 ka twinned sea-level jump could have catalyzed this worldwide (e.g. Turney & Brown 2007) and in the North Sea, the Storegga tsunami occurring ca. 8.1 ka (Rydgren & Bondevik 2015) could have had a similar impact (e.g. Dawson *et al.* 1990; Weninger *et al.* 2008). In the context of this book and chapter, it is the impact on preservation and taphonomy of the superimposed transgression events that should be noted. For the Storegga tsunami, this could be considered one of destruction, but for the sea-level 'jumping' in centuries before it appears to have improved preservation of surfaces (Hijma & Cohen 2011) and actual archaeological sites (Vos *et al.* 2015), at least in the southern North Sea nearshore paleovalley settings that were transgressed by the event.

Given the role SLIPs play in GIA-modeling in determining regional modeling outcomes and predicted paleogeographies, their collection and availability is particularly important. The sides of drowned paleovalleys in particular provide suitable circumstances for collecting series of SLIPs (Vis *et al.* 2015). Furthermore, it is useful to group offshore SLIPs for the North Sea by parent paleovalley system/drainage network, because valleys were transgressive pathways. Although an estimated 400 suitable SLIPs are presently available from the British, Belgian, Dutch, German and Danish North Sea sectors and coastal zones, these mainly cover the Middle and Late Holocene depth intervals (final 15–20 m of sea-level rise; e.g. Shennan *et al.* 2006). Data points for 14 ka to 7.5 ka, obtained from offshore areas where the North Sea is 20 m to 60 m deep, are much more scarce (uplifting coastal areas of Scotland and Norway excluded). Shennan *et al.* (2000) and Hazell (2008) report 15 to 20 such data points for the British sector, and some 20 more from the Dutch and German sectors collected in the 1970s. With recent archaeological dating campaigns in the various wind farm localities, the number of offshore collected SLIPs is now increasing on the British side. Vink *et al.* (2007) list another 60 offshore data points from the Dutch and German sectors combined (Fig. 7.11) and Alappat *et al.* (2010) add five more data points from the offshore Elbe paleovalley.

Besides collecting SLIPs from the offshore, typically from below 25 m depth, collecting them from the nearshore region is critical to cover the interval between 15 m and 25 m deep. All along the southern North Sea — the English coast, the Dutch-Belgian shores, the Frisian Islands (Waddenzee) — this is the depth range where 9 ka to 8 ka sea levels are projected, but where data points are particularly scarce. The reason is that deposits at this target depth in these coastal/nearshore environments are difficult to survey and core either from the sea or the land. From the sea, the difficulty arises firstly because at water depths of just a few meters seismic techniques investigating the target depth suffer from multiples of the water/sediment contact, and secondly because to reach the target transgressed surfaces, meters of recent shoreface and coastal sediment have to first be cored. From the land, depths of 15 m to 25 m are cored infrequently. Regionally along the North Sea this makes the nearshore area a 'white zone' for direct data on submerged landscapes, and has notably affected SLIP datasets too. Exceptions to this are areas with harbors and concentrations of economic and building activity, such as the Rotterdam harbor complex that overlies the Early Holocene valley of the Rhine and Meuse. Hijma and Cohen (2010) and Vos *et al.* (2015) added ten SLIPs from this critical 15-m- to 25-m-deep interval.

Relative sea-level rise, due to residual GIA and background tectonic subsidence combined, has raised the seawater surface of the North Sea further above the seabed by ca. 5 m to 8 m in the last 6 kyr, compared to a rise of 15 m to 35 m between 9 ka and 6 ka (lower values for the south than for the north, see Fig. 7.11). In the last 6000 years, sedimentary processes in the shallow sea (creating offshore sand-wave fields, shore-connected ridges, foreshore and beach bar systems, tidal inlets) controlled the preservation and taphonomy of drowned and buried landscapes of the nearshore zone (see 'Physical geography and modern environment', pages 150–152).

Outlook on data, mapping and reconstruction quality

Before discussing the submerged landscape inventory and taphonomic ramifications of the above, it should be noted that the overview presented here represents only the broadest picture, whereas dedicated original

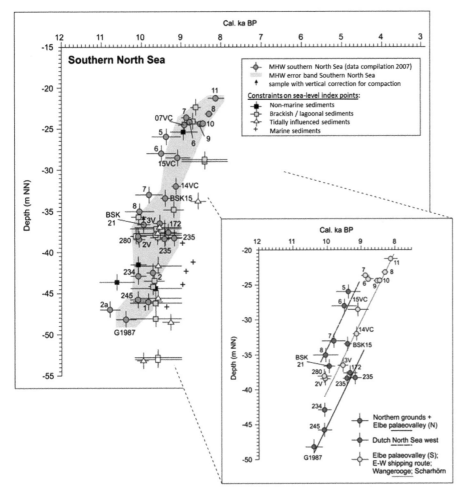

Fig. 7.11 Time-depth distribution and local mean high water (MHW) error band of sea-level index points obtained from basal peat data from the German Bight ('Elbe paleovalley') and Southern Bight ('Dutch North Sea west'), as compiled from various data sources. Inset shows breakdown by subregion. Early Holocene paleoshorelines of the same age in the Elbe area occur at greater depth than in the Southern Bight, reflecting differential GIA. Figure from Vink *et al.* (2007).

studies at subregional scale provide much more detail (e.g. Hijma *et al.* 2012; Tizzard *et al.* 2014). Furthermore, paleogeographic reconstruction inevitably involves a degree of uncertainty and interpolation, particularly where there are gaps in the geological record caused by erosion as well as uncertainties in the dating and spatio-temporal correlation of the parts of the record which are preserved — and the examples from the North Sea show this (Cohen *et al.* 2014). Inevitably, the gaps and uncertainties are greatest for the earlier parts of the record, generally reducing as the data get younger. Thus, for the Middle Pleistocene, paleogeographic and paleolandscape changes are only reconstructed on a basin scale, usually as rather generic highstand/lowstand scenarios with considerable uncertainties in the timing of events and exact position of features such as rivers and coastlines. For the Late Pleistocene and the post-Last

Glacial Maximum (LGM), the data allows time-stepped reconstructions at millennial intervals to be produced, ideally by combining information from GIA-modeling results and geological reconstructions, including those based on seismic data (e.g. Gaffney *et al.* 2007; 2009; van Heteren *et al.* 2014).

In the best-case scenario, detailed investigations combining multidisciplinary data and secure chronology can provide a more refined reconstruction which includes the identification of specific landform assemblages and also accounts for changes in erosion and deposition patterns (Bicket 2013; Cohen *et al.* 2014). For the Dutch sector, Erkens *et al.* (2014) assessed whether the current wealth of data (Figs. 7.2 & 7.3) would allow reconstructions of the submerged terrestrial geology offshore comparable to those for the onshore. It was concluded that the desired level of detail of paleogeographic reconstructions suitable for archaeological prediction could not be achieved without re-interpreting most of the existing data and carrying out additional fieldwork at considerable cost. In this light, it is good that ca. £60 million (approx. $73 million) is being spent on surveying and understanding the British part of Dogger Bank alone, as part of wind farm planning and development (Forewind consortium; www.forewind.co.uk).

Evidence of Submerged Landscapes on the Shelf

The North Sea holds a range of archaeological, sedimentary and geomorphological evidence for submerged prehistoric landscapes, i.e. drowned preserved former land surfaces, conserved and not substantially eroded or reworked by younger landscape and seabed-forming processes. This evidence has been found in different parts of the sea basin at a range of depths (both water and burial depth) reflecting patterns in preservation and taphonomy. Detection and reconstruction of buried elements of the submerged paleolandscape has, in recent years, been greatly aided by the provision of vast seismic data sets, often originally collected for hydrocarbon prospection (e.g. Gaffney et al. 2007; 2009; van Heteren et al. 2014). For a full review of the inventory of features composing the former landscapes in the North Sea, see Cohen et al. (2014). Reviews of the archaeological evidence can be found in Peeters and Momber (2014) and Roebroeks (2014).

Geomorphological evidence includes a range of glacial and periglacial landforms. The geological literature pays considerable attention to the North Sea's many subglacial tunnel valleys. These have been found across the sea basin, principally to the north of the Southern Bight, from all glaciations covering the central North Sea (Huuse & Lykke-Andersen 2000; Praeg 2003; Stewart et al. 2013). Upon deglaciation, the tunnel valleys developed into lakes, coastal embayments or local deeps in a shallow shelf sea and were accordingly filled with whatever sediments they trapped. Because of their transformed state, the tunnel valleys are of archaeological landscape relevance.

An example for which the postglacial archaeological potential has been explored is the Outer Silver Pit (Fig. 7.2; Fitch et al. 2005; Fitch 2011). These studies highlight that the shores of larger freshwater bodies would be attractive base-camp locations for terrestrial hunter-gatherer communities. Furthermore, tidal conditions were in play early on in the transgressive episode (Fig. 7.10; 10.5–10 ka BP), which established a depositional environment that could cap and protect former lake shores from erosion in later stages. Tunnel valleys are of further relevance as seismo-stratigraphic marker contacts that can be regionally traced through 3D seismic volumes and allow the distinction of Pleistocene sequences and associated relative ages. Other currently identified glacial/periglacial landforms include iceberg scours (particularly in the northern North Sea), till sheets with mega-scale glacial lineations (north of maximum ice limits), and moraines, ice-pushed ridge complexes and ice-marginal river valleys in the maximum ice-limit regions (Busschers et al. 2008; Moreau et al. 2012).

Preserved river valley floors (floodplains and channel fills) from periglacial and interglacial landscapes form a second major component of North Sea paleolandscapes. Some of these are particularly well-traced using 3D seismic data in the central North Sea in the vicinity of the Dogger Bank and Outer Silver Pit (Fig. 7.12; See Gaffney et al. 2007; 2009; van Heteren et al. 2014). Most of the mapped channels from this area are inferred to be of Late Glacial and Early Holocene age (where they connect to the Outer Silver Pit, which is itself of Devensian age), while older examples can also be seen in the data at greater depths (Fitch et al. 2005). Paleochannel fills and their channel belts have also been identified off the Netherlands (Hijma et al. 2012), northern Germany (Konradi 2000; Streif 2004), Belgium (Mathys 2009) and East Anglia (Bridgland et al. 1993; Bridgland & D'Olier 1995; Dix & Sturt 2011; Tizzard et al. 2014). Other identified interglacial features include lakes, marshes, floodplains, estuaries (with tidal channels), inland eolian dunes, and various nearshore sedimentary bodies (Hijma et al. 2010; Cohen et al. 2014; van Heteren et al. 2014). In general, basal parts of these features are better preserved because of

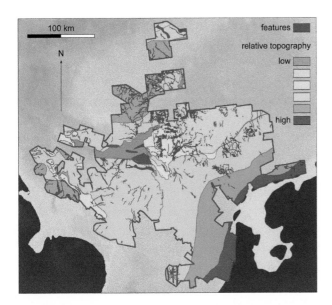

Fig. 7.12 Early Holocene paleolandscape reconstruction derived from 3D seismic data showing the area around the Dogger Bank, spanning the UK and Dutch sectors. Features in blue are mapped water bodies (rivers, estuaries and lakes); color coding shows relative topography. Van Heteren *et al.* (2014). Reproduced with permission from Cambridge University Press.

erosive truncations of their tops since transgression (tidal scour, wave ravinement — see 'Physical geography and modern environment', pages 150–152).

Over considerable areas offshore, because the Early Holocene transgression was rapid (Fig. 7.10) and rates of sea-level rise high (Fig. 7.11), the odds for preservation of patches of terrestrial landscape offshore may be better than in the coastal zone of the Middle Holocene (due to tidal scouring of inlet channels), especially where depositional terrestrial environments were transgressed. In general, Early Holocene valley surfaces would be expected to preserve better (intact soils) than the surfaces of their interfluve areas (ravinement decapitated). Moreover, even if the Early Holocene valley surface was transgressively eroded, the truncated valley deposit may have buried a Late Glacial precursor surface, so there would still be a chance that a paleolandscape surface is preserved and *in situ* archaeology encountered.

Some of the aforementioned paleolandscape elements have been identified from sedimentary (i.e. lithological,

biostratigraphical) rather than geomorphological evidence, generally obtained from core or borehole samples. While the geomorphological evidence, derived principally from seismic and to a lesser extent bathymetric data, is well suited to regional-scale landscape mapping, the sedimentary record is essential in providing paleoenvironmental (e.g. freshwater vs. marine sediment) and dating evidence that allows a more accurate paleolandscape reconstruction, and is crucial when trying to reconstruct change over time (van Heteren *et al.* 2014).

The most immediately obvious paleoenvironmental resource comprises submerged peats formed upon flooding of the terrestrial landscape (Arends 1833; Reid 1913). *In situ* basal peats are compacted peats, buried by tidal brackish deposits marking transgression (Jelgersma 1979; Vis *et al.* 2015). Throughout the central and southern North Sea, these date to the Late Glacial and Holocene depending on depth and geographical position (Fig. 7.11). They have been found at various localities including the Dogger Bank (Ward *et al.* 2006; Hazell 2008; Vink *et al.* 2007; Alappat *et al.* 2010; Hijma & Cohen 2010). Older examples have been reported, for instance, as part of the top of the Yarmouth Roads Formation (Zagwijn 1979; Funnell 1996). These older peats represent the remnants of the coastal plain transgressions from the Cromerian Complex (see Fig. 7.4) and have been correlated, based on stratigraphical position and contained palynology, with the onshore formations that contain *in situ* Paleolithic artifacts at Pakefield and Happisburgh (Parfitt *et al.* 2005; Wessex Archaeology 2009; Bicket 2013; Tizzard *et al.* 2014).

Overall, a range of other identified sedimentary deposits representing different environments and periods ranging from the Middle Pleistocene to the Holocene includes tidal flat, periglacial eolian, lagoonal/lacustrine, floodplain, channel infill, river-valley inland dune, marine, glacio-lacustrine and glacio-fluvial deposits (Cameron *et al.* 1992; Gatliff *et al.* 1994; Balson *et al.* 2002; Hijma *et al.* 2012; Cohen *et al.* 2014; Tizzard *et al.* 2014). Besides peat beds, it is negative relief landforms which host lakes and collect lacustrine fills that provide excellent paleoenvironmental records, which is an extra reason to map (Fig. 7.11) and target them for coring (e.g. Tizzard *et al.* 2014).

An aspect which remains under-researched is vegetation reconstruction. So far, pollen studies from these submerged environments are few and far between; hence, vegetation reconstructions are reliant on data from land which may be over 100 km away (Peeters & Momber 2014). As an illustration, the Late Glacial and Early Holocene vegetation developments in Doggerland's northerly interfluve landscapes, between the major Elbe (German Bight) and Rhine-Thames (Southern Bight) paleovalleys, could be expected to deviate from those in the far west of onshore northern England, the far south of onshore Netherlands and Germany, and the far east of Denmark and Sweden. For example, the timing of the invasion of a key tree species such as *Corylus* (hazel) into Doggerland is not yet directly established. *Corylus* is important because its nuts were a major Mesolithic food resource and charred remnants are often indicative of human activity or settlement. The appearance of *Corylus* in pollen records also marks the Preboreal-Boreal transition in the regional pollen zone scheme. Direct correlation to onshore areas such as the eastern Netherlands (van Geel *et al.* 1980) would place this transition at ca. 10.25 ka (ca. 9125 ±90 ^{14}C BP based on radiocarbon dates from the eastern Netherlands), but it might be a few centuries later in the Dogger Bank region since it lies some 300 km further north.

The aforementioned deposits and landforms are the evidence of the former landscape, and provide fossil material indicative of the various paleoenvironments of the time (interglacial vs. glacial flora and fauna, terrestrial vs. aquatic species, freshwater vs. marine biota). In principle, this also applies to archaeology (e.g. harpoons encountered from wetland and coastal environments) although hunter-gatherers may also have moved resources between environments, especially in coastal and wetland environments (Sturt 2006; Waddington *et al.* 2007; Amkreutz 2013; Bell *et al.* 2013; Moree & Sier 2014; 2015; Peeters *et al.* 2014; 2015b). Concentrations of faunal remains appear to be in the same areas as the archaeological remains, coming principally from the Brown Bank and Eurogeul navigation channel (five nautical miles west of Rotterdam harbor), with additional material from the Zeeland Ridges, the Belgian coast and

off East Anglia (van Kolfschoten & van Essen 2004; Mol *et al.* 2006; Hublin *et al.* 2009; Peeters *et al.* 2009; SeArch 2014; Tizzard *et al.* 2014).

The vast majority of finds from the North Sea are paleontological, comprising a range of Pleistocene and Holocene fauna such as mammoth, reindeer, horse and bison. To date, thousands of fossil bones have been landed, principally in the Netherlands, by trawlers and dredgers (Glimmerveen *et al.* 2004; 2006; Mol *et al.* 2006; van der Plicht & Palstra 2016). On paleontological grounds, and for the younger assemblages by radiocarbon dating, the trawled bone finds fall into four broad periods (Fig. 7.4): Early Pleistocene, early Middle Pleistocene (Cromerian), Late Pleistocene (Weichselian) and Holocene, with the majority dating to the latter two intervals (van Kolfschoten & van Essen 2004). It is worth noting however that the Late Glacial Period, as with the archaeological record (discussed below), is under-represented with only two radiocarbon-dated finds falling within the period between the LGM and the earliest Holocene (ca. 20–10 ka) versus more than 50 radiometric dates from the pre-LGM and Early Holocene periods (Peeters & Momber 2014). In recent papers, the limits of radiocarbon dating of bone material (van der Plicht & Palstra 2016) and of shell material (Busschers *et al.* 2010) from within Late Pleistocene strata of the North Sea are under discussion.

Some areas and formations, based on collections over the past decades, have provided far greater numbers of fossils than others. For instance, van Kolfschoten and van Essen (2004) specifically identify the Yarmouth Roads Formation (Early-Middle Pleistocene deltaic to non-marine), the Brown Bank region (Fig. 7.9a) and the Rhine-Meuse paleovalley channel sands and floodplains (Fig. 7.9b) as the source of fossil remains (see also Ward & Larcombe 2008; Hijma *et al.* 2012). Where these fossils are trawled from the actual seabed, marine reworking and winnowing processes are typically considered to explain part of the apparent concentration of fossils at the sea floor (whether in the last 6000 years by wave action, or in earlier stages of marine transgression). That said, it must be remembered that each of these areas encompasses reworking fluvial channel environments as well as flood-plains, and that they formed over periods with oscillating climate and sea levels so that multiple phases of reworking are part of these units. Kuitems *et al.* (2015) provide an example from a sand extraction location off Rotterdam harbor as a local study, and Hijma *et al.* (2012) do so for the southern North Sea as a regional study. Local and smaller regional studies show that the context of finds from within subsurface units can only be revealed by detailed geological investigation. Doing so in more places, and also farther offshore, would be a step forward in improving assessments of taphonomy and paleogeography (e.g. Tizzard *et al.* 2014; Kuitems *et al.* 2015).

The evidence for hominin occupation of the North Sea presently comprises a variety of artifacts and bones trawled/dredged from the seabed. At present, the earliest known archaeological finds are Middle Paleolithic and include lithics from Area 240 (Fig. 7.3) off East Anglia (inferred date: Middle Saalian; ca. MIS 8–7: Tizzard *et al.* 2014) and the Zeeland Ridges off the Netherlands, the same area of the sea floor which has also provided a Neanderthal skull fragment (Hublin *et al.* 2009; Peeters & Momber 2014). The potential for earlier finds is suggested by trawled/dredged Early and Middle Pleistocene fauna and the presence of flint artifacts and footprints in Cromer Forest-bed sediments exposed on the lower beach and foreshore at Happisburgh, at the base of the eroding cliff (Ashton *et al.* 2014), of which extensions could also be preserved offshore, within reach of aggregate mining and wind-farm foundation activities (Wessex Archaeology 2009; Cohen *et al.* 2012; Bicket 2013; Ward *et al.* 2014).

Upper Paleolithic finds are rarer and comprise a barbed point from the Leman and Ower banks trawled up in 1931 (radiocarbon dated to 11,740 ±150 ^{14}C BP: Housley 1991) and a possible worked flint from the Viking-Bergen Bank in the far north of the sea basin (Long *et al.* 1986; Coles 1998; Peeters & Momber 2014). The majority of finds to date are Mesolithic and include human remains, lithic, bone and antler artifacts. Relatively many human remains, as well as some of the bone and stone implements, come from the De Stekels area south-west of the Brown Bank.

The Maasvlakte-Europoort area off Rotterdam has produced over 500 bone and antler implements including barbed points (Louwe Kooijmans 1975; Verhart 2004; Peeters & Momber 2014). Also the Eurogeul navigation channel has also produced bone artifacts dating to the Mesolithic (Glimmerveen *et al.* 2004; Mol *et al.* 2006). Most recently, excavations in the second Maasvlakte harbor extension (Yangztehaven) have led to the discovery of an *in situ* Mesolithic assemblage at ca. 20 m depth, recovering thousands of bone fragments (including many charred bones), charcoal, lithics, fish and plant materials (including charred plant materials; Moree & Sier 2014; 2015 and contributions therein). This material was found by systematic archaeological sampling, guided by the results of detailed geological mapping (Vos *et al.* 2012; Weerts *et al.* 2012; Borst *et al.* 2014) starting from established paleogeographical understanding of the geology and archaeology of the region (Louwe-Kooijmans 1975; 1980; 2005; Hijma & Cohen 2010; Vos *et al.* 2011; Vos 2015), including the taphonomy. That taphonomy for the isolated inland dune upland within the Rhine–Meuse wetland involves post-depositional processes such as decay/selection, horizontal and vertical displacement and diffusion/blurring of the archaeological assemblages which collect in the wetland zones at the foot of local uplands (Amkreutz 2013:75), and the way further depositional developments bury, seal, compact and further seal and bury the artifact-bearing strata as transgression proceeds (Vos *et al.* 2015). From a sand-extraction area just offshore of the *in situ* Mesolithic assemblage, human skull remains were found (Borst *et al.* 2014), for which radiocarbon dating produced an age matching that of the Yangtzehaven base camp (Weerts *et al.* 2015).

Finally, post-Mesolithic finds have also been recovered, in the form of Neolithic axes from the Brown and Dogger banks (Peeters *et al.* 2009; Peeters & Momber 2014). The last shallow parts of the southern North Sea and the Dogger Bank area drowned by 7.5 ka to 7.0 ka (Fig. 7.10), just prior to the Mesolithic to Neolithic transition in the Netherlands and England. At the present state of research it is unclear whether these Neolithic axes could indicate *in situ* sites on former islands in the North Sea.

Taphonomy

Taphonomic variables

In 'Quaternary background and paleogeographic framework' and 'Evidence of submerged landscapes on the shelf', links were made between paleogeographical circumstances and developments and the taphonomic variables that are summarized in this section. Similar to paleogeography, the background controls on landscape preservation — and therefore archaeological preservation and site taphonomy — are tectonics, glaciation and sea-level change. The continuous landscape changes over multiple glacial-interglacial cycles described above mean that landscape-formative and record-preserving processes and conditions varied spatially through the period of interest.

Over the 500,000 to 1 million years covered by the archaeology of the region, this has created a very complex and fragmentary geological and archaeological record. A first step to organise it is to consider what deposits should be considered locally to contain archaeology (either *in situ* or concentrated reworked assemblages) and what not, from a climate perspective and past hominin climatic/environmental tolerances. For the Middle Pleistocene time frame, preserved deposits relating to interglacial (or even early/late stage glacials before/after maximum cooling) land surfaces are most relevant, given the archaeological information on climate tolerances of the hominins occupying the North Sea area and immediate surroundings at that time (e.g. Pettitt & White 2012). From the Late Pleistocene onwards, archaeology can be expected to be associated also with preserved surfaces within deposits of colder periods, with boreal forest, steppe and steppe-tundra environments alternating in the uplands, with the Rhine-Meuse and Thames river valleys with associated vegetation running through them (Fig. 7.9b). Throughout the Holocene, the North Sea area from a climate perspective was habitable.

A second step is to consider the controls on spatial patterns of preservation, in order to characterize regions based on the time depth of the archaeology that might be encountered in the first meters below the seabed (e.g.

the first 10 m). Once again, these controls are tectonics, glaciation history and sea-level history primarily. These have been discussed above in relation to the Cromerian Complex of the Southern Bight (early Middle Paleolithic, pages 154–159). Crustal uplift/downwarping driven by long-term tectonics and glacio-isostasy is a taphonomic variable for younger deposits too, but becomes less influential with the increasingly direct impact of glaciation.

Sea level, glaciation, glacio-hydro-isostatic and climate history are strongly correlated in their cyclicity with ca. 100 kyr periodicity in the last 1 Myr. These correlations and interrelations make this second step of listing the taphonomic variables for the North Sea more complex than the first one. Also, for the North Sea, because of the vicinity to the Scandinavian ice-mass, tectonic basin longevity, and shelf size and gradient, this may work out differently compared to surrounding shelf areas and marine basins (i.e. English Channel, Baltic Sea). See Cohen and Lobo (2013) for a global perspective on the alternating morpho-sedimentary processes and human habitats on shelves.

With regard to tectonics, subsidence promotes sediment accumulation, which in turn affords preservation through burial. Since accumulation requires a supply of sediment as well as accommodation space for deposition, the location of depocenters represents a critical taphonomic variable. The depositional pattern also means that deposits are stacked vertically with younger ones overlying their older counterparts. Given the huge supply initially provided by the Eridanos River, the majority (ca. 80%) of the Quaternary succession — in southern depocenters over 500 m, towards the north over 1000 m thick — comprise shallow marine pro-deltaic and deltaic deposits of essentially this river system (Cameron *et al.* 1992; Funnell 1996). In contrast, on the basin shoulders where subsidence is minimal, preservation tends to be poor owing to reworking of unburied sediments. In special situations though, preservation is possible, and the minimal subsidence results in a complex mix of deposits of different ages located at roughly the same level.

This latter situation characterizes the Cromerian Complex sediments of East Anglia which are a composite of coastal highstand deposits of different ages interspersed with discontinuous transgressive and fluvial sediments respectively laid down during periods of sea-level rise and fall. Preservation of interglacial deposits here was enabled firstly by sediment supply at times with prolonged slow rates of sea-level rise which promoted coastal aggradation. Secondly, relatively consistent paleogeography promoted superposition of coastlines during successive interglacials (see Fig. 7.7a). Thirdly, restricted extents of ice sheets prevented direct glacial erosion and when glaciation did occur in MIS 12/the Anglian (Fig. 7.4), it buried the coastal plain deposits under a thick till sheet rather than eroding them, owing to waterlogged ice-front conditions (see Fig. 7.7b and Cohen et al. 2012).

It should be realized that rates of GIA (see 'Late Glacial and Holocene', pages 161–164) are an order of magnitude greater than long-term tectonic subsidence rates of the North Sea basin (Kooi et al. 1998) and that glaciations of the central North Sea have left positive topographic features that need multiple glacial-cycles to be topographically leveled out or sink away in sedimentation-filling subsidence-created accommodation space in the immediate surroundings. Besides controlling rates of subsidence and relative sea-level rise in times of postglacial transgression, during the part of the glacial interval that the North Sea floor was terrestrial, GIA in tandem with background tectonics affects the morphology of terraced valleys (Maddy et al. 2000). It is also part of the reason why considerable valley incision is seen over the tectonic depocenters (Fig. 7.5), as features in the 3D seismics from the central North Sea confirm (Fig. 7.12). The continued effects of tectonics and isostasy mean that one should not expect age-equivalent deposits to be situated at the same depths.

At times of upwarping towards the glacial maximum, GIA amplifies the depth of incision, whereas during deglaciation it amplifies the thickness of the eventual valley fill as the area subsides (Busschers et al. 2007; Hijma et al. 2012). Furthermore, the lower reaches of a tributary system tend to grade to the larger trunk river that they join, which has more stream power and erosive capacity. Submerged reaches of the Thames, Medway and Scheldt in the vicinity of the Axial Channel through the Dover Strait (Fig. 7.8b, Figs. 7.9a,b) thus have

steeper gradients and vertically more pronounced terrace staircases than inland reaches. Submerged terraces have been identified offshore from bathymetric and seismic data (e.g. Bridgland et al. 1993; Bridgland & D'Olier 1995; Hijma et al. 2012; Fig. 7.9a), though care should be taken in attempting correlations between onshore and offshore terrace systems (Bates et al. 2007). These two glaciation-related particularities of the river systems of the North Sea — GIA effects and proglacial pathway inheritance — could have bearings on the taphonomy of preserved surfaces and their contained sites (e.g. Pettitt & White 2012).

A further taphonomic variable for river valleys, directly related to the tectonic setting and incisional vs. aggradational regimes, is the substrate composition of valley floor and valley side. These differ greatly between the seabed over depocenters off the northwestern Netherlands (Fig. 7.5) where the substrate is mainly sand (apart from the occasional patch of glacial gravels), and the seabed between Flanders and Sussex in the Southern Bight where gravelly sand (including abundant flint) occurs. This availability of gravel-sized lithologies in former terrestrial and coastal landscapes would affect tool-making hominins, particularly in the Paleolithic and could be an additional control on site distribution (Cohen et al. 2012; Bicket 2013).

Following the increased intensity of climate oscillations, starting 1 Ma and becoming more direct by ca. 450 ka, glaciations and sea-level oscillations increasingly overprinted the taphonomic contribution of tectonics. The direct impact of glaciation is generally regarded as negative, resulting in the erosion and reworking of the pre-glacial sedimentary record. This is most notable in the relatively rarity of in situ interglacial landscapes compared to glacial and periglacial deposits since they tend to be reworked or destroyed by the subsequent glaciation (Cohen et al. 2014). Large glaciation tends to be destructive. Exceptions can of course occur and certainly did in East Anglia (see pages 154–159), but only immediately south of the margins of the glaciation at maximum limit. This also means that the preservation of post-LGM (i.e. Late Pleistocene/Holocene) deposits is enhanced relative to earlier periods as they have not yet been affected by subsequent glaciation. On the other hand, erosional features created by glaciation can form sinks (e.g. tunnel valleys, kettle holes) in which later interglacial or postglacial deposits can accumulate. Water-laid sequences accumulated in such depressions could provide excellent paleoenvironmental information, but the presence of archaeological evidence in these sequences is unlikely unless lakeshore/bankside sediments are also preserved (see also 'Evidence of submerged landscapes on the shelf', pages 166–169).

Outside the ice margins, coversand/loess deposition, enhanced in rate and extent by periglacial conditions, can provide a protective blanket for archaeological sites and associated landscapes. This has been implicated in the potential differential preservation of Paleolithic sites in Britain relative to their continental European counterparts during MIS 6 to MIS 4 (Hijma et al. 2012). The effects are however unclear for the exposed North Sea landscape as it may have been the source, rather than a sink, of the eolian sediment present in the surrounding uplands (see also Antoine et al. 2009).

Sea-level change affects preservation by shifting the erosive effects of waves and tides across the continental shelf. That said, the actual degree of impact is also dependent on the rapidity of sea-level change (at the rates seen in the Late Glacial and Early Holocene, with accelerations due to meltwater pulsing), the nature of the coastline in question and the local sedimentary regime (Flemming 2004). For instance, back-barrier lagoons and the inner parts of estuaries tend to experience lower energy regimes compared to exposed outer coasts, while significant deposition driven by fluvial supply and/or longshore transport can promote preservation through burial. Growth of 'basal' peat induced by rising sea level can also provide a further means of protecting surfaces in the run up to true transgression (submergence of the peat surface), when the submerged peat is buried by mud and compacts. In particular, submerged deltaic deposits from transgressive stages in the valleys of larger rivers may contain and cover intact terrestrial surfaces with in situ archaeological remains. This is the case over large areas in the Rhine-Meuse paleovalley offshore, nearshore and onshore (Hijma & Cohen 2011; Hijma et al. 2012; Vos et al. 2015) and would also apply

to the Elbe valley (Figge 1980; Vink et al. 2007; Alappat et al. 2010). Transgressive deposition could be especially efficient in burying valley floors in areas where tributary valleys joined (Vis et al. 2015). This appears to hold for some of the nearshore reaches of the smaller estuaries surrounding the Thames Estuary, such as the Medway, which has preserved submerged basal sequences from the Early Holocene along its fringes (Devoy 1979).

Of the taphonomic variables discussed above, some of them combine in peculiar ways. Postglacial marine transgressions, for example, use valley systems as the pathways to enter the North Sea. In the southern North Sea they have gradients controlled by the erosional history of the Channel through the Dover Strait, whereas in reaches upstream (i.e. north) of the tectonic hinge line, in the North Sea basin tectonic depocenter (Figs. 7.5, 7.7 & 7.9), subsidence rates and sediment delivery control gradients (see also Hijma et al. 2012). This affects the pacing of transgressions into these valleys, and thus the preservation potential of Upper Paleolithic and Mesolithic archaeology. As this is known to occur abundantly below the coastal plain and, though more patchily in preservation, also in the nearshore, it is expected also to be present in the offshore Rhine-Meuse and Thames LGM-inherited valley systems (Fig. 7.9b).

Since transgression also occurred for sub-periods within the Early Glacial in this area (towards the MIS 5c and MIS 5a highstands, sea-level curve of Fig. 7.9a), positive taphonomic effects may apply to depositional surfaces with potential Neanderthal sites that occur within the Brown Bank Formation (in the area where this unit preserved, Fig. 7.9b). Tizzard et al. (2014) identified such an Early Glacial transgressive valley fill unit (their Unit 4), and a related unit (Unit 5) that overlies the flanking platform area from which 33 Middle Paleolithic handaxes were dredged. Where these handaxes are attributed a (late) MIS 7 age (Tizzard et al. 2014; Unit 3b), the depositional architecture and taphonomy suggest a protective role for MIS 5c/a transgressive deposits for the platform created at the MIS 7/6 transition, to explain the preserved situation today. This may apply to other areas off East Anglia and in the Thames Estuary (e.g. Dix & Sturt 2011).

Mapping archaeological potential

Since key tectonic and sedimentary controls (i.e. drainage and coastline configurations) appear to have remained largely stable prior to MIS 12 (see Fig. 7.4), a larger part of the early Middle Pleistocene landscape and associated archaeology may have been preserved within the subsiding depocenter off the northwestern Netherlands (at the top of the Yarmouth Roads Formation), and at equivalent depth onshore in the Netherlands. However, this will be deeply buried and hard to access, particularly in the central part of the basin. An exception may be the areas where ice-pushed ridges have been built by displacing blocks of Yarmouth Roads Formation, for instance, at the limit of the MIS 6 glaciation (Moreau *et al.* 2012). Onshore in the Netherlands, it is from quarry exposures in such Saalian ice-pushed ridges that Middle Paleolithic archaeology (e.g. Rhenen industry), is accessible from formations that otherwise occur tens of meters below the modern coastal plain surface (Stapert 1987).

After MIS 12, the preserved record becomes oriented (aligned) to drainage networks (periglacial fluvial and proglacial fluvial), whereas the main patch of record preserved from the Cromerian Complex is a fragmented coastal plain. From the younger period, the Holocene would hold preserved equivalents of such a coastal plain situation (Early Holocene: offshore; Middle and Late Holocene: onshore). In some places we would expect comparable Last Interglacial deposits (onshore) and Early Glacial deposits (offshore: regressive, interrupted by transgressions, see Fig. 7.9a). From the remaining periods it is expected that mainly mixed-age lag deposits characterize the North Sea floor (Hijma *et al.* 2012).

Across much of the basin, fluvial systems/paleochannels, owing to their ubiquity in the geomorphological record and their ability to preserve through aggradation, infilling or floodplain deposition may hold evidence that ranges in its taphonomic condition from *in situ* to reworked (Hosfield 2007; Ward & Larcombe 2008; Cohen *et al.* 2012). Superficial marine reworking will be widespread, deep marine reworking by tidal channels localized, and Holocene reworking will often affect deposits that had also experienced Late Pleistocene fluvial (periglacial rivers, Fig. 7.9b) and

marine (Last Interglacial and Early Glacial, Fig. 7.9a) reworking cycles.

Attempts have been made to use understanding of these processes combined with offshore geological and geomorphological evidence to indicate or predict where archaeological deposits are preserved. Goodwyn *et al.* (2010), working at the level of the UK continental shelf, based predictions on paleogeographic reconstructions coupled with evidence of known paleolandscape features (e.g. paleochannels, submerged forests). Simply put, high-potential areas were sub-aerially exposed during lowstands and have known evidence of paleolandscapes. Low-potential areas were never exposed, or were ice-covered, and presently comprise bedrock overlain by seabed sediment; for instance, the erosional Southern Bight. This approach does go some way to addressing shelf-scale taphonomy, but when considering an individual sea basin, suffers from a lack of geological detail and chronological depth. The input sea-level reconstructions are solely based on the post-LGM (Brooks *et al.* 2011) and given the scale of analysis, little consideration is given to the aforementioned differences in glaciation, sea level, sedimentation, change of drainage systems, GIA and tectonics which result in preservation varying in both space and time.

A more regional-scale conceptual approach, developed by Ward and Larcombe (2008), identified the taphonomic processes at play in particular landscape settings/environments and accordingly assigned a preservation rating to each. For example, floodplains were assigned a high rating because burial by fine-grained flood deposits or within peat-infilling depressions (i.e. channels and basins) was considered to promote preservation of archaeological and paleoenvironmental remains. Conversely, open sandy coasts were regarded as low potential owing to significant reworking by waves (see Ward & Larcombe 2008: table 1 for full list of landscape settings and ratings). In conjunction with information on post-LGM paleogeographical change, Ward and Larcombe (2008) applied these ratings to selected areas of the central and southern North Sea: the Dogger, Brown and Norfolk banks. This accordingly identified areas of high potential around and under these features predominantly where artifacts or fossil bones had previously been found, or where Pleistocene and Holocene peat, fluvial, estuarine or floodplain deposits had previously been identified. However, the absence of detailed geological information prevented spatial localization of areas (or depths) of potential beyond generalized zones (e.g. east of the Brown Bank) or geological formations (e.g. the Elbow Formation).

Similarly, Flemming (2002; 2004) discussed particular landscape settings (e.g. estuaries and river valleys, peat layers, former archipelagos) in which preservation was more/less likely. As in Ward and Larcombe (2008), the lack of high-resolution geological data hindered the assignment of potential beyond large-scale or wide-ranging classifications, such as that the large areas of the central North Sea around the Dogger and Brown banks were of high potential (Flemming 2002). Gaffney et al. (2007; 2009) and Fitch (2011) on the other hand, attempted to map archaeological potential by combining assigned archaeological potential (based on the likelihood of occupation and preservation of a given landscape feature; a similar approach to Ward & Larcombe 2008) with paleolandscape maps derived from 3D seismic data. This mapping identified zones of higher preservation potential centered on the shores of the Outer Silver Pit and around a series of former channels running north of it (see for example Fitch et al. 2007: fig. 9.7). A drawback to this is that it is presently restricted to the Early Holocene landscapes around the Dogger Bank and Outer Silver Pit in the UK sector. Also, extensive archaeological and geological sampling, which could test the model and refine the paleolandscape reconstruction, remains to be done.

Nevertheless, the use of seismic data in conjunction with geological data from cores or boreholes has demonstrably allowed progression from shelf-scale and conceptual assessments of preservation potential. Effectively, detailed geological mapping, particularly if combined with accurate chronological information, affords the possibility to identify the elements of the paleolandscape which have been preserved in a given area and also equally importantly, identify what has not been preserved. Accuracy is of course contingent on the survey and sampling density and, crucially, the availability of secure

chronological information (see van Heteren et al. 2014). Such studies are presently limited to small regions within the basin; for instance, off East Anglia (Tizzard et al. 2014) and the Netherlands (Kuitems et al. 2015). The single example that has attempted to go (successfully) from geological mapping to archaeological assessment and sampling is the discovery of the Yangtzehaven site in the Rotterdam Maasvlakte harbor extension (see 'Evidence of submerged landscapes on the shelf' pages [xxx]–[xxx]). Part of the secret of the success may be the presence of transgressive deposits protecting the surfaces hosting the archaeology. This suggests that mapping of valley networks with geophysics (swath bathymetry, shallow seismics) and coring the uppermost meters of the seabed to identify whether transgressive units are present between 'terrestrial substrate' and 'active seabed' may be a useful and informative exercise. This transgressive unit need not be a basal peat everywhere but may well be muds, similar to the floodplains in the Ward and Lacombe (2008) model. While examples for this are mostly drawn from the Dutch and British sectors of the North Sea, this would also hold for the Belgian sector (paleovalleys of the rivers Scheldt and IJzer) and for the German and Danish sectors in the German Bight (major paleovalleys of the Elbe and Weser, and those of their tributaries).

An important issue to consider is the extent to which occasional success in local projects (e.g. a single gravel extraction area or selected construction site) can be utilized at regional-scale (e.g. a wind-farm planning zone) or the scale of entire sectors of the North Sea. Similarly, evaluating to what degree success is due to the relative vicinity to coastal land (with known finds and established taphonomy) matters if the suitability of methodologies for areas further offshore is to be assessed. The drowned landscapes and archaeology of the nearshore and offshore may well have much in common even in taphonomy, but not with respect to amounts of useful prior information.

The level of mapping and sampling of the greater part of the North Sea is insufficient to go beyond relatively broad predictions of archaeological preservation potential (Bicket 2013). Nonetheless, the research done to date shows the potential of what can be achieved with sufficient resources. At the time of writing, work is underway to develop an offshore version of the Netherlands' Indicative Map of Archaeological Values (IKAW) for the Dutch sector, using as its basis the geological mapping described above (Erkens *et al.* 2014; Ward *et al.* 2014). This map identifies zones with higher or lower potential for intact Upper Paleolithic and Mesolithic landscapes. However, when destructive activities such as aggregate mining are planned, additional site-specific research will always be necessary because the current data density does not allow for detailed prediction. Procedures and experience from combined geotechnical, geological and archaeological surveying in future wind-farm areas (e.g. Cotterill *et al.* 2015) are another form of present research that could be utilized in other parts of the North Sea in the future.

Conclusion

The North Sea basin is an important region for submerged landscape research. It potentially holds some of the earliest evidence for hominin occupation of northwestern Europe and figures strongly in addressing questions on hominin migration and responses to environmental change through the last 500,000 years. From the Early Holocene, it contains the Doggerland area which is known to have been transgressed while occupied by Mesolithic people and forms a contact zone between northerly and central European Mesolithic populations. Importantly, there is confirmation that archaeological, faunal, paleoenvironmental and paleolandscape evidence has been preserved on the continental shelf. However, the evidence base is fragmentary, temporally discontinuous and often chronologically mixed — unsurprising given the Quaternary climatic, paleoenvironmental and paleogeographic changes that have affected the study area and their influence on taphonomic processes. Even so, the extant evidence is a powerful indication that the North Sea was habitable, almost certainly a focus of settlement and dispersal and, crucially, can be effectively studied.

The oceanographic conditions and size of the modern North Sea, coupled with the discontinuous and often buried nature of its Quaternary depositional record, mean that geological approaches to submerged landscape reconstruction and archaeological inventorying are essential (Bicket 2013; Ward *et al.* 2014; Vos *et al.* 2015; this chapter). In turn, the effectiveness of these approaches and the level to which the submerged landscape and its contents can be effectively resolved depend on two things: firstly, the taphonomic processes which have either helped preserve or destroy large parts of the evidence; and secondly, the density of geological sampling and geophysical survey. Certain regions will have better preservation than others, and are more suitable for certain sampling and survey techniques than others. It is important to lay out surveying and sampling campaigns in such a way that they address the questions of taphonomy besides more basic lithological characterization of the sub-surface. Hereto pages 170–174 summarized taphonomic variables and their use in mapping archaeological potential at local, regional and super-regional scale. Sections on 'Quaternary background and paleogeographic framework' and 'Evidence of submerged landscapes on the shelf' (see pages 152–169) supported that assessment by providing an overview of geological history and landform inventory, highlighting links with taphonomy throughout the text.

Recent decades have seen considerable advances relevant to submerged landscape research, in terms of data quantities, technical developments in geological and archaeological surveying and conceptual developments in dealing with the data at new resolutions and integrating it with knowledge acquired from the surrounding onshore regions. Conceptual approaches are now well established (e.g. Flemming 2004; Bailey & Flemming 2008) and research and management frameworks covering the North Sea have been developed (e.g. Peeters *et al.* 2009; Ward *et al.* 2014). Paleolandscape reconstructions have also become more accurate and detailed (e.g. compare Coles 1998 with Gaffney *et al.* 2007; 2009; Hijma *et al.* 2012), and now afford the possibility to go beyond the shelf-scale geological reports and papers that were for many years state of the art (e.g. Jelgersma 1979; Zagwijn 1979; Cameron *et al.* 1992; Bridgland & D'Olier 1995; Funnel 1996).

Together with improved reconstruction comes better understanding of taphonomic processes and hence of the preservation potential of archaeological material. However, detailed studies are still the exception rather than the rule and issues remain outstanding regarding chronology and correlation. Furthermore, the ability to take the geological mapping to the level where it can pinpoint archaeological deposits is still in its infancy and in the North Sea to date has only been demonstrated in a few instances.

A further step up in the quantity and quality of data is required to make such techniques work further offshore (see 'Mapping archaeological potential', pages 173–174). The research done to date on the submerged landscape of the North Sea provides a secure foundation on which future work can build and also provides examples that can point the way for other less well-studied shelves. Insight into geological developments governing formation and preservation of the Quaternary record and its taphonomy allows regionalization of what archaeology can be expected to be encountered where (this chapter, see also Bicket 2013), aiding landscape-based management of submerged archaeological heritage.

Wind-farm planning activities in the Dogger Bank and Outer Silver Pit areas in the UK sectors may be the next activity to reveal offshore North Sea Mesolithic discoveries. This Mesolithic potential should extend into the Elbe paleovalley in the German Bight and the Danish sector south-west of the Skagerrak. In this region it connects to the rich submerged Mesolithic archaeology in the straits connecting the Skagerrak with the Baltic Sea (e.g. Fischer 1995; Pedersen *et al.* 1997).

Regarding Middle Paleolithic and older archaeology, aggregate extraction in the UK sector of the southern North Sea is expected to continue to provide occasional discoveries and surveying opportunities (such as Tizzard *et al.* 2014). Late Middle Paleolithic archaeology (Neanderthals) and associated paleontology is expected to be found in the Belgian sector and the south-west of the Dutch sector, in the offshore continuations of the Last Interglacial Rhine and Meuse valleys. From before the Holocene, what could or should be present between the Dogger Bank, East Anglia, Rhine and Elbe in the central North Sea is the greatest unknown.

Data sources/Useful links

Belgium

SeArch Project: www.sea-arch.be/en
Flemish Hydrography: www.vlaamsehydrografie.be
Flanders Marine Institute (VLIZ): www.vliz.be/en

Denmark

Geological Survey of Denmark and Greenland (GEUS): www.geus.dk/geuspage-uk.htm
Danish Geodata Agency: eng.gst.dk

Germany

Federal Maritime & Hydrographic Agency (BSH): www.bsh.de/de/index.jsp
Federal Institute for Geosciences & Natural Resources (BGR): www.bgr.bund.de

Norway

Norwegian Geological Survey (NGU): www.ngu.no/en/node
Norwegian Mapping Authority (Kartverkert): www.kartverket.no/en/

The Netherlands

TNO Geological Survey of the Netherlands: www.dinoloket.nl/en
Royal Netherland Navy Hydrographic Service: www.defensie.nl/english/topics/hydrography
RCE Cultural Heritage Agency (Rijksdienst Cult. Erfgoed): www.culturalheritageagency.nl/en
Maasvlakte 2 program: www.20metersunderwater.nl
Netherlands Enterprise Agency (public datasets on wind-farm zones): offshorewind.rvo.nl
Facebook public outreach site displaying numerous photos of beach finds of bones from the artificial beach of the Rotterdam harbor extension: www.facebook.com/maasvlakte.strandvondsten

UK

British Geological Survey (BGS): www.bgs.ac.uk

UK Hydrographic Office (UKHO): www.gov.uk/government/organisations/uk-hydrographic-office

British Oceanographic Data Centre: www.bodc.ac.uk

The Crown Estate: www.thecrownestate.co.uk

Historic England / English Heritage: www.historicengland.org.uk www.english-heritage.org.uk

General

EMODnet initiative: www.emodnet.eu

INQUA Commission Marine Processes: www.inqua.org/cmp/index.html

Acknowledgments

Our colleague Dr. Sytze van Heteren (TNO Geological survey of the Netherlands) is thanked for his constructive comments and suggestions on an earlier draft of this chapter. Editors Nicholas C. Flemming and Anthony Burgess are thanked for their guidance, support and patience throughout the manuscript production.

References

Alappat, L., Vink, A., Tsukamoto, S. & Frechen, M. 2010. Establishing the Late Pleistocene–Holocene sedimentation boundary in the southern North Sea using OSL dating of shallow continental shelf sediments. *Proceedings of the Geologists' Association* 121:43-54.

Allen, J. R. L. 2000. Morphodynamics of Holocene salt marshes: a review sketch from the Atlantic and Southern North Sea coasts of Europe. *Quaternary Science Reviews* 19:1155-1231.

Amkreutz, L. W. S. W., 2013. *Persistent Traditions. A Long-term Perspective on Communities in the Process of Neolithisation in the Lower Rhine Area (5500-2500 cal BC).* Ph.D thesis. University of Leiden. Sidestone Press Dissertations: Leiden.

Anthony, D. & Leth, J. O. 2002. Large-scale bedforms, sediment distribution and sand mobility in the eastern North Sea off the Danish west coast. *Marine Geology* 183:247-263.

Antoine, P., Rousseau, D-. D., Moine, O. *et al.* 2009. Rapid and cyclic aeolian deposition during the Last Glacial in European loess: a high-resolution record from Nussloch, Germany. *Quaternary Science Reviews* 28:2955-2973.

Arends, F. 1833. *Physische Geschichte der Nordsee-Küste und deren Veränderungen durch Sturmfluthen seit der Cymbrischen Fluth bis jetzt.* H. Woortman: Emden.

Ashton, N., Lewis, S. G., De Groote, I. *et al.* 2014. Hominin footprints from Early Pleistocene deposits at Happisburgh, UK. *PLoS ONE* 9(2): e88329. doi:10.1371/journal.pone.0088329.

Bailey, G. N. & Flemming, N. C. 2008. Archaeology of the continental shelf: Marine resources, submerged landscapes and underwater archaeology. *Quaternary Science Reviews* 27:2153-2165.

Balson, P., Butcher, A., Holmes, R. *et al.* 2002. *North Sea Geology. Technical report Produced for Strategic Environmental Assessment — SEA2 & SEA3.* UK Department of Trade and Industry: London.

Barne, J. H, Robson, C. F., Kaznowksa, S. S., Doody, J. P. & Davidson, N. C. (eds.) 1995a. *Coasts and Seas of the United Kingdom. Region 5 North-east England: Berwick-upon-Tweed to Filey Bay.* Joint Nature Conservation Committee: Peterborough.

Barne, J. H, Robson, C. F., Kaznowksa, S. S., Doody, J. P. & Davidson, N. C. (eds.) 1995b. *Coasts and Seas of the United Kingdom. Region 6: Eastern England: Flamborough Head to Great Yarmouth.* Joint Nature Conservation Committee: Peterborough.

Barne, J. H., Robson, C. F., Kaznowksa, S. S. & Doody, J. P. (eds.) 1997. *Coasts and Seas of the United Kingdom. Region 4 South-east Scotland: Montrose to Eyemouth.* Joint Nature Conservation Committee: Peterborough.

Barne, J. H, Robson, C. F., Kaznowksa, S. S., Doody, J. P., Davidson, N. C. & Buck, A.L. (eds.) 1998. *Coasts and Seas of the United Kingdom. Region 7 South-east England: Lowestoft to Dungeness.* Joint Nature Conservation Committee: Peterborough.

Bates, M. R., Bates, C. R. & Briant, R. M. 2007. Bridging the gap: a terrestrial view of shallow marine sequences and the importance of the transition zone. *Journal of Archaeological Science* 34:1537-1551.

Beets, D. J. & van der Spek, A. J. F. 2000. The Holocene evolution of the barrier and the back-barrier basins of Belgium and the Netherlands as a function of late Weichselian morphology, relative sea-level rise and sediment supply. *Netherlands Journal of Geosciences* 79:3-16.

Behre, K.-E. 2004. Coastal development, sea-level change and settlement history during the later Holocene in the Clay District of Lower Saxony (Niedersachsen), northern Germany. *Quaternary International* 112:37-53.

Bell, M., Warren, G., Cobb, H. *et al.* 2013. The Mesolithic. In Ransley, J. Sturt, F., Dix, J. Adams, J. & Blue, L. (eds.) *People and the Sea: a Maritime Archaeological Research Agenda for England.* CBA Research Report 171. pp. 30-49. Council for British Archaeology: York.

Bicket, A. 2013. *Audit of Current State of Knowledge of Submerged Palaeolandscapes and Sites. English Heritage Project No. 6231.* Wessex Archaeology: Salisbury.

Borst, W., Weerts, H., Vellinga, T., & Otte, A. 2014. Monitoring programme for MV2, Part IV — archaeological and palaeontological finds. *Terra et Aqua* 135:5-16.

Böse, M., Luthgens, C., Lee, J. R. & Rose, J. 2012. Quaternary glaciations of northern Europe. *Quaternary Science Reviews* 44:1-25.

Bridgland, D. R., & D'Olier, B. 1995. The Pleistocene evolution of the Thames and Rhine drainage systems in the southern North Sea Basin. *Geological Society London, Special Publications* 96:27-45.

Bridgland, D. R., D'Olier, B., Gibbard, P. L. & Roe, H. M. 1993. Correlation of Thames terrace deposits between the Lower Thames, eastern Essex and the submerged offshore continuation of the Thames-Medway valley. *Proceedings of the Geologists Association* 104:51-57.

Briggs, K., Thomson, K. & Gaffney, V. 2007. A geomorphological Investigation of submerged depositional features within the Outer Silver Pit, Southern North Sea. In Gaffney V., Thomson, K. & Fitch, S. (eds.) 2007. *Mapping Doggerland: The Mesolithic Landscapes of the Southern North Sea.* pp. 43-59. Archaeopress: Oxford.

Brooks, A. J., Bradley, S. L., Edwards, R. J. & Goodwyn, N. 2011. The palaeogeography of Northwest Europe during the last 20,000 years. *Journal of Maps* 7:573-587.

Busschers, F. S., Kasse, C., van Balen, R. T. *et al.* 2007. Late Pleistocene evolution of the Rhine-Meuse system in the southern North Sea basin: imprints of climate change, sea-level oscillation and glacio-isostacy. *Quaternary Science Reviews* 26:3216-3248.

Busschers, F. S., van Balen, R. T., Cohen, K. M. *et al.* 2008. Response of the Rhine-Meuse fluvial system to Saalian ice-sheet dynamics. *Boreas* 37:377-398.

Busschers, F. S., Wesselingh, F., Kars, R. H. *et al.* 2010. Radiocarbon dating of Late Pleistocene marine shells from the southern North Sea. *Radiocarbon* 56:1151-1166.

Cameron, T. D., Crosby, A., Balson, P. S. *et al.* 1992. *UK Offshore Regional report: The Geology of the Southern North Sea.* HMSO for the British Geological Survey: London.

Carr, S. J. 2004. The North Sea basin. In Ehlers, J. & Gibbard, P. L. (eds.) *Quaternary Glaciations — Extent and Chronology. Part 1 Europe.* pp. 261-270. Elsevier: Amsterdam.

Carr, S. J., Holmes, R., van der Meer, J.M. & Rose, J. 2006. The Last Glacial Maximum in the North Sea Basin: micromorphological evidence of extensive glaciation. *Journal of Quaternary Science* 21:131-153.

Chang, T. S., Flemming, B. W., Tilch, E., Bartholomä, A. & Wöstmann, R. 2006. Late Holocene stratigraphic evolution of a back-barrier tidal basin in the East Frisian Wadden Sea, southern North Sea: transgressive deposition and its preservation potential. *Facies* 52: 329-340.

Clark, C. D., Hughes, A. L. C., Greenwood, S. L., Jordan, C. & Sejrup, H. P. 2012. Pattern and timing of retreat of the last British-Irish Ice Sheet. *Quaternary Science Reviews* 44:112-146.

Cloetingh, S., Ziegler, P. A., Beekman, F. *et al.* 2005. Lithospheric memory, state of stress and rheology:

neotectonic controls on Europe's intraplate continental topography. *Quaternary Science Reviews* 24:241-304.

Cohen, K. M. & Lobo, F. J. 2013. Continental shelf drowned landscapes: Submerged geomorphological and sedimentary record of the youngest cycles (special issue editorial). *Geomorphology* 203:1-5.

Cohen, K. M., MacDonald, K., Joordens, J. C. A, Roebroeks, W. & Gibbard, P. L. 2012. The earliest occupation of north-west Europe: a coastal perspective. *Quaternary International* 271:70-83.

Cohen, K. M., Gibbard, P. L. & Weerts, H. J. T. 2014. North Sea palaeogeographical reconstructions for the last 1 Ma. *Netherlands Journal of Geosciences* 93: 7-29.

Coles, B. J. 1998. Doggerland: a speculative survey. *Proceedings of the Prehistoric Society* 64:45-81.

Cotterill, C., James, L., Long, D., Forsberg, C-. F., Tjelta, T. I. & Mulley S. 2015. Dogger Bank and the case of the complex stratigraphy (abstract). *Quaternary Research Association Annual Discussion Meeting.* 6th–8th January 2015, Edinburgh, p. 1.

Dawson, A. G., Smith, D. E. & Long, D. 1990. Evidence for a tsunami from a Mesolithic site in Inverness, Scotland. *Journal of Archaeological Science* 17:509-512.

Devoy, R. J. N. 1979. Flandrian sea level changes and vegetational history of the lower Thames Estuary. *Philosophical Transactions of the Royal Society of London B: Biological Sciences* 285:355-407.

Dix, J. K & Sturt, F. C. 2011. *The Relic Palaeo-landscapes of the Thames Estuary.* Report prepared for the Marine Aggregate Levy Sustainability Fund.

Dong, P. & Guzzetti, F. 2005. Frequency-size statistics of coastal soft-cliff erosion. *Journal of Waterway, Port, Coastal and Ocean Engineering* 131:37-42.

Ehlers, J. & Gibbard, P. L. (eds.) 2004. *Quaternary Glaciations - Extent and Chronology. Part I: Europe.* Elsevier: Amsterdam.

EMODnet 2014. Seabed substrate map. EMODnet Geology website: www.emodnet-geology.eu/.

Erkens, G., Hijma, M. P., Peeters, J., van Heteren, S., Marges, V. & Vonhögen-Peeters, L. 2014. *Proef Indicatieve Kaart Archeologische Waarden (IKAW) Noordzee. Deltares Report 1206731.*

Figge, K. 1980. Das Elbe-Urstromtal im Bereich der Deutschen Bucht (Nordsee). *Eiszeitalter und Gegenwart* 30: 203-211.

Fischer, A. 1995. An entrance to the Mesolithic world below the ocean. Status of ten years' work on the Danish sea floor. In Fischer, A. (ed.) *Man and Sea in the Mesolithic.* pp. 371-384. Oxbow Books: Oxford.

Fitch, S. E. J. 2011. *The Mesolithic landscape of the southern North Sea.* Ph.D Thesis. University of Birmingham, UK.

Fitch, S., Thomson, K., & Gaffney, V. L. 2005. Late Pleistocene and Holocene depositional systems and the palaeogeography of the Dogger Bank, North Sea. *Quaternary Research* 64:185-196.

Fitch, S., Gaffney, V. L. & Thomson, K. 2007. The archaeology of North Sea palaeolandscapes. In Gaffney, V., Thomson, K. & Fitch, S. (eds.) *Mapping Doggerland: The Mesolithic Landscapes of the Southern North Sea.* pp. 105-118. Archaeopress: Oxford.

Flemming, N. C. 2002. *The Scope of Strategic Environmental Assessment of North Sea areas SEA3 and SEA2 in Regard to Prehistoric Archaeological Remains.* UK Department of Trade and Industry: London.

Flemming, N. C. (ed.) 2004. *Submarine Prehistoric Archaeology of the North Sea.* CBA Research Report 141. Council of British Archaeology: York.

Fruergaard, M., Andersen, T. J., Nielsen, L. H., Johannessen, P. N., Aagaard, T. & Pejrup, M. 2015. High-resolution reconstruction of a coastal barrier system: impact of Holocene sea-level change. *Sedimentology* 62:928-969.

Funnell, B. M. 1995. Global sea-level and the (pen-)insularity of late Cenozoic Britain. In Preece, R. C. (ed.) *Island Britain: a Quaternary Perspective.* pp. 3-13. Geological Society: London.

Funnel, B. M. 1996. Plio-Pleistocene palaeogeography of the southern North Sea Basin (3.75-0.60 Ma). *Quaternary Science Reviews* 15:391-405.

Gaffney, V., Thomson, K. & Fitch, S. (eds.) 2007. *Mapping Doggerland: The Mesolithic Landscapes of the Southern North Sea.* Archaeopress: Oxford.

Gaffney V., Fitch, S. & Smith, D. 2009. *Europe's Lost World: the rediscovery of Doggerland.* CBA Research Report 160. Council for British Archaeology: York.

Gatliff, P. W., Richards, P. C., Smith, K. *et al.* 1994. *UK Offshore Regional Report: The Geology of the Central North Sea.* HMSO for the British Geological Survey: London.

Gibbard, P. L. 1988. The history of the great northwest European rivers during the past three million years. *Philosophical Transactions of the Royal Society, London* B318:559-602.

Gibbard, P. L. 1995. The formation of the Strait of Dover. In Preece, R. C. (ed.) *Island Britain: a Quaternary Perspective.* pp. 15-26. Geological Society: London.

Gibbard, P. L. 2007. Palaeogeography: Europe cut adrift. *Nature* 448:259-260.

Gibbard, P. L. & Cohen, K. M. 2008. Global chronos-tratigraphical correlation table for the last 2.7 million years. *Episodes* 31:243-247.

Gibbard, P. L. & Cohen, K. M. 2015. Quaternary evolution of the North Sea and the English Channel. *Proceedings of the Open University Geological Society* 1:65-75.

Gibbard, P. L., Turner, C. & West, R. G. 2013. The Bytham river reconsidered. *Quaternary International* 292:15-32.

Glimmerveen, J., Mol, D., Post, K. *et al.* 2004. The North Sea project: the first palaeontological, palyno-logical and archaeological results. In Flemming, N. C. (ed.) *Submarine Prehistoric Archaeology of the North Sea.* CBA Research Report 141. pp. 43-52. Council of British Archaeology: York.

Glimmerveen, J., Mol, D. & van der Plicht, H. 2006. The Pleistocene reindeer of the North Sea — initial palaeontological data and archaeological remarks. *Quaternary International* 142-143:242-246.

Goodwyn, N., Brooks, A.J. & Tillin, H. 2010. *Water-lands: Developing Management Indicators for Sub-merged Palaeoenvironmental Landscapes.* Report prepared under the Marine Aggregates Levy Sustainability Fund (Ref No. MEPF 09/P109). ABP Marine Environmental Research.

Gupta, S., Collier, J. S., Palmer-Felgate, A. & Potter, G. 2007. Catastrophic flooding origin of shelf valley systems in the English Channel. *Nature* 448:342-345.

Gyllencreutz, R. 2005. Late Glacial and Holocene pale-oceanography in the Skagerrak from high-resolution grain size records. *Palaeogeography, Palaeoclimatology, Palaeoecology* 222:344-369.

Hazell, Z. J. 2008. Offshore and intertidal peat deposits, England — a resource assessment and development of a database. *Environmental Archaeology* 13:101-110.

Hijma, M. P., & Cohen, K. M. 2010. Timing and magnitude of the sea-level jump preluding the 8200 yr event. *Geology* 38:275-278.

Hijma, M. P., & Cohen, K. M. 2011. Holocene transgression of the Rhine river mouth area, The Netherlands/Southern North Sea: palaeogeography and sequence stratigraphy. *Sedimentology* 58:1453-1485.

Hijma, M. P., van der Spek, A. J. F. & van Heteren, S. 2010. Development of a mid-Holocene estuarine basin, Rhine-Meuse mouth area, offshore The Netherlands. *Marine Geology* 271:198-211.

Hijma, M. P., Cohen, K. M., Roebroeks, W., West-erhoff, W. E. & Busschers, F. S. 2012. Pleistocene Rhine-Thames landscapes: geological background for hominin occupation of the southern North Sea region. *Journal of Quaternary Science* 27:17-39.

Hijma, M. P., Engelhart, S. E., Törnqvist, T. E., Horton, B. P., Hu, P. & Hill, D. F. 2015. A protocol for a geological sea-level database. In Shennan, I., Long, A. J. & Horton, B. P. (eds.) *Handbook of Sea-Level Research.* pp. 536-553. John Wiley & Sons: Chichester.

Hosfield, R. T. 2007. Terrestrial implications for the maritime geoarchaeological resource: A view from the Lower Palaeolithic. *Journal of Maritime Archaeology* 2:4-23.

Houmark-Nielsen, M. & Kjær, K. H. 2003. Southwest Scandinavia, 40–15 kyr BP: palaeogeography and environmental change. *Journal of Quaternary Science* 18:769-786.

Housley, R. A. 1991. AMS dates from the Late Glacial and early Postglacial in North-west Europe: A review. In N. Barton, Roberts A. J. & Roe, D. A (eds.) *The Late Glacial in North-West Europe. Human Adaption and Environmental Change at the End of the Pleistocene.* CBA Research Report 77. pp. 25-39. Council for British Archaeology: London.

Hublin, J-. J., Weston, D., Gunz, P. *et al.* 2009. Out of the North Sea: the Zeeland Ridges Neandertal. *Journal of Human Evolution* 57:777-785.

Huthnance, J. M. 1991. Physical oceanography of the North Sea. *Ocean & Shoreline Management* 16:199-231.

Huuse, M. & Lykke-Andersen, H. 2000. Overdeepened Quaternary valleys in the eastern Danish North Sea: morphology and origin. *Quaternary Science Reviews* 19:1233-1253.

Jelgersma, S. 1979. Sea-level changes in the North Sea Basin. In Oele, E., Schüttenhelm, R. T. E. & Wiggers, A. J. (eds.) *The Quaternary History of the North Sea. Acta Universitatis Upsaliensis Symposia Universitatis Upsaliensis Annum Quingentesimum Celebrantis* 2:233-248.

Konradi, P. B. 2000. Biostratigraphy and environment of the Holocene marine transgression in the Heligoland Channel, North Sea. *Bulletin of the Geological Society of Denmark* 47:71-79.

Kooi, H., Johnston, P., Lambeck, K., Smither, C. & Molendijk, R. 1998. Geological causes of recent (~100 yr) vertical land movement in the Netherlands. *Tectonophysics* 299:297-316.

Kuitems, M., van Kolfschoten, Th., Busschers, F. & De Loecker, D. 2015. The Geoarchaeological and Palaeontological research in the Maasvlakte 2 sand extraction zone and on the artificially created Maasvlakte 2 beach — a synthesis. In Moree, J. M. & Sier, M. M. (eds.) *Interdisciplinary Archaeological Research Programme Maasvlakte 2, Rotterdam. BOORrapporten 566.* pp. 351-398. Gemeente Rotterdam / Rijksdienst voor het Cultureel Erfgoed & Port of Rotterdam: Rotterdam.

Laban C. 1995. *The Pleistocene Glaciations in the Dutch Sector of the North Sea.* Ph.D thesis, University of Amsterdam, the Netherlands.

Laban, C. & van der Meer, J. J. M. 2011. Pleistocene Glaciation in The Netherlands. In Ehlers, J., Gibbard, P. L. & Hughes P. D. (eds.) *Quaternary Glaciations— Extent and Chronology. A Closer Look. Developments in Quaternary Sciences vol. 15.* pp. 247-260. Elsevier: Amsterdam.

Lambeck, K. 1995. Late Devensian and Holocene shorelines of the British Isles and the North Sea from models of glacio-hydro-isostatic rebound. *Journal of the Geological Society* 152:437-448.

Lambeck, K., Purcell, A., Funder, S., Kjær, K. H., Larsen, E. & Moller, P. 2006. Constraints on the Late Saalian to early Middle Weichselian ice sheet of Eurasia from field data and rebound modelling. *Boreas* 35:539-575.

Lee, J. R., Busschers, F. S. & Sejrup, H. P. 2012. Pre-Weichselian Quaternary glaciations of the British Isles, The Netherlands, Norway and adjacent marine areas south of 68°N: implications for long-term ice sheet development in northern Europe. *Quaternary Science Reviews* 44:213-228.

Lekens, W. A. H., Sejrup, H. P., Haflidason, H., Knies, J., & Richter, T. 2006. Meltwater and ice rafting in the southern Norwegian Sea between 20 and 40 calendar kyr BP: Implications for Fennoscandian Heinrich events. *Paleoceanography* 21 PA3013 doi:10.1029/2005PA001228.

Long, D., Wickham-Jones, C. R. & Ruckley, N. A. 1986. A flint artefact from the northern North Sea. In Roe, D. A. (ed.) *Studies in the Upper Palaeolithic of Britain and North West Europe.* BAR International Series (No. 296). pp. 55-62. British Archaeological Reports: Oxford.

Louwe Kooijmans, L. P. 1975. Benen jacht- en visgerei uit de Midden-Steentijd, gevonden op de Maasvlakte (circa 7000 v.Chr.). In Louwe Kooijmans, L. P. & Sarfatij, H. & Verhoeven, A. (eds.) *Archeologen werken in Zuid-Holland, Opgravingen en vondsten uit de laatste 15 jaar.* p. 11. Rijksmuseum van Oudheden/Leiden tentoonstellingscatalogus: Leiden.

Louwe Kooijmans, L. P. 1980. Archaeology and coastal change in the Netherlands. In Thompson, F. H. (ed.) *Archaeology and Coastal Change.* pp. 106-133. The Society of Antiquaries of London: London.

Louwe Kooijmans, L. P. 2005. Jagerskamp in de moerassen. De donken bij Hardinxveld. In Louwe Kooijmans, L. P., van den Broeke, P. W, Fokkens, H & van Gijn, A. (eds.) *Nederland in de prehistorie.* pp. 183-186. Bert Bakker: Amsterdam.

Lunkka, J. P. 1994. Sedimentation and lithostratigraphy of the North Sea Drift and Lowestoft Till Formations in the Coastal cliffs of northeast Norfolk, England. *Journal of Quaternary Science* 9:209-233.

Maddy, D., Bridgland, D. R. & Green, C. P. 2000. Crustal uplift in southern England: evidence from the river terrace records. *Geomorphology* 33:167-181.

Martinius, A. W. & van den Berg, J. H. 2011. *Atlas of Sedimentary Structures in Estuarine and Tidally-Influenced River Deposits of the Rhine-Meuse-Scheldt System*. EAGE Publications bv: Houten.

Mathys, M. 2009. *The Quaternary Geological Evolution of the Belgian Continental Shelf, Southern North Sea*. Ph.D Thesis. Ghent University, Belgium.

Meijer, T. & Preece. R. C. 1995. Malacological evidence relating to the insularity of the British Isles during the Quaternary. In Preece, R. (ed.) *Island Britain: a Quaternary Perspective*. Geological Society London Special Publication 96:89-110.

Milne, G. A. 2015. Glacial isostatic adjustment. In Shennan, I., Long, A. J. & Horton, B. P. (eds.) *Handbook of Sea-Level Research*. pp. 421-438. John Wiley & Sons: Chichester.

Mol, D., Post, K., Reumer, J. W. F. *et al.* 2006. The Eurogeul - first report of the palaeontological, palynological and archaeological investigations of this part of the North Sea. *Quaternary International* 142-143:178-185.

Moreau, J., Huuse, M., Janszen, A., van der Vegt, P., Gibbard, P. L. & Moscariello, A. 2012. The Glaciogenic Unconformity of the Southern North Sea. *Geological Society London Special Publications* 368:99-110.

Moree, J. M. & Sier, M. M. (eds.) 2014. *Twintig meter diep! Mesolithicum in de Yangtzehaven — Maasvlakte te Rotterdam. Landschapsontwikkeling en bewoning in het Vroeg Holoceen. BOORrapporten 523*. Bureau Oudheidkundig Onderzoek Rotterdam (BOOR): Rotterdam.

Moree, J. M. & Sier, M. M. (eds.) 2015. Twenty metres deep! The Mesolithic Period at the Yangtze Harbour site — Rotterdam Maasvlakte, the Netherlands. Early Holocene landscape development and habitation. In Moree, J. M. & Sier, M. M. (eds.) *Interdisciplinary Archaeological Research Programme Maasvlakte 2, Rotterdam. BOORrapporten 566*. pp. 7-350. Gemeente Rotterdam / Rijksdienst voor het Cultureel Erfgoed & Port of Rotterdam: Rotterdam.

Murton, D. K. & Murton, J. B. 2012. Middle and Late Pleistocene glacial lakes of lowland Britain and the southern North Sea Basin. *Quaternary International* 260:115-142.

Neill, S. P., Scourse, J. D. & Uehara, K. 2010. Evolution of bed shear stress distribution over the northwest European shelf during the last 12,000 years. *Ocean Dynamics* 60:1139-1156.

Oost, A. P. 1995. *Dynamics and Sedimentary Developments of the Dutch Wadden Sea with a Special Emphasis on the Frisian Inlet: a Study of the Barrier Islands, Ebb-tidal Deltas, Inlets and Drainage Basins*. Ph.D Thesis. Utrecht University, the Netherlands.

Otto, L., Zimmerman, J. T. F, Furnes, G. K., Mork, M., Sætre, R. & Becker, G. 1990. Review of the physical oceanography of the North Sea. *Netherlands Journal of Sea Research* 26:161-238.

Overeem, I., Weltje, G. J., Bishop-Kay, C. & Kroonenberg, S. B. 2001. The Late Cenozoic Eridanos delta system in the Southern North Sea Basin: a climate signal in sediment supply? *Basin Research* 13:293-312.

Paramor, O. A. L., Allen, K. A., Aanensen, M. *et al.* 2009. *MEFEPO North Sea Atlas*. University of Liverpool: Liverpool.

Parfitt, S. A., Barendregt, R. W., Breda, M. *et al.* 2005. The earliest record of human activity in northern Europe. *Nature* 438:1008-1012.

Parfitt, S. A., Ashton, N. M., Lewis, S. G., *et al.* 2010. Early Pleistocene human occupation at the edge of the boreal zone in northwest Europe. *Nature* 466:229-233.

Pawley, S. M., Bailey, R. M., Rose, J. *et al.* 2008. Age limits on Middle Pleistocene glacial sediments from OSL dating, north Norfolk, UK. *Quaternary Science Reviews* 27:1363-1377.

Pedersen, J. B. T., Svinth, S. & Bartholdy, J. 2009. Holocene evolution of a drowned melt-water valley in the Danish Wadden Sea. *Quaternary Research* 72:68-79.

Pedersen, L. D., Fischer, A. & Aaby, B. (eds.) 1997. *The Danish Storebaelt since the Ice Age: Man, Sea and Forest*. Storebaeltsforbindelsen: Copenhagen.

Peeters, H., Murphy, P. & Flemming, N. C. (eds.) 2009. *North Sea Prehistory Research and Management Framework (NSPRMF) 2009*. English Heritage and RCEM: Amersfoort.

Peeters, J., Busschers, F. S., & Stouthamer, E. 2015a. Fluvial evolution of the Rhine during the last interglacial-glacial cycle in the southern North Sea basin: A review and look forward. *Quaternary International* 357:176-188.

Peeters, J. H. M. & Cohen, K. M. 2014. Introduction to North Sea submerged landscapes and prehistory. *Netherlands Journal of Geosciences* 93:3-5.

Peeters, J. H. M. & Momber, G. 2014. The southern North Sea and the human occupation of northwest Europe after the Last Glacial Maximum. *Netherlands Journal of Geosciences* 93:55-70.

Peeters, J. H. M., Brinkhuizen, D. C., Cohen, K. M. *et al.* 2014. Synthese - Mesolithicum in de Yangtzehaven-Maasvlakte te Rotterdam. In Moree, J. M. & Sier, M. M. (eds.) *Twintig meter diep! Mesolithicum in de Yangtzehaven — Maasvlakte te Rotterdam. Landschapsontwikkeling en bewoning in het Vroeg Holoceen. BOORrapporten 523.* pp. 287-318. Bureau Oudheidkundig Onderzoek Rotterdam (BOOR): Rotterdam.

Peeters, J. H. M., Brinkhuizen, D. C., Cohen, K. M. *et al.* 2015b. Synthesis — The Mesolithic in the Yangtzeharbour, Maasvlakte, Rotterdam. In Moree, J. M. & Sier, M. M. (eds.) 2015. *Twenty metres deep! The Mesolithic Period at the Yangtze Harbour site. BOORrapporten 566 (Interdisciplinary Archaeological Research Programme Maasvlakte 2, Rotterdam).* pp. 287-318. Port of Rotterdam Harbour Authority: Rotterdam.

Peeters. J., Busschers, F. S., Stouthamer, E. *et al.* 2016. Sedimentary architecture and chronostratigraphy of a late Quaternary incised-valley fill: A case study of the late Middle and Late Pleistocene Rhine system in the Netherlands. *Quaternary Science Reviews* 131(A):211-236.

Peltier, W. R. 2004. Global glacial isostasy and the surface of the ice-age Earth: The ICE-5G (VM2) model and GRACE. *Annual Review of Earth and Planetary Sciences* 32:111-149.

Peltier, W. R., Shennan, I., Drummond, R. & Horton B. 2002. On the postglacial isostatic adjustment of the British Isles and the shallow viscoelastic structure of the Earth. *Geophysical Journal International* 148:443-475.

Pettitt, P. & White, M. 2012. *The British Palaeolithic: Human Societies at the Edge of the Pleistocene World.* Routledge: Abingdon/New York.

Praeg, D. 2003. Seismic imaging of mid-Pleistocene tunnel-valleys in the North Sea Basin — high resolution from low frequencies. *Journal of Applied Geophysics* 53:273-298.

Reid, C. 1913. *Submerged Forests.* Cambridge University Press: Cambridge.

Rieu, R., van Heteren, S., van der Spek, Ad. J. F. & De Boer, P. L. 2005. Development and preservation of a mid-Holocene tidal-channel network offshore the western Netherlands. *Journal of Sedimentary Research* 75:409-419.

Roe, H. M. & Preece, R. C. 2011. Incised palaeo-channels of the late Middle Pleistocene Thames: age, origins and implications for fluvial palaeogeography and sea-level reconstruction in the southern North Sea basin. *Quaternary Science Reviews* 30:2498-2519.

Roebroeks, W. 2014. Terra incognita: The Palaeolithic record of northwest Europe and the information potential of the southern North Sea. *Netherlands Journal of Geosciences* 93:43-53.

Roep, Th. B., Holst, H., Vissers, R. L. M., Pagnier, H. & Postma, D. 1975. Deposits of southward-flowing, Pleistocene rivers in the Channel region, near Wissant, NW France. *Palaeogeography, Palaeoclimatology, Palaeoecology* 17:289-308.

Rohling, E. J., Grant, K., Bolshaw, M. *et al.* 2009. Antarctic temperature and global sea level closely coupled over the past five glacial cycles. *Nature Geoscience* 2:500-504.

Rose, J. 1994. Major river systems of central and southern Britain during the Early and Middle Pleistocene. *Terra Nova* 6:435-443.

Rose, J. 2009. Early and Middle Pleistocene landscapes of eastern England. *Proceedings of the Geologists' Association* 120:3-33.

Rose, J., Candy, I., Moorlock, B. S. P. *et al.* 2002. Early and early Middle Pleistocene river, coastal and neotectonic processes, southeast Norfolk, England. *Proceedings of the Geologists' Association* 113: 47-67.

Rydgren, K. & Bondevik, S. 2015. Moss growth patterns and timing of human exposure to a Mesolithic tsunami in the North Atlantic. *Geology* 43:111-114.

SeArch Project 2014. *Map showing an overview of the archaeological findings.* SeArch website: www.sea-arch.be. Last accessed March 2017.

Sejrup, H. P., Larsen, E., Haflidason, H. *et al.* 2003. Configuration, history and impact of the Norwegian Channel Ice Stream. *Boreas* 32:18-36.

Sejrup, H. P., Nygård, A, Hall A. M. & Haflidason. H. 2009. Middle and Late Weichselian (Devensian) glaciation history of south-western Norway, North Sea and eastern UK. *Quaternary Science Reviews* 28:370-380.

Shennan, I., Lambeck, K., Flather, R. *et al.* 2000. Modelling western North Sea palaeogeographies and tidal changes during the Holocene. In Shennan, I. & Andrews, J. (eds.) *Holocene Land-Ocean Interaction and Environmental Change around the North Sea.* Geological Society London Special Publication 166:299-319.

Shennan, I., Bradley, S., Milne, G., Brooks, A., Bassett, S. & Hamilton, S. 2006. Relative sea-level changes, glacial isostatic modelling and ice-sheet reconstructions from the British Isles since the Last Glacial Maximum. *Journal of Quaternary Science* 21:585-599.

Shennan, I., Long, A. J. Horton, B. P. (eds.) 2015. *Handbook of Sea-Level Research.* John Wiley & Sons: Chichester.

Sier, M. J., Peeters, J., Dekkers, M. *et al.* 2015. The Blake Event recorded near the Eemian type locality - A diachronic onset of the Eemian in Europe. *Quaternary Geochronology* 28:12-28.

Smith, A. J. 1985. A catastrophic origin for the palaeo-valley system of the eastern English Channel. *Marine Geology* 64:65-75.

Smith, D. E., Harrison, S., Firth, C. R. & Jordan, J. T. 2011. The early Holocene sea level rise. *Quaternary Science Reviews,* 30:1846-1860.

Stapert, D. 1987. A progress report on the Rhenen industry (Central Netherlands) and its stratigraphical context. *Palaeohistoria* 29:219-243.

Steffen, H. & Wu, P. 2011. Glacial isostatic adjustment in Fennoscandia — a review of data and modeling. *Journal of Geodynamics* 52:169-204.

Stewart, M. A., Lornegan, L. & Hampson, G. 2013. 3D seismic analysis of buried tunnel valleys in the central North Sea: morphology, cross-cutting generations and glacial history. *Quaternary Science Reviews* 72:1-17.

Streif, H. 2004. Sedimentary record of Pleistocene and Holocene marine inundations along the North Sea coast of Lower Saxony, Germany. *Quaternary International* 112:3-28.

Stringer, C. 2006. *Homo Britannicus: the incredible story of human evolution.* Penguin Books: London.

Sturt, F. 2006. Local knowledge is required: a rhythm-analytical approach to the late Mesolithic and early Neolithic of the East Anglian Fenland, UK. *Journal of Maritime Archaeology* 1:119-139.

Sturt, F., Garrow, D. & Bradley, S. 2013. New models of North West European Holocene palaeogeography and inundation. *Journal of Archaeological Science* 40:3963-3976.

Teller, J. T., Leverington, D. W. & Mann, J. D. 2002. Freshwater outbursts to the oceans from glacial Lake Agassiz and their role in climate change during the last deglaciation. *Quaternary Science Reviews* 21:879-887.

Tizzard, L., Baggaley, P. A. & Firth, A. J. 2011. Seabed Prehistory: investigating palaeolandsurfaces with Palaeolithic remains from the southern North Sea. In Benjamin, J., Bonsall, C., Pickard, C. & Fischer, A. (eds.) *Submerged Prehistory.* pp. 65-74. Oxbow: Oxford.

Tizzard, L., Bicket, A. R., Benjamin, J. & De Loecker, D. 2014. A Middle Palaeolithic site in the southern North Sea: investigating the archaeology and palaeogeography of Area 240. *Journal of Quaternary Science* 29:698-710.

Törnqvist, T. E., Rosenheim, B. E., Hu, P. & Fernandez, A. B. 2015. Radiocarbon dating and calibration. In Shennan, I., Long, A. J. & Horton, B. P. (eds.) *Handbook of Sea-Level Research.* pp. 349-360. John Wiley & Sons: Chichester.

Toucanne, S., Zaragosi, S., Bourillet, J. F. *et al.* 2009. Timing of massive 'Fleuve Manche' discharges over the last 350 kyr: insights into the European ice-sheet oscillations and the European drainage network from MIS 10 to 2. *Quaternary Science Reviews* 28:1238-1256.

Turney, C. S. M. & Brown, H. 2007. Catastrophic early Holocene sea-level rise, human migration and the Neolithic transition in Europe. *Quaternary Science Reviews* 26:2036-2041.

van den Berg, J. H., Jeuken, M. C. J. L. & Van der Spek, A. J. F. 1996. Hydraulic processes affecting the morphology and evolution of the Westerscheldt Estuary. In Nordstrom, K. F. & Roman, C. T. (eds.) *Estuarine Shores: Evolution, Environments and Human Alterations.* pp. 157-184. John Wiley & Sons: Chichester.

van der Molen, J. 2002. The influence of tides, wind and waves on the net sand transport in the North Sea. *Continental Shelf Research* 22:2739-2762.

van der Plicht, J. & Palstra, S. W. L. 2016. Radiocarbon and mammoth bones: What's in a date. *Quaternary International* 406 (part B):246-251

van der Spek, A. J. F. 1996. *Large-scale Evolution of Holocene Tidal Basins in the Netherlands.* Ph.D thesis. Utrecht University, the Netherlands.

van Geel, B., Bohncke S. J. P. & Dee, H. 1980. A palaeoecological study of an upper Late Glacial and Holocene sequence from 'De Borchert', The Netherlands. *Review of Palaeobotany and Palynology* 31:367-392, 397-448.

van Heteren, S., Meekes, J. A. C., Bakker, M. A. J. *et al.* 2014. Reconstructing North Sea palaeolandscapes from 3D and high-density 2D seismic data: An overview. *Netherlands Journal of Geosciences* 93:31-42.

van Kolfschoten, T. & van Essen, H. 2004. Palaeozoological heritage from the bottom of the North Sea. In Flemming, N. C. (ed.) *Submarine Prehistoric Archaeology of the North Sea.* CBA Research Report 141. pp. 70-80. Council of British Archaeology: York.

van Vliet-Lanoë, B., Laurent, M., Bahain, J. L. *et al.* 2000. Middle Pleistocene raised beach anomalies in the English Channel: regional and global stratigraphic implications. *Journal of Geodynamics* 29:15-41.

van Vliet-Lanoë, B., Vandenberghe, N., Laurent, M. *et al.* 2002. Palaeogeographic evolution of northwestern Europe during the Upper Cenozoic. *Geodiversitas* 24: 511-541.

Verhart, L. B. M. 2004. The implications of prehistoric finds on and off the Dutch coast. In Flemming, N.
C. (ed.) *Submarine Prehistoric Archaeology of the North Sea.* CBA Research Report 141. pp. 57-61. Council of British Archaeology: York.

Vink, A., Steffen, H., Reinhardt, L. & Kaufmann, G. 2007. Holocene relative sea-level change, isostatic subsidence and the radial viscosity structure of the mantle of northwest Europe (Belgium, the Netherlands, Germany, southern North Sea). *Quaternary Science Reviews* 26:3249-3275.

Vis, G. J., Cohen, K. M., Westerhoff, W. E. *et al.* 2015. Paleogeography. In Shennan, I., Long, A. J. & Horton, B.P. (eds.) *Handbook of Sea-Level Research.* pp. 514-535. John Wiley & Sons: Chichester.

Vos, P. C. 2015. Compilation of the Holocene palaeogeographical maps of the Netherlands. In Vos, P. C. *Origin of the Dutch Coastal Landscape: Long-term Landscape Evolution of the Netherlands During the Holocene, Described and Visualized in National, Regional and Local Palaeogeographical Maps Series.* pp. 50-79. Barkhuis Publishing: Deltares / Groningen.

Vos, P. C. & van Heeringen R. M. 1997. Holocene geology and occupation history of the province of Zeeland. *Mededelingen Nederlands Insitituut voor Toegepaste Geowetenschappen* 59:5-109.

Vos, P., Bazelmans, J., Weerts, H. & van der Meulen, M. (eds.) 2011. *Atlas van Nederland in het Holoceen: Landschap en bewoning vanaf de laatste IJstijd tot nu.* Bert Bakker: Amsterdam (in Dutch).

Vos, P. C., de Kleine, M. P. C. & Rutten, G. L. 2012. Efficient stepped approach to site investigation for underwater archaeological studies. *The Leading Edge* 31:940-944.

Vos, P. C., Bunnik, F. P. M., Cohen, K. M. & Cremer, H. 2015. A staged geogenetic approach to underwater archaeological prospection in the Port of Rotterdam (Yangtzehaven, Maasvlakte, the Netherlands): A geological and palaeoenvironmental case study for local mapping of Mesolithic lowland landscapes. *Quaternary International* 367:4-31.

Waddington, C., Bailey, G. & Milner, N. 2007. Howick: discussion and interpretation. In Waddington, C. (ed.) *Mesolithic Settlement in the North Sea Basin. A Case Study from Howick, North-east England.* pp. 189-202. Oxbow: Oxford.

Ward, I. & Larcombe, P. 2008. Determining the preservation rating of submerged archaeology in the postglacial southern North Sea: a first-order geomorphological approach. *Environmental Archaeology* 13:59-83.

Ward, I., Larcombe, P. & Lillie, M. 2006. The dating of Doggerland — postglacial geochronology of the southern North Sea. *Environmental Archaeology* 11:207-218.

Ward, I., Larcombe, P., Firth, A. & Manders, M. 2014. Practical approaches to management of the marine prehistoric environment. *Netherlands Journal of Geosciences* 93:71-82.

Weerts, H., Otte, A., Smit, B. *et al.* 2012. Finding the needle in the haystack by using knowledge of Mesolithic human adaptation in a drowning delta. *eTopoi: Journal for Ancient Studies* 3:17-24.

Weerts, H. J. T., Borst, W. G., Smit, B. I., Smits, E., van der Plicht, J. & van Tongeren. O. F. R. 2015. Epilogue. Mesolithic human skull fragments of the Maasvlakte 2 artificial beach. In Moree, J. M. & Sier, M. M. (eds.) *Interdisciplinary Archaeological Research Programme Maasvlakte 2, Rotterdam. BOORrapporten 566.* pp. 399-413. Gemeente Rotterdam / Rijksdienst voor het Cultureel Erfgoed & Port of Rotterdam: Rotterdam.

Weninger, B., Schulting, R., Bradtmöller, M. *et al.* 2008. The catastrophic final flooding of Doggerland by the Storegga Slide tsunami. *Documentia Praehistorica* 35:1-24.

Wessex Archaeology 2009. *Seabed Prehistory: Site Evaluation Techniques (Area 240) Existing Data Review.* Archaeology Data Service: York.

White, M. J. & Schreve. D. C. 2000. Island Britain—Peninsular Britain: palaeogeography, colonisation and the Lower Palaeolithic settlement of the British Isles. *Proceedings of the Prehistoric Society* 66:1-28.

Zachos, J., Pagani, M., Sloan, L., Thomas, E. & Billups, K. 2001. Trends, rhythms, and aberrations in global climate 65 Ma to present. *Science* 292:686-693.

Zagwijn, W. H. 1979. Early and Middle Pleistocene coastlines in the southern North Sea Basin. In Oele, E., Schüttenhelm, R. T. E. & Wiggers, A. J. (eds.) *The Quaternary History of the North Sea. Acta Universitatis Upsaliensis Symposia Universitatis Upsaliensis Annum Quingentesimum Celebrantis* 2: 31-42.

Zagwijn, W. H. 1983. Sea-level changes in The Netherlands during the Eemian. *Geologie en Mijnbouw* 62:437-450.

Chapter 8

Northern North Sea and Atlantic Northwest Approaches

Sue Dawson,[1] Richard Bates,[2] Caroline Wickham-Jones[3] and Alastair Dawson[3]

[1] Geography, School of Social Sciences, University of Dundee, Scotland, UK
[2] Department of Earth Sciences and Scottish Oceans Institute, University of St. Andrews, Fife, Scotland, UK
[3] Department of Archaeology, School of Geosciences, University of Aberdeen, St Mary's, Scotland, UK

Introduction

This section provides a general overview of existing knowledge and information relating to the submerged prehistoric landscape and archaeology of the continental shelf of the northern North Sea and Atlantic Northwest Approaches (Fig. 8.1). Research in this area (apart from in a few isolated cases) is still in its infancy and this document serves primarily to highlight potential. For this region we must be aware of the significant impact that Late Glacial and Holocene changes in relative sea level (RSL) have had, not only on the coastal landscape, but also across the continental shelf area that has undoubtedly supported human settlement throughout large parts of the period under consideration.

The sea area of the northern North Sea and Atlantic Northwest Approaches roughly encircles Scotland, spans the North Sea from Scotland to southern Norway, and runs west towards Rockall. For significant periods in the past this area not only formed a land bridge that allowed prehistoric populations to migrate between existing landmasses, but it also formed a hidden continent that is likely to have supported its own cultural groupings (Gaffney *et al.* 2009). To the south was a land that has now been well documented with its own topography, including hills, rivers, marshland and lakes, about which we are only just beginning to understand the detail (Gaffney *et al.* 2007). Based on simple bathymetric reconstructions this land likely extended far to the north. Recent research has dubbed the submerged North Sea terrestrial landscapes 'Doggerland' (Coles 1998) and this name will be used here.

Archaeological evidence from the land suggests that the archaeological potential of this seabed area should be

Fig. 8.1 Northwest Scotland and its position in relation to the North Sea and adjacent countries. Image by R. Bates.

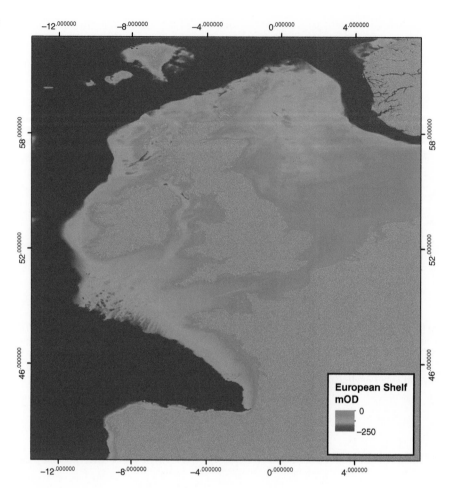

taken seriously, and this is supported by isolated finds of artifacts as well as finds of animal bone (see 'Known Submerged Finds', pages 204–205).

Earth Sciences and Sources of Data

Geomorphological background

The geomorphological pattern of the Scottish coastline follows that of the UK generally, showing a strong influence of broad geological variation with generally older and more resistant rocks to the north and west, and younger weaker rocks on the east coast. Local geological variation combined with the legacy of past glacial erosion and the dominant influence of the prevailing westerly climate has resulted in a highly complex shoreline.

The north and west coast today is dominated by erosion on a highly irregular coastline that bears the strong marks of glacial erosive sequences during and subsequent to the Last Glacial Maximum (LGM). The west coast is characterized by over-deepened fjords behind narrow and shallow inlets or sills, and over the highlands and islands to the far north and west by a landscape that has been planed clean by ice scour. Within both these types of landscapes, sediment deposition is minor and usually occurs in small isolated pockets that have been protected not only from the scour of ice but also from subsequent erosion by the strong westerly climate.

Scotland shows an unusually high tidal range, with stormy conditions a frequent occurrence due to its exposure to the strong westerly winds and North Atlantic swell. The east coast experiences both the tidal movements and the influence of waves in relatively shallow waters of the North Sea to create currents and the associated movement of sediment in the nearshore zone. In addition, the central northern North Sea area links to the Norwegian coastal zone and a renewed area of interest for the preservation of sediments. The interplay of the Atlantic and North Sea forces is most strongly manifest along the north coast, for example in the Pentland Firth where extreme currents are channeled between the mainland and the Orkney Isles.

While erosion is a feature around most of the coastline, deposition of sediment tends to occur more widely on the east coast, where the legacy of glaciation in the form of glaciogenic sediments (such as sands and gravels) is found both in the nearshore zone and offshore. The east coast also features the terminus of most Scottish major rivers that supply the majority of modern sediment to the shelf. The heads of these estuaries are the main locations for salt marshes, although due to the strong North Sea current regime and large wave fetch, the marshes are not as extensive as their counterparts in Wales and England.

Across all coastlines the complex history of eustatic and isostatic change is manifest by particular geomorphological features. For example, onshore raised beaches are testament to high sea levels in the past (see pages 195–200 for specific descriptions) and flooded lagoons over previous marsh and coastal lakes. Offshore, submerged cliff lines and platforms are accompanied by drowned river valleys, for example in the Firth of Forth.

Data sources: BGS seabed sediments and Quaternary sheets 1:250,000

Relatively high-resolution mapping vector and raster data for the coast of Scotland is held by the Ordnance Survey (OS), the Crown Estates and the British Geological Survey (BGS). Very high-resolution mapping using LiDAR (Light Detection and Ranging) has not been accomplished in a systematic manner around the Scottish coast to the extent that it has in Wales and England, with small surveys acquired on the Outer Hebrides, Orkney and around the urban centers of Glasgow and Edinburgh. This information is held with the commissioning bodies: Scottish Natural Heritage, Historic Environment Scotland and the Scottish Parliament. High-resolution vertical and oblique aerial photographs are held by the Royal Commission on Ancient and Historic Monuments for Scotland. Data on wave and tidal conditions are held by the United Kingdom Hydrographic Office (UKHO) and also by the British Oceanographic Data Centre (BODC).

The BGS collaborated with its opposite numbers in the Netherlands and Norway during the 1980s and 90s to produce a series of seabed sediment maps for the UK continental shelf at a scale of 1:250,000. These maps, and the associated cores, are an essential tool for assessing the prehistoric archaeological potential and sensitivity of areas of the sea floor. They provide classification of surface sediments by grain size, thickness of active marine sediments, thickness of Holocene deposits, standard cross-sections, information on tidal currents, sand waves and sand ripples, carbonate percentage, and other items of information which vary from sheet to sheet. Some sheets, but not all, include copious technical notes, sections, core profiles, and analysis of sources, references, and comments on the various facies. All sheets show positions of platforms and pipelines at date of publication. Notes on some of the most relevant sheets follow (from north to south). This analysis refers only to the geological, sedimentary, and taphonomic conditions relevant to primary occupation in the area, and the preservation of sites.

There have been numerous studies (e.g. Bates *et al.* 2013) of coastal change within the context of archaeological potential and potential loss for the strip of land that is intertidal and just above high-water mark. Significant work has been done on a number of sectors around the Scottish coast by Scottish Coastal Archaeology and the Problem of Erosion (SCAPE) (Dawson 2007).

Background Bedrock and Quaternary Geology

Bedrock geology

The north and west of the area is situated on continental crust that is part of the Eurasian tectonic plate. The region is divided by a number of globally significant sutures that separate ancient terrains with Archaen (>2500 Ma) rocks separated by the Moine Thrust from Proterozoic rocks to the east. The oldest rocks consist mainly of quartzofeldspathic gneisses, ultrabasic to acid intrusives and metasediments of the Lewisian Complex with Proterozoic rocks consisting of psammites and pelitic schists. Closer to shore and onshore, the Lewisian is overlain by a cover sequence of predominantly fluvial sandstones, conglomerates and shallow marine sediments. There is a strong dominance of northeast–southwest trending sutures that cross the seazone inherited from events during the Caledonian orogeny (470–410 Ma). This regional weakness is exploited throughout geological history with the most recent glacial activity reinforcing the deep Faroe–Shetland Channel. Subsequent to the Caledonian orogeny, local extension and strike-slip tectonics allowed depositional centers to form extended sequences of sediment during the Mid to Late Devonian, most notably in the Orcadian basin. During the Late Paleocene to Early Eocene the region experienced uplift associated with volcanic extrusions and extensive intrusive activity with associated extension. The Miocene to Pliocene is marked by shallow shelf deposition with small regional uplift toward the Holocene lowstand.

While the North Sea also sits within the Eurasian plate, its bedrock geology is dominated by structural features associated with a complex pattern of down-faulted basins separated by uplifted platforms formed through the Mesozoic and Cenozoic. Sediment cover is thin in the northern sector apart from sediments in marginal basins and within the fault-controlled uplift area of the Inner Moray Firth basin. To the south the Forth Approaches basin represents a Carboniferous extensional setting along northeast–southwest aligned structures with sediments thickening to the depositional centers offshore. A major graben, the Viking Graben, dominates the subsurface structure along the southeast Norwegian coast.

Interpretation of the regional geology is supported by legacy geological data for this area, which is prolific from offshore exploration activity in the North Sea over the last 40 years. Exploration for oil and gas and the formation of archives of geophysical acoustic data and cores have been carried out in all countries bordering the northern North Sea. Examples of this data exist as both 3D seismic survey blocks (Fig. 8.2) and 2D seismic lines (Fig. 8.3). In addition to these data is a program of regional seismic data in the UK acquired by the BGS in the form of single airgun, boomer and sparker profiles.

While these data exist, they were not acquired, and have not necessarily been processed, with the near-surface Quaternary geology in mind. However, it is likely that reprocessing the data, in a similar manner to that which has yielded such successful results for the southern North Sea (Gaffney et al. 2007), would yield significant new information of prehistoric archaeological significance.

Quaternary geology

Two BGS monographs summarize the Quaternary geology, seabed sediments and bathymetry of the area. 1) The geology of the Malin-Hebrides sea area (Fyfe et al. 1994) and 2) the geology of the Hebrides and West Shetland shelves, and adjacent deepwater areas (Stoker et al. 1993). The Malin-Hebrides area has a varied offshore topography. Deep water occurs in the Inner Sound, east of Raasay, at 316 m, the deepest recorded on the UK continental shelf. The topography of the sea floor has been altered by Quaternary ice sheets which have eroded the weaker sedimentary rock rather than the igneous and metamorphic rocks, thus leading to the over-deepening of the sedimentary basins. These basins have been infilled with Quaternary sediments of considerable thicknesses. The presence of seabed sediments is related to the strength of the tidal streams which have swept and eroded much material. The geology of the Hebrides and West Shetland shelves summarizes the Quaternary sediments, seabed sediments and bathymetry to the west of the Outer Hebrides. The coastal geomorphology is of a drowned landscape due to the interaction of eustatic processes and isostatic changes in relative sea level since deglaciation. Below sea level there is a northward increase in the gradient of the coastal slope. West of the Uists (Outer Hebrides) the 20-m isobath occurs up to 10 km offshore and there is no well-defined break of slope. The shelf edge and slope comprise a marked break of slope at the Geikie Escarpment with a descent into the Rockall Trough at ca. −700 m.

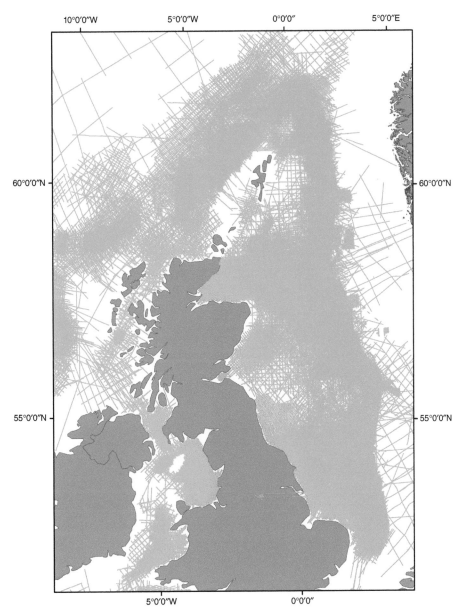

Fig. 8.2 3D seismic survey blocks acquired as part of hydrocarbon exploration licenses. Image by R. Bates.

Bathymetry

The continental shelf to the west and north of Scotland consists of a west to northwest dipping shelf intersected by a series of deep troughs (depths greater than 150 m) out to the shelf break of slope at 100 km to 200 km from the present coastline. Large topographic variation resulting from ice-scoured bedrock of geologically controlled erosion results in a bathymetry of upstanding reefs set with occasional pockets of fine-grained sediments. In contrast to this, the east coast consists of a gently sloping

(to 50 m depth) and undulating platform extending toward the center of the North Sea where depths increase to over 100 m. This slope is broken by a series of moraine ridges and sand banks. The nearshore on all coastlines is marked by shore-parallel shelf margins that extend out from the present coast to 3 km to 12 km at depths of 10 m to 30 m.

The offshore bathymetry across the continental shelf is available in vector and raster form as DigBath250 from the BGS. This covers the region of interest and extends

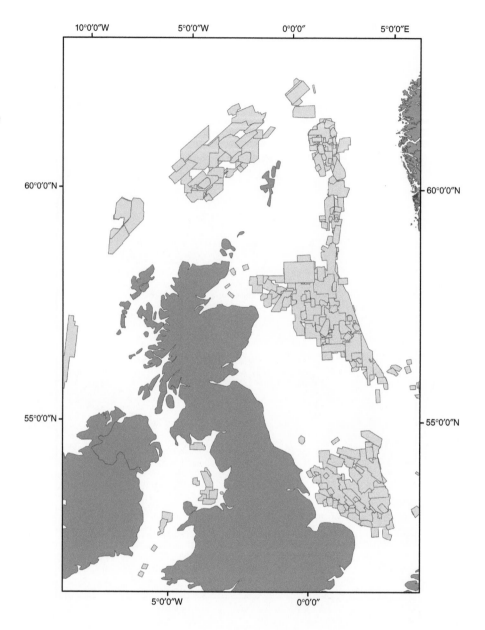

Fig. 8.3 Example of 2D seismic exploration lines acquired as part of hydrocarbon exploration licenses. Holdings for metadata are with the BGS and geological surveys of Norway, Denmark and the Netherlands. Image by R. Bates.

to include the whole of the UK continental shelf and slope and the northern North Sea across to Norway and Denmark, compiled from data gathered from the BGS, the UKHO, Flemish Hydrographic Survey, German Hydrographic Office, Netherlands Hydrographic Office, Norwegian Hydrographic Service and the Royal Danish Administration of Navigation and Hydrography.

A continuing program of shelf multibeam survey at higher resolution is ongoing by the UKHO and is supplemented by both commissioned survey for offshore development and speculative academic-related survey. Much has been made recently of SeaZone TruDepth points data sets for investigation of archaeological potential and this holds great promise for regional appraisal. Of particular note in regards to high-resolution data acquisition are the recent surveys undertaken by Marine Scotland in collaboration with various government partners and academic institutes. An example of the recent multibeam data sets is shown in Fig. 8.4 for the area to the south of Skye and east of Soay. From the

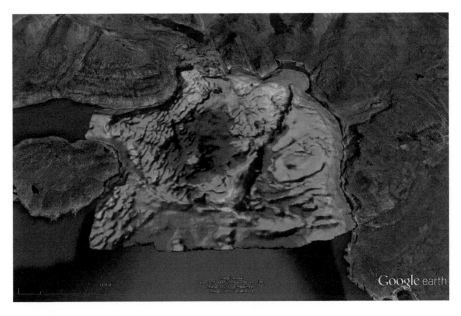

Fig. 8.4 Multibeam sonar bathymetry map of Loch Scavaig, south of the island of Skye on the west coast of Scotland. Color image shows depth variation from red (shallow) to blue (deep). The high-resolution bathymetry highlights moraine ridges that show ice advance limits together with bedrock signatures such as the dipping layers of the Torridonian Sandstones on the west side of the map. Image by R. Bates.

high-resolution data the bedrock geology and the moraine records resulting from deglaciation are readily apparent.

Post-Last Glacial Maximum Climate, Sea Level and Paleoshorelines

During the timescale covered by this chapter, that is the period from the Last Glacial Maximum onwards (a period of some 20,000 years), the climate and environment across northern Doggerland varied. While precise information relating to vegetation and ground conditions is still lacking, it is possible to reconstruct more detail for climate. For much of this period the surrounding landmasses were almost entirely glaciated and this increased the importance of the available land as a potential homeland for the prehistoric populations who were well established in the area. It was the impact of the isostatic adjustment of the landmasses on either side, due to the weight of glacial ice, together with the eustatic changes in oceanic waters that resulted in the emergence of a landmass of such size during the glacial maximum and for several millennia afterwards.

The submergence of this area took place at the start of the Holocene, but it was not a uniform process. To reconcile the detail of submergence it is necessary to know ice retreat and detailed sea-level changes that are not yet documented in great detail. As the ice melted and retreated across the continental shelves, land rebounded in certain areas of Scotland and Norway but global ice melt led to increases in the eustatic sea level and an overall submergence (Lambeck 1995). The submergence was greatest in the central North Sea and the area around the periphery of the last Scottish ice sheet such as Orkney, Shetland and the Outer Hebrides. Here, final submergence of coastal areas may not have taken place until ca. 4000 years ago (Bates *et al.* 2013). According to most authorities, the western extent of the Late Devensian ice sheet reached as far as the edge of the continental shelf (Fig. 8.5) (Hughes *et al.* 2011).

During the maximum extent of glaciation, regional sea level was ca. 120 m lower than present (Fairbanks 1989). However, owing to the loading of the last British ice sheet on the underlying lithosphere (crust), the position of relative sea level during the Last Glacial Maximum around the former ice sheet was much higher. This was due to the glacio-isostatic depression of the crust beneath the ice, the amount of depression increasing towards the western Highlands where the ice was thickest. During the

Fig. 8.5 Ice extent at the maximum of the last Scottish ice sheet. Clark *et al.* (2012). Reproduced with permission from Elsevier.

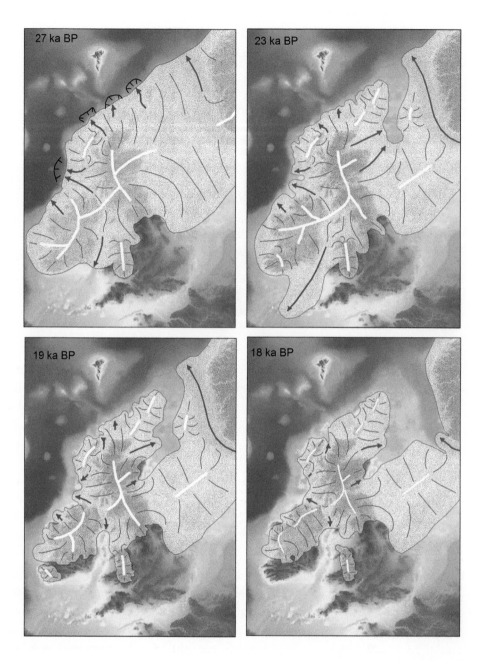

LGM, the sea floor beyond the maximum limit of the ice and across a belt several hundred kilometers in width was raised vertically. This was due to a compensatory radial outward movement of sub-crustal material from beneath the lithosphere underneath the ice sheet (Peltier 1998). In such areas during the glacial maximum therefore, the sea level was lowered on the one hand but on the other, the crust beneath the sea floor was raised, causing a regional shallowing and increased exposure of the continental

shelf. Figure 8.6 shows a schematic paleogeography off western Scotland. Areas of glacio-marine sedimentation, possible land areas and over-deepened basins are noted. The position of the shelf break is seen to the north-west of the Outer Hebrides.

As the ice sheet started to melt, regional sea level started to rise as a result of the melting of ice sheets worldwide. At the same time, as the British ice sheet started to retreat and thin, the land underneath the ice

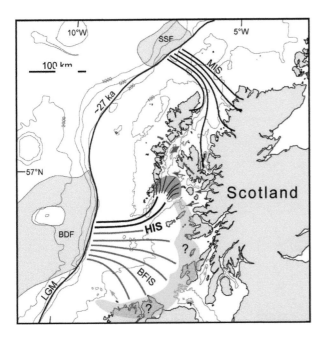

Fig. 8.6 Paleoglaciological reconstruction of ice-sheet configuration at maximum extent. LGM limit ca. 27 ka (MIS 2) (Bradwell *et al.* 2008). HIS = Hebrides Ice Stream (this study); MIS = Minch Ice Stream (Bradwell *et al.* 2008); BFIS = Barra-Fan Ice Stream (Scourse *et al.* 2009; Dunlop *et al.* 2010). Black lines are ice-stream flow lines inferred from this study; gray lines are hypothesized ice-sheet flow lines for the rest of the ice-stream catchment (not proven). Dark blue shading is ice-stream onset zone (deduced from this study); light blue shadings are hypothesized (glaciologically plausible) ice-stream onset zones for the rest of the catchment. Thick blue dashed line is the probable location of ice divide at time of HIS operation. Orange arrows show ice-sheet flow direction inferred from seabed drumlins (Dunlop *et al.* 2010; Ó Cofaigh *et al.* 2012). Blue arrows show ice-sheet flow direction from other evidence within the study area; these probably relate to later flow events (or possibly earlier). Offshore gray shaded areas are trough-mouth fans; SSF = Sula Sgeir Fan; BDF = Barra Donegal Fan (after Stoker 1995). Thin gray lines are bathymetric contours with values in meters.

sheet started to rise. The rate of vertical rise of the land surface beneath the ice varied regionally with the greatest rates of rise in areas where the ice was formerly thickest. By contrast, smaller amounts of vertical rebound took place towards the edge of the ice sheet where the ice thicknesses were much less. Beyond the ice margin, as sub-crustal material started to migrate back into areas underneath the lithosphere underlying the former ice

sheet, the areas of ocean floor started to sink (Lambeck *et al.* 1996).

Accompanying the eastward retreat and thinning of the ice sheet was a rapid rise in regional sea level. For the most part the early stages of ice-sheet decay (deglaciation) were characterized by a rise of (glacio-eustatic) sea level. Thus, one might imagine that as the ice sheet retreated eastwards towards the present Scottish mainland the rapidly rising sea level was immediately flooding into land areas exposed by the melting ice. During the early stages of regional deglaciation there were no new land areas exposed for possible Paleolithic habitation.

This complex interplay between the global eustatic component of sea-level rise and the rebounding of the land with the removal of the ice sheets leads to a variable pattern of relative sea-level changes around the Scottish coast and in the central North Sea. For Scotland, this trend, of rising sea level outpacing the rate of rise of the land surface, continued until the end of the Younger Dryas (Loch Lomond Stadial ca. 11–10 ka) period. For archaeology, the end of the Younger Dryas is highly significant since it is at this stage that parts of the Scottish coastline close to the center of glacio-isostatic rebound in the western Highlands started to experience a rate of crustal rebound faster than the rate of rise in regional sea level (graphs for a range of sites across Scotland demonstrate this variability — Fig. 8.7). Farther west and north (i.e. towards the periphery of the UK landmass) however, throughout the Western Isles and Orkney and Shetland, the rate of crustal rebound remained insufficient to exceed the rate of rise in sea level caused by an increasing volume of sea water in the Global Ocean. Such areas, therefore, continued to experience net submergence, as did the land area in the central North Sea. The central North Sea upon deglaciation opened up an exceptional area of dry land, an area known as Doggerland (Coles 1998).

Late Glacial shoreline isobase map

The history of relative sea level in Scotland during the Holocene (last 10,000 years) is complex. The area around Oban in the west had greater thickness of ice than areas of the Outer Hebrides, the north coast and the Northern

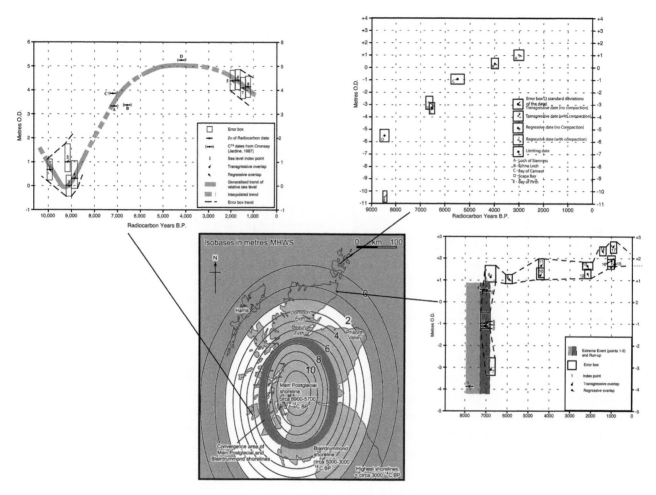

Fig. 8.7 Relative sea-level change graphs for selected areas across the north and west coast of Scotland. Modified from Jordan *et al.* (2010).

Isles. This led to varying amounts of isostatic rebound and therefore the various positions in the landscape where we see relict shorelines today. For the Oban area, this translates to visible shorelines dated to ca. 10 ka up to 10 m Ordnance Datum (OD). However, the same Late Glacial shoreline is well below present sea level in the areas of the islands of Coll, Tiree, and Islay, and the Solway Firth coastline. Predicted shorelines for the Outer Hebrides and Orkney suggest they are located between 20 m and 30 m below present (Fig. 8.8).

Shoreline uplift isobases (in meters OD) for the Main Late Glacial shoreline of Younger Dryas age have been extrapolated to 40-m water depth (below OD) based on various sources (Fig. 8.8). The best estimate of regional

eustatic sea level for the Younger Dryas is between −40 m and −45 m OD. Thus, the shoreline isobases are plotted to −40 m but no further. The coastal areas shown inside (and above) the 0 m shoreline contour indicate those areas where any coastal settlement archaeology of early Holocene age would be preserved at or above present day sea level. Outside of the 0 m isobase contour, any existing Mesolithic coastal settlement archaeology could occur below sea level across those areas of seabed shallower than the contours indicated. Figure 8.9 shows a GIS (Geographic Information System) reconstruction of dry land areas around Scotland's west coast that would have existed at the start of the Holocene interglacial and Fig. 8.10, areas around Orkney.

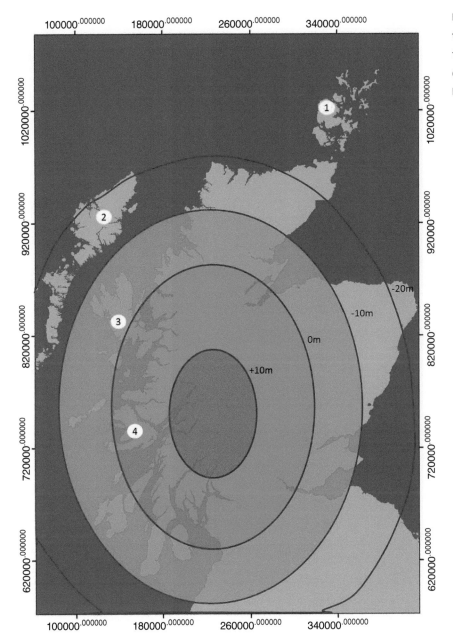

Fig. 8.8 Shoreline uplift isobases (m OD) for the Main Late Glacial Shoreline. Locations for key locations in text, (1) Orkney Islands; (2) Outer Hebrides; (3) Skye; (4) Mull. Image by R. Bates.

Holocene Relative Sea-Level Changes

Smith *et al.* 2006 and Fig. 8.7 show the most recent analysis of shorelines around the Scottish coast with the production of empirical models, based upon shoreline altitude data. The isobase maps show how the sequences of Holocene shorelines reflect variations in the relationship between isostatic movements and sea-surface change with increasing distance from the center of the Late Devensian ice sheet, with differences of as much as 10 m in elevation between shorelines of the same age. Thus, towards the heads of estuaries (such as the Forth, Tay and Clyde) close to the center of uplift, shorelines are uplifted and easily identified. On the peripheries of the Scottish uplift center, for example the Outer Hebrides to the west (including the Isle of Harris, Fig. 8.7), shorelines

Fig. 8.9 GIS reconstruction of additional dry land areas around Scotland's west coast and Outer Hebrides which would have existed at the start of the Holocene. Image by R. Bates.

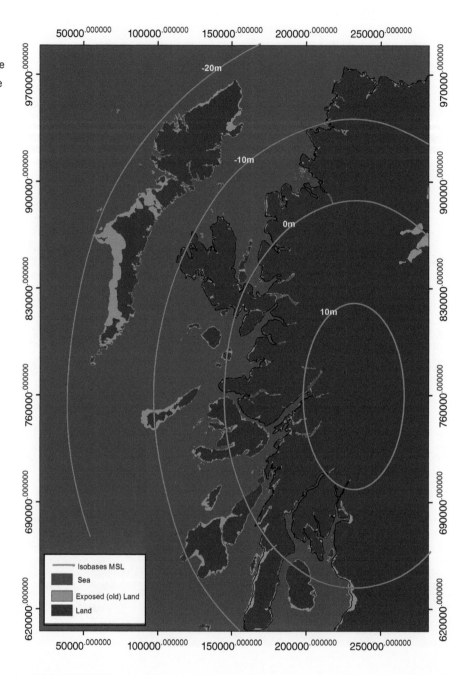

are now buried below present coastal sediments. Thus, some areas of the sea floor may have been dry land 10,000 to 12,000 years ago, flooded by rising sea level around 7000 years ago, and then exposed again a few thousand years later by the isostatic uplift of the land.

Along the east coast of Scotland shallow coastal waters range from the inland heads of the Firths where the land uplift is 8 m to 10 m out to the headlands of

Wick and Fraserburgh on the zero-isobase. This means that the rate of RSL change varies radically at different points on the coast. Edinburgh has been continuously uplifted more rapidly than the rising sea for the last 15,000 years, so that the sea level has undergone relative lowering throughout that time. At Lerwick, Shetland, however, the land has been sinking continuously since the ice melted so that the RSL has been continuously rising.

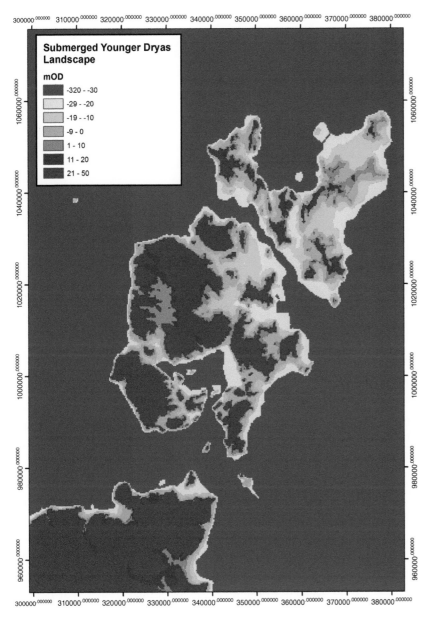

Fig. 8.10 GIS reconstruction of dry land areas around the Orkney Islands which would have existed at the start of the Holocene. Image by R. Bates.

Cromarty and Ythan are interesting because their rate of uplift has been closely similar to the sea-level rise, with a relative fall of sea level from 14 ka to 9 ka dropping below the present coastline, rising again to a maximum above the present coastline at about 5 ka and then dropping gradually to the present level. To the east of the present coastline, exposure of land across extensive areas of the entire North Sea area allows for a greater potential occupation area than hitherto realized (Gaffney *et al.* 2007; 2009).

Off the east coasts of Shetland, Fair Isle and Orkney, one would expect to find submerged caves or materials trapped in gullies and cracks in the bedrock. Various combinations of floating sea ice, rocky shelters, terrestrial mammals such as reindeer, or marine mammals such as seals, walrus, otter, and cetaceans, would have been present, depending on the exact local topography and conditions. Andrews *et al.* (1990) indicate scattered small islands emerging from the sea-ice typically a few tens of kilometers across. Their estimate of exposed land is conservative and they note that the areas of exposed land may have been more extensive than they show on their maps. This suggests a complex terrain of sheltered sea almost totally protected from North Atlantic storms, dotted with islands, covered with floating sea ice, and bordered to the west by the grounded Scottish ice cap.

Models of Postglacial Isostatic Adjustment

Rheological models of glacio-hydro-isostatic rebound, based upon geophysical data and modeled to fit the available sea-level data, have been developed by Lambeck *et al.* (1996) and Peltier *et al.* (2002). Models of British ice-sheet behavior since the LGM have also been developed. The model ICE-4G (VM2) is the model used to constrain the RSL pattern for the British Isles, and gives the best fit to the observational data (Peltier *et al.* 2002). Nevertheless, there are discrepancies between model predictions and the observational data sets for some areas of Scotland, e.g. Orkney (Bates *et al.* 2013) and Wick (Dawson & Smith 1997), both in the north of Scotland. Models of glacio-hydro-isostatic sea-level change have also been developed for the Irish Sea basin as the original British Isles models failed to represent the changes observed along the Irish Sea margins since deglaciation (Lambeck & Purcell 2001).

Relative Sea-Level Changes

Rockall

The Rockall Plateau is an extensive shallow-water area to the west of the British Isles and separated from the Scottish mainland by the 3000-m-deep Rockall Trough. The trough is a southwest-trending basin which lies between the mainland and the Rockall Plateau. The small pinnacle of Rockall Island is the only area to appear now subaerially on the plateau. A possibly wave-cut platform around the island exists at ca. 110 m. Beyond the island a break of slope at ca. 180 m depth and dredged beach

material suggest subaerial erosion during a Holocene sea-level stillstand (Roberts 1975). The continental margin north-west and west of the British Isles consists of a broad continental shelf, a narrow continental slope and a broader continental rise. Between latitudes 55°N and 60°N, an inner and an outer shelf are separated by the Outer Hebrides. A shallow basin separating the islands of Rona, Flannan and St. Kilda from the Outer Hebrides is the only area of relief on the outer shelf.

St. Kilda

The St. Kilda archipelago (57°49'N, 08°35'W) lies 64 km to the west of the Outer Hebrides towards the edge of the western Scottish continental shelf. The shelf slopes gently westward from the Outer Hebrides to a depth of 120 m to 140 m around St. Kilda over a distance of ca. 100 km. There is little sediment deposition on the shelf, and mainly erosional landforms. Near-vertical and cliff-like, the marginal slope exhibits a pronounced break of slope with the western margin fronted by a near-level surface at ca. 120 m depth (Sutherland 1984). The cliffed coastline of the islands and stacks that comprise the archipelago plunges to depths of 40 m. This shallows at −120 m to −80 m and then to −40 m around the base of the clifflines around the islands. This upper erosional surface culminates in a clear marine erosional platform at ca. −40 m. Planation of the bedrock surfaces and cliff formation must have occurred during periods of sea-level stability. It is suggested that during the peak of the Late Devensian ice sheet, sea levels around St. Kilda were around 120 m lower than present. The marine planation surfaces that occur between −70 m and −40 m are thought to have been formed during the last period of Northern Hemisphere glaciation, the Loch Lomond Stadial (Younger Dryas, 11–10 ka), when local RSL may have stood at around −40 m (Sutherland 1984).

Outer Hebrides

The sedimentary evidence for Holocene RSL fluctuations in the Outer Hebrides is somewhat limited. Radiocarbon dating of intertidal peats can be indirectly related to a sea level at least as low as the occurrence of the peat/organic deposits. Thus, a date of ca. 5700 BP was obtained for intertidal deposits in Benbecula (Ritchie 1998), which correlate to a sea level at least 5 m lower than the present. Work on the isles of Lewis and Harris by Jordan (2004) shows that the Mid–Late Holocene RSL record can be summarized as comprising two main events occurring between 5500±60 BP and 4500±100 BP, and between 3000±80 BP and 820±50 BP. Rising sea levels throughout the Holocene controlled the onshore movement of vast amounts of sediment from the extensive and fairly shallow coastal shelf to the west, which in turn has developed into the modern beach and dune systems and the machair grassland that fringes the western Atlantic seaboard of the whole island chain. The sea-level index points (SLIPs) identified by Jordan (2004) are thought to be representative of coastal barrier formation as opposed to direct shoreline development. However, as barrier formation is closely related to the relative movement of the sea, it is believed that both the Main Postglacial Transgression and the later Blairdrummond Shoreline (Smith *et al.* 2000) are recorded in the sediments. One site in particular, Northton on the Isle of Harris, has also recorded an Early/Mid Holocene (7370±80 BP) storm surge event that was imprinted upon the rising RSL trend for the area. Sediments from this event have a similar sedimentological signature to deposits from the storm surge that occurred in January 2005 across the southern areas of the Outer Isles.

Northwest Scotland mainland

The northwest of Scotland has been subject to extensive study of RSL change during the last twenty years (Peacock 1970; Shennan *et al.* 1994; 1995; 2000; 2005). Peacock (1970) defined the local marine limit around Arisaig at ca. 41 m OD from terraces and banks of shingle and sand interpreted as raised beaches at Sunisletter. The marine limit occurs between Upper Loch Dubh (sill elevation 36.48 m OD) and Cnoc Pheadir (sill elevation 42.5 m OD). Rapid RSL fall continued until at least 14,000 cal BP. The relative sea level then rose and reached no higher than ca. 10 m during the Holocene.

Northeast Scotland and the Northern Isles

South of Fraserburgh the situation changes because the Scandinavian ice sheet extended across the Norwegian Trench onto the UK shelf as far west as the Greenwich meridian, and a substantial dryland area of tundra was exposed between the Scandinavian grounded ice and the sea-lake. This dry land continued widening to the south and east over the southern North Sea basin, and was continuous with the land which is now Germany and France. The River Elbe discharged across this shelf to the north, and probably other rivers drained the landscape, as in northern Russia today. The combination of extensive land to the south, the proximity of the ice sheets, the large sea-lake, and the scattered islands projecting from the floating ice suggest a complex topography which could have supported humans exploiting sea mammals, fishing and land mammals. The tidal range is of the order of 3 m to 4 m along much of the coast, and tidal streams in the north-east area, where the tide flows through the gaps in the Orkney–Shetland ridge, have velocities of 2 m/sec to 3 m/sec. This area is co-extensive with a seabed of bare bedrock with very little sediment. In the Moray Firth the currents drop to 1.0 m/sec to 1.2 m/sec, with a further peak of 1.8 m/sec around Fraserburgh. In the southern half of the Moray Firth this is associated with 10 m to 20 m thickness of Quaternary sediments. Further offshore the currents over the whole area are of the order of 0.8 m/sec to 1.0 m/sec (Blackham *et al.* 1985). Where the currents have exposed the bedrock, artifacts would only survive trapped in gullies or caves. Over much of the area the currents have been strong enough to winnow out fine mud or clay, but this process would leave lithic artifacts in place.

Taphonomy and Potential for Archaeological Site Survival

In a glaciated area such as the submerged landscape of northern Doggerland, specialized conditions such as ice scour, glacial erosion, frost shattering, and normal subaerial erosion processes all have to be taken into account when considering the survival of archaeological material. In view of the work of Pitulko *et al.* (2004) it is also important to consider the effect of sea water rising over archaeological deposits in permafrost, which can result in the good preservation of artifacts.

An extreme event which needs to be taken into account is the Storegga Submarine Slide, which occurred off the coast of Norway at 7200 BP (about 8100 cal BP), and caused a tsunami which has been detected in coastal sediments on land on the east coast of Scotland (Dawson *et al.* 1988; Smith *et al.* 2004). At the date of the submarine landslide the sea level was still approximately 20 m to 30 m below present, and Dogger Bank was a promontory connected to north Germany, while the land bridge from the Netherlands to the Humber coast had recently been inundated. The tsunami wave locally may have penetrated several hundred meters inland, with a run-up of 1 m to 2 m in open areas, and much greater in enclosed lochs. Long and Holmes (2001) suggest that the human impact would have been small, due to the low population (*op.cit.* p.365), though this is disputed by Weninger *et al.* (2008). The impact may have been greater on the north shore of the Dogger Bank, if people were living there, and in Elbe Estuary.

Large areas of the Scottish shelf have been dry land for considerable periods in the last 700,000 years — the period of human (Paleolithic) settlement in Britain. England and Wales have a good record of early settlement sites on land, particularly in the south, but there are no absolutely-dated Paleolithic sites in Scotland so far, though artifact finds from three sites are significant. Typologically-specific lithics from Fairnington (Saville 2004), Howburn (Ballin *et al.* 2010a) and Kilmelfort (Saville & Ballin 2009) have been interpreted as providing evidence for human activity in the Late Glacial Period between 14,000–11,000 years ago. Environmental and osteological evidence suggests that the submerged landscape of the northern North Sea has, at times, been suitable for human settlement and it is possible that surviving Paleolithic sites from the 'Scottish sector' of the seabed still survive (Long *et al.* 1986), while comparable sites on land have been destroyed or buried by the actions of the Last Glacial ice sheet, which blanketed mainland Scotland.

The relative sea-level history of Scotland during the Holocene (last 10,000 years) is complex and the net result of this is that in some areas the present seabed has been dry land within the last 10,000 years. As this is the period within which Scotland has a comprehensive record of human activity, it is likely that these areas were once settled. They offer the possibility that submerged archaeological sites may be preserved.

Perhaps the best-known submerged landscape is that around the archipelago of Orkney, where the sea did not reach present levels until about 4000 years ago (Flemming 2003; Bates *et al.* 2013), but another area lies to the west of the Western Isles (Wickham-Jones & Dawson 2006), and there are small localized areas elsewhere, for example around Coll, Tiree and Islay. Although there is no specific data on RSL rise for some of these areas, it is assumed that sea level reached roughly its present level between 3000 and 5000 years ago, meaning that any submerged archaeological sites are likely to relate to Mesolithic or Early Neolithic settlement. Interestingly, both Orkney and the Western Isles stand out from the rest of Scotland in that they have little evidence on land for Mesolithic settlement. Mesolithic sites are few and far between in Orkney and lacking, with the exception of a few dates on anthropogenic deposits, e.g. at Northton (Gregory *et al.* 2005; Simpson *et al.* 2006), in the Western Isles. Given the importance of coastal resources in the Mesolithic and the apparent concentration of sites around Scotland's coastlands, this may be significant as an indication that evidence for the first 5000 years of human settlement in these areas is lying in the present offshore area. Recent field research in Orkney suggests the possible preservation of stone structures relating to the Neolithic on the seabed (Bates *et al.* 2013), and Benjamin (2010) has proposed a model for investigating the potential for submerged archaeology in other parts of Scotland, which gives practitioners the opportunity to test different techniques.

Large parts of the seabed sediments west of mainland Scotland towards the Outer Hebrides are not of marine origin but are submerged terrestrial deposits and deposits resulting from the erosion and re-deposition of material by glaciers and ice sheets. Sands, gravels, silts, clays and organic-rich deposits are referred to as Marine Aggregate

Deposits (MAD) (Wenban-Smith 2002). The potential for preservation of archaeological remains within these sediments depends upon the depositional and post-depositional processes on the offshore landscape prior to inundation. Many areas will have been subject to repeated glaciations and marine inundation since the peak of the ice sheet at ca. 22 ka. Material will have been transported, remixed and reworked. Rising water levels may favor the preservation of associated intercalated organic deposits. Once buried by fine-grained material, these may be more resistant to the effects of aerial exposure during marine regression (Wenban-Smith 2002). Evidence offshore for estuarine clays and silts, littoral and estuarine peats and silt-rich floodplain deposits is likely to provide good preservation potential for archaeological material, (e.g. Clachan Old Harbour, Raasay, Inner Sound: Richardson & Cressey 2007; Ballin *et al.* 2010b).

Many sites are likely to be deeply buried. Reconstruction of the conditions which may have buried archaeological sites and facilitated their re-discovery has recently been improved by new analytical techniques. Praeg (2003) for example, has used seismic imaging to detect buried glacial tunnels under modern sediments. Fitch *et al.* (2005) have re-interpreted extensive sub-bottom seismic records to detect the changes in sediment characteristics indicating buried river valleys. This technique has exposed a wide meandering river system draining northwards from the north-east flank of the Dogger Bank, and it is now being tested on other parts of the UK shelf, including the northern North Sea. Detailed reconstruction of the prehistoric topography of the submerged shelf is important as it may then be 'populated' using modeling based on the known locations of prehistoric sites in similar landscapes (Lakes *et al.* 1998; Fitch 2011). This allows construction of a hypothetical settlement pattern that may be tested, e.g. through diving, coring and remote sensing as appropriate. In this way, a more accurate map of potential sites can be drawn up.

Potential discovery hotspots in northern Doggerland cannot be listed exhaustively at this stage because of the lack of research in the area, but see the following section 'Potential Example Areas for Future Work' for a preliminary discussion of some areas. The steps needed to create high-resolution local sensitivity maps can however be identified. In principle the key factors that increase the potential for both early human settlement, and archaeological survival, are:

'fossil' estuaries and river valleys;

valleys, depressions, or basins with wetland or marsh deposits;

nearshore creeks, mudflats, and peat deposits;

'fossil' archipelago topographies where sites would have been sheltered by low-lying islands as the sea level rose;

niche environments in present coastal zones, wetlands, intertidal mudflats, lochs, and estuaries;

caves and rock shelters in re-entrant bays;

deposits of sediments formed within, or washed into rocky gullies and depressions.

Potential Example Areas for Future Work

The coastlands and islands contain some of the earliest recorded settlement in the area, dating back to the early ninth millennium BC and relating to the Mesolithic, or to Late Stone Age hunters who settled in Scotland after the end of the Last Glacial (Saville 2008). Several of Scotland's earliest dated archaeological sites come from the coast and islands of the north and west (Mithen & Wicks 2008; Hardy & Wickham-Jones 2009; Mithen *et al.* 2015). This undoubtedly reflects an element of bias in that archaeologists have long been attracted to the wealth of archaeological material surviving here, but it also reflects the importance of the area to an early population who were reliant upon water-based transport and who were attracted to the rich resources of the coastal lands and islands. Not only did the coast offer concentrations of marine and terrestrial food in terms of fish, shellfish, seabirds, mammals, nuts, roots and berries, it also had other advantages such as the shelter afforded by the many rock shelters and caves, and easy access to boat travel in inshore waters (Wickham-Jones 2014).

Initial study of these island areas suggests good prospects for the conservation of submarine prehistoric remains. They are likely to have been settled from earliest times though evidence for Paleolithic occupation on land has been affected by ice action during the Late Devensian glaciation which ended ca. 14,000 years ago. A complex history of RSL change, however, means that parts of the seabed have been exposed as dry land during and since this period. These areas were suitable for human settlement from early on and they may well preserve the record of that settlement. Some of these submerged areas remained exposed well into recent times (4 ka) and are thus likely to have been settled through the Mesolithic and into the Neolithic. Indeed the lack of Mesolithic sites in the Western Isles, Orkney and Shetland is notable and probably to be explained by this history of sea-level change. This dearth of terrestrial sites means that any archaeological sites to be found on the submerged landscape would be particularly important. Furthermore the terrestrial record can now be shown to be biased. The changing landscape means that terrestrial sites must be interpreted not so much as coastal in nature but rather in line with the topography created by lower relative sea levels.

Nevertheless not all archaeological sites will have survived submergence by the sea. In general, the lack of detailed research in an archaeological context means that hotspots cannot at present be mapped. However, as oil and gas exploration continues around Scotland and Scandinavia, data from coring and seismic survey ahead of major pipeline installation may yield valuable sources of information and this remains an area of future potential.

Known Submerged Finds

Compared to the southern North Sea where there are abundant faunal remains and occasional artifacts, the fossils from the bed of the northern North Sea are of more fragmentary distribution. The species of mammal recorded from the Scottish North Sea are (in order of abundance of fossils) reindeer (*Rangifer tarandus*), bison (*Bison* sp.), musk-ox (*Ovibos moschatus*), woolly mammoth (*Mammuthus primigenius*), red deer (*Cervus elaphus*), and some woolly rhinoceros (*Coelodonta antiquitatis*) (Flemming 2003).

With regard to Scotland, finds of artifacts are limited to a single worked flint from vibrocore number 60+01/46 obtained as part of a BGS program in the UK shelf some 150 km north-east of Lerwick, near the Viking Bank, in a water depth of 143 m (Long *et al.* 1986). It is possible that this find came from an area of dry land and is thus to be regarded as a submerged indication of prehistoric occupation in a beach environment.

Many intertidal peat deposits and examples of sub-merged woodland have been noted along the western coastal stretches of the Hebrides, though few have been accurately recorded or studied. The result is considerable evidence for submergence in the last few thousand years. Recent storm activity in the Outer Hebrides has uncovered many new exposures of intertidal peats and on-going studies include those on Coll (Dawson *et al.* 2001), and Raasay (Dawson 2009; Ballin *et al.* 2010b). These provide evidence of a slowly rising RSL with stillstands of sufficient length to permit the growth of woodland. Conditions like this would have permitted the local Mesolithic inhabitants to settle in the vicinity of the (now-submerged) coastline.

At Clachan Old Harbour on the south coast of Raasay there is a deposit of submerged peat and tree roots. Much of this has been destroyed by recent digging for fuel, though this digging has now stopped. There is anecdotal evidence for the removal of stone tools from here and when the site was visited by archaeologists in the summer of 2001, a single stone flake was recovered. Since then, further work, including excavation by CFA Archaeology Ltd, has resulted in the recovery of a small assemblage of worked stone, indicative of activity which has been dated to the Early Mesolithic (Ballin *et al.* 2010b). Dawson's work has shown that this site relates to a slightly lower stillstand in RSL that lasted long enough, probably 500–1000 years, to allow the growth of woodland (Dawson 2009).

In 1991, a scallop boat dredged a gold torc from the seabed near the Shiant Isles (Cowie 1994). This artifact is Bronze Age in date and assumed to relate to loss at sea, whether deliberate or accidental. During the Bronze Age the deliberate deposition of valuable objects in water was a common phenomenon. The characteristics of the Sound of Shiant mean that this artifact could have travelled here from some distance, but the find is also an indication that similar prehistoric material might occur elsewhere on the seabed.

In 1981, a group of divers recovered a gold arm ring from the seabed near to Ruadh Sgeir at the north end of the Sound of Jura (Graham-Campbell 1983). This artifact has been dated to the Viking period, probably tenth century AD, and is assumed to have resulted from a loss at sea. Again, it signifies the potential of the seabed to yield prehistoric remains that reflect to maritime trade and travel rather than direct settlement.

Areas of high potential

There is great likelihood of finds relating to the Mesolithic (10–6 ka) and Neolithic (6–4 ka) periods on the shallower parts of the Scottish shelf (down to ca. −45 m) in areas where the conditions for site preservation can be met. There is also a high possibility of finds relating to the Paleolithic period, prior to the Mesolithic, especially on lower stretches of the Scottish shelf and in the central northern North Sea across to Norway, though it is difficult to pinpoint hotspots for this.

Areas of high potential include the waters around Orkney and Shetland, as well as the Western Isles. Research around Orkney has provided local detail of past RSL change (Bates et al. forthcoming) and indicates the probability that stone-built structures relating to the Neolithic may have survived on the seabed (Bates et al. 2013).

Research around Shetland has focused on the period ca. 8.2 ka related to the Storegga tsunami event, and there is less detail of RSL change extending into more recent periods of human settlement. Nevertheless, paleogeographic models suggest the presence of a considerable submerged landscape around Shetland which may well hold the key to little-known periods of settlement here such as the Mesolithic. The potential archaeological importance of this submerged landscape is high.

Recently, work to investigate the potential for submerged archaeology has been initiated around the Western Isles (Benjamin & Hale 2012). Relative sea-level change suggests a considerable submerged landscape here.

Further potential locations for the survival of archaeological material on the seabed include the shelf to the west of the Hebrides; the Hawes Bank and seabed around Coll and Tiree; and between and around Islay, Jura, Colonsay and Oronsay. More specific locations include parts of the Rum and Canna coastline, sheltered inlets and reaches to the east of the Hebrides, and sheltered inlets around Skye. Recent research at the University of Ulster, Coleraine, has highlighted the previous existence of a low-energy strait with various islands between the north Irish coast and the south Hebrides in the Early Holocene (Cooper et al. 2002) thus confirming the importance of this area as another potential archaeological hotspot.

Conclusion and Outlook

Submerged prehistoric archaeology comprises a considerable resource for Scotland, a resource that, unlike other parts of Britain, is relatively unexplored to date.

The development of increasingly sophisticated detection methods, mapping, and underwater excavation means that the recovery of archaeological information from the northern North Sea is increasingly likely. While research has started, it is still patchy and underfunded. Some networks have already been set up by those working elsewhere, (e.g. the Submerged Landscapes Archaeological Network – submergedlandscapes.wordpress.com) and there is considerable scope for Scotland to benefit from the experience of those already working in the field.

A systematic approach is needed in order to catalogue areas that need investigation. These areas should be subject to targeted investigations following a wide area assessment for paleolandscape reconstruction. Specific features of pilot projects should include:

increased data on relative sea-level change, especially for island groups;
detailed mapping of Scotland showing the coast at various dates and based on evidence, not modeling;
predictive modeling for submerged site survival, including 3D modeling derived from energy and aggregate industry (third party) sources;

survey for submerged sites in locations with high potential;

a database of submerged paleoenvironmental information.

References

Andrews, I. J., Long, D., Richards, P. C. *et al.* 1990. *United Kingdom Offshore Regional Report: the Geology of the Moray Firth.* HMSO for the British Geological Survey: London.

Ballin, T. B., Saville, A., Tipping, R. & Ward, T. 2010a. An Upper Paleolithic flint and chert assemblage from Howburn Farm, South Lanarkshire, Scotland: first results. *Oxford Journal Of Archaeology* 29:323-360.

Ballin, T. B, White, R., Richardson, P. & Neighbour, T. 2010b. An Early Mesolithic stone tool assemblage from Clachan Harbour, Raasay, Scottish Hebrides. *Lithics* 31:94-104.

Bates, M. R., Nayling, N., Bates, R., Dawson, S., Huws, D. & Wickham-Jones, C. 2013. A multidisciplinary approach to the archaeological investigation of a bedrock-dominated shallow marine landscape: an example from the Bay of Firth, Orkney, UK. *International Journal of Nautical Archaeology* 42:24-43.

Bates, R., Bates, M. Dawson, S. & Wickham-Jones C.R. forthcoming. Marine Geophysical Survey of the Loch of Stenness, Orkney. In Edmonds, M. (ed.) *Heart of Neolithic Orkney World Heritage Site Geophysical Survey.* Oxbow Books: Oxford.

Benjamin, J. 2010. Submerged prehistoric landscapes and underwater site discovery: Re-evaluating the 'Danish Model' for international practice. *Journal of Island & Coastal Archaeology* 5:253-270.

Benjamin, J. & Hale, A. 2012. Marine, maritime or submerged prehistory? Contextualizing the prehistoric underwater archaeologies of inland, coastal and off-shore environments. *European Journal of Archaeology* 15:237-256.

Blackham, A, M., Field, D. & Cowling, G. 1985. *The North Sea Environmental Guide.* Oilfield Publications: Ledbury.

Bradwell, T., Stoker, M. S., Golledge, N. R. *et al.* 2008. The northern sector of the last British Ice Sheet: maximum extent and demise. *Earth-Science Reviews* 88:207-226.

Clark, C. D., Hughes, A. L. C., Greenwood, S. L., Jordan, C. & Sejrup, H. P. 2012. Pattern and timing of retreat of the last British-Irish Ice Sheet. *Quaternary Science Reviews* 44:112-146.

Coles, B. J. 1998. Doggerland: a speculative survey. *Proceedings of the Prehistoric Society* 64:45-81.

Cooper, J. A. G., Kelley, J. T., Belknap, D. F., Quinn, R. & McKenna, J. 2002. Inner shelf seismic stratigraphy off the north coast of Northern Ireland: new data on the depth of the Holocene lowstand. *Marine Geology* 186:369-387.

Cowie, T. 1994. A Bronze Age gold torc from the Minch. *Hebridean Naturalist* 12:19-21.

Dawson, A. G., Long, D. & Smith, D. E. 1988. The Storegga slides: evidence from eastern Scotland for a possible tsunami. *Marine Geology* 82:271-276.

Dawson, A. G., Dawson, S., Mighall, T. M., Waldmann, G., Brown, A. & Mactaggart, F. 2001. Intertidal peat deposits and early Holocene relative sea-level changes, Traigh Eileraig, Isle of Coll, Scottish Hebrides. *Scottish Journal of Geology* 37:11-18.

Dawson, S. 2009. Relative sea level changes at Clachan Harbour, Raasay. In Hardy, K. & Wickham-Jones, C. R. (eds.) *Mesolithic and Later Sites Around the Inner Sound, Scotland: the Work of the Scotland's First Settlers Project 1998–2004.* Scottish Internet Archaeological Reports 31.

Dawson, S. & Smith, D. E. 1997. Holocene relative sea-level changes on the margin of a glacio-isostatically uplifted area: an example from northern Caithness, Scotland. *The Holocene* 7:59-77.

Dawson, T. 2007. *A Review of the Coastal Zone Assessment Surveys of Scotland, 1996-2007: Methods and Collected Data.* SCAPE Trust and Historic Scotland: Edinburgh.

Dunlop, P., Shannon, R., McCabe, M., Quinn, R. & Doyle, E. 2010 Marine geophysical evidence for ice sheet extension and recession on the Malin Shelf: New evidence for the western limits of the British-Irish Ice Sheet. *Marine Geology* 276:86-99.

Fairbanks, R. G. 1989. A 17,000-year glacio-eustatic sea level record: influence of glacial melting rates on the Younger Dryas event and deep-ocean circulation. *Nature* 342:637-642.

Fitch, S. 2011. *The Mesolithic Landscape of the Southern North Sea*. Ph.D thesis. University of Birmingham, UK.

Fitch, S., Thomson, K. & Gaffney, V. 2005. Late Pleistocene and Holocene depositional systems and the palaeogeography of the Dogger Bank, North Sea. *Quaternary Research* 64:185-196.

Flemming, N. C. 2003. *The Scope of Strategic Environmental Assessment of Continental Shelf Area SEA4 in Regard to Prehistoric Archaeological Remains*. Department of Trade and Industry: London.

Fyfe, J. A., Long, D. & Evans, D. 1994. *UK Offshore Regional Report: The Geology of the Malin-Hebrides sea area*. HMSO for the British Geological Survey: London.

Gaffney, V., Thomson, K. & Fitch, S. (eds.) 2007. *Mapping Doggerland: The Mesolithic Landscapes of the Southern North Sea*. Archaeopress: Oxford.

Gaffney, V., Fitch, S. & Smith, D. 2009. *Europe's Lost World: The Rediscovery of Doggerland*. CBA Research Report 160. Council of British Archaeology: York.

Graham-Campbell, J. 1983. A Viking-age gold arm-ring from the Sound of Jura. *Proceedings of the Society of Antiquaries of Scotland* 113:640-642.

Gregory, R. A., Murphy, E. M., Church, M. J., Edwards, K. J., Guttmann, E. B. & Simpson, D. D. A. 2005. Archaeological evidence for the first Mesolithic occupation of the Western Isles of Scotland. *The Holocene* 15:944-950.

Hardy, K. & Wickham-Jones C. R. (eds.) 2009. *Mesolithic and Later Sites Around the Inner Sound, Scotland: the Work of the Scotland's First Settlers Project 1998–2004*. Scottish Archaeological Internet Reports 31.

Hughes, A. L. C., Greenwood, S. L. & Clark, C. D. 2011. Dating constraints on the last British-Irish Ice Sheet: a map and database. *Journal of Maps* 7:156-183.

Jordan, J. T. 2004. *Holocene Coastal Change in Lewis and Harris, Scottish Outer Hebrides*. Ph.D Thesis. Coventry University, UK.

Jordan, J. T., Smith, D. E., Dawson, S. & Dawson, A. G. 2010. Holocene relative sea-level changes in Harris, Outer Hebrides, Scotland, UK. *Journal of Quaternary Science* 25:115-134.

Lake, M. W., Woodman, P. E. & Mithen, S. J. 1998. Tailoring GIS software for archaeological applications: an example concerning viewshed analysis. *Journal of Archaeological Science* 25:27-38.

Lambeck, K. 1995. Late Devensian and Holocene shorelines of the British Isles and the North Sea from models of glacio-hydro-isostatic rebound. *Journal of the Geological Society* 152:437-448.

Lambeck, K. & Purcell, A. P. 2001. Sea-level change in the Irish Sea since the Last Glacial Maximum: constraints from isostatic modelling. *Journal of Quaternary Science* 16:497-506.

Lambeck, K., Johnston, P., Smither, C. & Nakada, M. 1996. Glacial rebound of the British Isles — III. Constraints on mantle viscosity. *Geophysical Journal International* 125:340-354.

Long, D. & Holmes, R. 2001. Submarine landslides and tsunami threat to Scotland. *In Proceedings of the International Tsunami Symposium (Session 1, numbers 1-12)*. 7th – 10th August 2001, Seattle, USA, pp. 355-366.

Long, D., Wickham-Jones, C. R. & Ruckley, N. A. 1986. A flint artefact from the northern North Sea. In Roe, D. A. (ed.) *Studies in the Upper Palaeolithic of Britain and North West Europe*. BAR International Series (No. 296). pp. 55-62. British Archaeological Reports: Oxford.

Mithen, S. J. & Wicks, K. 2008. Inner Hebrides Archaeological Project — Fiskary Bay, Coll. *Discovery and Excavation in Scotland* 9:36.

Mithen, S., Wicks, K., Pirie, A. *et al.* 2015. A Lateglacial archaeological site in the far north-west of Europe at Rubha Port an t-Seilich, Isle of Islay, western Scotland: Ahrensburgian-style artefacts, absolute dating and geoarchaeology. *Journal of Quaternary Science* 30:396-416.

Ó Cofaigh, C., Dunlop, P. & Benetti, S. 2012. Marine geophysical evidence for Late Pleistocene ice sheet extent and recession off northwest Ireland. *Quaternary Science Reviews* 44:147-159.

Peacock, J. D. 1970. Some aspects of the glacial geology of west Inverness-shire. *Bulletin of the Geological Survey of Great Britain* 33:43-56.

Peltier, W. R. 1998. Postglacial variations in the level of the sea: Implications for climate dynamics and solid-Earth geophysics. *Reviews of Geophysics* 36:603-689.

Peltier, W. R., Shennan, I., Drummond, R. & Horton B. 2002. On the postglacial isostatic adjustment of the British Isles and the shallow viscoelastic structure of the Earth. *Geophysical Journal International* 148:443-475.

Pitulko, V. V., Nikolsky, P. A, Yu. Girya, E. *et al.* 2004. The Yana RHS Site: Humans in the Arctic before the Last Glacial Maximum. *Science* 303:52-56.

Praeg, D. 2003. Seismic imaging of mid-Pleistocene tunnel-valleys in the North Sea Basin — high resolution from low frequencies. *Journal of Applied Geophysics* 53:273-298.

Richardson, P. & Cressey, M. 2007. Churchton Bay, Raasay: excavation, palaeoenvironmental assessment, survey and evaluation. *Discovery and Excavation in Scotland* 8:121.

Ritchie, W. 1998. Borve. In Gordon, J. E. & Sutherland, D. G. (eds.) *Quaternary of Scotland (Geological Conservation Review Series).* pp. 429-431. Chapman & Hall: London.

Roberts, D. G. 1975. Marine geology of the Rockall Plateau and Trough. *Philosophical Transactions of the Royal Society of London (Series A Mathematical and Physical Sciences)* 278:447-509.

Saville, A. 2004. The material culture of Mesolithic Scotland. In Saville, A. (ed.) *Mesolithic Scotland and its Neighbours: the Early Holocene Prehistory of Scotland, its British and Irish Context, and some Northern European Perspectives.* pp. 185-220. Society of Antiquaries of Scotland: Edinburgh.

Saville, A. 2008. The beginning of the later Mesolithic in Scotland. In Sulgostowska, Z. & Tomaszewski, A. J. (eds.) *Man — Millennia — Environment: Studies in Honour of Romuald Schild.* pp. 207-213. Polish Academy of Sciences, Institute of Archaeology and Ethnology: Warsaw.

Saville, A. & Ballin, T. B. 2009. Upper Palaeolithic evidence from Kilmelfort Cave, Argyll: A re-evaluation of the lithic assemblage. *Proceedings of the Society of Antiquaries of Scotland* 139:9-45.

Scourse, J. D., Haapaniemi, A. I., Colmenero-Hidalgo, E. *et al.* 2009. Growth, dynamics and deglaciation of the last British-Irish ice sheet: the deep-sea ice-rafted detritus record. *Quaternary Science Reviews* 28:3066-3084.

Shennan, I., Innes, J. B., Long, A. J. & Zong, Y. 1994. Late Devensian and Holocene relative sea-level changes at Loch nan Eala, near Arisaig, northwest Scotland. *Journal of Quaternary Science* 9:261-283.

Shennan, I., Innes, J. B., Long, A. J. & Zong, Y. 1995. Late Devensian and Holocene relative sea-level changes in northwestern Scotland: new data to test existing models. *Quaternary International* 26:97-123.

Shennan, I., Lambeck, K., Horton, B. P. *et al.* 2000. Late Devensian and Holocene records of relative sea-level changes in northwest Scotland and their implications for glacio-hydro-isostatic modelling, *Quaternary Science Reviews* 19:1103-1136.

Shennan, I., Hamilton, S., Hillier, C. & Woodroffe, S. 2005. A 16,000-year record of near-field relative sea-level changes, northwest Scotland, United Kingdom. *Quaternary International* 133-134:95-106.

Simpson, D. D. A., Murphy, E. M. & Gregory, R. A. 2006. *Excavations at Northton, Isle of Harris.* BAR British Series (No. 408). British Archaeological Reports: Oxford.

Smith, D. E., Cullingford, R. A. & Firth, C. R. 2000. Patterns of isostatic land uplift during the Holocene: evidence from mainland Scotland. *The Holocene* 10:489-501.

Smith, D. E., Shi, S., Cullingford, R. A. *et al.* 2004. The Holocene Storegga Slide tsunami in the United Kingdom. *Quaternary Science Reviews* 23-24:2291-2321.

Smith, D. E., Fretwell, P. T., Cullingford, R. A. & Firth, C. R. 2006. Towards improved empirical isobase models of Holocene land uplift from mainland Scotland, UK. *Philosophical Transactions of the Royal Society (Series A)* 364:949-972.

Stoker, M. S. 1995. The influence of glacigenic sedimentation on slope-apron development on the continental margin off Northwest Britain. In Scrutton, R. S.,

Stoker, M. S., Shimmield, G. B. & Tudhope, A.W. (eds.) *The Tectonics, Sedimentation and Palaeoceanography of the North Atlantic Region. Geological Society London Special Publication* 90:159-177.

Stoker, M. S., Hitchen, K. & Graham, C. C. 1993. *The Geology of the Hebrides and West Shetland Shelves, and Adjacent Deep-water Areas.* British Geological Survey Offshore Regional Report for the HMSO: London.

Sutherland, D. G. 1984. The submerged landforms of the St Kilda archipelago, western Scotland. *Marine Geology* 58:435-442.

Wenban-Smith, F. 2002. *Palaeolithic and Mesolithic Archaeology on the Seabed: Marine Aggregate Dredging and the Historic Environment.* Wessex Archaeology: Salisbury.

Weninger, B., Schulting, R., Bradtmöller, M. *et al.* 2008. The catastrophic final flooding of Doggerland by the Storegga Slide tsunami. *Documentia Praehistorica* 35:1-24.

Wickham-Jones, C. R. & Dawson, S. 2006. *The Scope of Strategic Environmental Assessment of North Sea Area SEA7 with regard to Prehistoric and Early Historic Archaeological Remains.* UK Department of Trade and Industry: London.

Wickham-Jones C. R. 2014. Coastal adaptions. In Cummings, V., Jordan, P. & Zvelebil, M. (eds.) *Oxford University Handbook of the Archaeology and Anthropology of Hunter-Gatherers.* pp. 695-711. Oxford University Press: Oxford.

Chapter 9

Paleolandscapes of the Celtic Sea and the Channel/La Manche

R. Helen Farr,[1] Garry Momber,[2] Julie Satchell[2] and Nicholas C. Flemming[3]

[1] Southampton Marine and Maritime Institute, Archaeology, University of Southampton, Highfield, UK
[2] Maritime Archaeology Trust, National Oceanography Centre, Southampton, UK
[3] National Oceanography Centre, Southampton, UK

Introduction

This chapter provides an overview of the work that has been undertaken in the region of the English Channel and Celtic Sea area. Whilst not exhaustive, it contains background information about the environment, geology and paleogeography of the region alongside a summary of current research, data sets and knowledge of submerged prehistoric sites. Bibliographic references of key texts and URLs are provided to sources of data.

Earth Sciences and Sources of Data

Modern coastline, best sources of high-resolution data

The Channel/La Manche and Celtic Sea area encompasses the south and south-west coast of the UK, the English Channel and the coastal waters of northern France (Fig. 9.1). The modern coastline consists of a

Fig. 9.1 Map indicating area subject to study. Julian Whitewright; Maritime Archaeology Trust. Reproduced with permission.

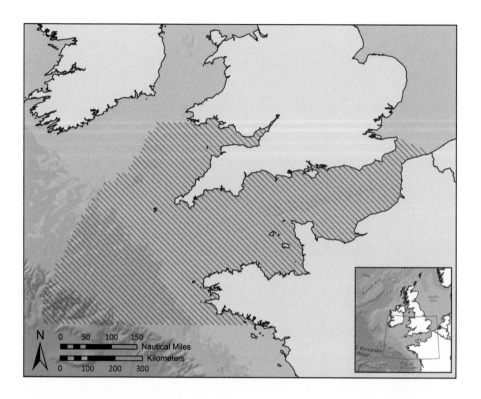

wide range of coastal units from hard and soft rock cliffs, estuaries and barrier beaches.

This region extends from Dover and Calais in the east, to St. David's Head in the north-west, and expands out across the Celtic Sea to the continental shelf edge, Brest and the Cotentin Peninsula. As such this region incorporates the Channel/La Manche, the Solent and Isle of Wight, the Cornish Peninsula, southern Wales, and the north coast of France. Water depths are generally shallower in the eastern channel (<50 m), deepening to 50 m to 200 m further west, and dropping to 1000 m towards the outer continental shelf boundary (UKMMAS 2010) (Fig. 9.2). The eastern Channel is characterized by shelving sand, shingle and pebble beaches interspersed with cliffs, whilst the western coastline is predominantly rocky. Areas of intertidal sediment and salt marsh can be found in bays and inlets, especially those in the Solent (Chichester, Langstone, and Portsmouth harbors, and Southampton Water), Poole Harbour and the Bristol Channel and Milford Haven, St. Malo, the Seine Estuary, and along the Normandy river estuaries.

Data sources

There are many multidisciplinary assessments of the modern coastline that may be useful for the investigation of coastal change in this region. This is probably one of the most intensively studied sea areas in the world with a substantial number of marine research centers.

On the French coast these include La Station de biologie marine de Concarneau, La Station biologique de Roscoff (SBR), Boulogne and IFREMER (Institut français de recherche pour l'exploitation de la mer), Brest. National coastal monitoring programs include the SOMLIT project (Service d'Observation du Milieu Littoral) which co-ordinates the monitoring activities carried out by marine stations on French coasts in order to understand the impact of global change on coastal zones. The French processing and archiving facility CERSAT (Centre ERS d'Archivage et de Traitement), the Coriolis Centre for *in situ* oceanographic data, and Sextant, a server for georeferenced marine data, provide access to databases of gridded and vector data produced by IFREMER and its partners. Most of these data are fully accessible to the public.

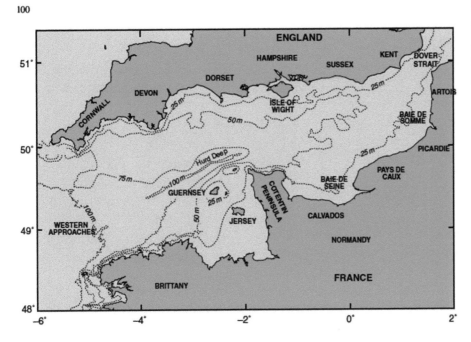

Fig. 9.2 Bathymetric contours in the Channel/La Manche and the Western Approaches. Jasmine Noble-Shelley; Maritime Archaeology Trust, after Paphitis (2000). Reproduced with permission.

SISMER (Systèmes d'Informations Scientifiques pour la Mer) is the national oceanographic data center of France, and archives French oceanographic data from 1968 onwards including physical, chemical oceanography and geophysical data.

Across the Channel, the southern British coast is well mapped by the Ordnance Survey (OS) and the British Geological Survey (BGS) and many data sets are available as high-resolution vector and raster data. Over the last decade, English Heritage has undertaken a series of projects to characterize the nature of the English coast including Rapid Coastal Zone Assessment Surveys (RCZAS) and the Historic Seascape Characterization (HSC). These expand historic landscape characterization into the marine environment. Additional regional surveys include *Charting Progress 2: the State of UK Seas* (UKMMAS 2010), which assessed the coastal environment and marine ecosystem around Britain, and the Marine Aggregates Levy Sustainability Fund (ALSF) Regional Environmental Characterizations (REC) — a series of regional coastal surveys of Britain's submerged habitats and heritage. RECs of the south coast are available for 2007–2010.

The Marine Aggregate Industry Protocol (MAI) for reporting finds of archaeological interest was set up in 2005, and is run in partnership with the British Marine Aggregate Producers Association (BMAPA), The Crown Estate and Historic England, and is implemented by Wessex Archaeology. Unexpected finds by offshore industries is covered by the Offshore Renewables Protocol for Archaeological Discoveries (ORPAD), established in 2010 and run by The Crown Estate and Wessex Archaeology. Since their creation, over 2000 objects have been reported.

The Channel Coastal Observatory (CCO) is the data management center for the regional coastal monitoring programs of England. It has freely accessible data for the Channel and Celtic Sea region including aerial photography, LiDAR (Light Detection and Ranging) data, topographic data, hydrogeographic data, photogrammetric data, seabed mapping, sediment distribution data, cliff lines and wave and tidal information. This resource enables modern coastal dynamics to be searched and mapped both by region and date.

The Standing Conference on Problems Associated with the Coastline (SCOPAC) sediment transport study (Carter *et al.* 2004) covers the coastline of central/southern England between Lyme Regis, Dorset and Shoreham-by-Sea, West Sussex. SCOPAC includes strategic regional coastal monitoring programs, analysis of coastal processes and a bibliographic database. The sediment transport study is regionally searchable and provides an overview of the coastal sediment types involved in transport and the transport mechanism involved.

Offshore wave and tidal data is available from the United Kingdom Hydrographic Office (UKHO), the Proudman Oceanographic Laboratory (POL) (now the National Oceanography Centre (NOC) — Liverpool), and the British Oceanographic Data Centre (BODC) — a national facility for preserving and distributing marine data.

The PISCES project (Partnerships Involving Stakeholders in the Celtic Sea Eco-System) was a three-year project funded by the European Commission to unite people across different sectors and countries for the sustainable management of the marine ecosystem. It provides interactive maps of the present-day Celtic Sea marine environment. Partnership projects provide routes for managing multi-disciplinary research on regional coastal problems.

Further studies of coastal geomorphology and erosion modeling for coastal management include the European-wide EUROSION and Response projects that looked to integrate EU climate change policy (2003–06) and the UK Futurecoast project.

A number of current EU projects researching and looking to manage the effects of coastal change are located within the INTERREG IVA European Regional Development Fund (ERDF) program. These have built on early INTERREG projects such as MESSINA (2003–06), EMDI (2004–07), and IMCORE (2008–11) and include:

FLOODCOM — led by Essex County Council
3i — Integrated Coastal Zone Management led by Delft University of Technology
Flood Aware, led by Provincie Zeeland
CC2150 — Coastal Communities 2150 and Beyond, led by the UK Environment Agency
PRISMA — led by Waterwegen en Zeekanaal NV
Arch-Manche — led by the Maritime Archaeology Trust (Hampshire and Wight Trust for Maritime Archaeology).

Further resources

EMODnet: www.emodnet.eu
IFREMER: wwz.ifremer.fr
IFREMER (Boulogne-sur-Mer region) wwz.ifremer.fr/manchemerdunord/Environnement/LER-Boulogne-sur-Mer
SBR: www.sb-roscoff.fr
CERSAT: cersat.ifremer.fr
SISMER: www.ifremer.fr/sismer/index_FR.htm
BGS: www.bgs.ac.uk/opengeoscience/home.html
Ordnance Survey (Scotland, Wales, England, Isle of Man): www.ordnancesurvey.co.uk
RCAZS report: historicengland.org.uk/advice/planning/marine-planning/rczas-reports/
CCO: www.channelcoast.org
MAI /ORPAD: www.arcgis.com/home/item.html?id=952ee00d3ff3459a91c6f821c0f13ab3
SCOPAC: www.scopac.org.uk/sedimenttransport.htm
NOC (POL): www.pol.ac.uk
BODC: www.bodc.ac.uk
PISCES: ec.europa.eu/environment/life/project/Projects/index.cfm?fuseaction=search.dspPage&n_proj_id=3281
EUROSION: www.eurosion.org
The RESPONSES Project: www.responsesproject.eu
Futurecoast: www.coastalwiki.org/coastalwiki/FUTURECOAST_project,_UK
INTERREG: www.interreg-messina.org
Arch-Manche project: archmanche.maritimearchaeologytrust.org
IMCORE: www.imcore.eu
CEFAS: www.cefas.co.uk/cefas-data-hub/research-and-publications-list/?page=168
UKHO charts: www.admiralty.co.uk/charts
Charting Progress 2: webarchive.nationalarchives.gov.uk/20141203181034/ http://chartingprogress.defra.gov.uk/

Wetlands, deltas, marshes, lagoons, coastal lakes

Much of the offshore Channel and Celtic Sea area is dominated by high-energy, scour environments. However low-energy coastal formations such as wetlands, marshes

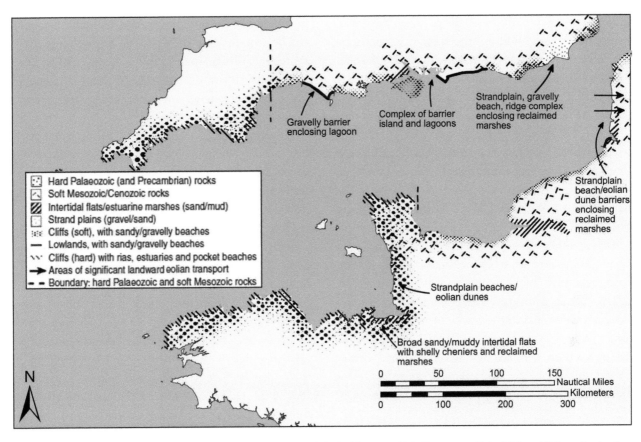

Fig. 9.3 General characterization of coastal geological and geomorphological features. Note the areas of intertidal mud flats and estuarine marsh. Jasmine Noble-Shelley; Maritime Archaeology Trust after Paphitis (2000). Reproduced with permission.

and lagoons exist in specific sheltered locations and provide areas of potential preservation of archaeological material. There are extensive wetlands, lagoons, marshes and mudflats on the French coast east of St. Malo, around Mont-Saint-Michel in Normandy, around the mouth of the River Seine, the Somme Estuary and on the west-facing coasts of Pas-de-Calais. The coastline of Brittany was formed where the Armorican Massif reaches the sea. Here, hard headlands and cliff lines are interspersed with ria inlets that track the course of submerged paleovalleys. These now form sheltered inlets with marshes and lagoons (Prigent *et al.* 1983). Figure 9.3 shows the distribution of geomorphological features including the wetland areas across the region.

Wetlands and mudflats in England include the Severn Estuary, Dartmouth and Kingsbridge estuaries, the Tamar, Falmouth, Fowey estuaries and the Taw-Torridge Estuary, Christchurch and Poole harbors along with the harbors and waterways around the Solent. Wetland areas

have been the focus of particular prehistoric archaeological interest. This includes work by the West Coast Palaeolandscapes Survey (Fitch & Gaffney 2011) and the Severn Estuary Levels Research Committee (SELRC). SELRC formed in 1985 because of the increasing awareness of the potential of the estuary wetlands for preserving archaeology. The Mesolithic site at Goldcliff and findings at Uskmouth, both on the Severn Estuary, including preserved Mesolithic human footprints (Bell 2007) highlight the excellent preservation within this wetland environment. In the Solent, there has been a long history of research. Many local surveys have been undertaken by the Maritime Archaeology Trust (formerly the Hampshire and Wight Trust for Maritime Archaeology), Wessex Archaeology and the New Forest Council, as well as various projects by the University of Southampton, the University of Bournemouth and the Isle of Wight County Council. Collaborative projects have mapped the effects of dredging and coastal development on these environments (e.g. Hodson & West 1972), as well as monitoring change. Work on the submerged landscape at Bouldnor Cliff addressed the formation of the Solent in relation to the archaeological, sedimentological and paleoenvironmental studies of submerged Holocene deposits (Momber *et al.* 2011).

The RESPONSE Project, a three-year project supported by the LIFE financial instrument of the European Community, launched in December 2006 to investigate the risks from climate change on both sides of the Channel. This built on the Coastal Change, Climate and Instability EC LIFE project that concluded in 2000.

Further information on the geology of southern England and intertidal estuary environments including a comprehensive bibliography can be found online on Ian West's University of Southampton website (www.southampton.ac.uk/~imw/). In addition, a UK country-wide survey of intertidal and offshore peat deposits has been compiled by English Heritage/Historic England into one searchable database organized by region (Hazell 2008). Reports are available from the website.

Further resources

- RESPONSE Project Publications: www.coastalwight. gov.uk/RESPONSE_webpages/r_e_publications.htm
- Peat database: content.historicengland.org.uk/content/ docs/research/peat-database-nsea.pdf
- MAT surveys: maritimearchaeologytrust.org/intertidal
- Wessex Archaeology surveys: www.wessexarch.co.uk/ projects/hampshire/portsmouth/langstone/index.html
- WCPS: www.dyfedarchaeology.org.uk/lostlandscapes/ WCPStechnical.pdf
- SELRC: www.selrc.org.uk/archaeology_bibliography .html
- Geology of the Wessex Coast of Southern England: www.southampton.ac.uk/~imw/

Coastal geomorpho-dynamics, erosion, accumulation

Rates and patterns of erosion vary across the region depending on the susceptibility of the coastal topography and geology to weathering. Exposure of the coastline

to the large Atlantic fetch has resulted in complex coastal geomorphology. Rugged outcrops along hard rock shorelines are often interspersed with sheltered, sediment-filled bays, inlets or estuaries. Where the underlying bedrock is softer, wide bays are backed by lines of eroding cliffs.

Towards the eastern end of the Channel, dune systems have developed along low-lying alluvial coastlines. These are particularly notable around the estuary mouths of the Seine and the Somme, the west coast of the Contentin Peninsula and along the south-facing coasts from Kent to West Sussex. In the adjacent county of Hampshire a complex of natural harbors contains a rich archaeological and sediment archive (Collins & Ansell 2000; Wenban-Smith & Horsfield 2001).

The intermittent exposures of soft cliffs along the Channel from Kent and the Pas-de-Calais in the east, to Lyme Bay and the Contentin Peninsula in the west, are the source of sediments that are transported to form spits such as Chesil Beach, and sandy headlands like that at Dungeness. Sediment also accumulates at the mouth of the major rivers such as the Somme and Seine. These sedimentary features can protect archaeological deposits beneath them or in the areas they shelter. Conversely, erosion of land that produces the sediment can have a negative impact on the cultural resource. Cliff erosion often manifests itself in the form of landslides or rotational slumping that carries material into the sea. The south-east facing cliffs of the Isle of Wight form the largest coastal landslip complex in Europe.

The west Cornish Peninsula is geologically similar to the Armorican Massif. It is a rocky, eroded, high-energy coast exposed to Atlantic storms. Despite this, coastal prehistoric sites have been preserved in the sheltered bays and protected inlets etched into the rocky headlands. For a map of the Armorican sites see Prigent *et al.* (1983: fig. 1). The incised valleys provide effective shelters from the dominant westerly waves and storms. The area around Vannes in southern Brittany is particularly significant as it is rich in archaeological Neolithic and Mesolithic monuments that extend into the intertidal zone and underwater; the stone circles of Er Lannic in the Golfe du Morbihan being the most visible example (www.stonepages.com/france/erlannic.html). In addition, prehistoric sites and hundreds of fish traps are found in embayments and estuaries around Brittany and Normandy. These are subject to study and recording by the Centre de Recherche en Archéologie, Archéosciences, Histoire (CReAAH), a network of organizations that are actively researching threats to the coastal heritage around the coastline of northwest France. Many of these archaeological features are dateable and can be used as indicators of coastal change.

The north coasts of Cornwall, Devon and Somerset contain protective environments in sheltered inlets where they are covered by the fluvial sediments. Submerged forests and Mesolithic material have been found off Westward Ho!, Porlock and Hele Bay; however the steep valleys of the north Devon coast and the broad catchments of Exmoor above can lead to devastating flash floods causing severe coastal erosion (e.g. Lynmouth in 1952).

For further information on local geomorphology, erosion and accumulation, reference should be made to the French Geological Survey (Bureau de Recherches Géologiques et Minières — BRGM), and the British Geological Survey in addition to the RCZAs and RECs for the region of interest. In the Solent region, SCOPAC also maps erosion and sediment transport.

Further resources
- BRGM: www.brgm.eu/content/digital-data-services
- SCOPAC: www.scopac.org.uk/sedimenttransport.htm
- RCZAS reports: historicengland.org.uk/advice/plan ning/marine-planning/rczas-reports
- www.cefas.co.uk/cefas-data-hub/research-and-publicat ions-list/?page=132
- CReAAH: www.creaah.univ-rennes1.fr
- BGS: www.bgs.ac.uk/data/home.html

Solid geology

The two dominant geologic processes acting in this region are the east–west extensional forces caused by the break-up of Europe from America, and the north–south compressional forces caused by the Alpine orogeny. The fault-bounded Western Approaches Trough forms

a sedimentary basin between the Armorican Massif in Brittany and the Cornubian Massif in Cornwall (Evans 1990). The planation surface of the sea floor in this area and that of the Celtic Sea is thought to date to the Mid Tertiary. At depths of 38 m to 69 m, submerged cliffs and terrace features dating to the Miocene and Pliocene have been eroded into the southern offshore region of Cornwall and Devon (Evans 1990).

Between 1 Ma and 400 ka, Britain was joined to the mainland of Europe at all times, despite glacially controlled fluctuations of sea level (Gibbard 1988; Stringer 2006). To the east of our study area, a high chalk plateau, the Weald–Artois anticline provided a solid bedrock link between southern Kent and northern France, even at times of interglacial high sea level. Some time between 400 ka and 200 ka this chalk ridge was eroded, and at subsequent times of interglacial high sea level, Britain and Ireland were cut off from Europe. The geology of the eastern Channel, the Weald–Artois anticline (Dover Strait) and southern Bight are well presented in Hijma et al. (2011).

The French Geological Society (Société géologique de France) has a variety of collated resources including the *Geology of France* journals which cover both sub-surface and surface geology, and 1:50,000 maps. The BRGM offers access to its geoscientific and environmental data through its InfoTerre portal.

The BGS has undertaken research into bedrock geology and seabed sediments across the region. This work is available as a series of publications and maps (1:250,000 and 1:65,000). A searchable database (GeoIndex) is also available online, comprising both onshore and offshore geological and geophysical data. 3D geology models (50-m grid) are available for terrestrial regions. The Strategic Environment Assessment program (SEA), funded by the Department for Business, Energy and Industrial Strategy (DBEIS), provides a shelf-wide resource with summaries of the geology of eight designated regions around the United Kingdom.

Further regionally specific geological reviews are numerous. The bedrock geology of the Channel and Celtic Sea is summarized by Evans (1990) for the Western Approaches, and Hamblin et al. (1992) for the UK sector of the Channel. The geology of the Bristol Channel

Fig. 9.4 Cap Lévi, Fermanville, showing the granite headland that is protecting the Paleolithic site at 20 m depth on the eastern side. Jasmine Noble-Shelley after Garry Momber; Maritime Archaeology Trust. Source of image Google. Reproduced with permission.

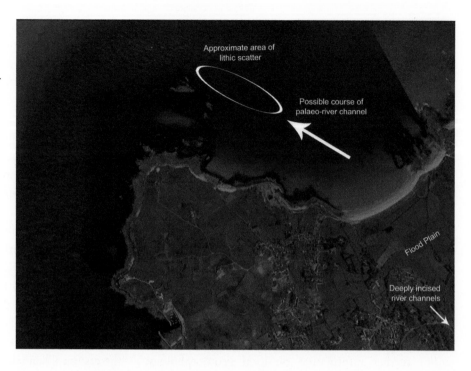

is outlined by Tappin *et al.* (1994) and that of the Solent by Dix (2001), Velegrakis (2000), and Velegrakis *et al.* (1999). Gibbard (1988) describes the background of tectonic processes that have formed the depressions of the English Channel and the North Sea over many millions of years and influenced the courses of the great rivers flowing onto and across the continental shelf. The shelf valley systems in the Channel have been studied by Gupta *et al.* (2008). A special issue of the *Journal of Quaternary Science* was published in 2003 devoted to the Quaternary history of the English Channel (Gibbard & Lautridou 2003). Further details of regional and local geology of the coast and the seabed are given by a number of papers in that journal (e.g. Bourillet *et al.* 2003; Lagarde *et al.* 2003; Reynaud *et al.* 2003).

Understanding of local geological features is necessary for assessment of potential archaeological preservation. The lee of rocky outcrops can provide protection for paleoterrestrial features and anthropogenic material. One famous archaeological example of this within this region is the Fermanville Paleolithic site (Scuvée & Veragne 1988; Cliquet *et al.* 2011). Prehistoric artifacts have been found in 20 m of water in the protective shadow of the Cap Lévi–Biéroc granite outcrop, east of Cherbourg. The

submerged occupation site is shielded from storm damage and as such has withstood climatic variations for over 60,000 years (Fig. 9.4).

Further resources
– The French Geological Survey: www.geosoc.fr
– BRGM: www.brgm.eu/content/digital-data-services
– The BGS provide free access to some of their data, online data holdings: www.bgs.ac.uk/opengeoscience/home.html
– www.bgs.ac.uk/data/mapViewers/home.html
– SEA program: www.gov.uk/guidance/offshore-energy-strategic-environmental-assessment-sea-an-overview-of-the-sea-process

Bathymetry

The Western Approaches are located on the north-west corner of the European continental shelf where water depths are generally less than 300 m. Around the coastlines of England and Wales, the main features of the bathymetry are nearshore water depths generally less than 50 m (see Fig. 9.2). The deepest areas (>100 m) are found offshore on the south coast to the west of Start Point, and also on the west coast off the westernmost tip of Wales. Off the west coast, the Celtic Sea is characterized by a deep (100–200 m) channel running north–south, and off the south coast the western half of the Channel is characterized by a fairly deep (100 m) central channel which runs (and shallows) in a west–east direction. West of Normandy, the continental shelf drops gently towards the Channel Islands. As such, these islands were accessible from mainland Europe for large parts of the Pleistocene. This is evidenced by sites like La Cotte de St. Brelade on the island of Jersey, indicating that prehistoric hominins exploited the landscape when sea levels were lower (Callow & Cornford 1986; Scott *et al.* 2014).

The bathymetry of the Channel and Celtic Sea provides further information on the potential for survival of archaeological remains. By identifying deeper areas it is possible to target those sites that are less likely to be affected by hydrodynamic processes, which can scour and disturb archaeological remains. The depth of the archaeological resource also affects the biological and chemical conditions influencing the preservation of sites. Archaeological material can survive for hundreds of thousands of years in the right taphonomic context if undisturbed or exposed by human action or changes in hydrodynamic patterns.

Maritime and hydrographic data sources

There are a range of publicly accessible data sets for this region. The most readily available and wide-ranging are the satellite-derived combined topography and bathymetry data sets (ETOPO1) from the National Oceanic and Atmospheric Administration (NOAA) National Geophysical Data Center. ETOPO1 provides a 1 arcminute global relief model of topography and bathymetry. An equally extensive database is that from General Bathymetric Chart of the Oceans (GEBCO) providing a 30 arcseconds gridded bathymetry. However, this data set was designed to target the deep oceans, and as such, data from the continental shelves is limited.

A number of higher resolution data sets generated from single-beam echo sounder (SBES), multibeam echo sounder (MBES) and bathymetric LiDAR surveys also cover parts of the region. Some of the main data sources are listed here:

– Maritime Coastguard Agency (MCA) Hydrographic Data. These data sets are not directly available for download, but it is possible to obtain agreements between organizations to access them for research purposes. Data for the length of southern England is available through the Channel Coastal Observatory;
– Service hydrographique et océanographique de la Marine (SHOM), the French hydrographic agency, provides access to bathymetric data at varying resolutions. Data is available for the whole Golfe de Gascogne (Bay of Biscay), the Manche (Channel) region and the Atlantic coast;
– The national Litto3D program, maps the coastal zone at higher resolution, aiming for a seamless onshore/offshore survey combining LiDAR and bathymetric data;
– IFREMER manages national oceanographic data including SISMER. Amongst other forms of oceanographic data, SISMER manages geophysical data;
– SeaZone Solutions Ltd is a commercial organization affiliated with the United Kingdom Hydrographic Office data. Bathymetric data includes MBES data, gridded SBES data and digitized Admiralty data. The data must be purchased;
– MEDIN (Marine Environmental Data & Information Network) contains some publicly available alternative sources of bathymetric data;
– REC projects were funded through the Marine ALSF from 2002–2011. RECs include large-scale bathymetric seismic and acoustic surveys. Sub-bottom data and core log material were collected in collaboration with the marine aggregates industry. Those of interest within the Channel and Celtic Sea zone include the offshore Arun River (Gupta *et al.* 2008; Wessex Archaeology 2008a,b), eastern English Channel and the Severn Estuary (MoLAS 2007).

The commercial sector provides an additional source of bathymetric data. Offshore surveys required for the renewable energy sectors or offshore infrastructure are making their data available through the Cowrie Data Management System.

Small localized data sets within the research region include:

New Forest District Council, north shore, Western Solent, 2006;

Hampshire and Wight Trust for Maritime Archaeology/Maritime Archaeology Trust;

University of Southampton;

EMODnet (European Marine Observation and Data Network) hydrography portal has access to digital terrain models and metadata for selected maritime basins across Europe including the Channel and Celtic Seas and Western Approaches.

For a fuller account of available geophysics data across the UK see Dix and Sturt (2013).

Further resources

NOAA: www.ngdc.noaa.gov/ngdc.html
GEBCO: www.gebco.net
MCA: www.mcga.gov.uk
SHOM portal: www.shom.fr/les-services-en-ligne/portail -datashomfr/
IFREMER: wwz.ifremer.fr
SeaZone: www.seazone.com
MEDIN: www.oceannet.org
Somerset CC: www.swheritage.org.uk/somerset-archives
ASLF and wind data: www.marinedataexchange.co.uk/
CCO: www.channelcoast.org
EMODnet: www.emodnet-hydrography.eu

Vertical earth movements

The dominant influence on vertical earth movement in this region is glacio-isostatic adjustment (GIA) (see Chapter 2, pages 20–24). The complex relationship between the rates of rebound and subsidence, and the global eustatic rise in the postglacial are key factors in determining the varying impacts of the sea on the land in different regions today. The process is long term and led to the separation of the British Isles from continental Europe around 7500 years ago (Flemming 1998; Bradley et al. 2009; 2011).

There has been much work modeling glacio-isostatic adjustment across the region. High-resolution GIAs are in constant development. The most recent GIA, 11 ka time slab for the British Isles is reproduced in Fig. 9.5.

The south coast of England and northern Brittany are subsiding from postglacial readjustment. On the French Armorican coast numerous megalithic passage graves and other prehistoric remains are preserved in the intertidal zone down to the low-tide limit where the tidal range is very large (5–10 m) (Giot 1968; Prigent et al. 1983). Discussion of preserved sites can be seen below.

Pleistocene and Holocene sediment thickness on the continental shelf

The unconsolidated sediment cover in the Channel is relatively thin when compared with the North Sea, which is a net area of accumulation. There are recent sediments in shallow water including peat deposits dating to the last postglacial transgression. The sediment cover is plotted on the BGS Bottom Sediment map series (BGS, 1:250,000). The thickest sediment deposits are in the over-deepened troughs of the Hurd Deep, the Fosse de L'île Vierge and the Fosse d'Ouessant. The Hurd Deep analyzed by Evans (1990: 75–76) is a narrow graben (Antoine et al. 2003), which has been successively infilled at periods of low sea level and then partially scoured out by tidal currents during and after marine transgressions. It is located far to the south of any phase of ice scour. The incision reaches a maximum depth of 240 m with a sediment infill thickness of 80 m, while the surrounding sea bed is quite flat at a depth of 70 m to 90 m, with a sediment cover often less than 0.5 m.

The Channel river system is an erosive landscape and a relatively sediment-starved part of the Northwest European Shelf (Gupta et al. 2008); strong tidal streams have led to scouring of sediments. Large areas of the Bristol Channel, Pembrokeshire and north Devon coasts have been eroded of Quaternary sediments (Tappin et al. 1994).

Data sources

Much of the sediment thickness data can be found summarized regionally within the Marine ALSF REC projects. Offshore geological records include seismic profiles, bathymetry and backscatter data, core or borehole and grab samples. The offshore geological records are

11,000 BP

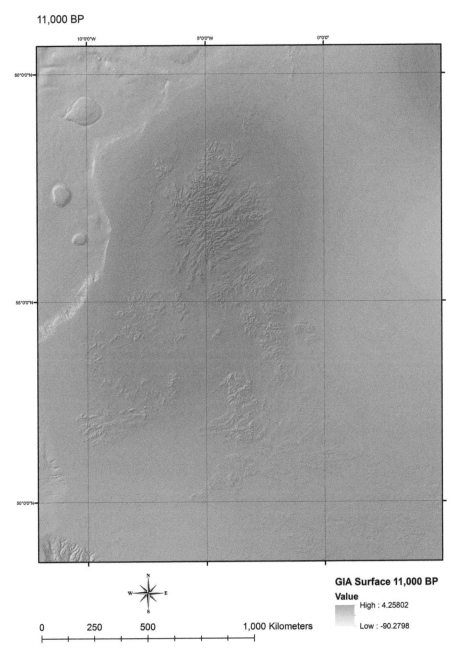

Fig. 9.5 GIA model at 11,000 BP,
superimposed on a map of the British
Isles. Courtesy of Fraser Sturt (2015). Map
produced in part from GEBCO 08
(www.gebco.net) data. Reproduced with
permission.

GIA Surface 11,000 BP
Value
High : 4.25802
Low : -90.2798

held by the BGS along with some borehole data. The BGS Offshore Regional Reports for Cardigan Bay and the Bristol Channel (Tappin *et al.* 1994) and the western English Channel and Western Approaches (Evans 1990) contain data for this region. Borehole data is collected during port and offshore development projects. The MAREMAP project holds seabed mapping data across the UK. Industry data from the UK offshore oil and gas companies can be searched via the UK Oil and Gas Data

website. The University of Southampton holds various boomer sub-bottom data sets from within the Channel and Celtic Sea region.

On a European scale, EMODnet has collated national data sets for habitat mapping that contains bathymetry, oceanographic and geologic seabed data and the Geo-Seas project collates offshore geological data from national providers. There is some overlap with the BGS archives.

SISMER is the designated national oceanographic data center for France. It is based at IFREMER, Brest. Local bathymetric data is being collected in the region for archaeological purposes by the Association pour le développement de la recherche en archéologie maritime (ADRAMAR). Surveys by the vessel *L'André Malraux* (launched in 2012) by Le Département des recherches archéologiques subaquatiques et sous-marines (DRASSM) is collecting data for archaeological purposes around French coastal waters, although this is usually focused on shipwreck research.

Further resources

– BGS: www.bgs.ac.uk/GeoIndex/offshore.htm
– UK Oil and Gas Data: www.ukoilandgasdata.com
– ASLF data: www.marinedataexchange.co.uk/aggregates
  -data.aspx
– MAREMAP: www.maremap.ac.uk/index.html
– EMODnet: www.emodnet-geology.eu
– Geo-Seas: www.geo-seas.eu
– SISMER: www.ifremer.fr/sismer/index_FR.htm
– DRASSM: www.culture.gouv.fr/fr/archeosm/archeo
  som/drasm.htm
– ADRAMAR: www.adramar.fr

Post-LGM Climate, Sea Level, and Paleoshorelines

The Last Glacial cycle has been studied intensively for global eustatic sea-level change and coastal evolution (see for example Lambeck 1991; 1993a,b; Lambeck & Chappell 2001; van Andel & Davies 2003; Waller & Long 2003) and the regional ice loading and isostatic response (Lambeck 1995; Peltier *et al.* 2002; Shennan & Horton 2002; Shennan *et al.* 2002; 2006) although no definitive reconstructions exist for this region as a whole (Westley *et al.* 2013). Paleolandscape reconstructions have been limited to areas where 3D seismic data has been available (Gaffney *et al.* 2007). Within the Channel and Celtic Sea region this includes parts of the English Channel including Poole Harbour and Christchurch Bay.

General climatic conditions and changes after the LGM

Sea-level research and modeling has been a rapidly growing field of research in the last decade, and for an introduction to literature focusing on fluctuations during the Last Glacial cycle see Rohling *et al.* (1998; 2008), Lambeck and Chappell (2001), and Siddall *et al.* (2003; 2010). Van Andel and Davies (2003) have published a multidisciplinary analysis of the climatic fluctuations during Marine Isotope Stage 3 (MIS 3), approximately 60 ka to 20 ka, and the consequent effects on the distribution of Neanderthal and anatomically modern humans (AMH), leading up to the LGM. During MIS 3 the Greenland ice core data GISP2 (Meese *et al.* 1997; Johnsen *et al.* 2001) show rapid fluctuations of temperature of the order of 5–10°C every few thousand years, the so-called Dansgaard-Oeschger oscillations (van Andel & Davies 2003: 58). The sequence of calculations and plotted maps, correlated with summaries of known major archaeological sites, provides an interesting analysis. By integrating archaeological data from the seabed, these maps and calculations can be used as a starting point to identify potential areas of interest for locating submerged habitation sites at times of low sea level. For an early example of this approach see Louwe Kooijmans (1970–71).

Reconstructions of the last glacial ice sheet (Clark *et al.* 2012) show at its greatest extent that it covered Scotland, Ireland and northern England and stretched west across to the Atlantic shelf edge at the LGM. Southern England avoided glaciation, and the ice had retreated from the Celtic Sea and southwest Wales by ca. 19 ka. Murton and Lautridou (2003) cite a number of authors to indicate the probable limits of continuous permafrost at different dates (Fig. 9.6). Most models show continuous permafrost barely extended south of the English coastline. Even the coldest estimates show Cornwall and Brittany free of continuous permafrost at the peak of the LGM. Thus the western Channel and Western Approaches may have served as a refuge for Paleolithic communities, highlighting the importance of the finds that have been uncovered from submerged sites such as the Fermanville, Biéroc, Paleolithic site (Scuvée & Veragne 1988; Cliquet *et al.* 2011), Bouldnor Cliff

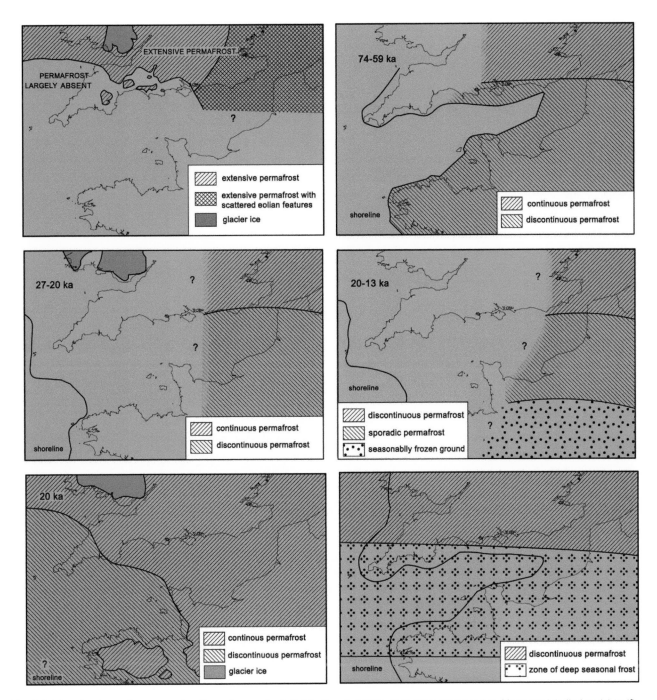

Fig. 9.6 Map showing various reconstructions of permafrost distribution during the Devensian/Weichselian. Maps are labelled as (a) to (f) from left to right and from top to bottom as follows (a) Last Glacial Period, based on Williams (1969); (b) 74–69 ka, based on Huijzer & Vandenberghe (1998); (c) 27–20 ka, based on Huijzer & Vandenberghe (1998); (d) 20–13 ka, based on Huijzer & Vandenberghe (1998); (e) 20 ka, based on van Vliet-Lanöe (2000); (f) 12.9–11.5 ka (Younger Dryas), based on Isarin (1997). After Murton & Lautridou (2003, fig.2). Adapted by Jasmine Noble Shelley, Maritime Archaeology Trust.

(Momber *et al.* 2011) or the artifacts recovered from paleochannels in the paleo-Arun tributary of the Channel River (www.wessexarch.co.uk/book/export/html/1200).

Evolution of sea level and coastline since the LGM

Following the LGM ca. 22,000 years ago, the climate continued to oscillate. Conditions ameliorated as deglaciation continued, reaching a peak in the Woodgrange or Windermere Interstadial at ca. 12.9 ka to 14.7 ka. This was followed by the colder harsher conditions of the Younger Dryas/GS-1 (ca. 12.9–11.5 ka) which then led into the ameliorated dry climate of the Holocene, Boreal regime. By ca. 6 ka to 7 ka there was a shift into the Atlantic regime which brought warmer and wetter conditions once more.

The process of sea-level rise that resulted from the thawing ice following the LGM has been subject to several studies (see for example Lambeck 1991; 1993a,b; 1995; Shennan & Horton 2002; Shennan *et al.* 2002; 2006; Waller & Long 2003; Bradley *et al.* 2009; 2011). Shelf-scale paleogeographic reconstructions (Fig. 9.7) have been undertaken using a combination of high-resolution bathymetry and glacio-isostatic modeling. However due to local variations in ice-sheet extent and deglaciation processes, and the need to account for differing rates of sedimentation and erosion, reconstructions of the sea-level evolution varies from region to region and there are many uncertainties (Brooks & Edwards 2006).

Britain became an island approximately 7500-7000 years ago, in the latter part of the Mesolithic (ca. 11000–6000 years ago). Within the Channel region, Waller and Long (2003) show rapid sea-level rise in the Early Holocene that led to marine transgression. This rapid sea-level rise began to decline ca. 6800 years ago as evidenced by a change from minerogenic to organogenic sedimentation. This was characterized by Scaife as part of the EU LIFE project, 'Coastal Change, Climate and Instability', and interpreted by Momber when looking at sea-level rise and formation of the Solent (Momber 2011: 132).

Notwithstanding regional earth movements and hydro-isostatic readjustment, global eustatic sea level reached within a few meters of the present level around 6 ka to 5 ka. Following this time, relative sea levels are dominated by isostatic realignment. Lambeck (1993a,b) models relative sea-level curves for different parts of the British coast indicating submergence over the last 6000 years along the Channel coast. Waller and Long (2003: 357) compare curves of sea-level indicators for the south coast, indicating a general submergence of the order of 4 m to 5 m in the last 6000 years. Carter (1988: 263) cites tide gauge data from a number of authors to illustrate a range of submergence rates from 1.7 to 5.4 mm/year along the south coast.

The appreciation of coastal geomorphology and dynamics in the consideration of paleocoastlines has a scholarly history in this region. In 1948, Steers reviewed the coastal geomorphology along the whole length of the English coast, describing historical changes, alluviation, the growth of Dungeness, erosion of parts of the coastline, the existence of drowned forests and rates of cliff retreat. These processes are reviewed in a more technical way by Carter (1988). Cracknell (2005) conducted a useful literature review when considering the impact of storm floods and coastal erosion during the last 2000 years. This theme is being developed with the EU, INTERREG IVa, Arch-Manche Project which is using artistic representations, archaeology and paleoenvironmental material to indicate change.

Climate conditions on the shelf

Review of the available literature and research associated with the human occupation of northern France and southern Britain, and associated climate changes, has highlighted how essential it is to understand the changing landscape for human occupation. This provides context for the often sparsely scattered archaeological remains that have been discovered to date.

Four typical and repeated phases of activity and occupations can be identified that are directly associated with each glacial cycle:

1 Migration into the British Isles. After a phase of abandonment, people migrated across the shelf to re-occupy parts of Britain (see Clark *et al.* (2012) for discussion

11,000 BP

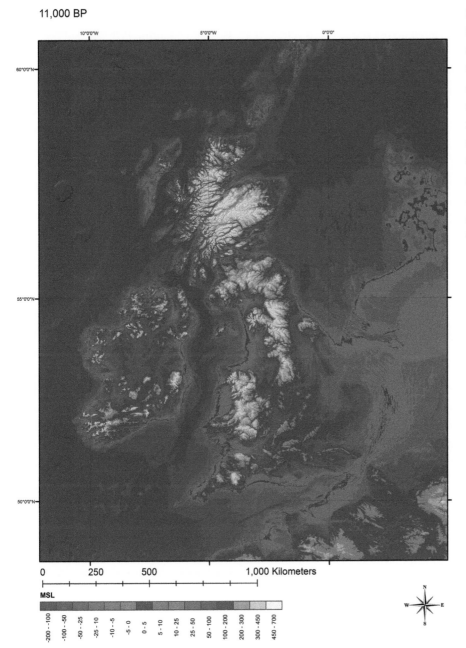

0   250   500   1,000 Kilometers

MSL

-200 - -100 | -100 - -50 | -50 - -25 | -25 - -10 | -10 - -5 | -5 - 0 | 0 - 5 | 5 - 10 | 10 - 25 | 25 - 50 | 50 - 100 | 100 - 200 | 200 - 300 | 300 - 450 | 450 - 700

Fig. 9.7 Paleogeographic reconstructions from 11,000 BP. Courtesy of Fraser Sturt, (2015). Map produced in part from Ordnance Survey Digimap, SeaZone solutions and GEBCO 08 (www.gebco.net) data. ©Crown Copyright/database right 2015. An Ordnance Survey/EDINA supplied service. ©Crown Copyright/SeaZone Solutions. All rights reserved. Licence no. 052006.001 31 July 2011. Not to be used for navigation. Additional data courtesy of the Channel Coastal Observatory.

on the retreat of the ice sheet). Occupation may have been short-lived; however, if it were sustained, a full breeding population must have crossed the shelf, or inter-breeding must have continued with the rest of Europe;

2 Mainland British Isles are occupied. The shelf area may have been exposed as dry land during climate amelioration, geomorphology of the Channel allowing, but, conversely, must have been abandoned if the shelf were inundated by high sea levels in the interglacial. Stringer (2006: 300) argues that Britain was a peninsula continuously during this phase. If this was the case, the shelf itself may have been occupied. The major determinant of the length of occupation period is geology and inundation; for a more recent discussion of this see Hijma *et al.* (2011);

3 Climate deterioration, ice increase and population decline. The population may have retreated across the shelf, or died out *in situ*. In these conditions occupation of the shelf is possible, but did not necessarily occur;

4 Cold periods, maximum shelf area exposed as periglacial tundra environment. Inuit-type occupation of the shelf would be possible, but did not necessarily occur.

Evidence for Submerged Terrestrial Landforms and Ecology

The change in relative sea level across southern England and northern France has caused submergence of important archaeological sites. These include intertidal Mesolithic footprints in clay below a peat dating from 6260±90 BP (CAR-1178) in the Severn Estuary (Bell 2007); Bronze Age walls in the shallow water between the Isles of Scilly (Crawford 1927; Thomas 1985; Garrow & Sturt 2011); numerous intertidal sites in the Solent such as the Bronze Age occupation and burial evidence from Langstone Harbour (Allen & Gardiner 2000); Neolithic trackways from Wootton Quarr (Loader *et al.* 1997) and submerged occupation sites associated with Bouldnor Cliff and the northwest Solent (Momber *et al.* 2011). In France, archaeological evidence includes the Middle Paleolithic site of La Mondrée, near Fermanville, Cherbourg (Cliquet *et al.* 2011); the submerged Neolithic standing stones in the Golfe du Morbihan (Cassen *et al.* 2011); the intertidal Neolithic monuments around the Brittany coast such as the passage grave on Kernic Beach at Plouescat; the standing stones on the archipelago of Mullein (Giot 1968; Prigent *et al.* 1983); and the many fish traps set at different altitudes on the beaches of Brittany and Normandy currently being investigated by CReAAH. Yet the archaeological record for northwest Europe extends much further back to around one million years (Stringer 2006), and the now-submerged continental shelf would have been part of the terrestrial landscape for most of the Pleistocene, as evidenced by

recent discoveries at Happisburgh (Ashton *et al.* 2014). It is clear that an understanding of the submerged terrestrial landscape is required, both for the recognition of the potential location of sites, but also to provide a fuller understanding of the regional landscape and ecology. Archaeological remains are likely to have existed across the whole of the Channel and Celtic Sea area at some time in the last 700,000 years. However, the potential survival of material from the prehistoric period is dependent on taphonomy, sedimentation, erosion and ongoing coastal/marine geomorphological processes (Dix *et al.* 2008).

Submerged river valleys

For the Mesolithic period the association between wetlands, estuaries and human occupation is well-established. Bailey (2004) has made the case that Paleolithic cultures at least as far back as the Last Interglacial were capable of exploiting marine resources, and it is noteworthy that both La Mondrée off Fermanville, Boxgrove and Pakefield are situated in coastal plain environments, even if there is no direct evidence of exploiting marine species for food. Sites such as Pontnewydd (Green 1984; Aldhouse-Green *et al.* 2012), La Cotte de St. Brelade (Callow & Cornford 1986; Scott *et al.* 2014) and Paviland (Aldhouse-Green 2000; Jacobi & Higham 2008) are deposits in caves, and such deposits in a sheltered location underwater could survive, especially if consolidated in rockfalls. Many of the Paleolithic sites discovered on land are associated with river gravels. Thus the potential for pre-LGM sites in the Channel and Celtic Sea is influenced by possible association both with rivers and/or coastlines, and also with caves. The pattern of rivers at any given date, the associated shoreline, and the intermediate wetlands and marshes, are important indicators of the probability of human occupation and the survival of remains. The submerged archaeological sites of La Mondrée, Bouldnor Cliff and the finds from the paleo-Arun are all associated with paleochannels that have survived the transgression. Work to detect and preserve the prehistory of the sea floor in the Channel and Celtic Sea area depends upon our understanding of the river drainage pattern. Work by Gibbard (1988), Gibbard and Lautridou (2003), Antoine *et al.* (2003), Lericolais *et al.* (2003) and Hijma *et al.* (2011) has addressed this challenge in the Channel. Research focusing on the Solent River as a significant tributary of the Channel river system can be found in Dix (2001).

The rivers, tributary junctions, deltas, braids, and over-deepened channels which are seen today on the Channel floor have been influenced by multi-cycle glacial retreats and re-advances of the sea level over the fluvial and deltaic features. In the Channel, the shoreline at the eastern terminal, prior to the erosion of the Weald–Artois Ridge, reached approximately from Brighton to Le Crotoy (Bates *et al.* 2003: 325, fig. 4A). The ridge was over 100 km wide, providing a permanent land connection even at times of high sea level. It has been suggested that the lowering of the chalk barrier at the Dover Strait could have been a catastrophic rupture (Gupta *et al.* 2008). If so, the erosive forces would have been one of the greatest influences on the relict seabed as the southern North Sea drained. For a discussion of this proposition see Gibbard (1988: 588-591), Hamblin *et al.* (1992: 80-81), Stringer (2006: 162-163) and Hjima *et al.* (2011). However, the evidence for a catastrophic flood is not convincing. When the level of the glacio-lacustrine water body trapped in the southern North Sea between the ice sheet to the north and the Dover Ridge to the south overtopped the ridge, the flow out would be identical to the input into the lake. The erosion may have been rapid, combined with frost shattering, but need not have been catastrophic, and the progressive widening and lowering of the sill may have been the result of several different events. Work in the Bay of Biscay has dated distal sediments associated with discharge from the North Sea that suggest that the rapid flooding that created the Channel river system was initiated during MIS 12 (Toucanne *et al.* 2009). For further review of chronology and paleogeography see Dix and Sturt (2013).

The gradient of the Channel River from the Dover Strait to the shelf edge is of the order of 1:4000. This is comparable with the present-day Rhine from the border of Switzerland to the North Sea at 1:2000, which in prehistoric times meandered through massive bends and oxbow lakes. Given the southward flow of the Thames and Rhine through the Channel, and the addition of tributaries from the Somme, Seine, and southern English rivers, one would therefore expect frequent changes of channel course, with meanders and braids. The incised and submerged valleys of the various rivers on the floor of the English Channel are shown on the BGS Bottom Sediment Maps and are analyzed by Hamblin *et al.* (1992: 78-79). Some, but not all, of the archaeological material within the landscape deposits would have been redeposited or reworked as the watercourses migrated.

Descriptions or maps of known seabed-submerged terrestrial features

Whilst submerged forests are known in several locations, including in the Solent and Severn Estuary, known submerged terrestrial features are still relatively rare, but are likely to increase given the recent proliferation of seabed mapping, and the coverage and availability of large industry data sets. Recent winter storms (2014) exposed new features and preserved paleolandscapes (Ashton *et al.* 2014).

Paleoclimate and faunal indicators

Peat deposits dating between 12,650 to 7000 years ago exist in the Dover Strait area, the Solent, and within the inlets along the south coast of Devon, Cornwall, Normandy and Brittany. The La Mondrée site is also associated with peat from before the LGM. Prehistoric archaeological discoveries in the Channel and Celtic Sea are frequently associated with submerged peat and drowned forests. While there is circumstantial evidence that prehistoric settlements, especially in the Mesolithic, were often close to coasts, rivers and marshes, this association should not be exaggerated. More logically, if a settlement is in a location which is a wetland, or which subsequently becomes a wetland, with peat formation, the archaeological materials are likely to become embedded in the cohesive sediments, and thus to survive. In addition to this, peat provides preserved evidence for paleoclimate. Detailed reports are found within site reports (e.g. Bouldnor Cliff 2012) and within RECs and Historic England's peat database (Hazell 2008).

Further resources

Some records are held in online European or global databases; for example the European Pollen Database and NEOTOMA.

– European Pollen Database: www.europeanpollendata base.net/
– Neotoma Paleoecology Database: www.neotomadb .org/
– Peat database: content.historicengland.org.uk/content/ docs/research/peat-database-nsea.pdf/

Taphonomy and Potential for Archaeological Site Survival

Factors favorable for the survival of archaeological strata in the original area of deposition can include:

– Very low beach gradient and offshore gradient so that wave action is attenuated and is constructional in the surf zone;
– Minimum fetch so that wave amplitude is minimal, wavelength is short, and wave action on the seabed is minimal;
– Original deposit embedded in peat or packed lagoonal deposits to give resistance and cohesion during marine transgression. Drowned forests and peat are good indicator environments;
– Deposits in a cave or rock shelter, where roof falls, accumulated debris, concretions, breccia, conglomerate formation and indurated wind-blown sand can all help to secure the archaeological strata;
– Local topography that contains indentations, re-entrants, bays, estuaries, beach-bars, lagoons, nearshore islands, or other localized shelter from dominant wind fetch and currents at the time of transgression of the surf zone;
– Frozen ground or permafrost enclosing archaeological deposit at time of inundation;
– Braided river pattern or deltaic islands, which can provide numerous lee environments protected locally from wave action from the west and south-west.

The factors above are those which promote survival of the original deposit *in situ*. However, if an archaeological deposit is buried under 5 m to 10 m of mud or sand it is unlikely to be discovered, except in very unusual circumstances. Thus, in the absence of major industrial excavation or dredging, the final requirements for survival and discovery are:

low-net modern sediment accumulation rate so that the artifacts are not buried too deeply;
no fields of sand waves or megaripples over the site;
a slight change in oceanographic conditions so that the site is being gently eroded to expose deposits when visited by archaeologists;
absence of heavy and continuous erosion which could remove the deposit completely;
absence of accumulation of successive layers of sediment during successive glacial cycles which would bury the archaeological material completely.

Oceanographic conditions, wind, waves, and currents

The Channel and Celtic Sea wave climate is dominated by westerly winds from the Atlantic, and the open-ocean swell waves generated by the fetch of many thousands of kilometers to the west and south-west. Within the Channel itself, winds from other directions are only blowing over a few hundred kilometers of sea and can generate only shorter period waves of limited amplitude. Wave data are summarized conveniently by Draper (1991) in atlas format, with successive maps showing the distribution of the significant wave height which is exceeded for different percentages of the time for spring, summer, autumn, winter, and the average for the whole year. More sophisticated data, and local data, can be obtained from databases and numerical models, including wave period and directional spectra, but the climatic data on wave height are usefully presented by Draper. This provides a general picture of the extent to which different parts of the seabed are exposed to wave action and possible erosion.

As would be expected, maximum wave height is found in the open Atlantic, decreasing to the east from the shelf edge. At the shelf edge the significant wave height

(H_s) exceeds 5 m for 10% of the year. This decreases to 4 m around the Irish southern coast, Pembroke, Land's End, and Brittany, and drops progressively to 1.5 m to 2.0 m at Dover. There are sheltered lee areas to the east of Start Point and Torbay, and east of Portland Bill and Cherbourg, where the 10% annual exceedance H_s drops to 0.5 m to 1.0 m.

Since winter is the season when storm waves are likely to be highest, and with the longest period, so that they interact with the seabed in deeper water, it is important to consider the wave climate specifically in this season. In winter H_s exceeds 6 m at the shelf edge for 10% of the time. The 10% winter exceedance H_s then drops to 5 m on the exposed western headlands, and decreases steadily further east in the Channel, dropping to 1.5 m to 2.0 m in the Dover Strait.

Maximum bed stress is caused by a combination of waves and tidal currents. While shelter from waves in areas of limited fetch is generally a favorable condition for the survival of seabed archaeological remains, the exceptionally high current velocities and tidal amplitudes in the English Channel mean that erosion may continue in locations that are fetch-limited. Thus the Solent is more eroded and scoured than one would expect from its sheltered wave climate, and tidal current gyres generated by headlands such as Portland Bill also promote bed stress and erosion.

Hamblin *et al.* (1992: 87) summarize the tidal current environment in the Channel and Celtic Sea. Greater detail can be obtained from Meteorological Office models or commercial models available to support offshore operations in the area. Some of the highest tidal amplitudes in the world occur in this area, with over 11 m in the Severn Estuary and a similar amplitude in the Golfe de Saint-Malo. In the English Channel, Bristol Channel, and southern Irish Sea the amplitude is generally more than 2 m. Tidal amplitude around the Isle of Wight and the Solent is in the range of 2 m to 5 m.

Tidal current velocities are greatest around the Pembroke coast, in the Severn Estuary, around Land's End, the Channel Islands, the central Channel between the Isle of Wight and Cherbourg, and in the narrowest part of the Dover Strait. This pattern of currents and associated bottom stress results in the bed-load parting of the central

Channel, and the resulting accumulation of sediments on the outer shelf, the eastern Channel and the southern North Sea. Massive tidal sand ridges accumulate on the French side of the Dover Strait. While the strong currents make archaeological work on the seabed difficult to carry out, the winnowing effect of currents is tending to reveal deposits which may contain prehistoric remains. In the extreme cases, archaeological materials will be eroded completely, and the context destroyed.

Data sources

There are a variety of data sources on modern oceanographic conditions. The UKHO holds information on tides and currents around the British Isles as does the NOC — Liverpool (formerly the POL). The CCO hold tidal stream information. Finally, historic tide gauge records for the UK are accessible via the BODC.

Further resources
– UKHO charts: www.admiralty.co.uk/charts
– NOC (Liverpool): www.pol.ac.uk/
– BODC: www.bodc.ac.uk/

Areas with the potential for discovery of archaeological material

The potential for the discovery of archaeological material within the Channel/Celtic Sea area is considered below in light of the oceanographic, climatic and geomorphological parameters outlined above. The area has been split into four zones, each with distinct geographical characteristics (Fig. 9.8).

Zone A: Celtic Sea, shelf margin, and Western Approaches up to the Start–Cotentin Ridge

Kenyon and Stride (1970), Johnson *et al.* (1982) and Hamblin *et al.* (1992: 88) describe the impact of bed-load parting on the sediment transport in the central English Channel. From an area between the Isle of Wight to Cherbourg the sediments tend to migrate in both directions on a west–east axis.

Fig. 9.8 Image showing the four zones which have been divided to show their potential for preservation of prehistoric archaeology. Julian Whitewright; Maritime Archaeology Trust. Reproduced with permission.

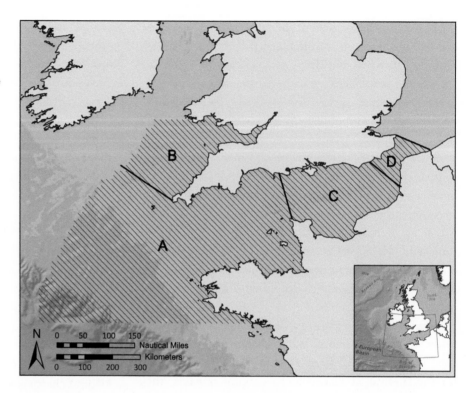

Bourillet *et al.* (2003: 261) describe the flow of sediments westwards through the dune fields or banks of the Celtic Sea into the valleys and canyons of the Biscay Margin. The origin of the Celtic Banks is complex and unresolved (Bourillet *et al.* 2003: 262). They occur between the 120-m and 200-m isobaths, and might be regarded as too deep to have been exposed during glaciations. However, the probable eustatic lowering of sea level to 130 m to 150 m, combined with the slight isostatic uplift outside the glaciated area, indicates that the coastline would have been well to the west of the present 120-m isobaths (see also the model outputs of Lambeck 1995, and Shennan *et al.* 2002). River valleys in this area are not deeply incised (Antoine *et al.* 2003: 230) and the rivers flowing from western Normandy, Cornwall, and Brittany did not flow into the Channel River or join to each other. They made separate courses to the shelf margin (Antoine *et al.* 2003).

Oppenheimer (2006) shows that much of the later Upper Paleolithic re-occupation of Britain may have taken place on the western margins of Europe. It has been demonstrated above that the Celtic Sea and western Channel would have formed either a refugium

or an exit route whenever the climate in the British Isles deteriorated and the ice advanced during earlier glaciations. Thus, in principle, this is an area where important archaeological deposits might have occurred that would shed considerable light on the understanding of the occupation of northwest Europe and the British Isles. During the transgression, however, it was exposed to the maximum force of the Atlantic waves and is now partly covered in sand waves (Evans 1990; Bourillet *et al.* 2003). The western part of Zone A is therefore an area where the chances of discovery of artifacts are low, and the costs of work would be extremely high.

The archaeological potential along the coastline is more favorable for preservation. There are submerged sites within estuaries and ria systems that have now been flooded. An area that is particularly rich is the Golfe du Morbihan, which has many Late Mesolithic and Neolithic monuments. Er Lannic is now isolated in the gulf but was occupied by people before it was turned into an island by rising sea levels. It contains two large, adjoining stone circles, one of which is now totally submerged. Much of the gulf saw inundation during the early periods of settlement; therefore we would expect to find more material underwater. Indeed, recent acoustic survey has recorded submarine Neolithic standing stones on the northern edge of the gulf (Cassen *et al.* 2011).

Along the coastline at Quiberon, a Mesolithic site dating to around ca. 6200 cal BC was threatened by erosion (Marchand & Dupont 2014). It was a shell midden and was excavated by the Centre national de la recherche scientifique (CNRS) and CReAAH. On the north coast of Brittany, at Le Penthièvre, Côtes d'Armor, France, Pleistocene sea-level changes during the three last interglacial/glacial cycles and associated Paleolithic occupations have been studied by the University of Rennes (Daire *et. al.* 2012; 2013).

Zone A is potentially important because it may have acted as a route for migrations and as a refugium, but in water deeper than 100 m the potential for either survival of archaeological deposits, or the possibility of finding and studying them, is low. In areas shallower than 100 m, and especially in coastal waters shallower than 50 m around the Channel Islands (Sebire 2005), Brittany, Cornwall (Crawford 1927) and Devon, there is a higher chance of finding prehistoric archaeology, probably in association with peat deposits (Evans 1990: 11).

Further resources
– CReAAH:www.creaah.univ-rennes1.fr
– CNRS: www.cnrs.fr

Zone B: The Bristol Channel and Severn Estuary

The southern Irish Sea was analyzed in the report on the prehistoric potential of SEA6 (Flemming 2005). The northern borders of the Channel and Celtic Sea in the open-sea zone are similar, with paleoglacial features (Tappin *et al.* 1994). The geology of the Bristol Channel is also reviewed by Tappin *et al.* (1994). The Irish Sea has not so far revealed submerged prehistoric sites of any period, which is odd, given the preservation of many geomorphological features of periglacial origin on the sea floor (see Chapter 10, this volume).

There are many proven Paleolithic and Mesolithic sites in Wales, and one would expect some materials from these

periods to survive in the Bristol Channel, particularly in sheltered embayments and areas that have been subject to deposition in the past. Mesolithic footprints have been found stratified within postglacial marine sediments on the intertidal foreshore at Uskmouth and Magor Pill, near Newport, Gwent (Bell *et al.* 2004; Bell 2007). Human interference with the coastline along the Severn Estuary is affecting the hydrodynamic regime and inducing erosion in areas where previously there was none. These areas of localized scour have the potential to uncover more archaeological evidence as the sediments continue to be removed.

Zone C: The central English Channel from the Start–Cotentin Ridge to Beachy Head

This zone is the area of bed-load divergence (Johnson *et al.* 1982), and hence maximum regional and local erosion. It also includes the confluence of the major submerged river channels converging on the Hurd Deep, the Seine and Somme submerged valleys, the Northern Paleovalley, the Arun River extension, and the Solent River. Many handaxes have been recovered from the river terrace deposits around the UK (Wymer 1999), a number of which have been found in trawls from coastal waters (Wessex Archaeology 2003). The braided river valleys in the English Channel are a unique feature of this part of the continental shelf. Given that so many terrestrial prehistoric deposits in Britain are found in river gravels and on abandoned river terraces it is relevant to consider whether the gravels in the area of the submerged rivers might contain similar remains. In general, where tidal currents or wave action have winnowed out the fines and left a lag gravel, the larger stone tools are likely to remain, albeit after some disturbance. If they have also been eroded or rounded beyond recognition, then no archaeological signal is likely to be detected. However, the general circumstance of the bed-load parting towards the center of the channel makes it likely that nearsurface archaeological sites will be exposed or eroded, rather than buried.

In the coastal regions of Zone C, more shelter is afforded by the land resulting in some locally sheltered and protected environments. This zone is potentially highly prospective for prehistoric archaeology especially where there are well-preserved river valleys infilled with sediments. A paleochannel that has become exposed is found at Bouldnor Cliff, off the Isle of Wight. This was a low-lying river flood plain in the western Solent, surrounded by hills, and possibly, with lakes. It was occupied during the Mesolithic around 8100 years ago when sea level was in the order of 10 m lower. As the Flandrian transgression progressed, the water rose and a layer of brackish sediments was deposited. They covered and protected the landscape including fine organic material and archaeological remains. Around 3500 years ago a new marine channel, the Solent, was formed as the sea level rose. This ran across the Holocene deposits, perpendicular to the initial waterway. Consequently, the sediments laid down during the transgression were exposed by the new current and washed away to leave thin strips of material along the edges of the waterway (Fig. 9.9). This example demonstrates potential for well-preserved paleolandscapes buried within the submerged river valleys. The distribution and concentrations of anthropogenic material within these landscapes can only be identified by visual inspection or sampling once the archaeological horizon is exposed.

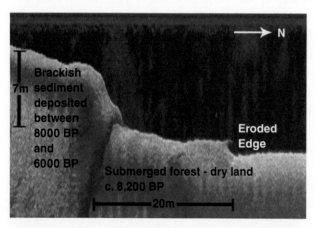

Fig. 9.9 Side-scan sonar image across the submerged cliff at Bouldnor. The underlying geology is exposed to the right. The peat-covered woodland bench is in the middle, while the sediment deposited over the landscape during the transgression is seen on the left. The area is now subject to erosion from right to left. Jasmine Noble-Shelley after Garry Momber; Maritime Archaeology Trust. Reproduced with permission.

Just offshore from Cap Lévi, east of Cherbourg, on the south side of the Channel, the site of La Mondrée contains a sediment sequence within a river-drainage system that has remained undisturbed for tens of thousands of years. Over 2500 worked flints were recovered after the initial discovery in the 1970s (Scuvée & Veragué 1988). The archaeological material is lying on the seabed and below the surface in stratified deposits that were laid in a hollow or paleochannel. Excavation in 2002 and coring in 2010 revealed over a meter of sediment remaining undisturbed beneath the surface (Cliquet et al. 2011). The sediments were interlaced with organic material that indicated human occupation in a flood plain environment during a climatic downturn ca. 70,000 years ago during MIS 5a. The layering of the archaeological material suggests there was ongoing or repeated occupation in the same area as the deposits built up across the paleochannel. The surface finds show there has been deflation of the seabed but the stratified deposit within the channel still remains intact despite fluctuating sea levels and climatic oscillations.

Zone D: The eastern Channel

This zone is dominated by the strong tidal currents through the Dover Strait, and the accumulation of marine sands. The BGS Bottom Sediment charts show active features such as megaripples and sand ribbons consisting of coarse and medium sand. In view of the active modern sediment bedforms and the strong tidal currents, this is not a prospective area for the survival and accessibility of prehistoric deposits. The tidal current regime and the active sand banks of the eastern Channel are described by Reynaud et al. (2003).

Hamblin et al. (1992) report peat beds detected in bore holes and cores, with peat at 36 m and 37 m dated at 10,530 and 9920 years old (Hamblin et al. 1992: 81), and a borehole at Cap Gris Nez showing peat with ages 12,650 and 8250 years old (Hamblin et al. 1992: 81). The occurrence of peat is a positive indicator, but if it is buried under thick layers of modern marine sand it is unlikely that any investigation could be made to search for prehistoric indicators. However, as with areas A, B and C, the potential does remain to

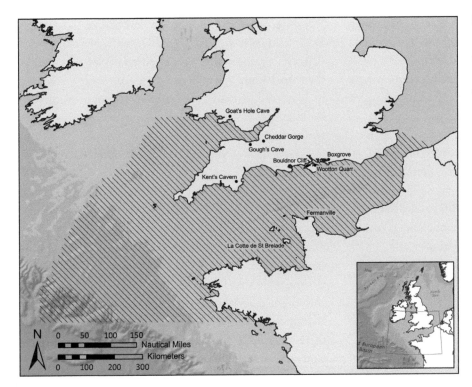

Fig. 9.10 Distribution of Paleolithic and Mesolithic sites bordering the Celtic and Channel Sea area coastline. Jasmine Noble-Shelley; Maritime Archaeology Trust. Reproduced with permission.

find archaeological material within the sheltered channels that have become filled with protective sediments before becoming submerged as sea level has risen.

General conclusions about possibility of site preservation

This section brings together the factors tending to favor occurrence, preservation, and accessibility of submerged prehistoric materials in the Channel and Celtic Sea. We have not tried to apply all the discriminatory factors to all areas or features in the region, but it is apparent that they do correlate well with the actual archaeological sites already found, and the assessment made of the potential in the Zones A–D (see Fig. 9.10). The unknown factor is the extent to which archaeological deposits protected from the initial stages of inundation may have survived the subsequent strong tidal currents and scour which may have occurred in the Bristol Channel, the central English Channel, the Western Approaches or the Dover Strait.

In view of the work of Pitulko *et al.* (2004) it is important to consider the effect of sea water rising over prehistoric archaeological deposits in permafrost, which would indicate the possibility of good preservation of artifacts. Although other factors also apply, for example river scour, frost shattering, and normal sub-aerial erosion processes, the critical period for survival of an archaeological deposit is the time when the surf zone starts to impact on the site, the ensuing few hundred years as the sea level rises over the site, and coastal shallow water waves are breaking over it, or washing into a cave mouth. The literature on these processes has been reviewed by Dix and Westley (2006). Favorable factors for survival of archaeological strata in the original area of deposition are outlined above.

This analysis demonstrates that survival or destruction of an archaeological deposit, whether originally inland or on the coast, depends acutely upon the local topography within a few hundred meters or a few kilometers of the site. Generalized coarse resolution maps tend to omit the details which show the necessary local topographic clues. The BGS 1:250,000 maps, although they are primarily designed to present sediment data, map isobaths at 10-m intervals and therefore provide a more accurate representation of topography than the Admiralty Charts. Additional high-resolution swath bathymetry and sub-bottom profiling would be valuable in detecting probable sites. It is no coincidence that the most prolific area in Europe of proven submerged Mesolithic sites is between the islands of the Danish archipelago, where many hundreds of sites have been mapped and sampled by the National Museum of Denmark's Institute of Maritime Archaeology, and the Danish Nature Agency, assisted by amateur divers (e.g. Skaarup & Grøn 2004). Further submerged Baltic sites have been discovered in sheltered waters off the coast of northern Germany (Lübke 2001; 2002). The Bouldnor Cliff and La Mondrée sites associated with protected paleochannels are textbook examples of preservation by local topography.

Potential Example Areas for Future Work

The archaeological record for the British Isles extends back almost 1,000,000 years and across at least six glacial cycles. In the course of each interglacial the environment on the British mainland ameliorated making it habitable. As the climate warmed, the ice sheets melted causing a rise in sea level. This process would have taken thousands of years during which time the Channel and Celtic Sea area would have been suitable for occupation. The research for this report has identified several sources of evidence that demonstrate why the Channel and Celtic Sea area would have been occupied on many occasions by *Homo heidelbergensis*, *Homo sapiens neanderthalensis* and *Homo sapiens sapiens*.

The archaeological material from Boxgrove, Clacton, Swanscombe, Purfleet, Crayford and the Cotentin Peninsula (notably at La Mondrée) was associated with deposits that were laid at times when sea level was rising or falling and was not at its maximum height. Large parts of the Channel would have been dry and habitable, potentially for long periods, before rising waters or a deteriorating climate forced people to move.

Middle Paleolithic activity during cold phases of the glacial oscillation is found in La Cotte de St. Brelade in Jersey (Callow & Cornford 1986; Scott *et al.* 2014), Harnham near Salisbury (Bates *et al.* 2014) and La

Mondrée, France (ca. 250–45 ka)(Scuvée & Verague 1988; Cliquet *et al.* 2011). This demonstrates that Middle Paleolithic people had developed strategies that enabled them to endure a harsher climate. An ability to survive in the cold extends the window of opportunity for exploitation of Channel and Celtic Sea areas by *Homo sapiens neanderthalensis*.

The arrival of varying Upper Paleolithic technologies to mainland Britain may have followed a similar pattern to that which has been postulated during the Lower and Middle Paleolithic. The concept of a steady colonization by different peoples is endorsed by studies of mitochondrial DNA dispersal which identified a distinct western 'Celtic' population that originated from the Basque region and another from refugia to the east. The colonizers from the south migrated along the Atlantic Margin and reached Britain first. Throughout this process there may well have been populations occupying the continental shelf, which in turn acted as a springboard into the British Isles.

The final and current interglacial, the Holocene, marks the start of the Mesolithic (ca. 11–6 ka). Sea levels rose about 30 m during this transgression and large areas of the Channel and Celtic Sea were finally inundated. The Mesolithic came to a close about the same time as sea level reached comparable levels to those we see today. Coastal and riverine resources were exploited extensively during the Mesolithic. Europe's northwest peninsula was severed for the last time when the English Channel met the North Sea as the Mesolithic Age was drawing to a close. Water transport was now necessary to reach the British Isles.

References

Aldhouse-Green, S. (ed.) 2000. *Paviland Cave and the 'Red Lady'. A Definitive Report*. Western Academic & Specialist Press: Bristol.

Aldhouse-Green, S, Peterson, R. & Walker, E. A. (eds.) 2012. *Neanderthals in Wales: Pontnewydd and the Elwy Valley Caves*. National Museum Wales Books and Oxbow Books: Bristol.

Allen, M. J. & Gardiner, J. 2000. *Our Changing Coast; a Survey of the Intertidal Archaeology of Langstone*

Harbour, Hampshire. CBA Research Report 124. Council for British Archaeology: York.

Antoine, P., Coutard, J-P., Gibbard, P., Hallegouet, B., Lautridou, J-P. & Ozouf, J. C. 2003. The Pleistocene rivers of the English Channel region. *Journal of Quaternary Science* 18:227-243.

Ashton, N., Lewis, S. G., De Groote, I. *et al.* 2014. Hominin footprints from Early Pleistocene deposits at Happisburgh, UK. *PLoS ONE* 9(2): e88329. doi:10.1371/journal.pone.0088329.

Bailey, G. 2004. World prehistory from the margins: the role of coastlines in human evolution. *Journal of Interdisciplinary Studies in History and Archaeology* 1:39-50.

Bates, M. R., Keen, D. H. & Lautridou, J-P. 2003. Pleistocene marine and periglacial deposits of the English Channel. *Journal of Quaternary Science* 18:319-337.

Bates, M. R., Wenban-Smith, F. F., Bello, S. M. *et al.* 2014. Late persistence of the Acheulian in southern Britain in an MIS 8 interstadial: evidence from Harnham, Wiltshire. *Quaternary Science Reviews* 101:159-176.

Bell, M. 2007. *Prehistoric Coastal Communities. The Mesolithic in Western Britain*. CBA Research Report 149. Council for British Archaeology: York.

Bell, M., Allen, J. R. L., Buckley, S., Dark, P. & Nayling, N. 2004. Mesolithic to Neolithic coastal environmental changes: excavations at Goldcliff East, 2003 and research at Redwick. *Archaeology in the Severn Estuary* 14:1-26.

Bourillet, J-F., Reynaud, J-Y., Baltzer, A. & Zaragosi, S. 2003. The 'Fleuve Manche': the submarine sedimentary features from the outer shelf to the deep-sea fans. *Journal of Quaternary Science* 18:261-282.

Bradley, S. L., Milne, G. A., Teferle, F. N., Bingley, R. M. & Orliac, E. J. 2009. Glacial isostatic adjustment of the British Isles: new constraints from GPS measurements of crustal motion. *Geophysical Journal International* 178:14-22.

Bradley, S. L., Milne, G. A., Shennan, I. & Edwards, R. 2011. An improved glacial isostatic adjustment model for the British Isles. *Journal of Quaternary Science* 26:541-552.

Brooks, A. & Edwards, R. 2006. The development of a sea-level database for Ireland. *Irish Journal of Earth Sciences* 24:13-27.

Callow, P. & Cornford, J. M. (eds.) 1986. *La Cotte de St. Brelade 1961–1978: Excavations by C.B.M. McBurney*. Geobooks: Norwich.

Carter, R. W. G. 1988. *Coastal Environments: An Introduction to the Physical, Ecological and Cultural Systems of Coastlines*. Academic Press: London.

Carter, D., Bray, M. & Hooke, J. 2004. *Solent Sediment Transport Study 2004*. Available at www.scopac.org.uk/sediment-transport.html.

Cassen, S., Baltzer, A., Lorin, A., Fournier, J. & Sellier, D. 2011. Submarine Neolithic stone rows near Carnac (Morbihan), France: Preliminary results from acoustic and underwater survey. In Benjamin, J., Bonsall, C., Pickard, C. & Fischer, A. (eds.) *Submerged Prehistory*. pp. 99-110. Oxbow Books: Oxford.

Clark, C. D., Hughes, A. L. C., Greenwood, S. L., Jordan, C. & Sejrup, H. P. 2012. Pattern and timing of retreat of the last British-Irish Ice Sheet. *Quaternary Science Reviews* 44:112-146.

Cliquet, D., Coutard, S., Clet, M. *et al.* 2011. The Middle Palaeolithic underwater site of La Mondrée, Normandy, France. In Benjamin, J., Bonsall, C., Pickard, C. & Fischer, A. (eds.) *Submerged Prehistory*. pp. 111-128. Oxbow Books: Oxford.

Collins, M. & Ansell, K. (eds.) 2000. *Solent Science: A Review. Proceedings in Marine Science, 1*. Elsevier: Amsterdam.

Cracknell, B. E. 2005. *"Outrageous Waves": Global Warming and Coastal Change in Britain through Two Thousand Years*. Phillimore: Chichester.

Crawford, O. G. S. 1927. Lyonesse. *Antiquity* 1: 5-14.

Daire, M-Y., Lopez-Romero, E., Proust, J-N., Regnauld, H., Pian, S. & Shi, B. 2012. Coastal changes and cultural heritage (1): Assessment of the vulnerability of the coastal heritage in Western France. *Journal of Island and Coastal archaeology* 7:168-182.

Daire, M-Y., Dupont, C., Baudry, A. *et al.* (eds.) 2013. *Ancient Maritime Communities and the Relationship Between People and Environment along the European Atlantic Coasts / Anciens peuplements littoraux et relations homme/milieu sur les côtes de l'Europe atlan-tique*. Proceedings of the HOMER 2011 Conference, Vannes (France), 28th September – 1st October 2011. British Archaeological Reports International Series 2570. Archaeopress: Oxford.

Dix, J. K. 2001. The Pleistocene geology of the Solent River system. In Wenban-Smith, F. F. & Horsfield, R. T. (eds.) *Palaeolithic Archaeology of the Solent River - Proceedings of the Lithic Studies Society day meeting (Lithic Studies Society Occasional Papers No. 7)*. 15th January 2000, Southampton, pp. 7-14.

Dix, J. & Sturt, F. 2013. Marine geoarchaeology and investigative methodologies. In Ransley, J., Sturt, F., Dix, J., Adams, J. & Blue, L. (eds.) *People & the Sea: a Maritime Archaeological Agenda for England*. CBA Research Report 171. pp. 1-9. Council for British Archaeology: York.

Dix, J. & Westley, K. 2006. Archaeology of continental shelves: a submerged pre-history. In Newell, R. C. & Garner, D. J. (eds.) *Marine Aggregate Dredging: Helping to Determine Good Practice. Proceedings of the Marine Aggregate Levy Sustainability Fund (ALSF) conference*. September 2006, Bath, pp. 90-91.

Dix, J., Quinn, R. & Westley, K. 2008. *Re-assessment of the Archaeological Potential of Continental Shelves* [data-set]. Archaeology Data Service: York.

Draper, L. 1991. *Wave Climate Atlas of the British Isles*. HMSO: London.

Evans, C. D. R. 1990. *UK Offshore Regional Report: The Geology of the Western English Channel and its Western Approaches*. HMSO for the British Geological Survey: London.

Fitch, S. & Gaffney, V. 2011. *West Coast Palaeolandscapes Survey (Project No. 1997)*. University of Birmingham.

Flemming N. C. 1998. Archaeological evidence for vertical movement on the continental shelf during the Palaeolithic, Neolithic and Bronze Age periods. In Stewart, I. S. & Vita-Finzi, C. (eds.) *Coastal Tecton-ics*. Geological Society, London, Special Publications 146:129-146.

Flemming, N. C. 2005. *The Scope of Strategic Environ-mental Assessment of Irish Sea Area SEA6 in regard to Prehistoric Archaeological Remains*. UK Department of Trade and Industry: London.

Gaffney V., Thomson, K. & Fitch, S. (eds.) 2007. *Mapping Doggerland: The Mesolithic Landscapes of the Southern North Sea*. Archaeopress: Oxford.

Garrow, D. & Sturt, F. 2011. Grey waters bright with Neolithic argonauts? Maritime connections and the Mesolithic-Neolithic transition within the 'western seaways' of Britain, c. 5000–3500 BC. *Antiquity* 85:59-72.

Gibbard, P. L. 1988. The history of the great northwest European rivers during the past three million years. *Philosophical Transactions of the Royal Society, London* B318:559-602.

Gibbard, P. L. & Lautridou, J-P. 2003. The Quaternary history of the English Channel: an introduction. *Journal of Quaternary Science* 18:195-199.

Giot, P. R. 1968. La Bretagne au péril des mers holocènes. In Bordes, F. & de Sonneville-Bordes, D. (eds.) *La Préhistoire, problèmes et tendances*. pp. 203-208. CNRS: Paris.

Green, H. S. 1984. *Pontnewydd Cave. A Lower Palaeolithic Hominid Site in Wales: the First Report*. National Museum of Wales: Cardiff.

Gupta, S., Collier, J., Palmer-Felgate, A., Dickinson, J., Bushe, K. & Humber, S. 2008. *Submerged Palaeo-Arun and Solent Rivers: Reconstruction of Prehistoric Landscapes* [data-set]. Archaeology Data Service: York.

Hamblin, R. J. O., Crosby, A., Balson, P. S. *et al.* 1992. *UK Offshore Regional Report: The Geology of the English Channel*. HMSO for the British Geological Survey: London.

Hazell, Z. J. 2008. Offshore and intertidal peat deposits, England — a resource assessment and development of a database. *Environmental Archaeology* 13:101-110.

Hijma, M. P., & Cohen, K. M. 2011. Holocene transgression of the Rhine river mouth area, The Netherlands/Southern North Sea: palaeogeography and sequence stratigraphy. *Sedimentology* 58:1453-1485.

Hodson, F. & West, I. M. 1972. Holocene deposits of Fawley, Hampshire and the development of Southamp-ton Water. *Proceedings of the Geologists' Association* 83:421-441.

Huijzer, A. S. & Vandenberghe, J. 1998. Climatic reconstruction of the Weichselian Pleniglacial in northwestern and central Europe. *Journal of Quaternary Science* 13:391-417.

Isarin, R. F. B. 1997. Permafrost distribution and temperatures in Europe during the Younger Dryas. *Permafrost and Periglacial Processes* 8:313-333.

Jacobi, R. M. & Higham, T. F. G. 2008. The 'Red Lady' ages gracefully: new ultrafiltration AMS determinations from Paviland. *Journal of Human Evolution* 55:898-907.

Johnsen, S. J., Dahl-Jensen, D., Gundestrup, N. *et al.* 2001. Oxygen isotope and palaeotemperature records from six Greenland ice-core stations: Camp Century, Dye-3, GRIP, GISP2, Renland and NorthGRIP. *Journal of Quaternary Science* 16:299-307.

Johnson, M. A., Kenyon, N. H., Belderson, R. H. & Stride, A. H. 1982. Sand transport. In Stride, A. H. (ed.) *Offshore Tidal Sands: Processes and Deposits.* pp. 58-94. Chapman and Hall: London.

Kenyon, N. H. & Stride, A. H. 1970. The tide-swept continental shelf sediments between the Shetland Isles and France. *Sedimentology* 14:159-173.

Lagarde, J. L., Amorese, D., Font, M., Laville, E. & Dugué, O. 2003. The structural evolution of the English Channel area. *Journal of Quaternary Science* 18:201-213.

Lambeck, K. 1991. Glacial rebound and sea-level change in the British Isles. *Terra Nova* 3:379-389.

Lambeck, K. 1993a. Glacial rebound of the British Isles—I. Preliminary model results. *Geophysical Journal International* 115:941-959.

Lambeck, K. 1993b. Glacial rebound of the British Isles—II. A high resolution, high-precision model. *Geophysical Journal International* 115:960-990.

Lambeck, K. 1995. Late Devensian and Holocene shorelines of the British Isles and the North Sea from models of glacio-hydro-isostatic rebound. *Journal of the Geological Society* 152:437-448.

Lambeck, K. & Chappell, J. 2001. Sea level change through the last glacial cycle. *Science* 292: 679-686.

Lericolais, G., Auffret, J-P. & Bourillet, J. F. 2003. The Quaternary Channel River: Seismic stratigraphy of its palaeo-valleys and deeps. *Journal of Quaternary Science* 18:245-260.

Loader, R., Westmore, I. & Tomalin, D. 1997. *Time and Tide: An Archaeological Survey of the Wootton-Quarr Coast.* Isle of Wight Council: Isle of Wight.

Louwe Kooijmans, L. P. 1970–71. Mesolithic bone and antler implements from the North Sea and from the Netherlands. *Berichten van de Rijksdienst voor het Oudheidkundig Bodemonderzoek* 20-21: 27-73.

Lübke, H. 2001. Eine hohlendretuschierte Klinge mit erhaltener Schäftung vom endmesolithischen Fundplatz Timmendorf-Nordmole, Wismarbucht, Mecklenburg-Vorpommern. *Nachrichtenblatt Arbeitskreis Unterwasserarchäologie* 8:46-51.

Lübke, H. 2002. Steinzeit in der Wismarbucht: ein Überblick. *Nachrichtenblatt Arbeitskreis Unterwasserarchäologie* 9:75-87.

Marchand G. & Dupont C. 2014. Maritime hunter-gatherers of the Atlantic Mesolithic: current archaeological excavations in the shell levels of Beg-er-Vil (Quiberon, Morbihan, France). *Mesolithic Miscellany* 22(2):3-9

Meese, D. A., Gow, A. J., Alley, R. B. *et al.* 1997. The Greenland Ice Sheet Project 2 depth-age scale: Methods and results. *Journal of Geophysical Research* 102:26411-26423.

MoLAS (Museum of London Archaeology Service) 2007. *Severn Estuary: Assessment of Sources for Appraisal of the Impact of Maritime Aggregate Extraction* [data-set]. Archaeology Data Service: York.

Momber, G. 2011. Changing landscapes. In Momber, G., Tomalin, D., Scaife, R., Satchell, J. & Gillespie, J. (eds.) *Mesolithic Occupation at Bouldnor Cliff and the Submerged Prehistoric Landscapes of the Solent.* CBA Research Report 164. pp. 168-178. Council for British Archaeology: York.

Momber, G., Tomalin, D., Scaife, R., Satchell, J. & Gillespie, J. (eds.) 2011. *Mesolithic Occupation at Bouldnor Cliff and the Submerged Prehistoric Landscapes of the Solent.* CBA Research Report 164. Council for British Archaeology: York.

Murton, J. B. & Lautridou, J-P. 2003. Recent advances in the understanding of Quaternary periglacial features of the English Channel coastlands. *Journal of Quaternary Science* 18:301-307.

Oppenheimer, S. 2006. *The Origins of the British: A genetic detective story*. Constable: London.

Paphitis, D., Velegrakis, A. F. & Collins, M. B. 2000. Residual circulation and associated sediment dynamics in the eastern approaches to the Solent. In Collins, M. & Ansell, K. (eds.) *Solent Science: a Review. Proceedings in Marine Science, 1*. pp. 107-109. Elsevier: Amsterdam.

Peltier, W. R., Shennan, I., Drummond, R. & Horton B. 2002. On the postglacial isostatic adjustment of the British Isles and the shallow viscoelastic structure of the Earth. *Geophysical Journal International* 148: 443-475.

Pitulko, V. V., Nikolsky, P. A, Yu. Girya, E. *et al.* 2004. The Yana RHS Site: Humans in the Arctic before the Last Glacial Maximum. *Science* 303:52-56.

Prigent, D., Vissent, L., Morzadec-Kerfourn, M. T. & Lautridou, J-P. 1983. Human occupation of the submerged coast of the Massif Armoricain and postglacial sea level changes. In Masters, P. M. & Flemming, N. C. (eds.) *Quaternary Coastlines and Marine Archaeology*. pp. 303-324. Academic Press: London and New York.

Reynaud, J-Y., Tessier, B., Auffret, J-P. *et al.* 2003. The offshore Quaternary sediment bodies of the English Channel and its Western Approaches. *Journal of Quaternary Science* 18:361-371.

Rohling, E. J., Fenton, M., Jorissen, F. J., Bertrand, P., Ganssen, G. & Caulet, J. P. 1998. Magnitudes of sea-level lowstands of the past 500,000 years. *Nature* 394:162-165.

Rohling, E. J., Grant, K., Hemleben, Ch. *et al.* 2008. High rates of sea-level rise during the last interglacial period. *Nature Geoscience* 1:38-42.

Scott, B., Bates, M., Bates, R. *et al.* 2014. A new view from La Cotte de St. Brelade, Jersey. *Antiquity* 88:13-29.

Scuvée, F. & Verague, J. 1988. *Le gisement sous-marin du Paléolithique moyen de l'anse de La Mondrée à Fermanville, Manche*. CEHP-Littus: Cherbourg

Sebire, H. 2005. *The Archaeology and Early History of the Channel Islands*. Tempus: Stroud.

Shennan, I. & Horton, B. 2002. Holocene land- and sea-level changes in Great Britain. *Journal of Quaternary Science* 17:511-526.

Shennan, I., Peltier, W. R., Drummond, R. & Horton, B. 2002. Global to local scale parameters determining relative sea-level changes and the post-glacial isostatic adjustment of Great Britain. *Quaternary Science Reviews* 21:397-408.

Shennan, I., Bradley, S., Milne, G., Brooks, A., Bassett, S. & Hamilton, S. 2006. Relative sea-level changes, glacial isostatic modelling and ice-sheet reconstructions from the British Isles since the Last Glacial Maximum. *Journal of Quaternary Science* 21:585-599.

Siddall, M., Rohling, E. J., Almogi-Labin, A. *et al.* 2003. Sea-level fluctuations during the last glacial cycle. *Nature* 423:853-858.

Siddall, M., Rohling, E. J., Blunier, T. & Spahni, R. 2010. Patterns of millennial variability over the last 500 ka. *Climate of the Past* 6:295-303.

Skaarup, J. & Grøn, O. 2004. *Møllegabet ll: A Submerged Mesolithic Settlement in Southern Denmark*. British Archaeological Reports International Series 1328. Archaeopress: Oxford.

Steers, J. A. 1948. *The Coastline of England and Wales*. Cambridge University Press: Cambridge.

Stringer, C. 2006. *Homo Britannicus: the incredible story of human evolution*. Penguin Books: London.

Tappin, D. R., Chadwick, R. A., Jackson, A. A., Wingfield, R. T. R. & Smith, N. J. P. 1994. *UK Offshore Regional Report: The Geology of Cardigan Bay and the Bristol Channel*. HMSO for the British Geological Survey: London.

Thomas, C. 1985. *Exploration of a Drowned Landscape: Archaeology and History of the Isles of Scilly*. Batsford: London.

Toucanne, S., Zaragosi, S., Bourillet, J. F. *et al.* 2009. Timing of massive 'Fleuve Manche' discharges over the last 350 kyr: insights into the European ice-sheet oscillations and the European drainage network from MIS 10 to 2. *Quaternary Science Reviews* 28:1238-1256.

UKMMAS, (UK Marine Monitoring and Assessments Strategy Community) 2010. *Charting Progress 2: An Assessment of the State of UK Seas*. DEFRA: UK.

van Andel, T. H. & Davies, W. (eds.) 2003. *Neanderthals and Modern Humans in the European Landscape during the Last Glaciation*. McDonald Institute for Archaeological Research: Cambridge.

van Vliet-Lanoë, B. 2000. Permafrost extent during the Last Glacial Maximum (20 ka BP). *French IGBP-WCRP News Letter* 10:38-43.

Velegrakis, A. 2000. Geology, geomorphology and sediments of the Solent system. In Collins, M. & Ansell, K. (eds.) *Solent Science: A Review. Proceedings in Marine Science, 1.* pp. 21-43. Elsevier: Amsterdam.

Velegrakis, A. F., Dix, J. K. & Collins M. B. 1999. Late Quaternary evolution of the upper reaches of the Solent River, Southern England, based on marine geophysical evidence. *Journal of the Geological Society, London* 156:73-87.

Waller, M. P. & Long, A. J. 2003. Holocene coastal evolution and sea-level change on the southern coast of England: a review. *Journal of Quaternary Science* 18:351-359.

Wenban-Smith, F. F. & Horsfield, R. T. (eds.) 2001. *Palaeolithic Archaeology of the Solent River - Proceedings of the Lithic Studies Society day meeting (Lithic Studies Society Occasional Papers No. 7).* 15th January 2000, Southampton.

Wessex Archaeology 2003. *Artefacts from the Sea (ALSF Report).* Wessex Archaeology: Salisbury.

Wessex Archaeology 2008a. *Seabed Prehistory: Gauging the Effects of Marine Aggregate Dredging Round 2 Final Report — Volume II: Arun (ref. 57422.32).* Wessex Archaeology: Salisbury.

Wessex Archaeology 2008b. *Seabed Prehistory: Gauging the Effects of Marine Aggregate Dredging Round 2 Final Report — Volume III: Arun additional grabbing (ref. 57422.33).* Wessex Archaeology: Salisbury.

Westley, K., Bailey, G., Davies, W. *et al.* 2013. The Palaeolithic. In Ransley, J., Sturt, F., Dix, J., Adams, J. & Blue, L. (eds.) *People & The Sea: A Maritime Archaeological Agenda for England.* CBA Research Report 171. pp. 10-29. Council for British Archaeology: York.

Williams, R. B. G. 1969. Permafrost and temperature conditions in England during the last glacial period. In Péwé T. L. (ed.) *The Periglacial Environment.* pp. 399-410. McGill-Queen's University Press: Montreal.

Wymer, J. 1999. *The Lower Palaeolithic Occupation of Britain (Vols. 1 & 2).* Trust for Wessex Archaeology (in association with English Heritage): Salisbury.

Chapter 10
Irish Sea and Atlantic Margin

Kieran Westley[1] and Robin Edwards[2]

[1] School of Geography and Environmental Sciences, Ulster University, Coleraine, Northern Ireland, UK
[2] School of Natural Sciences, Trinity College Dublin, Republic of Ireland

Introduction

This section provides an overview of the Irish Sea and Atlantic Margin areas. This covers the entirety of the Irish Sea basin dividing the island of Ireland from Great Britain and the continental shelf bounding the north, south and west of Ireland between the Celtic Sea and Northwest Approaches. It encompasses the shelf areas of the Republic of Ireland, Northern Ireland, southwest Scotland, northwest England, Wales and the Isle of Man (Fig. 10.1).

The following sub-sections will describe the modern physical environment (coastal geomorphology, bathymetry, seabed substrate) of the study area before outlining the past environmental changes it experienced since the Last Glacial Maximum (LGM) around 20,000 years ago. These consisted primarily of extensive glaciation followed by spatially and temporally variable patterns of relative sea level (RSL) change resulting in locally varying extents and rates of shelf exposure

and flooding. It is highly likely that both past and present conditions have exerted a strong control on the formation and preservation of submerged archaeological landscapes.

Earth Sciences

Modern coastline and physical processes

The modern coastline of the study area can be subdivided into broad zones with similar geological and geomorphological characteristics. The southwestern, western and northern shores of Ireland are extensively cliffed and rock-dominated, while its eastern and southeastern coasts are typified by unconsolidated glacial sediment and fewer rocky headlands. All shorelines are incised by bays and estuaries, with those on its northwestern, western and southwestern sides typically forming long and large indentations (e.g. Bantry Bay, Lough Swilly), creating an irregular coastline. Those on the east coast

Fig. 10.1 Overview map showing the Irish Sea and Atlantic Margin study area with key place-names marked.

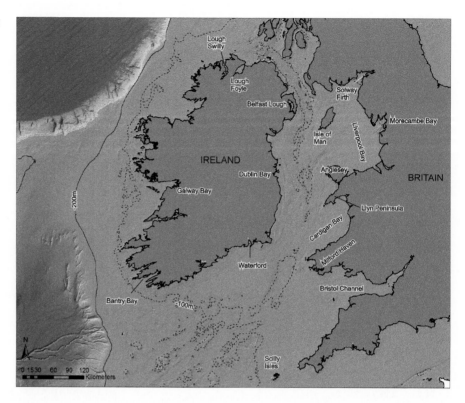

consist of wide shallow embayments frequently with barrier and spit features (e.g. Dublin Bay) resulting in an open linear shoreline. The western and northwestern coasts are high-energy environments exposed to the full force of Atlantic swells and storms. However, the east coast is much lower energy, receiving only ca. 20% of the wave energy of the Atlantic shores (Devoy 2008). Nonetheless, waves still remain more influential in shaping coastal morphology than tides (e.g. large sand or gravel beaches rather than extensive intertidal flats). Large proportions of the coast are therefore made up of wave-dominated forms including cliffs, headlands, beaches and barriers. A number of the larger sandy beaches are also characterized by extensive dune systems (Carter 1991; Cooper 2006; Devoy 2008). Low-energy coastal formations such as wetland, marshes and lagoons are of limited extent in Ireland due to the predominance of high-wave states and exposed shorelines. Typically, low-energy deposits are situated within the most sheltered portions of bays and estuaries, reflecting local variations in hydrodynamics and morphology. Whilst small examples

of these environments are dispersed around the Irish coast (e.g. Curtis & Sheehy-Skeffington 1998), more extensive deposits are found in areas such as Waterford Harbour, Bantry Bay, Galway Bay, Donegal Bay, Lough Swilly, Lough Foyle, Belfast Lough and Strangford Lough. These constitute the largest examples of lower energy areas.

The western coast of Great Britain is similarly diverse. In the northeast Irish Sea (i.e. Liverpool and Morecambe bays) the coast is generally low-lying with beaches, dunes and extensive low-energy formations such as intertidal flats. Tidal ranges are higher than on the western side of the Irish Sea, promoting the development of more extensive intertidal flat and salt-marsh environments (e.g. the Dee and Mersey estuaries and the Solway Firth) (Barne et al. 1996a). Rocky shores and cliffs are relatively sparse and restricted to the southern coast of the Isle of Man, southeast Scotland and parts of northwest England. To the south, the diverse Welsh coastline comprises extensive sandy beaches interspersed among long stretches of rock cliffs and headlands. A number of beaches have associated dune systems, while the estuaries (e.g. Milford Haven, Dyfi Estuary) are characterized by salt marshes and in some cases barrier and spit formation (Barne *et al.* 1995). A particularly large area of low-energy marsh and intertidal mudflats is located within the sheltered embayment formed by the Bristol Channel/Severn Estuary, reflecting its unusually extreme macrotidal conditions (Allen *et al.* 1987; Allen 1990). In general, the bedrock headlands on the British coast are larger than those on the opposing Irish side, creating prominent peninsulas (e.g. Llŷn Peninsula) separated by wide shallow bays (e.g. Cardigan Bay) which contrast to the more linear shore and less indented bays of eastern Ireland (Knight 2001).

Coastal geomorpho-dynamics vary considerably around the study area as a result of differences in wave energy (e.g. Atlantic versus inner Irish Sea coasts), geology (e.g. bedrock versus unconsolidated glacial sediment), topography/bathymetry (e.g. steep versus gently sloping/shallowing) and antecedent patterns of coastal change (e.g. differential isostatic uplift). Tidal currents are also particularly strong where topographic and bathymetric form constricts water bodies such as the Bristol, North and St. George's channels (ABPMer 2008). High-energy rocky shorelines (e.g. large zones of the western and northern coasts of Ireland) are scoured by constant wave and tidal action, and the only sedimentary deposits are typically high-energy sand and gravel beaches which accumulate where sediment supply and topography permit. Active erosion and retreat is visible where unconsolidated glacial sediment lies directly at the modern shoreline. This is particularly apparent along the eastern and southeastern coasts of Ireland, northern Isle of Man and northwest Wales where soft cliff-lines have been formed due to the easily erodible nature of the glacial deposits. The numerous beaches and dune fields dotted across the study area are the product of wave and wind action combined with extensive sources of glacial sediment, either derived from coastal erosion or reworked from the continental shelf (Barne *et al.* 1995; 1996a). Generally, it is the sheltered bays and lower energy Irish Sea coast where most accumulation would be expected. Long stretches of the coast have also been anthropogenically modified, for instance by sea defenses or extensive land reclamation in the eighteenth and nineteenth centuries. Examples include Belfast Lough, Lough Foyle, Wexford Harbour, the Shannon Estuary, Swansea Bay, and Morecambe Bay.

Data sources

High-resolution vector and raster mapping data (e.g. 1:50,000 scale and smaller) is held by national mapping agencies: the Ordnance Survey of Ireland (OSI: Republic of Ireland), Land & Property Services (LPS: Northern Ireland) and the Ordnance Survey (Scotland, Wales, England, Isle of Man). All organizations also hold high-resolution vertical aerial photos. These data form part of countrywide mapping data sets rather than specific coastal resources. Data including LiDAR (Light Detection and Ranging) and vertical aerial photos covering the coastal strip of England are also held by the Channel Coastal Observatory (CCO). Oblique aerial photographic surveys of the coast were conducted in 2003 for the Republic of Ireland (Office of Public Works (OPW)) and 2006 for Northern Ireland (Department of Environment Northern Ireland (DOENI) now the Department of Agriculture, Environment and Rural Affairs (DAERA)). Wave and tidal data are held in the Republic of Ireland by the Marine Institute (MI), in Northern Ireland by the Agri-Food & Biosciences Institute (AFBI) and for UK sites on both sides of the Irish Sea by the British Oceanographic Data Centre (BODC).

Text overviews of modern physical conditions and coastal geomorphology for Northern Ireland and eastern Irish Sea can be found in Barne *et al.* (1995; 1996a,b; 1997a,b) as part of the series "Coasts and Seas of the United Kingdom", produced by the Joint Nature Conservation Committee (JNCC). An overview of the Irish coastline, including its physical geography, geology, geomorphology and sea-level history, is provided in the *Coastal Atlas of Ireland* (Kozachenko *et al.* forthcoming). Physical conditions and their effect on seabed sediment are also dealt with on a broad scale by the regional offshore geology reports produced by the British Geological Survey (BGS) (Evans 1990; Fyfe *et al.* 1994; Tappin *et al.* 1994; Jackson *et al.* 1995). For the UK sector, the *Atlas of UK Marine Renewable Energy Resources* contains maps of modeled wave and tidal conditions (ABPMer 2008). Several exercises have been undertaken in coastal classification, primarily for coastal management or erosion modeling, such as the UK-level Futurecoast project (Burgess *et al.* 2004). Within the study area, there have been numerous studies of coastal change and published references that are too numerous to mention comprehensively in this report. These range from regional-scale overviews (e.g. Carter *et al.* 1987; Carter 1991; Knight 2002; May & Hansom 2003) to site-specific studies (e.g. Pye & Neal 1994; Moore *et al.* 2009). Good starting points are the aforementioned JNCC volumes and the May and Hansom (2003) overview. Finally, the Marine Environmental Data & Information Network (MEDIN) is a useful portal, providing links to organizations holding marine and coastal data for the UK, including all those already mentioned.

– Ordnance Survey of Ireland: www.osi.ie
– Land and Property Services: www.finance-ni.gov.uk/land-property-services-lps
– Ordnance Survey (Scotland, Wales, England, Isle of Man): www.ordnancesurvey.co.uk
– Agri-Food and Biosciences Institute: www.afbini.gov.uk
– Department of Agriculture, Environment and Rural Affairs (DAERA): www.daera-ni.gov.uk/
– Marine Institute: www.marine.ie/Home/
– Republic of Ireland Oblique Coastal Survey:
– www.coastalhelicopterview.ie/imf5104/imf.jsp?site=Helicopter
– Joint Nature Conservation Committee: jncc.defra.gov.uk
– Futurecoast: www.coastalwiki.org/coastalwiki/FUTURECOAST_project,_UK
– British Oceanographic Data Centre: www.bodc.ac.uk
– MEDIN Portal: www.oceannet.org
– UK Atlas of Marine Renewable Energy Resources: www.renewables-atlas.info
– Channel Coastal Observatory: www.channelcoast.org

Solid geology

The Quaternary sequence in the study area is underlain by a variety of rock types reflective of its geological past with different areas representing past episodes of climate change, volcanic activity, tectonic movements, sea-level fluctuation and erosion. The text below is a simplified summary based on the BGS and JNCC overviews (e.g. Jackson *et al.* 1995; Barne *et al.* 1996a) and online geological mapping resources (e.g. EMODnet). It provides only the broadest-scale overview and does not take into account local-scale variations. The reader is advised to consult the references and data sources for more detailed information and on- and offshore geological maps.

The North Channel between northeast Ireland and Scotland is floored by a sequence of mudstones, sandstone and halites deposited during the Triassic (250–203 million years ago (Ma)) by water-lain, eolian and evaporative processes. These differ from the coastal bedrock onshore, which in northeast Ireland consists either of Tertiary basalts (ancient lava flows) or Cretaceous (ca. 144–65 Ma) limestone (deposited in a shallow marine setting). However, the extensive chalk and basalt beds are not present on the eastern side of the North Channel, and are replaced by a complex series of deposits chiefly made up of Devonian (ca. 408–360 Ma) and Carboniferous (ca. 360–286 Ma) sedimentary rocks (sandstones, siltstones, conglomerates, limestone), and metamorphosed Dalriadan (ca. 800–600 Ma) schists. To the south of the Southern Uplands Fault, Silurian (ca. 438–408 Ma) sedimentary rocks (e.g. shale, mudstone, graywacke) can be found on both sides of the North Channel, but only extend offshore immediately off the Northern Ireland coast.

A broad divide runs approximately north–south through the central Irish Sea and Isle of Man separating it into western and eastern basins. The western basin is floored predominantly by Carboniferous mudstone and limestone which extend onshore in the vicinity of Dublin. Exceptions to this include nearshore areas where the aforementioned Silurian rocks are present, the Peel basin where Permo-Triassic (ca. 286–213 Ma) sedimentary rocks subcrop beneath the Quaternary sequence, and off Anglesey where Precambrian (ca. 600 Ma) schists, gneisses and igneous rocks are present. The eastern basin by contrast, is floored predominantly by younger Triassic mudstones and sandstones similar to those in the North Channel. Permo-Triassic sandstones also form the dominant coastal bedrock in Liverpool Bay with occasional Carboniferous sedimentary outcrops such as shale, limestone and sandstone. The Isle of Man itself is primarily Ordovician (ca. 505–438 Ma) slate with localized coastal outcrops of Carboniferous limestone and Triassic sandstone.

The southern sector of the Irish Sea, separating Ireland from Wales, is floored mainly by Triassic sandstones, siltstones and mudstones with the exception of Jurassic (ca. 213–144 Ma) mudstone and siltstone in Cardigan Bay. Jurassic sedimentary rocks also form the bedrock in the Bristol Channel. Onshore, however, the sequence is very different as both sides of the Irish Sea consist of older rocks. Northwest Wales, including Anglesey, comprises mainly Precambrian gneiss, schist and granite with some slate, the coast of west Wales is primarily Silurian and Ordovician slates and sandstones, while the south coast of Wales is generally Carboniferous limestone and shale. The eastern Irish coast too consists mainly of Ordovician mudstones and sandstones with some Precambrian gneiss and Carboniferous limestone in the extreme southeast.

The Celtic Sea basin is bounded by extensive Devonian sandstones with some Carboniferous limestone and sandstone on the south coast of Ireland and on the southwest coast of England by Devonian slates and sandstones and Carboniferous shales and sandstones, with occasional igneous intrusive granites (e.g. Land's End and the Scilly Isles). However, under the Quaternary cover of the seabed, the bedrock is younger with extensive areas mapped as Cretaceous chalk/limestone or Tertiary sedimentary rocks.

Unlike the Irish Sea, the Atlantic shelf off western Ireland is less well studied from a geological perspective, and lacks regional overviews such as the JNCC or BGS reports. Consequently, the seabed bedrock sequence is less well constrained. Large-scale maps produced for the EMODnet/OneGeology Europe databases suggest Devonian sandstone and Carboniferous sandstone and limestone off southwest Ireland; Precambrian gneiss or schist off western Ireland; and Carboniferous sedimentary rocks, Silurian metamorphic rocks, Precambrian schist and intrusive Devonian igneous rock off northwest Ireland. This ties in with the coastal bedrock in western Ireland which is predominantly sandstone and shale in the south-west; schist, gneiss and granite in the north-west

and limestone with localized metamorphic and igneous outcrops in the central part.

Data sources

National geological institutes are the Geological Survey of Ireland (GSI: Republic of Ireland), Geological Survey of Northern Ireland (GSNI: Northern Ireland), and British Geological Survey (BGS: England, Wales and Scotland). Note that while onshore data for Northern Ireland is held by the GSNI, offshore geology datasets are held by the BGS. The most comprehensive sources on offshore geology are the aforementioned BGS regional reports (e.g. Jackson *et al.* 1995). All the above institutes have online mapviewers showing data sources, extents, and in some instances the actual data are available for download (e.g. bedrock geology, surficial geology). Text overviews for Ireland can also be found in Sleeman *et al.* (2004), Mitchell and Ryan (2001), and Holland and Sanders (2009). For the UK, regional overviews include Mitchell (2004) for Northern Ireland and Howells (2008) for Wales. Many areas are also covered by geological maps down to 1:50,000 or 1:10,000 scale with accompanying memoirs (e.g. Wilson & Manning 1978), while the JNCC volumes have descriptive summaries of coastal geology.

- Geological Survey of Ireland digital data holdings: www.gsi.ie/Publications+and+Data/Digital+Data/ Available+Digital+Data.htm
- British Geological survey online data holdings: www.bgs.ac.uk/opengeoscience/home.html
- Geological Survey of Northern Ireland data: www.bgs.ac.uk/gsni/data/
- BGS Geology of Britain mapviewer: www.bgs.ac.uk/ discoveringGeology/geologyOfBritain/viewer.html

Bathymetry, sources of bathymetric data and digital archives

The Atlantic continental shelf to the west, north-west and south of Ireland consists of a broad, relatively level surface that slopes gradually down to the shelf break, situated up to 100 km to 200 km off the present coast. By contrast, off the northeast and east coast of Ireland, there is much greater topographic variation. A narrow (kilometers to low tens of kilometers) inshore/coastal platform drops off into a series of bathymetric deeps — the Rathlin Trough off the north coast, the North Channel between Northern Ireland and Scotland, and the Western Trough and St. George's Channel between Ireland and Wales — which collectively, are referred to as the Celtic Trough. This reaches widths of 30 km to 70 km and depths of 60 m to 120 m (Jackson *et al.* 1995). Within, and on the margins of, the Celtic Trough, isolated deeps incise even further into the seabed. Beaufort's Dyke for example has a maximum depth of 312 m (Callaway *et al.* 2011). On the eastern side of the Irish Sea, the seabed rises up to form a broad coastal platform such that most of Liverpool and Cardigan bays are characterized by water depths of

<50 m. To the south, the Celtic Sea shelf extends over 300 km to 400 km from the present coastline. Its overall morphology is broad and gently sloping, though there are a series of massive ridges inshore of the shelf break in depths of 100 m to 180 m (Fig. 10.2).

Data sources

A range of regional data sources are available for the Irish Sea and Atlantic Margin in addition to the global GEBCO and European EMODnet datasets. These include data digitized from nautical charts or obtained from single-beam echo sounder (SBES) surveys (10s of meters resolution) e.g. SeaZone Solutions Ltd or the BGS DigBath 1:250,000 scale vector map of bathymetric contours. Higher resolution (down to 1 m) multibeam

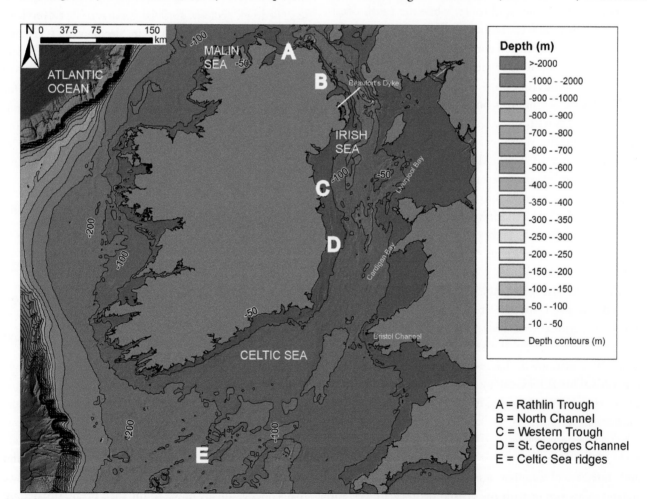

Fig. 10.2 Bathymetry within the Irish Sea and Atlantic Margin. The bathymetric Digital Terrain Model data products have been derived from the EMODnet Hydrography portal (www.emodnet-hydrography.eu).

echo sounder (MBES) and inshore bathymetric LiDAR surveys also cover large parts of the Irish Sea and Atlantic shelves.

Within Republic of Ireland territorial waters (i.e. Atlantic Margin and western Irish Sea), the MI and GSI are currently undertaking the Integrated Mapping For the Sustainable Development of Ireland's Marine Resource (INFOMAR) program. This is one of the largest seabed mapping projects in the world and, to date, has mapped several hundred thousand square kilometers of Ireland's seabed (Dorschel *et al.* 2011). Deeper shelf areas have been extensively mapped using MBES, with inshore bays and nearshore areas being filled in by MBES and bathymetric LiDAR during the current project phase, which is scheduled for completion by the end of 2016. These data are freely accessible for download via the INFOMAR website. Raw MBES data have also been archived with, and are downloadable from, the United States National Ocean and Atmospheric Administration (NOAA). At the time of writing the survey is incomplete, with the main gaps to be filled in the western Irish Sea and nearshore Atlantic Margin (Fig. 10.3).

Within UK waters, the Maritime and Coastguard Agency (MCA) has overall responsibility for the Civil Hydrography Programme (CHP), an ongoing exercise aimed at updating nautical charts. A Memorandum of Understanding (MoU) exists between various UK bodies collecting high-resolution bathymetric data (e.g. the UK Hydrographic Office (UKHO), BGS, JNCC, MCA) to exchange data and avoid duplicating survey efforts (Fig. 10.3).

In Northern Irish waters, cross-border partnerships between the MCA, MI and local agencies (NIEA, AFBI) include the Joint Irish Bathymetric Survey (JIBS: completed in 2008; Quinn *et al.* 2009) and Ireland, Northern Ireland and Scotland Hydrographic Survey (INIS Hydro) program (which also covers Scottish waters; ongoing at time of writing). Both have collected data under the auspices of the CHP. Mapping within the Irish Sea is much more limited, with only planned surveys in Cardigan Bay and isolated blocks in Liverpool Bay and around Anglesey (Van Landeghem *et al.* 2009a). CHP-collected data are available (at no cost under an Open Government License) via the UKHO INSPIRE

data portal. A limited quantity of data are freely available from other portals, such as the CCO, which holds some inshore MBES and SBES data for English waters. High-resolution data from commercial projects (e.g. offshore wind farms, seabed cables and pipelines) also exist within the study area, and access generally requires contacting the relevant developer or surveyor. However, data access is improving via the Crown Estate Marine Data Exchange portal, which archives survey data and reports from offshore renewable projects on the UK shelf.

- British Geological Survey bathymetry: www.bgs.ac.uk/products/digbath250/
- INFOMAR webpage: www.infomar.ie
- INIS Hydro webpage: www.inis-hydro.eu
- UK Hydrographic Office: www.gov.uk/government/organisations/uk-hydrographic-office
- Maritime and Coastguard Agency: www.gov.uk/government/organisations/maritime-and-coastguard-agency
- UK Pan-Governmental Hydrographic data sources: www.gov.uk/guidance/share-hydrographic-data-with-maritime-and-coastguard-agency-mca
- UKHO INSPIRE Portal: aws2.caris.com/ukho/mapViewer/map.action
- Crown Estate Marine Data Exchange: www.marinedataexchange.co.uk
- NOAA Bathymetry data portal: maps.ngdc.noaa.gov/viewers/bathymetry/

Pleistocene and Holocene sediment thickness on the continental shelf

The most detailed overviews of offshore geology are the BGS regional reports which cover the UK sector only. In the study area these are: Malin-Hebrides Sea (Fyfe *et al.* 1994), Irish Sea (Jackson *et al.* 1995), Cardigan Bay–Bristol Channel (Tappin *et al.* 1994) and Western Approaches (Evans 1990). Much of the Pleistocene and Holocene geological information within these has also been summarized in Strategic Environmental Assessment (SEA) reports by Flemming (2005) for the Irish Sea, and Wickham-Jones and Dawson (2006) for western Scotland.

Fig. 10.3 Coverage of INFOMAR, CHP and other UK, pan-governmental MoU organizations (BGS, Ministry of Defence, CEFAS — Centre for Environment, Fisheries and Aquaculture Research, JNCC, Natural England, Crown Estate) multibeam bathymetry surveys in the Irish Sea and on the Atlantic Margin. Note that coverage is representative of the time of writing (2013–2016), and updates can be obtained from the INFOMAR and UKHO INSPIRE data portals. See text for weblinks.

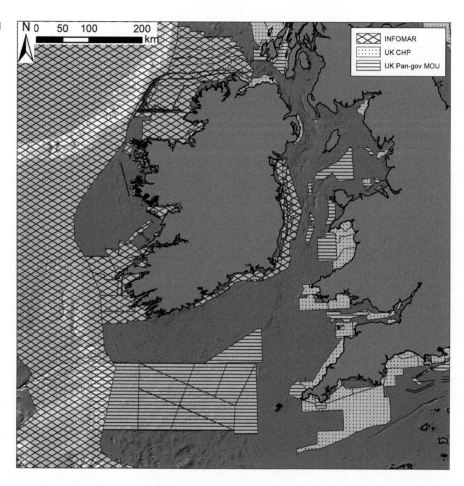

The Irish Sea preserves variable thickness of Quaternary sediment with the thickest deposits (up to 300 m) found in the Western Trough versus areas of thin or absent Quaternary deposits concentrated off northeast Ireland and Anglesey. The platforms flanking the trough are estimated to have 50 m or less thickness with the exception of infilled depressions east of the Isle of Man which have localized thicknesses up to 200 m (Jackson *et al.* 1995). To the south, in St. George's Channel, the deeply infilled central trough continues (generally 100–200 m with localized thicknesses up to 375 m) and is flanked by platforms with <50 m sediment thickness. Large areas absent of Quaternary sediment are mapped within and at the mouth of the Bristol Channel (Tappin *et al.* 1994) (Fig. 10.4).

Nowhere in the Irish Sea basin preserves a complete Quaternary succession, and all areas have some evidence of erosion surfaces (Fig. 10.4). This is unsurprising since

the basin was probably glaciated on multiple occasions during the Pleistocene, and probably also experienced multiple marine transgressions, either due to isostatic depression, or interglacial sea-level highstands. The oldest Pleistocene sediments range in age from possible pre-MIS 12 (Marine Isotope Stage 12) (Bardsey Loom Formation) to MIS 12 (Caernarfon Bay Formation) buried over 100 m deep in St. George's Channel (Tappin *et al.* 1994; Jackson *et al.* 1995). These have been interpreted as fluviatile/shallow marine and glacigenic respectively. Glacial (including subglacial, glaciomarine and glaciolacustrine sediments) constitute a large component of the Pleistocene sequence. These include Early Saalian marine/glaciomarine deposits (St. George's Formation), Saalian and Weichsalian tills (Cardigan Bay Formation Upper and Lower Till members) and Late Weichsalian glaciomarine and glacio-lacustrine deposits (Western Irish Sea Formation) (Tappin *et al.* 1994; Jackson *et al.*

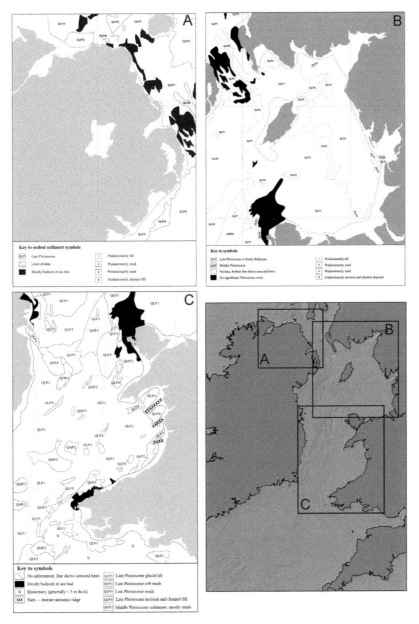

Fig. 10.4 Quaternary seabed sediment distribution in the Irish Sea. From Barne *et al.* (1995; 1996a; 1997b), based on BGS data, reproduced with permission of the JNCC.

1995). The most recent unit — the Surface Sands Formation — has some evidence of sub-aerial deposition, particularly in its lower member: Sediment Layer 2 (SL2). Much of this is located in the eastern Irish Sea, though boreholes from the Kish Bank on its western side show intertidal muds stratified below modern marine sands (Jackson *et al.* 1995).

Tidal streams exert a strong influence on sediment distribution in the Irish Sea, with powerful currents funneled through the North and St. George's channels creating areas of bed-load parting. These zones of net erosion are characterized by gravelly substrates, in contrast to the areas of slack tide situated in between where muddy sediments have accumulated: the western and eastern Irish Sea mud belts located off northeast Ireland and northwest England respectively. The variation in accumulation versus erosion is also illustrated by the thickness of the Seabed Sands Formation, which ranges from absence (resulting in the western Irish Sea or Cardigan Bay Upper Till Formation outcropping at the sea floor) to over 40 m thick (Tappin *et al.* 1994; Jackson *et al.* 1995).

Quaternary sediment thicknesses for the Malin-Hebrides area are reduced compared to the Irish Sea, generally less than 20 m with localized increases up to 50 m. The thickest deposit reportedly consists of the Jura Formation: Late Pleistocene/Early Holocene marine-deposited sediment. However, the remainder of the sediments are mapped as undifferentiated or thin Quaternary (Fyfe *et al.* 1994). This may partly be a product of strong tidal currents, which sweep across the north coast of Ireland at speeds of 1 to >2 m/sec at spring tides. Consequently, large areas are interpreted as sandy gravel or gravel (which may represent outcropping Pleistocene till), or rock outcrop. This certainly appears to be the case for the North Channel which is subject to particularly strong tidal currents which have scoured out much of its sedimentary (including Quaternary) cover (Fyfe *et al.* 1994). The recently acquired JIBS dataset has also highlighted the dynamic conditions on this shelf, displaying numerous large (meters to 10s of meters-scale) bedforms at depths over 100 m, exposed rock or gravel around headlands and cliffs and numerous upstanding rock pinnacles (Quinn *et al.* 2009).

For the Celtic Sea, Evans (1990) reports a break between Late Pliocene/Early Pleistocene Little Sole Formation and Late Pleistocene/Early Holocene sediments (the Melville Formation, Layers A and B) with the hiatus attributed to glacial erosion. The exceptions are deep sequences along the continental slope which were probably too deep to be exposed during lowstands (but which could preserve evidence pertaining to glacial and paleo-oceanographic change) and infilled depressions on the shelf such as the Hurd Deep and southern end of the Celtic Trough. The infill in the former is between 80 m to 137 m thick while that in the latter is up to 200 m thick and includes pre-LGM units (Evans 1990). The Melville Formation forms the area of massive sand ridges on the south-west margin of the shelf but thins to the north and east. The other Holocene layers in the study area — Layers B and A — are respectively regarded as a thin lag deposit created by marine transgression and a subsequent post-transgression/marine deposit reworked from pre-existing sediment and deposited under modern conditions (Evans 1990).

In addition to the overviews, there are also localized studies which examine seabed geology (though not necessarily sediment thickness specifically) at a finer resolution through some combination of grab, core, seismic and bathymetric data. These tend to focus on specific research questions, such as sea-level change (e.g. Kelley *et al.* 2006; Roberts *et al.* 2011), past glaciation (e.g. Gallagher *et al.* 2004; Callaway *et al.* 2011), sediment dynamics (e.g. Van Landeghem *et al.* 2009b), seabed mapping (Wheeler *et al.* 2001; McDowell *et al.* 2005) or habitat mapping (e.g. Callaway *et al.* 2009; Plets *et al.* 2012).

On the Atlantic shelf off Ireland, the thickness and origin of seabed sediment are less well-constrained and detailed overviews do not exist. Extant localized studies have focused more on the shelf surface (e.g. glacial bedform mapping: Benetti *et al.* 2010; Dunlop *et al.* 2010; Ó Cofaigh *et al.* 2012b) than the Quaternary sequence. Rough seabed substrate maps are available from habitat mapping projects (e.g. EUSeaMap). These show extensive areas of coarse-grained sediment off northwest Ireland with increasing sand content in large indented bays (e.g. Donegal Bay, Lough Swilly), but provide little information for the rest of the shelf. It is likely though that coverage will improve in the near future because of the INFOMAR seabed mapping program.

Data sources

Offshore geological records include seismic profiles, bathymetry and backscatter data, grab samples and core/borehole samples. Useful derived products include substrate, sediment thickness and geological maps and vertical profiles or sections.

Offshore geological records for the UK sector of the Irish Sea (including Northern Ireland) are held by the BGS, while those for the Republic of Ireland are held by the GSI. At present, records searchable and available online for Republic of Ireland waters consist of survey locations, bathymetry and backscatter data (both derived from MBES), and grab samples and seabed classification maps (derived from backscatter and ground-truth samples). The location of seismic profile surveys conducted by the GSI/MI are the same as the MBES surveys (i.e. a 3.5 kHz profiler was run alongside the MBES), but the actual profiler data must be obtained directly from the GSI. Core/borehole logs must also be obtained direct from the GSI.

The BGS offers a searchable web portal (the Geo-Index) and identifies the location of acoustic surveys, grab samples and core/borehole data. Some data has been digitized and is freely accessible, but some records must be obtained direct from the BGS. Some metadata on both acoustic and sampling surveys is available via the EU-SEASED and EU-CORE portals. Another web portal is provided by the MAREMAP project, a collaborative venture between several UK institutions to collate and make accessible seabed mapping data. At present, this web portal offers basic geological information, such as 1:250,000 scale seabed sediment maps covering the entire UK.

Maps of basic substrate and wave/tidal energy at the seabed have been produced by biological habitat mappers because these represent some of the physical input parameters for broad-scale modeling of seabed habitats. In isolation, these data do not provide direct evidence for submerged prehistoric landscapes. However, they give an indication of modern conditions which could be important for assessing preservation potential (e.g. areas of stripped bedrock versus zones of accumulated fine sediment) while the original sources of the data (e.g. borehole records, bathymetric measurements) could be reused for submerged landscape purposes. Key projects include UKSeaMap (covering the UK sector of the study area) and the EUSeaMap (covering the Irish sector of the study area). At present, the maps have best coverage for the well-mapped areas in the UK sector and support the above description of seabed geology (unsurprising since they are based on the original BGS datasets).

Key text overviews for the UK sector come from the BGS's Offshore Regional Reports (Evans 1990; Fyfe et al. 1994; Tappin et al. 1994; Jackson et al. 1995). These contain descriptions of seabed geology along with paper maps and sections. Equivalent reports do not exist for Republic of Ireland waters; the recently published seabed atlas for the Irish shelf (Dorschel et al. 2011) deals primarily with deepwater (>200 m) areas.

Numerous geophysical and geotechnical surveys have also taken place in the study area for commercial/industrial purposes. However, the data and associated metadata remain dispersed among various commercial organizations and are therefore difficult to access. There are links on the UK Government's Department for Business, Energy & Industrial Strategy website to organizations which may be willing to supply the data at cost. Another portal which also provides a catalogue of information (including seismics) collected by the oil and gas industry is the UK Oil and Gas Data site; similar data and reports from offshore renewable projects are archived at the Crown Estate Marine Data Exchange.

- Geological Survey of Ireland digital data holdings: www.gsi.ie/Publications+and+Data/Digital+Data/Available+Digital+Data.htm
- British Geological survey offshore data: www.bgs.ac.uk/GeoIndex/offshore.htm
- MAREMAP: www.maremap.ac.uk/index.html
- Department for Business, Energy & Industrial Strategy data access page: og.decc.gov.uk/en/olgs/cms/data_maps/data_release/data_release.aspx
- UK Oil and Gas Data: www.ukoilandgasdata.com
- UKSeaMap 2010: jncc.defra.gov.uk/page-2117
- EUSeaMap: jncc.defra.gov.uk/page-5020

Post-LGM Climate and Sea-Level Change

Climate change and ice-sheet evolution

The Irish Sea and Atlantic Margin experienced repeated Pleistocene glaciations and complex associated patterns of sea-level change and continental shelf exposure driven by the interplay between global eustasy and regional isostasy (Fig. 10.5). As on the rest of the Northwest Shelf, the pattern of ice growth/decay and sea-level change is best documented for the Weichselian/MIS 2. Evidence for preceding glacial periods is fragmentary, with only isolated sites surviving Last Glacial erosion even in onshore areas.

Fig. 10.5 Retreat of the British–Irish ice sheet at selected time steps covering the LGM and postglacial. Note that this is only one scenario of change. Clark *et al.* (2012a: fig 18); image courtesy of Chris Clark, reproduced with permission of Elsevier.

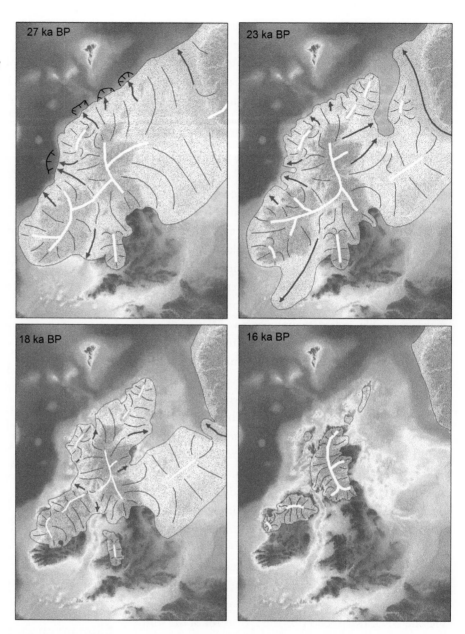

Earlier studies reconstructed a largely terrestrially-based British-Irish ice sheet with parts of Scotland (e.g. Caithness and the Shetland and Orkney Isles) and southern Ireland regarded as ice-free (e.g. Bowen *et al.* 1986; 2002). However, the latest research indicates a much larger ice sheet with approximately one-third of its extent on the continental shelf and reaching the shelf edge (Clark *et al.* 2012a). Importantly, the entirety of the study area (Irish Sea basin and Atlantic Margin including the Celtic Sea) is now believed to have been ice-covered during the last glacial cycle (Fig. 10.5).

The extensive growth of ice over the British Isles during the Last Glacial probably began after ca. 35 ka. This ice is believed to have reached the Atlantic shelf edge by 29 ka to 27 ka and is estimated to have reached its maximum extent across the British Isles by 27 ka. At this point it covered all of Ireland and most of Britain with the exception of southern and central England. Offshore, the Atlantic Margin, Irish Sea and most of the North Sea, where it was confluent with Scandinavian ice, were also covered (Greenwood & Clark 2009; Scourse et al. 2009; Ballantyne 2010; Clark et al. 2012a; Ó Cofaigh et al. 2012a).

Retreat of the ice sheet is currently believed to have been asynchronous and non-monotonic. Ice started to retreat from the Atlantic shelf edge from 26 ka; however, a short-lived advance of ice from the Irish Sea to the Scilly Isles took place between 25 ka and 23 ka before rapidly retreating again. By 19 ka, the Celtic Sea, southeast Ireland and southwest Wales were ice-free. The British and Irish ice sheets had started to 'unzip' such that the southern Irish Sea was deglaciated. Retreat from the shelf edge continued with ice margins moving closer to modern shorelines, though with large ice lobes still situated off the west and north-west of Ireland. Continued retreat saw the south of Ireland ice-free by 18 ka, and an independent Welsh ice cap. Parts of the coastal fringes of northwest Ireland were also by now ice-free. From 18 ka to 17 ka on, the marine-based ice sheets broke up even further and by 16 ka the North Sea, Irish Sea and Atlantic Margin were completely deglaciated and the remaining ice was located on land across Scotland and the north of Ireland. Finally, by 15 ka, the formerly vast ice sheet was restricted to upland/highland ice and minor caps on some of the Scottish isles and northwest Ireland (Hughes et al. 2011; Clark et al. 2012a; see also Fig. 10.5).

Within the overarching pattern of advance and retreat, evidence also suggests periodic re-advances between 19 ka and 15 ka. For instance, the Armoy moraine across northeast Ireland records a renewed incursion of Scottish ice post-dating the ice-sheet maxima, while several sites on the northeast coast of Ireland also record local expansion of Irish ice (McCabe & Dunlop 2006). Stillstands and possible oscillations in ice retreat can also be inferred from sequences of moraines preserved on the shelf off northwest Ireland (Ó Cofaigh et al. 2012b). Note however, that it is still not certain whether these re-advances were local events or part of a pan-ice sheet response to climate changes in the wider North Atlantic, such as Heinrich events (Clark et al. 2012b). It is also worth noting that, at the time of writing, the British-Irish ice sheet is the subject of a major research project (BRITICE-CHRONO), which includes geophysical survey and geotechnical sampling of the continental shelf. It is therefore possible that the above pattern of ice sheet growth and decay will be refined in the near future.

Deglaciation continued, and peaked during the Bølling–Allerød/GI-1 Interstadial (ca. 12.9–14.7 ka). As in the rest of northwest Europe, climate then deteriorated following this warm period. This period — the Younger Dryas/GS-1 interval (12.9–11.5 ka) — was characterized by cold arid conditions and limited mountain glaciation. Following this, rapid warming resumed at the onset of the Holocene reaching, and then exceeding (by ca. 1–3°C), modern temperatures during the climatic optimum between ca. 9 ka to 4 ka (Bell & Walker 2005).

The extension of ice onto the shelf has been comprehensively demonstrated by high-resolution seafloor mapping which shows glacial features (e.g. moraines, drumlins) preserved on the Irish Sea and Atlantic shelves (e.g. Van Landeghem et al. 2009a; Dunlop et al. 2010; Ó Cofaigh et al. 2012b). Onshore evidence, such as subglacial bedforms running into the coastal zone, is also indicative of ice sheets extending off the present coastline (e.g. Greenwood & Clark 2009). However, the majority of the offshore geomorphological evidence is not directly dated, and therefore most of the timing and pattern of changes described above currently relies on dated measurements of fluctuations in ice rafting (seen as a proxy of ice growth/decay) from deep-sea cores (Scourse et al. 2009). Moreover, the level of mapping is still variable, with some areas better mapped and studied than others (see section on bathymetry, pages 245–247).

Data sources

Most of the information on climate and ice-sheet development comes from the published literature, such as the references above. Additional sources include the BRITICE database, which contains maps of onshore glacial features and Geographic Information System (GIS) layers for the British ice sheet, and the BRITICE-CHRONO database, which contains dates relevant to reconstructing the last glacial ice advance/retreat over the British Isles.

– BRITICE: www.shef.ac.uk/geography/staff/clark_chris/britice
– BRITICE-CHRONO (database and project): www.britice-chrono.group.shef.ac.uk/

Sea-level and paleogeographic change

Given the uncertainties in ice-sheet extent and chronology, its effects on RSL are still debated. This is exacerbated by spatial and temporal variations in the distribution of datable evidence of past sea level, in the form of Sea-Level Index Points (SLIPs) and limiting dates. Summary reviews of RSL change in Britain and Ireland are available in Shennan *et al.* (2006a) and Edwards and Craven (2017). There are at present far fewer data points from the continental shelf and for the pre-Holocene than from onshore areas (Shennan *et al.* 2002; Brooks & Edwards 2006). It is therefore not possible at present to produce an accurate and consistent paleogeographic reconstruction for the study area using the datable evidence alone.

An alternative approach is to derive shelf-scale reconstructions of paleogeographic change from glacio-isostatic adjustment (GIA) models. These calculate the Earth's crustal response to changing ice and water loads and, in conjunction with glacio-eustatic data, offer the ability to model past RSL change. While there are still uncertainties in these models, not least in terms of the input earth and ice models (see for example the differences between Bradley *et al.* (2011) and Kuchar *et al.* (2012)), they represent the best available means of reconstructing regional- to continental-scale paleogeography. The most recent models (Brooks *et al.* 2008; Bradley *et al.* 2011; Kuchar *et al.* 2012) have indicated variable patterns across the Irish Sea and Atlantic Margin relating to regional differences in ice loading and the timing and pattern of deglaciation. Where the ice was thickest, the crust was depressed by up to several tens of meters below contemporary sea level. The resultant pattern of rebound as the ice retreated then varied across the study area based on the local thickness of ice, length of glaciation and timing of retreat. Generally,

the greatest rebound concentrated in the northern half of the Irish Sea basin and the northeastern fringe of the Atlantic Margin (i.e. northeast Ireland, southwest Scotland and northwest England) due to the combined effects of thicker, longer-duration and local ice loading, and closer proximity to the major center of regional ice loading over Fennoscandinavia. Rebound decreased to the south and south-west with increasing distance from the former centers of loading (Brooks *et al.* 2008; Bradley *et al.* 2011; Fig. 10.6). Rebound rates were up to several mm/year during the postglacial and were, across much of the northern half of the Irish Sea, capable of offsetting the global eustatic rise for part of the postglacial.

The following general description of sea-level and paleogeographic change after 20 ka is based on Brooks

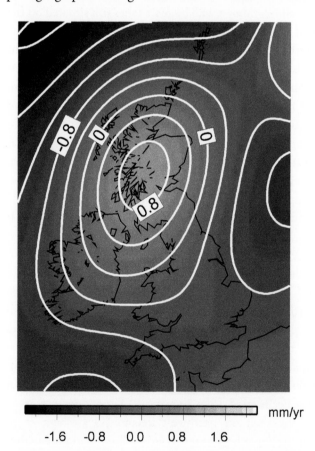

Fig. 10.6 Modeled present-day vertical isostatic uplift of the British Isles. Note the centre of uplift over Scotland and the northern Irish Sea basin, and the trend of decreasing uplift to the south. From Bradley *et al.* (2011: fig. 9b); reproduced with permission from John Wiley & Sons.

et al. (2008) (Fig. 10.7). Two additional modeling studies are available for the region: Bradley *et al.* (2011), which uses the Brooks *et al.* (2008) ice model with an updated eustatic term; and Kuchar *et al.* (2012), which uses alternative ice models produced by the numerical ice-sheet modeling of Hubbard *et al.* (2009). Whilst the general patterns of RSL change produced by these models are similar, differences exist in the precise values of former RSL, and these can be taken as an informal indication of potential model uncertainty.

The highest simulated RSL are attained in south-west Scotland (Knapdale) and the northeast of Ireland (northern Antrim) at 20 ka owing to significant isostatic depression in this area. At Knapdale, an RSL highstand of +40 m is simulated, with a corresponding highstand in northern Antrim at around +30 m. Despite rising eustatic (global) sea levels, rapid isostatic rebound in this part of the study area resulted in falling RSL until the start of the Holocene, although this trend was briefly interrupted around 14.5 ka by Meltwater Pulse 1a (MWP-1A) (Deschamps *et al.* 2012). At both sites, the Holocene is characterized by RSL rise from a lowstand of between – 6 m (Knapdale) and –10 m (northern Antrim) but, owing to the ongoing isostatic rebound in this area, simulated

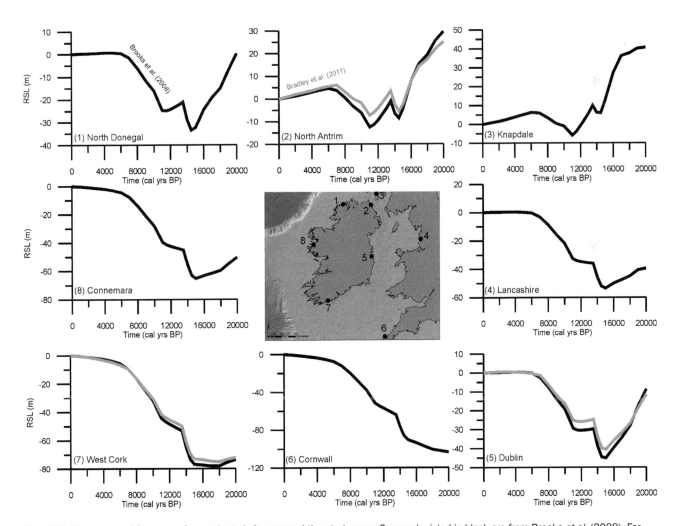

Fig. 10.7 GIA-modeled RSL curves from selected sites around the study area. Curves depicted in black are from Brooks *et al.* (2008). For comparative purposes, curves from Bradley *et al.* (2011) have been plotted in gray for three of the sites: (2) northern Antrim; (5) Dublin; and (7) West Cork. Data courtesy of Tony Brooks and Sarah Bradley.

RSL peaked in the middle part of the Holocene around 5 m to 6 m above present, before falling to present-day levels.

In contrast, the lowest simulated RSL are attained in the southernmost portion of the study area (Cornwall) where simulated RSL at 20 ka was more than 100 m below present. Simulated RSL then rose continuously during deglaciation and throughout the Holocene with rates of rise diminishing around 6 ka, reflecting the general cessation of global ice melt by this time. In the Bradley et al. (2011) model, the revised eustatic curve extends ice melt resulting in a slight shift in the simulated RSL curves (Fig. 10.7). This general pattern of continuous RSL rise is mirrored in southwest Ireland (west Cork) although in this instance, local isostatic depression causes a slight initial RSL fall and an extended lowstand around –80 m.

At intermediate sites such as Lancashire, Dublin, and Connemara, a similar pattern of simulated RSL change to that of west Cork is apparent, with differences in the magnitude of the initial RSL fall and the extreme depths of the lowstands reflecting the overall geometry of ice loading within the region. In all cases, simulated RSL fall is terminated by MWP-1a, after which RSL rises from the lowstands of –50 m (Lancashire), –45 m (Dublin) and –65 m (Connemara), attaining its modern position by around 6000 years ago in the case of the former two locations, and around 1000 years ago in the case of the latter. Significantly, simulated RSL does not rise at a uniform rate after MWP-1a, and all three intermediate sites show an extended interval of RSL stillstand or slow rise between ca. 13.5 ka and 11 ka, at depths of around 30 m (Lancashire and Dublin) and 40 m (Connemara). It should be noted that all sites within this region have simulated RSL histories that significantly depart from the generalized eustatic global sea-level curve, due to GIA effects.

When combined with modern bathymetry (and notwithstanding errors caused by post-transgression erosion and deposition), simulated RSL can be used to provide approximate maps of paleogeographic change (Edwards & Brooks 2008; Brooks et al. 2011; Sturt et al. 2013; Fig. 10.8). These suggest that Late Glacial shelf exposure on the Atlantic Margin was relatively small, of the order of a few kilometers to less than 30 km around

Ireland. Larger areas were exposed in the eastern Irish Sea, such that most of Liverpool Bay, Cardigan Bay and the Bristol Channel were sub-aerial. Even so, it is apparent that the majority of the central Irish Sea, Celtic Sea and western Atlantic Margin were not exposed sub-aerially. The extent of exposed shelf is predicted to have remained relatively stable until the onset of more rapid sea-level rise from ca. 16 ka to 15 ka. As land in the south and west was starting to flood, however, the northern sector of the study area, which had been isostatically depressed, was now rising above contemporaneous sea level. This lowstand, however, was relatively shallow (<ca. –30 m) and short lived. Consequently, by the Holocene (ca. 11–10 ka), exposed shelves were limited to fringing strips around the modern coast, and infilled areas that are now modern bays, such as the Bristol Channel. By the Mid Holocene, modern paleogeography was largely attained. Importantly, the modeled paleogeographies suggest that there was never a postglacial terrestrial connection linking Britain and Ireland. This contrasts with previous research which had suggested a land bridge between southeast Ireland and southwest Wales created by the migration of a glacial forebulge from the shelf edge back towards the center of isostatic depression around 11 ka to 10 ka (Wingfield 1995).

In addition to the GIA-based scenario, an alternative interpretation of RSL change (the 'glaciomarine' hypothesis) around the Irish Sea basin has been developed from onshore field evidence, namely raised Late Pleistocene marine sediment, primarily from the north of Ireland (McCabe et al. 2007; Clark et al. 2012b). This initially agrees with the GIA models in that both reconstruct high sea levels immediately following deglaciation (ca. +30 m above present sea level for the field evidence versus +15–30 m for the GIA models: see Edwards et al. 2008). Thereafter, they differ considerably with the field evidence used to infer a rapid drop to a –15 m lowstand below present sea level at 20 ka, then a rapid rise above present by 19 ka. The period from 19 ka to 14 ka, when the GIA models predict continuously falling RSL, is interpreted as a series of highstand (up to +20 m) 'sawtooth' oscillations resulting from rapid variations in isostatic depression/uplift driven by rapid re-advance and retreat of the British-Irish ice sheet. Following this, RSL

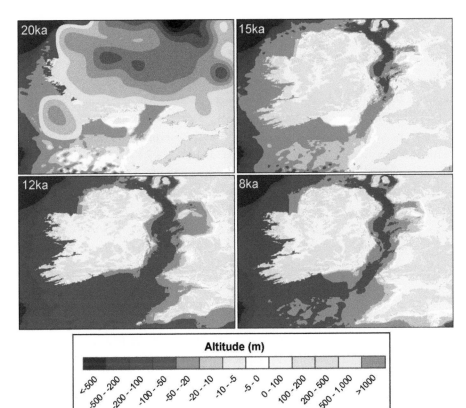

falls to a lowstand of ca. 30 m at 13.5 ka before rising to present levels. Proponents of this interpretation assert that such rapid changes are simply not reproduced by the GIA models (McCabe *et al.* 2007; McCabe 2008; Clark *et al.* 2012b). Interestingly, the supporting evidence seems to be restricted to Ireland, with the Welsh and Celtic Sea evidence more supportive of lower (though not necessarily exactly the same) sea level as predicted by the GIA models (Knight 2001; McCarroll 2001; Scourse & Furze 2001; Roberts *et al.* 2006). From a submerged landscape perspective, the implications of the glaciomarine hypothesis are that post-LGM sub-aerial landscape exposure may have been relatively short, and even more limited in extent than predicted by the GIA models. Both viewpoints agree that Ireland had become separated from Britain no later than 15 ka (Brooks *et al.* 2011; Montgomery *et al.* 2014).

The debate is presently unresolved and highlights the need for additional evidence based on dated lowstand SLIPs and limiting dates. At the time of writing, new RSL data from submerged contexts intended to fill this gap is currently being generated by the Natural Environment Research Council (NERC)-funded Sea Level Minima project (Cooper 2012) and should therefore be available in the near future. It can also be expected that the models will undergo refinement as new evidence on the lateral extent and chronology of the British-Irish ice sheet comes to light (e.g. Clark *et al.* 2012a; Ó Cofaigh *et al.* 2012a,b). Therefore, the main conclusion at present is that we can expect continental shelves to have been exposed to varying degrees during the Late Pleistocene and Early–Mid Holocene and that the degree and timing of exposure varied around the study area, but the precise pattern has yet to be conclusively reconstructed.

Data sources

Databases of all known Sea-Level Index Points (SLIPs) and limiting dates for Britain and Ireland are currently held at the University of Durham and Trinity College (Dublin) respectively and are described in Brooks and Edwards (2006) and Shennan and Horton (2002).

Several generations of GIA models covering the British Isles have been developed over the past two decades (e.g. Lambeck 1995; Lambeck & Purcell 2001; Shennan *et al.* 2006a,b; Brooks *et al.* 2008; Bradley *et al.* 2011). The models periodically undergo revision and attempt to take into account new information on local RSL change, regional ice extents and global factors such as eustatic volume changes and gravitational fluctuations when these become available. Lambeck (1995), Shennan *et al.* (2002), Brooks *et al.* (2011) and Sturt *et al.* (2013) also provide time-stepped sequences of paleogeographic maps based on the models.

Sea-Level Minima project: sealevelminima.weebly.com

Shelf climate

Environmental conditions on the shelf varied over the period of exposure due to changes in global climate and fluctuations in the position of the British-Irish ice sheet. Prior to the LGM, the British Isles had experienced multiple alternations between cold glacial and warm interglacial stages with limited evidence for shorter stadial or interstadial episodes within the longer glacial/interglacials (Mitchell & Ryan 2001).

By 29 ka to 23 ka, large parts of the shelf were under ice and nearshore regions presumably characterized by extensive sea ice. As the ice retreated between 19 ka and 15 ka, climates became periglacial. O'Connell *et al.* (1999) for example estimate that winter temperature in southwest Ireland during the final phase of deglaciation (ca. 16.8–15 ka) was probably lower than –20°C and not more than 5°C during the warmest month. This cold period was ended by rapid warming at the onset of the GI-1 interstadial (ca. 14.7 ka). Chironomid-based reconstructions from northwest Britain and Ireland are consistent in showing a July temperature rise from ca. 6–7°C to ca. 12–14°C, close to modern values for all the study sites. Interstadial temperatures remained relatively warm, though with short centennial-scale cold oscillations of a few degrees. The interstadial was followed by renewed cooling during the GS-1 stadial (ca. 12.9–11.5 ka) and is reflected by a fall to ca. 6–8°C for July temperatures (Brooks & Birks 2000; Lang *et al.* 2010; Watson *et al.* 2010; van Asch *et al.* 2012). The stadial in turn was terminated by a rapid Holocene warming to near present-day temperatures (Fig. 10.9). This pattern of climate change is broadly similar to the rest of northwest Europe and is substantiated by a range of proxies including beetles (Coope *et al.* 1998), pollen (O'Connell *et al.* 1999) and stable isotopes from sites within the study area (e.g. Diefendorf *et al.* 2006) and regionally from the Greenland ice cores (e.g. Rasmussen *et al.* 2006). Through the Early–Mid Holocene, climate

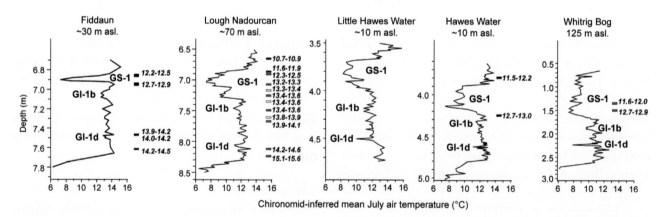

Fig. 10.9 Reconstructed July temperatures from chironomid data from sites across the study area. Fiddaun is in western Ireland, Lough Nadourcan is in northwest Ireland, Haweswater is in northwest England, and Whitrig Bog is in southeast Scotland. All curves clearly show the cold GS-1 Stadial intervening between the warmer Holocene and GI-1 Interstadial. Dates in calibrated years (ka) are shown on the right-hand side of each temperature curve. From van Asch *et al.* (2012: fig. 7); image courtesy of Nelleke van Asch; reproduced with permission from Elsevier.

remained broadly stable (except for a short, rapid cold episode at 8.2 ka) and potentially slightly warmer than the modern Atlantic-influenced climate that prevails throughout the British Isles.

All the above evidence has come from presently onshore regions rather than the continental shelf. However, given the wide-ranging nature of the climate changes (e.g. similar changes recorded across the British Isles, limited extent of shelf exposure relative to terrestrial areas), we can expect shelf climatic conditions to have been very similar. Precise confirmation would however require assessment of suitable paleoclimate proxies obtained from the shelf itself.

Shelf paleoenvironments (floral evidence)

Amongst the key data sources for reconstructing past climate changes are fossil plant evidence including macro- and microfossils typically from stratified dated sequences accumulating in depressions, lakes, bogs and peat. High rainfall and the existence of many depressions (some of which were created by glaciation, e.g. kettle holes) means that there are numerous onshore sites scattered across the study area that are suitable for the preservation of paleobotanical evidence. This is particularly true of Ireland where peat and bog expanded considerably during the Holocene.

Many such deposits have undergone extensive investigation over the past century (e.g. Erdtman 1928; Jessen 1949; Godwin 1975; Lowe *et al.* 1995). Consequently, the Late Pleistocene and Holocene are reasonably well constrained in terms of chronology (both radiocarbon and dendrochronological) and pattern of vegetation change, which clearly record the climatic changes described above (Walker 1995; Pilcher *et al.* 2008). The initial colonizing vegetation reaching western Britain and Ireland during initial deglaciation consisted of cold-adapted open herb-shrub communities (e.g. *Artemesia*). Climate warming during the GI-1 Interstadial (ca. 14.7–12.9 ka) then saw transitions to low scrub and heathland (e.g. *Salix*) and then to juniper (*Juniperus*) woodland. Some sites also record short-lived colder oscillations within the interstadial. By the end of the warm interval, open scrub and grasslands had returned. These were in turn

replaced by open tundra/alpine species (e.g. *Artemesia*) during the GS-1 Stadial (ca. 12.9–11.5 ka). The Holocene warming was then typified by forest development, firstly open juniper and birch woodland which then gave way to increasing quantities of hazel (*Corylus*) and pine (*Pinus*). These in turn gave way to increasing oak (*Quercus*) and elm (*Ulmus*). A subsequent alder (*Alnus*) expansion occurred around ca. 7 ka to 8 ka while mixed hazel-oak-elm-alder forests developed in lowlands, with birch and pine restricted to highland areas (see overviews in Walker 1995; Mitchell & Ryan 2001; Coxon 2008).

Obviously, there were local variations in the timing and pattern of vegetation change depending on local factors such as shelter, rainfall, soil development, and paleogeography. For example, Innes *et al.* (2004) document a delayed Early Holocene birch expansion on the Isle of Man relative to the rest of the northern Irish Sea region which they attribute to soil instability and RSL rise severing the terrestrial connection between the island and the mainland. Mitchell (2006) points out that birch and juniper were extirpated in Ireland during GS-1 but survived locally in Britain and France, thus permitting rapid migration back into Ireland during the Early Holocene. In addition, Mitchell (2006) also points out that the flooding of the Irish Sea acted as a filter to re-colonizing vegetation, with numerous species such as beech (*Fagus*) and lime (*Tilia*) reaching Britain but not Ireland.

The numerous peats, bogs and lakes around Ireland and western Britain mean that the extant pollen and plant macrofossil records are derived overwhelmingly from terrestrial rather than submerged contexts. The exceptions are the Pleistocene deposits from Cork Harbour (a nearshore estuarine location in an enclosed bay rather than fully offshore (Dowling & Coxon 2001)), nearshore cores in the Menai Strait (Roberts *et al.* 2011), Cardigan Bay (Haynes *et al.* 1977), and various intertidal peats and forests dotted around the shores of the study area (e.g. Bell 2007; Simpson 2008; Wilson & Plunkett 2010). Owing to glaciation, the pre-LGM evidence is restricted to a series of disconnected sites with nowhere preserving a complete or relatively long undisturbed sequence (Coxon 1993).

Shelf paleoenvironments (faunal evidence)

Both Britain and Ireland have numerous sites with faunal data which help constrain the pattern of Quaternary climate change, for instance with alternating sequences or isolated episodes of cold- versus warm-adapted species (or even periods of extinction) confirming the multiple glacial/interglacial and stadial/interstadial cycles described previously (see overviews in Mitchell & Ryan 2001; Stringer 2006). The evidence ranges from large mammals to mollusks, insects (generally beetles (*coleoptera*) and midges (chironomids)), and microfauna (e.g. foraminifera). These latter types of data have proved particularly useful in creating higher resolution and numerical estimates of past climate change within the study area (e.g. Brooks & Birks 2000; Bedford *et al.* 2004; Lang *et al.* 2010; van Asch *et al.* 2012).

The critical issue regarding the faunal evidence from the study area concerns the difference between the Irish and British records. A key feature of Ireland's Late Glacial and Holocene fauna is its impoverished nature compared to Britain and the rest of Europe (Davenport *et al.* 2008). Many large species found in the latter areas did not reach Ireland, such as mammoth, elk, aurochs and horse. Some species, such as wild boar, arrived only in the Early Holocene, while others, such as red deer and wolf, are restricted to short intervals during the Late Glacial. This also contrasts with MIS 3, during which all the species occupying Britain (except for humans and woolly rhino) have been found in Ireland (Woodman *et al.* 1997; Woodman 2008). The evidence for these faunas comes from a variety of sites including caves, bogs, infilled lakes and for the Holocene, archaeological sites.

Faunal remains from the shelf in the study area are surprisingly sparse especially when compared to the vast assemblages recovered from the North Sea. Known finds consist of a mammoth tusk dredged from Waterford (presently on display in the National Museum in Dublin), a possible mammoth humerus dredged from Galway Bay, and a mammoth mandible dredged from Holyhead Harbour (Adams 1878; Savage 1966). Some material has also been found in deeply buried peats excavated during harbor construction or within intertidal peat and forests. For example, red deer, boar and giant deer (*Megaloceros*) were excavated from Belfast Harbour (Praeger 1896), giant deer and aurochs from Heysham Harbour (Hazell 2008), and red deer, aurochs and brown bear from Liverpool Docks (Huddart *et al.* 1999a). Examples of faunal remains from intertidal peat and forest include reindeer from Roddan's Port (Singh 1963), red deer from the Shannon Estuary (O'Sullivan 2001), red deer, aurochs and dog from Formby Point (including bones and footprints) (Huddart *et al.* 1999b), and red deer, aurochs, boar and roe deer from the Severn Estuary (Bell 2007). The waterlogged conditions at intertidal peats and forests provide favorable conditions for the preservation of microfossils (both floral and faunal) and insects, and there are examples of research which have utilized such sites for paleoenvironmental reconstruction (e.g. Bell 2007; Simpson 2008; Whitehouse *et al.* 2008b).

There are several possible reasons for the lack of fully submerged faunal evidence, not least the lack of research and lack of known submerged land surfaces with organic preservation. It is certainly possible that more faunal remains exist in museum stores (particularly for specimens recovered during the nineteenth century), but to date little attempt has been made to investigate this. It may also be that the nature of the fishing activity and techniques (e.g. less bottom trawling?) within the Irish Sea and Atlantic Margin differs to that of the North Sea, though again, this has yet to be investigated.

Data sources

The best starting point for paleoenvironmental data is the published literature, such as the above or similar references which deal with the British or Irish Quaternary. For example, Bell (2007) has a gazetteer of intertidal/submerged peats and forest which also describes archaeological finds and faunal remains from each site. Supplementing this are databases or data archives which include the global or European-scale examples (e.g. European Pollen Database, NEOTOMA, BugsCEP: see Chapter 6 this volume). For Ireland, an Irish Pollen Database (IPOL) containing metadata (location, chronology, published reference) from over 475 sites has recently been compiled. Additional useful sources of information or lists of contacts can be obtained from research associations or networks such as the Quaternary Research Association (QRA), Irish Quaternary Association (IQUA), Irish Palaeoecology and Environmental Archaeology Network (IPEAN) and the Association for Environmental Archaeology (AEA). Both the QRA and IQUA have produced useful field guides which cover the onshore part of the study area (e.g. Whitehouse *et al.* 2008a).

- Association for Environmental Archaeology: envarch.net
- Quaternary Research Association: www.qra.org.uk
- Irish Palaeoecology and Environmental Archaeology Network: www.ipean.ie
- Irish Pollen Site Database: www.ipol.ie
- Irish Quaternary Association: www.iqua.ie

Submerged Landscape Evidence

Paleochannels

The most obvious exposed depressions visible on bathymetric rather than sub-bottom data are the Celtic Trough and its associated enclosed deeps (e.g. Beaufort's Dyke). In addition, seismic profiles show that some of these are partly infilled and have also identified other sets of fully buried paleovalleys or incisions across the Irish Sea and Bristol Channel (Eyles & McCabe 1989; Tappin *et al.* 1994; Jackson *et al.* 1995; Fitch & Gaffney 2011).

The Celtic Trough itself does not seem to be controlled by the underlying geological structure, and therefore has been interpreted as the end product of multiple erosive episodes during glacial maxima and subsequent sea-level rise since MIS 12 at least (Tappin *et al.* 1994; Jackson *et al.* 1995). Interpretation of the smaller incised deeps varies. Devoy (1995) points out that they could be polygenetic; formed by (pre-LGM) lowstand rivers and later modified by glacial and deglacial processes. Tappin *et al.* (1994) and Jackson *et al.* (1995) attribute them to kettle holes. Eyles and McCabe (1989) describe the network of channels off the eastern Irish coast as subglacial tunnel valleys, a view supported for Beaufort's Dyke at least by Callaway *et al.* (2011). In addition to these large deeps, there are several sets of

smaller anastomosing channels off Anglesey and the Llŷn Peninsula which are interpreted as former braided rivers on sub-aerial outwash (Tappin *et al.* 1994; Jackson *et al.* 1995). Fitch and Gaffney (2011) meanwhile distinguish several types of buried channels in Liverpool Bay including thin ephemeral end-glacial drainage channels, tunnel valleys, a potential small Holocene delta system and several Holocene channels and fluvial floodplains. In the Bristol Channel, Fitch and Gaffney (2011) also made a distinction between wide Pleistocene channels and smaller infilled Holocene channels.

Paleovalleys mapped by sub-bottom or bathymetric data have also been identified outside the Irish Sea basin. These include the Celtic Sea shelf off Waterford (Gallagher 2002; Gallagher *et al.* 2004), the Bann Estuary (Quinn *et al.* 2010; see Fig. 10.10) and Courtmacsherry Bay (Devoy *et al.* 2006). The west coast of Ireland lacks mapped paleovalleys, though this may stem from a lack of research rather than an absence of evidence. Various interpretations have been proposed for these features. The Bann paleovalley off the north coast lies directly opposite the modern Bann Estuary and could represent its lowstand extension (Quinn *et al.* 2010). Off the south coast, both the Waterford and Courtmacsherry paleovalleys are interpreted as former glaciofluvial channels (Gallagher 2002; Gallagher *et al.* 2004; Devoy *et al.* 2006). With the exception of the Waterford paleochannel, detailed investigation of the chronology, geomorphology and evolution of these features is lacking at the time of writing.

The general picture is that paleovalleys are known across the study area, with most examples concentrating in the Irish Sea where most research has been done. Many of these incisions, particularly the larger and deeper examples, are attributed to glacigenic rather than fluvial processes, though examples of the latter are visible, particularly on higher resolution seismic data (e.g. Fitch & Gaffney 2011; see Fig. 10.11). Presumably though, if sub-aerially exposed following ice retreat and not fully infilled, then the incised features would have formed natural watercourses or lake basins. Studies reconstructing paleodrainage patterns and linking the offshore paleovalleys to presently onshore river systems have yet to take place.

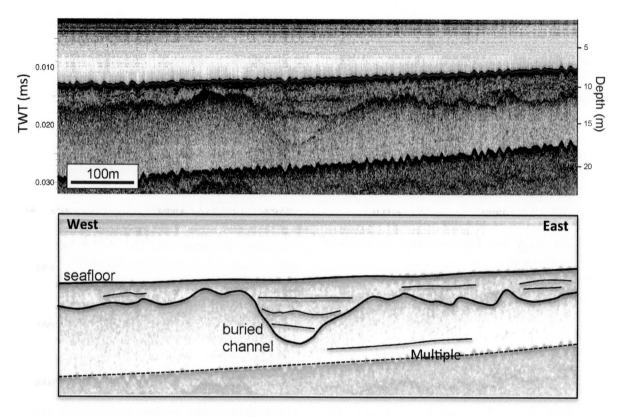

Fig. 10.10 Possible infilled paleochannel imaged by CHIRP sub-bottom profiler off the Bann Estuary, Northern Ireland. From Quinn *et al.* (2010); image courtesy of R. Quinn. Reproduced with permission.

Fig. 10.11 (a) Distribution of offshore infilled paleochannels as mapped from 3D seismic data in Liverpool Bay; (b) and (c) horizontal 2D timeslices through the 3D data volume showing the paleochannels as they appear on the 3D seismic data. Fitch & Gaffney (2011); data archived at the Archaeological Data Service: archaeologydataservice.ac.uk; images (b) and (c) courtesy of V. Gaffney.

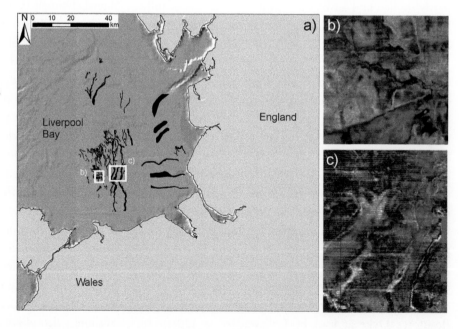

Glacial features

Within the study area, arguably the most widespread geomorphic paleolandscape features are glacial in origin, for example moraines and drumlins (e.g. Dunlop *et al.* 2010; Ó Cofaigh *et al.* 2012b). These are found across the study area, including the Irish and Celtic seas and Atlantic Margin, with the best examples identified from high-resolution multibeam surveys.

In the central Irish Sea, Van Landeghem *et al.* (2009a) describe moraines (including both ribbed and de Geer varieties), flutes, eskers, drumlins and iceberg scour marks (Fig. 10.12). These are interpreted as evidence of the decay of a grounded ice sheet which had previously advanced south-west towards the central Irish Sea basin. In the northern Irish Sea, an extensive submerged drumlin field can be seen along the coast of northeast Ireland (UKHO/AFBI unpublished data). At the time of writing, a full analysis has yet to be conducted to determine its implications for ice-sheet growth and decay. Other identified glacial features in the central part of the Irish Sea basin include rôches moutonnées, pingos and ice-wedge polygons (Jackson *et al.* 1995). Further south, linear ridges perpendicular to the shore of Cardigan Bay (locally known as 'sarnau') are believed to be moraines, and ice-wedge casts are reported from the outer Severn Estuary (Tappin *et al.* 1994).

On the Atlantic Margin to the north-west of Ireland, subglacial bedforms (including drumlins and moraines) confirm that grounded ice advanced to the continental shelf edge (Fig. 10.13). Extensive iceberg scour marks distal to the outermost moraines attest to initial retreat by large calving events followed by a slower recession with no calving (Benetti *et al.* 2010; Dunlop *et al.* 2010; Ó Cofaigh *et al.* 2012b). These bedforms also demonstrate two directions of ice flow and retreat, with ice advancing north-west out of Ireland (and subsequently retreating back towards the south-east: Benetti *et al.* 2010; Ó Cofaigh *et al.* 2012b) and converging with ice flowing west out of Scotland (again retreating back in the opposite direction: Dunlop *et al.* 2010). Additional large moraines to the west of Ireland confirm ice extension to the shelf edge in this direction as well (Sejrup *et al.* 2005).

The southern part of the study area has more limited evidence of glacial features, possibly due to the short-lived extension of ice into this area (Ó Cofaigh *et al.* 2012a). Off southeast Ireland, Gallagher *et al.* (2004) identify two boulder ridges which are interpreted as evidence of a former submerged ice margin. The maximum extent of ice across the shelf is not presently known, though it does seem to have reached as far south as the Isles of Scilly (Hiemstra *et al.* 2006) and perhaps even the shelf edge, creating a set of subglacial bedforms (esker ridges and transverse moraines) which were previously interpreted as a set of relict paleotidal sand banks (Praeg 2012).

Without supporting evidence, submerged glacial features in isolation should not be assumed to provide evidence of formerly sub-aerial landscapes. This is because they can be created underwater by ice grounded on the seabed. Good examples of this are seen in Benetti *et al.* (2010) and Van Landeghem *et al.* (2009a) where iceberg scours in association with subglacial features (e.g. moraines, drumlins) are clear evidence that the ice margin was grounded in a watery environment (either marine or proglacial lake respectively).

Paleoshoreline features

Verifiable submerged shoreline features are not widespread, though this may be due more to a lack of research than lack of evidence. For example, recently acquired multibeam bathymetric data from the north and northeastern coasts of Ireland have revealed consistent sets of breaks in slope or low cliff lines cut into rocky substrates which could be submerged shorelines (Quinn *et al.* 2010; Westley *et al.* 2011; Fig. 10.14). Such features would have been invisible on earlier low-resolution bathymetry. Thébaudeau *et al.* (2013) used this high-resolution data set to map more than 500 submerged sub-horizontal platform and terrace features along the northern Irish coast between Rathlin Island and Lough Swilly. They explored the genesis of these features by driving a geomorphological model of rock shore platform development with a suite of possible RSL curves for the region. The resulting profiles produced some platform-like features but failed to generate the number and size of sub-horizontal platforms identified

Fig. 10.12 Glacial features (drumlins, flutes, eskers, moraines and iceberg scours) on the Irish Sea shelf off northwest Wales. Van Landeghem *et al.* (2009a: fig. 6); image courtesy of K. Van Landeghem; reproduced with permission from John Wiley & Sons.

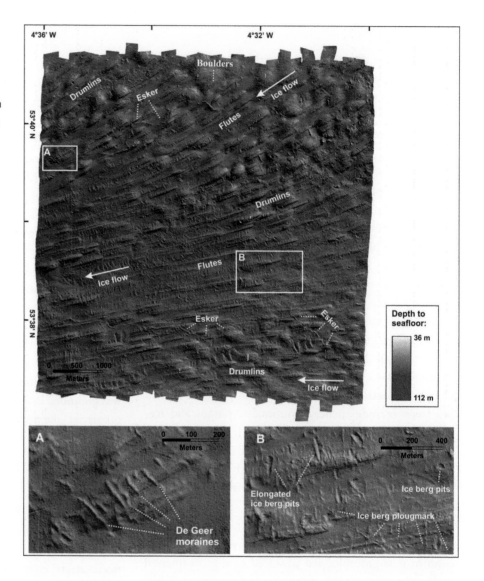

in the mapping exercise. The authors concluded that many of the rock-cut platforms in the study area are likely inherited from one or more earlier phases of RSL, illustrating the likelihood of destruction or reworking of deposits in these high-energy settings, and the challenges associated with developing a reliable chronology in such contexts.

It is likely that similar features exist elsewhere in the Irish Sea basin and Atlantic Margin, but these will remain undetectable in the absence of high-resolution swath bathymetry. As higher quality data become available (e.g. the INFOMAR program) more detailed mapping of seafloor features will become possible. However, not all former shorelines will be preserved on the seabed, as they

can also be eroded or buried during transgression. This latter category will be invisible on bathymetric data but could be detectable by sub-bottom surveys. Examples of this have been presented by Kelley *et al.* (2006) for Belfast Lough and Fitch and Gaffney (2011) for Liverpool Bay.

Aside from the geomorphological evidence, there is limited sedimentary evidence of deposits indicative of paleoshorelines. Kelley *et al.* (2006) for example interpret a sand deposit cored from −30 m in Belfast Lough as a lowstand beach on the basis of its shell assemblage and distinction from under- and overlying mud layers (interpreted as glaciomarine and estuarine muds respectively). The BGS offshore geology overviews note several instances of relict intertidal or nearshore

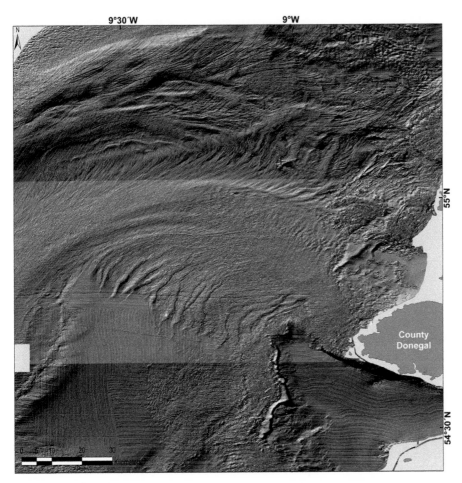

Fig. 10.13 Glacial moraines on the Atlantic Margin created by ice retreating from the shelf edge into northwest Ireland. High-resolution bathymetric data collected by the INFOMAR program; image from Ó Cofaigh *et al.* (2012b, fig. 2); image courtesy of P. Dunlop; reproduced with permission from Elsevier.

deposits within the SL2 member of the Surface Sands Formation, a deposit formed during the Late Glacial and postglacial transgression. These include Early Holocene sediments in the northeast Irish Sea (depths ranging from 20 m to 50 m), intertidal muds buried under the Kish Bank (western Irish Sea; ca. −30 m to −35 m), peaty silts indicative of reed swamp off north Wales (ca. −22 m; >9.2 ka) and shallow water, possibly intertidal, deposits

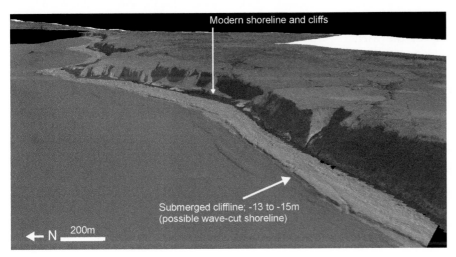

Fig. 10.14 Possible wave-cut shoreline cut into chalk bedrock at 13 m to 15 m depth off the north coast of Ireland. See Thébaudeau *et al.* (2013) for detailed explanation of this type of feature. Modern cliffs are ca. 100 m high. Bathymetric data is from the JIBS multibeam dataset; terrestrial DEM and draped aerial photos are from the Land & Property Service (Northern Ireland). Reproduced with permission.

in the north Celtic Sea (−123 m; ca. 11 ka) (Tappin *et al.* 1994; Jackson *et al.* 1995; see also Devoy 1995; Wingfield 1995).

In general, evidence for submerged shorelines is discontinuous across the study area. The above examples provide only local indications, which, given the spatio-temporal variability in isostatic adjustment (and hence RSL change) should not be extrapolated across wide areas unless there is compelling and substantive evidence to do so.

Former terrestrial environments

Evidence of former terrestrial environments takes a variety of forms ranging from deposits with significant preservation of organic remains (e.g. peat), to sands, silts or gravels which are often only recognizable as terrestrial after analysis of their microfossil assemblages (e.g. freshwater versus marine sediments). Of these, it is the deposits with large terrestrial organic remains that are most easily recognized without detailed scientific analysis and therefore have been frequently recorded in the study area, either in core/borehole investigations accompanying engineering or scientific projects, deep excavations or naturally exposed on the seabed or in the intertidal zone. Formerly terrestrial deposits lacking large organics have been less frequently recorded and therefore it is possible they have been sampled but not recognized due to insufficient analysis. The following discussion will therefore focus primarily on peat deposits and forest remains.

Cores and boreholes have sampled terrestrial deposits on both sides of the Irish Sea basin. In the Menai Strait, organic remains, including peat with wood, were recovered at various depths down to ca. 25 m and have been dated to the Late Pleistocene/Early Holocene (ca. 13–8 ka) (Roberts *et al.* 2011). Wingfield (1995) notes that borehole investigations in the estuaries and coastal lowlands of Lancashire (southeast Liverpool Bay) have recovered interbedded sequences of freshwater marsh and shallow marine deposits from −16 m upwards and post-dating 9.3 ka, while Hazell (2008) records numerous boreholes along the coast of Liverpool Bay reaching Holocene peat at depths up to 17 m. In Belfast Lough,

borehole investigations accompanying harbor works and land reclamation have sampled peats from as deep as 12 m and as early as 11.5 ka (Carter 1982; Brooks & Edwards 2006). Though many records are undated, those which are generally fall within the Early–Mid Holocene.

Similar sites can be found on the Atlantic Margin coast. For instance, silts containing pollen and freshwater diatoms were found in a borehole from Bantry Bay (southwest Ireland), and are interpreted as a lakebed deposit. These were recovered from a depth of ca. 57 m and were assigned (presumably on the basis of the pollen assemblage, as no details were provided in the original paper) to the Allerød/GI-1 Interstadial (ca. 13–14 ka) (Stillman 1968). At Courtmacsherry Bay (southern Ireland), layers of peat up to 7 m deep and 7.5 ka BP in age have been cored under saltmarsh deposits (Devoy *et al.* 2006). Recent engineering investigations in the west of Ireland (Sruwaddacon Bay) recovered peat samples from depths of 5.6 m to 10.7 m below the modern seabed. At time of writing, the samples are undated (RPS Group 2010).

In addition to the post-LGM evidence, at least two sites have submerged paleolandscape remains from earlier periods. Deep within the infill of the Celtic Trough, layers of peat were identified within the clay, sand and gravel of the Bardsey Loom Formation at a depth of 120 m below the seabed (itself located at ca. 100 m) and are believed to predate the MIS 12 glaciation (Tappin *et al.* 1994). In Cork Harbour (southwest Ireland), boreholes show buried intertidal, estuarine and mudflat sediments at ca. −17 m to −20 m, which, on the basis of pollen, appear to date from a pre-LGM temperate stage, potentially the MIS 11 or MIS 9 interglacials (Dowling *et al.* 1998; Dowling & Coxon 2001).

Larger expanses of formerly terrestrial landscapes have also been revealed by dredging, deep coastal excavation, or by natural erosion. In Belfast, nineteenth and early twentieth century harbor works exposed large vertical sections which reached several meters below low water. In these, a distinct peat bed, with tree remains, was recorded ca. 8 m to 9 m below high water and stratified beneath estuarine deposits (Praeger 1892; Charlesworth & Erdtman 1935). Though not radiocarbon dated, its floral and faunal assemblages and other dated peat

Fig. 10.15 Exposure of Holocene (ca. 7.4–6.5 ka) intertidal peat, Portrush (Northern Ireland). The peat underlies a section of beach extending both under the water from where the person is standing and under the seawall. It is generally covered by sand and only exposed after winter storms. Photo courtesy of K. Westley.

samples from Belfast suggest an Early Holocene age. Peat was also found in similar works on the eastern side of the Irish Sea at Heysham Port, Birkenhead and Barry Docks (Reid 1913; Bell 2007; Hazell 2008).

The most widespread expanses of formerly terrestrial landscape are found in the intertidal zone of beaches and estuaries, where they are naturally exposed by erosion of the overlying sediment (Fig. 10.15). Some of the best-studied examples come from the Severn Estuary, where numerous exposures appear at low tide. At the Goldcliff locality, extensive archaeological and paleoenvironmental surveys have provided considerable evidence of the past environment and Mesolithic human activity (Bell 2007). Outside of the Severn Estuary, similar deposits can be found around the eastern Irish Sea coast and the Isle of Man. For example, Bell (2007) maps 75 sites across northwest England and Wales. Hazell (2008) contains even more sites than Bell (2007) due to a wider geographical coverage, for example with 77 sites mapped around Liverpool Bay alone. However, unlike the latter, it provides no coverage for Wales. Sites on the Isle of Man are referred to by Roberts *et al.* (2006).

Ireland also has numerous reported intertidal peats and submerged forests. For example, Charlesworth (1963) and Mitchell (1976) map 28 and 33 locations respectively of 'submerged peat or forest', though they do not provide detailed information on each site. Most sites occur within the more sheltered bays and loughs (e.g. Strangford Lough: McErlean *et al.* (2002), Shannon Estuary: O'Sullivan (2001)) though their presence outside these areas should not be entirely ruled out as demonstrated by a peat bed on a wave-exposed sandy beach at Portrush on the north coast (Wilson *et al.* 2011) (see also Fig. 10.15). As with the British sites, the vast majority are Holocene in date, with a few Late Pleistocene examples (Carter 1982). In some cases, similar beds can be traced into shallow water offshore, either as extensions of the intertidal layers (e.g. Portrush: Westley *et al.* 2014) or isolated submerged deposits (e.g. Breen 1993; Westley 2015).

In general, evidence for submerged prehistoric landscapes can be found across the study area (Fig. 10.16). The extant offshore evidence tends to favor glacial features, with some paleovalleys and a handful of evidence for shoreline features or deposits. Organic remains relating to the former terrestrial landscape (i.e. peat and forest) are numerous, but cluster overwhelmingly in the intertidal zone with a few nearshore occurrences.

Data sources

The majority of data relating to submerged landscapes within the study area is disparate and spread across the published literature. There are however a few exceptions: the Waterlands project produced a GIS layer collating all recorded paleolandscape evidence for UK waters

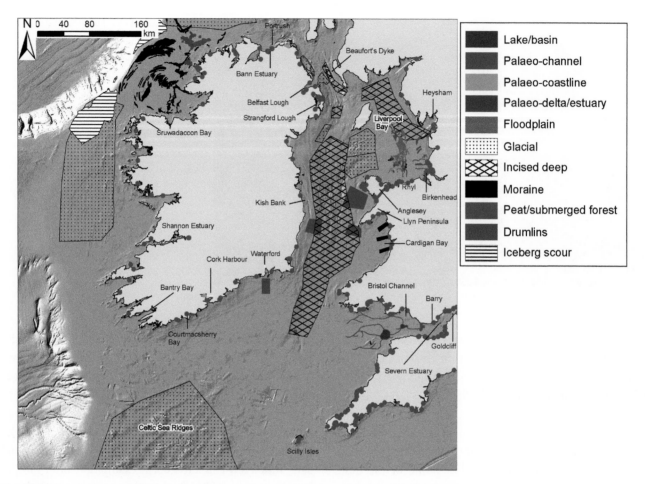

Fig. 10.16 Summary map compiled from multiple sources, showing extant sites with intertidal and submerged paleoenvironmental deposits or geomorphic features. Names of sites mentioned in the text are shown, see text for source references. Note that for some areas (e.g. central Irish Sea), precise data on the location of individual features (e.g. paleochannels, moraines) was not available. Therefore zones have been used to denote wide areas in which these paleolandscape features have been found.

(Goodwyn *et al.* 2010). Intertidal and submerged peats from around the English coast (relevant counties for the study area are Cumbria, Lancashire, Merseyside, Gloucestershire, Somerset, City of Bristol and the north coasts of Devon and Cornwall) are summarized in the Historic England Coastal Peat database (Hazell 2008). Bell (2007) contains a gazetteer of intertidal peats and submerged forests for western Britain (chiefly Wales and the adjacent areas of England). Some sites (principally those with accurate dating, elevation and location data) are contained within the British and Irish sea-level databases (Shennan & Horton 2002; Brooks & Edward 2006).

– Waterlands project hosted on the Marine ALSF Navigator Portal: www.marinealsf-navigator.org.uk/tree
– Historic England Intertidal and Coastal Peat database:
– content.historicengland.org.uk/content/docs/research/peat-database-nsea.pdf

Taphonomy and Potential for Archaeological Site Survival

Three main controls affect taphonomic conditions and the archaeological potential of the Irish Sea and Atlantic

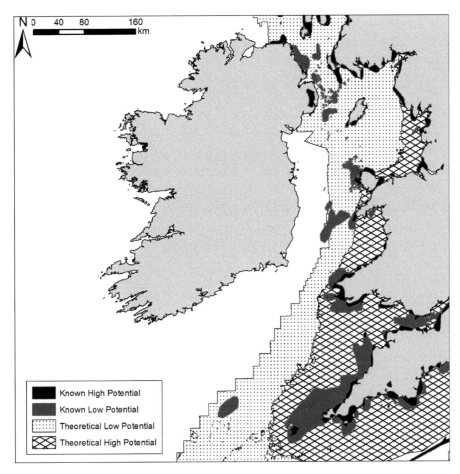

Known High Potential
Known Low Potential
Theoretical Low Potential
Theoretical High Potential

Fig. 10.17 Submerged landscape potential of the Irish Sea and Atlantic Margin as mapped by the Waterlands project. Goodwyn *et al.* (2010). Reproduced with permission. Note that this project only mapped areas within UK territorial waters.

Margin. Firstly, extensive glaciation; secondly, past sea-level change; and thirdly, high-energy marine conditions. The impact of the last two is well illustrated by Goodwyn *et al.* (2010) (see also Fig. 10.17).

Extensive glaciation has meant that much of the pre-LGM record has been removed, and what is left is highly fragmented. Preservation should not be entirely ruled out, as shown by the occasional survival of pre-Last Glacial paleoenvironmental and archaeological data. Onshore examples include the Mid Pleistocene Neanderthal site at Pontnewydd, Wales (Stringer 2006) and the MIS 3 organic deposits at Aghnadarragh, Northern Ireland (McCabe *et al.* 1987), both of which survived glaciation. However, the likelihood is that discoveries from these periods will probably be limited compared to the postglacial and Early Holocene. This is particularly apparent in that the vast majority of identified formerly terrestrial land surfaces (e.g. intertidal peats and forest) are

Holocene in date. The identified submerged pre-LGM deposits tend to be found through borehole investigations which have been able to penetrate the deep overlying sediment cover (e.g. Bardsey Loom Formation: Tappin *et al.* 1994; Cork Harbour: Dowling & Coxon 2001). Until recently there were no confirmed *in situ* Paleolithic sites or finds in Ireland. Dowd and Carden (2016) have reported that a brown bear patella recovered from the Alice and Gwendoline Cave in County Clare during 1903 shows clear cut marks of butchering, and recent radiocarbon dating provides a date of 12,590 cal BP to 12,810 cal BP. This pushes back the earliest human occupation of Ireland by 2500 years. The cave is about 25 km inland from the Atlantic coast. Ice models show that the ice cap had retreated north of this point (Fig 10.5), but the access route by which humans arrived in Ireland is completely obscure. Paleolithic material is preserved at several sites on the British side of the Irish

Sea, and thus further discoveries in Ireland, or offshore, are possible, and both Paleolithic and Mesolithic materials could be preserved underwater. If the pattern of past sea-level change, as reconstructed by extant data and models is correct (e.g. Brooks *et al.* 2011), then large parts of the Atlantic Margin and Irish Sea shelf may not have been sub-aerially exposed after the LGM. This applies primarily to the northern sector of the study area (northeast Ireland, southwest Scotland), where lowstands may have been less than −20 m to −30 m resulting in maximum shelf exposure of the order of a few kilometers. Exposed shelf extents increased to the south where the postglacial lowstand could have fallen to −80 m to −90 m. Maximum shelf exposure could then have increased to several tens of kilometers. Even so, bathymetry is such that Brooks *et al.* (2011) predict that the Irish coastline was never more than 30 km from its present position. In addition, the depth of the Celtic Trough means that it was probably not sub-aerial and formed a water body (most likely a marine seaway, but possibly a glacio-lacustrine lake during deglaciation) dividing Britain from Ireland. Conversely, the wider platforms on the eastern Irish Sea coast imply that large areas were exposed in Liverpool and Cardigan bays and the Bristol Channel. Together, this implies that large parts of the Irish Sea and Atlantic Margin shelves, especially those towards the shelf break, were probably not available for human occupation. This is reflected in Goodwyn *et al.* (2010) (see also Fig. 10.17) whereby the largest areas of theoretical high potential are located in the eastern Irish Sea where the exposed shelf is reconstructed to have been at its most extensive.

The high-energy conditions that prevail around much of the study area mean that many areas have suffered erosion during and after transgression, creating zones of scoured bedrock or lag gravels. For example, much of the Celtic Sea seabed is a thin lag deposit reworked from pre-existing deposits (Evans 1990). Survival potential would therefore be expected to be low with the exception of infilled depressions which may have collected and protected material. Preservation of large-scale landscape features (e.g. glacial landforms: Benetti *et al.* 2010) is clear on the western shelf off Ireland. However, whether this indicates also the survival of fine-grained organic and archaeological deposits is presently unknown. The open

and exposed conditions here imply that survival potential is low except for localized exceptions (see Westley 2015 for an example of local archaeological and paleoenvironmental preservation on the high-energy north coast). The Irish Sea probably has a range of preservation states stemming from variable energy conditions, including low-energy environments. For example, areas of erosion in the constricted St. George's and North channels have low potential for landscape survival, but the more heavily buried shallow shelves on the flanks of the Celtic Trough have a higher potential. Across both western Britain and Ireland, the most immediately prospective areas are probably the sheltered bays and estuaries which have clear evidence in the form of intertidal peats and forests that submerged landscapes (including archaeological and paleoenvironmental evidence) can survive. This is clearly mapped by Goodwyn *et al.* (2010) (see also Fig. 10.17) in that areas of known high potential are located close to the coast where there are recorded preserved intertidal landscapes, and areas of known low potential are located where strong currents have cleared the overlying sediment so that bedrock is exposed at the seabed.

Potential Areas for Future Work

Regions with least and highest chance of site survival

Assuming that the present GIA models are correct, then deeper areas (>40–50 m in the north-east and down to >90–100 m in the south-west) were probably not exposed during the Last Glacial and therefore are of low potential. Areas of exposed bedrock or modern sediment directly over bedrock will have no *in situ* archaeological and paleoenvironmental remains (at best, reworked remains) and are also of low potential. The only attempt to classify or identify these environments in the study area is the Waterlands project (Goodwyn *et al.* 2010). This classifies the majority of the western Irish and Celtic Seas as theoretical or known low potential with their eastern margins (e.g. Liverpool Bay, Cardigan Bay, Bristol Channel) defined as theoretical high potential. This analysis however, does not cover the shelf inside Republic of Ireland waters. In addition, local attempts have been made to classify potential on the basis of seabed and sub-seabed evidence for landscape preservation (e.g. Westley *et al.* 2011 for northeast Ireland; Fitch & Gaffney 2011 for Liverpool Bay and the Bristol Channel). These show that smaller 'pockets' of higher potential can exist within larger areas of low potential.

Prioritization of potential sites

On the basis of this overview, prioritized areas consist of sheltered inshore waters, bays and estuaries due to a) likelihood of exposure during the post-LGM, and b) lower energy conditions facilitating preservation. Within these environments, additional priority could be given to areas with known paleoenvironmental preservation e.g. inter- or subtidal peats and forests (see examples in Westley *et al.* 2014; Westley 2015).

In terms of correlation with terrestrial archaeology, there are several issues to consider. The first relates to the distribution of archaeological evidence, namely the paucity of evidence for an Irish Paleolithic with only one recently confirmed site (Dowd & Carden 2016). Although it is still not known whether this dearth is a genuine absence, any studies investigating Paleolithic submerged landscapes might be advised to prioritize the eastern side of the study area. Within Ireland itself, prioritization could also be based on the distribution of the extant Early Mesolithic, particularly the lack of such evidence in southwest Ireland. Whether this relates to the area having the deepest lowstand is presently uncertain. Research could therefore be directed here to determine whether this is the key factor in the lack of evidence. The east and northeast coasts are closest to Britain and therefore the logical place to examine evidence for the earliest colonization (Early Mesolithic and potentially Paleolithic?) of Ireland assuming that the early colonists moved in from this direction.

Any place where paleoshorelines and submerged coastal areas are present could provide useful information on the coastal/maritime adaptations of either the Mesolithic or Paleolithic; topics which are poorly understood at present. For example stable isotope measurements from a site within the study area

(Kendrick's Cave, north Wales: Richards *et al.* 2005) indicate Late Paleolithic marine resource use. Artifactual evidence for this (e.g. specialized tools, faunal evidence of marine exploitation) however, is still lacking and is presumably due to the submergence of the relevant coastal sites. A final area worth considering is a prioritization of the Celtic Sea/Western Approaches area as a potential postglacial Atlantic seaboard colonization route into Ireland and western Britain, not only for people but for flora and fauna as well (e.g. Mitchell 2006; Davenport *et al.* 2008). This however would be tempered with the likelihood of low preservation potential inferred for much of this area (Fig. 10.17).

At this broad-scale shelf overview level, precise areas cannot be prioritized any further without analysis of existing and future seabed data. Approaches for going beyond this include targeting areas on the basis of archaeological/paleoenvironmental questions (such as discussed above) or extant data availability. The extant data can then be used to assess paleolandscape extents (i.e. the degree of exposure during lower sea level) and preservation conditions (i.e. bedrock outcropping at seabed versus buried layers). Regarding the issue of extant data availability, as described above, there is a considerable amount of accessible marine scientific data for both Irish and UK waters. It therefore makes sense to first assess the usefulness of this data from a submerged archaeological landscape perspective before targeting areas of no/little data with new surveys (e.g. Fitch & Gaffney 2011; Westley *et al.* 2011). Once the data have been assessed and high-potential areas identified, then further refinement of priorities (e.g. specific areas within large bays) can be done through collection and analysis of additional high-resolution bathymetric and seismic profiles (e.g. Boomer, CHIRP, Pinger), and *in situ* sampling and observation (e.g. cores, ROV, divers).

Conclusion and Outlook

The Irish Sea and Atlantic Margin represent a considerable area spanning a range of distinct coastal environments, from the indented and wave-dominated rocky coast of western Ireland to the intertidal mudflats of western Britain. The shelf too is highly variable and includes broad expanses in the Celtic Sea, bathymetric trenches in the Irish Sea and narrower fringes around the west and north coasts of Ireland.

From the perspective of submerged landscape taphonomy, the key characteristics of the study area are its glacial and RSL history and the modern physical conditions. Its glacial history means that preservation of pre-LGM evidence is likely to be rare, particularly in the northern and western parts of the study area (as is the case for presently onshore areas). Its RSL history means that the size of exposed shelf was less than in other parts of the Northwest Shelf, restricting the areas available for human settlement, especially if this was delayed towards the terminal Pleistocene and Early Holocene when rising sea levels were rapidly flooding the exposed landscape. Finally, regarding modern physical conditions, much of the study area has high-energy environments ranging from Atlantic waves and storms to strong tidal streams at the constricted entrances to the Irish Sea. Consequently, large expanses of seabed are probably of low potential for landscape and archaeological preservation. On a shelf scale, these factors together give a rather negative outlook for submerged archaeological landscape research.

However, using a finer scale of analysis, we can see that the physical environment of the coast and shelf at present, and in the past, is sufficiently varied that pockets of high potential can, and do, exist. This is exemplified by the sheltered bays and nearshore regions where deep excavations, geotechnical investigations and natural erosion have uncovered fragments of the past landscape, mainly dating to the Holocene but with a handful of earlier examples. Further offshore, there is clear evidence that large glacial geomorphological features have survived, though the preservation of widespread postglacial landscape evidence remains a more open question. Precise mapping of these pockets and zones of high potential will only be possible through detailed analysis of seabed geophysical and geotechnical data, from both extant and future surveys. It is likely that, as research progresses and we advance from the broad-scale to a level of detail which can pick out the locations of highest potential, the above view of landscape survival will change for the better.

At present, there are still numerous gaps in knowledge, not least with respect to archaeological evidence or even faunal remains from the seabed, in contrast to the North and Baltic seas, and the English Channel. That said, this sea basin is less well-studied than those mentioned above and future research aided by the considerable and ever-increasing quantity of seabed data will hopefully redress this situation.

References

ABPMer 2008. *Atlas of UK Marine Renewable Energy Resources*. Technical Report prepared for the UK Department of Business, Enterprise and Regulatory Reform (BERR). Available at www.renewables-atlas.info/.

Adams, A. L. 1878. On the recent and extinct Irish Mammals. *Scientific Proceedings of the Royal Dublin Society* 2:45-86.

Allen, J. R. L. 1990. The Severn Estuary in southwest Britain: its retreat under marine transgression, and fine-sediment regime. *Sedimentary Geology* 66:13-28.

Allen, J. R. L. & Rae, J. E. 1987. Late Flandrian shoreline oscillations in the Severn Estuary: a geomorphological and stratigraphical reconnaissance. *Philosophical Transactions of the Royal Society* B315:185-230.

Ballantyne, C. K. 2010. Extent and deglacial chronology of the last British-Irish Ice Sheet: implications of exposure dating using cosmogenic isotopes. *Journal of Quaternary Science* 25:515-534.

Barne, J. H., Robson, C. F., Kaznowksa, S. S. & Doody, J. P. (eds.) 1995. *Coasts and Seas of the United Kingdom. Region 12 Wales: Margam to Little Orme*. Joint Nature Conservation Committee: Peterborough.

Barne, J. H., Robson, C. F., Kaznowska, S. S., Doody, J. P. & Davidson, N. C. (eds.) 1996a. *Coasts and Seas of the United Kingdom. Region 13 Northern Irish Sea: Colwyn Bay to Stranraer, including the Isle of Man*. Joint Nature Conservation Committee: Peterborough.

Barne, J. H., Robson, C. F., Kaznowska, S. S., Doody, J. P., Davidson, N. C. & Buck, A. L. (eds.) 1996b. *Coasts and Seas of the United Kingdom. Region 11*

The Western Approaches: Falmouth Bay to Kenfig. Joint Nature Conservation Committee: Peterborough.

Barne, J. H., Robson, C. F., Kaznowska, S. S., Doody, J. P., Davidson, N. C. & Buck, A. L. (eds.) 1997a. *Coasts and Seas of the United Kingdom. Region 14 South-west Scotland: Ballantrae to Mull.* Joint Nature Conservation Committee: Peterborough.

Barne, J. H., Robson, C. F., Kaznowska, S. S., Doody, J. P., Davidson, N. C. & Buck, A. L. (eds.) 1997b. *Coasts and Seas of the United Kingdom. Region 17 Northern Ireland.* Joint Nature Conservation Committee: Peterborough.

Bedford, A., Jones, R. T., Lang, B., Brooks, S. & Marshall, J. D. 2004. A Late-glacial chironomid record from Hawes Water, northwest England. *Journal of Quaternary Science* 19:281-290.

Bell, M. 2007. *Prehistoric Coastal Communities. The Mesolithic in Western Britain.* CBA Report 149. Council for British Archaeology: York.

Bell, M. & Walker, M. J. C. 2005. Late Quaternary Environmental Change. *Physical and Human Perspectives* (2nd Ed.). Pearson: Harlow.

Benetti, S., Dunlop, P. & Ó Cofaigh, C. 2010. Glacial and glacially-related features on the continental margin of northwest Ireland mapped from marine geophysical data. *Journal of Maps* 6:14-29.

Bowen, D. Q., Rose, J., McCabe, A. M. & Sutherland, D. G. 1986. Correlation of Quaternary glaciations in England, Ireland, Scotland and Wales. *Quaternary Science Reviews* 5:299-340.

Bowen, D. Q., Phillips, F. M., McCabe, A. M., Knutz, P. C. & Sykes, G. A. 2002. New data for the Last Glacial Maximum in Great Britain and Ireland. *Quaternary Science Reviews* 21:89-101.

Bradley, S. L., Milne, G. A., Shennan, I. & Edwards, R. 2011. An improved glacial isostatic adjustment model for the British Isles. *Journal of Quaternary Science* 26:541-552.

Breen, C. 1993. The 'Pearl', Tran a Fearla, Allihies. *Excavations Bulletin 1993: Summary Accounts of Archaeological Excavations in Ireland.* Wordwell: Dublin.

Brooks, A. & Edwards, R. 2006. The development of a sea-level database for Ireland. *Irish Journal of Earth Sciences* 24:13-27.

Brooks, A. J., Bradley, S. L., Edwards, R. J., Milne, G. A., Horton, B. & Shennan, I. 2008. Postglacial relative sea-level observations from Ireland and their role in glacial rebound modelling. *Journal of Quaternary Science* 23:175-192.

Brooks, A. J., Bradley, S. L., Edwards, R. J. & Goodwyn, N. 2011. The palaeogeography of Northwest Europe during the last 20,000 years. *Journal of Maps* 7:573-587.

Brooks, S. J. & Birks, H. J. B. 2000. Chironomid-inferred Late-glacial air temperatures at Whitrig Bog, South-east Scotland. *Journal of Quaternary Science* 15:759-764.

Burgess, K., Hay, H. & Hosking, A. 2004. Futurecoast: predicting the future coastal evolution of England and Wales. *Journal of Coastal Conservation* 10:65-71.

Callaway, A., Smyth, J., Brown, C. J., Quinn, R., Service, M. & Long, D. 2009. The impact of scour processes on a smothered reef system in the Irish Sea. *Estuarine, Coastal and Shelf Science* 84:409-418.

Callaway, A., Quinn, R., Brown, C. J., Service, M., Long, D. & Benetti, S. 2011. The formation and evolution of an isolated submarine valley in the North Channel, Irish Sea: an investigation of Beaufort's Dyke. *Journal of Quaternary Science* 26:362-373.

Carter, R. W. G. 1982. Sea-level changes in Northern Ireland. *Proceedings of the Geologists' Association* 93:7-23.

Carter, R. W. G. 1991. *Shifting Sands. A Study of the Coast of Northern Ireland from Magilligan to Larne.* HMSO: Belfast.

Carter, R. W. G., Johnston, T. W., McKenna, J. & Orford, J. D. 1987. Sea-level, sediment supply and coastal changes: examples from the coast of Ireland. *Progress in Oceanography* 18:79-101.

Charlesworth, J. K. 1963. *Historical Geology of Ireland.* Oliver & Boyd: Edinburgh.

Charlesworth, J. K. & Erdtman, G. 1935. Postglacial section at Milewater Dock, Belfast. *Irish Naturalists' Journal* 5:234-235.

Clark, C. D., Hughes, A. L. C., Greenwood, S. L., Jordan, C. & Sejrup, H.P. 2012a. Pattern and timing of retreat of the last British-Irish Ice Sheet. *Quaternary Science Reviews* 44:112-146.

Clark, J., McCabe, A. M., Dowen, D. Q. & Clark P. U. 2012b. Response of the Irish Ice Sheet to abrupt climate change during the last deglaciation. *Quaternary Science Reviews* 35:100-115.

Coope, G. R., Lemdahl, G., Lowe, J. J. & Walkling, A. 1998. Temperature gradients in northern Europe during the last glacial-Holocene transition (14-9 ^{14}C kyr BP) interpreted from coleopteran assemblages. *Journal of Quaternary Science* 13:419-433.

Cooper, A. 2012. Late glacial sea level minima in the Western British Isles: observed and modelled. *Quaternary International* 279-280 (XVIII Inqua Congress Abstracts):97-98.

Cooper, J. A. G. 2006. Geomorphology of Irish estuaries: inherited and dynamic controls. *Journal of Coastal Research Special Issue No.39 – Proceedings of the 8th International Coastal Symposium (ICS 2004)*:176-180.

Coxon, P. 1993. Irish Pleistocene biostratigraphy. *Irish Journal of Earth Sciences* 12:83-105.

Coxon, P. 2008. Landscapes and environments of the last glacial-interglacial transition: a time of amazingly rapid change in Ireland. In Davenport, J., Sleeman, D. & Woodman, P. (eds.) *Mind the Gap: Postglacial Colonization of Ireland. Special Supplement to the Irish Naturalists' Journal.* 29:45-62.

Curtis, T. G. F. & Sheehy Skeffington, M. J. 1998. The salt marshes of Ireland: An inventory and account of their geographical variation. *Proceedings of the Royal Irish Academy* 98B:87-104.

Davenport, J., Sleeman, D. & Woodman, P. (eds.) 2008. *Mind the Gap: Postglacial Colonization of Ireland. Special Supplement to the Irish Naturalists' Journal.* 29:1-138.

Deschamps, P., Durand, N., Bard, E. *et al.* 2012. Ice-sheet collapse and sea-level rise at the Bølling warming 14,600 years ago. *Nature* 483:559-564.

Devoy, R. J. N. 1995. Deglaciation, Earth crustal behaviour and sea-level changes in the determination of insularity: a perspective from Ireland. In Preece, R. (ed.). *Island Britain: a Quaternary Perspective. Geological Society Special Publication No. 96*:181-208.

Devoy, R. J. N. 2008. Coastal vulnerability and the implications of sea-level rise for Ireland. *Journal of Coastal Research* 24:325-341.

Devoy, R. J. N., Nichol, S. L. & Sinnott, A.M. 2006. Holocene sea-level and sedimentary changes on the south coast of Ireland. *Journal of Coastal Research Special Issue No.39 – Proceedings of the 8th International Coastal Symposium (ICS 2004)*:146-150.

Diefendorf, A. F., Patterson, W. P., Mullins, H. T., Tibert, N. & Martini, A. 2006. Evidence for high-frequency late Glacial to mid-Holocene (16,800 to 5500 cal yr B.P.) climate variability from oxygen isotope values of Lough Inchiquin, Ireland. *Quaternary Research* 65:78-86.

Dorschel, B., Wheeler, A. J, Monteys, X. & Verbruggen, K. 2011. *Atlas of the Deep-Water Seabed: Ireland.* Springer: Amsterdam.

Dowd, M. & Carden, R. F. 2016. First evidence of a Late Upper Palaeolithic human presence in Ireland. *Quaternary Science Reviews* 139:158:163.

Dowling, L. A & Coxon, P. 2001. Current understanding of Pleistocene temperate stages in Ireland. *Quaternary Science Reviews* 20:1631-1642.

Dowling, L. A., Sejrup, H. P., Coxon, P. & Heijnis, H. 1998. Palynology, aminostratigraphy and U-series dating of marine Gortian interglacial sediments in Cork Harbour, southern Ireland. *Quaternary Science Reviews* 17:945-962.

Dunlop, P., Shannon, R., McCabe, M., Quinn, R. & Doyle, E. 2010. Marine geophysical evidence for ice sheet extension and recession on the Malin Shelf: New evidence for the western limits of the British Irish Ice Sheet. *Marine Geology* 276:86-99.

Edwards, R. & Brooks, A. 2008. The island of Ireland: drowning the myth of an Irish land-bridge? In Davenport, J., Sleeman, D. & Woodman, P. (eds.) *Mind the Gap: Postglacial Colonisation of Ireland. Special Supplement to the Irish Naturalists' Journal.* 29:19-34.

Edwards, R. & Craven, K. 2017. Relative sea-level change around the Irish Coast. In Coxon, P., McCarron, S. & Mitchell, F. (eds.) *Advances in Quaternary Science: The Irish Quaternary.* pp. 181-216. Springer.

Edwards, R., Brooks, A., Shennan, I., Milne, G. & Bradley, S. 2008. Reply: Postglacial relative sea-level observations from Ireland and their role in glacial rebound modelling. *Journal of Quaternary Science* 23:821-825.

Erdtman, G. 1928. Studies in the postarctic history of the forests of north-western Europe. *Geologiska Föreningens Förhandlingar* 50:123-192.

Evans, C. D. R. 1990. *UK Offshore Regional Report: The Geology of the Western English Channel and its Western Approaches.* HMSO for the British Geological Survey: London.

Eyles, N. & McCabe, A. M. 1989. Glaciomarine facies within subglacial tunnel valleys: the sedimentary record of glacioisostatic downwarping in the Irish Sea Basin. *Sedimentology* 36:431-448.

Fitch, S. & Gaffney, V. 2011. *West Coast Palaeolandscapes Survey (Project No. 1997).* University of Birmingham.

Flemming, N. C. 2005. *The Scope of Strategic Environmental Assessment of Irish Sea Area SEA6 in regard to Prehistoric Archaeological Remains.* UK Department of Trade and Industry: London.

Fyfe, J. A., Long, D. & Evans, D. 1994. *UK Offshore Regional Report: The Geology of the Malin-Hebrides Sea Area.* HMSO for the British Geological Survey: London.

Gallagher, C. 2002. The morphology and palaeohydrology of a submerged glaciofluvial channel emerging from Waterford Harbour onto the nearshore continental shelf of the Celtic Sea. *Irish Geography* 35:111-132.

Gallagher, C., Sutton, G. & Bell, T. 2004. Submerged ice marginal forms in the Celtic Sea off Waterford Harbour, Ireland. Implications for understanding regional glaciation and sea-level changes following the last glacial maximum in Ireland. *Irish Geography* 37:145-165.

Godwin, H. 1975. *History of the British Flora: a Factual Basis for Phytogeography* (2nd Ed.). Cambridge University Press: Cambridge.

Goodwyn, N., Brooks, A.J. & Tillin, H. 2010. *Waterlands: Developing Management Indicators for Submerged Palaeoenvironmental Landscapes.* Report prepared under the Marine Aggregates Levy Sustainability Fund (Ref No. MEPF 09/P109). ABP Marine Environmental Research.

Greenwood, S.L. & Clark, C. D. 2009. Reconstructing the last Irish Ice Sheet 2: a geomorphologically-driven model of ice sheet growth, retreat and dynamics. *Quaternary Science Reviews* 28:3101-3123.

Hazell, Z. J. 2008. Offshore and intertidal peat deposits, England – a resource assessment and development of a database. *Environmental Archaeology* 13:101-110.

Haynes, J.R., Kiteley, R.J., Whatley, R.C. & Wilks, P.J. 1977. Microfaunas, microfloras and the environmental stratigraphy of the Late Glacial and Holocene in Cardigan Bay. *Geological Journal* 12:129-158.

Hiemstra, J.F., Evans, D. J. A., Scourse, J. D., McCarroll, D., Furze, M. F. A. & Rhodes, E. 2006. New evidence for a grounded Irish Sea glaciation of the Isles of Scilly, UK. *Quaternary Science Reviews* 25:299-309.

Holland, C. H. & Sanders, I. S. (eds.) 2009. *The Geology of Ireland* (2nd Ed). Dunedin Academic Press: Edinburgh.

Howells, M.F. 2008. *Regional Geology Guide: Wales.* British Geological Survey: London.

Hubbard, A, Bradwell, T, Golledge, N. *et al.* 2009. Dynamic cycles, ice streams and their impact on the extent, chronology and deglaciation of the British–Irish ice sheet. *Quaternary Science Reviews* 28:758-776.

Huddart, D. Gonzales, S. & Roberts, G. 1999a. The archaeological record and mid-Holocene marginal coastal palaeoenvironments around Liverpool Bay. *Journal of Quaternary Science* 14:563-574.

Huddart, D., Roberts, G. & Gonzales, S. 1999b. Holocene human and animal footprints and their relationships with coastal environmental change, Formby Point, NW England. *Quaternary International* 55:29-41.

Hughes, A. L. C., Greenwood, S. L. & Clark, C. D. 2011. Dating constraints on the last British-Irish Ice Sheet: a map and database. *Journal of Maps* 7:156-183.

Innes, J. B., Chiverrell, R. C., Blackford, J. J. *et al.* 2004. Earliest Holocene vegetation history and island biogeography of the Isle of Man, British Isles. *Journal of Biogeography* 31:761-772.

Jackson, D. I., Jackson, A. A., Evans, D., Wingfield, R. T. R., Barnes, R. P. & Arthur, M. J. 1995. *UK Offshore Regional Report: the Geology of the Irish Sea.* HMSO for the British Geological Survey: London.

Jessen, K. 1949. Studies in Late Quaternary deposits and flora-history of Ireland. *Proceedings of the Royal Irish*

Academy Section B: Biological, Geological & Chemical Science 52B:85-290.

Kelley, J. T., Cooper, J. A. G., Jackson, D. W. T., Belknap, D. F. & Quinn, R. J. 2006. Sea-level change and inner shelf stratigraphy off Northern Ireland. *Marine Geology* 232:1-15.

Knight, J. 2001. Glaciomarine deposition around the Irish Sea basin: some problems and solutions. *Journal of Quaternary Science* 16:405-418.

Knight, J. (ed.). 2002. *Field Guide to the Coastal Environments of Northern Ireland*. International Coastal Symposium 2002. University of Ulster: Coleraine.

Kozachenko, M., Devoy, R. & Cummins, V. forthcoming. *Shorelines: the Coastal Atlas of Ireland*. Cork University Press: Cork.

Kuchar, J., Milne, G., Hubbard, A. *et al.* 2012. Evaluation of a numerical model of the British-Irish ice sheet using relative sea-level data: implications for the interpretation of trimline observations. *Journal of Quaternary Science* 27:597-605.

Lambeck, K. 1995. Late Devensian and Holocene shorelines of the British Isles and the North Sea from models of glacio-hydro-isostatic rebound. *Journal of the Geological Society* 152:437-448.

Lambeck, K. & Purcell, A.P. 2001. Sea-level change in the Irish Sea since the Last Glacial Maximum: constraints from isostatic modelling. *Journal of Quaternary Science* 16:497-506.

Lang, B., Brooks, S. J., Bedford, A., Jones, R. T., Birks, H. J. B. & Marshall, J. D. 2010. Regional consistency in Lateglacial chironomid-inferred temperatures from five sites in north-west England. *Quaternary Science Reviews* 29:1528-1538.

Lowe, J. J., NASP Members 1995. Palaeoclimate of the North Atlantic seaboards during the Last Glacial/Interglacial transition. *Quaternary International* 28:51-61.

May, V. J. & Hansom, J. D. 2003. *Coastal Geomorphology of Great Britain*. Geological Conservation Review Series No. 28. Joint Nature Conservation Committee: Peterborough.

McCabe, A. M. 2008. Comment: Postglacial relative sea-level observations from Ireland and their role in glacial rebound modelling. *Journal of Quaternary Science* 23:817-820.

McCabe, A. M. & Dunlop, P. 2006. *The Last Glacial Termination in Northern Ireland*. Geological Survey of Northern Ireland: Belfast.

McCabe, A. M., Coope, G.R., Gennard, D. E. & Doughty, P. 1987. Freshwater organic deposits and stratified sediments between Early and Late Midlandian (Devensian) till sheets, at Aghnadarragh, County Antrim, Northern Ireland. *Journal of Quaternary Science* 2:11-33.

McCabe, A. M., Cooper, J. A. G. & Kelley, J. T. 2007. Relative sea-level changes from NE Ireland during the last glacial termination. *Journal of the Geological Society, London* 164:1059-1063.

McCarroll, D. 2001. Deglaciation of the Irish Sea Basin: a critique of the glaciomarine hypothesis. *Journal of Quaternary Science* 16:393-404.

McDowell, J. L., Knight, J. & Quinn, R. 2005. High-resolution geophysical investigations seaward of the Bann Estuary, Northern Ireland coast. In FitzGerald, D. M. & Knight, J. (eds.) *High Resolution Morphodynamics and Sedimentary Evolution of Estuaries*. pp. 11-31. Springer.

McErlean, T., McConkey, R. & Forsythe, W. 2002. *Strangford Lough: An Archaeological Survey of the Maritime Cultural Landscape*. Blackstaff Press: Belfast.

Mitchell, F. J. G. 2006. Where did Ireland's trees come from? *Proceedings of the Royal Irish Academy* 106B:251-259.

Mitchell, G. F. 1976. *The Irish Landscape*. Collins: London.

Mitchell, G. F. & Ryan, M. 2001. *Reading the Irish Landscape*. Townhouse: Dublin.

Mitchell, W. I. (ed.) 2004. *The Geology of Northern Ireland*. Geological Survey of Northern Ireland: Belfast.

Montgomery, W. I., Provan, J., McCabe, A. M., & Yalden, D. W. 2014. Origin of British and Irish mammals: disparate post-glacial colonisation and species introductions. *Quaternary Science Reviews* 98:144-165.

Moore, R. D., Wolf, J., Souza, A. J. & Flint, S. S. 2009. Morphological evolution of the Dee Estuary, Eastern Irish Sea, UK: a tidal asymmetry approach. *Geomorphology* 103:588-596.

Ó Cofaigh, C., Telfer, M. W., Bailey, R. M. & Evans, D. J. A. 2012a. Late Pleistocene chronostratigraphy and ice sheet limits, southern Ireland. *Quaternary Science Reviews* 44:160-179.

Ó Cofaigh, C., Dunlop, P. & Benetti, S. 2012b. Marine geophysical evidence for Late Pleistocene ice sheet extent and recession off northwest Ireland. *Quaternary Science Reviews* 44:147-159.

O'Connell, M., Huang, C. C. & Eicher, U. 1999. Multidisciplinary investigations, including stable-isotope studies of thick Late-glacial sediments from Tory Hill, Co. Limerick, western Ireland. *Palaeogeography, Palaeoclimatology, Palaeoecology* 147:169-208.

O'Sullivan, A. 2001. *Foragers, Farmers and Fishers in a Coastal Landscape: an Intertidal Archaeological Survey of the Shannon Estuary*. Royal Irish Academy: Dublin.

Pilcher, J., Plunkett, G. & Whitehouse, N. J. 2008. Palaeoecology of the North of Ireland – history and overview. In Whitehouse, N. J., Roe, H. M., McCarron, S. & Knight, J. (eds.) *North of Ireland Field Guide*. pp. 29-40. Quaternary Research Association: London.

Plets, R., Clements, A., Quinn, R. & Strong, J. 2012. Marine substratum map of the Causeway Coast, Northern Ireland. *Journal of Maps* 8:1-13.

Praeg, D. 2012. New evidence on glaciation of the Celtic Sea: results from GLAMAR. *Quaternary International* 279-280 (XVIII Inqua Congress Abstracts):385.

Praeger, R. L. 1892. The Irish Post-Glacial estuarine deposits. *The Irish Naturalist* 1:138-141.

Praeger, R. L. 1896. A submerged pine-forest. *The Irish Naturalist* 5:155-160.

Pye, K. & Neal, A. 1994. Coastal dune erosion at Formby Point, north Merseyside, England: causes and mechanisms. *Marine Geology* 119:39-56.

Quinn, R., Forsythe, W., Plets, R. *et al.* 2009. *Archaeological applications of the Joint Irish Bathymetric Survey (JIBS) data – Phase 2*. Final Report submitted to the Heritage Council.

Quinn, R., Plets, R., Clements, A. *et al.* 2010. *Archaeological Applications of the Joint Irish Bathymetric Survey (JIBS) data – Phase 3*. Final Report submitted to the Heritage Council.

Rasmussen, S. O., Andersen, K. K., Svensson, A. M. *et al.* 2006. A new Greenland ice core chronology for the last glacial termination. *Journal of Geophysical Research - Atmospheres* 111:D06102.

Reid, C. 1913. *Submerged Forests*. Cambridge University Press: Cambridge.

Richards, M. P., Jacobi, R., Cook, J., Pettitt, P.B. & Stringer, C. B. 2005. Isotope evidence for the intensive use of marine foods by Late Upper Palaeolithic humans. *Journal of Human Evolution* 49:390-394.

Roberts, D. H., Chiverrell, R. C., Innes, J. B. *et al.* 2006. Holocene sea levels, Last Glacial Maximum glaciomarine environments and geophysical models in the northern Irish Sea Basin, UK. *Marine Geology* 231:113-128.

Roberts, M. J., Scourse, J. D., Bennell, J. D., Huws, D. G., Jago, C. F. & Long, B. T. 2011. Late Devensian and Holocene relative sea-level change in North Wales, UK. *Journal of Quaternary Science* 26:141-155.

RPS Group. 2010. *Corrib Onshore Pipeline. Environmental Impact Statement*. Unpublished report prepared for An Bord Plenála, the Department of Communications, Energy & Natural Resources and Department of Environment, Heritage & Local Government, Republic of Ireland.

Savage, R. J. G. 1966. Irish Pleistocene mammals. *Irish Naturalists' Journal*. 15:117-130.

Scourse, J. D. & Furze, M. F. A. 2001. A critical review of the glaciomarine model for Irish sea deglaciation: evidence from southern Britain, the Celtic shelf and adjacent continental slope. *Journal of Quaternary Science* 16:419-434.

Scourse, J. D., Haapaniemi, A. I., Colmenero-Hidalgo, E. *et al.* 2009. Growth, dynamics and deglaciation of the last British–Irish ice sheet: the deep-sea ice-rafted detritus record. *Quaternary Science Reviews* 28:3066-3084.

Sejrup, H. P., Hjelstuen, B. O., Torbjørn Dahlgren, K. I. *et al.* 2005. Pleistocene glacial history of the NW European continental margin. *Marine and Petroleum Geology* 22:1111-1129.

Shennan, I. & Horton, B. 2002. Holocene land- and sea-level changes in Great Britain. *Journal of Quaternary Science* 17:511-526.

Shennan, I., Peltier, W. R., Drummond, R. & Horton, B. 2002. Global to local scale parameters determining relative sea-level changes and the post-glacial isostatic adjustment of Great Britain. *Quaternary Science Reviews* 21:397-408.

Shennan, I., Bradley, S., Milne, G., Brooks, A., Bassett, S., & Hamilton, S. 2006a. Relative sea-level changes, glacial isostatic modelling and ice-sheet reconstructions from the British Isles since the Last Glacial Maximum. *Journal of Quaternary Science* 21:585-599.

Shennan, I., Hamilton, S., Hillier, C. *et al.* 2006b. Relative sea-level observations in western Scotland since the Last Glacial Maximum for testing models of glacial isostatic land movements and ice-sheet reconstructions. *Journal of Quaternary Science* 21:601-613.

Simpson, D. 2008. Greyabbey Bay, Strangford Lough: marine, non-marine and terrestrial micro- and macro-fossils. In Whitehouse, N.J., Roe, H. M., McCarron, S. & Knight, J. (eds.) 2008. *North of Ireland Field Guide.* pp. 155-167. Quaternary Research Association: London.

Singh, G. 1963. Pollen-analysis of a deposit at Roddans Port, Co. Down, N. Ireland, bearing reindeer antler fragments. *Grana Palynologica* 4:466-474.

Sleeman, A., McConnell, B. & Gatley, S. 2004. *Understanding Earth Processes, Rocks and the Geological History of Ireland.* Geological Survey of Ireland: Dublin.

Stillman, C. J. 1968. The post glacial change in sea level in southwestern Ireland: new evidence from freshwater deposits on the floor of Bantry Bay. *Scientific Proceedings of the Royal Dublin Society* A3(11):125-127.

Stringer, C. 2006. *Homo Britannicus: the incredible story of human evolution.* Penguin Books: London.

Sturt, F., Garrow, D. & Bradley, S. 2013. New models of North West European Holocene palaeogeography and inundation. *Journal of Archaeological Science* 40:3963-3976.

Tappin, D. R., Chadwick, R. A., Jackson, A. A., Wingfield, R. T. R. & Smith, N. J. P. 1994. *UK Offshore Regional Report: The Geology of Cardigan Bay and the Bristol Channel.* HMSO for the British Geological Survey: London.

Thébaudeau, B., Trenhaile, A. S., & Edwards, R. J. 2013. Modelling the development of rocky shoreline profiles along the northern coast of Ireland. *Geomorphology* 203:66-78.

van Asch, N., Lutz, A. F., Duijkers, M. C. H., Heiri, O., Brooks, S.J. & Hoek, W. Z. 2012. Rapid climate change during the Weichselian Lateglacial in Ireland: Chironomid-inferred summer temperatures from Fiddaun, Co. Galway. *Palaeogeography, Palaeoclimatology, Palaeoecology* 315-316:1-11.

Van Landeghem, K. J. J., Wheeler, A. J., & Mitchell, N. C. 2009a. Seafloor evidence for palaeo-ice streaming and calving of the grounded Irish Sea Ice Stream: Implications for the interpretation of its final deglaciation phase. *Boreas* 38:119-131.

Van Landeghem, K. J. J., Uehara, K., Wheeler, A. J., Mitchell, N. C. & Scourse, J. D. 2009b. Post-glacial sediment dynamics in the Irish Sea and sediment wave morphology: Data–model comparisons. *Continental Shelf Research* 29:1723-1736.

Walker, M. J. C. 1995. Climatic changes in Europe during the Last Glacial/Interglacial Transition. *Quaternary International* 28:63-76.

Watson, J. E., Brooks, S. J., Whitehouse, N. J., Reimer, P. J., Birks, H. J. B. & Turney, C. 2010. Chironomid-inferred late-glacial summer air temperatures from Lough Nadourcan, Co. Donegal, Ireland. *Journal of Quaternary Science* 25:1200-1210.

Westley, K. 2015. Submerged archaeological landscape investigation, Eleven Ballyboes, Republic of Ireland. *International Journal of Nautical Archaeology* 44:243-257.

Westley, K., Quinn, R., Forsythe, W. *et al.* 2011. Mapping submerged landscapes using multibeam bathymetric data: a case study from the north coast of Ireland. *International Journal of Nautical Archaeology* 40:99-112.

Westley, K., Plets, R. & Quinn, R. 2014. Holocene paleogeographic reconstructions of the Ramore Head area, Northern Ireland, using geophysical and geotechnical data: Paleo-landscape mapping and archaeological implications. *Geoarchaeology* 29:411-430.

Wheeler, A. J., Walshe, J. & Sutton, G. D. 2001. Seabed mapping and seafloor processes in the Kish, Burford,

Bray and Fraser Banks area, south-western Irish Sea. *Irish Geography* 34:194-211.

Whitehouse, N. J., Roe, H. M., McCarron, S. & Knight, J. (eds.). 2008a. *North of Ireland Field Guide.* Quaternary Research Association: London.

Whitehouse, N. J., Watson, J. & Turney, C. 2008b. Roddans Port Late Glacial site. In Whitehouse, N. J., Roe, H. M., McCarron, S. & Knight, J. (eds.). *North of Ireland Field Guide.* pp. 168-173. Quaternary Research Association: London.

Wickham-Jones, C. R., & Dawson, S. 2006. *The Scope of Strategic Environmental Assessment of North Sea Area SEA7 with regard to Prehistoric and Early Historic Archaeological Remains.* UK Department of Trade and Industry: London.

Wilson, H.E. & Manning, P. I. 1978. *Geology of the Causeway Coast.* HMSO: Belfast.

Wilson, P. & Plunkett, G. 2010. Age and palaeoenvironmental significance of an inter-tidal peat bed at Ballywoolen, Bann estuary, Co. Londonderry. *Irish Geography* 43:265-275.

Wilson, P., Westley, K., Plets, R. & Dempster, M. 2011. Radiocarbon dates from the inter-tidal peat bed at Portrush, County Antrim. *Irish Geography* 44: 323-329.

Wingfield, R. T. R. 1995. A model of sea-levels in the Irish and Celtic Seas during the end-Pleistocene to Holocene transition. In Preece, R. (ed.). *Island Britain: a Quaternary Perspective.* Geological Society Special Publication 96:209-242.

Woodman, P. 2008. Mind the gap or gaps? In Davenport, J., Sleeman, D. & Woodman, P. (eds.) *Mind the Gap: Postglacial Colonization of Ireland. Special Supplement to the Irish Naturalists' Journal.* 29:5-18.

Woodman, P., McCarthy, M. & Monaghan, N. 1997. The Irish Quaternary Fauna Project. *Quaternary Science Reviews* 16:129-159.

Chapter 11

The Iberian Atlantic Margin

Delminda Moura, Ana Gomes and João Horta

Universidade do Algarve, Faculdade de Ciências e Tecnologia, Centro de Investigação Marinha e Ambiental (CIMA), Faro, Portugal

Introduction

The Iberian Atlantic Margin is a passive margin and may be divided into six sectors determined by the geomorphic context (Fig. 11.1): (i) Northern margin, (ii) Northwestern margin: northern Nazaré canyon, (iii) Western margin: southern Nazaré canyon to Cape Sines, (iv) Western margin between Cape Sines and Cape St. Vicente, (v) Southern margin between Cape St. Vicente and Cape Santa Maria and, (vi) Gulf of Cádiz. In this chapter we start with a general overview of the Iberian Atlantic shelf, and then consider each region in turn. The high-resolution imagery used here to show the sectors discussed in the text was provided courtesy of the Instituto Geográfico Português (IGP) through an agreement with ESRI (Environmental Systems Research Institute) Portugal (www.esri.com). In addition, a global 30 arcsecond grid was generated by combining quality-controlled ship depth soundings with interpolation between sounding points guided by satellite-derived gravity data. The General Bathymetric Chart of the Oceans (GEBCO) consists of an international group of experts who work on the development of a range of bathymetric data sets and data products, including gridded bathymetric data sets, the GEBCO Digital Atlas, the GEBCO World Map and the GEBCO Gazetteer of Undersea Feature Names (www.bodc.ac.uk/data/online_delivery/gebco/). The hydrographic information shown in maps was obtained at www.puertos.es/en-us/oceanografia/Pages/portus.aspx and www.ieo.es, and snirh.apambiente.pt for Spain and Portugal respectively.

The Atlantic coastal zones expose crystalline and sedimentary rocks displaying very different coastal morphologies, ranging from quite straight sectors to very crenulated coastlines, from high cliffs to wide portions of low-lying land. Along the cliffed coast, pocket beaches and embayed beaches occur, mainly associated with stream mouths. Capes and peninsulas are important morphological features influencing longshore drift that in some localities is in the opposite direction to the

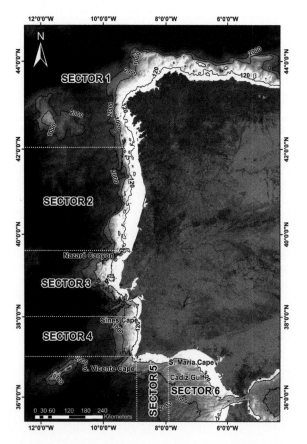

Fig. 11.1 Iberian Atlantic Margin showing the six sectors discussed in the text. Solid lines are bathymetric contours. The −120 m bathymetric contour limits to seaward the portion of the continental shelf exposed to aerial conditions during the Last Glacial Maximum (LGM). The striking manner in which the Nazaré canyon traverses the shelf is clearly distinguishable.

main axis of wave propagation. The most extensive Meso-Cenozoic basins in the Iberian Atlantic Margin are the Basque-Cantabrian basin in northern Spain, and the Lusitanian and Algarve basins, in the center and south of Portugal respectively. The morphology of the Atlantic Iberia continental shelf ranges from almost smooth to canyoned and very irregular, from narrow to wide, with variable slopes and shelf breaks at different depths, which are sometimes difficult to distinguish. The shelf sedimentary cover represents a balance between the sedimentary fluvial input and the redistribution by waves and currents. This is highly dynamic on the north and west coasts, which are exposed to massive ocean swells

and storm waves, whereas the southern sector is a low-energy environment. Several morphological features such as marine abrasion surfaces are preserved on the Atlantic Iberian continental shelf and some publications estimate their age by comparing their depth with the global sea-level curves. After the Last Glacial Maximum (LGM) the paleoenvironmental evolution is well known and very similar along the Iberian Atlantic Margin consisting of common aspects: (i) rapid sea-level rise up to 6 ka BP, (ii) main infilling phase of estuaries, rias and proto-coastal lagoons between 8 ka BP and 5 ka BP with the landward migration of the coastline, (iii) deceleration of sea-level rise after 5 ka BP, leading to the genesis of barrier islands protecting coastal watersheds and, (iv) infilling of coastal environments by fluvial sediments mainly after 3 ka BP sometimes leading to seaward coastline migration.

Geomorphological Framework of the Iberian Atlantic Margin

Northern margin

Coastal geomorphology

The northern Iberian Atlantic Margin in Spain (Sector 1, Figs. 11.1 and 11.2) comprises two main geological contexts: (i) The Basque-Cantabrian basin composed mostly of Mesozoic rocks, (mainly) Cretaceous carbonate and detritus sedimentary rocks (Agirrezabala *et al.* 2002) and, (ii) Coruña and Galicia where outcrops are mainly crystalline Paleozoic rocks. 1:50,000 scale geological maps can be accessed from the web page of the Instituto Geológico Y Minero de España (www.igme.es/internet/default.asp).

The littoral of Galicia (northwest Spain) is a high-energy irregular coast exposing mainly granite and metamorphic rocks from the Paleozoic. It is a crenulated coastline with pocket beaches between headlands, embayed beach barriers and marine embayments (rias). Several islands and archipelagos occur on the Galician Margin (Fig. 11.2), and the rias near their mouths are sometimes protected from wave attack by small islands (Gomez-Gesteira *et al.* 2011). The rias were previously

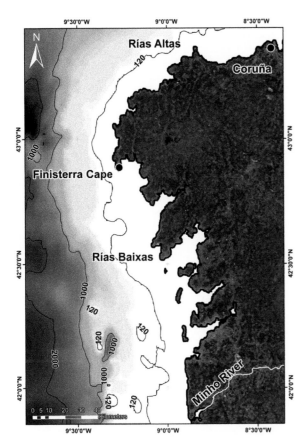

Fig. 11.2 Northern Sector (north of the Iberian Atlantic Margin to Minho River – please see text). Note the crenulated coast at the Galician Margin due to the marine embayments (rias).

incised fluvial valleys that were flooded during the Holocene transgression. The rias north of Cape Finisterra are named Rias Altas and are backed by higher coastal cliffs (reaching 200 m; Alonso & Pagés 2007) than the Rias Baixas to the south from Cape Finisterra (Fig. 11.2) where several coastal wetlands occur. The Rias Baixas on the Galician western coast are generally funnel-shaped and the depths in the central axes may reach 60 m in the outer zone (seaward) and some of them have tectonic control (Garcia-Gil *et al.* 2000; Méndez & Vilas 2005). The northern coast of the Iberian Atlantic Margin has been uplifted in part due to isostatic rebound and that uplift is higher in Cantabria relative to Galicia (Alonso & Pagés 2007). Between Cape Finisterra and Coruña several raised beaches, which range in age between 100 ka to 73 ka, are particularly well preserved up to 3 m above

the present mean sea level (MSL) whereas raised beaches dated as 71 ka occur up to 6 m above MSL in Cantabria.

Continental shelf

The continental shelf bordering the Cantabrian Sea is narrow, between 7 km and 20 km at Cape Matxitxako and off the Oria Estuary respectively (Uriarte *et al.* 2004), with the shelf break located between 180 m and 245 m water depth (Ercilla *et al.* 2008) and it receives the sedimentary contribution of numerous rivers. These sediments discharged from the major estuaries intersect an almost continuous belt of rocky substrate, the structure of which largely controls the shelf morphology and slope (Galparsoro *et al.* 2010). Several submerged paleofluvial channels and nine marine terraces between 37 m and 92 m water depth were sculpted during past low sea levels and were identified in this shelf sector (Galparsoro *et al.* 2010). The location of the submarine canyons intercepting the continental slope with heads at the outer shelf near the break is tectonically controlled (Fernàndez-Viejo *et al.* 2014). The Cantabrian shelf is dominated by erosional processes leading to the absence of sedimentary cover on the outer shelf. The scarcity of recent sediments, despite the numerous fluvial discharges onto the shelf, is probably due to its narrowness associated with the frequent drops of mean sea level during the Quaternary when rivers discharged directly onto the continental slope (Ercilla *et al.* 2008) as well as due to the energetic marine conditions. In contrast to the poor sedimentary cover on the overall continental shelf in this sector, a mud patch up to 7 m thick occurs in the eastern part of the Cantabrian Margin due to favorable geomorphological conditions acting as a trap mechanism for the suspended particles (Jouanneau *et al.* 2008).

The Galician continental shelf is relatively narrow, ranging between 25 km and 40 km. The shelf break occurs at 160 m to 180 m and several rock outcrops appear along the inner and outer portions of the shelf (Dias *et al.* 2002). The Galician shelf is characterized by small coastal embayments flooded during the Holocene transgression (rias) (Bernárdez *et al.* 2008a) that act as sediment traps (Prego 1993). The Galicia Mud Patch (GMP) is a noticeable sedimentary body corresponding to the underwater deposition of the sediments mainly from the Minho and Douro rivers, the main sediment suppliers to the shelf (Bernárdez *et al* 2008b). The GMP is roughly parallel to the current shoreline and lies 121 m below the sea surface. Its genesis relates to the abundant supply of fluvial sediments mainly during flood events and with favorable hydrographic conditions for fine-sediment accumulation at the mid shelf (Martins *et al.* 2007; Lantzsch *et al.* 2009). Moreover, the effect of the Minho freshwater discharge is so important to the Rias Baixas circulation that it can reverse the normal gradient of salinity (Sousa *et al.* 2013). A palimpsest deposit 50 cm thick, occurring on the Galician outer shelf was probably the result of strong bottom currents acting under storm conditions at 2850 cal BP at the Sub Boreal–Sub Atlantic transition (González-Álvarez *et al.* 2005). The Galician shelf morphological evolution during the Late Quaternary was the result of the highly energetic hydrodynamic regime and the low sediment accumulation, particularly during the Holocene sea-level rise when the sediments were trapped at the river mouths (Lantzsch *et al.* 2010).

Northwestern margin: northern Nazaré canyon

Coastal geomorphology

In this sector of the Iberian Atlantic Margin between the Minho River and the Nazaré canyon (Fig. 11.3) the Paleozoic rocks (mainly granites) crop out, composing the higher cliffs at the northernmost part, north of the Douro Estuary. This sector includes the Portuguese Western Meso-Cenozoic Fringe (PWMCF) which consists of a set of carbonate plateaux and hills up to 618 m (Ramos *et al.* 2012). With the exception of the northern littoral cliffs and a few Mesozoic carbonate cliffs of the PWMCF (e.g. Cape Mondego and Nazaré), the coast is low. South of the estuary of the Douro River, the Aveiro lagoon and the São Jacinto Dunes (natural reserve with 700 ha) are noteworthy morphological units in this sector north of Figueira da Foz. Digital and paper geological maps at several scales may be purchased through the web page of the Laboratório Nacional de Energia e Geologia (www.lneg.pt/lneg/).

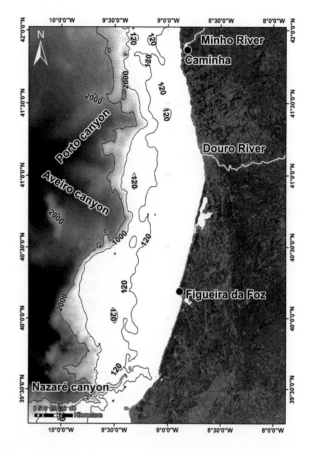

Fig. 11.3 Sector of the Iberian Atlantic Margin between the Minho River and the Nazaré canyon. Note the relatively straight coastal sector between the Douro Estuary and Nazaré, with Cape Mondego near Figueira da Foz being the most conspicuous promontory.

Continental shelf

The continental shelf width ranges between 30 km and 60 km with the shelf break at 160 m depth and it receives a significant sedimentary input from several rivers (Dias & Nittrouer 1984) including the Douro River (see National System of Water Resources Information at snirh.pt/snirh/download/Douro_hoje.pdf) which is the major contributor with 79% of the total annual sediment supply to the shelf (Oliveira *et al.* 2002). A horst system of the rocky substrate forms several plateaus with elevations up to 20 m, which run parallel to the shelf break, thereby creating an obstacle to sediment exportation off-shelf (Dias *et al.* 2002).

The shelf sedimentary characteristics result from both the shelf topography and the effectiveness of the strong hydrodynamic regime, leading to the wide distribution of the sediments along the platform and therefore to a thin sedimentary cover (Dias *et al.* 2002). The coarser sediments are deposited near the coast and subsequently transported southward by longshore drift, whereas the finer sediments are exported to the outer shelf (Dias *et al.* 2002). The deposits of these fine sediments occur mainly on the Douro and Minho mud patches at 120 m depth (Oliveira *et al.* 2002). Sedimentation rates in the Minho and Douro mud patches are 0.10 cm/year to 0.23 cm/year and 0.17 cm/year to 0.4 cm/year respectively (Drago *et al.* 2000). Waves are the main erosional agent of these muddy sediments during winter because shear velocities exceed the erosion critical value, whereas the wave shear values are not strong enough to remobilize the sediment during summer (Vitorino *et al.* 2002).

Several relict sediments on the inner shelf testify to paleocoastlines. Between Caminha and Figueira da Foz the shelf shows a mean width of 43.7 km and depth ranges between 130 m to 200 m. The main physiographic aspects are the submarine canyons of Porto, Aveiro and Nazaré. The latter is a morphological, sedimentological and biological barrier and marks the limit of this sector at the southernmost margin. The submarine canyon of Nazaré is a 'gouf' canyon (head near the coastline) developed along the Nazaré Fault. Its head is located at 50 m depth and it acts as a trap for sediments from longshore drift (Oliveira *et al.* 2007; Schmidt *et al.* 2001).

Western margin: southern Nazaré canyon to Cape Sines

Coastal geomorphology

The coastline is characterized by quite straight-cliffed sectors with different geographical orientations bounded by capes between Nazaré and Cape Sines (Fig. 11.4). These capes have a fundamental role in the coastal circulation patterns as described below. Pocket beaches and bay beaches occur adjacent to rivulet mouths interrupting coastal cliffs with heights up to 100 m exposing carbonate rocks and marls from the Jurassic and Cretaceous. The most conspicuous feature occurring in this sector is the Peniche Peninsula (Fig. 11.4), a small island until the twelfth century, which is currently connected to

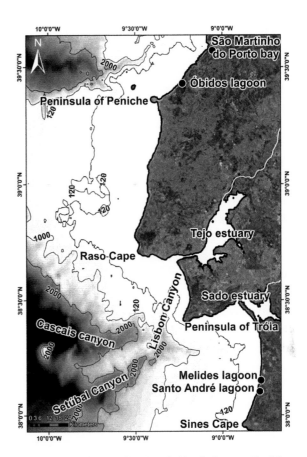

Fig. 11.4 Sector of the Iberian Atlantic Margin from south of the Nazaré canyon to Cape Sines, containing two important estuaries (the Tejo and the Sado) as discussed in the text. Note the irregular shelf dissected by several submarine canyons in this sector.

the mainland through a tombolo due to the silting of the mouth of the São Domingos River (Dias 2004; Oliveira *et al.* 2011). Therefore, the island could have been occupied during low sea levels and shelf exposure, as happened during the LGM. In addition, there are natural caves that could have been used as shelters during the Paleolithic, similar to the Furninha Cave (Bicho & Cardoso 2010). The Furninha Cave is a natural karstic cave located in the southern face of the coastal cliff of the Peniche Peninsula. That cave is filled by eolian sands (9 m thick) enclosing remains of Quaternary fauna (including *Homo sapiens* bones) and was occupied from the Middle Paleolithic to the end of the Chalcolithic (Raposo 1995; Cardoso 2006). The São Martinho do Porto Bay and the Óbidos coastal lagoon are particular coastal environments located between Nazaré and the Peniche Peninsula, the latter one with intermittent artificial communication with the ocean.

The coastal sectors between Cape Raso and Cape Espichel (which lies between the Tejo and Sado estuaries) and between Cape Espichel and Cape Sines enclose two littoral arcs, each of them containing an estuary, the Tejo and Sado estuaries respectively. The Tejo River is the longest of the Iberian Peninsula with the third largest catchment area. The lower Tejo valley is incised into the Tejo sedimentary basin of tertiary sediments up to 1200 m thick (Vis *et al.* 2006). The coastal plain of the Raso-Espichel littoral arc becomes narrow southward and is backed by cliffs on top of which are accumulated eolian sands deposited during the Holocene. In this littoral arc, the Albufeira coastal lagoon (near the Sado Estuary), with depths of up to 15 m and a major axis oblique to the coastline, has a flooded surface of 1.3 km² and is separated from the ocean by a fixed sand barrier (Ferraz *et al.* 2005). The Tróia Peninsula protects the Sado Estuary within the Espichel–Sines littoral arc from wave attack and the river discharges show a strong seasonality (Moreira *et al.* 2009). The Santo André and Melides coastal lagoons cover an area of 2.5 km² and 0.4 km² and have a maximum depth of 4 m and 2 m, respectively. The Santo André lagoon is formed by a main lagoon body and by two smaller elongated bodies oriented north–south, while the Melides lagoon has an elongated shape with the major axis oblique to the shoreline, oriented NE–SW (Pires *et al.* 2011). The southernmost part of this arc, Cape Sines, exposes limestones from the Jurassic that are planed off by a marine abrasion platform at 15 m to 20 m (Carvalho *et al.* 2000).

Continental shelf

This sector is limited to the north and south by the Nazaré canyon and Cape Sines respectively and characterized by the occurrence of several submarine canyons (Fig. 11.4), which intersect the very irregular continental shelf (Lastras *et al.* 2009). The Nazaré canyon is the most extensive gouf-type canyon in Europe: 170 km in length between the −60-m bathymetric contour line (0.5 km off the coast) and the Iberian Abyssal Plain (Andrade 1938; Oliveira *et al.* 2011). It is not connected with any river

and is of tectonic origin. Adjacent to the Nazaré canyon, sandy, silty and muddy sediments show high contents both of $CaCO_3$ and organic matter, the latter associated mainly with the finer sediment (Mil-Homens *et al.* 2006). The Nazaré canyon is an active zone of sedimentation and deposition of the Western Iberian Margin (Schmidt *et al.* 2001; van Weering *et al.* 2002).

This section of the continental shelf is irregular, cut by the Lisbon, Setúbal and Cascais submarine canyons (Gomes 2000). The shelf break occurs mostly at depths greater than 200 m and the shelf width is highly variable, ranging from 15 km to 70 km between the Nazaré and Lisbon canyons. The width of the inner portion of the continental shelf increases southward from 50 km to 90 km. Near Lisbon, the shelf width is 30 km narrowing close to the heads of the Lisbon and Setúbal submarine canyons (Mougenot 1988; Gomes 2000). It is covered by sedimentary bodies such as beaches and littoral ridges inherited from past coastlines as well as muddy sediments from river mouths. This sedimentary cover points to the effectiveness of the shelf in trapping sediments from fluvial discharges and preventing their transfer to deeper zones (Jouanneau *et al.* 1998). In contrast, northward between the Peniche Peninsula and Cape Raso, the sediments are scarce due to the lack of major rivers and the highly energetic marine climate, and the substrate crops out at the inner shelf between 50 m (northern sector) and 90 m (southern sector). North-west of the Peniche Peninsula there is the Berlengas archipelago, composed of small islands and islets towards the edge of the continental shelf. The largest island (Berlenga Grande) is the only one that is inhabited and is of a granitic nature. There, two beach levels of Pleistocene age were identified. Several caves and bays have developed, controlled by the pattern and distribution of granite fracture (Bicho & Cardoso 2010).

At Cape Raso, the Sintra eruptive massif extends to the continental shelf leading to a very irregular submarine morphology. In addition, submerged paleocliffs with sub-horizontal abrasion platforms at the foot also occur. The morphology of the mid shelf, down to −140 m, is structurally controlled, favoring the development of particular sedimentary environments southward from the Peniche Peninsula such as in the Lourinhá valley and the graben of the Ericeira Sea. Elongated rocky ridges with very inclined slopes occur at between −70 m and −40 m. North and south of the Ericeira valley two sub-sectors can be defined on the outer shelf, with the northern sector having a gradient of 2° and remarkable width, extending to a depth of 390 m. The southern sector shows a steeper slope and there are several submarine ridges resulting from differential erosion (Vanney & Mougenout 1981; Badagola *et al.* 2006). Southward from Ericeira, the shelf break approaches the coast and is shallower (−170 m) near Cape Raso.

Between Cape Raso and the Setúbal canyon, the continental shelf is narrow, showing its minimum width of 3 km near Cape Espichel. It extends to a depth of 180 m and the major morphological features are the estuaries of the Tejo and Sado rivers, the submarine canyons of Cascais (8 km length), Setúbal (25 km length) and its tributary to the Lisbon canyon (Matos *et al.* 2006). The Sado Estuary has a direct connection with the Setúbal canyon (Garcia *et al.* 1997). The sedimentary cover is mainly influenced by the Tejo and Sado sedimentary input. Off the Tejo Estuary there occurs a muddy and silty body up to 15 m thick, spread over an area of 560 km² indicating a sedimentation rate of 0.38 cm/year to 1.2 cm/year (Drago *et al.* 2000). The Sado ebb delta is a submerged sandy body, which extends up to 5 km from the mouth of the gorge and covers an area of 47 km² (Brito *et al.* 2005). Off the Sado Estuary, several slope ruptures were interpreted as paleocoastlines between 60 ka and 11 ka (Brito *et al.* 2010). Between the submarine Setúbal canyon and Cape Sines, the continental shelf width is less than 25 km and the continental shelf break at a depth between 160 m and 180 m is poorly defined (Matos *et al.* 2006).

Western margin between Cape Sines and Cape St. Vicente

Coastal geomorphology

Southward from Cape Sines, the southwestern Portuguese coast (Fig. 11.5) extends as a rocky cliffed coast exposing mainly shales and graywackes from the Paleozoic, where beaches occur only at the mouths of rivers and rivulet. The coastline, roughly straight, is NE–SW orientated.

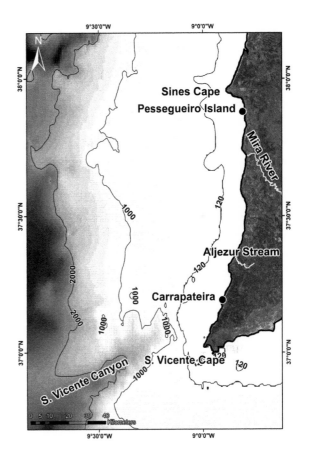

Fig. 11.5 Iberian Atlantic Margin between Cape Sines and Cape St. Vicente, where the Mira and Aljezur rivers are the only waterways contributing to the sedimentary input to the shelf.

Cliffs up to 100 m high have experienced neotectonic uplift (Plio-Pleistocene) as evidenced by raised beaches and paleocliffs (Figueiredo *et al.* 2009). The Mira and Aljezur estuaries are the main interruptions on this cliffed coast.

Continental shelf

Southward from Cape Sines, which is an effective obstacle to northward longshore drift, the continental shelf is narrow, less than 25 km wide (only 10 km near Cape St. Vicente), and quite atypical, formed by uneven surfaces, without well-defined physiographic provinces, except in southern Carrapateira. This is mainly due to the thick sequence of Neogene sediments, which fossilize and preserve the paleorelief. The shelf break, roughly defined, is located at between 160 m and 180 m depth. Several levels of marine abrasion were recognized at −90 m to −97 m and −111 m to −125 m corresponding to Quaternary sea levels. Pessegueiro Island, 250 m from the coastline is formed by eolianites overlying Paleozoic rocks accumulated during the Late Pleistocene (Carvalho *et al.* 2000) probably during the LGM when the shelf was exposed to strong winds. Several trackways of *Cervus elaphus* can be observed in the eolianites of Pessegueiro Island in at least six stratigraphic horizons (Neto de Carvalho 2009).

The Setúbal canyon (Fig. 11.4) transfers sediments from the Tejo and Sado rivers to the abyssal plain. Thus, the shelf southward from Cape Sines shows a sparse sedimentary cover because in addition to the reduced sedimentary load carried by rivulets (with the exception of the Mira and Aljezur rivers), the energetic marine climate is efficient in dispersing sediments. The St. Vicente canyon at the southern limit of the southwest Iberian Atlantic Margin is not associated with any drainage net, and is of tectonic origin.

Southern margin between Cape St. Vicente and Cape Santa Maria

Coastal geomorphology

This sector extends between Cape St. Vicente and the Ria Formosa coastal lagoon near Faro, from where the coast-line changes from NW–SE to SW–NE (Fig. 11.6). The coastal zone is mainly a cliffed one exposing carbonate rocks, uncohesive sands and sandstones. In the eastern part of this sector are located the lowlands of the coastal plain and the coastal lagoon system. Between Sagres (near Cape St. Vicente) and Lagos, limestone, dolomite and marl from the Jurassic and to a less extent from the Cretaceous, form the highest coastal cliffs of this sector, up to 60 m in Cape St. Vicente where the hardest rocks from the Jurassic crop out. Cliffs are quite abrupt with marine caves at the foot, several of them with apertures only accessible by diving. In the central part of the sector (between Lagos and Olhos de Água) the cliffs expose mainly Miocene fossiliferous calcarenite. These rocks are extremely karstified leading to a crenulated coastline with headlands, pocket beaches, caves and numerous stacks. Moreover, Pliocene and Pleistocene fluvio-marine sediments fill the paleovalleys that are incised in the Miocene Formation and the Lagos-Portimão Carbonate Formation (Pais *et al.* 2012). Due to the tectonic behavior of the Algarve basin, rocky cliffs grade laterally into sandy and sandstone cliffs eastward from Olhos de Água. Here, the exposed sedimentary sequence assumes a key role in the establishment of the Plio-Pleistocene boundary of the Algarve region where it is possible to identify the Gauss-Matuyama geomagnetic inversion (Moura &

Fig. 11.6 Southeastern sector of the Iberian Atlantic Margin. Note the crenulated coastline between Cape St. Vicente and Olhos de Água exposing Miocene karstified carbonate rocks. Between Olhos de Água and Quarteira, coastal cliffs expose sands and sandstones from the Pliocene and Pleistocene.

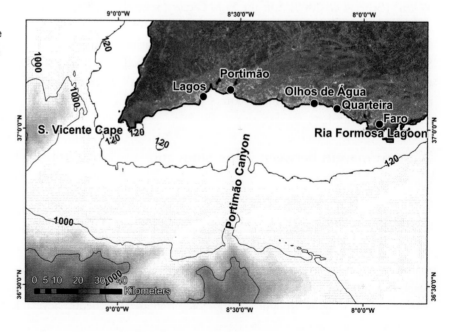

Boski 2009). An upper surface attributed to the marine abrasion during the Late Pliocene sea-level highstand cuts across all the formations previously referred to, creating a littoral plane the elevation of which decreases eastward from Sagres. Major rivers draining into this coastal sector have estuaries that were filled during the Holocene.

Continental shelf

The continental shelf is narrow (8–28 km) and the shelf break is between 110 m and 150 m depth, with the St. Vicente and Portimão submarine canyons being the major morphological features. Other canyons of lesser dimensions are also observable on the edge of the shelf, where the morphology is relatively simple, but their connections to the emerged river valleys are obliterated by sedimentary infill. To the east of the Portimão canyon, which is of tectonic origin (Lopes *et al.* 2006), a progradational sedimentary body developed, which preserved the Miocene paleorelief, having been deposited during the Plio-Quaternary (Vanney & Mougenot 1981). Roughly parallel relief near Lagos at −105 m and −110 m is probably formed by cemented paleocoastal ridges which could have been generated during a sea-level lowstand (Vanney & Mougenot 1981). At a moderate water depth (ca. 30 m) paleobarriers, spit bars and beaches composed of lithified sands of diverse granulometries occur parallel to the current coast testifying to paleoquaternary shores eastward from Lagos (Infantini *et al.* 2012). Eastward from Quarteira, a prograding sedimentary body has a remarkable thickness of 500 m that covers the continental shelf, showing sedimentary discontinuities and fossil valleys off the Faro region (Vanney & Mougenot 1981). In a general way, the sediments covering the Algarve shelf (southern Portugal, Figs. 11.6 and 11.7) are finer than the ones in the northern and western Iberian shelf. However, two coarse deposits were observed, one of them near Cape St. Vicente (pebbly sand at 125 m depth), the other one near Olhos de Água at a mean depth of 37 m (bioclastic sand) (Dias *et al.* 1980).

Gulf of Cádiz

Coastal geomorphology

The Gulf of Cádiz lies between Cape Santa Maria (Portugal) and the Gibraltar Strait (Spain). The coastline is bay shaped between Cape Santa Maria and the Guadalquivir Estuary, backed mainly by coastal lagoons and estuaries of which the Guadiana, the Tinto-Odiel and the Guadalquivir (Fig. 11.7) are the most important and contribute large amounts of sediment to the

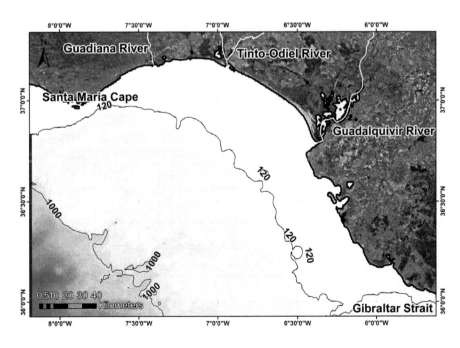

Fig. 11.7 Gulf of Cádiz showing the main rivers that supply sediments to the shelf (Guadiana, Tinto–Odiel and Guadalquivir). Note the narrowness of the continental shelf off Cape Santa Maria.

continental shelf. Between the Guadalquivir Estuary and the Gibraltar Strait, the coastline is roughly orientated NW–SE and is a mixed sandy-rocky coast with several embayments fed by sediments from short rivers draining the western Betic ranges (Del Rio *et al.* 2002).

Continental shelf

Since the Miocene, the central and western sectors of the Gulf of Cádiz have been subsiding, favoring the deposition of thick sedimentary sequences, thus creating a depocenter during the Pliocene and Quaternary. In contrast, the southeastern sector maintained an elevation structurally controlled during the Quaternary. Here, thin deposits were formed, some of them later eroded (Maldonado & Nelson 1999). The continental shelf width ranges between 30 km and 40 km and has a slope of 0.2° to 0.32° (Hernández-Molina *et al.* 2002). Continental shelf evolution was greatly influenced by tectonics in the Gulf of Cádiz leading to the formation of fault-bounded basins in the northwest sector, whereas a compressive regime in the southeastern sector was responsible for the formation of fault-bounded blocks and therefore the absence of significant sedimentary basins (Maldonado & Nelson 1999). During the Quaternary, the tectonic regime has been differentiated in south Iberia into several sectors due to its complexity. An extensional regime associated with an effective subsidence produced a large space in the northwest sector of the Gulf of Cádiz, which could accommodate thick sedimentary sequences, with contributions from the Guadalquivir and Guadiana rivers (Maldonado & Nelson 1999). The Gulf of Cádiz circulation is influenced by the water mass exchange between the Mediterranean Sea and the Atlantic Ocean through the Gibraltar Strait and is highly dynamic (Hernández-Molina *et al.* 2002).

Marine Climate

The Iberian Atlantic coast experiences two main types of marine climate: (i) the northwest and west coasts, which are well exposed both to wind waves generated locally and to Atlantic swell and, (ii) the south and southeast coasts, which are in a sheltered position relative to the very energetic conditions from the Atlantic Ocean. Waves approaching the coast are modified, refracted, reflected and dissipate or concentrate energy, according to the underwater topography and coastal morphology. The Iberian Atlantic Margin experiences a semi-diurnal mesotidal tidal regime. Online information concerning the marine climate of the Portuguese coast may be accessed at the Instituto Hidrográfico web page (www.hidrografico.pt/) and Instituto Português do Mar e da Atmosfera (www.ipma.pt/pt/).

Northwestern coast

The Cantabrian coast is dominated by the north-west storms from the Atlantic (Arteaga Cardineau & González 2005). Swell waves represent 23.5% of the occurrence frequency with a main direction from the north-west, whereas wind waves have an occurrence of 7.9% from WNW and 7.6% from NNW (Lechuga *et al.* 2012). During summer, the wave peak period (T_p) is lower than 10 s and the significant wave height (H_s) is 1.5 m during 75% of the year. During winter, the T_p and H_s are 13 s and 2 m respectively for 50% of occurrences (www.puertos.es). However, during storms, waves can reach up to 10 m in height. The tidal regime is semidiurnal with a mean tidal range of 1.65 m and 4.01 m during neap and spring tides respectively (www.puertos.es). The Galician coast of northwest Spain is a typical mesotidal regime with a wave-dominated coast. Swell waves from the north-west have H_s of 1.7 m to 3.2 m and T_p of 9.2 s to 12 s, but wave heights may be as high as 18 m during storms (Bao *et al.* 2006). Nevertheless, in spite of the fact that H_s may be as high as 8 m during winter, a generally low-energy marine climate allows fine sediments to accumulate at the mid shelf off the Douro Estuary (Fig. 11.3) during the summer months (the Douro Mud Patch) (Vitorino *et al.* 2002).

Western coast

Swell waves generated far away in the Atlantic approach mainly from the north-west (77.3% of occurrence), followed by the western waves (20% of occurrences), southwestern (2.4%) and southern (0.2%) (Costa *et al.*

2001). The presence of capes and peninsulas leads to the occurrence of shadow zones where the hydrodynamic conditions are protected from the predominant waves from the north-west, but they are very exposed to storm conditions from the south-west. Significant wave height ranges from 1 m to 3 m and the period from 7 s to 15 s. On this coast, storms are associated with the south-west climate and do not exceed 3% of occurrences (Costa *et al.* 2001). Tides are semidiurnal with an average range of 2.8 m to 2.9 m. Additionally, the wave refraction against the major headlands induces a northward longshore drift, leading to the formation of extensive littoral ridges anchored to the mainland at its southern part and aggrading northward. However, the resultant transport generated by this complex circulation pattern is from south to north, in spite of the waves approaching from the north-west for most of the year. Wave refraction induced by Cape Sines produces a northward littoral current and therefore northward sedimentary transport.

Southern coast between Cape St. Vicente and Cape Santa Maria

In the southernmost sector, between Cape St. Vicente and Faro, the mean tidal range is 2.8 m during spring tides and 1.3 m during neap tides, with a maximum tidal range of 3.5 m (www.hidrografico.pt/). Mean tidal range at the Gulf of Cádiz is 2.02 m. The Algarve coast has a low-energy regime when compared to the west and north regions. Storm conditions are associated with western and southwestern wind waves and affect mainly the southern coast between Cape St. Vicente and Cape Santa Maria. Conditions from the south-east (10% of occurrences) cause waves with low fetch generated in the Gibraltar Strait. The mean of significant wave height is 1 m in the southernmost coastal sector, being lower than 1 m during 68% of the year. Waves higher than 3 m represent only 2% of the occurrence frequency in a year. Only 4% of the record at the southern margin shows mean periods greater than 7 s and, during 63% of the year the mean period ranges between 3 s and 5 s. Peak period is strongly associated with the wave direction from two dominant sectors: (i) SW–W (71% of the year), (ii) SE (23% of the year). This difference in

wave period is caused by the origin of wave generation and hence the fetch distance over which the waves have traveled. Higher periods are associated with waves from the south-west and west whereas lower values are related to the short fetch from the south-east. Stormy conditions (significant wave height >3.5 m) are associated with SW–W conditions and represent ca. 10 days during winter on average (Costa *et al.* 2001). In the Gulf of Cádiz, the waves are mainly from the south-west (associated with stormy conditions) and the south-east (with 10% of occurrences) near Huelva. Here, 75% of the recorded waves have H_s lower than 0.5 m. Near Cádiz, waves approach mainly from the north-west and as in Huelva, this is a low-energy coastal sector. This coast has a mesotidal regime and waves come mainly from the west and south-west leading to longshore drift towards the east and south-east. The changing of the coastline orientation to NW–SE after the Guadalquivir mouth influences the wave angle to the coast decreasing southward and, near the Gibraltar Strait, littoral currents and longshore drift are weak (Del Rio *et al.* 2002). The significant wave height is 0.6 m and during storms may reach 1.5 m (Benavente *et al.* 2005). The circulation over the inner shelf in the northern sector of the Gulf of Cádiz is the consequence of counteracting flows on the outer shelf and continental slope. A warm inshore current counter to the offshore upwelling circulation extends westward from the inner sector of Gulf of Cádiz (Sánchez *et al.* 2006).

Quaternary Evolution

The Quaternary is characterized by frequent climatic changes inducing eustatic sea-level changes. Two interglacial periods are identified as responsible for eustatic mean sea levels above the present one: Marine Isotopic Stage (MIS) 11 and the sub stage MIS 5e. However, heights attained above MSL are widely regarded as often due to local causes such as crustal movements and changes in the sedimentary budget. Moreover, sites so far from ice sheets are not influenced as much by isostatic rebound, compared to high-latitude environments. The Iberian Atlantic Margin extends through a large latitudinal zone and displays a huge variety of morphotectonic contexts. While behavior of the MSL during the Holocene is well understood, this is not the case for the Pleistocene due to several factors: (i) the MSL was mostly below the present one and therefore the geomorphological evidence is now submerged, (ii) several marine terraces and shore platforms are difficult to date due to a lack of datable elements and, (iii) the geological evidence of MSL higher than the present has been rapidly destroyed due to the intense rate of coastal erosion along almost the entire Iberian Atlantic Margin.

During the last interglacial (MIS 5e) the mean sea level was higher than the present as testified by beach deposits along much of the Iberian Atlantic Margin, such as the beach deposit identified on the west coast at Magoito and São Julião (northern Lisbon) at heights of 8 m to 10 m, attributed to MIS 5e (Pereira *et al.* 2006). Several shore platforms along the Iberian Atlantic Margin ranging between 3 m' and 9 m above the present MSL were attributed to MIS 5e (Moura *et al.* 2006). By the end of the Pleistocene, after the warm conditions of MIS 5e, an extreme cold event (the LGM) occurred at 18 ka BP to 20 ka BP within the Last Glacial Period (LGP) which contains MIS 4, MIS 3 and MIS 2. MIS 3 was previously assumed to be a warm phase, but is now accepted as a transitional phase of extreme climatic instability (Bard 2002) and, therefore, belongs to the LGP together with MIS 4 and MIS 2 (van Meerbeeck *et al.* 2009), during which surface water temperature at the Iberian Atlantic Margin was between 5°C and 12°C lower than at present (Abreu *et al.* 2003). Several cold events occurring between 70 ka and 25 ka are typified by an increase of terrigenous input onto the Portuguese Margin. At the northern part of the Portuguese Margin, the frequent transport by turbidity currents was probably due to a humid climate and high fluvial input between 70 ka and 30 ka (Baas *et al.* 1997). Continental shelf slope breaks at 80 m, 75 m to 65 m and 45 m to 35 m depth, were interpreted as marine abrasion surfaces carved during MIS 4 and MIS 3 between 60 ka and 25 ka at the southwest margin (Brito *et al.* 2010). Several spit bars and/or dunes submerged at the Algarve shelf were reported by Vanney and Mougenot (1981). Those carbonate sandy (up to 80% $CaCO_3$) bodies are karstified and therefore should

have been exposed during the LGM (Infantini *et al.* 2012).

At 18 ka BP the coastline was close to the shelf break between −130 m and −140 m (Dias *et al.* 2000) and the polar front reached northern Portugal (McIntyre *et al.* 1976; Molina-Cruz & Thiede 1978). Changes in temperature during the LGP provoked successive changes between *Quercus* forest and steppe in the Iberian Peninsula (Goñi *et al.* 2000). During the LGM a strong hydrodynamic regime provoked the accumulation of coarse sediment in the shoreface zone, whereas the finer particles were exported off-shelf in Galicia (Lantzsch *et al.* 2009). Between 15 ka BP and 14.5 ka BP a transgressive pulse led to a sea-level rise from −120 m to −100 m, and afterwards to −40 m during the warm climatic event (Bølling–Allerød) after 13 ka BP (Dias *et al.* 2000). The coarse sedimentary facies off Galicia were overlaid by finer sediments deposited in shallow water during the rapid MSL rise after 13 ka cal BP on the outer shelf, where they have been preserved (Lantzsch *et al.* 2009). At the estuary of the Rio Minho a fluvial facies started to accumulate at 13.4 ka cal BP fossilizing the bedrock paleorelief (Araújo *et al.* 2005). However, an important cold phase occurred at 11.5 ka BP (the Younger Dryas), due to the intense influx of melt waters into the Atlantic. This led to a new phase of enhanced river flow and higher erosional capacity when MSL was 40 m to 60 m below present levels (Dias *et al.* 2000). This temporary return to glacial conditions was recorded on the Portuguese Margin by an increased input of terrigenous silt and sand (Cascalho *et al.* 1994). The Younger Dryas is a meltwater event equivalent to Heinrich events in the central North Atlantic, but only a few icebergs reached the Portuguese Margin. Minor sediment winnowing occurred on the northern part of the Portuguese Margin during the Younger Dryas. Holocene sediments have a typical interglacial signature, with a dominance of biogenic over terrigenous sedimentation (Baas *et al.* 1997).

After the Younger Dryas, the rate of sea-level rise was high during the Early Holocene up to 6 ka BP to 5 ka BP and as a response wide estuaries began to form once sediments were no longer efficiently exported to the shelf. In Galicia, sea level was 25 m to 30 m lower than present (Méndez & Vilas 2005) and rose rapidly in the beginning of the Holocene transgression (between 8 ka cal BP and 6 ka cal BP). After this rapid transgressive pulse, subsequent deceleration led to the infilling of valleys, and several complexes of beach-barrier wetlands started to form at 5.7 ka cal BP. Rates of sedimentation in the estuaries range between 0.28 mm/year at 5.7 ka cal BP and 2.60 mm/year after 1 ka cal BP (Bao *et al.* 2006). The Ria of Vigo, a completely exposed terrestrial feature during the LGM with an erosional profile in the main channel, was flooded during the Holocene transgression, and a sedimentary depocenter 57 m thick was situated in the central part of the ria (García-García *et al.* 2005). At the Rio Minho Estuary the first marine influence was identified at 9,020 cal BP in sediments overlying the previous fluvial facies (Araújo *et al.* 2005). The large tidal environments of the lower Tejo valley started to form after 12 ka cal BP as a response to sea-level rise, substituting the pre-Holocene transgression braided fluvial system, and marine transgression attained its maximum at 7 ka cal BP (Vis *et al.* 2008). Similarly, the lower estuary of the Sado River and the Mira Estuary (southwest coast), were completely flooded between 7 ka BP and 5 ka BP (Freitas *et al.* 2003; Brito *et al.* 2010). Mean sea level was ca. 20 m lower than the present at 8 ka BP (Boski *et al.* 2002) on the Algarve eastern coast and rose ca. 0.85 m/century (ca. 30 times more than today) until 6.5 ka BP when it reached −15 m to −20 m. Between that time and 5 ka BP, sea-level rise slowed down (0.25 m/century) and sea level attained a level similar to the present (Boski *et al.* 2002). At the Guadiana Estuary, about 80% of the Holocene sequence accumulated with a rate of 80 cm/century during the first phase of the sea-level rise up to 6.7 ka cal BP (Boski *et al.* 2002). The fluvial facies corresponding to the LGP were overlaid by thick sequences of salt-marsh facies accumulated during the rapid rise of MSL up to 6 ka BP (Boski *et al.* 2002). The accumulation in the Guadiana River is a 16-m thick sedimentary sequence influenced by the estuary morphology. In fact, the infilling pattern of the estuaries depends on the balance between sedimentary input and the volume to be filled. A good example of the importance of the balance between sea-level rise velocity and sedimentary input is the Gulf of Cádiz. Two patterns of sedimentation during the Holocene highstand can be distinguished, due to different fluvial inputs and wave energy. The lack of major rivers west of the Guadiana River (Portuguese–Spanish boundary) and a moderate-to-high energy-wave climate lead to the dominance of erosive processes, whereas, eastward from the Guadiana River, depositional systems prevail due to the contribution of a larger fluvial supply (Lobo *et al.* 2004). The Late Pleistocene–Holocene sedimentary body records the eustatic transgression (18–6 ka BP) through transgressive sequences overlapped by highstand facies (after 6 ka BP), which support current coastal environments (Gutierrez-Mas *et al.* 1996). During the last eustatic maximum, sea level reached 2 m above the present level leading to the generation of estuaries and sea inlets which persisted in the Gulf of Cádiz until Roman times (Zazo *et al.* 1994).

The deceleration of the MSL rise after 6 ka BP to 5 ka BP led to the genesis of extensive mud flats, ria-like environments, coastal lagoons and barrier systems throughout the Iberian Atlantic Margin (Zazo *et al.* 1994; Freitas *et al.* 2003; Andrade *et al.* 2004; González-Villanueva *et al.* 2009; Lantzsch *et al.* 2009). Coastal dune generation was favored during cold and dry phases of the Pre-Boreal (9.5–9 ka BP) and Boreal (8.7–7.5 ka BP) along the Portuguese coast (e.g. Soares & Sousa 2003; 2006; Moura *et al.* 2007). Epipaleolithic shell middens are preserved in the Magoito and São Julião eolianites near Lisbon (Soares & Sousa 2003).

Past Landscapes and Their Preservation

Landscapes on the continental shelves experienced polygenic evolution (marine and sub-aerial) depending on sea-level changes, and landscapes and materials under water may be better preserved than the ones exposed to aerial weathering. However, the success of preservation depends on several environmental variables such as marine climate, rate of burial, and frequency and rate of sea-level rise and fall. Intrinsic properties of the landscape and factors such as chemical processes and mechanical cohesive strength are also important to the potential for preservation. The effect of erosion and transport

TABLE 11.1 COMPARISON OF THE POTENTIAL PRESERVATION OF UNDERWATER LANDSCAPES AT THE IBERIAN ATLANTIC MARGIN, BASED ON MARINE CLIMATE, MORPHOLOGY AND SEDIMENTARY PROCESSES.

SECTOR (see Fig. 11.1)		Fluvial Sedimentary Input	Energy	Shelf Sedimentary Coverage	Erosional vs. Depositional Processes	Preservation Potential
1 Northern Margin Cádiz	Cantabria	High	High	Poor	Erosion> Deposition	Low Exception: The 7-m mud patch at the easternmost part
	Galicia	High	High	Thin	Erosion> Deposition	Low Exceptions: 1) coastal embayments that were flooded during the Holocene transgression 2) mid-shelf
2 Northwestern between Minho River and Nazaré canyon		High	High	Thin	Erosion> Deposition	Low
3 Southern Nazaré canyon to Sines		Low northward of Cape Raso	High to Moderate at the sites in the sheltered position conferred by capes and peninsulas	Thin Between the Peniche Peninsula and Cape Raso	Deposition> Erosion southward of Cape Raso	Moderate Mainly southward of the Cabo da Roca, the Berlengas archipelago (submerged caves and shelters) and Peniche Peninsula (submerged caves and shelters)
		High southward of Cape Raso		Thick southward of Cape Raso to Cape Sines		
4 Western between capes Sines and St. Vicente		Low	High	Poor	Erosion> Deposition	Poor
5 Southern Margin between capes St. Vicente and Santa Maria		Moderate	Low	Variable	Deposition> Erosion	High
6 Gulf of Cádiz		High	Low	Thick	Deposition> Erosion	High

by waves and currents may be attenuated if a rapid burial occurs. Therefore, the balance between energy, sedimentary input and tectonics should be taken into account when the aim is to evaluate the preservation potential of underwater remains. Tectonics plays an important role because subsidence and uplift favors burial by deepening the basin of deposition and exhumation (due to exposure to wave attack) respectively. A summary of the potential preservation of remains at the Iberian Atlantic Margin is shown in Table 11.1 considering two main scenarios: (i) erosive shelf conditions experiencing high-energy levels and low sedimentary inputs and (ii) constructive shelf conditions showing high sedimentary supply and low hydrodynamics. Collapsed and filled marine caves, which are common in karstic landscapes of carbonate shores, are assumed to be excellent traps to preserve archaeological remains. In addition, eolianites and beachrocks of cemented sands provide relatively soft material to be excavated and sliced and are therefore favorable to various structures such as wells or shelters and are capable of withstanding marine abrasion for a considerable period. Nonetheless, where the continental shelf is very narrow (e.g. 3 km near Cape Espichel) the vestiges of human occupation during the LGM are probably and inevitably exposed close the current coastline.

Data Sources

Sistema Nacional de Informação de Recursos Hídricos (SNIRH): snirh.pt/

Instituto Español de Oceanografía, Boya Océano-meteorológica Augusto González de Linares, Santander: www.ieo-santander.net/datosoceanografi cos/boya-agl.php

Instituto Hidrográfico Marinha Portuguesa: www. hidrografico.pt/

Instituto Geológico y Minero de España: www.igme.es/ internet/default.asp

Instituto Português do Mar e da Atmosfera: www. ipma.pt/pt/

Laboratório Nacional de Energia e Geologia (LNEG): www.lneg.pt/lneg/

Puertos del Estado. Gobierno de España, Ministerio del Fomento: www.puertos.es/

Atlas da Água (SNIRH): snirh.apambiente.pt

Ministerio de Agricultura y Pesca, Alimentación y Medio Ambiente. Gobierno de España: www. mapama.gob.es/

Map service by Esri: www.esri.com

British Oceanographic Data Centre — Natural Environ-mental Research Council. General Bathymetric Chart of the Oceans (GEBCO): www.gebco.net

Acknowledgements

The authors would like to thank the Fundação para a Ciência e a Tecnologia (FCT) [The Foundation for Science and Technology] for their financial support through numerous research projects, the data from which has underpinned much of what has been stated in this chapter.

References

Abreu, L. de, Shackleton, N. J., Schönfeld, J., Hall, M. & Chapman, M. 2003. Millennial-scale oceanic climate variability off the western Iberian margin during the last two glacial periods. *Marine Geology* 196:1-20.

Agirrezabala, L. M., Owen, H. G. & García-Mondéjar, J. 2002. Syntectonic deposits and punctuated limb rotation in an Albian submarine transpressional fold (Mutriku village, Basque-Cantabrian basin, northern Spain). *Geological Society of America Bulletin* 114:281-297.

Alonso, A. & Pagés, J. L. 2007. Stratigraphy of Late Pleistocene coastal deposits in Northern Spain. *Journal of Iberian Geology* 33:207-220.

Andrade, C. F. 1938. *Os Vales submarinos portugueses e o diastrofismo das Berlengas e Estremadura.* Casa Portuguesa: Lisbon.

Andrade, C., Freitas, M. C., Moreno, J. & Craveiro, S.C. 2004. Stratigraphical evidence of Late Holocene barrier breaching and extreme storms in lagoonal sediments of Ria Formosa, Algarve, Portugal. *Marine Geology* 210:339-362.

Araújo, M. F., Lobato, M. F., Cruces, A. & Drago, T. 2005. Paleoenvironmental geochemical patterns in the Holocenic evolution of Minho Estuary. In Freitas, M. C. & Drago, T. (eds.) *Iberian Coastal Holocene Paleoenvironmental Evolution — Coastal Hope Conference.* 24th – 29th July 2005, Lisbon, pp. 8-10.

Arteaga Cardineau, C. & González, J. A. 2005. Natural and human erosive factors in Liencres Beach spit and dunes (Cantabria, Spain). *Journal of Coastal Research* (Proceedings of the 2nd Meeting in Marine Sciences. 6th – 8th May 2004, Valencia — Special Issue 49):70-75.

Baas, J. H., Mienert, J., Abrantes, F. & Prins, M. A. 1997. Late Quaternary sedimentation on the Portuguese continental margin: climate-related processes and products. *Palaeogeography, Palaeoclimatology, Palaeoecology* 130:1-23.

Badagola, A., Rodrigues, A., Terrinha, P. & Veiga, L. 2006. Geomorphological characterization of the Estremadura Spur Continental shelf. In Balbino, J. (ed.) *VII Congresso Nacional de Geologia*. 29th June – 13th July 2006, Universidade de Évora, Pólo de Estremoz, Portugal, pp. 377-380.

Bao, R., Alonso, A., Delgado, C. & Pagés, J. L. 2006. Identification of the main driving mechanisms in the evolution of a small coastal wetland (Traba, Galicia, NW Spain) since its origin 5700 cal yr BP. *Palaeogeography, Palaeoclimatology, Palaeoecology* 247:296-312.

Bard, E. 2002. Climate shock: Abrupt changes over millennial time scales. *Physics Today* (Dec):32-38.

Benavente, J., Borja, F., Gracia, F. J. & Rodriguez, A. 2005. Introduction to the Gulf of Cadiz Coast. In Gracia, F. J. (ed.) *Geomorphology of the South-Atlantic Spanish Coast. Field trip guide A4. VI International Conference on Geomorphology*. 7th – 11th September 2005, Zaragoza (Spain), pp. 1-12.

Bernárdez, P., González-Álvarez, R., Francés, G., Prego, R., Bárcena, M. A. & Romero, O. E. 2008a. Late Holocene history of the rainfall in the NW Iberian peninsula — Evidence from a marine record. *Journal of Marine Systems* 72:366-382.

Bernárdez, P., González-Álvarez, R., Francés, G., Prego, R., Bárcena, M. A. & Romero, O. E. 2008b. Palaeoproductivity changes and upwelling variability in the Galicia Mud Patch during the last 5000 years: geochemical and microfloral evidence. *The Holocene* 18:1207-1218.

Bicho, N. & Cardoso, J. L. 2010. Paleolithic occupations and lithic assemblages from Furninha Cave, Peniche (Portugal). *Zephyrus* 66:17-38.

Boski, T., Moura, D., Veiga-Pires, C. *et al.* 2002. Postglacial sea-level rise and sedimentary response in the Guadiana Estuary, Portugal/Spain border. *Sedimentary Geology* 150:103-122.

Brito, P., Terrinha, P., Rebêlo, L. & Monteiro, H. 2005. Deltaic sedimentary structure interpreted from high resolution seismic data: Sado estuary, Portugal. In Freitas M. C. & Drago, T. (eds.) *Iberian Coastal*

Holocene Paleoenvironmental Evolution — Coastal Hope Conference. 24th – 29th July 2005, Lisbon, pp. 15-16.

Brito, P., Terrinha, P., Duarte, H., Rebêlo, L., Ferraz M. & Costas, S. 2010. Submerged paleo-coastlines on the continental shelf offshore the Sado estuary. In *Geosciences On-line Journal e-Terra*, vol. 12 (5) *VIII Congresso Nacional de Geologia*, 9th – 16th July, Braga, Portugal, pp. 1-4 (in Portuguese).

Cardoso, J. L. 2006. The Mousterian complex in Portugal. *Zephyrus* 59:21-50.

Carvalho, G. S., Gomes, F. V. & Pinto, F. T. (eds.) 2000. *A zona costeira do Alentejo*. Associação EUROCOAST-PORTUGAL: Porto.

Cascalho, J., Magalhães, F., Dias, J. M. A. & Carvalho, A. G. 1994. Sedimentary unconsolidated cover of the Alentejo continental shelf (first results). *Gaia: Revista de Geociências* 8:113-118.

Costa, M., Silva, R. & Vitorino, J. 2001. Contribuição para o estudo do clima de agitação Marítima na costa portuguesa. In *Proceedings of 2as Jornadas Portuguesas de Engenharia Costeira e Portuária*. Associação Internacional de Navegação: Sines (Portugal).

Del Rio, L., Benavente, J., Gracia, F. J. *et al.* 2002. The quantification of coastal erosion processes in the South Atlantic Spanish coast: Methodology and preliminary results. In Gomes, F. V., Pinto, T. & das Neves, L. (eds.) *6th International Conference Littoral 2002*. 22nd – 26th September, Porto, Portugal, pp. 1-9.

Dias, J. M. 2004. A história da evolução do litoral português nos últimos vinte milénios. In Tavares, A. A., Tavares, M. J. F. & Cardoso, J. L. (eds.) *Evolução Geohistórica do Litoral Português e Fenómenos Correlativos: Geologia, História, Arqueologia e Climatologia*. pp. 157-170. Universidade Aberta: Lisbon.

Dias, J. M. A. & Nittrouer, C. A. 1984. Continental shelf sediments of northern Portugal. *Continental Shelf Research* 3:147-165.

Dias, J. M. A., Monteiro, J. H. & Gaspar, L. C. 1980. Potencialidades em cascalhos e areias da plataforma continental portuguesa. *Comunicações dos Serviços Geológicos de Portugal* 66:227-240.

Dias, J. M. A., Boski, T., Rodrigues, A. & Magalhães, F. 2000. Coast line evolution in Portugal since the Last

Glacial Maximum until present — a synthesis. *Marine Geology* 170:177-186.

Dias, J. M. A., Gonzalez, R., Garcia, C. & Diaz-del-Rio, V. 2002. Sediment distribution patterns on the Galicia-Minho continental shelf. *Progress in Oceanography* 52:215-231.

Drago, T., Jouanneau, J. M., Weber, O., Naughton, F. & Rodrigues, A. 2000. Environmental factors controling the depositional facies charateristics of Minho, Douro and Tejo mud patches. In Dias, J. A. & Ferreira, Ó. (eds.) *3ʳᵈ Symposium on the Iberian Atlantic Margin*. 25ᵗʰ – 27ᵗʰ September 2000, Faro, Portugal, pp. 217-218.

Ercilla, G., Casas, D., Estrada, F. *et al.* 2008. Morphosedimentary features and recent depositional architectural model of the Cantabrian continental margin. *Marine Geology* 247:61-83.

Fernàndez-Viejo, G., López-Fernández, C., Domínguez-Cuesta, M. J & Cadenas, P. 2014. How much confidence can be conferred on tectonic maps of continental shelves? The Cantabrian-Fault case. *Scientific Reports* 4:article 3661.

Ferraz, M., Silva, E., Cruces, A. *et al.* 2005. Environmental characterization of the Albufeira Lagoon (Portugal) at micro timescale using a multidisciplinar approach. In Freitas M. C. & Drago, T. (eds.) *Iberian Coastal Holocene Paleoenvironmental Evolution — Coastal Hope Conference*. 24ᵗʰ – 29ᵗʰ July 2005, Lisbon, pp. 54-57.

Figueiredo, P. M., Cabral, J. & Rockwell, T. 2009. Actividade tectónica Plio-Plistocénica no SO de Portugal: Novos dados adquiridos por observações de campo e trincheiras. In Flor Rodríguez, G., Gallastegui, J., Flor Blanco, G. & Martín Llaneza, J. (eds.) *Nuevas Contribuciones al Margen Ibérico Atlántico, 6° Simposio sobre el Margen Ibérico Atlántico*. 1ˢᵗ – 5ᵗʰ December 2009, Oviedo (Spain), pp. 1-4.

Freitas, M. C., Andrade, C., Rocha, F. *et al.* 2003. Lateglacial and Holocene environmental changes in Portuguese coastal lagoons 1: the sedimentological and geochemical records of the Santo André coastal area. *The Holocene* 13:433-446.

Galparsoro, I., Borja, A., Legorburu, I. *et al.* 2010. Morphological characteristics of the Basque continental shelf (Bay of Biscay, northern Spain); their implications for Integrated Coastal Zone Management. *Geomorphology* 118:314-329.

Garcia, C., Dias, J., Oliveira, A., Jouanneau, J. & Rodrigues, A. 1997. Turbid plumes connected with Tagus and Sado rivers: processes and dinamic. In Hernández-Molina, F. J. & Vázquez, J. T. (eds.) *2ⁿᵈ Symposium on the Atlantic Iberian Continental Margin*. 17ᵗʰ – 20ᵗʰ September 1997, Cádiz, Spain, pp. 133-134.

García-García, A., García-Gil, S. & Vilas, F. 2005. Quaternary evolution of the Ría de Vigo, Spain. *Marine Geology* 220:153-179.

Garcia-Gil, S., Durán, R. & Vilas, F. 2000. High resolution seafloor mapping of the Ría de Pontevedra (Galicia, NW Spain). In Dias, J. A. & Ferreira, Ó. (eds.) *3ʳᵈ Symposium on the Iberian Atlantic Margin*. 25ᵗʰ – 27ᵗʰ September 2000, Faro, pp. 47-48.

Gomes, A. A. T. 2000. Evidências geomorfológicas de alguns processos responsáveis pela evoluçã quaternária do canhão de Setúbal. *Ciências da Terra (UNL)* 14:213-222.

Gomez-Gesteira, M., Beiras, R., Presa, P. & Vilas, F. 2011. Coastal processes in northwestern Iberia, Spain. *Continental Shelf Research* 31:367-375.

Goñi, M. F. S., Turon, J-L., Eynaud, F. & Gendreau, S. 2000. European climatic response to millennial-scale changes in the atmosphere-ocean system during the Last Glacial Period. *Quaternary Research* 54:394-403.

González-Álvarez, R., Bernárdez, P., Pena, L. D. *et al.* 2005. Paleoclimatic evolution of the Galician continental shelf (NW of Spain) during the last 3000 years: from a storm regime to present conditions. *Journal of Marine Systems* 54:245-260.

González-Villanueva, R., Pérez-Arlucea, M., Alejo, I. & Goble, R. 2009. Climatic-related factors controlling the sedimentary architecture of a Barrier-Lagoon complex in the context of the Holocene transgression. *Journal of Coastal Research* (Proceedings of the 10ᵗʰ International Coastal Symposium, 13ᵗʰ – 18ᵗʰ April, Lisbon — Special Issue 56):627-631.

Gutierrez-Mas, J. M., Hernández-Molina, F. J. & López-Aguayo, F. 1996. Holocene sedimentary dynamics on the Iberian continental shelf of the Gulf of Cádiz (SW Spain). *Continental Shelf Research* 16:1635-165.

Hernández-Molina, F. J., Somoza, L., Vazquez, J. T. 2002. Quaternary stratigraphic stacking patterns on the continental shelves of the southern Iberian Peninsula: their relationship with global climate and palaeoceanographic changes. *Quaternary International* 92:5-23.

Infantini, L., Moura, D. & Bicho, N. 2012. Utilização de ferramentas SIG para o estudo da morfologia submersa da Baía de Armação de Pêra. In Almeida, A. C., Bettencourt, A. M., Moura, D., Monteiro-Rodrigues, S. & Alves, M. I. C. (eds.) *Environmental Changes and Human Interaction along the Western Atlantic Edge.* pp. 227-241. APEQ: Coimbra.

Jouanneau, J-M., Garcia, C., Oliveira, A., Rodrigues, A., Dias, J. M. A. & Weber, O. 1998. Dispersal and deposition of suspended sediment on the shelf off the Tagus and Sado estuaries, SW Portugal. *Progress in Oceanography* 42:233-257.

Jouanneau, J-M., Weber, O., Champilou, N. *et al.* 2008. Recent sedimentary study of the shelf of the Basque country. *Journal of Marine Systems* 72:397-406.

Lantzch, H., Hanebuth, T. J. J., Bender, V. B. & Krastel, S. 2009. Sedimentary architecture of a low-accumulation shelf since the Late Pleitocene (NW Iberia). *Marine Geology* 259:47-58.

Lantzsch, H., Hanebuth, T. J. J. & Henrich, R. 2010. Sediment recycling and adjustment of deposition during deglacial drowning of a low-accumulation shelf (NW Iberia). *Continental Shelf Research* 30:1665-1679.

Lastras, G., Arzola, R. G., Masson, D. G. *et al.* 2009. Geomorphology and sedimentary features in the Central Portuguese submarine canyons, Western Iberian margin. *Geomorphology* 103:310-329.

Lechuga, A., De la Peña, J. M., Antón, A. I. & Díez, G.F. 2012. Behavior of the beaches on the north of Spain, is global warming involved? In Lynett, P. & Smith, J. M. (eds.) *Proceedings of 33rd Conference on Coastal Engineering.* 1st – 6th July 2012, Santander, Spain, pp. 1-10.

Lobo, F. J., Sánchez, R., González, R. *et al.* 2004. Contrasting styles of the Holocene highstand sedimentation and sediment dispersal systems in the northern shelf of the Gulf of Cadiz. *Continental Shelf Research* 24:461-482.

Lopes, F. C., Cunha, P. P., & Le Gall, B. 2006. Cenozoic seismic stratigraphy and tectonic evolution of the Algarve margin (offshore Portugal, southwestern Iberian Peninsula). *Marine Geology* 231:1-36.

Maldonado, A. & Nelson, C. H. 1999. Interaction of tectonic and depositional processes that control the evolution of the Iberian Gulf of Cadiz margin. *Marine Geology* 155:217-242.

Martins, V., Dubert, J., Jouanneau, J-M. *et al.* 2007. A multiproxy approach of the Holocene evolution of shelf-slope circulation on the NW Iberian continental shelf. *Marine Geology* 239:1-18.

Matos, M., Santos, A. & Echol, C. 2006. Superficial sediments mapping of Portuguese Continental Shelf between Cabo da Roca and Cabo de Sines: their evolution in the last 80 years. In Balbino, J. (ed.) *VII Congresso Nacional de Geologia.* 29th June – 13th July 2006, Universidade de Évora, Pólo de Estremoz, Portugal, pp. 409-412.

McIntyre, A., Kipp, N. G., Be, A. W. H. *et al.* 1976. Glacial North Atlantic 18,000 years ago: a CLIMAP reconstruction. *Geological Society America Memoirs* 145:43-76.

Méndez, G. & Vilas, F. 2005. Geological antecedents of the Rias Baixas (Galicia, northwest Iberian Peninsula). *Journal of Marine Systems* 54:195-207.

Mil-Homens, M., Stevens, R. L., Abrantes, F. & Cato, I. 2006. Heavy metal assessment for surface sediments from three areas of the Portuguese continental shelf. *Continental Shelf Research* 26:1184-1205.

Molina-Cruz, A. & Thiede, T. 1978. The glacial eastern boundary current along the Atlantic Eurafrican continental margin. *Deep Sea Research* 25:337-356.

Moreira, S., Freitas, M., Araújo, M. *et al.* 2009. Contamination of intertidal sediments — the case of Sado Estuary (Portugal). *Journal of Coastal Research* (10th International Coastal Symposium, 13th – 18th April, Lisbon — Special Issue 56):1380-1384.

Mougenot, D. 1988. *Geologie de la marge Portugaise (vol. 1).* Ph.D thesis. Université Pierre et Marie Curie, Paris.

Moura, D. & Boski, T. 2009. Plio-Pleistocene Boundary: Olhos de Água outcrop. In Gomes, A., Boski, T. &

Moura, D. (eds.) *VII reunião do Quaternário Ibérico - Field Guide.* 5th – 9th October 2009, Universdade do Algarve, Portugal, pp. 30-34.

Moura, D., Albardeiro, L., Veiga-Pires, C., Boski, T. & Tigano, E. 2006. Morphological features and processes in the central Algarve rocky coast (south Portugal). *Geomorphology* 81:345-360.

Moura, D., Veiga-Pires, C., Albardeiro, l., Boski, T., Rodrigues, A. L. & Tareco, H. 2007. Holocene sea level fluctuations and coastal evolution in the central Algarve (southern Portugal). *Marine Geology* 237:127-142.

Neto de Carvalho, C. 2009. Vertebrate tracksites from the Mid-Late Pleistocene eolianites of Portugal: the first record of elephant tracks in Europe. *Geological Quarterly* 53:407-414.

Oliveira, A., Rocha, F., Rodrigues, A. *et al.* 2002. Clay minerals from the sedimentary cover from the Northwest Iberian shelf. *Progress in Oceanography* 52:233-247.

Oliveira, A., Santos, A. I., Rodrigues, A. & Vitorino, J. 2007. Sedimentary particle distribution and dynamics on the Nazaré canyon system and adjacente shelf (Portugal). *Marine Geology* 246:105-122.

Oliveira, A., Palma, C. & Valença, M. 2011. Heavy metal distribution in surface sediments from the continental shelf adjacent to Nazaré canyon. *Deep-Sea Research Part II: Topical Studies in Oceanography* 58:2420-2432.

Pais, J., Cunha, P., Pereira, D. *et al.* 2012. *The Paleogene and Neogene in Western Iberia (Portugal). A Cenozoic Record in the European Atlantic Domain.* Springer: Heidelberg.

Pereira, A. R., Neves, M., Trindade, J., Borges, B., Angelucci, D. E. & Soares, A. M. 2006. Carbonate dunes and related deposits in Estremadura (Portugal). Sea-level changes and neotectonics. In Pereira, A. R., Trindade, J., Garcia, R. & Oliveria, S (eds.) *Dinâmicas Geomorfológicas. Metodologias. Aplicação.* pp. 165-178. Associação Portuguesa de Geomorfólogos: Lisbon (in Portuguese).

Pires, A. R., Freitas, M. C., Andrade, C. *et al.* 2011. Morphodynamics of an ephemeral tidal inlet during a life cycle (Santo André Lagoon, SW Portugal). *Journal of Coastal Research* (Proceedings of ICS2011, 9th – 14th May 2011, Szczecin, Poland — Special Issue 64):1565-1569.

Prego, R. 1993. General aspects of carbon biogeochemistry in the ría of Vigo, northwestern Spain. *Geochimica et Cosmochimica Acta* 57:2041-2052.

Ramos, A. M., Cunha, L. & Cunha, P. P. 2012. Cartografia geomorfológica aplicada ao ordenamento do território area da Figueira da Foz-Nazaré (Portugal central). *Revista Geonorte* 3:1433-1449.

Raposo, L. 1995. Ambientes, territorios y subsistencia en el Paleolítico Medio de Portugal. *Complutum* 6:57-77.

Sánchez, R. F., Mason, E., Relvas, P., da Silva, A. J. & Peliz, Á. 2006. On the inner-shelf circulation in the northern Gulf of Cádiz, southern Portuguese shelf. *Deep-Sea Research Part II: Topical Studies in Oceanography* 53:1198-1218.

Schmidt, S., Stigter, H.C de. & van Weering, T. C. E. 2001. Enhanced short-term sediment deposition within the Nazaré Canyon, north-east Atlantic. *Marine Geology* 173:55-67.

Soares, A. M. M. & Sousa, A. C. 2003. Aeolianites of the coastal region of Lisbon — a contribution to their dating. In Vilas, F., Rubio, B., Diez, J. B. *et al.* (eds.) *Special Volume on the 4th Symposium on the Atlantic Iberian Continental Margin.* 7th – 10th July, Vigo, *Thalassas* 19:180-181.

Soares, A. M. M., Moniz, C. & Cabral, J. 2006. The consolidated dune of Oitavos (west of Cascais — Lisbon Region). Its dating by radiocarbon method. *Comunicações Geológicas* 93:105-118 (in Portuguese).

Sousa, M. C., Vaz, N., Alvarez, I. & Dias, J. M. 2013. Effect of Minho estuarine plume on Rias Baixas: numerical modeling approach. *Journal of Coastal Research* (Proceedings of the 12th International Coastal Symposium, 8th – 12th April 2013, Plymouth, England – Special Issue 65):2059-2064.

Uriarte, A., Belzunce, M. J. & Solaun, O. 2004. Characteristics of estuarine and marine sediments. In Borja, Á. & Collins, M. (eds.) *Oceanography and Marine Environment of the Basque Country.* pp. 273-282. Elsevier: Amsterdam.

van Meerbeeck, C. J., Renssen, H. & Roche, D. M. 2009. How did Marine Isotope Stage 3 and Last

Glacial Maximum climates differ? Perspectives from equilibrium simulations. *Climate of the Past* 5:33-51.

van Weering, T. C. E., Stigter, H. C. de, Boer, W. & Haas, H de. 2002. Recent sediment transport and accumulation on the NW Iberian margin. *Progress in Oceanography* 52:349-371.

Vanney, J. & Mougenot, D. 1981. La Plate-forme continentale du Portugal et les provinces adjacentes: Analyse géomorphologique. *Memória dos Serviços Geológicos de Portugal* 28.

Vis, G., Kasse, C. & Vandenbergh, J. 2006. The Holocene fluvio-deltaic sedimentary evolution of lower-Tagus river, Portugal. In Balbino, J. (ed.) *VII Congresso Nacional de Geologia.* 29th June – 13th July 2006, Universidade de Évora, Pólo de Estremoz, Portugal, pp. 611-613.

Vis, G., Kasse, C. & Vandenbergh, J. 2008. Late Pleistocene and Holocene relative sea-level curve and palaeogeography of the lower Tagus valley (Portugal). *Proceedings of 4th Annual Conference, IGCP Project 495-Quaternary Land-Ocean Interactions: Driving Mechanisms and Coastal Responses.* 27th October – 1st November 2008, Faro, Portugal, pp. 106-108.

Vitorino, J., Oliveira, A., Jouanneau, J-M. & Drago, T. 2002. Winter dynamics on the northern Portuguese shelf. Part 2: bottom boundary layers and sediment dispersal. *Progress in Oceanography* 52:155-170.

Zazo, C., Goy, J-L., Somoza, L. 1994. Holocene sequence of sea-level fluctuations in relation to climatic trends in the Atlantic-Mediterannean linkage coast. *Journal of Coastal Research* 10:933-945.

Chapter 12
The Western Mediterranean Sea

Miquel Canals,[1] Isabel Cacho,[1] Laurent Carozza,[2] José Luis Casamor,[1] Galderic Lastras[1] and Anna Sànchez-Vidal[1]

[1] GRC Geociències Marines, Dept. de Dinàmica de la Terra i de l'Oceà, Universitat de Barcelona, Barcelona, Spain
[2] UMR 5602 Géode Géographie de l'Environnement, Maison de la Recherche de l'Université du Mirail, Toulouse, France

Introduction

The study area mainly includes the continental shelves of France and Spain and partly Italy, three North African countries (Algeria, Morocco and Tunisia), and Gibraltar and Monaco. Overall, the total shelf area in the western Mediterranean down to the 150-m isobath is 208,000 km^2 within a total basin area of 1,343,620 km^2. There is a major difference in the degree of available information from EU member states and North African countries.

The sea basin area extends west to east from the Gibraltar Strait to the Sicily Strait, and north to south from the Ligurian Sea and the Golfe du Lion to North Africa. The main rivers are the Rhône and the Ebre, opening into the Golfe du Lion and the Golfo de Valencia, respectively. The widest shelves develop off these two river mouths, with 120 km and 70 km width,

respectively (Fig. 12.1). Locally, large submarine canyons are incised into the shelf, thus segmenting it and isolating some of these segments from sediment fluxes carried by littoral drift currents. The western Mediterranean Sea includes several archipelagos of which the largest ones are, from west to east, the Baleares (Balearics Islands) and the Corsica and Sardinia block. The sea area between Sardinia and Sicily represents the transition from the western Mediterranean Sea to the Ligurian and Ionian seas (Fig. 12.1).

The climate of the western Mediterranean region is characterized by hot, dry summers, two wetter seasons in fall and spring, and cool and relatively wet winters (UNEP/MAP 2003), with an overall temperature and precipitation gradient from north to south. Where there is a significant sediment input to the continental shelf, it tends to form coastal prisms that mostly cover the inner shelf and part of the middle shelf, down to 80 m

Submerged Landscapes of the European Continental Shelf: Quaternary Paleoenvironments, First Edition.
Edited by Nicholas C. Flemming, Jan Harff, Delminda Moura, Anthony Burgess and Geoffrey N. Bailey.

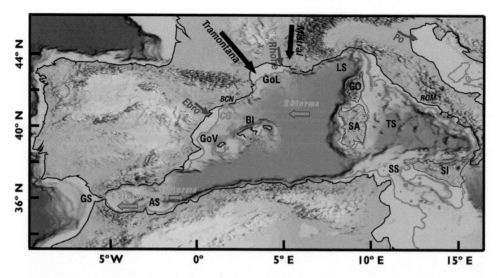

Fig. 12.1 General map of the western Mediterranean Sea and adjacent landmasses. The orange line marks the 150-m isobath which corresponds approximately to the shelf edge. The strongest winds (black arrows) and the direction of main storms (red arrows) in the various areas are also shown. Red asterisks indicate the location of submarine caves with evidence of prehistoric remains and paleo-sea level indicators. Blue arrows represent main rivers. AS = Alboran Sea; BI = Balearic Islands; BCN = Barcelona; CC = Cosquer Cave; CG = Cova del Gegant; CO = Corsica Island. CT = Cueva del Tesoro; GC = Gibraltar caves. GoL = Golfe du Lion; GoV = Golfo de Valencia; GS = Gibraltar Strait; LS = Ligurian Sea; = MC = Mallorca caves; ROM = Rome; SA = Sardinia; SI = Sicily; SS = Sicily Strait; TS = Tyrrhenian Sea. University of Barcelona.

depth. The outer shelf generally is of relict character and not covered by postglacial sediments, so that its topography preserves the coastal landscape of the Last Glacial Maximum.

The rocks forming the shoreline and the shelf basement are limestone in many locations, which favored the formation of large caves. Some of these caves (e.g. Cosquer Cave; Clottes *et al.* 1992) have already provided excellent finds in terms of prehistoric studies, while other caves (e.g. Mallorca caves; Dorale *et al.* 2010) hold a unique record of paleo-sea levels.

The survival of coastal and submerged prehistoric sites is partly determined by the regional and local wave climate and the extent to which extreme waves strike the shore. Wind-wave research is well developed in the Mediterranean Sea (e.g. Ardhuin *et al.* 2006; Bertotti & Cavaleri 2009), based on numerical models of wave generation, and on satellite measurements of waves (see 'Sources of oceanographic data', pages 317-322).

Earth Sciences and Sources of Data

Modern coastline: best sources of high-resolution data

High-resolution inland and coastline maps and data sets are available for the EU member states. These data are collected and distributed by state agencies and institutes, namely the national geographic institutes of France (www.ignfi.fr/) and Spain (www.ign.es/), and the National Cartographic Portal of Italy (www.pcn. minambiente.it/). Some regional administrations have cartographic institutes which also release high-quality data in various formats, including classic topographic maps, orthophotomaps and aerial photographs. An example is Institut Cartogràfic i Geològic de Catalunya (ICGC — www.icgc.cat/).

Color aerial photographs of the coastline are potentially useful, and have been acquired by different ministries (e.g. Ministerio de Agricultura y Pesca, Alimentación y Medio Ambiente in Spain — www.mapama. gob.es/) for a variety of purposes. Such photographs are useful when there are extensive areas of interspersed land and sea such as in marshes, lagoons, and over shallow rock terraces, and they usually have some penetration into clear waters down to limited depths.

LiDAR (Light Detection and Ranging) projects are ongoing in the relevant EU countries. Depending on water conditions, topo/bathymetric LiDARs are able to penetrate down to a depth of a few tens of meters into the water, thus providing an opportunity to merge LiDAR information with multibeam bathymetric data over the continental shelf. The submerged coastal fringe is usually a critical area in terms of topographic/bathymetric coverage, as survey vessels cannot enter very shallow waters, which results in bathymetry data gaps which could be covered by LiDAR data. Furthermore, LiDAR produces high-resolution information from the emerged coast, thus providing an excellent opportunity to integrate land and seafloor data. Instituto Geográfico Nacional (IGN — pnoa.ign.es/presentacion) runs the LiDAR mapping program in Spain.

The above-mentioned base maps depict coastal geomorphic elements and include, wherever relevant, information on wetlands, deltas, marshes, lagoons and other coastal features of interest. Historical maps and aerial photographs usually are also available through those institutions. These documents are useful to extract information on coastal morphodynamics, erosion and sediment accumulation. Furthermore, state and regional organizations and agencies dealing with coastal management (e.g. the Directorate General for Coasts of the Spanish Ministry for the Environment) hold huge data sets that could be of relevance, though extracting the right information may require a tremendous data-mining effort. Furthermore, some of these agencies might be, in practice, reluctant to release data sets that they may consider as 'critical' or 'sensitive' for social, economic or environmental reasons.

Western Mediterranean rivers show strong seasonal fluctuations and are often ephemeral in character. The regime of the vast majority is torrential, and the main events leading to the transport of significant volumes of sediment to the shoreline are flash floods following heavy rains typical of the Mediterranean climate. The two main rivers in the study area, the Rhône and Ebre, in France and Spain respectively, form large deltas and through geological time have built the two widest shelves (see page 301).

Apart from the low-lying deltaic areas and coastal stretches down-current of the main river mouths (e.g. the shores of the Golfe du Lion and a large part of the Valencia coastline down-current of the Rhône and Ebre mouths, respectively), cliffy coasts with elongated beaches a few kilometers long and pocket beaches characterize most coastlines in this sector. The largest watersheds excluded, the majority of river basins are small to very small, which results from a highly compartmentalized drainage system because of the geological and topographical complexities of the emerged landmasses (see, for instance, fig. 1 in Arnau *et al.* 2004). The water collecting capability of these watersheds is rather small and is not continuous through time, which leads to many sediment-depleted and sediment-starved continental shelf segments that have been identified offshore.

Solid geology

Coastal geological maps are available in France, Italy and Spain as produced by national geological institutes and, on some occasions, regional geological institutes. National institutes are Instituto Geológico y Minero de España (IGME — www.igme.es), Bureau de Recherches Géologiques et Minières (BRGM — www.brgm.fr) in France, and Istituto Superiore per la Protezione e la Ricerca Ambientale (ISPRA — www.isprambiente.gov.it) in Italy. The scale of the maps is usually 1:50,000, 1:250,000 and smaller scales. Digital versions of these maps are becoming more common, thus contributing to the implementation of the INSPIRE Directive on infrastructures for spatial information in Europe. At the European scale there are a variety of initiatives for searching national geological maps (see Chapter 4, this volume).

Figure 12.2 shows a general geological map of the western Mediterranean region, illustrating that most outcrops along the coastal area consist of Tertiary and

Fig. 12.2 Geological map of the western Mediterranean region. The geology of North Africa is not included as there is no integrated map available to a uniform standard and with sufficient resolution. Adapted from Kirkaldy (1963: fig. 2) with permission from Blandford Press.

younger sedimentary rocks. The exceptions to this are Corsica, Sardinia and a few other areas. Young volcanic rocks appear mainly in Sicily and some of the Aeolian Islands off Italy within the Tyrrhenian Sea.

Some administrative regions also have their own geological surveys (e.g. ICGC in Catalonia), which often produce maps that are more detailed than those from national geological surveys.

Geological mapping *sensu stricto* of the shelf floor is in general less advanced, despite some efforts to carry out dedicated cartographic projects, such as Fondos Marinos (FOMAR) in Spain, started in 1972 and later abandoned before completion of all the intended work. FOMAR led to the production of several sets of thematic maps at 1:200,000 in several continental shelf segments, like those off Catalonia, Murcia and Andalucía (Fig. 12.3). In France, BRGM published some maps, *Cartes de Géologie Marine*, with disparate objectives (e.g. sedimentology, structural geology, general geology) but so far all of them are related to the French Atlantic shelf.

Bathymetry

As for the bathymetric mapping of the continental shelf, four main data sources exist for the EU member states opening into the western Mediterranean Sea:

1 Hydrographic institutes such as Service hydrographique et océanographique de la Marine

(SHOM) in France and Instituto Hidrográfico de la Marina in Spain;
2 Specific mapping programs like ESPACE in Spain (Acosta *et al.* 2012) (Fig. 12.4);
3 *Ad hoc* maps made to address specific problems, generally of local extent, which often are commissioned by government agencies and carried out by private companies;
4 Bathymetric data and maps collected and produced by research groups within the frame of research projects.

Part of the cartographic results is publicly available in the form of map collections and other publications. Hydrographic institutes' maps usually take the form of nautical charts and, in general, they cannot be considered as high resolution since they are not multibeam bathymetry maps *per se* and largely rely on traditional depth-sounding methods. The soundings plotted on published sheets are also widely spaced, not reflecting the details of topography. Specific mapping programs for the continental shelf usually publish and disseminate the maps they produce, though often the coverage is incomplete and the published documents have less resolution than the maximum achievable from raw multibeam data. Derivative products from multibeam data sets usually include backscatter maps, slope gradient maps, and 3D visualizations. After the processing of raw data, GIS (Geographic Information System) and imaging techniques are commonly applied, which can

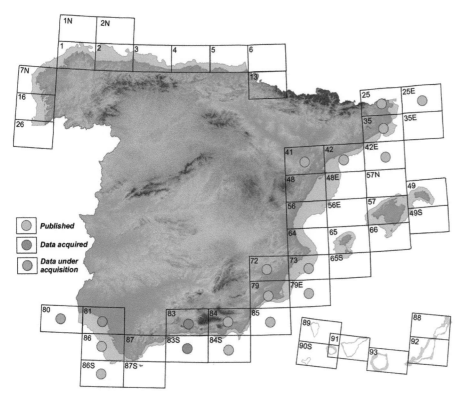

Fig. 12.3 Latest available status of the Spanish FOMAR program for geological mapping of the continental shelf. Continental shelf width in blue color. University of Barcelona, after www.igme.es and cuarzo.igme.es/sigeco.

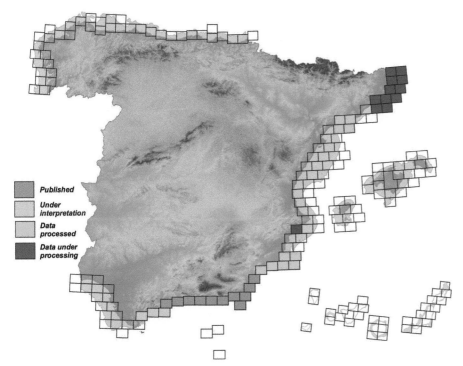

Fig. 12.4 Latest published status of the Spanish ESPACE program for the multibeam mapping of the continental shelf. University of Barcelona, after various sources.

greatly enhance the graphic and visual appearance of the final maps and images. To different extents, and with variable coverage, these data exist for French, Spanish and Italian continental shelves, and for rare North African locations. Because of their original purpose, *ad hoc* maps can be very high resolution but often are difficult to find. Finally, research groups tend to publish their cartographic products in the open literature and in reports of limited circulation. One common problem is the limited size and resolution of the published images, which impedes searching for small-sized, specific targets. There are, however, some exceptions such as the *Morpho-bathymetric Chart of the Gulf of Lion* (Fig. 12.5) published in large format, though with limited resolution as its scale is 1:250,000 (Berné *et al.* 2001).

In almost all these cases what would best accommodate the search for prehistoric evidence is not only having the final map products, but also being able to access the

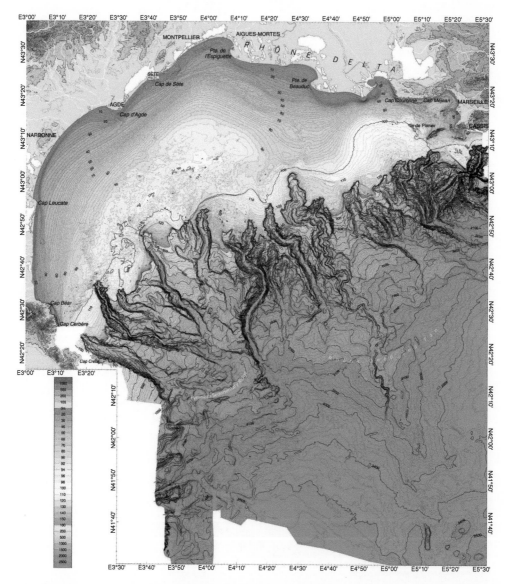

Fig. 12.5 Bathymetric map of the Golfe du Lion including both multibeam (mostly beyond the shelf edge) and single-beam (mostly over the continental shelf) data. Berné & Gorini (2005). Reproduced with permission from Elsevier.

raw multibeam bathymetry data, ideally accompanied by side-scan sonar, LiDAR, direct observations and other data types.

During the 1980s, the Spanish Ministry of Public Works carried out an extensive inventory of sand resources over the Mediterranean continental shelf, including numerous thematic maps which could be of interest for prehistoric landscape research.

The *Morpho-bathymetry of the Mediterranean Sea* at 1:2,000,000, published under the auspices of the Commission Internationale pour l'Exploration Scientifique de la Méditerranée (CIESM) and Institut Français de Recherche pour l'Exploitation de la Mer (IFREMER) (MediMap Group 2007), with contributions from many research groups from different nations, covers the entire Mediterranean Sea, though there are significant gaps in some areas (Fig. 12.6). This chart aims at compiling on a voluntary basis all available multibeam data sets from the Mediterranean Sea, though with limited resolution

(500-m grid size). However, most of the information in its latest version is in deep waters beyond the shelf edge.

Vertical earth movements

The general tectonic setting of the western Mediterranean region is shown in Fig. 12.7. The main tectonic element is the mountainous system resulting from the Alpine orogeny that extends from the Alps to the Apennines and then to the North African chains of the Tell, Kabyles and Rif to enter into southern Spain where it forms the Betic Chains which extend to the Baleares. The different segments of the Alpine orogenic systems are still tectonically active, though to varying degrees. The northwestern and central western Mediterranean area is the most tectonically stable in the entire domain.

According to the *European-Mediterranean Seismic Hazard Map* of Giardini *et al.* (2003), high to moderately high hazard occurs mainly along the Algerian coast in

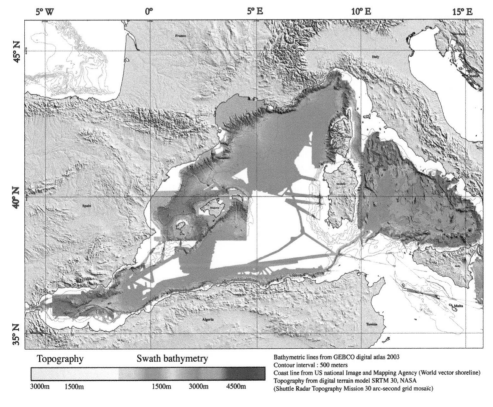

Fig. 12.6 Morpho-bathymetry of the Mediterranean Sea after the compilation carried out by the MediMap transnational group, as compiled by IFREMER and published jointly with CIESM. Loubrieu *et al.* (2007). Reproduced with permission from CIESM.org.

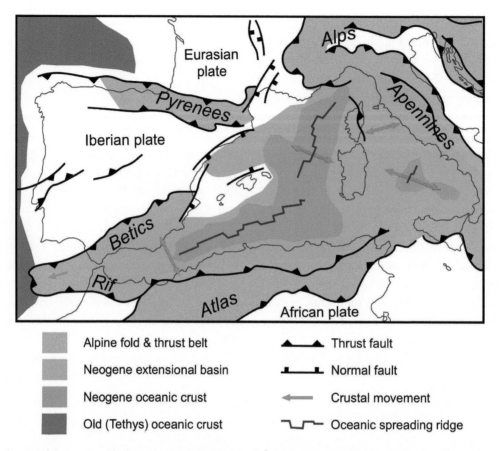

Fig. 12.7 Tectonic map of the western Mediterranean region, where the European plate system including the Iberian plate and the African plate collide, with subduction along the steep North African Margin. Oceanization (light blue) took place in the deepest parts of the Algero-Balearic basin and the Tyrrhenian Sea. Woudloper. https://upload.wikimedia.org/wikipedia/commons/9/92/Tectonic_map_ Mediterranean_EN.svg. Used under CC BY-SA 1.0 (http://creativecommons.org/licenses/by-sa/1.0).

North Africa, the coasts of eastern Andalucía, Murcia and south Alicante in Spain, and around the Ligurian Sea between France and Italy (Fig. 12.8).

Sea-level change is a measure of the relative shift in position of the sea surface relative to the adjacent land (see the more general discussion in Chapter 2, pages 20-24). Jointly with thermal expansion/contraction of ocean waters, the exchange of water mass between continental ice sheets and oceans, upon which vertical land movements driven by active tectonic processes are superimposed, is the principal process contributing to sea-level change on glacial timescales, by changing ocean volume, deforming the ocean basins and their margins, and modifying the gravitational field. The change at a given location of sea surface relative to land at a given time compared to its position at the present time is represented in the following formulation (Lambeck & Purcell 2005):

$$\Delta\zeta_{rsl}(\phi, t) = \Delta\zeta_{esl}(t) + \Delta\zeta_I(\phi, t) + \Delta\zeta_T(\phi, t) \quad (12.1)$$

The expression on the left represents the change of relative sea level (RSL) in terms of position co-ordinates (ϕ) and time (t), where Δ is the Laplacian operator. The first term on the right hand side of the equation represents the ice-volume equivalent sea level contribution; the second term, the glacio-hydro-isostatic contribution (including the response of the sea surface to the changing ice load —glacio-isostatic, and the response to concomitant change in water load — hydro-isostatic); and the last term, the tectonic contribution.

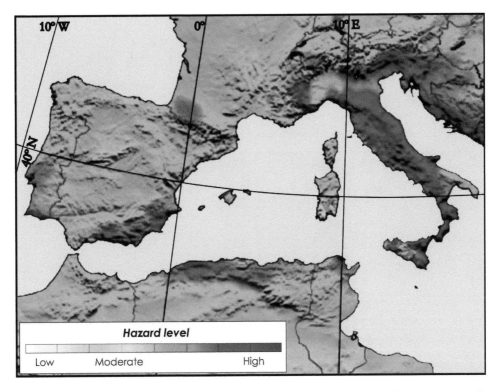

Fig. 12.8 Seismic hazard map of the western Mediterranean region. Modified from Giardini *et al.* (2003) with permission from European Seismological Commission and International Correlation Program.

In the case of the Mediterranean Sea, the dominant ice sheets contributing to the isostatic components of sea-level change are the European ice sheet (Scandinavia, Barents–Kara, and Eurasia) and the western hemisphere ice sheets (Laurentide, Cordilleran, Arctic Canada, and Greenland). For the northwestern Mediterranean (Lambeck & Purcell 2005), the glacio-isostatic contribution at 12 ka was approximately –10 m to –11 m while at 6 ka it was –2.5 m to –3 m (Fig. 12.9a,b).

The hydro-isostatic contribution at 12 ka ranged from 0 m in the Baleares to 16 m in the Golfe du Lion, while at 6 ka it ranged from –5 m in the Baleares to 2 m in the Golfe du Lion (Fig. 12.9c,d).

Lambeck and Purcell (2005) estimated that the relative sea-level change, including both isostatic terms and the Ice-Equivalent Sea Level (ESL) contributions (Fig. 12.10), varied from –150 m in the Baleares to –120 m in the Golfe du Lion at 20 ka, from –64 m in the Baleares to –48 m in the Golfe du Lion at 12 ka, and from –8 m in the Baleares to –2 m in the Golfe du Lion at 6 ka, calculated with equivalent levels at these epochs (i.e. 20,

12 and 6 ka) of –142 m, –54 m and 0 m, respectively. In their study, tectonic contributions are assumed to be either absent or corrected for.

With the onset of broad sea-level stability at 6 ka, coastal prehistoric sites and classical ports showing relative submersion or uplift can be linked to two different geological factors in addition to the residual ongoing glacio-isostatic adjustment (GIA) processes: (1) tectonic mobility; and/or (2) sediment compaction (Flemming 1969; Flemming & Webb 1986; Marriner & Morhange 2007). The comparison of observed sea-level indicators (e.g. geomorphological markers of paleo-sea levels, coastal archaeological data and sedimentary core analysis) and GIA predictions allows estimating the vertical tectonic displacement for a given area. Such an exercise has been made, for example, for the Italian coast (Lambeck *et al.* 2011).

Long-term vertical earth movements (or long-term coastal displacement rates) in the Mediterranean are inferred from the present elevation of markers of the Last Interglacial highstand Marine Isotope Stage

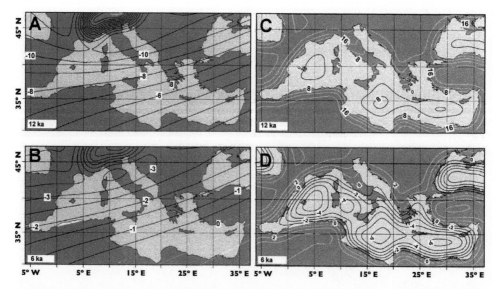

Fig. 12.9 Glacio-isostatic contribution to sea-level change in the Mediterranean Sea at 12 ka (A); and at 6 ka (B) from the total global ice sheets, including Antarctica and mountain glaciers. Hydro-isostatic contribution to sea-level change in the Mediterranean Sea at 12 ka (C); and at 6 ka (D), corresponding to the decay of the global ice sheets, with cessation of melting at 6.8 ka. Contours in meters. Black contours denote negative values, and white contours positive values. Modified from Lambeck and Purcell (2005) with permission from Elsevier.

(MIS) 5.5 (124±5 ka), which occurs typically between 3 m and 10 m above present sea level, at sites where the geological and seismic evidence points to stability. In some parts of southern Italy, Holocene uplift rates are greater (2 mm/year) with respect to these long-term uplift rates inferred from the MIS 5.5 highstand, whereas in northeastern Italy there is subsidence at rates of 0.5 to 1 mm/year. Compaction of the younger sediments accounts for higher estimated subsidence rates from the younger sediments (4-0 ka cal BP) than the average based on the older Holocene (11-6 ka cal BP), which are consistent with the rates inferred from the MIS 5.5 markers (Lambeck *et al.* 2011).

In contrast to the Italian coast, tectonic activity is generally negligible in the northwestern Mediterranean Sea, and the region is considered as tectonically relatively stable. For example, the Last Interglacial shoreline in the Provençal coast can be found at elevations that differ by no more than a few meters from the present sea level (Conchon 1975). Thus, Pirazzoli (2005) considers that even the rate of 0.3 mm/year of linear sea-level rise during the last 5,000 years accurately reconstructed for the Provençal coast from underwater identification and sampling of remains of biogenic rims built by the coralline rhodophyte *Lithophyllum lichenoides* (Morhange *et al.* 2001) is too high to be ascribed to tectonics. Other species, like *Lithophyllum byssoides*, also form continuous rims in some places along the coasts of the western Mediterranean and therefore hold potential to further investigate sea-level fluctuations (Ballesteros *et al.* 2014).

In southern France, the partially drowned Paleolithic Cosquer Cave is one of the best Mediterranean examples of human occupation of the continental margin and postglacial sea-level rise (see section below on 'Submerged terrestrial landforms and ecology', and Chapter 12 Annex). The entrance, 37 m below sea level, was submerged around 7000 years ago (Sartoretto *et al.* 1995). Painted horses dated at ~18,000 BP have been partially eroded at current sea level, testifying to the absence of any sea-level oscillation higher than present (Marriner & Morhange 2007).

On the rocky coasts of the northwestern Mediterranean basin, the above mentioned *Lithophyllum lichenoides* rim remains may be preserved for millennia when submerged in a rising sea environment. These remains can be used as biological indicators of recent sea-level variations.

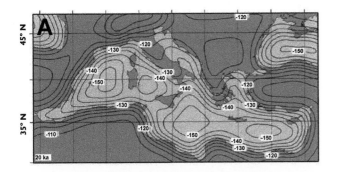
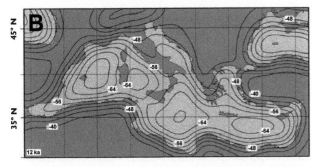
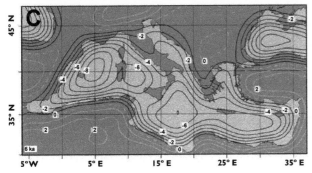

Fig. 12.10 The total predicted relative sea-level change in the Mediterranean Sea at 20 ka, 12 ka and 6 ka (A, B and C, respectively), for the nominal earth and ice model. Modified from Lambeck and Purcell (2005) with permission from Elsevier.

The study of *Lithophyllum lichenoides* biogenic littoral rims in the coasts of Var and Bouches du Rhône, west of Marseille in southern France, shows that relative sea level rose about 1.6 m in the area during the last 4500 years without exceeding the present datum. The rate of sea-level rise was 0.4 mm/year between 4500 and 1500 years ago and slowed down to 0.2 mm/year from 1500 years ago to the present (Laborel *et al.* 1994). Regions at the periphery of the above zone (including French and Spanish Catalonia) were also surveyed, but a weaker development of *Lithophyllum* rims and bad preservation

of algal remains led to unconvincing dates. This is similar to observations in the ancient harbor of Marseille of marine fauna fixed to archaeological structures as well as of bio-sedimentary units (Morhange *et al.* 2001)

In spite of the view that the area is tectonically stable, the study of peat formations from the eastern limit of the Rhône Delta yields a relative subsidence of between 1.5 m and 0.5 m prior to 2260 BP after data from the rocky coast to the east of Marseille (Vella & Provansal 2000). Two periods of rapid relative sea-level rise (about 2 mm/year) were identified, separated by a period of stability (Fig. 12.11). The difference is hypothetically related to tectonic-subsidence movements, that is, subsidence due to an excessive sedimentary overload, or tectonic movements that were the result of fault reactivation at the edges of the delta (Vella & Provansal 2000). Contrary to the commonly held view, subsidence in the Rhône Delta was of limited magnitude at least over the last 6000 years (Vella *et al.* 2005).

A number of paleocoastal indicators consisting of beach ridges and beachrocks have been identified on the Barcelona continental shelf at depths down to 104 m and linked to short Late Pleistocene–Holocene stillstands. Two sets of indicators have been related to transgressions that followed MIS 4 and MIS 2 (Liquete *et al.* 2007) (see 'Submerged terrestrial landforms and ecology', pages 314-317).

Further south, in the Ebre Delta, the relative mean sea-level curve for the last 7000 years may be fitted to a notably linear subsidence rate of 1.75 mm/year relative to present sea level (Sornoza *et al.* 1998). In other areas, particularly in the southeast of Spain, both positive and negative relative mean sea-level variations have been reported for recent millennia, related to tectonic uplift and subsidence. In La Marina area of Alicante and in Cope Basin in Murcia, uplift rates of 0.023 and up to 0.038 mm/year were obtained by Zazo *et al.* (2003) for the last 130 kyr. In the southeastern tip of Spain, in Almeria, a relative sea-level lowering of 0.8 m has been documented for the last 2000 years, with relative variations ranging from −0.8 m to 0.8 m in the last 6000 years. For the entire Holocene, estimated changes of relative mean sea level do not exceed 1.3 m there (Goy *et al.* 2003). However, in Campo de Dalías, within the

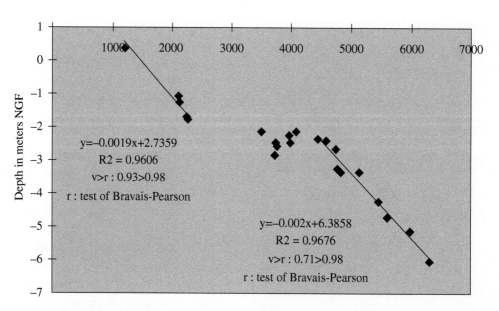

Fig. 12.11 Relative sea-level rise during the periods 6295 BP to 4450 BP, and 2260 BP to 1200 BP, based on the study of peat formations from the eastern limit of the Rhône Delta. Vella and Provansal (2000). Reproduced with permission from Elsevier.

same general area, Zazo *et al.* (2003; 2013) identified 27 Pleistocene raised marine terraces placed up to 82 m above sea level on the up-thrown block of a fault, which represent a 0.046 mm/year average uplift rate. Proxy indicators could not detect these rates of vertical tectonic displacement of the order of 20 mm/kyr to 40 mm/kyr over historic periods of 2 kyr to 3 kyr.

Pleistocene and Holocene sediment thickness on the continental shelf

The most extensive data on Holocene sediment thickness over the Mediterranean continental shelf of Spain are those in the FOMAR map series (Fig. 12.3), though additional sources of information, generally of local character, also exist. The main thicknesses of the Holocene deposits are off river mouths and down-current following the littoral drift, and tend to be close to the shoreline. Holocene sediments do not cover large parts of the outer shelf.

In the Golfe du Lion, Got and Aloisi (1990) addressed the Holocene sedimentation in a quantitative way. As for the Spanish Mediterranean shelf, the outer shelf of the Golfe du Lion is essentially relict. Lobo *et al.* (2014) have provided useful summary maps and discussion of

the underlying geology of the Spanish Mediterranean shelf, with descriptions of fluvial influence, canyons, escarpments, and Late Quaternary deposits. Examples are provided of seismic sub-bottom stratigraphic profiles.

The lack of significant postglacial sediment accumulation along large stretches of the outer continental shelf of the western Mediterranean Sea, which was the coastal area closest to the Last Glacial Maximum (LGM) shoreline, offers an excellent opportunity for paleolandscape reconstruction and the detection of prehistoric archaeological material (see '1 Potential areas for future work', pages 323-326).

Scattered in the scientific literature are data on modern sedimentation rates, but as previously stated, the highest rates are determined by the proximity of river mouths.

Post-LGM Climate, Sea Level and Paleoshorelines

General climatic conditions

The LGM finished about 18 ka cal BP (Mix *et al.* 2001) but the rapid warming of the last deglaciation did not start in the Mediterranean region until 15 ka cal BP (Cacho et al. 2001). During these 3 kyr, the climate in the Mediterranean region severely deteriorated, with colder temperatures than during the LGM and extremely arid conditions (Cacho et al. 2001; González-Sampériz et al. 2006). This was the time when the Heinrich 1 (H1) event took place in the North Atlantic, which led to a rapid reorganization of the thermohaline circulation in the North Atlantic Ocean and related changes in the hydrography and temperature patterns at intermediate and lower latitudes (Broecker 2006). Lake and pollen records from northwest Iberia highlight this as the most arid period of the last 30 kyr (González-Sampériz et al. 2006; Morellón et al. 2009). Major changes in surface and deepwater properties from the Catalano-Balearic basin have been documented for this period with strong changes in the mode of operation of the deepwater cell in the Golfe du Lion (Cacho et al. 2006; Frigola et al. 2008).

The deglacial warming that began at about 15 ka cal BP lasted about 5 kyr. The total warming during this transition in the surface temperatures of the Mediterranean Sea was about 8°C (Cacho et al. 2001) but this transition was interrupted by the pronounced cooling event of the Younger Dryas (12.9-11.7 ka cal BP).

Evolution of sea level and coastline since the LGM

For submerged prehistoric research some of the newest and most interesting results about past sea levels come from the analysis of speleothems from caves on the island of Mallorca. Overall, these speleothem-rich caves present excellent conditions to investigate past sea-level changes. The caves were flooded and emptied on several occasions by glacio-eustatic sea-level oscillations throughout the Middle and Late Quaternary. Encrustations of calcite or aragonite record the water level of each flooding event, as shown by Dorale et al. (2010) (Fig. 12.12). Although models predict an almost constant sea-level rise of about 0.2 mm/year (i.e. 60 cm of sea-level rise over the past 3000 years) around Mallorca during the Late Holocene, caused by hydro-isostatic subsidence of the basin (Mitrovica & Milne 2002; Peltier 2004) (Fig. 12.12h), the actively accreting speleothems from the

island (Fig. 12.12c,d) demonstrate instead that relative sea level has been stable for the last ~2800 years (Dorale et al. 2010; Tuccimei et al. 2010). The estimate of post-LGM GIA and the global ice-volume equivalent change of sea level determined from far-field data sites (Lambeck et al. 2014) confirms that the sea level approached its present level about 4000 years ago, and never exceeded present sea level during the last 6000 years. During short-term observations of relative sea level at any specific location, allowance needs to be made for the small Mediterranean tidal amplitude, which is seldom more than 10 cm (Tsimplis et al. 1995) in this area, and for seasonal and barometric factors, which can cause short-term relative changes of sea level of up to 30 cm to 50 cm.

Dorale et al. (2010) presented results from speleothems in coastal caves on Mallorca, and these show that the western Mediterranean sea level was ~1 m above modern sea level ~81,000 years ago during MIS 5a. The authors state that though these findings seemingly conflict with the eustatic sea-level curve of far-field sites, they corroborate an alternative view that MIS 5a was at least as ice-free as the present, and they challenge the prevailing view of MIS 5 sea-level history and certain facets of ice-age theory. This is an important result in terms of estimating the preservation potential of Mallorca caves for prehistoric research, as the maximum relative sea level for the last 130,000 years was never more than 1 m above present sea level (Fig. 12.13).

Global models (Waelbroeck et al. 2002), the SPECMAP (SPECtral MAPping project) curve derived from the Atlantic Ocean (Imbrie et al. 1984; 1990), and the Red Sea curve (Siddall et al. 2003), jointly with the results of Shackleton (2000), suggest that at a larger scale the same situation could apply for at least the last 200,000 years (Fig. 12.14).

For the isostatic component of relative sea level variations see 'Vertical earth movements', pages 307-312.

Broad classification of the climate conditions on the shelf

The entire region of the western Mediterranean Sea has a Mediterranean climate, though with a marked north–south gradient in terms of both temperature and

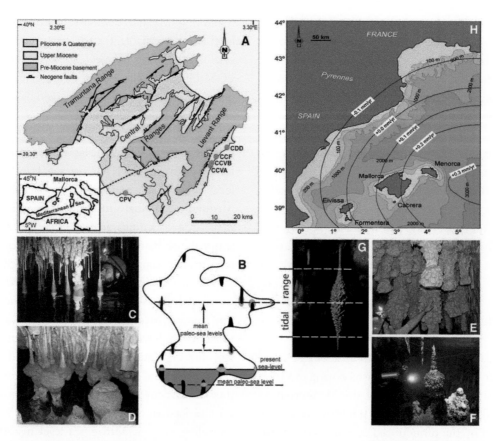

Fig. 12.12 Encrusted speleothems at various levels in caves from Mallorca. (A) Geological map of Mallorca (after Tuccimei *et al.* 2006) and location of sampled caves (red dots); (B) Schematic cross-section through a coastal cave in Mallorca showing multiple carbonate encrustation levels; (C and D) Present-day and paleolevels of encrusted speleothems related to higher (E), and lower (F), sea-level stands; (G) Typical morphology for tidal range-related carbonate encrustation (size of speleothem, 20 cm); (H) Bathymetric map of the western Mediterranean region and predicted present-day rate of sea-level change due to glacio-isostatic adjustment. Adapted from Mitrovica and Milne (2002), courtesy of Dorale *et al.* (2010). Reproduced with permission from AAAS.

precipitation (see 'Introduction', pages 301-302). Wave climate is characterized by rather short wavelength wind-waves. Western winds mostly have an effect in the Alboran Sea, while eastern storms are the most energetic ocean process over the shelf along the Spanish coast opening to the east. In the Golfe du Lion, the northern winds *tramontana* and *mistral* may generate high seas over the open shelf mainly in winter months (Fig. 12.1). The wave base does not reach the outer shelf, uncovered by Holocene sediment, during storms, and the shallowest areas protected by sea cliffs are not severely affected by northern winds in the area. It is suspected that the outer shelf is cleaned of fine sediment by currents associated with Dense Shelf Water Cascading (DSWC)

events (see 'Taphonomy and potential for archaeological site survival', pages 317-322).

Submerged Terrestrial Landforms and Ecology

Submerged river valleys

Though numerous buried channels are known to exist on western Mediterranean shelves, as imaged by seismic reflection profiles, to our best knowledge no systematic mapping of these features has been carried out so far. Traditionally, these buried valleys have been attributed to

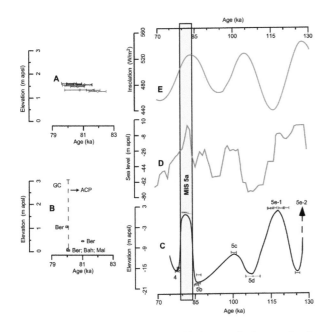

Fig. 12.13 Comparison between the Mallorca and other sea-level estimates. (A) Elevation of MIS 5a encrusted speleothems from Mallorca; (B) Other deposits from tectonically stable locations such as the Bahamas (Bah), Hearty (1998); the Atlantic Coastal Plain (ACP), Muhs et al. (2002); Wehmiller et al. (2004); Bermuda (Ber), Ludwig et al. (1996); Vacher and Hearty (1989), Grand Cayman (GC) Coyne et al. (2007), and Mallorca (Mal) Hearty (1987); (C) Sea-level reconstruction for Mallorca. Elevations and U/Th ages of encrusted speleothems throughout MIS 5 and at the onset of MIS 4 are shown. Ages and 2s error bars are color-coded by sample; blue-colored ages are obtained from Tuccimei et al. (2006); (D) The reconstructed ocean water $\delta^{18}O$ scaled as sea level. Shackleton (2000); (E) 60°N June insolation. Berger & Loutre (1991). The vertical yellow bar denotes the timing of the peak MIS 5a sea level recorded at Mallorca, and shows a good correlation with 60°N June insolation and the reconstructed ocean water $\delta^{18}O$ scaled as sea level. Modified from Dorale et al. (2010) with permission from AAAS.

fluvial incision during sea-level lowstands and transitional periods. It is also thought that submarine canyons that are deeply incised into the continental shelf represent the offshore continuation of terrestrial river valleys. This is the case of Blanes canyon opposite the Tordera River and La Fonera canyon opposite the Ter River in Catalonia (northeast Spain), and especially of many submarine valleys and land barrancos in some islands, such as Corsica. Collina-Girard has published on the drainage

patterns of rivers and karstic water flow on the shelf around Marseille, and adjacent to the Cosquer Cave (Collina-Girard 1992; 1999, see also Chapter 12 Annex, this volume).

Seabed submerged terrestrial features

The continental shelves of the western Mediterranean Sea are particularly rich in submerged terrestrial and coastal landforms that formed during past sea-level lowstands, with the last imprint attributed to the LGM. Paleo-shorelines and other relict coastal features such as deltas and dune systems are known to occur in a number of places, and specially in sediment-depleted and sediment-starved shelf stretches, such as the continental shelf of the Alboran Sea, Murcia, Alicante, north Catalonia and the Baleares in Spain; the southern Roussillon, Alpes Maritimes and Corsica in France; and also on the Italian shelves of the western Mediterranean including the Sicily Strait. However, many of the best data are still to be published at high resolution and, in some places, have yet to be acquired. The collection of very high-resolution bathymetric data would be extremely beneficial at specific locations where old low-resolution data have shown tremendously promising evidence of flooded paleocoastal landscapes.

Even in shelf areas with high sediment input offshore of the largest river mouths, the outer shelf is usually relict in character, with the sea floor depicting a large variety of now-submerged terrestrial and coastal landforms. This situation could be explained by a number of factors including i) longshore transport of sediment, which prevents it from reaching the outer shelf in significant volumes, and ii) hydrodynamic processes such as DSWC events which remove the loose sediment on the shelf up to sand sizes thus keeping it 'clean' and avoiding the sediment masking of paleolandscapes.

Coastal, submarine and mixed caves (i.e. those with underwater access though not fully flooded) are known to occur in many of the coastal cliffs, especially if made of limestone as in Gibraltar, Andalucía, Murcia, Alicante, Castelló, Tarragona, Barcelona and Girona (Garraf and Montgrí massifs), the Baleares, Provence-Alpes-Côte d'Azur and so on. Some of the caves that

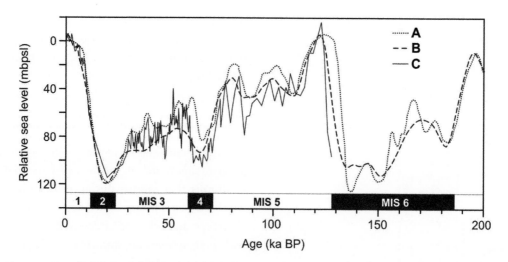

Fig. 12.14 Relative sea-level curves estimated from δ¹⁸O records. Vertical scale is in meters with respect to present sea level (zero value). A refers to the Waelbroeck *et al.* (2002) sea-level curve that corresponds to a global model; B refers to the SPECMAP curve, Imbrie *et al.* (1984; 1990), that derives from the Atlantic Ocean; C refers to the Siddall *et al.* (2003) curve originating from the Red Sea. Modified from Liquete *et al.* (2007).

are close to the shoreline have been adapted for tourist visits (e.g. Artà, Drac and some others in Mallorca) or diving excursions (e.g. Medes Islands off Girona). Others have been closed because of the danger they represent for inexperienced divers or because of underground water pollution, as they are located in karstic massifs (e.g. La Falconera south of Barcelona). In some areas, like Mallorca, the abundance of large caves at the shoreline or at a short distance from it suggests the existence of caves with underwater access or fully submarine still to be discovered. For a more detailed discussion of the role of submarine karst, see also Chapter 12 Annex, this volume.

Regional paleoclimate and vegetation indicators, peat, pollen, organics

Vegetation cover in the western Mediterranean borderlands changed drastically during the Quaternary glacial-interglacial cycles but also at millennial timescales (Moreno *et al.* 2014). Pollen records from both lake and marine sediments support the dominance of arid and cold conditions during the Last Glacial Period (Pons & Reille 1988; González-Sampériz *et al.* 2006; Fletcher &

Sánchez-Goñi 2008). These conditions were particularly extreme during millennial-scale Heinrich events, when major changes in the North Atlantic oceanography occurred, and had a major impact on the climatology and oceanography of the western Mediterranean region (Cacho *et al.* 1999; 2000; Allen & Huntley 2000; Sánchez-Goñi *et al.* 2002; Moreno *et al.* 2005; Sierro *et al.* 2005; Frigola *et al.* 2008). The severe climate conditions in the western Mediterranean during Heinrich events are further supported by evidence of major reduction in lake levels according to geochemical and sedimentological parameters from lake sediments (Morellón *et al.* 2009).

The last deglaciation in the western Mediterranean region involved a rapid improvement in forest cover, suggesting a parallel increase in both temperature and humidity (Carrión *et al.* 1998; Fletcher & Sánchez-Goñi 2008), and also in lake levels (Morellón *et al.* 2009), with corresponding changes in vegetation dynamics and fire regimes (Pérez-Sanz *et al.* 2013). Deglacial warming was interrupted during the Younger Dryas, an event that involved a major change in the oceanography of the North Atlantic (Hughen *et al.* 2000), and a return to quasi-glacial conditions in Europe and the western Mediterranean (Cacho *et al.* 2001). Pollen records from the Iberian Peninsula and ultra-high resolution speleothem records from the southern Pyrenees indicate a return to relatively arid conditions (González-Sampériz et al. 2006). The latter also show a progressive shift towards humid conditions in parallel with the resumption of the North Atlantic Meridional Overturning Circulation (Bartolomé et al. 2015).

Holocene pollen records show a new transition, this time from humid towards more arid conditions by the Middle Holocene, although the precise timing of this transition and the associated internal characteristics show subregional differences (Carrión et al. 2007; Pérez-Obiol et al. 2011; Sadori et al. 2011). Other paleoclimatic archives such as coastal deposits (Zazo et al. 2008) and peat records (Martínez-Cortizas et al. 2009) further support such a Middle Holocene climate transition in the western Mediterranean region. Holocene climate variability has also been detected in records from North Africa, which point to a strong climate control on African cultural adaptation (Mercuri et al. 2011).

Climate variability and marine productivity

Climate oscillations during the Quaternary induced major changes in faunal distribution both in marine and terrestrial environments. One of the most productive areas in the western Mediterranean Sea is the Alboran Sea. Paleoceanographic studies focused on millennial-scale climate variability of the Last Glacial Period have documented the succession of high–low productivity events in relation to the Dansgaard-Oeschger (D-O) climate variability. High productivity conditions were detected during warm interstadial events and have been attributed to the enhancement of low-pressure systems over the western Mediterranean region (Moreno et al. 2004). This pattern is also reflected in the phytoplankton composition with high concentrations of the small Gephyrocapsa and Emiliania huxleyi during warm interstadials (Colmenero-Hidalgo et al. 2004). The reduction of coccolithophore productivity was particularly noticeable during the cold stadials, directly resulting from Heinrich events that drove cold and less saline Atlantic waters through the Gibraltar Strait and caused strong surface stratification at least in the western part of the western Mediterranean Sea (Ausín et al. 2015a). Preservation of diatoms is very rare in the western Mediterranean due to the undersaturation of silica. However, high preservation of diatoms occurred during the cooling event of the Younger Dryas in the Alboran Sea, which points to very high primary production in the area at that time (Bárcena et al. 2001). Intense changes in the planktonic foraminifera assemblages have also been described in relation to both orbital and millennial timescale climate changes but their signal has mostly been interpreted in terms of water temperature oscillations (Sbaffi et al. 2001; Pérez-Folgado et al. 2003). In the Holocene a major change in the planktonic foraminifera has been described at 8 ka BP, with a shift from a dominance by Neogloboquadrina pachyderma and Globigerina bulloides to a dominance of Globorotalia inflata with G. bulloides, which has been interpreted as a marker of the onset of the modern oceanographic conditions in the basin including distinct geostrophic fronts such as the ones that currently control productivity patterns in the Alboran Sea (Rohling et al. 1995). This improvement in primary productivity at around 8 ka BP has also been detected in the coccolithophore assemblage, which shows significant Holocene centennial-millennial scale variability in productivity linked to different dominant phases of the North Atlantic Oscillation (Ausín et al. 2015b).

Taphonomy and Potential for Archaeological Site Survival

Sources of oceanographic data

Time series of atmospheric and oceanographic parameters including wind speed and direction, air temperature and atmospheric pressure, wave height, wave period and direction, and derived data, are available for the region, including a few series of up to 20 years for European waters (Table 12.1). A selection of statistics can be provided for each location, which may include averaged and extreme values of wind speed and wave height (Fig. 12.15), duration statistics, periods and directions (e.g. scatter diagrams of wind or wave height and period).

TABLE 12.1 SOURCES OF METOCEAN DATA SETS FOR WESTERN MEDITERRANEAN WATERS.

Country	Buoys	Period	Institution	Data Source
Spain	15 buoys	1985–Present	Puertos del Estado	www.puertos.es
	6 buoys	1985–2012*	XIOM (Xarxa d'Instruments Oceanogràfics i Meteorològics, Generalitat de Catalunya)	Nothing further available
France	19 buoys	1990–Present	CANDHIS (Centre d'Archivage National de Données de Houle In Situ)	candhis.cetmef.developpement-durable.gouv.fr
Italy	14 buoys	2002–Present	Italian national data buoy network	www.telemisura.it

*The XIOM network was disbanded due to budget cuts.

Time series of current speed and direction are available at discrete periods (1–2 years) in a few locations in European waters. For instance, current profiles through the water column have been monitored on the Têt pro-delta (northwest Mediterranean Sea) at 28 m depth in 2004–2005, or in the Venice Gulf at the *Acqua Alta* oceanographic platform at 16 m depth (www.ve.ismar.cnr.it/piattaforma/) using upward looking Acoustic Doppler Current Profilers (ADCPs). Regional and local administrations, and research groups, may own current-speed and current-direction data sets from specific coastal and shelf areas. Obtaining these data would require a considerable effort of data mining for which the help of local experts would be required.

The recently created European Marine Observation and Data Network (EMODnet — www.emodnet.eu/) also integrates marine data resources including physical conditions as monitored by fixed stations, Argo buoys, profilers or gliders (Fig. 12.16).

While having real data and long time-series statistics of the atmospheric and oceanographic parameters in an area of interest is certainly useful, many key parameters can be modeled through computer simulations.

There are also several online atlases that provide wind and wave data obtained through high-resolution numerical models. For example, within the Medatlas project an extensive atlas of the wind and wave conditions in the Mediterranean Sea was completed

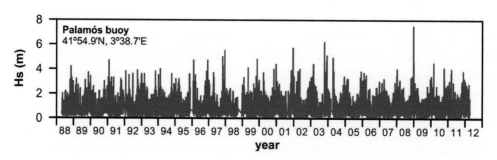

Fig. 12.15 Significant wave height (H_s) time-series recorded by the Palamós (northeast of Spain) buoy from 1988 to 2012. University of Barcelona.

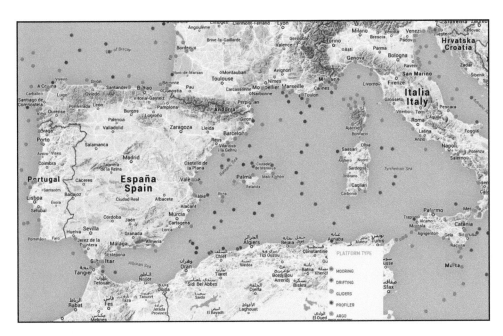

Fig. 12.16 Screenshot of EMODnet showing the distribution of moorings (including metocean buoys), drifting buoys, gliders, profilers and Argo floats in the western Mediterranean Sea since 1950. An on-line query option in the EMODnet website allows access in real time to the meta data of every component of the observational network, including types of data lists. Time-series data from the different private and public organizations is also available to download from the EMODnet website (www.emodnet.eu/).

in 2004 (Medatlas Group 2004). In addition, there have been various efforts through both national and multi-national projects to develop a subregional and Mediterranean-scale monitoring and forecasting capacity. The Mediterranean ocean Forecasting System (MFS) has been producing forecasts since September 1999, and is continuously updated and improved (Pinardi *et al.* 2003). Each day a 10-day forecast is released. MFS is the Mediterranean component of the European Marine Core Service (MyOcean2 project), which offers a unique access to oceanographic products through an online catalogue (marine.copernicus.eu). Furthermore, several Mediterranean national marine forecasting systems are nested into MFS in the so-called MONGOOS (Mediterranean Oceanography Network for the Global Ocean Observing System — www.mongoos.eu) subsystems network, which aims to coordinate and consolidate the existing operational oceanography systems in the Mediterranean Sea (former MOON and MedGOOS alliances). Detailed forecasts for the western Mediterranean are also available through the joint effort of Puertos del Estado and the

Agencia Estatal de Meteorología (the State Meteorological Agency) of Spain, or made *ad hoc* for particularly relevant events (Fig. 12.17).

Time series of sea level at several tidal gauges are available for European waters. The European Sea-Level Service (ESEAS) was an international collaboration of organizations in 23 countries that made the initial step of bringing together the formerly scattered sea-level data and research in Europe. ESEAS aimed at developing into a major research infrastructure for all aspects related to sea level, be they in the field of climate change research, natural hazards or marine research. However, it has been inactive since about 2008. An alternative is the Permanent Service for Mean Sea Level (PSMSL) of the Federation of Astronomical and Geophysical Data Analysis Services (FAGS), with impressive data coverage made from about 2000 stations worldwide for which monthly and annual mean sea-level information is available (Fig. 12.18). However, the number of stations with at least 40 years of revised local reference data is much lower, with only two of them on the northernmost shores of the western

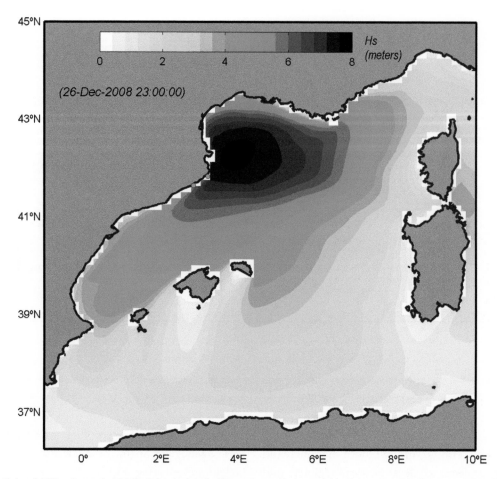

Fig. 12.17 Spatial variability of wave height H_s (m) calculated with an atmospheric model in the western Mediterranean Sea from a violent sea storm (locally named the "storm of the century") which hit the continental shelf and coastline of north Catalonia and western Golfe du Lion on the 26th December 2008. Modified from Sanchez-Vidal et al. (2012).

Mediterranean Sea and a few more in the Gibraltar area, and none along the coastline of North African countries. The relative sea-level trend where measurements exist in the western Mediterranean Sea is about 1 mm/year of rise, with an anomaly of more than 100 mm for 2013 relative to 1960–1990 (detrended) (see www.psmsl.org/pro ducts/trends/ and www.psmsl.org/products/anomalies/). PSMSL is based at the National Oceanography Centre (NOC) in Liverpool. More information is available at www.eea.europa.eu/data-and-maps/data/external/perma nent-service-for-mean-sea.

In addition, there are several historically important long sea-level records discussed in the literature, which are not available in a monthly and annual mean format,

or are not true mean sea levels (i.e. based perhaps on intermittent observations of high and low waters rather than on continuous records by a 'tide gauge' or 'tide pole' as now understood). More information is available at PSMSL through www.psmsl.org/data/longrecords/.

Areas of rapid erosion, and rapid accumulation of sediments since the LGM

Erosion is a common feature along most western Mediterranean low-lying shores, while rapid accumulation of sediments mostly occurs off river mouths and down-current of them. Holocene pro-deltas have developed

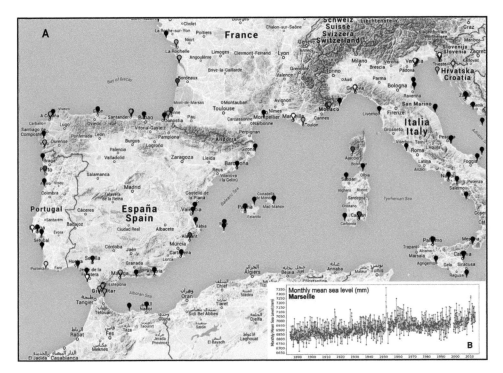

Fig. 12.18 (A) Map of the tide gauge stations in the western Mediterranean Sea. Each color shows a different series length; (B) Monthly mean sea-level data from the tide gauge of Marseille (France), obtained from the Permanent Service for Mean Sea Level. Red data must be used with caution (www.psmsl.org/data/obtaining/map.html). Reproduced with permission.

off the main rivers such as the Rhône, Ebre, Llobregat and others. Modern coastal sedimentary prisms develop over the inner shelf locally extending to the mid-shelf. Even in areas with high sediment input, such as the Golfe du Lion, the outer shelf is generally not covered by recent sediment and therefore the present-day seafloor topography corresponds to that of the LGM when the shelf was sub-aerially exposed and the shoreline was located at the current shelf edge (Fig. 12.19).

Some shelf segments are starved of sediment, either because they do not receive fluvial sediment inputs or because they are bounded by submarine canyons deeply incised into the shelf that capture the longshore sediment before it reaches them, or because of both. A good example of this situation is La Planassa shelf segment in between the submarine canyons of La Fonera and Blanes, in the North Catalan Margin. A narrow shelf, with low sediment input and numerous canyons also parallels most of the southern coast of France from Marseille to the Italian border.

Areas of maximum protection from wave and current damage

With the exception of the Gibraltar area and a few other places where there are topographic constrictions, such as prominent coastal promontories, currents in the western Mediterranean Sea are not as strong as in many North Atlantic settings. Islands always have a sheltered side that is protected from the dominant winds and seas. In the Baleares this is particularly the case of the south-facing coastline. Other coasts with cliffs that are downwind of the main airflows also may offer a good protection, as demonstrated for instance for the Cosquer Cave in southern France.

Other parameters to take into account in terms of protection from wave and current damage are the wave base (with the greatest depths being reached during storms) and the limits of the areas actually affected by the strongest winds. Exceptional records of waves of up to 10 m or larger have been measured at some coastal

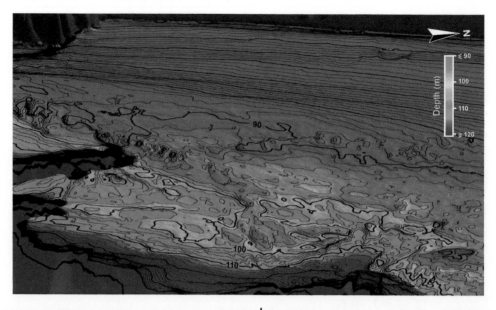

Fig. 12.19 Paleo-deltaic landscape on the outer shelf of the northwestern Mediterranean Sea. Note the steep paleo-delta front, lateral spits, semi-enclosed embayments or lagoons, distributary channels, and a coastline-parallel dune field. Also, observe the strong contrast between the inner and mid shelf that are covered by a modern littoral prism, and the uncovered outer shelf where the paleo-delta is. The image is constructed from low-resolution single depth soundings, which indicates the tremendous potential of multibeam mapping of the area, where strategically located sediment cores could be obtained for paleoclimatic reconstructions. University of Barcelona after single-beam data from SHOM.

locations in the northwestern Mediterranean Sea, but even during the largest storms significant wave heights are rarely more than 2 m high over the shelf, and this for short periods of time. Regarding the winds, for example, the *tramontana* does not go farther south than Minorca, with the Mallorca and Ibiza shelves remaining unaffected.

Tsunamis are potentially destructive waves that are most commonly triggered by earthquakes. The Algerian, south and southeast Spanish and Ligurian margins are the most tsunamigenic ones in the western Mediterranean basin according to both the seismic hazard map in Fig. 12.8 and historical records. Once generated, tsunamis propagate and can hit and damage coastlines that are far away from the source, as illustrated by the May 2003 event, which is the last to have occurred in the western Mediterranean Sea. That tsunami was triggered by the Boumerdès-Zemmouri earthquake in Algeria and first hit the Balearic coasts after less than 40 minutes, in the late evening of the 21st May, causing some infrastructural damage and sinking more than 100 boats. Resonance effects may increase the effects of tsunamis in sheltered

areas such as bays and harbors (Vela *et al.* 2014). On the other hand, tsunamis may uncover archaeological remnants that were buried under coastal and shelf sediments.

Coastal and wetland or lagoonal areas of prehistoric potential

Lagoonal areas mostly develop in association with the main deltas like the Rhône and Ebre and down-current of them. Lagoons with barriers and spits occur mainly along the coasts of Valencia, Languedoc-Roussillon and parts of north and west Sardinia.

General conclusions about site preservation

We consider the possibility of site preservation as rather high in most of the western Mediterranean area, as proven by some recent findings, such as the Cosquer Cave and caves in Mallorca, and by the identification of promising flooded landscapes at specific outer shelf locations (see following section).

Potential Areas for Future Work

Regions with least and highest chance of site survival

The triangle from Marseilles to Barcelona to the Baleares presently looks highly promising because of the co-occurrence of favorable conditions, also supported by good background information and previous findings, which are briefly summarized below:

1 Some of the best examples of coastal paleolandscapes known in the western Mediterranean Sea occur there (Figs. 12.19 & 12.20).
2 A number of coastal caves suggest a strong potential for new findings, as illustrated by the exceptional case of the Cosquer Cave (www.bradshawfoundation.com/cosquer/index.php) which was first discovered in 1985, although the prehistoric paintings were only reported and published in 1991. The Cosquer Cave is at Cape Morgiou in the Calanques, near Marseille, and despite partial destruction due to flooding by the sea, ranks

among the few where paintings on the rock have been preserved. More than 150 animal figures have been found, dated between 27,000 and 19,000 years ago. The main gallery can be accessed through a 175-m long tunnel, which starts at 37 m under the sea surface (Fig. 12.21). Another remarkable example is Cova del Gegant (the Giant's Cave) in Sitges, a few kilometers south of Barcelona, which hosts a Middle Paleolithic site including the presence of Neanderthals. After its discovery in 1954 and subsequent excavations, the most remarkable findings in Cova del Gegant have been a Neanderthal mandible (dated at 52.3 ± 2.3 ka by U/Th) and a Neanderthal lower incisor tooth from two different individuals (Daura *et al.* 2005; 2010; Rodríguez *et al.* 2011). This cave has also yielded fossils of Late Pleistocene large vertebrates and Middle Paleolithic artifacts. This cave is flooded at the mouth, but rises slightly in karstic tunnels further back. The archaeological deposits have been found just above present sea level.

3 Scattered accounts of accidental retrieval by fishermen of large mammal bones and teeth (Fig. 12.22) on

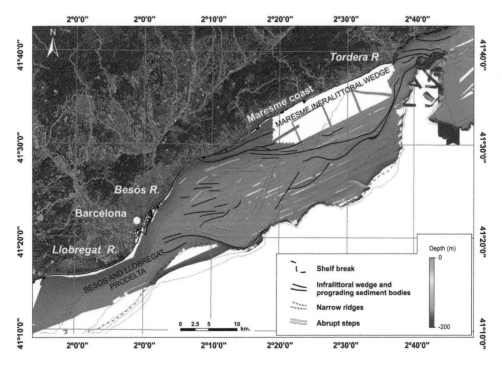

Fig. 12.20 Interpreted shaded relief map of the continental shelf of Barcelona depicting numerous paleoshorelines and beach rocks ("Narrow ridges" in the legend). Note the protruding rounded delta shape just east of Barcelona. University of Barcelona.

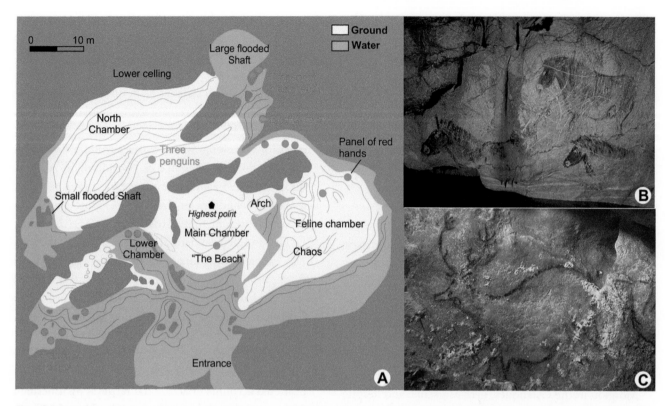

Fig. 12.21 (A) Plan of Cosquer Cave near Marseille, France; (B) Painting found in the Fresco of Horses. Note the water surface in the lower part of the image; (C) Painting found in the Lower Chamber. Map (redrawn) and images from www.culture.gouv.fr/fr/archeosm/en/fr-cosqu1.htm.

continental shelf areas where a potentially promising paleolandscape is known to occur. More than 15 bones and teeth (attributed to mastodon) have been dredged up during several years within an area about 500 m in diameter, which may suggest an accumulation due to prehistoric human action or a concentration effect because of natural conditions or events.

4 A good-to-excellent background information on shelf-floor topography, hydrodynamics, geology and sediment dynamics of this region, where the non-tidal character and limited geodynamic activity (no significant seismicity, no volcanism) enhances the preservation potential; the karstic nature of a large part of the coast and shelf adds to the potential of the area.

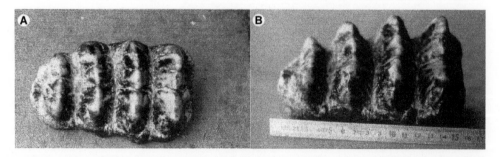

Fig. 12.22 (A and B) Photographs of mega-mammal teeth (attributed to a mastodon) fished up accidentally on the continental shelf off Girona, northeast Spain. The low quality of the images is due to their being taken with a mobile phone. Argo Maris Private Foundation.

5 A large number of land-based facilities and sea-going platforms and instruments, public and private, supported by and at the service of a large, multidisciplinary, transnational scientific community in the coastal nations (Fig. 12.23).

A second potentially promising area where background information in the form of, for instance, high-resolution multibeam bathymetric maps, has been made available recently is the continental shelf of the south and southeast of Spain, including Andalucía, Murcia and Alicante. The multibeam bathymetry has shown several, particularly conspicuous coast-parallel, cemented paleoshorelines with associated landforms (Fernández-Salas *et al.* 2003; 2015; Pinna 2013; Pinna *et al.* 2014) (Fig. 12.24). Furthermore, numerous coastal caves are known to occur in the area, some of which have yielded prehistoric archaeological objects and fossils, such as Cueva del Tesoro (Málaga).

In Gibraltar there has been a sequence of international projects led by the Gibraltar Museum to investigate the record of prehistoric sites, especially Gorham's Cave. Researchers from the Gibraltar Museum, the universities of York and Huelva and the National Oceanography Centre Southampton have collaborated in underwater investigations. An active focus of research was the submarine caves at a depth of 20 m to 25 m in the area known as Vladi's Reef to the south of the Rock. Radiocarbon dating of the finds from Gorham's Cave suggests that the region was a refugium allowing the survival of Neanderthals later than elsewhere (Finlayson *et al.* 2006 and references therein). GRC Geociències Marines from the University of Barcelona has also been involved in research about the last Neanderthals in Gibraltar and Europe (Tzedakis *et al.* 2007).

Insufficient background information exists on the North African coast to allow an informed assessment of the archaeological potential of the continental shelf.

Fig. 12.23 Modern survey vessel specially adapted for ROV inspection to 400 m of water depth, belonging to Argo Maris Private Foundation. Propulsion is by hull-mounted jets instead of propellers, which ensures safer operation at sea. Note the ROV hangar (in open position) to the left and small crane above it. The boat needs only three crew. This vessel is often used by GRC Geociències Marines from the University of Barcelona, which also uses other specially adapted vessels for high-resolution mapping, inspection and sampling of the continental shelf floor. Argo Maris Private Foundation.

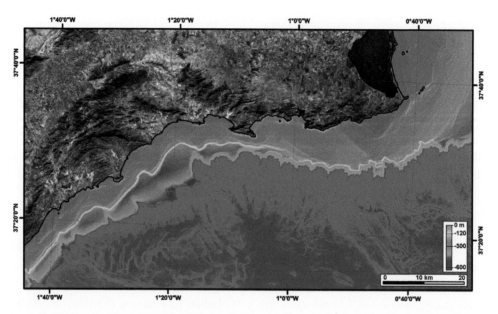

Fig. 12.24 Multibeam bathymetry map of the continental margin of southeast Spain. Note several continuous terrace levels over the continental shelf, which correspond to paleoshorelines. The 120-m isobath marks the approximate limit of the outer shelf. University of Barcelona, using a data set from Instituto Español de Oceanografía.

List of the institutions and names and contacts of scientists having special research-expertise in the example areas:

- Universitat de Barcelona: Miquel Canals, Isabel Cacho, Galderic Lastras, David Amblas, Joan Daura;
- Universitat de les Illes Balears: Joan Josep Fornós;
- Centre d'Arqueologia Subaquàtica de Catalunya (CASC): Gustau Vivar;
- Museo Nacional de Arqueología Subacuática (ARQUA): Iván Negueruela;
- CEFREM: Université de Perpignan: Xavier Durrieu de Madron;
- IFREMER: Gilles Lericolais;
- Maison de la Recherche de l'Université du Mirail, UMR 5602 Géode, Toulouse: Laurent Carozza;
- Département des recherches archéologiques subaquatiques et sous-marines (DRASSM): Michel l'Hour, Yves Billaud;
- Centro de Arqueología Subacuática (CAS) del Instituto Andaluz del Patrimonio Histórico;
- Centro de Arqueología Subacuática de la Comunidad Valenciana (CASCV);
- The Gibraltar Museum.

Major industrial activities in the area which tend to disturb seabed prehistoric sites

- Sand dredging for beach renourishment at specific locations, with extraction generally taking place on the inner shelf. Subsequent sand transfer by longshore currents and storms may lead to the increase of sediment accumulation locally.
- Bottom trawling over the continental shelf.
- Cable and pipeline trenching.
- Sewage dumping through pipes, which might be relevant at specific locations, generally on the inner shelf
- Anchoring zones off large harbors and in tourist spots frequently visited by yachts.

Recommendation

In order to promote the study of the Quaternary landscapes of continental shelf areas exposed during low sea-level stages, and assess their potential for hominin occupation, it would be advantageous to develop a mapping task force or correspondence group using consistent criteria, procedures and platforms to promote the acquisition of new high-resolution data and compile the existing cartographic products following a commonly agreed protocol and layout, and make those products available to approved users. Involving WebGIS experts would be necessary. The poor to non-existent background information on the North African continental shelves of the western Mediterranean is a gap that would require a sustained and devoted effort, including the pooling of resources and the involvement of institutions and experts (eventually including specialized training) from the relevant countries.

References

Acosta, J., Muñoz, A. & Uchupi, E. 2012. Systematic mapping of the Spanish continental margin. *Eos* 93:289-90.

Allen, J. R. M. & Huntley, B. 2000. Weichselian palynological records from southern Europe: correlation and chronology. *Quaternary International* 73-74:111-125.

Ardhuin, F., Bertotti, L., Bidlot, J. -R. *et al.* 2006. Comparison of wind and wave measurements and models in the Western Mediterranean Sea. *Ocean Engineering* 34:526-541.

Arnau, P., Liquete, C. & Canals, M. 2004. River mouth plume events and their dispersal in the northwestern Mediterranean Sea. *Oceanography* 17:22-31.

Ausín, B., Flores, J. A., Sierro, F. J. *et al.* 2015a. Coccolithophore productivity and surface water dynamics in the Alboran Sea during the last 25kyr. *Palaeogeography, Palaeoclimatology, Palaeoecology* 418:126-140.

Ausín, B., Flores, J. A., Sierro, F. J. *et al.* 2015b. Atmospheric patterns driving Holocene productivity in the Alboran Sea (Western Mediterranean): A multiproxy approach. *The Holocene* 25:583-595.

Ballesteros, E., Mariani, S., Cefali, M. E., Chappuis, E. & Terradas, M. 2014. *Manual dels hàbitats litorals de Catalunya*. Departament de Territori i Sostenibilitat: Barcelona.

Bárcena, M. Á., Cacho, I., Abrantes, F., Sierro, F. J., Grimalt, J. O. & Flores, J. A. 2001. Paleoproductivity variations related to climatic conditions in the Alboran Sea (western Mediterranean) during the last glacial-interglacial transition: the diatom record. *Palaeogeography, Palaeoclimatology, Palaeoecology* 167:337-357.

Bartolomé, M., Moreno, A., Sancho, C. *et al.* 2015. Hydrological change in southern Europe responding to increasing North Atlantic overturning during Greenland Stadial 1. *Proceedings of the National Academy of Sciences* 112:6568-6572.

Berger, A. & Loutre, M. F. 1991. Insolation values for the climate of the last 10 million years. *Quaternary Science Reviews* 10:297-317.

Berné, S. & Gorini, C. 2005. The Gulf of Lions: An overview of recent studies within the French 'Margins' Programme. *Marine and Petroleum Geology* 22:691-693.

Berné, S., Carré, D., Loubrieu, B., Mazé, J. -P. & Normand, A. 2001. *Carte morpho-bathymétrique du Golfe du Lion, 1:100,000.* 4 sheets. IFREMER: Brest, France.

Bertotti, L. & Cavaleri, L. 2009. Large and small scale wave forecast in the Mediterranean Sea. *Natural Hazards and Earth System Sciences* 9:779-788.

Broecker, W. S. 2006. Abrupt climate change revisited. *Global and Planetary Change* 54:211-215.

Cacho, I., Grimalt, J. O., Pelejero, C. *et al.* 1999. Dansgaard-Oeschger and Heinrich event imprints in Alboran Sea paleotemperatures. *Paleoceanography* 14:698-705.

Cacho, I., Grimalt, J. O., Sierro, F. J., Shackleton, N. & Canals, M. 2000. Evidence for enhanced Mediterranean thermohaline circulation during rapid climatic coolings. *Earth and Planetary Science Letters* 183:417-429.

Cacho, I., Grimalt, J. O., Canals, M. *et al.* 2001. Variability of the western Mediterranean Sea surface temperature during the last 25,000 years and its connection with the Northern Hemisphere climatic changes. *Paleoceanography* 16:40-52.

Cacho, I., Shackleton, N., Elderfield, H., Sierro, F. J. & Grimalt, J. O. 2006. Glacial rapid variability in deep-water temperature and $\delta^{18}O$ from the Western Mediterranean Sea. *Quaternary Science Reviews* 25:3294-3311.

Carrión, J. S., Munuera, M. & Navarro, C. 1998. The palaeoenvironment of Carihuela Cave (Granada, Spain): a reconstruction on the basis of palynological investigations of cave sediments. *Review of Palaeobotany and Palynology* 99:317-340.

Carrión, J. S., Fuentes, N., González-Sampériz, P. *et al.* 2007. Holocene environmental change in a montane region of southern Europe with a long history of human settlement. *Quaternary Science Reviews* 26:1455-1475.

Clottes, J., Beltrán, A., Courtin, J. & Cosquer, H., 1992. The Cosquer Cave on Cape Morgiou, Marseilles. *Antiquity* 66:583-598.

Collina-Girard, J. 1992. Présentation d'une carte bathymétrique au 1/25 000 du précontinent marseillais (au large de la zone limitée par la grotte Cosquer et l'habitat préhistorique de Carry-le-Rouet). *Géologie méditerranéenne* 19:77-87.

Collina-Girard, J. 1999. Les replats littoraux holocènes immergés en Provence et en Corse: implications eustatiques et néotectoniques. *Quaternaire* 10:121-131.

Colmenero-Hidalgo, E., Flores, J. A., Sierro, F. J. *et al.* 2004. Ocean surface water response to short-term climate changes revealed by coccolithophores from the Gulf of Cadiz (NE Atlantic) and Alboran Sea (W Mediterranean). *Palaeogeography, Palaeoclimatology, Palaeoecology*. 205:317-336.

Conchon, O. 1975. *Les formations quaternaries de type continental en Corse orientale*. Ph.D thesis. University of Paris VI (2 volumes).

Coyne, M. K., Jones, B. & Ford, D. 2007. Highstands during Marine Isotope Stage 5: evidence from the Ironshore Formation of Grand Cayman, British West Indies. *Quaternary Science Reviews* 26:536-559.

Daura, J., Sanz, M., Subirà, M. E., Quam, R., Fullola, J. M. & Arsuaga, J. L. 2005. A Neandertal mandible from the Cova del Gegant (Sitges, Barcelona, Spain). *Journal of Human Evolution* 49:56-70.

Daura, J., Sanz, M., Pike, A. W. G. *et al.* 2010. Stratigraphic context and direct dating of the Neandertal mandible from Cova del Gegant (Sitges, Barcelona). *Journal of Human Evolution* 59:109-122.

Dorale, J. A., Onac, B. P., Fornós, J. J. *et al.* 2010. Sea-level highstand 81,000 years ago in Mallorca. *Science* 327:860-863.

Fernández-Salas, L. M., Lobo, F. J., Hernández-Molina, F. J. *et al.* 2003. High-resolution architecture of late Holocene highstand prodeltaic deposits from southern Spain: the imprint of high-frequency climatic and relative sea-level changes. *Continental Shelf Research* 23:1037-1054.

Fernández-Salas, L. M., Durán, R., Mendes, I. *et al.* 2015. Shelves of the Iberian Peninsula and the Balearic Islands (I): Morphology and sediment types. *Boletín Geológico y Minero* 126:327-376.

Finlayson, C., Pacheco, F. G., Rodríguez-Vidal, J. *et al.* 2006. Late survival of Neanderthals at the southernmost extreme of Europe. *Nature* 443:850-853.

Flemming, N. C. 1969. *Archaeological Evidence for Eustatic Change of Sea Level and Earth Movements in the Western Mediterranean in the Last 2,000 Years*. Geological Society of America Special Paper 109.

Flemming, N. C. & Webb, C. O. 1986. Tectonic and eustatic coastal changes during the last 10,000 years derived from archaeological data. *Zeitschrift für Geomorphologie* 62:1-29.

Fletcher, W. J. & Sánchez-Goñi, M. F. 2008. Orbital- and sub-orbital-scale climate impacts on vegetation of the western Mediterranean basin over the last 48,000 yr. *Quaternary Research* 70:451-464.

Frigola, J., Moreno, A., Cacho, I. *et al.* 2008. Evidence of abrupt changes in Western Mediterranean Deep Water circulation during the last 50 kyr: A high-resolution marine record from the Balearic Sea. *Quaternary International* 181:88-104.

Giardini, D., Jiménez, M. -J. & Grünthal, G. (eds.) 2003. *European-Mediterranean Seismic Hazard Map, 1:5,000,000*. 1 sheet. European Seismological Commission and International Geological Correlation Program Project no 382: SESAME.

González-Sampériz, P., Valero-Garcés, B. L., Moreno, A. *et al.* 2006. Climate variability in the Spanish Pyrenees during the last 30,000 yr revealed by the El Portalet sequence. *Quaternary Research* 66: 38-52.

Got, H. & Aloisi, J. C. 1990. The Holocene sedimentation on the Gulf of Lions margin: a quantitative approach. *Continental Shelf Research* 10:841-855.

Goy, J. L., Zazo, C. & Dabrio, C. J. 2003. A beach-ridge progradation complex reflecting periodical sea-level and climate variability during the Holocene (Gulf of Almeria, Western Mediterranean). *Geomorphology* 50:251-268.

Hearty, P. J. 1987. New data on the Pleistocene of Mallorca. *Quaternary Science Reviews* 6:245-257.

Hearty, P. J. 1998. The geology of Eleuthera Island, Bahamas: A Rosetta Stone of Quaternary stratigraphy and sea-level history. *Quaternary Science Reviews* 17:333-355.

Hughen, K. A., Southon, J. R., Lehman, S. J. & Overpeck, J. T. 2000. Synchronous radiocarbon and climate shifts during the last deglaciation. *Science* 290:1951–1954.

Imbrie, J., Hays, J. D., Martinson, D. G. *et al.* 1984. The orbital theory of Pleistocene climate: support from a revised chronology of the marine $\delta^{18}O$ record. In Berger, A. L., Imbrie, J., Hays, J., Kukla, G. & Saltzman, B. (eds.) *Milankovitch and Climate* (Part 1). pp. 269-305. Reidel Publishing Company: Dordrecht.

Imbrie, J., Duffy, A., Mix, A. C. & Mcintyre, A. 1990. *SPECMAP Archive #1, IGBP PAGES/World Data Center for Paleoclimatology, Data Contribution Series # 90-001*. Brown University: Rhode Island.

Kirkaldy, J.F. 1963. *Minerals and Rocks*. Blandford Press: London.

Laborel, J., Morhange, C., Lafont, R., Le Campion, J., Laborel-Deguen, F. & Sartoretto, S. 1994. Biological evidence of sea-level rise during the last 4500 years on the rocky coasts of continental southwestern France and Corsica. *Marine Geology* 120:203-223.

Lambeck, K. & Purcell, A. 2005. Sea-level change in the Mediterranean Sea since the LGM: model predictions for tectonically stable areas. *Quaternary Science Reviews* 24:1969-1988.

Lambeck, K., Antonioli, F., Anzidei, M. *et al.* 2011. Sea level change along the Italian coast during the Holocene and projections for the future. *Quaternary International* 232:250-257.

Lambeck, K., Rouby, H., Purcell, A., Sun, Y. & Sambridge, M. 2014. Sea level and global ice volumes from the Last Glacial Maximum to the Holocene. *Proceedings of the National Academy of Sciences* 111:15296-15303.

Liquete, C., Canals, M., Lastras, G. *et al.* 2007. Long-term development and current status of the Barcelona continental shelf: A source-to-sink approach. *Continental Shelf Research* 27:1779-1800.

Lobo, F. J., Ercilla, G., Fernández-Salas, L. M. & Gámez, D. 2014. The Iberian Mediterranean shelves. In Chiocci, F. L & Chivas, A. R (eds.) *Continental Shelves of the World: Their Evolution During the Last Glacio-Eustatic Cycle (Geological Society Memoir No. 41)*. pp. 147-170. Geological Society: London.

Loubrieu, B., Mascle, J. & Medipmap Group 2007. Morpho-Bathymetry of the Mediterranean Sea. *CIESM & Ifremer Special Publications*, 1:3,000,000 scale.

Ludwig, K. R., Muhs, D. R., Simmons, K. R., Halley, R. B. & Shinn, E. A. 1996. Sea-level records at ~80 ka from tectonically stable platforms: Florida and Bermuda. *Geology* 24:211-214.

Marriner, N. & Morhange, C. 2007. Geoscience of ancient Mediterranean harbours. *Earth-Science Reviews* 80:137-194.

Martínez-Cortizas, A., Costa-Casais, M. and López-Sáez, J. A. 2009. Environmental change in NW Iberia between 7000 and 500 cal BC. *Quaternary International* 200:77-89.

MedatlasGroup 2004. *Wind and wave atlas of the Mediterranean Sea*. WEAO Research Cell.

MediMap Group 2007. *Morpho-bathymetry of the Mediterranean Sea, 1:2,000,000*. 2 sheets. Maps and Atlases. CIESM/IFREMER.

Mercuri, A. M., Sadori, L. & Uzquiano Ollero, P. 2011. Mediterranean and north-African cultural adaptations to mid-Holocene environmental and climatic changes. *The Holocene* 21:189-206.

Mitrovica, J. X. & Milne, G. A. 2002. On the origin of late Holocene sea-level highstands within equatorial ocean basins. *Quaternary Science Reviews* 21:2179-2190.

Mix, A. C., Bard, E. & Schneider, R. 2001. Environmental processes of the ice age: land, oceans, glaciers (EPILOG). *Quaternary Science Reviews* 20:627-657.

Morellón, M., Valero-Garcés, B., Vegas-Vilarrúbia, T. *et al.* 2009. Lateglacial and Holocene palaeohydrology in the western Mediterranean region: The Lake Estanya record (NE Spain). *Quaternary Science Reviews* 28:2582-2599.

Moreno, A., Cacho, I., Canals, M., Grimalt, J. O. & Sánchez-Vidal, A. 2004. Millennial-scale variability in the productivity signal from the Alboran Sea record, Western Mediterranean Sea. *Palaeogeography, Palaeoclimatology, Palaeoecology* 211:205-219.

Moreno, A., Cacho, I., Canals, M. *et al.* 2005. Links between marine and atmospheric processes oscillating on a millennial time-scale. A multi-proxy study of the last 50,000 yr from the Alboran Sea (Western Mediterranean Sea). *Quaternary Science Reviews* 24:1623-1636.

Moreno, A., Svensson, A., Brooks, S. J. *et al.* 2014. A compilation of Western European terrestrial records 60-8 ka BP: towards an understanding of latitudinal climatic gradients. *Quaternary Science Reviews* 106:167-185.

Morhange, C., Laborel, J. & Hesnard, A. 2001. Changes of relative sea level during the past 5000 years in the ancient harbor of Marseilles, Southern France. *Palaeogeography, Palaeoclimatology, Palaeoecology* 166:319-329.

Muhs, D. R., Simmons, K. R. & Steinke, B. 2002. Timing and warmth of the Last Interglacial period: new U-series evidence from Hawaii and Bermuda and a new fossil compilation for North America. *Quaternary Science Reviews* 21:1355-1383.

Peltier, W. R. 2004. Global glacial isostasy and the surface of the ice-age Earth: The ICE-5G (VM2) model and GRACE. *Annual Review of Earth and Planetary Sciences* 32:111-149.

Pérez-Folgado, M., Sierro, F. J., Flores, J. A. *et al.* 2003. Western Mediterranean planktonic foraminifera events and millennial climatic variability during the last 70 kyr. *Marine Micropaleontology* 48:49-70.

Pérez-Obiol, R., Jalut, G., Julià, R. *et al.* 2011. Mid-Holocene vegetation and climatic history of the Iberian Peninsula. *The Holocene* 21:75-93.

Pérez-Sanz, A., González-Sampériz, P., Moreno, A. *et al.* 2013. Holocene climate variability, vegetation dynamics and fire regime in the central Pyrenees: the Basa de la Mora sequence (NE Spain). *Quaternary Science Reviews* 73:149-169.

Pinardi, N., Allen, I. Demirov, E. *et al.* 2003. The Mediterranean Ocean Forecasting System: first phase of implementation (1998-2001). *Annales Geophysicae* 21:3-20.

Pinna, A. 2013. *Evidencias morfológicas de la evolución del nivel del mar en el margen continental entre el Mar Menor y el Golfo de Almería (Mediterráneo Occidental)*. Masters dissertation. University of Barcelona (unpublished).

Pinna, A., Lastras, G., Acosta, J., Muñoz, A. & Canals, M. 2014. The imprint of sea-level changes in the southeastern Iberian continental shelf, Western Mediterranean Sea. *Geophysical Research Abstracts* 16 EGU2014-601.

Pirazzoli, P. A. 2005. A review of possible eustatic, isostatic and tectonic contributions in eight late-Holocene relative sea-level histories from the Mediterranean area. *Quaternary Science Reviews* 24:1989-2001.

Pons, A. & Reille, M. 1988. The Holocene and Upper Pleistocene pollen record from Padul (Granada, Spain): a new study. *Palaeogeography, Palaeoclimatology, Palaeoecology* 66:255-263.

Rodríguez, L., García-González, R., Sanz, M. *et al.* 2011. A Neanderthal lower incisor from Cova del Gegant (Sitges, Barcelona, Spain). *Boletin de la Real Sociedad Española de Historia Natural. Sección Geológica* 105:25-30.

Rohling, E. J., Den Dulk, M., Pujol, C. & Vergnaud-Grazzini, C. 1995. Abrupt hydrographic change in the Alboran Sea (western Mediterranean) around 8000 yrs BP. *Deep Sea Research Part I: Oceanographic Research Papers* 42:1609-1619.

Sadori, L., Jahns, S. & Peyron, O. 2011. Mid-Holocene vegetation history of the central Mediterranean. *The Holocene* 21:117-129.

Sánchez-Goñi, M., Cacho, I., Turon, J. *et al.* 2002. Synchroneity between marine and terrestrial responses to millennial scale climatic variability during the Last Glacial Period in the Mediterranean region. *Climate Dynamics* 19:95-105.

Sanchez-Vidal, A., Canals, M., Calafat, A. M. *et al.* 2012. Impacts on the deep-sea ecosystem by a severe coastal storm. *PLoS ONE* 7(1):e30395.

Sartoretto, S., Collina-Girard, J., Laborel, J. & Morhange, C. 1995. Quand la grotte Cosquer a-t-elle été fermée par la montée des eaux? *Méditerranée* 82:21-24.

Sbaffi, L., Wezel, F. C., Kallel, N. *et al.* 2001. Response of the pelagic environment to palaeoclimatic changes in the central Mediterranean Sea during the Late Quaternary. *Marine Geology* 178:39-62.

Shackleton, N. J. 2000. The 100,000-year ice-age cycle identified and found to lag temperature, carbon dioxide, and orbital eccentricity. *Science* 289:1897-1902.

Siddall, M., Rohling, E. J., Almogi-Labin, A. *et al.* 2003. Sea-level fluctuations during the last glacial cycle. *Nature* 423:853-858.

Sierro, F. J., Hodell, D. A., Curtis, J. H. *et al.* 2005. Impact of iceberg melting on Mediterranean thermohaline circulation during Heinrich events. *Paleoceanography* 20:PA2019.

Sornoza, L., Barnolas, A., Arasa, A., Maestro, A., Rees, J. G. & Hernández-Molina, F. J. 1998. Architectural stacking patterns of the Ebro delta controlled by Holocene high-frequency eustatic fluctuations, delta-lobe switching and subsidence processes. *Sedimentary Geology* 117:11-32.

Tsimplis, M. N., Proctor, R. & Flather, R. A. 1995. A two-dimensional tidal model for the Mediterranean Sea. *Journal of Geophysical Research: Oceans* 100 (C8):16223-16239.

Tuccimei, P., Ginés, J., Delitala, M. C. *et al.* 2006. Last interglacial sea level changes in Mallorca island (Western Mediterranean). High precision U-series data from phreatic overgrowths on speleothems. *Zeitschrift für Geomorphologie* 50:1-21.

Tuccimei, P., Soligo, M., Ginés, J. *et al.* 2010. Constraining Holocene sea levels using U-Th ages of phreatic overgrowths on speleothems from coastal caves in Mallorca (Western Mediterranean). *Earth Surface Processes and Landforms* 35:782-790.

Tzedakis, P. C., Hughen K. A., Cacho, I. & Harvati K. 2007. Placing late Neanderthals in a climatic context. *Nature* 449:206-208.

UNEP/MAP (United Nations Environment Programme/Mediterranean Action Plan) 2003. *Riverine Transport of Water, Sediments, and Pollutants to the Mediterranean Sea.* UNEP MAP Technical Report Series 141. UNEP/MAP: Athens.

Vacher, H. L. & Hearty, P. 1989. History of stage 5 sea level in Bermuda: Review with new evidence of a brief rise to present sea level during Substage 5a. *Quaternary Science Reviews* 8:159-168.

Vela, J., Pérez, B., González, M. *et al.* 2014. Tsunami resonance in Palma bay and harbour, Majorca Island, as induced by the 2003 Western Mediterranean earthquake. *The Journal of Geology* 122:165-182.

Vella, C. & Provansal, M. 2000. Relative sea-level rise and neotectonic events during the last 6500 yr on the southern eastern Rhône delta, France. *Marine Geology* 170:27-39.

Vella, C., Fleury, T. -J., Raccasi, G., Provansal, M., Sabatier, F. & Bourcier, M. 2005. Evolution of the Rhône delta plain in the Holocene. *Marine Geology* 222-223:235-265.

Waelbroeck, C., Labeyrie, L., Michel, E. *et al.* 2002. Sea-level and deep water temperature changes derived from benthic foraminifera isotopic records. *Quaternary Science Reviews* 21:295-305.

Wehmiller, J. F., Simmons, K. R., Cheng, H. *et al.* 2004. Uranium-series coral ages from the US Atlantic Coastal Plain — the "80 ka problem" revisited. *Quaternary International* 120:3-14.

Zazo, C., Goy, J. L., Dabrio, C. J. *et al.* 2003. Pleistocene raised marine terraces of the Spanish Mediterranean and Atlantic coasts: records of coastal uplift, sea-level highstands and climate changes. *Marine Geology* 194:103-133.

Zazo, C., Dabrio, C. J., Goy, J. L. *et al.* 2008. The coastal archives of the last 15 ka in the Atlantic-Mediterranean Spanish linkage area: Sea level and climate changes. *Quaternary International* 181:72-87.

Zazo, C., Goy, J. L., Dabrio, C. J. *et al.* 2013. Retracing the Quaternary history of sea-level changes in the Spanish Mediterranean-Atlantic coasts: Geomorphological and sedimentological approach. *Geomorphology* 196:36-49.

Chapter 12

Western Mediterranean: Annex

Submerged Karst Structures of the French Mediterranean Coast: An Assessment

Yves Billaud

Ministère de la Culture/DRASSM, Marseille, France
CNRS, UMR 5204 Edytem, Université de Savoie, Le Bourget-du-Lac, France

Introduction

Numerous submerged caves in the limestone cliffs of southern France between the Bay of Marseille and the Bay of Villefranche have been surveyed and partially sampled in order to understand the circumstances of the earlier karstic systems as they were exposed during the Pleistocene, and then inundated during the last marine transgression. Terrestrial indicators and microfauna have been identified, and a few signs of human occupation, including the famous Cosquer Cave (Grotte Cosquer). Although much of the diving investigation was completed before 1980, there have been frequent modern studies to improve the understanding of

the origin of the major karst, and the effects of changing sea level and climate. Future research may still reveal more complete prehistoric deposits.

The Karst of Southeast France

The south coast of France includes several karstic massifs. From west to east these are as follows (Fig. 12a.1):

- from Perpignan to Narbonne, the extremity of the Corbières and the massif of La Clape;
- from Sète to Montpellier, the mountain of la Gardiole and the border of the Cévennes;

Submerged Landscapes of the European Continental Shelf: Quaternary Paleoenvironments, First Edition.
Edited by Nicholas C. Flemming, Jan Harff, Delminda Moura, Anthony Burgess and Geoffrey N. Bailey.
© 2017 John Wiley & Sons Ltd. Published 2017 by John Wiley & Sons Ltd.

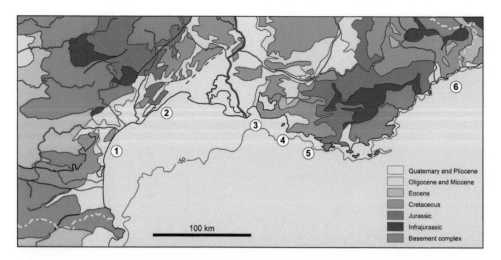

Fig. 12a.1 Karstic massifs of the French Mediterranean coast; (1) Corbières and La Clape; (2) Bas-Languedoc; (3) La Couronne; (4) Calanques of Marseille; (5) Var; (6) Pre-alpes of Grasse and Nice.

– from the east of Marseille to Toulon, the massif of la Couronne and the Calanques;
– from Antibes to the Italian frontier, the Pre-Alps of Grasse and Nice.

These massifs consist of Jurassic and Cretaceous limestone, and have been subjected to a multi-phase karstic evolution, starting in the Tertiary or even in the Cretaceous. In more recent periods, the major structural events were generated by the Messinian Salinity Crisis followed by the Pliocene marine transgression.

At 5.6 Ma the sea level was lowered to about −2000 m leading to the steady erosion of the margins of the Mediterranean basin. This erosion propagated towards the interior of the bordering lands through the cutting of deep canyons. The lowering of the base level and water table led in turn to the formation of important karstic networks. At 5.32 Ma the opening of the Gibraltar Strait caused the refilling of the basin and the drowning of the canyons and the deep karst. The response and adaptation of these outlet channels to the raised base level of drainage led to the formation of unique structures, especially the deep chimney-shafts (Mocochain & Clauzon 2010). During the Pleistocene the different glacio-eustatic cycles led to greater or lesser re-erosion of the Messinian canyons and to further reorganization of drainage patterns.

For the karst of southeast France, and in particular for those of the Calanques massif, there are numerous publications over a long period documenting the reconstruction of the karst (Blanc 1997; 2000; 2010).

In 1991, the publication of the discovery of Cosquer Cave, the only Paleolithic rock-painted cave known to now to the east of the Rhône, revived interest in the detailed study of the regional karst, and in the reconstruction of the process of marine transgression after the last glaciation. This cave is accessible at present only by diving through an entrance which opens at a depth of 37 m, and then by following a siphon ascending towards a half-flooded chamber (Fig. 12a.2). It records evidence of two periods of human habitation, during the Gravettian between 28,000 BP and 26,000 BP and in the Solutrean from 20,000 BP to 18,000 BP (Clottes *et al.* 2005).

The Calanques of Marseille

Ancient landscape

The islands in the sea around Marseille are the summits of a drowned karst landscape. At the Last Glacial Maximum (19,000 BP) the sea level was at −135 m (Lambeck & Bard 2000), thus exposing the whole of the continental shelf. A morpho-structural analysis applied to the bathymetric data from SHOM (Service hydrographique et océanographique de la marine) enables us to specify the morphology of this terrestrial landscape

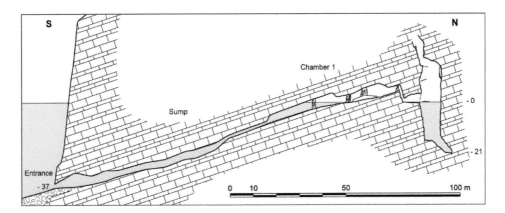

Fig. 12a.2 Projected elevation of Cosquer Cave. Billaud *et al.* (2014).

(see Collina-Girard 1992 — fig. 3). This vast calcareous plateau is shown to have been traversed by a network of valleys and shows a high density of mud-filled basins which could correspond to dolines (see Collina-Girard 1992 — fig. 2). The bathymetric chart with this interpretation (closed depressions, talus slopes, paleodrainage) has been edited and published at a scale of 1:50,000 (Collina-Girard 1996a).

The last rise of sea level

In the Marseille region, several chronological benchmarks allow us to sketch out a curve for the rising sea level:

– surrounding the lighthouse island of Planier, at a depth of –100 m, there is a fossil beach which has been dated to 13,850±200 BP (Collina-Girard 1996b);
– various littoral calcareous organisms dated to 9000 BP are still present at –55 m at the base of Grand Conglué Island;
– the drowning of the entrance to Cosquer Cave at –37 m is estimated to have happened at 7000 BP on the basis of radiocarbon dating of algal concretions (Sartoretto *et al.* 1995);
– the cardial Neolithic site on Riou, dated to 6000 BP, has revealed grindstones made from a rock which outcrops at present 25 m below present sea level;
– the fossil ridges of algal 'trottoir' between –1.25 m and the surface show that the sea level was already within this range by 2500 years ago (Collina-Girard & Monteau 2010).

The top and base of talus slopes and platforms have been interpreted, independently of the geology, as evidence of periods of relative stability during the period of rising sea level. These morphological structures are evident on navigational charts at depths of 50–55 m and 36 m, and as observed by divers, probably at –67 m, –55 m, –41 m, –36 m, –25 m, –16 m and –11 m (see Collina-Girard 1998; 1999 — fig. 6). These facts enable a time-depth curve to be constructed, which can be taken in parallel with that from the Rhône Delta.

Submerged Caves and Chambers

The approaches

Diving observations

Data on submerged caverns could be provided by several different sources and at different technical levels. Thus, one might imagine that the many divers active on the Mediterranean coast would be at least the most likely to discover and explore the caverns. However, it would appear that sports, or leisure, divers are being deterred by a number of accidents, some of which have been fatal: during the last decade alone there were fatalities in Cosquer Cave in 1991, the Grotte de la Tremie in 1994, and the Grotte du Méjean cliff in 2007. In contrast, dedicated 'spéléonautes', that is underground divers or speleological divers, notwithstanding their amateur status, currently conduct exploration and mapping which is of the highest standard in providing evidence of coastal karstification:

- the resurgence of Port-Miou has been explored for a distance of 2 km and to a depth of 223 m;
- the resurgence of the Bestouan has been explored for almost 3 km (Cavalera *et al.* 2010).

While most of these dives are primarily technical in seeking to penetrate as far as possible into the cave or cave complexes, they nevertheless provide some information on the morphology and sediments. Although there are issues of confidentiality in making this material publicly available, some details may be found in specialist speleological publications, club bulletins, on specialist web sites, and in official regional communications (e.g. cds13.free.fr and www.plongeesout.com/).

Freshwater research

In the same way, the discovery of submerged freshwater springs can lead to information on karst, especially regarding modern drainage patterns. This research is often carried out through an indirect approach by searching for visible signs on the surface, or with the help of ship-borne sensors which measure salinity and temperature (Gilli 2003).

This enquiry has been developed by the project COST 621 "Groundwater management of coastal karstic aquifers" (Tuplipano 2005). In the context of archaeological investigations of prehistoric settlement, freshwater springs play an important role, and there is a need to determine their existence and functioning in earlier periods.

Archaeological field projects

On the French Mediterranean coast archaeological studies of drowned caves are still rare, but they do show the potential of the topic. Leaving aside the relatively recent work in Cosquer Cave, most research was conducted from 1968–1976 by DRASM (Direction des Recherches Archéologiques Sous-Marines, which became DRASSM in 1996), thanks to the commissioning of a specialist research vessel, the *Archéonaute*. The research concentrated on the coastal sector of the Calanques of Marseille, and also, further east, the Bay of Villefranche.

The Calanques of Marseille (Bouches-du-Rhône)

In the Calanques, more than a dozen submerged caves were located in the course of the research by DRASSM (Fig. 12a.3). However, only four of them were thoroughly investigated. One cave produced a bone breccia and

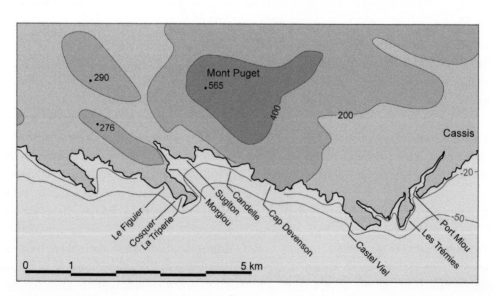

Fig. 12a.3 Main submerged caves of coastal sector of the Calanques of Marseille. Bonifay & Courtin (1998: fig. 3). Adapted with permission from National des Sociétés historique et scientifiques.

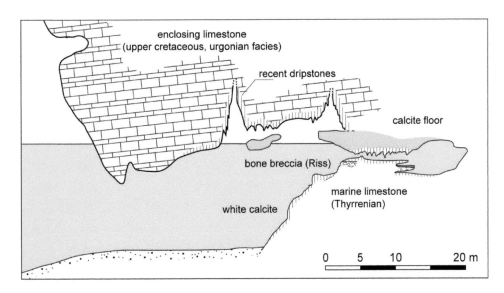

Fig. 12a.4 La Triperie Cave, Marseille. Bonifay & Courtin (1998 : fig. 5). Adapted with permission from National des Sociétés historique et scientifiques.

another some Paleolithic artifacts (Bonifay & Courtin 1998).

The Veyron Cave (Marseille)

In the Bay of Marseille, in front of the lighthouse on Île Planier, a bank between 20 m and 30 m depth shows karstic morphology with underground features, passages 5 m in diameter and two collapsed chambers. Signs of ancient sediments were quickly discovered (Froget 1972). A core taken in one of the collapsed chambers showed alternating layers of continental and marine sediments associated with the early Würm period, but there was no evidence of human occupation (Bonifay 1973).

The Sormiou 1 Cave (Marseille)

This cave opens at a shallow depth on the north side of the Calanque of Sormiou. The entrance arch gives access to a gallery that is 3 m wide, and 60 m long, descending to a depth of 17 m. A core taken 15 m from the entrance showed post-Würm marine deposits, and continental deposits probably of Rissian age (Courtin 1978). There was no discovery of either terrestrial faunal remains or of human occupation.

La Triperie Cave (Marseille)

The cave of La Triperie is situated in the same bay as Cosquer Cave, just a few dozen meters to the east.

It opens into a great arched entrance (with the roof at –6 m, and floor at –16 m), penetrating a huge cliff descending to a depth of 20 m (Fig. 12a.4). With a width of about 15 m and a lowest point at –20 m, the cave is passable for about 50 m into the cliff. Indurated sediments are preserved in about half of the cavity. Expeditions from 1974 to 1976, including the removal of blocks of stone extracted with a pick-axe, enabled the reconstruction of a generalized stratigraphy of this complex assemblage, unequally preserved in different parts of the cave, described from top to bottom as follows (Bonifay & Courtin 1998:19):

- Assemblage A: Würmian stalagmitic layers and breccias with a fauna of very small mammals;
- Assemblage B: marine deposits, probably of the Mid Tyrrhenian period;
- Assemblage C: terrestrial formations 10 m to 12 m thick, with dune sands, consolidated loess and bone breccia, probably from the end of the Riss glaciation;
- Assemblage D: calcareous marine deposits, probably Mid Pleistocene;
- Assemblage E: thick infill of pre-Riss stalagmitic layers, eroded.

The authors of the report stressed the importance of this chamber and also the need for new investigations to

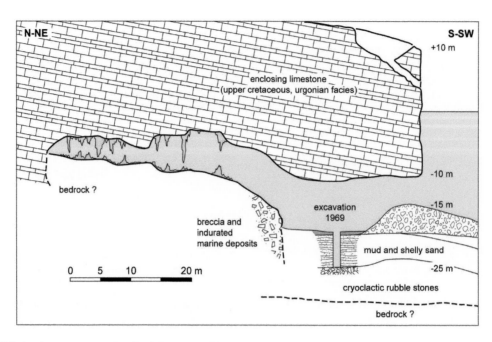

Fig. 12a.5 Les Trémies Cave, Marseille. Bonifay & Courtin (1998 : fig. 8). Adapted with permission from National des Sociétés historique et scientifiques.

obtain a precise age for the different deposits and to search for evidence of human occupation, which logically should occur there.

Cave of Les Trémies (Cassis)

The cave of Les Trémies is located at the extreme west of the Bay of Cassis. It owes its name to the nearby feed hopper for the products of a quarry. It is a vast cavern more than 20 m wide, opening at a depth of 10 m. It appears as a great hall with a floor that extends more than 60 m from the entrance, and a pile of rocks at the center (Fig. 12a.5). Several seasons of research (1968, 1969, 1971 and 1972) were devoted to the study of a complex of deposits with an important variety of sedimentary facies.

The unconsolidated sediments at the entrance (Assemblage I) were studied in detail using both surveys and coring. Consisting of marine sediments, they mark the arrival in the cave of the Versilian transgression. The base of the sequence is dated to 6000 BP, and it rests on continental bedrock rubble (Assemblage II). The oldest sediments (Assemblages IIIa to IIIg) are indurated. They show alternations of marine and continental deposits that are Würmian to pre-Rissian in age. Bones, flint

and even traces of hearths were discovered in the levels assigned to the Riss glaciation.

Other chambers and cavities

Several other caverns have been recorded in the Calanques sector of the coast. 100 m north of Cosquer, Fig Tree Cave has all the characteristics of a habitable cave with a vast entrance chamber (with a floor at –20 m descending to –30 m). It has not been the object of thorough survey. Other caverns cited in bibliographies but without details include: Castel-Vieil, Morgiou East, the Candelle, Cape Devenson and western Sugiton. Some of them contain significant infills which remain to be studied.

Cavities are also present in the islands. Thus at Île Flat, and Île Planier, located south-east of Marseille, a small cave opening at 10 m depth was reported. With a 3-m opening and stretching for a length of 30 m, it contains calcite floors and breccias.

The massif of La Couronne

The western limit of the Bay of Marseille is marked, around Carry-le-Rouet, by a karstified calcareous massif, the massif of La Couronne. However, few cavities have been identified except for the caves of Mona at Rouet and the caves of Cape Méjean (Bonifay & Courtin 1998 — fig. 2). The latter, of which there are three in total, open at a depth of 15 m to 20 m. Two of them show complex networks, although they remain only partially explored so far. One of them contains indurated bedding with microfauna (Courtin 1978:738).

Departement of Var

To the east of La Ciotat, in the department of Var, caves have similarly been reported (Courtin 1978). They are situated at a depth of 10 m to 15 m.

– Cave 1: on Fauconnière point: gallery of 5 m width and 5 m height, penetrating into the cliff a distance of 30 m, with an infill of indurated sediments;
– Cave 2: at a distance of 40 m from Cave 1, and parallel to it, it also penetrates into the headland and shows indurated sediment bedding;
– Cave 3: a further 100 m from the two previous caves. A large entrance arch and complex network which needs to be studied. Samples have been taken from the breccias and the floor of this cave;
– Cave at Défens point: several cavities of which one (known as Number 5) starts with an entrance of 8 m by 5 m, and continues with a gallery 30 m long containing infill.

Alpes Maritimes

Sector of the Bay of Villefranche
This sector was identified for exploration projects, conducted in 1969 and 1970 by the research ship *Archéonaute*, because of its proximity to the famous Paleolithic sites on land (Lazaret, Terra Amata, and others), providing a frame of reference (Bonifay 1970; 1973).

– Coral Cave: a cavity opening at a depth of 24 m, width 16 m, and a length of 60 m. A sample under the entrance arch showed a sequence of shell sands, 4.5 m thick, resting on a scree of huge blocks. It is proposed that the Würmian infill was redeposited during the later

marine transgression, during a probable stillstand at the level of the cave;
– Merou Cave: this cave opens at a depth of 24 m, and sediments have been sampled to a depth of 2 m;
– Agaraté Cave: opening at a depth of 10 m, has been the subject of a previous, unpublished, topographic survey.

Sector of the Gulf of Juan
A cavity has been the object of thorough investigation and a dozen important caves have been discovered. Two of them have been surveyed topographically (the Saint-Ferréol Cave and the Sainte-Marguerite Cave) and preliminary samples have been collected (Bonifay 1973:531).

– Huet Cave: this cave opens at a depth of 27 m, and was the subject of a probe to a depth of 3 m in unconsolidated sediments inside the entrance. In addition, sediments of indurated continental origin have been sampled. They consist of breccias containing carbon and fragments of flint.

Conclusion

The French Mediterranean coast exhibits several limestone massifs in which multiple karst phases with a long history have developed. The cavities that they conceal also contain sedimentary traps providing information about the evolution of human occupation and environment during the Quaternary, and particularly during the most recent glacio-eustatic cycles.

A provisional assessment shows that existing knowledge of the karsts is insufficient to fully evaluate the range of complex themes involved (speleological exploration, three-dimensional surveys, sea-level changes, freshwater resources, sedimentary infill, etc.). As regards underwater prehistoric archaeology, apart from Cosquer Cave, most research was conducted several decades ago, but it does show the potential of the region (important and significant sedimentary sequences, traces of human presence, terrestrial fauna, and so on).

The published results of these earlier investigations, notwithstanding the fact that they pose some difficulties, clearly provide the basis for new research and the potential for new results. Future investigations would benefit from technical advances in diving, and also from developments in archaeological methodology and improvements in dating methods.

References

Billaud, Y., Chazaly, B., Olive, M. & Vanrell, L. 2014. Acquisition 3D et documentation multiscalaire de la grotte Cosquer : une réponse aux difficultés d'accès et à une submersion inéluctable? *Karstologia* 64:7-16.

Blanc, J.-J. 1997. Géodynamique et histoire du karst: application au sud-est de la France. *Quaternaire* 8:91-105.

Blanc, J.-J. 2000. Les grottes du Massif des Calanques (Marseilleveyre, Puget, Archipel de Riou). Canevas tectonique, évolution et remplissages. *Quaternaire* 11:3-19.

Blanc, J.-J. 2010. Histoire des creusements karstiques et des surfaces d'érosion en Provence occidentale. *Physio-Géo* 4:1-26.

Bonifay, E. 1970. Antiquités archéologiques sous-marines. *Gallia préhistoire* 13:585-592.

Bonifay, E. 1973. Circonscription des antiquités préhistoriques sous-marines. *Gallia préhistoire* 16:525-533.

Bonifay, E. & Courtin, J. 1998. Les remplissages des grottes immergées de la région de Marseille. In Camps, G (ed.) *L'homme préhistorique et la mer. Proceedings of the 120th Congress National des Sociétés historiques et scientifiques.* pp. 11-29. Éditions du CTHS: Paris.

Cavalera, Th., Arfib, B. & Gilli, E. 2010. Port Miou et le Bestouan, les plus longs fleuves souterrains d'Europe. *Karstologia Mémoires* 19:246:247.

Clottes, J., Courtin, J. & Vanrell, L. (eds.) 2005. *Cosquer redécouvert.* Seuil: Paris.

Collina-Girard, J. 1992. Présentation d'une carte bathymétrique au 1/25,000 du précontinent marseillais (au large de la zone limitée par la grotte Cosquer et l'habitat préhistorique de Carry-le-Rouet). *Géologie méditerranéenne* 19:77-87.

Collina-Girard, J. 1996a. Préhistoire et karst littoral: la grotte Cosquer et les Calanques marseillaises (Bouches-du-Rhône, France). *Karstologia* 27:27-40. Please note that an English language version of this was published as: Collina-Girard J. 2004. Prehistory and coastal karst area: Cosquer Cave and the "Calanques" of Marseille. *Speleogenesis and Evolution of Karst Aquifers* 2(2):1-13. Available at www.speleogenesis.info/directory/karstbase/pdf/seka_pdf4500.pdf.

Collina-Girard, J. 1996b. *Bassin de Marseille, topographie du plateau continental.* Institut Géographique National: Aix-en-Provence/Paris.

Collina-Girard, J. 1998. Niveaux de stationnement marins observés cartographiquement et en plongée (0 à -60m) entre Marseille et Cassis (B.-du-Rh., France). In Camps, G (ed.) *L'homme préhistorique et la mer. Proceedings of the 120th Congress National des Sociétés historiques et scientifiques.* pp. 31-52. Éditions du CTHS: Paris.

Collina-Girard, J. 1999. Les replats littoraux holocènes immergés en Provence et en Corse: implications eustatiques et néotectoniques. *Quaternaire* 10:121-131.

Collina-Girard, J. & Monteau, R. 2010. Les Calanques de Marseille à Cassis. *Karstologia Mémoires* 19:244-245.

Courtin, J. 1978. Recherches Archéologiques sous-marines. *Gallia préhistoire* 21:735-746.

Froget, C. 1972. Découverte de formations quaternaires sous-marines au banc du Veyron (baie de Marseille). *Comptes Rendus de l'Academie des Sciences* 274:1352-1354.

Gilli, E. 2003. Les karsts littoraux des Alpes-Maritimes: inventaire des émergences sous-marines et captage expérimental de Cabbé. *Karstologia* 40:1-12.

Lambeck, K. & Bard, E. 2000. Sea-level change along the French Mediterranean coast for the past 30,000 years. *Earth and Planetary Sciences Letters* 175:203-222.

Mocochain, L. & Clauzon, G. 2010. Spéléogénèse et crise de salinité messinienne de Méditerranée. *Karstologia Mémoires* 19:52-53.

Sartoretto, S., Collina-Girard, J., Laborel, J. & Morhange, C. 1995. Quand la grotte Cosquer a-t-elle été fermée par la montée des eaux? *Méditerranée* 82:21-24.

Tulipano, L. (ed.) 2005. *COST Action 621. Groundwater Management of Coastal Karstic Aquifers.* EU: Luxembourg.

Chapter 13
The Central Mediterranean

Fabrizio Antonioli,[1] Francesco L. Chiocci,[2] Marco Anzidei,[3] Lucilla Capotondi,[4] Daniele Casalbore,[5] Donatella Magri[6] and Sergio Silenzi[7]

[1] ENEA, Roma, Italy
[2] Earth Science Dept. University La Sapienza, Rome, Italy
[3] INGV, Rome, Italy
[4] CNR, ISMAR, Bologna, Italy
[5] University La Sapienza, Rome, Italy
[6] Plant Biology Dept., University La Sapienza, Rome, Italy
[7] ISPRA, Rome, Italy

Introduction

Because of their geologically young age and complex geological setting, central Mediterranean shelves show a large variability in morphology (width, slope, unevenness), stratigraphy (different thickness of depositional bodies resulting from the last climatic/eustatic cycle) and sedimentology (shelf-mud offshore of the main river mouths, bioclastic sediment in under-supplied areas). The overall morphology usually encompasses a well-defined shelf break as deep as 120 m to 150 m; above this depth, a well-developed erosional unconformity cutting across older deposits is present, overlaid by the depositional sequence formed in the last 20,000 years during the last sea-level rise and the present highstand of sea level. The thickness of such a sequence is highly variable, from many tens of meters offshore of the main rivers (e.g. the Po and Tiber) or in tectonically active areas (e.g. the Ionian Sea), to a few decimeters on insular shelves or on uplifting regions with arid climates (e.g. Apulia and the volcanic islands of the Tyrrhenian Sea). Accordingly, the shelf is floored by detrital siliciclastic sediment in well-fed areas that show sandy shorelines and muddy shelves with a mud line located at about 20 m, and by bioclastic sand and silt in underfed areas.

The morphology of the shelf is profoundly conditioned by the tectonic and sedimentary setting. In stable/uplifting and underfed areas, bedrock may outcrop and create a complex setting of shoals and

Submerged Landscapes of the European Continental Shelf: Quaternary Paleoenvironments, First Edition.
Edited by Nicholas C. Flemming, Jan Harff, Delminda Moura, Anthony Burgess and Geoffrey N. Bailey.

paleo-headlands. In subsiding and well-fed areas the shelf is usually flat, with a slope of a few tenths of a degree.

Earth Sciences

Geodynamic setting of the central Mediterranean shelves

The central Mediterranean area is a complicated puzzle for geodynamic reconstruction. The main geodynamic factor controlling Mediterranean tectonics is the relative motion of Africa and Europe as a consequence of different spreading rates along the Atlantic oceanic ridge (e.g. Gueguen *et al.* 1998; Serpelloni *et al.* 2007). In this context, the geology of Italy can be traced from the Early Paleozoic Hercynian orogen, throughout the Mesozoic opening of the Tethys Ocean to the later closure and embayment of this ocean, during the Alpine and Apennine subductions (Scrocca *et al.* 2003). These latter formed two orogens (i.e. the Alps to the north, and the later Apennines along the Italian Peninsula and Sicily) that are the backbone of the geological and geomorphological setting of Italy (Fig. 13.1). There are very limited areas in Italy which are not part of these two thrust belts, notably the Apulian and Iblean forelands (Puglia region and southeast Sicily), a few areas on the Po Plain, and the island of Sardinia.

The eastern part of the central Mediterranean is dominated by the Dinarides (Fig. 13.1). These form a complex fold, thrust and imbricate belt which developed along the northeastern margin of the Adriatic Sea (e.g.

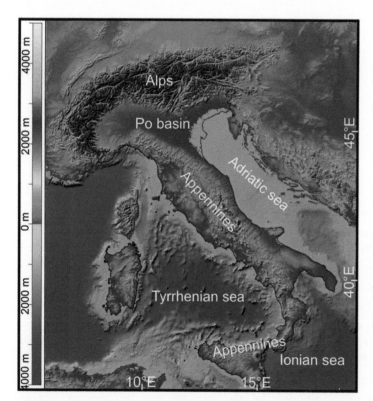
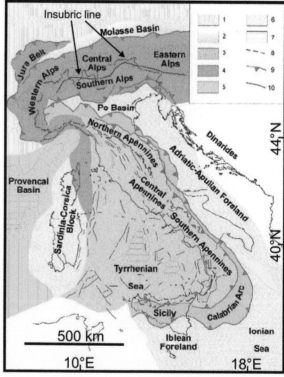

Fig. 13.1 On the left, a digital elevation model of Italy and surrounding seas (data from www.gebco.net/); note the change from light blue to dark blue marks the approximate limit of the continental shelf. On the right, a simplified tectonic map of Italy and surrounding regions, modified from Scrocca *et al.* (2003) with permission. (1) foreland areas; (2) foredeep deposits (delimited by the −1000-m isobath); (3) domains characterized by a compressional tectonic regime in the Apennines; (4) thrust belt units accreted during the Alpine orogenesis; (5) areas affected by extensional tectonics in response to the eastward roll-back of the west-directed Apennine subduction; (6) outcrops of crystalline basement (including metamorphic alpine units); (7) regions characterized by oceanic crust; (8) Apennine watershed; (9) thrust; (10) faults.

Dewey *et al.* 1973) or Apulia microplate (e.g. Dercourt *et al.* 1993). They may be considered as representing the 'southern branch' of the Alpine-Mediterranean orogenic belt, extending for about 700 km from the Southern Alps in the north-west to the Hellenides in the south-east. The main Dinaridic thrust-related deformations took place during either the Paleogene or the Miocene, when the complex tectonic structure of the region was formed (Korbar 2009). Fold, thrust and imbricate structures have mainly a northwest–southeast strike with a southwest-directed transport direction over most parts of the Dinarides, particularly in their central parts (e.g. Pamic *et al.* 1998).

The evolution of the Alpine chain is complex, but in broad terms it is the result of the subduction of Tethyan oceanic crust that resulted in a continent-to-continent collision between the European and African plates (Adria Promontory), and which occurred in several phases from the Middle Cretaceous to the Neogene. On the whole, the structure of the Alpine chain can be divided into two different structural and paleogeographic domains separated by the Insubric Line or Periadriatic Lineament, i.e. a series of very long faults running from west to east along the whole Alpine Arc (Fig. 13.1). The Northalpine domain, mainly made up of metamorphic rocks, is organized in folds and thrusts verging to the north, while in the Southalpine domain (or Southern Alps) the folds verge dominantly southward and they are made up mainly of sedimentary rocks.

The Apennines are a series of mountain ranges bordered by narrow coastlands that form the physical backbone of peninsular Italy for a total length of 1400 km (Fig. 13.1). The majority of geologic units of the Apennines are made up of marine sedimentary rocks (shales, sandstones, and limestones) which were mainly deposited on the southern margin of the Tethys Ocean in the Meso-Cenozoic. The basic tectonic evolution of the Apennines thrust belt-foredeep system may be considered as the result of the Late Oligocene-to-present north-eastward roll-back of a west-directed slab (Malinverno & Ryan 1986; Doglioni 1991). This mechanism is, in fact, the most plausible explanation for the progressive eastward migration of the thrust fronts, the foreland flexure (and consequent shift of the foredeep basins)

as well as the extensional processes along the internal Tyrrhenian back-arc basins (e.g. Faccenna *et al.* 2003). Paleo-reconstruction of the kinematics of the arc suggests up to 775 km of migration from the Late Oligocene up to the present along a transect from the Golfe du Lion to Calabria (Gueguen *et al.* 1998). The constant compression along the eastern margin induces the formation of great folds and pushes the Apennines against the Dalmatian coasts at a rate of 1 mm/year (Anzidei *et al.* 2001; 2005a; Serpelloni *et al.* 2005; 2007). The present-day foredeep basins are represented by the Po Plain and Adriatic Sea for the Northern Apennines, the Fossa Bradanica and Taranto valley for the Southern Apennines and the Catania-Gela Plain and Sicily Strait offshore for the Sicilian Apennines. Conversely, the western margin of the Apennines is characterized by extensional tectonics, with graben and half-graben structures paralleling the Tyrrhenian coast, which allowed the growth of several volcanic complexes, some of them still active.

The geology of Italy is described in a series of 132 geological maps at a scale of 1:100,000; their production started in 1877 (www.isprambiente.gov.it/it/cartografia/carte-geologiche-e-geotematiche/carta-geologica-alla-scala-1-a-100000). A completely new update of the national geological cartography is at present being carried out in within the framework of the CARG (Geological CARtography) project. This update (ongoing since 1988) involves over 60 institutions including CNR (Italian Research Council), university departments and research institutes, as well as local authorities, with the aim of producing 652 geological sheets on a scale of 1:50,000, fully covering Italian territory. To date some 40% of these maps have been completed or are in progress (www.isprambiente.gov.it/Media/carg/). Notably, the 1:50,000 guidelines foresee geological mapping of marine areas that were not considered in the 1:1,000,000 geological sheets.

Recently, a geological map of Italy at 1:1,250,000 has been produced integrating the cartographic products of the CARG project together with the 1:100,000 national geological cartography and geological maps from the literature (www.isprambiente.gov.it/en/projects/soil-and-territory/the-geological-map-of-italy-1-250000-scale/default; Fig. 13.2).

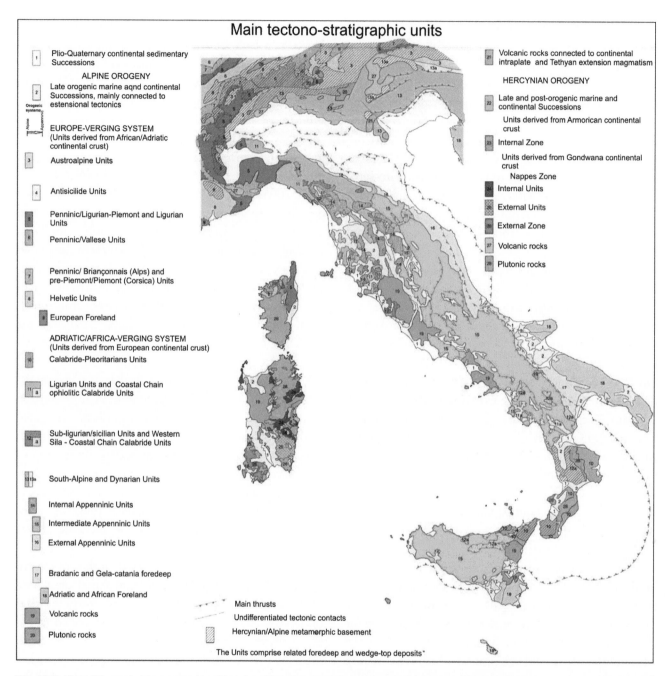

Fig. 13.2 Map of the main tectono-stratigraphic units of Italy from the geological map of Italy at 1:1,250,000 scale (www.isprambiente.gov.it/ en/projects/soil-and-territory/the-geological-map-of-italy-1-250000-scale/default). Reproduced with permission from ISPRA.

A relevant cartographic product is the Structural Model of Italy (Bigi *et al.* 1990). This model is compiled from gravimetric and aeromagnetic maps of Italy, as well as by detailed geo-structural maps of Italy at a scale of 1:500,000, where the Plio-Quaternary deposits of Alpine and Appennine foredeeps and the Plio-Quaternary sediment thickness are also shown. This publication constitutes the most important and fruitful integration of scientists with geological and geophysical backgrounds in a three-dimensional reconstruction of the anatomy of Italy and its geodynamic evolution.

In the 1970s, geological map sheets of the former Yugoslavia were published at a scale of 1:100,000 by the Federal Geological Institute Beograd, which also produced a schematic geological map at 1:500,000 (Federal Geological Institute Beograd 1970). More recently, geological and tectonic maps of the Alps and Dinarides were published by Pamic *et al.* (1998) and Schmid *et al.* (2004). In the last few years, a great effort has also been made recently to produce the first geological and morpho-tectonic map of the Mediterranean domain at a scale of 1:4,000,000 (Mascle & Mascle 2012).

Modern coastline

At present, the Italian GeoPortale Cartografico Nazionale (PCN; www.pcn.minambiente.it) represents the best available data source on the web concerning high-resolution cartography of Italy. The PCN is a data bank operated by the Italian Ministry for the Environment, Land and Sea, and it is based upon color and black and white digital orthophoto, digital elevation model (DEM), territorial and administrative data sources, place names (toponomastic), shoreline of seas and lakes, printed paper cartography of the Istituto Geografico Militare (IGM; www.igmi.org), which has been digitalized at a maximum scale of 1:1000, DEM by vector data, road network, and 3D models of the most relevant city centers. Moreover, vector data concerning hydrogeological risk, protected areas and land-use maps are available. All the data are referred to the WGS84 geographic coordinate system.

Recently, the Progetto Coste (Coast Project; www.pcn.minambiente.it/viewer/index.php?project=coste) has been added to the PCN with the aim of analyzing the state of coastal systems, advancing research, how to plan the best use of land, defense and adaptation, marine strategy, the prevention of erosion, and Integrated Coastal Zone Management (ICZM). Following the Piano Straordinario per il Telerilevamento, a high-resolution, remote-sensing project of the Italian Ministry for the Environment, Land and Sea, there will be a link to the PCN website for the LiDAR (Light Detection and Ranging) data: at present the data covers over 30% of Italian territory, including coastal areas and shorelines. By processing such data it will be possible to obtain high-resolution digital surface models and digital elevation models.

Wetlands, deltas, marshes, lagoons, coastal lakes

Several morphological types characterize the central Mediterranean coast. A detailed classification of such types, using a Geographic Information System (GIS) was performed by Ferretti *et al.* (2003) (www.santateresa. enea.it/wwwste/dincost/dincost1.html). Following the EU Water Framework Directive (ec.europa.eu/enviro nment/pubs/pdf/factsheets/water-framework-directive. pdf), Brondi *et al.* (2003) grouped the previous 12 coastal types of Ferretti *et al.* (2003) into six coastal geomorphological types (www.isprambiente.gov.it/files/ icram/tipi-geomorfologica.pdf/view) (Fig. 13.3). Their relative percentages are: 47% mountain coasts, 12% terrace coasts, 4% littoral plain, 4% torrent plain, 21% river plain and 12% dune plain.

Coastal geomorpho-dynamics, erosion, accumulation

During the Late Pleistocene and the Holocene, coastal areas were strongly modified, mainly by sea-level variation and by the impact of human activity on the coastal sectors. Land use is one of the most significant morpho-genetic factors of recent millennia. This is particularly evident in the Mediterranean area and in Italy, where coastal modification started in the first millennium BC (Pasquinucci *et al.* 2004). Such evidence supports the possibility of identifying areas for seabed prehistoric sites and landscapes. On the Italian coasts, beaches account for some 4000 km; of these, 42% are in a state of erosion (Aucelli *et al.* 2007). Further details are given in Table 13.1.

Not only are 42% of the Italian sandy coasts under-going erosion and coastal retreat, but also the parts of the coast classified as 'stable' are only prevented from retreating by coastal defense structures that, starting from 1907 with the 'Law for the Defense of Habitats from Coastal Erosion', completely altered the natural littoral

Fig. 13.3 Geomorphological types of the Italian coast, modified from Brondi *et al.* (2003) with permission from MEDCOAST. (A) Mountain coast and high rocky coast, where cliffs or pocket beaches with coarse sediments can be present; (B) Terraces, with flat emerged portion resulting from marine abrasion of a rocky substrate or from the deposition of sediments on the rocky substrate; (C) Littoral plain made up of terrace remnants, alluvial fan and alluvial plain bordered by hill or mountain chains; (D) Torrent plain, which corresponds to large, deep valleys and high coasts; (E) River plain, where regular alluvial plains display swamps, dunes, delta and lagoons; (F) Dune plain, covered by eolian or fluvial dunes, sometimes with interdune depressions, locally swampy.

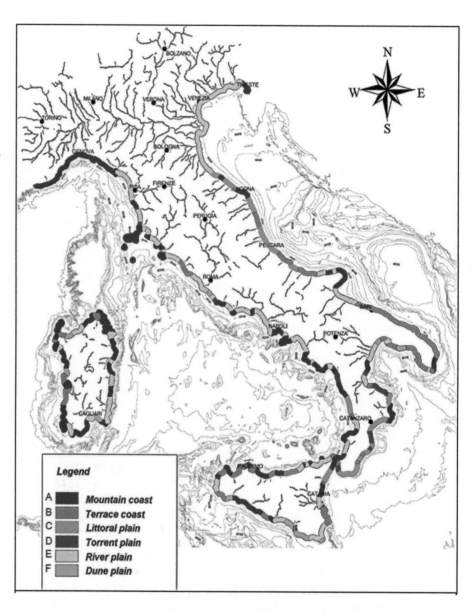

sedimentary processes, often inducing new erosional phenomena down-current of the structures.

Around the central Mediterranean there are many coastal plains threatened by potential erosive morphogenetic processes such as relative sea-level rise and increasing frequency of extreme events. For the Italian coast there is a specific study which identified 33 coastal plains with a high probability of flooding within the next 100 years (Lambeck *et al.* 2011). Along the coast of the Ligurian and Tyrrhenian seas, the areas undergoing strong modification are the Versilian Plain, the Ombrone Delta (at present this sector is losing 10 meters of unprotected

beach each year), the Orbetello lagoon, the Latium central area, the Pontina Plain and the mouths of the Volturno and Sele rivers.

Along the southern Adriatic Sea, the coastal lakes of Lesina and Varano are undergoing strong modification. Along the central and northern Adriatic Sea the greater part of the littorals from the Emilia Romagna region to the Slovenian border, including the Venice, Grado and Marano lagoons, is undergoing erosion. In Sardinia the most vulnerable areas are the wetlands of Cagliari and Oristano; many beaches in the north of the island are also eroding.

TABLE 13.1 LENGTH OF COASTS AND BEACHES, AND PERCENTAGE OF COAST BEING ERODED IN THE ITALIAN REGIONS. DATA AFTER AUCELLI *et al.* (2007).

Italian Region	Coastal length (km)	Sandy coasts (km)	Coast in erosion (km; %)
Liguria	466	94	31 (33)
Tuscany	442	199	77 (39)
Latium	290	216	117 (54)
Campania	480	224	95 (42)
Calabria	736	692	300 (43)
Sicily	1623	1117	438 (39)
Sardinia	1897	459	195 (42)
Basilicata	56	38	28 (74)
Apulia	865	302	195 (65)
Molise	36	22	20 (91)
Abruzzi	125	99	50 (50)
Marches	172	144	78 (54)
Emilia Romagna	130	130	32 (25)
Veneto	140	140	25 (18)
Friuli-Venezia	111	76	10 (13)
Italy	7569	3952	1681 (42)

In Sicily, the natural morpho-dynamic processes of the coastal sectors of Trapani and the Catania Plain have been compromised by land use and global changes. However, increasingly, erosion is affecting all the sandy beaches of the Italian littoral, where the dune systems have been strongly compromised or definitively destroyed.

Bathymetry

Sources of bathymetric data and digital archives are available from the Italian Navy's Italian Hydrographic Institute (1:100,000 scale everywhere, 1:50 to 1:25,000 in selected areas). Digital data are available from the General Bathymetric Database Chart of the Ocean — International Bathymetric Chart of the Mediterranean (GEBCO — IBCM) (1:250,000). Other sources of data are CARG and MAGIC (Marine Geohazards along the Italian Coasts), but these are not publicly available. The bathymetric data used for Fig. 13.4 were derived from GEBCO (www.gebco.net/data_and_products/gridded_bathymetry_data/). This database provides a bathymetric grid at 30 arcsecond intervals, with reference to geographical coordinates, latitude and longitude in the WGS84 datum. Bathymetry and topography were processed using the software Global Mapper 11 (www.globalmapper.com) in order to produce maps at different levels of detail (Fig. 13.5 and 13.6).

Vertical land movements

A variety of papers on vertical land movement along the Mediterranean coast, using archaeological and geomorphological markers, have been published during recent decades (Flemming & Webb 1986; Pirazzoli 1986; Lambeck *et al.* 2004; Antonioli *et al.* 2007; Anzidei *et al.* 2011a,b; 2013; Furlani *et al.* 2013; Pagliarulo *et al.* 2013). Lambeck *et al.* (2004), using reliable Last Interglacial and Holocene sea-level markers, assumed a nominal age of 124±5 ka and elevation, in the absence of tectonics, of 7±3 m above present sea level for this

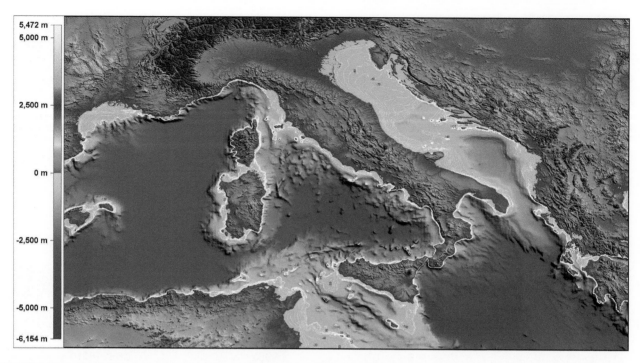

Fig. 13.4 The bathymetric data used in this map were derived from GEBCO. Isobaths between –130 m and –10 m are well defined in the Adriatic Sea.

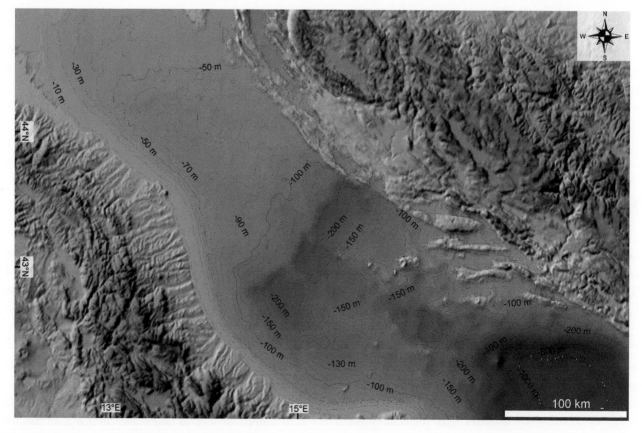

Fig. 13.5 Close-up of Fig. 13.4 for the central Adriatic Sea.

Fig. 13.6 Close-up of Fig. 13.4 for the Tyrrhenian Sea.

event. This elevation is higher than global estimates of this level because the Italian sites lie relatively close to the former ice margins, and the present Marine Isotope Stage (MIS) 5.5 shorelines in the Mediterranean may lie a few meters higher than for localities much further from the former ice margins (Potter & Lambeck 2004). These values are consistent with observations of tidal notches from Sardinia, southern Lazio, southern Campania and western Sicily, in areas believed to have been tectonically stable during the recent glacial cycles. Tidal notches are formed by solution and bioerosion of carbonatic rocks. Due to the low amplitude of tides in Italy (a mean of about 40 cm, with the exception of the northeastern Adriatic Sea), they are considered one of the best sea-level markers. Present-day tidal notches are always carved on stable carbonatic rock (with the exception of the Circeo promontory and Capri Island) by both dissolution and biological erosion. The vertical range of the notch in the central Mediterranean Sea (between 36 cm and 75 cm) seems to be related to local isostasy and tidal amplitude (Antonioli *et al.* 2013).

Ferranti *et al.* (2006) published a compilation of the MIS 5.5 highstand for 246 sites along the coasts of Italy, defining areas tectonically stable and areas with active vertical displacement since the Late Pleistocene (an extensive database of the latter is plotted in Table 13.1). Antonioli *et al.* (2009) and Ferranti *et al.* (2010) report more data on the altitude of MIS 5.5 in Tuscany and the northeastern Adriatic Sea and in other sites for the Holocene (Figs. 13.7 & 13.8).

To better understand relative sea-level changes during the Holocene, we have to consider that its variations are the sum of eustatic, glacio-hydro-isostatic and tectonic factors. The first is global and time dependent, while the other two also vary according to location. The glacio-hydro-isostatic factor along the central Mediterranean Sea coast was recently modeled and compared with field data at sites not affected by significant tectonic movements (Lambeck *et al.* 2004 (Fig. 13.9); 2011; Antonioli *et al.* 2007). The authors defined as tectonic factors all movements that are not eustatic or isostatic. Lambeck *et al.* (2011) electronic supplementary material provides:

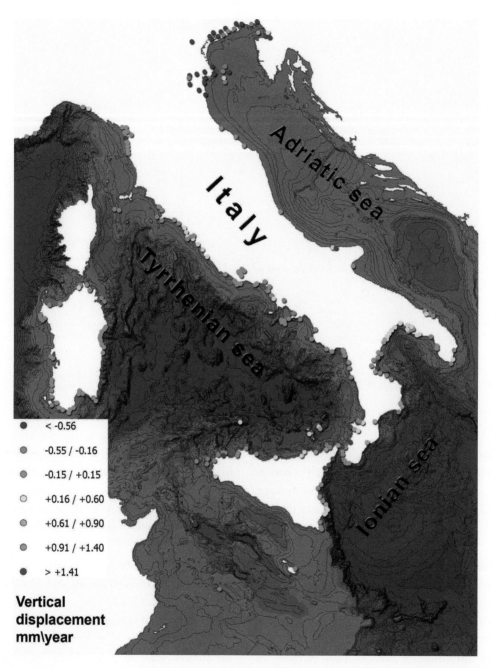

Fig. 13.7 Rates of vertical movement along the Italian coastlines for MIS 5.5 (125 ka) and the Holocene. Modified from Lambeck *et al.* 2011.

1) coordinates, time (ka cal BP) and predicted altitude (m) on 40 coastal sites dating back to 20 ka cal BP; 2) the database of the Holocene observational data; and 3) the database of Last Interglacial (125 ka) observational data. Tectonic uplift mainly occurs in the foredeep (Adriatic and Ionian coasts), as well as in the southern Tyrrhenian Sea. In contrast, subsidence occurs on many Italian coastal sites, strongly enhanced by sediment compaction and draining of water and organic-rich soils for land reclamation. Last Interglacial deposits do not outcrop along the Slovenian and Croatian coasts until Montenegro, and on the evidence of a submerged tidal

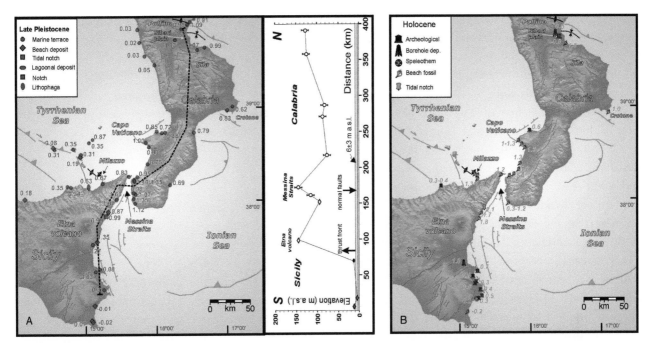

Fig. 13.8 Vertical displacement rates (mm/year) computed from the elevation of (A) Late Pleistocene and (B) Holocene markers plotted with main structures on a DEM of Calabria and eastern Sicily. Modified from Ferranti *et al.* 2010.

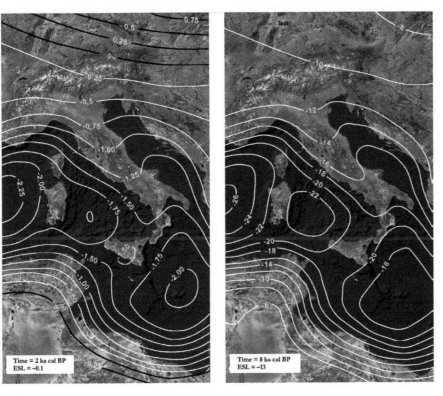

Fig. 13.9 Glacio-hydro-isostatic adjustment variations of sea level for the central Mediterranean for two distinct epochs (2 ka cal BP and 8 ka cal BP). The white contour refers to sea-level change. The ice-volume-equivalent sea level (ESL) values are given in meters (−0.1 m and −13 m at 2 ka cal BP and 8 ka cal BP respectively). Modified from Lambeck *et al.* (2004).

notch (dated to the Early Holocene, Furlani *et al.* 2011) that was found in all carbonate cliffs, it is supposed that this coastal area is still subsiding.

The well-known subsidence in the Venice area (north Adriatic Sea) occurred with a rate of ~2.3 mm/year to ~2.4 mm/year in the last 150 years (Camuffo & Sturaro 2004). The contribution from mean sea-level rise is 1.1 mm/year, while 0.3 mm/year is due to natural subsidence and 0.9 mm/year to anthropogenic subsidence (Carbognin *et al.* 2004). In the Emilia-Romagna region during recent decades the subsidence reached a peak of 70 mm/year in the Po Delta (Carminati & Martinelli 2002), with a subsidence due to land use of 10 mm/year to 30 mm/year (Bonsignore & Vicari 2000).

Along the Ligurian coast, in the Versilian Plain, several meters of Late Pleistocene and Holocene sediments indicate 1 mm/year of subsidence during the Pliocene–Quaternary (Tongiorgi 1978). Land reclamations during the last century increased subsidence rates along this plain (Mazzanti 1995; Nisi *et al.* 2003) and recent data show subsidence at rates of 2 mm/year for the Versilian Plain during recent decades, while in Lake Massaciuccoli these rates reached 7 mm/year for the same period (Palla *et al.* 1976; Auterio *et al.* 1978; Galletti Fancelli 1978; Antonioli *et al.* 2000; Nisi *et al.* 2003).

Volcanoes

In the central Mediterranean region, several volcanoes are present with ages and activity spans from a few millions of years ago to the present (www.volcano.si.edu and www.ingv.it) (Fig. 13.10). Their character and genesis vary according to the complex geodynamics of the Mediterranean region; a large number of volcanoes are located along the eastern and southern Tyrrhenian Margin and in the Sicily Strait. Mount Etna lies on the

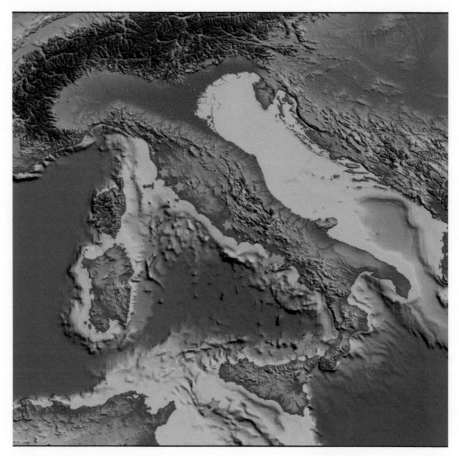

Fig. 13.10 Map of Italian volcanoes (marked with red stars) which have been active since the Holocene. Modified from www.volcano.si.edu/ and www.ingv.it. Volcano age generally decreases from north to south and some, such as Stromboli in the Aeolian Islands and Etna in Sicily, are still active.

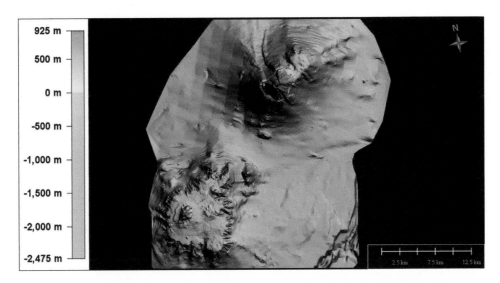

Fig. 13.11 DTM of the Panarea and Stromboli volcanoes. The DTMs are at 2.5-m and 5-m grid size at Panarea and Stromboli respectively. Local bathymetry in the area surrounding the Panarea archipelago (the islets of Panarelli, Dattilo, Lisca Nera, Bottaro, and Lisca Bianca) is at 0.5-m grid size, while regional bathymetry is at 10-m grid size.

Sicilian coast of the Ionian Sea. Besides these, there are additional large volcanoes, such as Marsili and Palinuro, rising from the sea floor of the southern Tyrrhenian Sea.

During activity, these volcanoes underwent significant vertical surface movements (Esposito *et al.* 2009); on coasts and islands these caused very fast relative sea-level changes. Coastal archaeological remains, mainly Roman, in a few cases prompted contemporary written accounts of such relative sea level (RSL) changes due to vertical land movements, well before the instrumental era. Particularly significant are the Phlegrean Fields (Naples) with the Temple of Serapis in Pozzuoli, the submerged Roman town of Baia, and the small Roman harbor of Basiluzzo in the Aeolian Islands (Tallarico *et al.* 2003; Anzidei *et al.* 2005b). As an example, in this latter area, a very high resolution DEM, which includes topography and bathymetry, is available (Fabris *et al.* 2010; Fig. 13.11). Given the continuous subsidence of nearby Panarea and the presence of prehistoric settlements on the island, it is possible that archaeological remains could also exist, submerged by up to tens of meters in this area as in other islands of the Aeolian archipelago, as in the case of Lipari (Roman age remains at −12.7 m; see www.regione.sicilia.it/beniculturali/archeologiasottomarina/).

Pleistocene and Holocene Sediment Thickness on the Continental Shelf

The continental shelves are shallow-water (<150 m) flat areas, at present usually characterized by silty-muddy sedimentation. The Italian shelves belong to the temperate climate zone and are wave-dominated, because of the narrow tidal range (microtidal regime) and the medium-regime of marine currents with speed of tens of cm/sec on average (IIM 1982). The shelves are mainly fed by rivers, so the deposits are usually siliciclastic, with subordinate extra-tropical carbonate intrabasinal deposits. The latter are characterized by temperate assemblages and are located on under-fed areas such as the very wide shelves (Tuscan, Adriatic, southern Sicily) or insular shelves (Tuscan, Pontine, Campanian, Aeolian, Egadi and Tremiti archipelagos among others).

Given their relatively young geological age, Italian shelves are extremely narrow and steep if compared with older margins. Present-day morphology is mainly the result of Plio-Pleistocene basin-ward margin progradation episodes, which partially or completely covered the uneven pre-existing substrate (Tortora *et al.* 2001).

Geodynamics controls the overall morphology of the shelves, with the Tyrrhenian and Adriatic shelves being the two end-members; the former lies on the margin of a basin undergoing oceanization and is thus affected by extensional tectonics, while the latter lies on an epicontinental sea which is the superficial expression of the foredeep basin of the Apennines to the west, and of the Dinarides to the east, and is thus affected by compressive tectonics and loading subsidence, both migrating over time. During the Pleistocene, shelf deposition was mainly controlled by high-frequency (fourth- and fifth-order, ca. 100 kyr and 20 kyr) depositional cycles that occur as multi-stacks of shelf deposits (e.g., Chiocci 2000; Ridente *et al.* 2009) similar to those observed in other Mediterranean areas (e.g. Somoza *et al.* 1997; Tesson *et al.* 2000). In detail, the stratigraphic architecture and physiography of the shelves varies considerably over short lateral distances in relation to the complex geology and tectonics of Italy (Martorelli *et al.* 2014). Some Italian coasts are characterized by uplift (the southern Tyrrhenian, southern Adriatic and Ionian margins); others are relatively stable (most of the eastern Tyrrhenian, from Tuscany to northern Calabria) or subsiding (northern Adriatic) (Ferranti *et al.* 2006).

The sediment supply and dispersal vary in the different areas according to inland climate and hydrogeological setting. Thus the shelves facing onto northern-central Italy are commonly fed by large rivers (the Po, Tiber, Arno, Ombrone, Volturno, and Sele being the principal ones), while those in southern Italy are fed by small rivers with a torrential regime or ephemeral streams. The complex interplay in time and space between these factors controlled and controls the character of depositional systems and favors or hinders the formation and preservation of lowstand, transgressive and highstand deposits on the shelf. On the whole, the complete succession of Late Quaternary systems rarely occurs. For the last depositional sequence, the falling-stage and lowstand systems are mainly confined to the very outer shelf and upper continental slope; transgressive-systems are often absent or represented by a thin lag of coarse, bioclastic sediment, while the highstand systems are usually (but not always) well developed, producing

a muddy wedge that covers the inner-middle shelf and thins towards the shelf break (Martorelli *et al.* 2014).

As far as source data is concerned, the Plio-Quaternary sediment thickness (Fig. 13.12) in Italy and in surrounding seas can be derived from the structural model of Italy at a scale of 1:500,000 published from Bigi *et al.* (1990) or from IBCM for all the Mediterranean basin at 1:1,000,000 scale (www.ngdc.noaa.gov/mgg/ibcm/ibcmsedt.html). The maximum thickness of Plio-Quaternary sediments is found in the foredeep areas of the Po Plain, the Adriatic Sea and southwest Sicily, with values up to 7 km. Other main depocenters are located in the Tyrrhenian intraslope basins off Campania, Calabria and Sicily.

As far as Pleistocene and Holocene sediment thickness is concerned, comprehensive maps do not exist, apart from the Adriatic Sea. Here, a series of thematic maps at a scale of 1:1,000,000 show the thickness of the postglacial deposits as well as the isochrones of the Quaternary sediment (Fig. 13.13; www.isprambiente.gov.it/Media/carg/index_marine.html). These latter reach values >1.4 s twt (two-way travel time) (more than 1000 m) in the northern and central Adriatic, while they rapidly decrease towards the Golfo di Trieste and south of the Gargano promontory. The lowstand deposits show a main depocenter of about 250 m immediately north of the Mid Adriatic Depression (MAD) (Fig. 13.13), where it records 150 km of shelf progradation due to the Po Delta and prodelta systems, while they thin to a minimum of a few meters on the southern flank of the MAD (Ridente & Trincardi 2005). The thickness of transgressive deposits increases from the northern Adriatic towards the south, reaching a maximum value of 90 m offshore of Pescara.

The highstand deposits show three main depocenters located: 1) offshore of the Po river mouth; 2) elongated parallel to the coast between Ancona and Vieste, and 3) south-east of the Gargano promontory, where the depocenter has been interpreted as a subaqueous delta, prograding far away from and not directly connected to any river source (Cattaneo *et al.* 2003).

In the case of the Tyrrhenian continental shelves, the reference sources are mainly related to the results

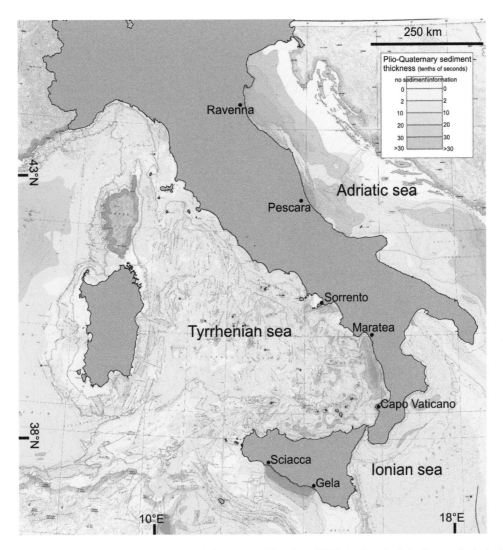

Fig. 13.12 Plio-Quaternary sediment thickness around the Italian coasts (data from IBCM charts, adapted with permission), displaying three main depocenters (see text for detail).

of 1) scientific projects, such as 'Oceanografia e Fondi Marini' (Sardinia, Liguria, Tuscany, Latium and Calabria continental shelves; Aiello *et al.* 1978; Arca *et al.* 1979; Corradi *et al.* 1984; Bartolini *et al.* 1986; Chiocci 1994), 'Mare del Lazio' (Fig. 13.14) (Latium continental shelf; Chiocci & La Monica 1996), 'Geologia dei margini continentali' (northwestern part of Sicilian shelf; Agate & Lucido 1995) and the EU INTERREG project 'Beachmed' (Liguria shelf; www.beachmed.it); 2) technical reports describing prospecting for relict sand on the continental shelf to find sources for beach replenishment (part of the Tuscan continental shelf, unpublished data);

and 3) the explanatory notes for the 1:50,000 national geological sheets in the framework of the CARG project (e.g. Foce del Fiume Sele or Salerno sheet for the Campanian shelf, www.isprambiente.gov.it).

For the Tyrrhenian shelves, postglacial deposits usually have depocenters (value >40–50 m) located in a close relationship with river mouths, or slightly displaced where an interaction between the river plume and the geostrophic currents is present, such as with the Tiber (Chiocci 1994). Transgressive deposits (5–10 m) are present locally and their depocenters are found in front of small river mouths and in protected topographic

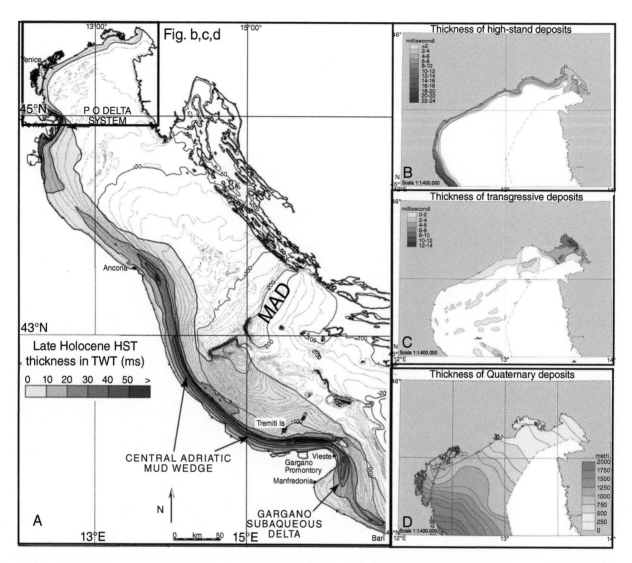

Fig. 13.13 (A) Thickness and depocenter distribution of the Late Holocene highstand mud-wedge in the Adriatic Sea. Modified from Cattaneo *et al.* (2003) and reproduced with permission from Elsevier; on the right (B, C, D), thickness of highstand, transgressive and Quaternary deposits. Data from 1:50.000 geological map of Italian sea 'Venezia sheet' (www.isprambiente.gov.it/Media/carg/index_marine.html). Reproduced with permission from ISPRA.

lows, but mainly along the narrow and steep shelf such as in Calabria (Chiocci *et al.* 1989; Tortora *et al.* 2001; Martorelli *et al.* 2010). These deposits can be organized in elongated bodies parallel to the coast or in linear depocenters normal to the coast, representing the sediment infilling of inner shelf paleoriver incision. Highstand deposits commonly have depocenters in front of the main river mouth, where they can reach a thickness of 30 m to 40 m; these depositional bodies rapidly thin towards the shelf edge (Fig. 13.14).

Post-LGM Climate, Sea Level, and Paleoshorelines

General climatic conditions and change since LGM. Mediterranean Sea Surface Temperature (SST) patterns

The Mediterranean Sea is a semi-enclosed basin with the Gibraltar Strait providing the only connection to the open

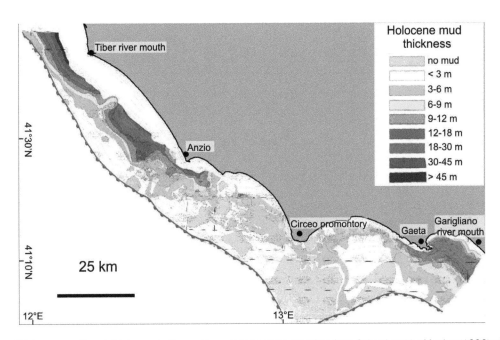

Fig. 13.14 Map of Holocene sediment thickness in the southern Latium shelf. Modified from Chiocci and La Monica (1996) with permission; note the main depocenters are located offshore of the main rivers draining this sector (the Tiber to the north and the Garigliano to the south).

ocean. The hydrodynamics of the basin are controlled by the North Atlantic water inflow, wind regime, and climate of the surrounding lands (Pinardi & Masetti 2000). Climate variability is related to the geographic position of the region located on the transition between high- and low-latitude influences. Paleoceanographic research demonstrates that during the Last Glacial interval, the Mediterranean Sea became a sensitive region, responding to North Atlantic rapid climatic changes, including Dansgaard-Oeschger (D-O) and Heinrich events (Rohling *et al.* 1998; Cacho *et al.* 1999; Sangiorgi *et al.* 2003; Sierro *et al.* 2005). In addition, it seems that the African monsoon extended at certain times over the Mediterranean basin and has been responsible for an increase in precipitation leading to increased freshwater input, reduced deepwater ventilation and deposition of sapropel layers (Tzedakis 2007). Various efforts have been made to estimate Quaternary SSTs in the Mediterranean Sea but most of these studies have considered single-core rather than multi-core analysis on a basin-wide scale (Kallel *et al.* 1997; Sangiorgi *et al.* 2003; Geraga *et al.* 2005; Di Donato *et al.* 2008). Additional problems are related to the different adopted methods (e.g. size fraction of faunal counting, use of different proxies) and

to the age quality control (robust age model and high temporal resolution), which prevents us from performing correlations between different cores in the same time slice. The principal attempts across the entire basin are reported in Table 13.2 and the results are illustrated in Figs. 13.15 and 13.16. They are related to well-constrained time intervals: the LGM and the Holocene Climatic Optimum.

Evolution of sea level and coastline since the LGM

Holocene relative sea-level change along the central Mediterranean coast is the sum of eustatic, glacio-hydro-isostatic, and tectonic factors. As pointed out above, the first is global and time dependent, while the latter two also vary with location. A large collection of Holocene data for Italy, using different markers well connected with sea level, was published by Lambeck *et al.* (2004). This large observational data set was compared with the predicted curves, leading to quantitative estimates of vertical tectonic movements (if present), identification of tectonically stable coastal areas and paleogeographic

TABLE 13.2 PRINCIPAL RECONSTRUCTION OF SST PERFORMED IN THE CENTRAL MEDITERRANEAN SEA.

Work	Time interval	Proxies	Web site/map
Molina-Cruz & Thiede (1978)	LGM (18 ka BP)	planktonic foraminifera census counts	
Thunell (1979)	LGM (18 ka BP) eastern Mediterranean sea	planktonic foraminifera census counts	
CLIMAP 1981: *Climate Long-range Investigation, Mapping, and Prediction*	LGM (18 ka BP) defined as maximum extent of continental glaciers	planktonic foraminifera census counts	www.ncdc.noaa.gov/ paleo/climap.html
Antonioli *et al.* (2004)	LGM (22 ±2 ka cal BP) and Holocene Climatic Optimum (8 ±1 ka cal BP).	Oxygen isotopic composition of calcareous planktonic foraminifera	CLIMEX MAP 2004, see Figs. 13.15 and 13.16
Hayes *et al.* (2005)	LGM (23–19 ka cal BP)	planktonic foraminifera census counts	
MARGO working group (2009): *Multiproxy Approach for the Reconstruction of the Glacial Ocean surface*	LGM interval (23–19 ka cal BP)	transfer functions (based on planktonic foraminifera, diatom, dinoflagellate cyst and radiolarian abundances) and geochemical palaeothermometers (alkenones and planktonic foraminifera Mg/Ca)	www.geo.uni-bremen.de/ ~apau/margo/margo_ EN.html

reconstructions for the central Mediterranean Sea at different epochs. These predictions were then extended to the whole Mediterranean (Lambeck & Purcell 2005). A complete review of the Holocene data, including the files of the predicted sea-level data for 40 Italian sites, has recently been published (see supplementary material in Lambeck *et al.* 2011). Significant differences can be observed in stable coastal areas if different sites are considered: as an example, sea level at 10 ka cal BP varies along the Italian coasts between −48 m and −35 m, whereas at about 8 ka cal BP it ranges between −19 m and −11 m. The Mesolithic and Neolithic paleoshorelines predicted from Lambeck *et al.* (2011) for Sardinia and the central Mediterranean are drawn in Figs. 13.17 and 13.18, and other examples using the same method and model are published for the islands of Pianosa and Elba (Tuscany) in Antonioli *et al.* (2011). An example of temporal and spatial evolution of land/sea extension and paleoshoreline evolution since the LGM is reported in Furlani *et al.* (2013) for the area between southern Sicily and northern Africa.

Central Mediterranean wave climate

The survival of prehistoric sites in shallow seas depends partly upon the prevailing local wave climate. Enclosed between the storm belt of northern Europe and the tropical area of northern Africa, the Mediterranean Sea has a relatively mild climate on average, but substantial storms are possible, usually in the six-month winter (Cavaleri *et al.* 1991). The maximum measured significant wave height (H_s) can reach 10 m, but model estimates for some non-documented storms suggest larger values. For instance, even in the relatively small Adriatic Sea, an oceanographic tower, located 15 km offshore in 16 m water depth, suffered heavy damage up to 9 m above sea level (Cavaleri 2000).

In particular, the Mediterranean winter climate is dominated by the eastward movement of storms originating over the Atlantic and impinging upon the western European coasts (Giorgi & Lionello 2008). Furthermore, Mediterranean storms can be produced within the region in cyclogenetic areas such as the

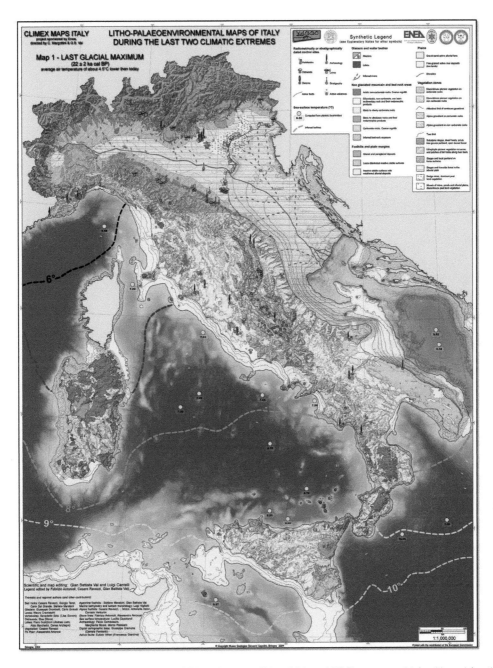

Fig. 13.15 Mean April–May sea-surface temperatures (SST) during the LGM at 22 ka cal BP. Figures are obtained by solving the paleotemperature equation of Shackleton (1974) (see Capotondi (2004) for details), using isotopic analyses of planktonic Foraminifera shells from selected marine sediment cores with well-constrained dates. Oxygen isotopic composition of water in the different basins of the Mediterranean Sea at time zero is from Pierre (1999). The figures indicate a cooling of approximately 7±2°C compared to the present day.

lee of the Alps, the Golfe du Lion and the Golfo di Genova (Lionello *et al.* 2006). In contrast, high pressure and descending air motions dominate over the Mediterranean region during the summer, leading to dry conditions, particularly over its southern part. The summer Mediterranean climate variability has been found to be connected with both the Asian and African monsoons (Alpert *et al.* 2006) and with strong geopotential

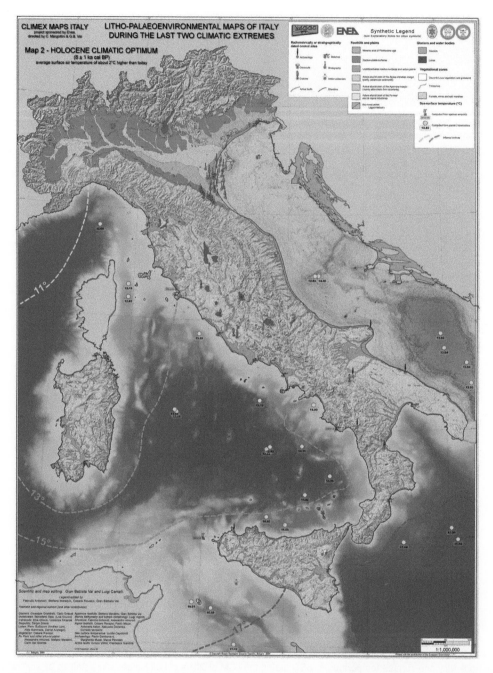

Fig. 13.16 Mean April–May sea-surface temperatures (SST) during the Holocene Climatic Optimum at 8 ka cal BP. Figures are obtained by the same methods as for those presented in Fig. 13.15.

blocking anomalies over central Europe (Trigo *et al.* 2006).

In addition to planetary-scale processes and teleconnections, the meteo-marine conditions of the Mediterranean area are strongly affected by local processes induced by the complex topography, coastline and vegetation cover of the region, which modulate the regional climate signal at small spatial scales (e.g. Lionello *et al.* 2006). The complex orography has, indeed, the effect of distorting large synoptic structures and produces

Fig. 13.17 Sardinia, Tyrrhenian Sea. DEM from GEBCO with paleoshorelines predicted from Lambeck *et al.* (2011) model. In red, the Neolithic coast (8 ka cal BP), and in yellow, the Mesolithic coast (11 ka cal BP). The coastal tectonic component in Sardinia is about zero (Ferranti *et al.* 2010).

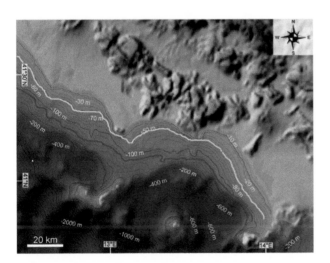

Fig. 13.18 Coast of southern Latium, Tyrrhenian Sea. DEM from GEBCO with paleoshorelines predicted from Lambeck *et al.* (2011) model. In red, the Neolithic coast (8 ka cal BP), in yellow, the Mesolithic coast (11 ka cal BP), and in blue, the LGM (22 ka cal BP). The tectonic component in this area is about zero. Ferranti *et al.* (2010).

local winds of sustained speed (for example the *bora*, *mistral–tramontana*, *scirocco* and *libeccio*, and the Etesian winds) which, over some areas of the region, blow up at almost any time of the year. Such wind regimes strongly determine wave-field dynamics in the sub-basins of the Mediterranean Sea. In the case of the central Mediterranean Sea, a major effort has been carried out recently to collect available information from buoys in a comprehensive atlas (Franco *et al.* 2004; www.isprambiente.gov.it/it/servizi-per-lambiente/stato-delle-coste/atlante-delle-coste). The latter gives an accurate description of the wave climate on sites but with poor extension and time resolution. A wind and wave atlas, MEDATLAS (MedatlasGroup 2004), of the entire Mediterranean Sea has been published by the WEAO's (Western European Armaments Organisation) Research Cell; this atlas represents the largest historical database of hydrographic measurements available for the Mediterranean.

Because of the interaction of processes at a wide range of spatial and temporal scales, even relatively minor modifications of the general circulation can lead to substantial changes in the Mediterranean climate and consequently in the meteo-marine regime. This makes the Mediterranean a region potentially vulnerable to climatic changes (e.g. Ulbrich *et al.* 2006). The Mediterranean region has, in fact, shown large climate shifts in the past (Luterbacher *et al.* 2006) and it has been identified as one of the most prominent hotspots in future climate change projections (Giorgi 2006). The distribution and intensity of wind fields over the Mediterranean Sea are expected to change during the twenty-first century planning horizon, as a result of anthropogenic climate change due to the enhanced greenhouse effect (IPCC 2007). One of the most comprehensive reviews of climate change projections over the Mediterranean region is reported by Ulbrich *et al.* (2006), based on a limited number of global and regional model simulations performed throughout the early 2000s. A number of papers have also reported regional climate-change simulations over Europe, including totally or partially the Mediterranean region (e.g. Giorgi & Lionello 2008). A robust and consistent picture of climate change over the Mediterranean emerges, consisting of a pronounced decrease in precipitation, especially in the warm season. A pronounced warming is also projected, with maximum temperatures in the summer season. Interannual variability is projected to mostly increase, especially in summer, which, together with the mean warming, would lead to a greater occurrence of extremely high temperature events.

Submerged Terrestrial Landforms and Ecology

Submerged river valleys

Submerged river valleys represent the prolongation of river drainage systems onto the continental shelf when it was sub-aerially exposed during the fall and successive rise of sea level before reaching its present-day highstand sea level. Paleovalleys are recognized in high-resolution seismic profiles through the discontinuity and different acoustic response with respect to the surrounding areas (Fig. 13.19), and the isopachs of transgressive deposits which often show local depocenters. An impressive example of a paleoriver drainage system has been found in the northern part of the Latium shelf between Ansedonia and Civitavecchia (see Chiocci & La Monica 1996; Tortora *et al.* 2001). These paleovalleys can be tracked from the coast down to 70 m to 80 m below sea level in alignment with the mouths of what are now very small rivers on the present coast, such as the Marta, Arrone, Fiora and Tafone (Fig. 13.19). These paleovalleys are represented by erosive features characterized by a width of a few thousand meters, a channel depth of a few

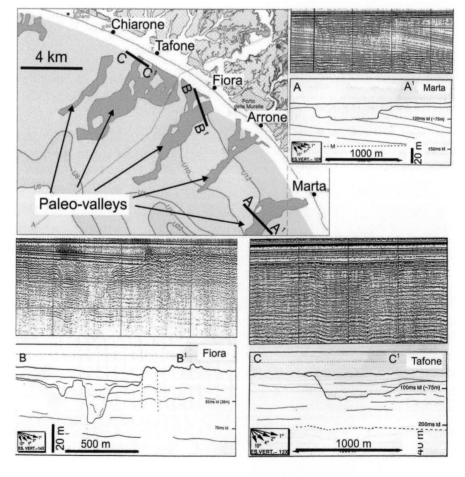

Fig. 13.19 Map of a reconstructed paleovalley on the northern Latium shelf. Data from the 1:50.000 geological map of 'Montalto di Castro' (www.isprambiente.gov.it/MEDIA/carg/lazio.html), with the location of high-resolution seismic profiles crossing paleovalleys of Marta (A–A[1]), Fiora (B–B[1]) and Tafone (C–C[1]) rivers. Reproduced with permission.

tens of meters and a variable shape, ranging from V- to U-shaped cross-sections (Fig. 13.19). These features affect the erosive unconformity related to the last lowstand of sea level (about 20 ka) and the paleovalleys are filled with deposits characterized by low transparency and short-term lateral variation of seismic attributes. The infilling material is often represented by fluvial deposits at the base of the paleovalleys overlain by lagoonal and then marine deposits. There is a poor correlation between the drainage system of present-day rivers and the width of paleovalleys, suggesting that paleoriver drainage behavior and runoff was very different from the present day during lowstand conditions. Similar features have also been found in the Golfo di Salerno (www.isprambiente.gov.it/Media/ carg/note_illustrative/467_Salerno.pdf), along the Liguria shelf in front of the Entella river mouth (Corradi et al. 1984) and in the southern Latium shelf (Chiocci & La Monica 1996). Here, submerged paleovalleys were recognized between the Circeo promontory and Gaeta, corresponding to present-day inactive small rivers (Fig. 13.14), also highlighting the difference between the lowstand and present-day drainage systems. Finally, it is noteworthy that there is a lack of seismic evidence in paleovalleys on the outer shelf corresponding to the larger rivers such as the Tiber or Po, where the progradation of these systems seems to continue undisturbed.

Submerged marine caves

On the limestone rocks outcropping along the Italian coastline (Fig. 13.20) there are well-developed karstic systems which were active during the sea-level lowstand in glacial periods. Thousands of submerged caves are known but they have only been partially studied. Two important books have been published describing in detail hundreds of sea-flooded caves, with sketches, sections and scientific (geological, biological and ecological) information (Alvisi et al. 1994 and Cicogna et al. 2003). Figs. 13.21 and 13.22 illustrate some flooded caves in Sardinia (Grotta dei Cervi, Fig. 13.21) and Riparo Blanc (central Tyrrhenian Sea), 100 km south of Rome (Fig. 13.22). The submerged caves are of interest for future research, including possible prehistoric occupation.

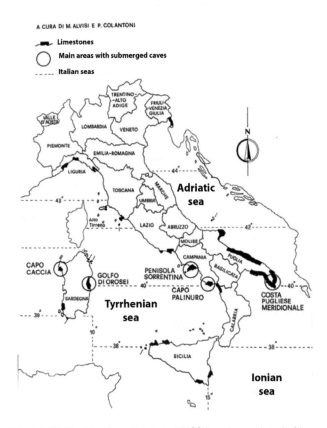

Italian coastal karst areas of interest

Fig. 13.20 Modified from Alvisi et al. (1994) and reproduced with permission from Istituto Italiano di Speleologia; black represents coastal areas of karstic limestone, where submerged caves are located.

Regional paleoclimate and vegetation indicators

A number of radiocarbon-dated pollen records from coastal areas and marine cores provide a reconstruction of the history of the coastal vegetation in the central Mediterranean regions from 22 ka cal BP to 5 ka cal BP (Fig. 13.23). Between 22 ka cal BP and 15 ka cal BP, a few pollen diagrams from marine cores from both the Tyrrhenian and Adriatic seas are available, whilst records from coastal sites are missing. The provenance of pollen grains extracted from marine sediments is not easily established, as the pollen grains of both coastal and inland vegetation may be transported into the sea by wind and rivers. In addition, marine records may be

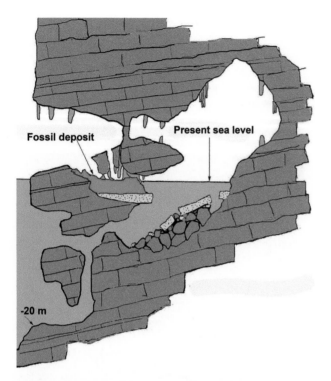

Fig. 13.21 Alghero, Sardinia. A partially submerged cave containing fossil deer (*Megaceros caziotii*). Today this cave is inaccessible due to sheer cliffs above and below the cave, descending to a depth of 40 m. Modified from Antonioli and Muccedda (1997).

biased by selective pollen accumulation and preservation, resulting in an over-representation of the saccate pollen of conifers and in reduced floristic richness. The distance from the coast, and the water depth of the coring sites, clearly affect the vegetation reconstruction. The deep-sea records of the Adriatic Sea (−300 m; Fig. 13.23) mostly represent the development of regional vegetation, as indicated by the high amounts of pollen from montane vegetation, corresponding to pollen inputs from far inland. In particular, both core KET 8216 and MD 90-917 (Rossignol-Strick *et al.* 1992; Combourieu-Nebout *et al.* 1998) show significant percentages of *Picea*, most probably from the mountain ranges of the Balkan Peninsula. The vegetational landscape of the Italian coasts was herb-dominated, as also indicated by the coeval lacustrine records from inland regions (Follieri *et al.* 1998; Allen *et al.* 2000). Excluding *Pinus*, the most important trees were deciduous *Quercus*, which could populate areas both at middle elevations and along the coasts. By contrast, modest frequencies of evergreen *Quercus* were probably from low elevation sites. Marine cores from above 300 m depth in the Adriatic Sea, which were relatively close to the ancient coast, show high percentages of *Juniperus*, together with abundant steppe and grassland elements, while deciduous trees were less abundant than in the coeval deep-sea records (Lowe *et al.* 1996).

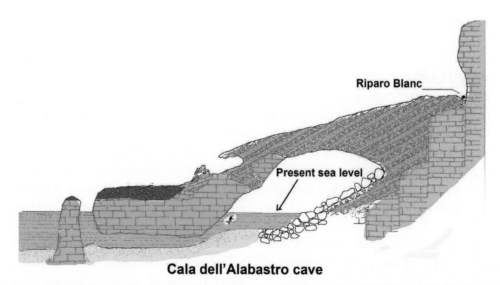

Cala dell'Alabastro cave

Fig. 13.22 The Cala dell'Alabastro cave (San Felice Circeo, Italy, black line). The arrow indicates a Mesolithic site (Riparo Blanc) containing the shell mounds of limpets (*Patella ferruginea*) dated by radiocarbon; the submerged cave is of interest for future research. Modified from Antonioli and Ferranti (1994).

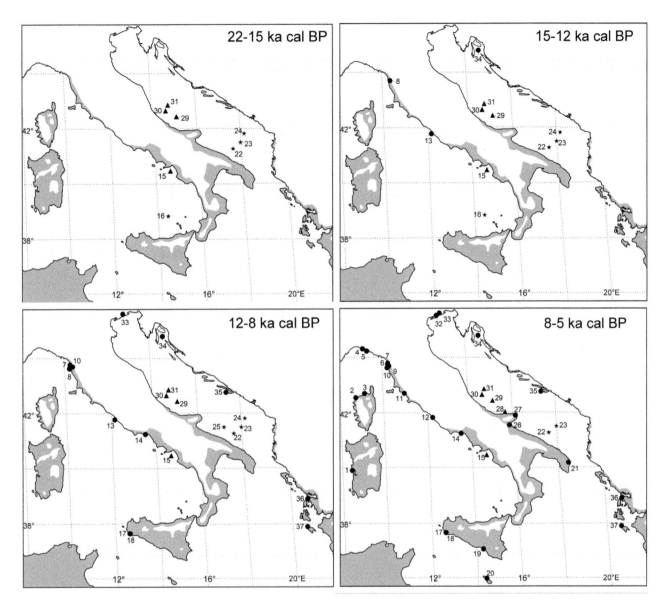

Fig. 13.23 Location of coastal and marine pollen records in Italy for the time interval 22 ka cal BP to 5 ka cal BP (dot = land site; triangle = marine site above 300 m; star = marine site below 300 m). The gray area represents the modern distribution of evergreen vegetation. (1) Mistras: Di Rita and Melis (2013); (2) Le Fango: Reille (1992); (3) Saleccia: Reille (1992); (4) Rapallo: Bellini *et al.* (2009); (5) Sestri Levante: Bellini *et al.* (2009); (6) Lago di Massaciuccoli: Colombaroli *et al.* (2007); (7) Versilian Plain: Bellini *et al.* (2009); (8) Arno S1: Amorosi *et al.* (2009); (9) Pisa Plain: Bellini *et al.* (2009); (10) Arno M1: Aguzzi *et al.* (2007); (11) Ombrone: Biserni and van Geel (2005); (12) Lingua d'Oca–Interporto: Di Rita *et al.* (2010); (13) Pesce Luna: Milli *et al.* (2013); (14) Vendicio: Aiello *et al.* (2007); (15) C106: Russo Ermolli and di Pasquale (2002); Di Rita and Magri (2009); (16) KET 8003: Rossignol-Strick and Planchais (1989); (17) Lago Preola: Magny *et al.* (2011); (18) Gorgo Basso: Tinner *et al.* (2009); (19) Biviere di Gela: Noti *et al.* (2009); (20) Burmarrad: Djamali *et al.* (2013); (21) Lago Alimini Piccolo: Di Rita and Magri (2009); (22) MD 90-917: Combourieu-Nebout *et al.* (1998); (23) KET 8216: Rossignol-Strick *et al.* (1992); (24) IN9: Zonneveld (1996); (25) SA03-1: Favaretto *et al.* (2008); (26) Lago Salso: Di Rita *et al.* (2011); Di Rita (2013); (27) Lago Battaglia: Caroli and Caldara (2007); Caldara *et al.* (2008); (28) RF93-30: Oldfield *et al.* (2003); (29) RF93-77: Lowe *et al.* (1996); (30) PAL94-8: Lowe *et al.* (1996); (31) CM92-43: Asioli *et al.* (2001); (32) Ca' Fornera: Miola *et al.* (2010); (33) Concordia Sagittaria: Miola *et al.* (2010); (34) Lake Vrana: Schmidt *et al.* (2000); (35) Malo Jezero: Jahns and van den Bogaard (1998); (36) Lake Voulkaria: Jahns (2005); (37) Alikes Lagoon: Avramidis *et al.* (2012).

Between 15 ka cal BP and 12 ka cal BP, a general increase in woody vegetation is recorded at all sites, including two coastal basins close to the present estuaries of the Arno and Tiber rivers respectively (Amorosi *et al.* 2009; Milli *et al.* 2013). Excluding *Pinus*, the main arboreal component was deciduous *Quercus*, accompanied by *Tilia*, *Corylus*, *Ulmus* and other deciduous trees in low frequencies. A continuous presence of evergreen oaks is recorded especially along the Tyrrhenian coasts, suggesting that the Late Glacial increase of tree populations near the sea involved both the deciduous and evergreen elements of the woodland. In general, a moderate decrease of the arboreal vegetation is recorded during the Younger Dryas event, which is much better recorded in lacustrine cores from inland (Magri 1999; Allen *et al.* 2000; Drescher-Schneider *et al.* 2007; Di Rita *et al.* 2013).

Between 12 ka cal BP and 8 ka cal BP (Fig. 13.23), a general development of woodland is recorded, in response to increased water availability. Regional variations in the floristic composition of the vegetation may be explained by the proximity of refuge areas for trees and local physiographic and climatic conditions. For example, on the Versilian Plain and on the Ligurian coast, large amounts of *Abies* were present, probably from the slopes of nearby mountain ranges, and possibly also from the coastal plains (Mariotti Lippi *et al.* 2007; Bellini *et al.* 2009). On the north Adriatic coastal plain, significant amounts of *Picea* could reflect the presence of spruce in the alpine forelands (Miola *et al.* 2010). The coasts of Latium and Campania were characterized by mixed oak forest with sparse evergreen elements (Russo Ermolli & Di Pasquale 2002; Aiello *et al.* 2007; Milli *et al.* 2013). Along the Tyrrhenian coast, increasing amounts of *Alnus* and riparian plants point to the expansion of freshwater habitats as a consequence of changes in the groundwater table. The Adriatic coasts of the Balkan Peninsula were characterized by oak-dominated open woodlands, accompanied by significant amounts of *Pistacia* in Greece (Jahns 2005). Similarly, in southwestern Sicily the landscape was characterized by open vegetation with Mediterranean evergreen shrubs, dominated by *Pistacia*. The development of woodland

was most likely hampered by climate aridity (Tinner *et al.* 2009).

Between 8 ka cal BP and 5 ka cal BP (Fig. 13.23), the development of the vegetation recorded by pollen diagrams in coastal sites was affected not only by plant population dynamics and climate change, but also by geomorphic processes connected to the sea-level rise and changes in the groundwater table (Di Rita *et al.* 2010). In the north Adriatic coast, pollen, plant macrofossils and non-pollen palynomorphs indicate the development of salt marsh plant communities around 6.7 ka cal BP in relation to the marine transgression (Miola *et al.* 2010). In the time interval 6.3 ka cal BP to 4 ka cal BP, a pollen record from the coastal Tavoliere Plain shows periodic fluctuations of a salt marsh, ascribed to a combination of slight sea-level fluctuations and changes in the precipitation/evaporation budget, connected to variations of the solar magnetic field and their influence on local atmospheric processes (Di Rita 2013). On the Tyrrhenian side of the Italian Peninsula, extensive freshwater basins along the coasts of Tuscany and Latium underwent significant variations of water depth, as indicated by changes in the extension of the *Alnus* carr (Di Rita & Magri 2012). On the southern Sicilian coast, the Neolithic cultural phase corresponds to the expansion of a close maquis, especially between 7 ka cal BP and 5 ka cal BP, which was not favorable for local agricultural practices and was only very slightly influenced by human activity (Noti *et al.* 2009; Tinner *et al.* 2009). The western coast of Greece experienced a reduction of *Pistacia* in favor of deciduous elements, suggesting a generally wetter climate (Jahns 2005).

On the whole, in the timespan 22 ka cal BP to 5 ka cal BP, an increasing number of pollen records describe the development of the vegetation on the central Mediterranean coasts (Di Rita & Magri 2012). Since the onset of the Holocene, the central Mediterranean coasts have hosted many diversified vegetational landscapes, correlated with regional and local physiographic and climate conditions, but in all cases the coastal vegetation appears very unstable and vulnerable to geomorphic processes and hydrological changes.

Areas of rapid erosion, and rapid accumulation of sediments since the LGM

Given the high variability of sediment supply, geological setting and shelf physiography, Italian shelves can be grouped in four end-members following the previous papers of Tortora *et al.* (2001) and Martorelli *et al.* (2014).

The first group is represented by wide shelves with high sediment supply, such as those developed in the Adriatic and central-northern Tyrrhenian Sea (part of the Latium and Tuscan shelves). These shelves are fed by large rivers (e.g. the Po, Tiber, Arno) or by several medium-sized rivers (e.g. the Apennine rivers for the central Adriatic). They are covered by a thick deposit of Holocene mud formed by the progradation of river deltas or the emplacement of mid-shelf mud belts (Figs. 13.13, 13.14; Amorosi & Milli 2001). Transgressive deposits are often lacking because the discharged river-mouth bed-load was trapped in coastal lagoons developing within fluvial valleys during sea-level rise, so that only a small amount of finer sediments were able to escape the lagoon, resulting in restricted units of fine sediment on the shelf.

The second group is represented by narrow shelves with high sediment supply, such as in the southern Tyrrhenian Sea, the western and northern Ionian Sea and the Ligurian Sea. These shelves are commonly characterized by steep slopes (>1°), limited width (few kilometers), and high sediment supply provided by ephemeral but powerful water courses, draining coastal ranges and transporting coarse-grained material into the sea during flash-flood events (e.g. Casalbore *et al.* 2011). The shelf is covered by a relatively large thickness of highstand mud that reaches the shelf break (Martorelli *et al.* 2010). In distinct contrast from the previous case, transgressive deposits are well developed, reaching their maximum thickness on Italian shelves, due to the lack of any major valley or coastal plain large enough to be able to store the sediment during the sea-level rise.

The third group is represented by shelves with a very low sediment supply and is typical of the Tyrrhenian islands, and of wide sectors of Sardinia, Apulia and Sicily (Fig. 13.24). These shelves are almost mud-free because of the lack of fluvial input and the high-energy

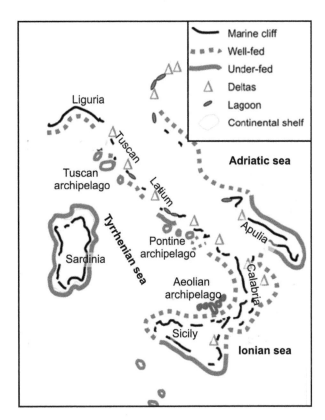

Fig. 13.24 Simplified map of the Italian continental shelf (in gray) with an indication of differences in the nature of sediment supply. Modified from Martorelli *et al.* (2014).

hydrodynamic regime. The sedimentation is mainly intrabasinal, made up of volcaniclastic and bioloclastic material. In such settings, submerged depositional terraces (Chiocci *et al.* 2004) are often the only depositional bodies overlying the bedrock on the shelf.

The fourth group is represented by wide shelves with low sediment supply and which have developed offshore of coastal lagoons or marshlands (such as in Tuscany, southern Latium, northern Apulia and Sardinia, Fig. 13.24). Here, the shelves are starved because fine sediments are trapped onshore, while beaches can be fed by longshore currents, creating narrow coastal prisms.

Conclusion

The central Mediterranean seas are very promising because of the presence of numerous partially sea-flooded

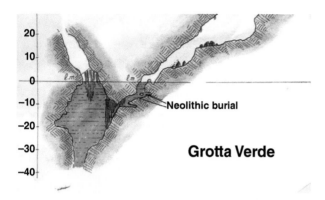

Fig. 13.25 The Grotta Verde (Sardinia, Italy) section of a Neolithic necropolis containing ceramics and skeletons at –9 m and anthropic evidence until –11 m. Modified from Antonioli *et al.* (1997) with permission from UCL.

caves, some of which have been shown in recent studies to contain dated prehistoric sites, especially in central and southern Italy. At the moment the only prehistoric site that has published data is the Grotta Verde in northwest Sardinia (Alghero), containing Neolithic-age burials with Cardial pottery (6.7–7.2 ka cal BP, Antonioli *et al.* 1997 Fig. 13.25) and the site of Pakostane in Croatia (Radic Rossi & Antonioli 2008), where some hand-made wooden remains were found *in situ* at –5 m and dated to the Neolithic, although this research is still a work in progress. Finally, we illustrate some interesting caves (Fig. 13.26) located in northern Sicily close to the coast and above sea level containing prehistoric remains, while there are unstudied sea-flooded caves between –40 m and –15 m located in the surrounding cliffs.

Fig. 13.26 Emerged caves on land, on or close to the modern coastline in northern Sicily containing prehistoric remains. Modified from Mannino and Thomas (2007) with permission from UCL.

References

Agate, M. & Lucido, M. 1995. Caratteri morfologici e sismostratigrafici della piattaforma continentale della Sicilia nord-occidentale. *Naturalista siciliana* 4:3-25.

Aguzzi, M., Amorosi, A., Colalongo, M. L. *et al.* 2007. Late Quaternary climatic evolution of the Arno coastal plain (Western Tuscany, Italy) from subsurface data. *Sedimentary Geology* 202:211-229.

Aiello, E., Bartolini, C., Gabbani, G. *et al.* 1978. Studio della piattaforma continentale medio tirrenica per la ricerca di sabbie metallifere: I) da Capo Linaro a Monte Argentario. *Bollettino Società Geologica Italiana* 97:495-525.

Aiello, G., Barra, D., De Pippo, T., Donadio, C., Miele, P. & Russo Ermolli, E. 2007. Morphological and palaeoenvironmental evolution of the Vendicio coastal plain in the Holocene (Latium, Central Italy). *Il Quaternario Italian Journal of Quaternary Sciences* 20:185-194.

Allen, J. R. M., Watts, W. A. & Huntley, B. 2000. Weichselian palynostratigraphy, palaeovegetation and palaeoenvironment: the record from Lago Grande di Monticchio, southern Italy. *Quaternary International* 73-74:91-110.

Alpert, P., Baldi, M., Ilani, R. *et al.* 2006. Relations between climate variability in the Mediterranean region and the tropics: ENSO, South Asian and African monsoons, hurricanes and Saharan dust. In Lionello, P., Malanotte-Rizzoli, P. & Boscolo, R. (eds.) *Mediterranean Climate Variability (Developments in Earth and Environmental Sciences series).* pp. 149-177. Elsevier: Amsterdam.

Alvisi, M., Colantoni, P. & Forti, P. 1994. *Memorie dell'Istituto Italiano di Speleologia 6 (Serie II).*

Amorosi, A. & Milli, S. 2001. Late Quaternary depositional architecture of Po and Tevere river deltas (Italy) and worldwide comparison with coeval deltaic successions. *Sedimentary Geology* 144:357–375.

Amorosi, A., Ricci Lucchi, M., Rossi, V. & Sarti, G. 2009. Climate change signature of small-scale parasequences from Lateglacial-Holocene transgressive deposits of the Arno valley fill. *Palaeogeography, Palaeoclimatology, Palaeoecology* 273:142-152.

Antonioli, F. & Ferranti, L. 1994. La Grotta sommersa di Cala dell'Alabastro (S. Felice Circeo). *Memorie della Società Speleologica Italiana* 6:137-142.

Antonioli, F. & Muccedda, M. 1997. La grotta dei Cervi a Punta Giglio. *Speleologia* 37:61-66.

Antonioli, F., Ferranti, L. & Lo Schiavo, F. 1997. The submerged Neolithic burials of the Grotta Verde at Capo Caccia (Sardinia, Italy) implication for the Holocene sea-level rise. *Memorie Descrittive del Servizio Geologico Nazionale* 52:329-336.

Antonioli, F., Improta, S., Nisi, M. F., Puglisi, C. & Verrubbi, V. 2000. Nuovi dati sulla trasgressione marina olocenica e la subsidenza della pianura versiliese (Toscana Nord-Occidentale). *Atti del Convegno: 'Le Pianure: Conoscenza e Salvaguardia. Il contributo delle scienze della terra'.* 8th – 10th November 1999, Regione Emilia-Romagna, pp. 214-218.

Antonioli F., Vai G. B. & Cantelli L. 2004. Litho-Palaeoenvironmental maps of Italy during the last two climatic extremes two maps 1:1.000.000. Explanatory notes edited by Antonioli F., and Vai G.B., 32° IGC publications: Bologna.

Antonioli, F., Anzidei, M., Auriemma, R. *et al.* 2007. Sea-level change during the Holocene in Sardinia and in the northeastern Adriatic (central Mediterranean Sea) from archaeological and geomorphological data. *Quaternary Science Reviews* 26:2463-2486.

Antonioli, F., Ferranti, L., Fontana, A. *et al.* 2009. Holocene relative sea-level changes and vertical movements along the Italian coastline. *Quaternary International* 206:102-133.

Antonioli, F., D'Orefice, M. & Ducci, S. *et al.* 2011. Palaeogeographic reconstruction of northern Tyrrhenian coast using archaeological and geomorphological markers at Pianosa island (Italy). *Quaternary International* 232:31-44.

Antonioli, F., Lo Presti, V., Anzidei, M. *et al.* 2013. Formazione di solchi di battente marini attuali sulle coste del Mediterraneo centrale. *Miscellanea INGV (no. 19) Riassunti del Congresso AIQUA 2013 L'ambiente Marino Costiero del Mediterraneo oggi e nel recente passato geologico. Conoscere per comprendere.* 19th – 21st June 2013, Naples, p.54.

Anzidei, M., Baldi, P., Casula, G. *et al.* 2001. Insights into present-day crustal motion in the central Mediterranean area from GPS surveys. *Geophysical Journal International* 146:98-110.

Anzidei, M., Baldi, P., Pesci, A. *et al.* 2005a. Geodetic deformation across the Central Apennines from GPS data in the time span 1999-2003. *Annals of Geophysics* 48:259-271.

Anzidei, M., Esposito, A., Bortoluzzi, G. & De Giosa, F. 2005b. The high resolution bathymetric map of the exhalative area of Panarea (Aeolian Islands, Italy). *Annals of Geophysics* 48:899-921.

Anzidei, M., Antonioli, F., Lambeck, K., Benini, A., Soussi, M. & Lakhdar, R. 2011a. New insights on the relative sea level change during Holocene along the coasts of Tunisia and western Libya from archaeological and geomorphological markers. *Quaternary International* 232:5-12.

Anzidei, M., Antonioli, F., Benini, A. *et al.* 2011b. Sea level change and vertical land movements since the last two millennia along the coasts of southwestern Turkey and Israel. *Quaternary International* 232:13-20.

Anzidei, M., Antonioli, F., Benini, A., Gervasi, A. & Guerra, I. 2013. Evidence of vertical tectonic uplift at Briatico (Calabria, Italy) inferred from Roman age maritime archaeological indicators. *Quaternary International* 288:158-167.

Arca, S., Carboni, A., Cherchi, S. *et al.* 1979. Dati preliminari sullo studio della piattaforma continentale della sardegna meridionale per la ricerca di placers. In *Atti Convegno Nazionale Progetto Finalizzato Oceanografia e Fondi Marini.* 5th – 17th October 1979, Rome, pp. 567-576.

Asioli, A., Trincardi, F., Lowe, J. J., Ariztegui, D., Langone, L. & Oldfield, F. 2001. Sub-millennial scale climatic oscillations in the central Adriatic during the Lateglacial: palaeoceanographic implications. *Quaternary Science Reviews* 20:1201-1221.

Aucelli, P. P. C., Aminti, P. L., Amore, C. *et al.* 2007. Lo stato dei litorali italiani. Gruppo Nazionale per la Ricerca sull'Ambiente Costiero. *Studi Costieri* 10:5-112.

Auterio, M., Milano, V., Sassoli, F. & Viti, C. 1978. Fenomeni di subsidenza nel comprensorio del consorzio di bonifica della Versilia. *Atti del Convegno: 'I problemi della subsidenza nella politica del territorio e della difesa del suolo'*. 9th – 10th November 1978, Pisa, pp. 65-82.

Avramidis, P., Geraga, M., Lazarova, M. & Kontopoulos, N. 2012. Holocene record of environmental changes and palaeoclimatic implications in Alykes Lagoon, Zakynthos Island, western Greece, Mediterranean Sea. *Quaternary International* 293:184-195.

Bartolini, C., Bernabini, M., Burragato F. & Maino, A. 1986. Rilievi per placers sulla piattaforma continentale del Tirreno centro-settentrionale. In *Rapporto Tecnico Finale del Progetto Finalizzato: Oceanografia e Fondi Marini Sottoprogetto 'Risorse minerarie'*. pp 97-117. CNR: Firenze.

Bellini, C., Mariotti-Lippi, M. & Montanari, C. 2009. The Holocene landscape history of the NW Italian coasts. *The Holocene* 19:1161-1172.

Bigi, G., Castellarin, A., Coli, M. *et al.* 1990. *Modello Strutturale d'Italia. CNR SELCA: Firenze.*

Biserni, G. & van Geel, B. 2005. Reconstruction of Holocene palaeoenvironment and sedimentation history of the Ombrone alluvial plain (South Tuscany, Italy). *Review of Palaeobotany and Palynology* 136:16-28.

Bonsignore, F. & Vicari, L. 2000. La subsidenza nella Pianura emiliano-romagnola: criticità ed iniziative in atto. *Atti del Convegno 'Le Pianure: Conoscenza e Salvaguardia. Il contributo delle scienze della terra'*. 8th – 10th November 1999, Regione Emilia-Romagna, pp. 119-121.

Brondi, A., Cicero, A. M., Magaletti, E. *et al.* 2003. Italian coastal typology for the European Water Framework Directive. In Özhan, E. (ed.) *Proceedings of the Sixth International Conference on the Mediterranean Coastal Environment Vol. 2 (MEDCOAST '03)*. 7th – 11th October 2003, Ravenna (Italy), pp. 1179–1188.

Cacho, I., Grimalt, J. O., Pelejero, C. *et al.* 1999. Dansgaard-Oeschger and Heinrich event imprints in Alboran Sea paleotemperatures. *Paleoceanography* 14:698-705.

Caldara, M., Caroli, I. & Simone O. 2008. Holocene evolution and sea-level changes in the Battaglia basin area (eastern Gargano coast, Apulia, Italy). *Quaternary International* 183:102-114.

Camuffo, D. & Sturaro, G. 2004. Use of proxy-documentary and instrumental data to assess the risk factors leading to sea flooding in Venice. *Global and Planetary Change* 40:93-103.

Capotondi, L. 2004. Marine Sea surface Temperature. In Antonioli, F. & Vai, G. B. (eds.) *Litho-Paleoenvironmental Maps of Italy During the Last Two Climatic Extremes. Map 1: Last Glacial Maximum Map 2: Holocene Climatic Optimum — Explanatory notes.* pp. 53-56. 32° IGC publications: Bologna.

Carbognin, L., Teatini, P. & Tosi, L. 2004. Eustacy and land subsidence in the Venice Lagoon at the beginning of the new millennium. *Journal of Marine Systems* 51:345-353.

Carminati, E. & Martinelli, G. 2002. Subsidence rates in the Po Plain, northern Italy: the relative impact of natural and anthropogenic causation. *Engineering Geology* 66:241-255.

Caroli, I. & Caldara, M. 2007. Vegetation history of Lago Battaglia (eastern Gargano coast, Apulia, Italy) during the middle-late Holocene. *Vegetation History and Archaeobotany* 16:317-327.

Casalbore, D., Chiocci, F. L., Scarascia Mugnozza, G., Tommasi, P. & Sposato, A. 2011. Flash-flood hyperpycnal flows generating shallow-water landslides at Fiumara mouths in Western Messina Straits (Italy). *Marine Geophysical Research* 32:257–271.

Cattaneo, A., Correggiari, A., Langone, L. & Trincardi, F. 2003. The late-Holocene Gargano subaqueous delta, Adriatic shelf: sediment pathways and supply fluctuations. *Marine Geology* 193:61-91.

Cavaleri, L. 2000. The oceanographic tower *Acqua Alta* — activity and prediction of sea states at Venice. *Coastal Engineering* 39:29-70.

Cavaleri, L., Bertotti, L & Lionello, P. 1991. Wind wave cast in the Mediterranean Sea. *Journal Geophysical Research — Oceans* 96:C6 10739-10764.

Chiocci, F. L. 1994. Very high-resolution seismics as a tool for sequence stratigraphy applied to outcrop scale-examples from eastern Tyrrhenian margin Holocene/Pleistocene deposits. *American Association of Petroleum Geologists Bulletin* 78:378-395.

Chiocci, F. L. 2000. Depositional response to Quaternary fourth-order sea-level fluctuations on the Latium margin (Tyrrhenian Sea, Italy). In Hunt, D. & Gawthorpe,

R. L. (eds.) *Sedimentary responses to forced regressions (Geological Society Special Publication)* 172:271-289.

Chiocci, F. L. & La Monica G. B. 1996. Analisi sismostratigrafica della piattaforma continentale. In *Il Mare del Lazio*. pp. 41-61. Università 'La Sapienza' di Roma: Lazio.

Chiocci, F. L., D'Angelo, S., Orlando, L. & Pantaleone, A. 1989. Evolution of the Holocene shelf sedimentation defined by high-resolution seismic stratigraphy and sequence analysis (Calabro-Tyrrhenian continental shelf). *Memorie Società Geologica* Italiana 48:359-380.

Chiocci, F. L., D'Angelo, S., & Romagnoli C. (eds.) 2004. *Atlante dei Terrazzi Deposizionali Sommersi lungo le coste italiane. Memorie Descrittive della Carta Geologica d'Italia* (vol. 58). Apat: Rome.

Cicogna Nike Bianchi, C. & Ferrari, G. 2003. *Grotte marine d'Italia 50 anni di ricerche*. Ministero dell'Ambiente: Rome.

CLIMAP 1981. *Climate Long-range Investigation, Mapping, and Prediction*. Available at www.ncdc.noaa.gov/paleo/climap.html.

Colombaroli, D., Marchetto, A. & Tinner W. 2007. Long-term interactions between Mediterranean climate, vegetation and fire regime at Lago di Massaciuccoli (Tuscany, Italy). *Journal of Ecology* 95:755-770.

Combourieu-Nebout, N., Paterne, M., Turon, J.-L. & Siani, G. 1998. A high-resolution record of the last deglaciation in the central Mediterranean Sea: Palaeovegetation and palaeohydrological evolution. *Quaternary Science Reviews* 17:303-317.

Corradi, N., Fanucci, F., Fierro, G., Firpo, M., Picazzo, M. & Mirabile, L. 1984. La piattaforma continentale ligure: caratteri, struttura ed evoluzione. In *Rapporto Tecnico Finale del Progetto Finalizzato: 'Oceanografia e Fondi Marini'*. pp. 1-34. CNR: Rome.

Dercourt, J., Ricou, L. E. & Vrielynck, B. (eds.) 1993. *Atlas Tethys Palaeoenvironmental Maps*. Gauthiers-Villars: Paris.

Dewey, J. F., Pitman, W. C., Ryan, W. B. F. & Bonnin, J. 1973. Plate tectonics and evolution of the Alpine system. *Geological Society of America Bulletin* 84:3137-3180.

Di Donato, V., Esposito, P., Russo-Ermolli, E., Scarano, A. & Cheddadi, R. 2008. Coupled atmospheric and marine palaeoclimatic reconstruction for the last 35 ka in the Sele Plain-Gulf of Salerno area (southern Italy). *Quaternary International* 190:146-157.

Di Rita, F. 2013. A possible solar pacemaker for Holocene fluctuations of a saltmarsh in southern Italy. *Quaternary International* 288:239-248.

Di Rita, F. & Magri, D. 2009. Holocene drought, deforestation, and evergreen vegetation development in the central Mediterranean: a 5500 year record from Lago Alimini Piccolo, Apulia, southeast Italy. *The Holocene* 19:295-306.

Di Rita, F. & Magri, D. 2012. An overview of the Holocene vegetation history from the central Mediterranean coasts. *Journal of Mediterranean Earth Sciences* 4:35-52.

Di Rita, F. & Melis R. T. 2013. The cultural landscape near the ancient city of Tharros (central West Sardinia): vegetation changes and human impact. *Journal of Archaeological Science* 40:4271-4282.

Di Rita, F., Celant, A. & Magri D. 2010. Holocene environmental instability in the wetland north of the Tiber delta (Rome, Italy): sea-lake-man interactions. *Journal of Paleolimnology* 44:51-67.

Di Rita, F., Simone, O., Caldara, M., Gehrels, W. R. & Magri, D. 2011. Holocene environmental changes in the coastal Tavoliere Plain (Apulia, southern Italy): A multiproxy approach. *Palaeogeography, Palaeoclimatology, Palaeoecology* 310:139-151.

Di Rita, F., Anzidei, A. P. & Magri, D. 2013. A Lateglacial and early Holocene pollen record from Valle di Castiglione (Rome): Vegetation dynamics and climate implications. *Quaternary International* 288:73-80.

Djamali, M., Gambin, B., Marriner, N. *et al.* 2013. Vegetation dynamics during the early to mid-Holocene transition in NW Malta, human impact versus climatic forcing. *Vegetation History and Archaeobotany* 22:367-380.

Doglioni C. 1991. A proposal for the kinematic modelling for W-dipping subductions — possible applications to the Tyrrhenian–Apennines system. *Terra Nova* 3:423-434.

Drescher-Schneider, R., de Beaulieu, J.-L., Magny, M. *et al.* 2007. Vegetation history, climate and human impact over the last 15,000 years at Lago dell'Accesa (Tuscany, Central Italy). *Vegetation History and Archaeobotany* 16:279-299.

Esposito, A., Anzidei, M., Atzori, S., Devoti, R., Giordano, G. & Pietrantonio G. 2009. Modeling ground deformations of Panarea volcano hydrothermal/geothermal system (Aeolian Islands, Italy) from GPS data. *Bulletin of Volcanology* 72:609–621.

Fabris, M., Baldi, P., Anzidei, M., Pesci, A., Bortoluzzi, G. & Aliani S. 2010. High resolution topographic model of Panarea Island by fusion of photogrammetric, Lidar and bathymetric digital terrain models. *The Photogrammetric Record* 25:382-401.

Faccenna, C., Jolivet, L., Piromallo, C. & Morelli, A. 2003. Subduction and the depth of convection in the Mediterranean mantle. *Journal of Geophysical Research* 108 (B2):2099.

Favaretto, S., Asioli, A., Miola, A. & Piva, A. 2008. Preboreal climatic oscillations recorded by pollen and foraminifera in the southern Adriatic Sea. *Quaternary International* 190:89-102.

Federal Geological Institute Beograd 1970. *Geological map SFR Yugoslavia 1:500.000*. Belgrade.

Ferranti, L., Antonioli, F., Mauz, B. *et al.* 2006. Markers of the last interglacial sea-level highstand along the coast of Italy: Tectonic implications. *Quaternary International* 145-146:30-54.

Ferranti, L., Antonioli, F., Anzidei, M., Monaco, C. & Stocchi, P. 2010. The timescale and spatial extent of recent vertical tectonic motions in Italy: insights from relative sea-level changes studies. *Journal of the Virtual Explorer* Vol. 36, paper 23.

Ferretti, O., Delbono, I., Furia, S. & Barsanti, M. 2003. Elementi di gestione costiera. Parte Prima. Tipi morfo - sedimentologici dei litorali italiani. *Rapporto Tecnico ENEA RT/2003/42/CLIM*. pp. 146.

Flemming, N. & Webb, C. O. 1986. Tectonic and eustatic coastal changes during the last 10,000 years derived from archaeological data. *Zeitschrift für Geomorphologie* 62:1-29.

Follieri, M., Giardini, M., Magri, D. & Sadori, L. 1998. Palynostratigraphy of the last glacial period in the volcanic region of central Italy. *Quaternary International* 47-48:3-20.

Franco, L., Piscopia, R., Corsini, S. & Inghilesi, R. 2004. *L'Atlante delle onde nei mari italiani*. APAT: Rome.

Furlani, S., Cucchi, F., Biolchi, S. & Odorico, R. 2011. Notches in the Northern Adriatic Sea: Genesis and development. *Quaternary International* 232:158-168.

Furlani, S., Antonioli, F., Biolchi, S. *et al.* 2013. Holocene sea level change in Malta. *Quaternary International* 288:146-157.

Galletti Fancelli, M. L. 1978. Ricerche sulla subsidenza della pianura pisana. Analisi polliniche di sedimenti quaternari della pianura costiera tra Pisa e Livorno. *Bollettino Società Geologica Italiana* 97:197-245.

Geraga, M., Tsaila-Monopolis, S., Ioakim, C., Papatheodorou, G. & Ferentinos, G. 2005. Short-term climate changes in the southern Aegean Sea over the last 48,000 years. *Palaeogeography, Palaeoclimatology, Palaeoecology* 220:311-332.

Giorgi, F. 2006. Climate change hot-spots. *Geophysical Research Letters* 33:L08707.

Giorgi, F. & Lionello, P. 2008. Climate change projections for the Mediterranean region. *Global and Planetary Change* 63:90-104.

Gueguen, E., Doglioni, C. & Fernandez, M. 1998. On the post-25 Ma geodynamic evolution of the western Mediterranean. *Tectonophysics* 298:259-269.

Hayes, A., Kucera, M., Kallel, N., Sbaffi, L. & Rohling, E. J. 2005. Glacial Mediterranean sea surface temperatures based on planktonic foraminiferal assemblages. *Quaternary Science Reviews* 24:999-1016.

IIM (Istituto Idrografico della Marina) 1982. *Atlante delle correnti superficiali dei mari Italiani*. Istituto Idrografico della Marina: Genoa.

IPCC 2007. *Climate Change 2007: Synthesis Report*. IPCC: Valencia. Available at www.ipcc.ch/pdf/ assessment-report/ar4/syr/ar4_syr.pdf

Jahns, S. 2005. The Holocene history of vegetation and settlement at the coastal site of Lake Voulkaria in Acarnania, western Greece. *Vegetation History and Archaeobotany* 14:55-66.

Jahns, S. & van den Bogaard, C. 1998. New palynological and tephrostratigraphical investigations of two salt lagoons on the island of Mljet, south Dalmatia, Croatia. *Vegetation History and Archaeobotany* 7:219-234.

Kallel, N., Paterne, M., Labeyrie, L., Duplessy, J.-C. & Arnold, M. 1997. Temperature and salinity records of the Tyrrhenian Sea during the last 18,000 years. *Palaeogeography, Palaeoclimatology, Palaeoecology* 135:97-108.

Korbar, T. 2009. Orogenic evolution of the External Dinarides in the NE Adriatic region: a model constrained by tectonostratigraphy of Upper Cretaceous to Paleogene carbonates. *Earth-Science Reviews* 96:296-312.

Lambeck, K. & Purcell, A. 2005. Sea-level change in the Mediterranean Sea since the LGM: model predictions for tectonically stable areas. *Quaternary Science Reviews* 24:1969-1988.

Lambeck, K., Antonioli, F., Purcell, A. & Silenzi, S. 2004. Sea-level change along the Italian coast for the past 10,000 yrs. *Quaternary Science Reviews* 23:1567-1598.

Lambeck, K., Antonioli, F., Anzidei, M. *et al.* 2011. Sea level change along the Italian coast during the Holocene and projections for the future. *Quaternary International* 232:250-257.

Lionello, P., Bhend, J., Buzzi, A. *et al.* 2006. Cyclones in the Mediterranean region: Climatology and effects on the environment. In Lionello, P., Malanotte-Rizzoli, P & Boscolo, R. (eds.) *Mediterranean Climate Variability (Developments in Earth and Environmental Sciences series).* pp. 325-372. Elsevier: Amsterdam.

Lowe, J., Accorsi, C. A., Bandini-Mazzanti, M. *et al.* 1996. Pollen stratigraphy of sediment sequences from lakes Albano and Nemi (near Rome) and from the central Adriatic, spanning the interval from oxygen isotope stage 2 to the present day. *Memorie dell'Istituto Italiano di Idrobiologia* 55:71-98.

Luterbacher, J., Xoplaki, E., Casty, C. *et al.* 2006. Mediterranean climate variability over the last centuries: A review. In Lionello, P., Malanotte-Rizzoli, P & Boscolo, R. (eds.) *Mediterranean Climate Variability (Developments in Earth and Environmental Sciences series).* pp. 27-148. Elsevier: Amsterdam.

Magny, M., Vannière, B., Calo, C. *et al.* 2011. Holocene hydrological changes in south-western Mediterranean as recorded by lake-level fluctuations at Lago Preola, a coastal lake in southern Sicily, Italy. *Quaternary Science Reviews* 30:2459-2475.

Magri, D. 1999. Late Quaternary vegetation history at Lagaccione near Lago di Bolsena (central Italy). *Review of Palaeobotany and Palynology* 106:171-208.

Malinverno, A. & Ryan W. B. F. 1986. Extension in the Tyrrhenian Sea and shortening in the Apennines as a result of arc migration driven by sinking of the lithosphere. *Tectonics* 5:227-245.

Mannino, M. A. & Thomas, K. D. 2007. New radiocarbon dates for hunter-gatherers and early farmers in Sicily. *Accordia Research Papers* 10:13-34.

MARGO 2009. *Multiproxy Approach for the Reconstruction of the Glacial Ocean surface.*

Mariotti Lippi, M., Guido, M., Menozzi, B. I., Bellini, C. & Montanari C. 2007. The Massaciuccoli Holocene pollen sequence and the vegetation history of the coastal plains by the Mar Ligure (Tuscany and Liguria, Italy). *Vegetation History and Archaeobotany* 16:267-277.

Martorelli, E., Chiocci, F. L. & Orlando, L. 2010. Imaging continental shelf shallow stratigraphy by using different high-resolution seismic sources: an example from the Calabro-Tyrrhenian margin (Mediterranean Sea). *Brazilian Journal of Oceanography* 58:55-66.

Martorelli, E., Falese, F. & Chiocci, F. L. 2014. Overview of Late Quaternary continental shelf deposits off the Italian peninsula. In Chiocci, F. L & Chivas, A. R. (eds.) *Continental Shelves of the World. Their Evolution During the Last Glacio-Eustatic Cycle.* Geological Society Memoirs No. 41:169-186.

Mascle, J. & Mascle, G. 2012. Geological and Morpho-Tectonic Map of the Mediterranean Domain. Published by the Commission For The Geological Map Of The World (CGMW) and UNESCO.

Mazzanti, R. 1995. Revisione e aggiornamento sui movimenti tettonici deducibili dalle dislocazioni nei sedimenti pleistocenici ed olocenici della Toscana costiera. *Studi Geologici Camerti (Special Issue)* 1:509-521.

MedatlasGroup 2004. *Wind and wave atlas of the Mediterranean Sea.* WEAO Research Cell.

Milli, S., D'Ambrogi, C., Bellotti, P. *et al.* 2013. The transition from wave-dominated estuary to wave-dominated delta: The Late Quaternary stratigraphic architecture of Tiber River deltaic succession (Italy). *Sedimentary Geology* 284-285:159-180.

Miola, A., Favaretto, S., Sostizzo, I., Valentini, G. & Asioli, A. 2010. Holocene salt marsh plant communities in the North Adriatic coastal plain (Italy) as reflected by pollen, non-pollen palynomorphs and plant macrofossil analyses. *Vegetation History and Archaeobotany* 19:513-529.

Molina-Cruz, A. & Thiede, T. 1978. The glacial eastern boundary current along the Atlantic Eurafrican continental margin. *Deep Sea Research* 25:337-356.

Nisi, M. F., Devoti, S., Gabellini, M., Silenzi, S., Puglisi, C. & Verrubbi, V. 2003. Acquisizione di dati territoriali per la valutazione del rischio da risalita del livello del mare in Versilia. *Studi Costieri* 6:91-131.

Noti, R., van Leeuwen, J. F. N., Colombaroli, D. *et al.* 2009. Mid- and late-Holocene vegetation and fire history at Biviere di Gela, a coastal lake in southern Sicily, Italy. *Vegetation History and Archaeobotany* 18:371-387.

Oldfield, F., Asioli, A., Accorsi, C. A. *et al.* 2003. A high resolution late Holocene palaeo environmental record from the central Adriatic Sea. *Quaternary Science Reviews* 22:319-342.

Pagliarulo, R., Antonioli, F. & Anzidei, M. 2013. Sea level changes since the Middle Ages along the coast of the Adriatic Sea: The case of St. Nicholas Basilica, Bari, Southern Italy. *Quaternary International* 288:139-145.

Palla, B., Cetti, T., Poggianti, M., Mengali, E. & Bartolini, A. 1976. *I movimenti verticali del suolo nella Pianura pisana dopo il 1920 dedotti dal confronto di livellazioni.* Provincia e Comune di Pisa: Pisa.

Pamic, J., Gusic, I. & Jelaska, V. 1998. Geodynamic evolution of the Central Dinarides. *Tectonophysics* 297:251-268.

Pasquinucci, M., Pranzini, E. & Silenzi, S. 2004. Evolucion Paleoambiental de los Puertos y Fondeaderos antiguos en el Mediterràneo occidental. In *Seminario ANSER, Variazioni del livello marino ed evoluzioni della costa toscana in epoca storica: opportunità di porti e approdi, vol. I*, pp. 87-102.

Pierre, C. 1999. The oxygen and carbon isotope distribution in the Mediterranean water masses. *Marine Geology* 153:41-55.

Pinardi, N. & Masetti, E. 2000. Variability of the large scale general circulation of the Mediterranean Sea from observations and modelling: a review. *Palaeogeography, Palaeoclimatology, Palaeoecology* 158:153-173.

Pirazzoli, P. A. 1986. Marine notches. In van de Plassche, O. (ed.) *Sea-level Research: a Manual for the Collection and Evaluation of Data.* pp. 361-412. Geo Books: Norwich.

Potter, E.-K. & Lambeck, K. 2004. Reconciliation of sea-level observations in the Western North Atlantic during the last glacial cycle. *Earth and Planetary Science Letters* 217:171-181.

Radic Rossi, I. & Antonioli, F. 2008. Preliminary considerations on the ancient port of Pakoštane (Croatia) based on archaeological and geomorphological research. In *European Association of Archaeologists 15th meeting 'From Antiquity to the Middle Ages'.* 16th – 21st September 2008, Malta, pp. 98-99.

Reille, M. 1992. New pollen-analytical researches in Corsica: the problem of *Quercus ilex* L. and *Erica arborea* L., the origin of *Pinus halepensis* Miller forests. *New Phytologist* 122:359-378.

Ridente, D. & Trincardi, F. 2005. Pleistocene 'muddy' forced-regression deposits on the Adriatic shelf: a comparison with prodelta deposits of the late Holocene highstand mud wedge. *Marine Geology* 222-223:213-233.

Ridente, D., Trincardi, F., Piva, A. & Asioli, A. 2009. The combined effect of sea level and supply during Milankovitch cyclicity: Evidence from shallow-marine $\delta^{18}O$ records and sequence architecture (Adriatic margin). *Geology* 37:1003-1006.

Rohling, E. J., Hayes, A., De Rijk, S., Kroon, D., Zachariasse, W. J. & Eisma, D. 1998. Abrupt cold spells in the northwest Mediterranean. *Paleoceanography* 13:316-322.

Rossignol-Strick, M. & Planchais, N. 1989. Climate patterns revealed by pollen and oxygen isotope records of a Tyrrhenian sea core. *Nature* 342:413-416.

Rossignol-Strick, M., Planchais, N., Paterne, M. & Duzer, D. 1992. Vegetation dynamics and climate during the deglaciation in the south Adriatic basin from a marine record. *Quaternary Science Reviews* 11:415-423.

Russo Ermolli, E. & di Pasquale, G. 2002. Vegetation dynamics of south-western Italy in the last 28 kyr inferred from pollen analysis of a Tyrrhenian Sea core. *Vegetation History and Archaeobotany* 11:211–219.

Sangiorgi, F., Capotondi, L., Combourieu Nebout, N. et al. 2003. Holocene seasonal sea-surface temperature variations in the southern Adriatic Sea inferred from a multiproxy approach. *Journal of Quaternary Science* 18:723-732.

Schmid, S. M., Fügenschuh, B., Kissling, E. & Schuster, R. 2004. Tectonic map and overall architecture of the Alpine orogen. *Eclogae Geologicae Helvetiae* 97:93-117.

Schmidt, R., Müller, J., Drescher-Schneider, R., Krisai, R., Szeroczyńska, K. & Barić, A. 2000. Changes in lake level and trophy at Lake Vrana, a large karstic lake on the Island of Cres (Croatia), with respect to palaeoclimate and anthropogenic impacts during the last approx. 16,000 years. *Journal of Limnology* 59:113-130.

Scrocca, D., Doglioni, C. & Innocenti, F. 2003. Constraints for an interpretation of the Italian geodynamics: a review. *Memorie Descrittive della Carta Geologica Italia* 62:15-46.

Serpelloni, E., Anzidei, M., Baldi, P., Casula, G. & Galvani, A. 2005. Crustal velocity and strain-rate fields in Italy and surrounding regions: new results from the analysis of permanent and non-permanent GPS networks. *Geophysical Journal International* 161:861-880.

Serpelloni, E., Vannucci, G., Pondrelli, S. et al. 2007. Kinematics of the Western Africa-Eurasia plate boundary from focal mechanisms and GPS data. *Geophysical Journal International* 169:1180-1200.

Shackleton, N. J. 1974. Attainment of isotopic equilibrium between ocean water and the benthonic foraminifera genus *Uvigerina*: Isotopic changes in the ocean during the Lastglacial. In Labeyrie, L. (ed.) *Variation du climat au cours du Pleistocène*. pp. 203-209. Centre Nationale de la Recherche Scientifique No. 219.

Sierro, F. J., Hodell, D. A., Curtis, J. H. et al. 2005. Impact of iceberg melting on Mediterranean thermohaline circulation during Heinrich events. *Paleoceanography* 20:PA2019

Somoza, L., Hernandez-Molina, F. J. & De Andres, J. R. 1997. Continental shelf architecture and sea-level cycles: Late Quaternary high-resolution sequence stratigraphy of the Gulf of Cadiz, Spain. *Geo-Marine Letters* 17:133-139.

Tallarico, A., Dragoni, M., Anzidei, M. & Esposito, A. 2003. Modeling long-term ground deformation cooling of a magma chamber: Case of Basiluzzo island, Aeolian Islands, Italy. *Journal of Geophysical Research* 108(B12):2568.

Tesson, M., Posamentier, H. W. & Gensous, B. 2000. Stratigraphic organization of Late Pleistocene deposits of the western part of the Golfe du Lion shelf (Languedoc shelf), western Mediterranean Sea, using high-resolution seismic and core data. *American Association of Petroleum Geologists Bulletin.* 84:119–150.

Thunell, R. C. 1979. Eastern Mediterranean Sea during the Last Glacial Maximum, an 18,000 years BP reconstruction. *Quaternary Research* 11:353-372.

Tinner, W., van Leeuwen, J. F. N., Colombaroli, D. et al. 2009. Holocene environmental and climatic changes at Gorgo Basso, a coastal lake in southern Sicily, Italy. *Quaternary Science Reviews* 28:1498-1510.

Tongiorgi, M. 1978. La subsidenza nelle basse pianure dell'Arno e del Serchio: una prima valutazione quantitativa dei fenomeni osservati in rapporto ai problemi della difesa del suolo. *Atti del Convegno: 'I problemi della subsidenza nella politica del territorio e della difesa del suolo'.* 9th – 10th November 1978, Pisa, pp. 7-14.

Tortora, P., Bellotti, P. & Valeri P. 2001. Late-Pleistocene and Holocene deposition along the coasts and continental shelves of the Italian peninsula. In Vai, G. B. & Martini, I. P. (eds.) *Anatomy of an Orogen: the Apennines and Adjacent Mediterranean Basins.* pp. 455-478. Kluwer Academic Publishers: Dordrecht.

Trigo, R., Xoplaki, E., Zorita, E. et al. 2006. Relations between variability in the Mediterranean region and mid-latitude variability. In Lionello, P., Malanotte-Rizzoli, P. & Boscolo, R. (eds.) *Mediterranean Climate Variability (Developments in Earth and Environmental Sciences series).* pp. 179-226. Elsevier: Amsterdam.

Tzedakis, P. C. 2007. Seven ambiguities in the Mediterranean palaeoenvironmental narrative. *Quaternary Science Reviews* 26:2042-2066.

Ulbrich, U., May, W., Li, L., Lionello, P., Pinto, J. G. & Somot, S. 2006. The Mediterranean climate change under global warming. In Lionello, P., Malanotte-Rizzoli, P. & Boscolo, R. (eds.) *Mediterranean Climate Variability (Developments in Earth and Environmental Sciences series)*. pp. 399-416. Elsevier: Amsterdam.

Zonneveld, K. A. F. 1996. Palaeoclimatic reconstruction of the last deglaciation (18-8 ka B.P.) in the Adriatic Sea region; a land-sea correlation based on palynological evidence. *Palaeogeography, Palaeoclimatology, Palaeoecology* 122:89-106.

Chapter 14

Physical Characteristics of the Continental Shelves of the East Mediterranean Basin, Submerged Settlements and Landscapes — Actual Finds and Potential Discoveries

Ehud Galili,[1,2] Yaacov Nir,[3] Dina Vachtman[4] and Yossi Mart[5]

[1] *Zinman Institute of Archaeology, University of Haifa, Haifa, Israel*
[2] *Israel Antiquities Authority, Israel*
[3] *Rehovot, Israel*
[4] *Statoil ASA, Harstad, Norway*
[5] *Recanati Institute of Maritime Studies, University of Haifa, Haifa, Israel*

Introduction and Background

The archaeological background

The south Levant coast is of importance as it is on one of the main pathways of early hominin dispersal out of Africa. Several Lower to Upper Paleolithic sites have been discovered on the south Levant coastal plain or close to it. Middle Paleolithic sites have been discovered in the red loam soils (locally termed *Hamra*) (Ronen 1977) in the Last Interglacial (i.e. Marine Isotope Stage

Submerged Landscapes of the European Continental Shelf: Quaternary Paleoenvironments, First Edition.
Edited by Nicholas C. Flemming, Jan Harff, Delminda Moura, Anthony Burgess and Geoffrey N. Bailey.
© 2017 John Wiley & Sons Ltd. Published 2017 by John Wiley & Sons Ltd.

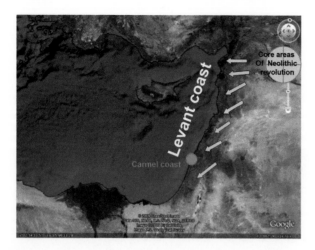

Fig. 14.1 The Levant coast and the core areas of the Neolithic Revolution. E. Galili (2015).

(MIS) 5e) beach deposits (Galili *et al.* 2007) and in coastal caves (Ronen *et al.* 2008). These finds suggest that there is a potential for discoveries of Paleolithic sites on the sea bottom. However, most of the submerged sites discovered so far are Neolithic. Very few Middle and Upper Paleolithic remains (scattered flint artifacts) have been recovered so far from the sea floor in Israel. The Levant coast (Fig. 14.1) is the closest marine environment to the core area of the Neolithic Revolution. The agro-pastoral-marine subsistence system, the so-called 'Mediterranean fishing village', emerged on this coast. It is the zone from which the coastal expansion of the European Neolithization process started. The westward-moving wave of Neolithic cultural and technological innovations along the Mediterranean coasts is one of the major developments in human history. At present, the settlements representing the first stages of that process are underwater. Until recently the data on the inundated Levantine coastal Neolithic communities were scarce. Because of sea-level rise, the coastal Neolithic sites were inundated by sea water and sediments, and were not available for research by conventional archaeological methods.

The geological/geomorphological background

Humans settling along the coastal plains during the Epipaleolithic and Neolithic periods commonly preferred

sites adjacent to flowing rivers supplying good quality water, one of life's essentials. The extensive global sea-level rise during the Late Pleistocene and Early Holocene covered the coastal plains of the Last Glacial Maximum (LGM) with sea water and sediment. These processes changed the landscape of the inundated plains, complicating the determination of the fluvial systems and the exploration of their significant paleoenvironments.

During Pleistocene sea-level lowstands, broad areas of the continental shelf were exposed to rainfall and water runoff, initiating the development of networks of fluvial channels with appropriately incised valleys which occupied the entire area of the LGM exposed shelf. During subsequent transgressions however, fluvial and estuarine deposition, as well as sediment reworking by waves and currents, filled and partly erased or buried these river networks (Dalrymple *et al.* 1994).

The continental shelf of the eastern Mediterranean is characterized by considerable variability. Starting from the wide expanses of the Nile Delta and southern Israel, the shelf narrows towards the north and it is relatively steep and narrow off Lebanon, Cyprus and Anatolia (Neev *et al.* 1976) (Figs. 14.2, 14.3, 14.4). Considering

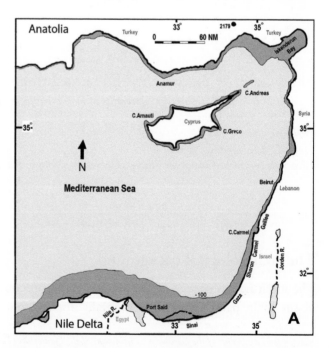

Fig. 14.2 The Levant, the eastern Mediterranean basin and location of the −100-m isobath. E. Galili (2014).

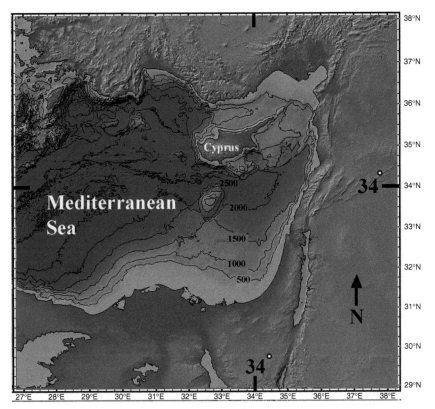

Fig. 14.3 The bathymetry of the eastern Mediterranean coast (source and courtesy of: www.geomapapp.org).

that the shelf was exposed sub-aerially during the LGM, and that the climate of the eastern Mediterranean was wetter than at present during the last cold spell of the Late Pleistocene, there are grounds to presume that most of the shelf area comprised fertile coastal plains. Rivers, which cross the present coastal plains, probably transected the LGM plains, and their elevated levées were probably targets for Epipaleolithic and Neolithic habitats.

Formation possesses of continental shelves during low sea stands

In order to understand how and where prehistoric settlements were located on the shelf, this chapter describes the processes that have formed the coast and continental shelf of the eastern Mediterranean since the Last Interglacial. It summarizes some of the physical characteristics of the coasts and shallow continental shelf of the eastern Mediterranean basin, discusses sea-level changes and the resulting coastal modifications and their possible impact

on ancient human societies. We also summarize briefly the main finds associated with submerged prehistoric settlements recovered so far, assess the potential for survival of sites and for the discovery of new ones. We propose a framework aimed at locating and studying additional submerged settlements.

It is generally understood that rivers play an important role in sediment delivery to the shelf, especially during sea-level lowstands (e.g. Suter & Berryhill 1985; Bartek *et al.* 1990; Coleman & Roberts 1990). It is also recognized that fluvial response to sea-level change has significant implications for sediment delivery and depositional geometries on the shelf and slope (Butcher 1990; Wescott 1993). River responses to base-level rise or drop have been reported in different, and sometimes contradictory, ways (see Blum & Törnqvist 2000 for a review). Talling (1998) recognized the importance of the thickness of highstand sedimentary deposits (i.e. the coastal prism) on the subsequent incision of the continental shelf. He assumed that if sea level does not fall below the shelf-slope break, the major valley incision

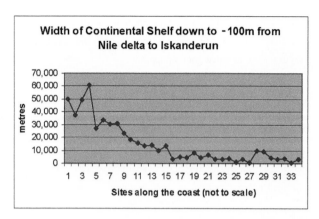

Fig. 14.4 Width of continental shelf down to −100 m from the Nile Delta to Iskenderun (Y. Nir). (1) 31 km west of Dumyat; (2) Dumyat; (3) Port Said; (4) west of Bardawil lagoon; (5) Mount Casius; (6) East Bardawil; (7) El Arīsh; (8) Haroubeh; (9) Rafah; (10) Gaza; (11) Ashdod; (12) Palmahim; (13) Yafo; (14) 'Atlit; (15) Cape Carmel; (16) 'Akko (Acre); (17) Akhziv; (18) Ras el Naqura; (19) Ras el Bayada; (20) Tyre; (21) 9 km north of Tyre; (22) 19 km north of Tyre; (23) 34 km north of Tyre; (24) Sidon; (25) 8 km north of Sidon; (26) 27 km north of Sidon; (27) Ras Beirut; (28) 49 km north of Beirut; (29) Tripoli; (30) 14 km north of Tripoli; (31) Tarsus; (32) Banias; (33) Ras el Hani; (34) Ras el Basit.

will occur where convex coastal prisms are deposited during highstands. However, when the shoreline drops below the shelf edge, headward erosion and knickpoint retreat concentrates the discharge and sediment load in a single channel, connecting the drainage basin to the depocenter on the shelf edge, thereby bypassing the stage of deposition on the exposed shelf. After sea-level fall, the inflection point works its way back up the river channel, rising to higher levels. This continues until the sea level starts to rise again, and the river then regrades itself to the new highstand. As the rate of sea-level rise increases towards the rise inflection point, the main shelf-bypass valley may become flooded and subsequently backfilled (Fagherazzi *et al.* 2004).

The rate of sea-level fall also has implications for the rate of deposition on the shelf and in fluvial valleys during the subsequent rise. Experimental study (van Heijst & Postma 2001) shows that a rapid sea-level fall accompanied by fast rates of horizontal shoreline migration results in large volumes of sediment deposition

compared to slow rates of sea-level fall. This in turn leads to an increased volume of sediment in transgressive systems tracts during the subsequent sea-level rise. The relative timing of the connection of one of the shelf canyons with the fluvial valley has a strong bearing on the final volume of the slope fan and the lowstand delta, because it determines the change from aggradation to degradation of the fluvial valley. Low rates of sea-level fall (i.e. long duration of fall) are most likely to produce large lowstand deltas, whereas high rates of sea-level fall (10 m/kyr, Cutler *et al.* 2003) will produce small ones. This emphasizes the importance of the timing of the connection relative to the sea-level cycle, which not only controls the lowstand-delta volume, but also affects its composition, i.e. the percentage of primary fluvial sediment relative to reworked shelf sediment. As a result, lowstand deltas that are formed during rapid rates of sea-level fall consist largely of shelf-derived material and with as little as 20% to 30% of fluvial sediment (van Heijst & Postma 2001). The rate of sea-level rise varies, but it is known, for example, that the Sea of Marmara was not marine during the Younger Dryas period, and that the present depth of the Dardanelles Strait is 87 m below present sea level, which reflects on the contemporaneous sea level (McHugh 2008). Presuming that the sea level stabilized ca. 6,000 years ago, the rate of sea-level rise in the Early Holocene was ca. 15 mm/year (Cutler *et al.* 2003; Galili *et al.* 2005). The rate of sea-level drop is more difficult to assess because the drawdown was associated with enhanced erosion, and the physical evidence has been partly eroded away.

Morphological features, such as submarine canyons that dissect the shelf-slope break and large sand ridges that are present on the shelf, would have affected channel incision and network development. Submarine canyons that convey sediments by gravity-driven processes (McAdoo *et al.* 2000) are occasionally shown to represent a favorable location for sub-aerial channel incision. Although it is realistic to suppose that channels will form by headward retreat once the canyon scarps are exposed to rainfall action, this does not necessarily imply that the major rivers crossing the shelf will discharge into them. Simulations indicate that the rivers are almost unaffected by the presence of canyons because the aggregation of the riverine network is mostly dictated by the local and upstream watershed distribution, and during sea-level fall, canyons that are still below sea level have negligible influence on sub-aerial channel incision and deltaic deposition (Fagherazzi *et al.* 2004). In other words, if the shelf topography is reworked during sea-level rise, the major rivers will usually excavate a new path on the shelf during sea-level retreat and at the latest stages of marine regression, and only if they are very close to a canyon will they change their course toward it. However, if previously excavated channels or canyons extend across the entire continental shelf, such channels will have a greater impact on the riverine platforms because of direct interaction with channel formation. The three-dimensional shape of the coastal plain and continental shelf system is also of fundamental importance for the development of the river network, as well as for the redistribution of alluvial sediment through erosion and deposition. Summerfield (1985) suggested that when the slope of the coastal plain is less than the slope of the adjacent offshore shelf, this leads to channel extension with deep incision during sea-level fall. Conversely, if the slope of the coastal plain is steeper than the shelf slope, this produces channel extension with progradation and aggradation. Mass wasting processes might have a critical role when sea level is lower, exposing the steep part of the continental edge.

Physical conditions

Climate

The climate characteristic of the eastern Mediterranean since (at least) the closure of the north-east marine passage to the Indian Ocean in the Late Miocene is a regime of winter rains and dry summers. Along the eastern section of northern Sinai and all along the Israeli coast, $CaCO_3$ solutions lithified (mostly during summer) the coastal dune belts into harder rock, an eolianite known by the local term *kurkar*. Similar ridges, though not so pronounced, occur on much of the coast of Libya and Egypt. Kurkar ridges form chains parallel to the shore, and are known to exist both inland and offshore in many

Levant coastal zones. Their present location reflects sea-level fluctuations and shoreline movement during the Pliocene and Quaternary.

The connection with global and regional pressure systems

The Levant coast is located in, and belongs to, the 'Mediterranean Climate' region. This type of climate is mainly the result of tropical 'highs' located in the 25°N to 30°N latitude. These 'highs' migrate northwards during winter, and southwards during summer. Other separate and independent pressure systems that have a significant influence on the region are the 'Red Sea' (during the spring and fall) and 'Persian Troughs' (mainly during the summer), and the 'Cyprus and the Siberian Lows' during winter. These climatic conditions result in relatively steady and moderate winter weather, while summer is characterized by a quiet, uniform regime and no rain.

Connection with landmasses

The southern Levant climatic conditions are influenced by pressure systems connected with landmasses: the Asian continent to the north and east, and the African to the south-west. The Mediterranean is cyclogenetic due to air masses that flow via Anatolia to the Aegean, and then to the Mediterranean proper. Therefore the 'lows' have some tendency to prolong their existence in the region. The Siberian 'high' results in low temperatures, and an easterly dry and cold winter wind. During the winter, the local weather is mainly governed by cyclonic systems moving from west to east. These cyclones create unstable conditions across the whole of the Levant.

Daily breezes along the southern Levant shores

Local wind breezes in the Levant are known by the Hebrew term *breeza*. The easterly breeze (*sharqiya* in Arabic) starts from WSW during morning hours having a speed of 1–5 knots. At noon, it rotates to a westerly wind with a somewhat increased speed of 3–8 knots. Further rotation to the NW is typical of the afternoon hours with speeds of 5–10 knots. At night, it continues its clockwise movement and is mostly easterly with just 1–5 knots until before dawn, when it reaches again 3–8 knots. These easterly fall and spring winds are the product of a moving 'low' along the Libyan desert to Egypt and Israel, the result of which are south-westerly and westerly waves in the central and northern Israeli coast.

Waves

The longest fetch of the south Levant shoreline is to the west. Therefore, the westerly winds create the highest waves. Long duration of these winds enables the formation of high-swell waves which have a significant effect on the beaches and sand transport in shallow water (Carmel *et al.* 1985).

In extreme cases, high and long swells reach the south Levant coast from a distance of ca. 2000 km from their zone of generation east of Sicily, and reach the southern Levant from azimuth 285° along a narrow sea strip (Carmel *et al.* 1985; Rosen 1998; Almagor 2002). Since winter weather in the eastern Mediterranean is determined by the passage of cyclonic systems moving eastwards, the direction of the maximal fetch is from the west; therefore this is also the directional origin of the highest waves.

Since most of the coasts of the southern Levant are characterized by gentle slopes, they are dissipative beaches, where the approaching wave loses its energy gradually. One of the most complete records of wave climate in the Levant is that maintained in Israel. The data base of the characteristics of the waves along the Israeli coast is held in a data bank containing measurements since the 1950s, and is stored in the data depository of the Israel Oceanographic and Limnological Research Institute (IOLR). The data have been collected since 1958 off Ashdod, Hadera and Haifa (Rosen 1998). The analysis of the directional distribution of the approaching waves shows that in all sea conditions the waves approach the Israeli coast from the sector of WSW through W to NNW. The highest waves arrive from the west, but the process of development and the termination of storms commonly takes place through gradual shifts in orientation. In many cases, the storm is initiated when it approaches from WSW, then the intensity increases

as the storm changes its orientation to the west. The storm ebbs as its orientation shifts to the north-west. Commonly, the waves at deep water off the Levant coast lack uniformity. Off the northern section of the Israeli coast their trend is further to the south-west, and the difference in orientation of the same storm off Ashdod and Haifa can reach 20°. However, the higher the significant wave height at deep water, the smaller the orientational difference between the waves off northern and southern Israel (Perlin & Kit 1999).

Currents

The hydrographic regime of the east Mediterranean Sea is constrained by water supply from a single source — the Atlantic Ocean that flows into the Mediterranean through the 14.3-km-wide Gibraltar Strait. As the Atlantic waters travel eastwards, they become warmer, and the warming has two contrary effects on the physical properties of the water. On the one hand, the warmer water becomes lighter; on the other, the increased temperature increases evaporation, the salinity increases and the water becomes denser. The fine balance between temperature and salinity controls the physical oceanography of the Mediterranean Sea.

The increased intensity of the monsoons in warmer years increases the intensity of the monsoon rains in East Africa, and consequently the flow of the Nile River upsurges, and that freshwater input enhances the stratification of the eastern Mediterranean and its euxinification. The effect of the wind on the surface water in the eastern Mediterranean, in the mixed layer above the pycnocline, is such that the currents flow predominantly eastwards. The Atlantic-Ionic Stream flows eastwards across the Gulf of Sirte and then along the continental slope of North Africa, and that orientation is maintained in the Mid Mediterranean Jet. The northward flowing Southern Levantine Current is derived from the Atlantic-Ionic Stream, and its extension, the Asia Minor Current, flows westwards along the Anatolian slope. The gyres in the eastern Mediterranean — the Mersa Matruh, Shikmona and Rhodes gyres — downwell the surface water to depth. Mediterranean currents (except the Mid Mediterranean one) flow along the distal continental shelf and upper slope, and they interact with the longshore current, which is derived from waves breaking along the shoreline (Rosentraub & Brenner 2007). Off North Africa and the southern Levant the longshore current is compatible with the regional currents, but off the northern Levant and southern Anatolia the two currents flow in opposite directions. The longshore current shapes the coasts of the southern Levant by carrying the sand from the Nile Delta and transporting it along the coasts of Sinai and Israel as far north as Haifa Bay. During storms, intense exchanges of water take place between the shelf and the open sea, which lead to the transport of sand from the shelf to the bathyal zone (Rosentraub & Brenner 2007). The currents in the Mediterranean can comprise sea currents and littoral currents.

Longshore currents are the predominant mechanism of sediment transport in the eastern Mediterranean, where the wave-induced current is very effective during sea storms, and the effect of local variation in wind direction and intensity during such storms is negligible. The velocities of the longshore currents during storms along the Israeli coast have never been measured, but Zviely *et al.* (2007) estimated them as high as 2 m/sec. In general, when the shore face slope is steep, the breaker zone is narrower and vice versa. The breaker zone in the easternmost Mediterranean is characterized by waves lower than 1 m during spring (April–June) and fall (October–November), and the breaker zone during these seasons extends to barely several tens of meters from the coastline. Swells of up to 2 m in height affect these beaches in the summer, and the breaker zone widens to ca. 200 m, to average water depths of 3 m. During winter storms, when wave height occasionally exceeds 5 m, the breaker zone is 350 m wide, and its width can reach 800 m during rare storms (Zviely *et. al.* 2007).

Rip currents are intermittently-spaced, shore-normal, offshore-flowing currents, generated by the shoreward transport of water by large waves, resulting first in longshore dispersal of the excess water and then shore-normal currents (*q.v.*). They flow across the breaker zone to the open sea. Commonly the rip currents remove the coastal sand and expose the underlying layer — clay or rock — and, in places, relics of shipwrecks or ancient settlements.

Tide currents are significant in the open sea, but their geomorphological signature on the coasts of the eastern Mediterranean Sea, where the tidal range averages 0.5 m or less, is negligible. The eastern Mediterranean is in the micro-tidal zone and the tidal cycles are diurnal and last ca. 11 hours.

Description of the Regions in the Study Area

The geomorphology of the continental shelf and the adjacent coastal zone of the easternmost Mediterranean is characterized by extreme differences between its landscape units. The sediment-starved zones of the shelf such as coastal El-Alamein or southern Cyprus, differ considerably from the sediment-rich domain of the littoral cell of the Nile or the Seyhan-Ceyhan Delta in Turkey. The steep continental margin and the narrow coastal belt of northern Israel, Lebanon, Syria, and large parts of southern Anatolia, stands in stark contrast to the wide coast and continental shelf of the gulfs of Iskenderun and Mersin, and their coastal plains. It seems that sediment supply rather than geological structure constrained the geographic distribution of the most extensive shelves. The persistent accumulation of sand dunes along many shores alongside shallow areas of continental shelf interfered with the seaward flow of rivers and thus led to the development of marshes along the coastal plains. These wetlands were repeatedly drained during times of political stability, and expanded again due to neglect during periods of instability (Braudel 1949). There is some evidence that similar geomorphological patterns prevailed also during events of lower sea level (Belknap & Mart 1999). The occurrence of these swamps could have restricted human habitation to high ground, such as sandstone ridges and river banks, along the present and the now-submerged coastal plains.

Libyan–Egyptian Mediterranean coast and shelf, and the Nile Delta

Already in the fifth century, Herodotus (ca. 484–425 BC) noticed the triangular shape of the sedimentary accretion at the mouth of the Nile River and coined the geographic term 'delta' for such features. The Nile has been flowing to the Mediterranean Sea since at least the end of the Miocene, so that the regime of eastward supply of quartz sand to the coasts of northern Sinai has been consistent during the last 5 Myr, and was interrupted only recently by human activities. The variations of sea level during that time span distributed a thick apron of sand along the coast and the shallow continental shelf. Some of the sand, which was carried onto the land by the combined activity of the breakers and the wind, accumulated and formed extensive fields of sand dunes. The accumulated sand smoothed the erosional features along the coastal domain, which were formed by the variability of the sea level and the resulting changes in the flow regime of the coastal rivers.

The sedimentary regime of the Nile and its delta has been constrained by several factors. Primarily the sedimentology is controlled by the flux of the river, which was estimated to be 150 million tons annually before the construction of the Aswan dams (Said 1981). This horizontal accretion is curtailed by the subsidence of the delta due to the sedimentary load (Stanley 1988) and by the wave-induced, eastward-flowing longshore current. The waves and the longshore currents graded the fluvial sediment.

Nile sediment supply to the south Levant coast

The Nilotic sand is carried by the longshore current, created by westerly winds and the associated wave climate (Almagor 1979; Goldsmith & Golik 1980; Almagor & Hall 1984; Almagor et al. 2000). This current transports the sandy fraction of the sedimentary load eastwards along the coast of northern Sinai. The effectiveness of the transport is derived from the acute angle between the prevailing western storms and the breakers that hit the east–west trending coast obliquely and thus convert much of the wave energy into eastward-flowing longshore currents. Large quantities of the sandy fraction are carried eastwards by the longshore currents while the clay fraction flows seawards and is deposited along

the distal continental shelf, slope and rise, to become the predominant constituent of the turbiditic deposits of the Levant basin (Venkatarathnam & Ryan 1971; Nir 1984). The abundance of smectitic clays (Nile lutite) along the shallow shelf and the coastal plain of the southern Levant suggests the possible sedimentological significance of counter-clockwise-flowing currents along the distal shelf, which could transport the Nilotic clays to the Levant coasts (Olausson 1960; Stanley et al. 1997; 1998). The sandy fraction of the Nilotic sedimentary load comprises predominantly medium to fine-grained quartz, derived from the Nubian sandstones of Upper Egypt, with minor quantities of basaltic minerals derived from the Ethiopian plateaux and transported by the Blue Nile. This Nilotic sand has been distributed along the entire continental shelf and the coastal plain of the Levant due to the variability of the sea levels since the Early Pliocene, but considerable quantities were carried onto the land to form sand dunes. Such dunes occur at present along northern Sinai and the coastal Levant, but their Plio-Pleistocene fossil equivalents are very abundant in the Nile littoral domain.

Eastern Mediterranean (Levant coast) — Alexandria to Anatolia

The eastern Mediterranean (Levant coast) is characterized by three different sedimentary regimes: 1) The Nile littoral cell, commencing east of Alexandria up to the Akko promontory in northern Israel (ca. 650 km long); 2) The Akko–Iskenderun (multi-littoral sedimentary cells: Akko–Ras el Naqoura, 20 km; Ras el Naqoura–RasBasit, ca. 420 km); 3) Alexandretta–Southern Anatolian coast (Lebanese coast, 25 km; Syrian coast, 193 km).

This part of the Mediterranean has a most important role in the development of prehistoric centers and sites.

The varying shelf width decreasing from south to north gives a range of conditions, gradients and contexts within which to find submerged sites. In all cases, occupation of the shelf would have moved westwards as sea level dropped, while during rising sea levels, sites would have to retreat landwards towards the present coastline.

The east Mediterranean is characterized, with the exception of the Nile, by a scarcity of large rivers. These are mostly of the seasonal stream type, with relatively large discharge rivers in northern Lebanon and Syria. As a result, the contiguous lands of Israel, Lebanon and Syria supply only small quantities of sediment to the beaches and to the section of the continental shelf relevant to this study. The Nile on the other hand (until the mid-1960s) supplied through the Dumyat (Damietta) and Rashid (Rosetta) tributaries a sediment yield of up to 120 million tons/year to 150 million tons/year, mostly composed of fines, i.e. silt and clay. The rest, ca. 10% to 15%, is of sand-sized particles (62–2000 microns). Inman and Jenkins (1984) on the other hand give much higher sand values, estimating the total annual sand yield ("when the river flowed freely, prior to 1961") to be about 30 million m^3 (equivalent to ca. 45 million tons, although it is hard to accept such a high figure).

The width of the continental shelf of the southeastern part of the Levantine basin is about 40 km to 50 km off northern Sinai, 25 km off Rafah, wedging out gradually to about 10 km off the Carmel Plain, and to just a few kilometers off southern Lebanon (Figs. 14.2, 14.3, 14.4). The surface water hydrological patterns of the currently submerged landscapes, during periods of low sea stand, can be identified by studying the sea bottom topography with multibeam sonar (see also section, 'Levantine known submerged terrestrial features', page 398). These include several submarine river canyons which incise the northern

section of the Israeli shelf. About 1000 m off the northern Carmel coast (Fig. 14.5) there are two adjacent parallel canyons (termed Adam Canyon) cutting the currently submerged kurkar ridge. The ridge summit is at about 14 m below sea level (BSL) and the bottom of the canyon is at 20 m BSL. These canyons represent an ancient river channel system of coastal streams, which drained the western slopes of Mount Carmel during periods of low sea level in the Last Glacial. Several coastal streams, including the largest one, Nahal (stream) Galim, joined together and crossed the western Carmel coast kurkar ridge, creating Adam Canyon. Off Haifa Bay, the submerged river channels of Nahal Kishon and Nahal Na'aman/Hilazon can be observed on the multibeam image (Figs. 14.6a,b). On the Galilee shelf the ancient channels of Nahal Yassf and Nahal Ga'aton are observed (Fig. 14.7). The Achziv mega canyon cuts the continental slope down to ca. 500 m (Fig. 14.8). The Lebanese and Syrian shelves and slopes are narrow, broken and incised by many deep and steep canyons.

Due to changes in the shoreline orientation from west–east along the Sinai shores, curving gradually to south–north along the Israeli shoreline, the amount of transported sand (influenced by the changing and decreasing potential of the longshore components), decreases sharply from around 500,000 m^3 in Gaza to ca. 150,000 m^3 per year, at the Carmel Head. The local sources of sediments for the coast and the shelf originate mostly from the coastal streams discharge and from

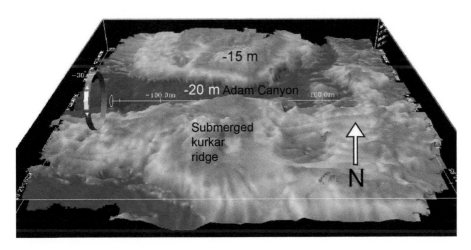

Fig. 14.5 Submerged river canyon (Adam Canyon) off the Carmel coast (multibeam image). Galili (2004).

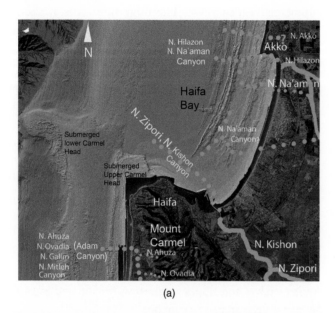

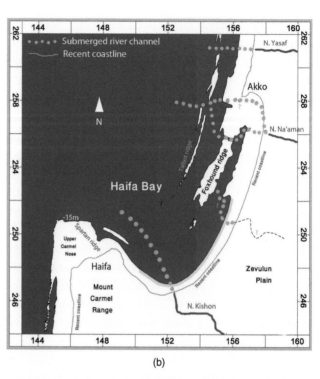

(a)   (b)

Fig. 14.6 (a) Submerged kurkar ridges and river channels on the shallow continental shelf of Haifa Bay. Modified by E. Galili, after Sade *et al.* (2006); (b) some 9000 years ago when sea level was ca. 15 m below present sea level. Modified by E. Galili, after Zviely (2006). Reproduced with permission from University of Haifa.

mollusk shells, and fossil carbonate grains coming from the eroded kurkar cliffs. The volume of the later is minute (Nir 1973; 1984).

Shelf conditions during the regressing shoreline of the Last Glacial Period would have been affected by the morphology of the paleocoastal plain, which for the most part should have resembled that of the present: various 'parallel-to-shore' kurkar ridges, terrestrial-alluvial clay in elongated basins between the ridges, and large sand-dune fields. There is no reason to expect great changes and differences of the Nile sediment discharge during the last 20,000 years, therefore we can assume that the present regime of two to three main sediment distribution patterns on the shelf existed throughout that period: a sand belt from the shoreline to about 25 m water depth; clayey-silt with some sand in the intermediate zone, and silty-clays and clay (mud: Nile lutite) in deeper water. Therefore, in order to understand the succession of the shelf sedimentary pattern during the past ca. 20,000 years, we may envisage a coastal sand

belt moving eastwards with the transgressing sea, with its speed depending on rates of sea-level rise. Whenever the rate of rise was slow, more imported sand accumulated along the coast and in sand dunes, while rapid sea-level rise resulted in thinner sand deposits. Some portion of that sand 'accompanied' the transgressing sea shore. The sedimentary column that one might expect on the present continental shelf beyond the present sand belt (ca. 20–25 m water depth: Nir 1973; 1984) comprises a muddy sea bottom which thickens westwards, covering sands, kurkar ridges, and sometimes alluvial clay deposits — all of varying thicknesses. Rates of clay deposition could have reached as high as 1 mm/year of unconsolidated mud (Nir 1973; 1984). Therefore, even if we assume a turbulent sea bottom that causes further transport to deeper water, we may expect very thick clay cover at the edge of the paleocoastal plain, thinning towards the present shoreline. If we assume that clay cover is a better medium for conserving archaeological and organic material than sand, the faster the shoreline retreats, the

Fig. 14.7 Submerged kurkar ridges and river channels on the shallow continental shelf of western Galilee. Modified by E. Galili after Sade *et al.* (2006), with permission from Geological Survey of Israel.

longer the clay deposition remains in place, resulting in better preservation conditions.

Littoral sand transport along the Israeli coast

A survey of recent littoral sand transport (LST) measurements along the littoral cell shows an ongoing decrease as the longshore currents move east and north up the coast. Inman and Jenkins (1984) and Inman (2002) estimated the rate of the eastward wave-induced LST at the Damietta eastern promontory of the Nile Delta at about 860,000 m^3/year. This rate decreases to about 500,000 m^3/year along the outer Bardawil lagoon sand bar at the northern Sinai coast. Further to the northeast, an updated estimate by Perlin and Kit (1999) shows that the average net LST along the southern Israeli coast decreases from 450,000 m^3/year at Ashkelon, to about 200,000 m^3/year at Ashdod. Moving north, the rate decreases to approximately 100,000 m^3/year at Tel Aviv,

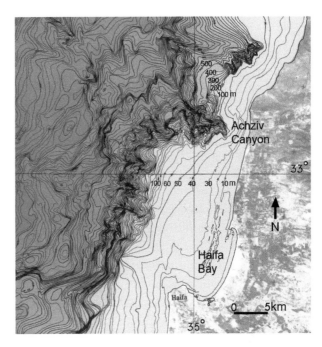

Fig. 14.8 Bathymetry of the Achziv mega canyon. Hall (1998). Adapted with permission from Geological Survey of Israel.

and diminishes to 60,000 m^3/year to 70,000 m^3/year at the south boundary of the northern Carmel coast just before reaching Haifa Bay, the northern end of the Nile littoral cell. Past this boundary, the lack of a sandy beach and the sharp curvature of the coast make accurate LST measurements difficult to obtain. An updated estimate of LST for the northern Carmel coast was carried out by Zviely (2006) based on Perlin and Kit's (1999) analysis, and on long-term sets of directional wave data collected between 1994 and 2004. The new estimate yielded an average net LST of sand to the north of 72,000 m^3/year, during the years 1994–2004, confirming the previous estimate of wave-induced LST in the vicinity of Haifa Bay.

Holocene coastal evolution of Haifa Bay region — a case study

Haifa Bay, on the northern coast of Israel, is a mega embayment and the most significant morphological structure along the southeastern Mediterranean coast. The bay is the northeastern end of the Nile littoral cell (Inman & Jenkins 1984), and constitutes the final depositional basin of the Nile-derived quartz sand (Emery & Bentor 1960; Emery & Neev 1960; Goldsmith & Golik 1980; Carmel et al. 1985; Perlin & Kit 1999). During the LGM about 20,000 years ago, the coastline in the bay area was located more than 20 km west of the present-day coast. Since then, and up to the beginning of the Holocene, sea level rose rapidly to about 40 m BSL (Fairbanks 1989; Bard et al. 1990; 1996; Lambeck & Bard 2000; Lambeck et al. 2002; 2004). At the beginning of the Holocene, at about 9500 cal BP to 9000 cal BP, marine sand started to enter the bay, and for the first time since the LGM, Haifa Bay assumed its present morphological character. The coastline was still 5.5 km west of the present coastline, and the kurkar ridges were still on land (Fig. 14.6b). The paleopatterns of surface runoff in this period can be traced on the recent sea bottom (also see paragraph 'Eastern Mediterranean (Levant coast) — Alexandria to Anatolia', pages 384-389) (Fig. 14.6a). The submerged Nahal Hilazon canyon crosses the submerged kurkar ridges in the north sector of the bay, west of Akko. Nahal Na'aman (recently joined to the Nahal Hilazon south-east of Akko) may have entered the central sector of Haifa Bay independently. It could have passed the now-submerged kurkar ridges by flowing north and joining Nahal Hilazon, or by flowing south and joining Nahal Kishon/Zipori (Fig 14.6a).

The sea invaded eastward and surrounded the kurkar ridges one after the other, and then inundated them all (Fig. 14.9). Sand continued to accumulate in the southern part of the bay. Only at about 8 ka to 7.15 ka did the flooding sea cross the present-day coastline, first in the south and later in the north. At around 6.8 ka to 6.6 ka the sea invaded the Zevulun Plain up to 2 km and the Qishon Estuary more than 4 km south-east of the present coastline. It is still unknown exactly when the sea reached its maximum penetration inland, but it seems that at around 4000 years ago sea level was approximately at the present level, and the coastline in Haifa Bay was up to 3 km east of the present-day coast in most of the Zevulun Plain, and 4.8 km in the Qishon Estuary (Fig. 14.9c). As marine sands accumulated in the bay and the rivers discharged sediment, the coastline migrated westward, and eolian dunes accumulated on top of the marine sand. To the east of the dunes, different wetland conditions developed, mainly along the rivers and streams (Zviely 2006).

The Carmel coast settlements, sea levels and coastal changes

Approximately 20,000 years ago, the Carmel coastal plain was wider and the coastline was some 10 km to the west. In the succeeding interglacial period, rising sea level reduced the area of coastal plains throughout the world and changed the paleoenvironments (Fig. 14.10). Prehistoric inhabitants had to cope with gradual changes in environmental conditions. Several submerged Neolithic settlements inundated in this period have been discovered on the northern Carmel coast (Fig. 14.11). The earlier a submerged prehistoric site on the Carmel coast is, the farther offshore it is located. Atlit-Yam is located 200 m to 400 m off the present shoreline at a depth of 8 m to 12 m. The submerged settlements Kfar Samir, Kfar Galim, Tel Hreiz, Megadim and Neve-Yam Pottery Neolithic (PN) sites are located 1 m to 180 m offshore at a depth of 0.2 m to 5 m. The bottom of a well, recovered and excavated in Atlit-Yam (Fig. 14.12) is at about 15.5 m depth. Hence the sea level was about 16 m BSL during the first stages of the well. The coastline was then located some 1 km west of today's coastline (Fig. 14.13). The distance of the well from the ancient shoreline was about 400 m to 800 m. During the Pottery Neolithic, sea level was ca. 10 m BSL and the coastline was some 600 m to the west with some islands at a distance of 1000 m to 1500 m offshore (Galili et al. 1988; 2005) (Fig. 14.14). The archaeological data enable a reconstruction of the rate of marine transgression from the tenth millennium BP to the present (Fig. 14.10). It can be concluded that there is a direct correlation between the constant rise in sea level and the abandonment of coastal settlements and translocation eastward. Sea level rose continuously from 9200 cal BP to 4000 cal BP, a time span comprising two main stages. Between 9200 cal BP and 7000 cal BP, sea level rose some 12 m, from 16 m to 4 m BSL, at a mean annual rate of ca. 5 mm/year to 6 mm/year. From 7000 cal BP to 4000 cal BP sea level rose an additional 4 m, from 4 m BSL to the present level, and the mean annual rate was ca. 1.33 mm/year. From ca. 4000 BP, sea level was relatively constant, with possible minor changes of less than the local tidal range (±0.25 m).

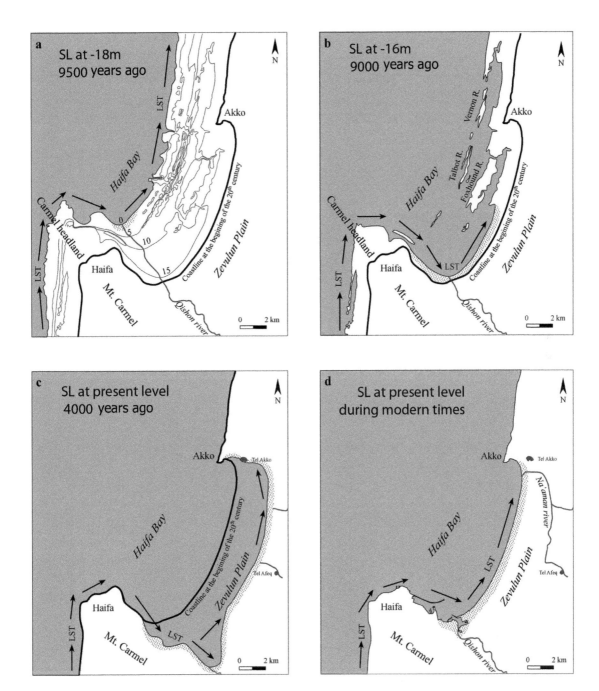

Fig. 14.9 Holocene coastal changes in Haifa Bay embayment: (a) about 9500 years ago, when sea level was ca. 18 m below present sea level; (b) about 9000 years ago, when sea level was ca. 15 m below present sea level; (c) about 4000 years ago, when sea level reached its present level; (d) the recent coastline in Haifa Bay (Zviely 2006). Reproduced with permission from University of Haifa.

Submerged settlements off the Carmel coast

The submerged settlements recovered off the Carmel coast are dated to the Pre-Pottery and the Pottery Neolithic period (Fig. 14.11). The sequence of settlements found and studied there is amongst the most perfectly preserved submerged prehistoric settlements found anywhere in the world. These settlements were uncovered as a result of intensive sand quarrying and construction

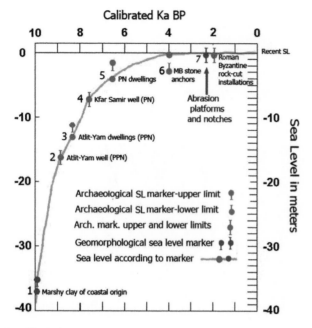

Fig. 14.10 Sea-level curve of the Carmel coast based on archaeological and geomorphological markers. E. Galili (2014).

of marine structures, resulting in massive erosion on the sea bottom along the Israeli coast. The finds enable a reconstruction of the material culture and the socio-economic system of the Neolithic inhabitants of the coastal plain. In addition, it was possible to reconstruct the paleoenvironmental changes and study their impact on coastal habitations during this important period. The prehistoric settlements submerged off the Carmel coast belong to two main entities: the Late Pre-Pottery Neolithic (PPNC), and the Late Pottery Neolithic (Wadi Rabah culture).

The Pre-Pottery Neolithic site of Atlit-Yam

The Atlit-Yam PPNC submerged village thrived some 9200 cal BP to 8500 cal BP. It is located in the north bay of Atlit, submerged at a depth of 8 m to 12 m, covering approximately 40,000 m^2. Its excavations revealed foundations of rectangular stone structures, round installations, a structure built of sandstone megaliths (Fig. 14.15), anthropomorphic stone steles, stone-built water wells (Fig. 14.12) and tens of hearths with charcoal remains. The degree of preservation is extraordinary. The tools found consist of stone, bone and flint artifacts including axes, spearheads, sickle blades and arrowheads. Sixty-five human skeletons buried in flexed positions were uncovered (Fig. 14.16) in and around structures. Organic remains include animal and fish bones, numerous charred and waterlogged seeds, tree branches and pollen grains.

Fig. 14.11 Location of the submerged prehistoric settlements off the Carmel coast. E. Galili (2010).

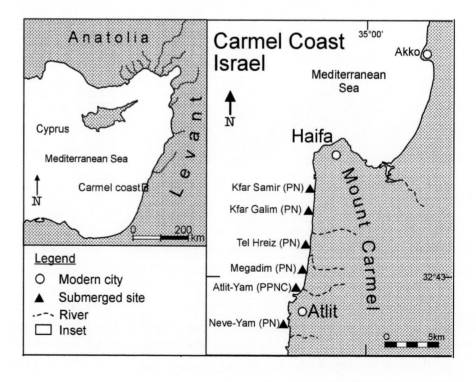

Fig. 14.12 Stone-built water well from Atlit-Yam. Photo courtesy of Itamar Grinberg.

Organic remains suggest that the village economy was complex, based on different food resources acquired through hunting, herding, fishing and farming (Galili et al. 1993; Galili 2004).

The Pottery Neolithic Wadi Rabah sites

Numerous remains from the Pottery Neolithic period were revealed in a narrow, almost continuous submerged belt (15 km long and 200 m wide), parallel to the present shoreline. They include five PN sites dated to the eighth millennium cal BP, from north to south: Kfar Samir, Kfar Galim, Tel Hreiz, Megadim and Neve-Yam, all situated at depths of 1 m to 5 m (Fig. 14.11) (Wreschner 1977; 1983; Raban 1983; Galili & Weinstein-Evron 1985; Galili et al. 1989; 1997; 1998; Horwitz et al. 2002). In these sites, stone and wood-built structures were found, as well as many stone, bone and flint artifacts, and numerous pottery items. Also found were stone-built graves containing human skeletons (at Neve-Yam), various types of features including pits (some containing charred and waterlogged plant remains), and animal bones. An example of such a site is the submerged settlement of Kfar Samir, located south of Haifa, 0.5 m to 5 m BSL (Fig. 14.11) (Galili & Weinstein-Evron 1985). Excavations and surveys there revealed paved floors, olive oil extraction facilities (Galili et al. 1997) and several water wells constructed of alternating layers of wooden branches and stones (Galili & Weinstein-Evron 1985) (Fig. 14.17). In addition, round pits containing plant

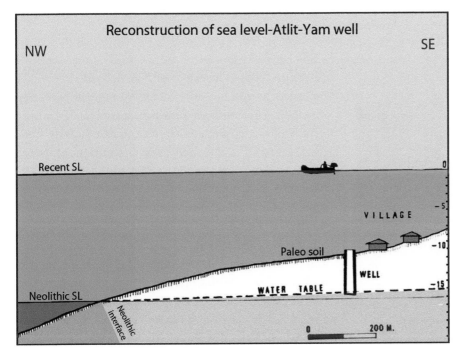

Fig. 14.13 Schematic cross-section of the Atlit-Yam region, with the water well and reconstruction of the Neolithic sea level. E. Galili (2010).

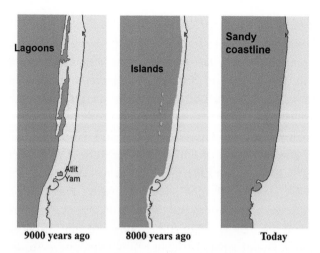

Fig. 14.14 Holocene coastal changes on the northern Carmel coast. E. Galili (2004).

remains (mainly broken olive stones and pulp) were found together with wooden bowls (Fig. 14.18), fragments of woven reed mats, and stone basins.

The Northern multi-littoral cells — Akko to Alexandretta

From Akko, the present northern limit of the Nile littoral cell, and northward to the Lebanese border at Rosh Haniqra, the beaches are beyond the present sedimentological influence of the Nile, and form a 20-km-long independent littoral cell (Boulos 1962; Emery & George 1963; Sanlaville 1970; 1982; Goedicke 1972; Sanlaville

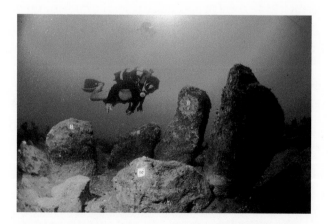

Fig. 14.15 Megalithic structure from Atlit-Yam submerged Pre-Pottery Neolithic C site. Photo courtesy of Itamar Grinberg.

Fig. 14.16 Human burial from Atlit-Yam. A. Zaid (1985). Reproduced with permission.

& Prieur 2005). The extent of Nile-derived sediment is indicated by specific suites of heavy minerals and clay minerals. The southern province coastal sand consists mainly of quartz, some calcium carbonate (mostly from local sources), and a very low proportion of heavy minerals. The mineral suite of the Lebanese and Syrian coast and shelf on the other hand differs from that of the Nilotic one and has more local carbonates, and quartz, mostly of Jurassic age sandstones originating in the mountain ranges of Lebanon and Syria. Figures 14.2, 14.3, and 14.4 show the various widths of the two littoral and shelf provinces: the Nile littoral cell, and the multi-littoral cells of Lebanon and Syria. In comparison, the Lebanese-Syrian cell is relatively narrow. As a result, the post-Last Glacial rate of sedimentation within this northern

(a)

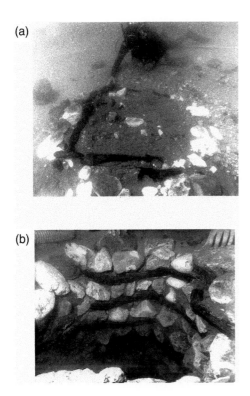

(b)

Fig. 14.17 Water well built of wooden branches and stones from Kfar Samir Pottery Neolithic site. E. Galili (1986).

area is relatively much smaller than that of the zone influenced by the Nile (and might show differences of orders of magnitude). Due to the fact that the Lebanese-Syrian coast was occupied through the ages by dense and very significant prehistoric settlements, the presumed underwater settlements would have a thinner cover of

Fig. 14.18 Wooden bowl from Kfar Samir Pottery Neolithic site. J. Galili (1983).

sand and mud. As a result, preservation conditions might be poor, but discovery could be easier.

Cyprus coast (the continental shelf down to 100 m water depth)

This section is based partly on 1:50,000 scale topographic maps of Cyprus, and Hall's bathymetric map (Nir 1993; 2010; Hall 1998). Cyprus, the second largest island in the Mediterranean with its ca. 770 km coastline, differs in its shelf characteristics from the southeastern Levantine basin continental shelves. The coastal plain varies in width and is mostly dependent on the proximity of the mountains to the shore: very narrow in the west, southwest and north, and somewhat wider elsewhere. In the southeastern region, and in the inter-mountainous Mesaoria Plain, the plains are much wider. In general, the Cyprus continental shelf is very narrow, never exceeding 5 km width (to the 100-m depth contour). In the two big bays at both ends of the Mesaoria Plain (Morfou Bay in the west and Famagusta Bay in the east), the 100-m depth contour is at ca. 3 km to 5 km offshore, while in other regions the 100-m depth contour is at 1 km to 1.5 km offshore only. Whenever the mountains are close to the shoreline the offshore gradients resemble those of the nearby land.

The four main sections of the continental shelf, defined from the shoreline out to 100 m water depth, can be summarized as follows: 1) Apostolos Andreas to Paphos — quite uniform shelf width, varying from 3 km to 5 km; 2) Paphos to Karavostasi (southern tip of Morfou Bay) — very narrow shelf, from few hundreds of meters to 1 km, at most 2 km; 3) Karavostasi to Cape Kormakitis — 3 km to 4 km width; 4) Cape Kormakitis to Apostolos Andreas — very narrow shelf along the northern underwater slopes of the elongated Kyrenia range.

The width of the Cyprus shelf thus varies from a few hundreds of meters to a maximum of 5 km. Being an island, its main coastal and shelf sediment source is of local origin, either transported from the nearby land, or from local production of sea fauna. As a result, the terrestrial sediment input to the sea around Cyprus is much lower than that of the Nile littoral cell. The land-to-sea sediment transport depends mostly on floods of seasonal streams which supply the beaches and the shallow shelf with coarse and fine sediments respectively. Although the nearby Troodos Mountain and the elongated chain of the Kyrenia range are high with steep gradients, the overall amount of sediment reaching the sea is relatively small. The result of these conditions is that the rate of shelf sediment deposition is low, and therefore preservation or survival of settlements of the Last Glacial is most problematic, due to the fact that sediment cover could have been too thin.

The physical characteristics of the coast are considerably different from those prevailing along the continental coast. The coastal and underwater profile in Cyprus is much steeper and most of the coastal regions are surrounded by steep erosional cliffs. Thus the chances of survival and preservation of coastal Neolithic settlements are rather slim. Such coastal sites that might have recently been on the sea bottom were probably eroded in the course of the postglacial sea-level rise and consequently no proven traces of such sites have yet been recovered. Any lithic materials or other robust artifacts would probably be wedged into crevices or cracks, or lying between larger rocks. This makes recovery even more unlikely. One possible site of early date is reported by Ammerman *et al.* (2011). However the flint artifacts reported there may have drifted from land.

The 100-m isobath indicates that Cyprus was never connected to the mainland of Turkey, thus human occupation of Cyprus must have depended on seacraft which could cross the wide (50–60 km) channel from southern Turkey. However, the Taurus Mountains and the Kyrenia range are intervisible on a clear day (Galili *et al.* 2015a; Ammerman & Davis 2013-14).

Discussion, Conclusions and Recommendations

The survival of submerged settlements off the Carmel coast

The submerged settlements found and studied off the Carmel coast are amongst the most perfectly preserved

submerged prehistoric settlements found anywhere in the world. It is therefore worth considering in detail how these sites survived ten millennia of wave attack exposed on a west-facing coast with a fetch of over 1000 km, and at times extreme wave heights greater than 10 m. The position of the Carmel coast between the straight sandy African coasts and the rocky indented European coasts provides perfect conditions for preservation and exposure of submerged landscapes, paleosols and settlements. These archaeological features survived and the organics are well preserved because of a unique combination of circumstances. To explain the burial, survival, and subsequent exposure of the prehistoric sites the following scenarios are proposed.

- The Carmel coast has been exposed to massive and continuous wave attack from the western fetch, and before the construction of the Nile dams and beach extraction there was a continuous northward-drifting cloak of sand that created stable beaches as far north as Haifa Bay.
- The kurkar ridges extend as far as 10 km seaward from the present shore. During the low sea stand of the LGM (ca. 20,000 years ago), rivers carrying fluvial sediments from the Carmel hills filled the troughs between the ridges. This sediment and water drained from the mountains created marshes and swamps in the shallow basins between the ridges.
- Climatic and geologic agents brought about the drying of these wetlands creating dry clayey surfaces and fertile soils.
- During the end of the Pre-Pottery Neolithic period population growth in the permanent inland settlements of the region, could have resulted in the over-exploitation of the surrounding natural resources, leading to an environmental crisis that forced people to look for new exploitable territories.
- The coastal region was attractive for occupation because of the availability of fertile soil and groundwater and, in addition, diverse coastal and marine resources. However, settling the coastal region required adaptation and overcoming new problems. These involved the development of natural resistance to new diseases, like malaria. Human ingenuity devised the invention of a man-made

permanent source of water — wells which reached the local groundwater. The newly arrived agriculturists merged with pre-existing coastal populations of hunters, gatherers, fishermen, co-evolving into the so-called 'Mediterranean fishing village'. These Neolithic agro-pastoral-marine subsistence systems consolidated the basis of the current Mediterranean subsistence system.

– Postglacial rising sea level gradually invaded the Neolithic coastal lands. The inundation was preceded by sea water flowing between gaps in the kurkar ridges and creating creeks and gullies between successive ridges.

– The continuous sea-level rise caused a loss of fertile land by the invading sea water and accumulations of wind-driven sand, the salinization of the water wells by sea water and eventually inundation of inhabited sites. That process eventually caused an abandonment of these coastal sites and a population shift inland. Fixed installations, structures, human burials, broken artifacts and other discarded materials were left behind, as well as small implements which were lost or deposited in wells before the sites were abandoned.

– Shortly before the sites were abandoned and inundated by the sea, the settlements were overlain with marine sand. It was brought from the south by longshore drift and driven ashore as wind-blown sand to accumulate a sediment layer several meters thick over the abandoned villages. This sediment blanket assured good preservation of structures, burials and organics.

– The outer kurkar ridges provided some protection of the inundated archaeological material from initial wave attack, but given the physical characteristics of the coast, many relics not embedded in the paleosol were washed away during the process of inundation. As the water depth increased, the outer kurkar ridges became submerged, and the waves ran over the crests, but the ridges still absorbed part of the wave energy.

– As sea level stabilized ca. 4000 years ago, thick drifts of sand moved northwards along the shore. The overlying sand actually provided further protection for the buried settlements. Also, shoaling waves on a low gradient tend to move sand shoreward, establishing an equilibrium gradient, and not eroding the seabed, provided that there is no net export of sand out of the area.

– The massive exposure of the covered and well-protected sites starts at the beginning of the twentieth century with intensified human interference with the coastal environment. Later, damming the Nile (the main source of sand to the south Levant coast), sand quarrying, and the construction of marine structures (breakwaters, etc.) resulted in a shortage of sand on the shore-face area.

– The forces tending to drive sand northwards are still in operation, but the supply of sand from the south has been reduced. Thus the sand cover of the sites often gets thinner with each storm. Even given the constructive nature of the waves in terms of onshore-offshore transport, the northward drift results in a reduced sand cover, and the waves then strike the beach directly, with less sand supply, thus causing net sand loss and erosion.

Since the 1960s, following each significant storm, systematic underwater rescue surveys have been conducted in the newly exposed areas. The surveys aimed at locating, documenting, removing to safety or preserving *in situ* any submerged prehistoric relics which were randomly exposed. The surveys, followed by rescue excavations and multi-disciplinary research, provided a jigsaw puzzle of data concerning the material culture, economy, subsistence, demography, environment and daily life of the Neolithic coastal dwellers. The finds shed light on previously unknown marine aspects of the Neolithic Revolution.

The Potential for Finding Paleolandscapes and Submerged Settlements — General Outlines

The following criteria and logical deductions help to identify the areas most likely to reveal submerged prehistoric deposits in the eastern Mediterranean:

– Most areas of continental shelves around the world, down to 120 m BSL, were dry land during prehistoric times and are potential areas for discovering paleolandscapes and traces of human existence;

- Most of the paleolandscape is overlaid by post-flooding marine sediments and sediments brought by rivers (in low places, basins and river deltas) or marine bio-structural biogenic rock or encrustations (in areas which were located on high topographic areas during flooding). These paleolandscapes are not visible on the surface and are not accessible to research unless the sediments are removed by natural or man-made erosion or by artificial excavations;

- Some of the paleolandscapes and submerged sites which were located on high topographic areas during the flooding were eroded by forces generated during the long-lasting, gradual postglacial sea-level rise;

- The geomorphological nature of shipwreck sites and submerged settlements differs, and so do the search and survey methods used to locate and study them. The origin of a shipwreck is from above the marine bottom sediments, while a submerged settlement originates in the paleolandscape under the marine bottom sediments. Shipwreck sites were created in the last 5000 years by a ship sinking through the water column and then settling on top of the ancient sediment. Submerged prehistoric sites were created by people living on the paleosol prior to the last sea-level rise, and these were then overlaid by accumulating marine sediments;

- Acoustic sub-surface techniques can be used to detect and reconstruct buried landscapes, but no technique available at present can efficiently detect anthropogenic material that is buried, and distinguish it from natural objects. Therefore the rate of discovery of prehistoric sites in areas of thick marine or terrestrial overlying sediments is slow, and tends to depend upon accidental industrial work such as dredging, pipe-laying or harbor excavation;

- The search for prehistoric sites on the continental shelf of the eastern Mediterranean, and especially on the Levant coast, requires strategic consideration of the effects of both depth and age of the site when selecting search targets. In general, sites that are at present on dry land were available for human occupation continually during the last 500 kyr. Such land sites can potentially contain signs of continuous occupation sequences starting in the Lower Paleolithic period, until recent times. Submerged sites, on the other hand, were not available for human occupation for all of this period. Due to fluctuations in sea level, the availability of occupational conditions at the currently submerged sites depends on the depth of the site below sea level and its age;

- The topographic pattern of the submerged regions (mainly the steepness) is also of importance when calculating the amount of dry land resources available for prehistoric occupation that were later flooded by the rising sea. The flatter the currently submerged continental shelf is, the more land was available for prehistoric occupation and the more land loss has occurred when sea level has risen. Conversely, the steeper the sea bottom is, the less dry land was available in prehistory;

- Analysis of the curve of global sea-level changes over the last 250 kyr (Cutler *et al.* 2003) indicates that the deeper a certain point is on the sea bottom, the less time it was on dry land (Galili 2017). In tectonically stable areas and those with minimum glacio-isostatic adjustment (GIA) like the Levant coast, locations that are currently at 5 m BSL, were on dry land for ca. 90 % of that period (ca. 225 kyr of the last 250 kyr).

- Locations currently at −50 m were on dry land for ca. 60% of that time (ca. 150 kyr of the last 250 kyr). Locations currently at −100 m were on dry land for ca. 20 % of that period (ca. 50 kyr of the last 250 kyr). Locations currently at −130 m were underwater all that time (the last 250 kyr). Thus in principle, when searching for *in situ* human occupations (rather than isolated artifacts) on the sea bottom, the deeper the search area is, the lower the chances of finding *in situ* prehistoric material. To deposit a considerable sequence of anthropogenic sediments and *in situ* archaeological layers (such as in the Paleolithic Tabun Cave on Mount Carmel), an occupation lasting tens of thousands of years is needed. Given the pattern of global sea-level changes, the existence of such long duration settlements on deeply submerged landscapes is less likely compared to locations on land that were never underwater.

– Submerged prehistoric sites that are at a depth greater than −50 m in the eastern Mediterranean must be older than 13,000 years. Such pre-Neolithic sites of hunter-gatherers were, most probably, not permanently occupied by sedentary settlements. Thus a limited amount of structures, features and traces that can be preserved and discovered underwater, are expected to be found in such depths. This is because such features are usually associated with sedentary settlements of the last 13 kyr. Nevertheless, Paleolithic sites may have existed almost anywhere on land and at any depth down to 130 m BSL. In northwestern European seas, a few sites with extensive accumulations of lithic artifacts have been discovered at 15 m to 20 m depth (see Chapter 7 and Chapter 9 this volume).

Given the priorities of research in the eastern Mediterranean, the points made in the previous paragraphs, and the increased costs of working in deeper water, the most promising areas for research are those which were dry land during the Epipaleolithic and Neolithic periods. These are, presently at 0 m to 40 m BSL, and it is this depth range and particularly the shallower end of the range that offers the best chances of finding *in situ* anthropogenic material. Submerged sites of Neolithic (and earlier) date are of particular importance in understanding the role of marine and coastal resources in Neolithic development and expansion.

Many sites in shallow water are exposed because of coastal and underwater erosion due to human activities and natural processes of erosion. In these circumstances, soon after exposure, these sites undergo rapid erosion of the deposits, and valuable archaeological material is lost, until the sand cover moves back over the site. Thus locating, investigating, protecting and monitoring shallow-water sites should be given high priority.

Decades of surveying and searching for submerged settlements on the Carmel coast indicate that most of the preserved sites are covered by sand most of the time and are then not visible. Sites were exposed every few decades for a short period of time (sometimes for a few days) and were covered again soon after exposure. Thus ongoing monitoring of promising areas is crucial for discovering submerged settlements in these conditions. Under-water rescue surveys aimed at discovering, documenting, salvaging and studying these sites should be carried out during the whole year in promising areas of the eastern Mediterranean, and over an extended period of years. Raising public awareness and relying on information provided by local offshore industries, fishermen, divers and members of the public in coastal areas is of great importance in contributing to the discovery and rescue of new material.

The Levant coast: survival of sites, promising areas for discovering submerged settlements and potential for future work

Due to the protection by overlying sand cover, there are probably additional submerged settlements preserved in underwater areas covered by thick sediment such as Haifa Bay, Iskenderun Bay, the Nile Delta, and offshore north Sinai. However, these sites are not accessible and the chances that they will be exposed are slim. Oil and gas exploration and excavation for pipeline entrenchment can reveal archaeological deposits, but it is only by on-site excavations and the provision of the necessary infrastructure that such sites can be properly identified and studied. Areas characterized by limited sand supply, such as the Galilee coast and most of the coasts of Lebanon and Syria, have suffered from intensive erosion during sea-level rise. Prehistoric sites on such coasts are probably not preserved. In areas where there is a protective overlying sand layer of 2 m to 3 m, and later erosion, there are good chances of preservation and discovery of submerged sites. Such is the situation in the Carmel coast, which is the region where all the submerged sites so far have been discovered. The following regions also have potential for the survival and exposure of submerged settlements and landscapes:

– Areas of known rapid coastal and underwater erosion: the Nile Delta, Ashkelon (Fig. 14.19), the Sharon cliffs, the Carmel coast, Cyprus coasts;
– Areas of known accumulation of sediments since the LGM: Haifa Bay, the Nile Delta, Iskenderun Bay;
– Areas of protection from wave and current damage: the leeside of kurkar islands and reefs in Yavne-Yam, Apollonia, Caesarea, Dor, Neve-Yam, Atlit and Achziv.

Fig. 14.19 Coastal erosion in Ashkelon, southern Israel. The blue (dark) line designates the location of the top of the coastal cliff during 1999; the dashed, bright, line designates the location of the top of the coastal cliff during 2013. E. Galili (2013).

Levantine known submerged terrestrial features

These include the following:

– Submerged kurkar ridges: in the Carmel coast and western Galilee. Submerged and flooded kurkar ridges in Egypt, Lebanon and Syria;
– Submerged river valleys in Israel, which were mentioned above (see sections on 'Eastern Mediterranean (Levant coast) — Alexandria to Anatolia', pages 384-387 and 'Holocene coastal evolution of Haifa Bay region — a case study', pages 387-388). These provide details on the surface-water hydrological patterns during periods of low sea stand: a) Adam Canyon off the Carmel coast (Fig. 14.5), is crossing the now-submerged kurkar ridge. It was formed during low sea level by several streams (Nahal Ahuza, Nahal Ovadia, Nahal Galim, and Nahal Mitleh) on the northern Carmel coast. These ephemeral streams were blocked by the eastern kurkar ridge (now on the coast) and flowed northward, towards the Nahal Ahuza Channel, where the kurkar ridge disappeared under the sediments. There they crossed this ridge by the now-submerged Adam Canyon (Galili *et al.* 2005); b) The Naaman

Canyon off Akko (Fig. 14.6a) (see Haifa Bay case study above, pages 387-388); c) Ein Sara Canyon (Fig. 14.7). This canyon was formed by the Nahal Ga'aton stream (presently running a few kilometers to the north), which flowed to the south when the sea level was lower (Galili & Eitam 1988) and possibly by Nahal Beit Ha'emeq; d) Achziv mega canyon, which was formed by Nahal Kziv and Nahal Sha'al (Fig. 14.8);
– Submerged paleosols along the Israeli coast: identified at Yavne-Yam, Caesarea south anchorage, Dor south anchorage and main bay, Neve-Yam, Atlit to Haifa, Shavei Ziyon, Achziv south (Fig. 14.20) (Galili & Inbar 1987).
– Archaeological remains in paleosols: found in Caesarea south, Neve-Yam, northern Carmel coast (from Neve-Yam to Haifa) and Shavei Ziyon (Galili 1985).

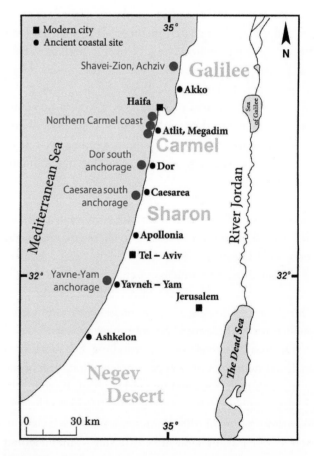

Fig. 14.20 Locations of paleosols found off the Israeli coast. Galili (1985).

Recommendation for a Mediterranean–European Collaboration

The proper understanding of the expansion of the Neolithic Revolution and the developments that may have preceded it, the development of Mediterranean subsistence and the role and time-depth of marine resources and fishing, as well as other cultural transitions in and before the Early Neolithic period, needs collaboration between European and Mediterranean institutions working together in the eastern Mediterranean. Underwater surveys aimed at locating, mapping and studying submerged settlements off the eastern Mediterranean and south European coasts should be carried out using available search methods (remote sensing, as well as traditional surveying), thus ensuring proper coverage of all areas where sites could be preserved and located. Priority should be given to searching in areas where the probability of finding sites is higher and in places where such sites are highly endangered. It is also essential to excavate, rescue and exploit the evidence from known selected sites and features which are at risk.

Oceanographic and Archaeological Data Sources

The following sources of data are relevant to the areas discussed in this chapter:

– Bathymetric chart of the eastern Mediterranean Sea (Hall 1994).
– Gas-pipe route survey bathymetric chart (Anon 2001) Sheets No. 1–9, 1:20,000.
– Reproduction and enhancement of British Admiralty maps of the Israeli continental shelf (Hall 1998).
– Hefa (Haifa). Chart 1585. 1:20,000 (Clarke 1998).
– Bay of Acre. Chart 1585. 1:45,400 (Mansell et al. 1863).
– Morphology and acoustic backscatter of the northern Israel continental margin, based on high-resolution multibeam sonar (Sade 2006).

– High bathymetry of the Mediterranean Sea off northern Israel (Sade et al. 2006).
– Morphology and sediments of the inner shelf off northern Israel (Atlit–Rosh Haniqra) (Eitam 1988; Eitam & Ben Avraham 1992).
– Government of Israel publications (Yalkut Hapirsumim) protected and declared antiquities sites. www.antiquities.org.il/gush_helka_heb.asp (in Hebrew).
– List of submerged settlements off the Israeli coast (the SPLASHCOS Viewer) (Jöns et al. 2016)
– Submerged settlements off the Israeli coast (Galili et al. 2015b; in press).

Author contributions

Yaacov Nir wrote the sections regarding the Nile littoral cell, the eastern Mediterranean (Alexandria to Iskenderun), the northern multi-littoral cells Akko to Alexandretta, Cyprus coast and contributed to the geological/geomorphological background, the physical conditions sections and the LST along the Israeli coast chapter. Yossi Mart wrote the sections concerning the Egyptian–Libyan Mediterranean coast and shelf and the Nile Delta, and contributed to the geological/geomorphological background, the physical conditions sections and the LST along the Israeli coast. Dina Vachtman wrote the section regarding formation possesses of continental shelves during low sea stands. Ehud Galili excavated and studied the submerged settlements, wrote the archaeological background, the sections about the Carmel coast settlements and sea-level changes, Holocene coastal evolution of the Haifa Bay region, submerged river channels on the continental shelf of northern Israel, submerged settlements off the Carmel coast, the discussion, conclusions and recommendations sections, the oceanographic and archaeological data sources, and edited the text.

Acknowledgments

We wish to thank the Israel Antiquities Authority and Haifa University for their institutional support, to Geoffrey N. Bailey and the SPLASHCOS project, to Nicholas C. Flemming for his useful comments, corrections and additions to the article and for editing it, to Josef Galili for the photography, archive management and the organization of the underwater excavations, to Itamar Grinberg for his photography, and to Dov Zviely for his useful help and providing the data about his Haifa Bay research.

References

Ammerman, A. J. & Davis, T. (eds.) 2013-14. *Island Archaeology and the Origins of Seafaring in the Eastern Mediterranean. Proceedings of the Wenner Gren Workshop held in memory of John D. Evans.* 19th – 21st October 2012, Reggio Calabria. *Eurasian Prehistory* vols. 10-11. Available online at: www.peabody.harvard.edu/files/19_Mannino_start.pdf

Ammerman, A. J., Howitt-Marshall, D., Benjamin, J. & Turnbull, T. 2011. Underwater investigations at the early sites of Aspros and Nissi Beach on Cyprus. p.263-271. In Benjamin, J., Bonsall, C., Pickard, C. & Fischer, A. (eds.) *Submerged Prehistory*. pp. 263-271. Oxbow Books: Oxford.

Anonymous 2001. *Gas pipe route survey bathymetric chart* (Sheets No. 1-9, scale 1:20,000). Oceana Marine Research Ltd (Ministry of National Infrastructure — Natural Gas Management): Rosh Haayin (Israel).

Almagor, G. 1979. Relict sediments of Pleistocene age on the continental shelf of northern Sinai and Israel. *Israel Journal of Earth Sciences* 28:70-76.

Almagor, G. 2002. The Mediterranean coasts of Israel. *Geological Survey of Israel Bulletin* 13:177-179.

Almagor, G. & Hall, J. K. 1984. *Morphology of the Mediterranean continental margin of Israel (a compilative summary and a bathymetric chart)*. Geological Survey of Israel: Jerusalem.

Almagor, G., Gill, D. & Perath, I. 2000. Marine Sand Resources Offshore Israel. *Marine Georesources and Geotechnology* 18:1-42.

Bard, E., Hamelin, B., Fairbanks, R. G. & Zinder, A. 1990. A calibration of the ^{14}C timescale over the past 30,000 years using mass spectrometric U-Th ages from Barbados corals. *Nature* 345:405-410.

Bard, E., Hamelin, B., Arnold, M. *et al.* 1996. Deglacial sea-level record from Tahiti corals and the timing of global meltwater discharge. *Nature* 382:241-244.

Bartek, L. R., Anderson, J. B. & Abdulah, K. C. 1990. The importance of overstepped deltas and 'interfluvial' sedimentation in the transgressive systems tract of high sediment yield depositional systems-Brazos-Colorado deltas, Texas. In Armentrout, J. M. & Perkins, B. F. (eds.) *Sequence stratigraphy as an exploration tool: Concepts and practices in the Gulf Coast (vol. 11).* pp. 59-70. GCSSEPM: Houston.

Belknap, D. & Mart, Y. 1999. Sea-level lowstand in the eastern Mediterranean: Late Pleistocene coastal terraces offshore northern Israel. *Journal of Coastal Research* 15:399-412.

Blum, M. D. & Törnqvist, T. E. 2000. Fluvial responses to climate and sea-level change: A review and look forward. *Sedimentology* (supplement s1) 47:2-48.

Boulos, I. 1962. *Carte de reconnaissance des côtes du Liban* (scale 1:150,000). Imp. Bassile Frères: Beirut.

Braudel, F. 1949. *La Méditerranée et le monde Méditerranéen à l'époque de Philippe II.* Armand Colin: Paris.

Butcher, S. W. 1990. The nickpoint concept and its implications regarding onlap to the stratigraphic record. In Cross, T. A. (ed.) *Quantitative Dynamic Stratigraphy.* pp. 375-385. Prentice-Hall: New York.

Carmel, Z., Imman, D. L. & Golik, A. 1985. Directional wave measurement at Haifa, Israel, and sediment transport along the Nile littoral cell. *Coastal Engineering* 9:21-36.

Clarke, J. P. 1998. *Hefa (Haifa) Chart 1585* (scale 1:20,000). United Kingdom Hydrographic Office: Taunton.

Coleman, J. M. & Roberts, H. H. 1990. Cyclic sedimentation of the northern Gulf of Mexico Shelf. In Armentrout, J. M. & Perkins, B. F. (eds.) *Sequence stratigraphy as an exploration tool: Concepts and practices in the Gulf Coast (vol. 11).* pp. 113-134. GCSSEPM: Houston.

Cutler, K. B., Edwards, R. L., Taylor, F. W. *et al.* 2003. Rapid sea-level fall and deep-ocean temperature change since the last interglacial period. *Earth and Planetary Science Letters* 206:253-271.

Dalrymple, R. W., Boyd, R. & Zaitlin, B. A. 1994. Preface. In Dalrymple, R. W., Boyd, R. & Zaitlin, B. A. (eds.). *Incised-valley systems: Origin and sedimentary sequences*. Society for Sedimentary Geology (Special Pub. 51).

Eitam, Y. 1988. *The shallow structure and the geological processes of the inner shelf off northern Israel in the Late Pleistocene*. Ph.D thesis. Tel Aviv University, Israel (in Hebrew).

Eitam, Y. & Ben Avraham Z. 1992. Morphology and sediments of the inner shelf off northern Israel. *Israel Journal of Earth Sciences* 41:27-44.

Emery, K. O. & Bentor, Y. 1960. The continental shelf of Israel. *Geological Survey of Israel Bulletin* 26:25-40.

Emery, K. O. & Neev, D. 1960. Mediterranean beaches of Israel. *Geological Survey of Israel Bulletin* 26:1-24.

Emery, K. O. & George, C. J. 1963. *The shores of Lebanon*. American University of Beirut: Beirut.

Fagherazzi, S., Howard, H. D. & Wiberg, P. A. 2004. Modeling fluvial erosion and deposition on continental shelves during sea level cycles. *Journal of Geophysical Research: Earth Surface* 109:F03010.

Fairbanks, R. G. 1989. A 17,000-year glacio-eustatic sea level record: influence of glacial melting rates on the Younger Dryas event and deep-ocean circulation. *Nature* 342:637-642.

Galili, E. 1985. *Clay exposures and archaeological finds on the sea bottom, between Haifa and Atlit*. Unpublished MA thesis (Dept. of Maritime Civilizations). University of Haifa, Israel (in Hebrew).

Galili, E. 2004. *Submerged settlements of the ninth to seventh millennia BP off the Carmel Coast*. Unpublished Ph.D thesis. Tel Aviv University, Israel (in Hebrew).

Galili, E. 2017. Prehistoric Archaeology on the Continental Shelf: A Global review. *The Journal of Island and Coastal Archaeology* 12:147-149.

Galili, E. & Eitam, Y. 1988. Young faulting in the shallow continental shelf: Evidence from northern Israel. *Proceedings of the 12th Annual Conference of the Israel Geological Society*. 21st – 24th February 1988, Ein Boqeq (Israel), pp. 30-31 (in English/Hebrew).

Galili, E. & Inbar, M. 1987. Underwater clay exposures along the Israeli coast, submerged archaeological remains and sea-level changes in the northern Carmel Coast. *Horizons in Geography* 22:3-34 (in Hebrew).

Galili, E. & Weinstein-Evron, M. 1985. Prehistory and paleoenvironments of submerged sites along the Carmel coast of Israel. *Paléorient* 11:37-52.

Galili, E., Weinstein-Evron, M. & Ronen, A. 1988. Holocene sea-level changes based on submerged archaeological sites off the northern Carmel coast in Israel. *Quaternary Research* 29:36-42.

Galili, E., Weinstein-Evron, M. & Zohary, D. 1989. Appearance of olives in submerged Neolithic sites along the Carmel coast. *Journal of the Israel Prehistoric Society* 22:95-97.

Galili, E., Weinstein-Evron, M., Hershkovitz, I. *et al.* 1993. Atlit-Yam: A prehistoric site on the sea floor off the Israeli coast. *Journal of Field Archaeology* 20:133-157.

Galili, E., Stanley, D. J., Sharvit, J. & Weinstein-Evron, M. 1997. Evidence for earliest olive-oil production in submerged settlements off the Carmel coast, Israel. *Journal of Archaeological Science* 24:1141-1150.

Galili, E., Sharvit, J. & Nagar, A. 1998. Nevé-Yam — underwater survey. *Excavations and Surveys in Israel* 18:35-36.

Galili, E., Zviely, D. & Weinstein-Evron, M. 2005. Holocene sea-level changes and landscape evolution on the northern Carmel coast (Israel). *Méditerranée* 104:79-86.

Galili, E., Zviely, D., Ronen, A. & Mienis, H. 2007. Beach deposits of MIS 5e high sea stand as indicators for tectonic stability of the Carmel coastal plain, Israel. *Quaternary Science Reviews* 26:2544-2557.

Galili, E., Sevketoglu, M., Salamon, A. *et al.* 2015a. Late Quaternary morphology, beach deposits, sea-level changes and uplift along the coast of Cyprus and its possible implications on the early colonists. In Harff, J., Bailey, G. & Lüth, F. (eds.) *Geology and Archaeology: Submerged Landscapes of the Continental Shelf*. Geological Society, London Special Publications 411:179-218.

Galili, E., Horwitz, L. K., Eshed, V., Rosen, B. & Hershkovitz, I. 2015b. Submerged prehistoric settlements off the Mediterranean Coast of Israel. *Skyllis* 181-204.

Galili, E., Horwitz, L. K., & Rosen, B. in press. Submerged Pottery Neolithic settlements off the Mediterranean coast of Israel: Evidence for subsistence activities, material culture and the emergence of separate graveyard. In Bailey, G. (ed.) *Under the Sea: Archaeology and Palaeolandscapes. Proceedings of the SPLASHCOS Final Conference.* 23rd – 27th September 2013, Szczecin, Poland.

Goedicke, T. R. 1972. Submarine canyons on the central continental shelf of Lebanon. In Stanley, D. J. (ed.) *The Mediterranean Sea — a natural sedimentation laboratory.* pp. 655-670. Dowden, Hutchinson and Ross: Stroudsburg (USA).

Goldsmith, V. & Golik, A. 1980. Sediment transport model of the southeastern Mediterranean coast. *Marine Geology* 37:147-175.

Hall, J. K. 1994. *Bathymetric chart of the eastern Mediterranean Sea (scale 1:625,000).* Geological Survey of Israel: Jerusalem.

Hall, J. K. 1998. *Israel Mediterranean coast bathymetric chart (scale 1:50,000, charts 1-5) Technical report TR-GSI/1/98.* Geological Survey of Israel: Jerusalem.

Horwitz, L. K., Galili, E., Sharvit, J. & Lernau, O. 2002. Fauna from five submerged Pottery Neolithic sites off the Carmel coast. *Journal of the Israel Prehistoric Society* 32:147-174.

Inman, D. L. 2002. Nearshore processes. In *McGraw-Hill Encyclopedia of Science and Technology* (9th Ed.) vol. 11:604-611. McGraw-Hill: New York.

Inman, D. L. & Jenkins, S. A. 1984. The Nile littoral cell and man's impact on the coastal zone of the southeastern Mediterranean. *Scripps Institution of Oceanography Series 84-31.* University of California: La Jolla.

Jöns, H., Mennenga, M. & Schaap, D. 2016. *The SPLASHCOS-Viewer. A European Information system about submerged prehistoric sites on the continental shelf. A product of the COST Action TD0902 SPLASHCOS (Submerged Prehistoric Archaeology and Landscapes of the Continental Shelf).* Accessible at splashcos-viewer.eu/

Lambeck, K. & Bard, E. 2000. Sea-level change along the French Mediterranean coast for the past 30,000 years. *Earth and Planetary Science Letters* 175:203-222.

Lambeck, K., Yokoyama, Y. & Purcell, T. 2002. Into and out of the Last Glacial Maximum: Sea-level change during Oxygen Isotope Stages 3 and 2. *Quaternary Science Reviews* 21:343-360.

Lambeck, K., Antonioli, F., & Silenzi, S. 2004. Sea-level change along the Italian coast for the past 10,000 yrs. *Quaternary Science Reviews* 23:1567-1598.

Mansell, A. L., Hull, T. A., & Christian, F. B. 1863. *Bay of Acre Chart 1585 (scale 1:45,400).* Hydrographic Office of the Admiralty: London.

McAdoo, B. G., Pratson, L. F. & Orange, D. L. 2000. Submarine landslide geomorphology, US continental slope. *Marine Geology* 169:103-136.

McHugh, C. M. G., Gurung, D., Giosan, L. *et al.* 2008. The last reconnection of the Marmara Sea (Turkey) to the World Ocean: a paleoceanographic and paleoclimatic perspective. *Marine Geology* 255:64-82.

Neev, D., Almagor, G., Arad, A., Ginzburg, A. & Hall, J. K. 1976. The geology of the southeastern Mediterranean. *Geological Survey of Israel Bulletin* Issue 68.

Nir, Y. 1973. Geological history of the recent and sub-recent sediments of the Israel Mediterranean shelf and slope (Report MG/73/2). Geological Survey of Israel: Jerusalem.

Nir, Y. 1984. *Recent sediments of the Israel Mediterranean continental shelf and slope.* Ph.D thesis. University of Gothenburg, Sweden.

Nir, Y. 1993. *The Coasts of Cyprus.* Geological Survey of Israel: Jerusalem.

Nir, Y. 2010. Cyprus. In Bird, E. C. (ed.) *Encyclopaedia of the World's Coastal Landforms.* pp. 841-848. Springer: Dordrecht.

Olausson, E. 1960. *Description of Sediment Cores From the Mediterranean and the Red Sea: Reports of the Swedish Deep Sea Expedition (vol. 8).* pp. 287-334. Elanders Boktryckeri: Gothenburg.

Perlin, A. & Kit, E. 1999. Longshore sediment transport on Mediterranean coast of Israel. *Journal of Waterway, Port, Coastal, and Ocean Engineering* 125:80-87.

Raban, A. 1983. Submerged prehistoric sites off the Mediterranean coast of Israel. In Masters, P. M. & Flemming, N. C. (eds.) *Quaternary Coastlines and Marine Archaeology.* pp. 215-232. Academic Press: London.

Ronen, A. 1977. Mousterian Sites in red loam in the coastal plain of Mount Carmel. *Eretz Israel* 13:183-190.

Ronen, A., Neber, A., Mienis, H. *et al.* 2008. *A Mousterian occupation on an OIS 5e shore near the Mount Carmel Caves, Israel.* In Sulgostowska, Z. & Tomaszewski, A. J. (eds.) *Man, Millennia, Environment.* pp. 197-205. Polish Academy of Sciences: Warsaw.

Rosen, D. S. 1998. *Assessment of marine environmental impacts due to construction of artificial islands on the coast of Israel. Progress report 4: Characterization of meteo-oceanographic climate in the study sector (Report H16/98).* Israel Oceanographic and Limnological Research: Haifa.

Rosentraub, Z. & Brenner, S. 2007. Circulation over the southeastern continental shelf and slope of the Mediterranean Sea: Direct current measurements, winds, and numerical model simulations. *Journal of Geophysical Research: Oceans* 112:C11001.

Sade, A. 2006. Morphology and acoustic backscatter of the Northern Israel Continental Margin based on high-resolution multibeam sonar. MA thesis. University of Tel-Aviv, Israel.

Sade, A., Hall, J. K., Golan, A. *et al.* 2006. *High bathymetry of the Mediterranean Sea off northern Israel (Report GSI/20/2006 and IOLR report H/44/2006).* Geological Survey of Israel: Jerusalem.

Said, R. 1981. *The Geological Evolution of the River Nile.* Springer-Verlag: New York.

Sanlaville, P. 1970. Les variations holocènes du niveau de la mer au Liban. *Revue de Geographie de Lyon* 45:279-304.

Sanlaville, P. 1982. Asia, Middle East, coastal morphology: Syria, Lebanon, Red Sea, Gulf of Oman, and Persian Gulf. In Schwartz, M. (ed.) *The Encyclopedia of Beaches and Coastal Environments.* pp. 98-102. Hutchinson Ross Publishing: Stroudsburg, Pennsylvania.

Sanlaville, P. & Prieur, A. 2005. Asia, Middle East, coastal ecology and geomorphology. In Schwartz, M. L. (ed.) *Encyclopedia of Coastal Science.* pp. 71-83. Springer: Dordrecht.

Stanley, D. J. 1988. Subsidence in the northeastern Nile Delta: Rapid rates, possible causes, and consequences. *Science* 240:497-500.

Stanley, D. J., & Galili, E. 1998. Clay mineral distributions to interpret Nile cell provenance and dispersal: III. Offshore margin between Nile Delta and northern Israel. *Journal of Coastal Research* 14:196-217.

Stanley, D. J., Mart, Y. & Nir, Y. 1997. Clay mineral distributions to interpret Nile cell provenance and dispersal: II. Coastal plain from Nile Delta to northern Israel. *Journal of Coastal Research* 13:506-533.

Suter, J. R. & Berryhill, H. L. 1985. Late Quaternary shelf-margin deltas, northwest Gulf of Mexico. *American Association of Petroleum Geologists Bulletin* 69:77-91.

Summerfield, M. A. 1985. Plate tectonics and landscape development on the African continent. In Morisawa, M. & Hack, J. T. (eds.) *Tectonic Geomorphology.* pp. 27-51. Allen and Unwin: Boston & London.

Talling, P. J. 1998. How and where do incised valleys form if sea level remains above the shelf edge? *Geology* 26:87-90.

van Heijst, M. W. I. M. & Postma, G. 2001. Fluvial response to sea-level changes: A quantitative analogue, experimental approach. *Basin Research* 13:269-292.

Venkatarathnam, K. & Ryan, W. B. F. 1971. Dispersal patterns of clay minerals in the sediments of the eastern Mediterranean Sea. *Marine Geology* 11:261-282.

Wescott, W. A. 1993. Geomorphic thresholds and complex response of fluvial systems — some implications for sequence stratigraphy. *American Association of Petroleum Geologists Bulletin* 77:1208-1218.

Wreschner, E. E. 1977. Neve-Yam. A submerged Late Neolithic settlement near Mount Carmel. *Eretz Israel* 13:260-271.

Wreschner, E. E. 1983. The submerged Neolithic village 'Neve-Yam' on the Israeli Mediterranean coast. In Masters, P. M. & Flemming, N. C. (eds.) *Quaternary Coastlines and Marine Archaeology.* pp. 325-333. Academic Press: London.

Zviely, D. 2006. *Sedimentological processes in Haifa Bay in context of the Nile littoral cell.* Ph.D thesis. University of Haifa, Israel (in Hebrew with English abstract).

Zviely, D., Kit, E. & Klein M. 2007. Longshore sand transport estimates along the Mediterranean coast of Israel in the Holocene. *Marine Geology* 238:61-73.

Chapter 15

Late Pleistocene Environmental Factors of the Aegean Region (Aegean Sea Including the Hellenic Arc) and the Identification of Potential Areas for Seabed Prehistoric Sites and Landscapes

Dimitris Sakellariou,[1] Vasilis Lykousis,[1] Maria Geraga,[2] Grigoris Rousakis[1] and Takvor Soukisian[1]

[1] *Institute of Oceanography, Hellenic Centre for Marine Research, Anavyssos, Greece*
[2] *Department of Geology & Geoenvironment, University of Patras, Greece*

Introduction

This chapter reviews the main environmental factors which have controlled the evolution of the submerged prehistoric landscape and therefore influenced prehistoric human migration and activity, and discusses the potential for the survival of prehistoric cultural and natural remains on the sea floor of the continental shelf in the Aegean region. For the purpose of this chapter, under the term 'Aegean region' we consider the Aegean Sea including the Ionian, Libyan and Levantine side of the Hellenic Arc (Fig. 15.1).

Submerged Landscapes of the European Continental Shelf: Quaternary Paleoenvironments, First Edition.
Edited by Nicholas C. Flemming, Jan Harff, Delminda Moura, Anthony Burgess and Geoffrey N. Bailey.
© 2017 John Wiley & Sons Ltd. Published 2017 by John Wiley & Sons Ltd.

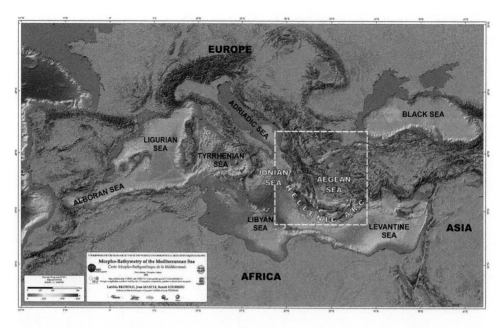

Fig. 15.1 Morpho-bathymetry map of the Mediterranean Sea, Brosolo *et al.* (2012), with location of the Aegean region and Hellenic Arc. Reproduced with permission from CGMW/UNESCO.

The overall geodynamic outline and the morpho-tectonic structure of the Aegean region provide the framework for discussion of the morphological evolution of the continental shelf. On the basis of the variable geomorphological configuration of the coastal and shelf areas we propose to divide the Aegean region into nine blocks, each of which has its own natural and morphological characteristics and evolution. A short description of the geological and tectonic background will provide the necessary basic information for the assessment of the natural resources available for human settlement. We use the available paleomorphological reconstructions of the exposed landmasses during the major low sea-level periods of the Late Pleistocene to define target areas for submerged landscape surveys. A brief description of the Last Glacial Maximum (LGM) and post-LGM climatic conditions as derived from numerous examples of paleo-oceanographic and paleoclimatic research provides the environmental framework. The present overall wind and wave climate in the eastern Mediterranean and Aegean seas enables us to identify the coastal areas protected from or exposed to wave erosion, and to assess the potential survival of shallow submerged prehistoric remains. It is certain that the wind and wave climate has changed

significantly over the course of the last 20,000 years. It is also reasonable to suggest that the relative effect of winds of a given strength from a given direction would be the same, allowing for changes in coastline.

The aforementioned environmental factors are used to describe the characteristics of the nine individual geomorphologic blocks of the Aegean region, with particular reference to the potential survival of natural and archaeological prehistoric remains on the shelf. We conclude with suggested potential example areas for future work on prehistoric submerged landscape and archaeology survey in the Aegean region.

Geodynamic Outline and Morpho-tectonics

The eastern Mediterranean domain (Fig. 15.1) is a unique environment in terms of active geodynamics, plate movements, ongoing geological processes, formation of new and increased relief, and destruction of older morphological structures.

The oceanic crust below the Ionian and Levantine sea basins represents the last remnant of the Tethys Ocean, a 200-million-year-old ocean, which disappeared because of the convergent motion between the Eurasian and the African-Indian continents. The collision of these two continental masses gave birth to some of the highest mountain ranges on Earth, like the Himalayas, Alps, Caucasus and others. Currently, the last remnant of this ocean is being consumed by the fast (5 mm/year) Africa–Eurasia convergence and associated north-northeastward subduction beneath the Hellenic Arc. Active subduction processes and ongoing migration of the continental and oceanic crustal blocks has led to the formation of a complicated geodynamic structure. Convergence between the African plate and the Hellenic fore-arc is nearly perpendicular along western Crete and highly oblique (30°) in the eastern Hellenic fore-arc as far as Rhodes (Fig. 15.2).

The curvature of the Hellenic plate boundary and consequent obliquity of convergence has increased systematically since the Middle/Late Miocene. This has been

attributed to several processes including rollback of the subduction interface induced by the slab-pull force of the downgoing African slab (Le Pichon & Angelier 1979; Angelier et al. 1982; Le Pichon 1982; Meulenkamp et al. 1988), gravitational body forces associated with over-thickened Alpine crust (e.g. Le Pichon et al. 1995; Jolivet 2001) and westward extrusion of the Anatolian block along the North Anatolian Fault (Taymaz et al. 1991; Le Pichon et al. 1995).

The boundaries of the actively deforming part of the Aegean region are defined by the following geotectonic features (Fig. 15.3):

1 The westward prolongation of the dextral, strike-slip North Anatolian Fault (NAF) into the Aegean Sea along the North Aegean Trough (NAT) marks the boundary between the Eurasian continent to the north and the deforming Aegean region to the south. The North Aegean Trough comprises a series of deep basins

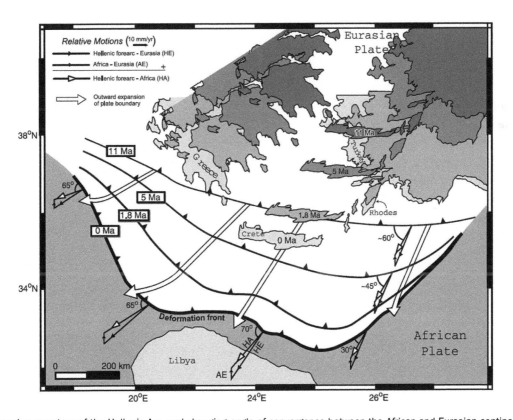

Fig. 15.2 Increasing curvature of the Hellenic Arc, and changing angle of convergence between the African and Eurasian continents along the Hellenic forearc since 11 Ma (Late Miocene), together with reconstructed paleogeography of the Aegean region using a Mercator projection of modern coastlines. Ten Veen and Kleinspehn (2002). Reproduced with permission from John Wiley & Sons.

Fig. 15.3 Main geotectonic boundaries and features of the Aegean region (Aegean Sea, Hellenic Arc and back-arc region) drawn on shaded relief map. Onshore morphology and offshore bathymetry are extracted from the *Morpho-Bathymetry Map of the Mediterranean Sea* (Brosolo *et al.* 2012). KF = Kephallinia Transform Fault; NAT = North Aegean Trough; NAF = North Anatolian Fault. Hellenic Volcanic Arc: Me = Methana; Mi = Milos; Th = Thera; Ni = Nisyros. Brosolo *et al.* (2012). Reproduced with permission from CGMW/UNESCO.

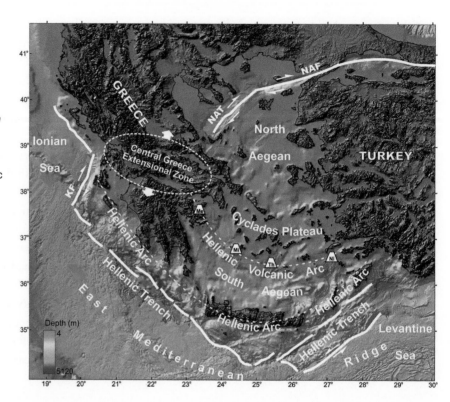

developed due to transtension along the westernmost segments of the NAF.

2 The Hellenic Trench, a series of deep elongate basins aligned along the Hellenic Arc defines the southeast, south and southwest boundary between the Aegean continental block and the East Mediterranean Ridge, the accretionary prism developed above the north-northeastward subduction of east Mediterranean oceanic crust.

3 The Kephallinia Transform Fault (KF), a SSW–NNE trending strike-slip fault, which separates the active part of the Hellenic Arc to the south-east from the inactive one to the north.

4 The Central Greece Extensional Zone, a diffuse boundary, an area of active extensional deformation, which accommodates the transfer of stress between the northern tip of the Kephallinia Fault and the southwestern tip of the North Anatolian Fault.

The main, ongoing processes that dominate the geodynamic evolution of the Hellenic region are: the westward extrusion of the Anatolia continental block along the North Anatolian Fault; the N–NNE subduction of the eastern Mediterranean lithosphere beneath the Hellenic Arc; the subsequent SSW–NNE extension of the Aegean back-arc region; the collision of northwest Greece with the Apulian block in the northern Ionian Sea north of the Kephallinia Fault; and the incipient collision with the Libyan promontory south of Crete (McKenzie 1970; 1978; Dewey & Şengör 1979; Le Pichon & Angelier 1979; Le Pichon *et al.* 1982; Meulenkamp *et al.* 1988; Mascle & Martin 1990; Meijer & Wortel 1997; Jolivet 2001; Armijo *et al.* 2004; Kreemer & Chamot-Rooke 2004) (Fig. 15.2).

The active, long-term, geodynamic processes, crustal movements and deformation of the Aegean region give birth to violent, short-term, catastrophic geological events. Build-up of stresses along the boundaries and in the interior of the Aegean region leads to brittle (and ductile) deformation in the upper crust, expressed with normal, reverse or strike-slip faulting. Vertical tectonics along major and minor faults create a puzzle of small-scale blocks which move vertically, either positively (uplift) or negatively (subsidence). Uplifting blocks are being eroded while subsiding blocks host sedimentary basins which receive the products of erosion. Submarine landslides are frequent along the steep, on/offshore, faulted slopes and are very commonly triggered by strong earth tremors.

Subduction of the eastern Mediterranean lithosphere beneath the Hellenic Arc and the subsequent melting of the downgoing sediments and rocks result in spectacular volcanoes and related volcanic structures in the southern Aegean Sea. The volcanoes are aligned along a 500 km-long by 40 km-wide concave arc, known as the Hellenic Volcanic Arc. Volcanic eruptions, on/offshore and coastal faulting, and submarine landslides have triggered numerous (large and small) tsunamis, some of which have caused devastation in many parts of the Aegean, Ionian and eastern Mediterranean coastlines. All of these processes have played a major role in the morphological evolution and configuration of the Aegean region's continental shelf and have modified the effect of Late Pleistocene sea-level fluctuations.

Geomorphology

The variety of active geological and tectonic processes in the Aegean region has created a puzzle of smaller blocks, each characterized by its own geomorphological evolution and configuration, which may be significantly different from the adjacent blocks. This is particularly the case for the coastal and shelf areas.

The southern and eastern parts of the Hellenic Arc, from the southern Peloponnese to Crete and to Rhodes, are characterized in general by fast, but not uniform, uplift, and are segmented by numerous faults which crosscut the general trend of the arc (Fig. 15.4). On the top of the uplifted blocks, the islands of the Hellenic Arc rise well above sea level and are separated by narrow but deep, faulted trenches. The shelf around the mostly rocky coasts of the islands is very narrow, and nowhere exceeds a few kilometers in width.

The western part of the Hellenic Arc, from Corfu in the north to the southern Peloponnese in the south, displays a similarly narrow shelf. The Ionian Margin hosts the deltaic plains of some of the largest rivers of Greece.

The shelf reaches its maximum width of up to 15 km to 20 km off the river deltas but in most areas the shelf width does not exceed 2 km to 5 km.

The Aegean Sea between the North Aegean Trough and the South Aegean basin includes the great majority of the numerous islands and islets which comprise the Aegean archipelago. Many, like the Cyclades Islands, are separated from each other by shallow shelves. Others, like the east and northeast Aegean Islands, the east Dodecanese and the north Sporades Islands, are joined to each other and to the adjacent landmasses of Greece or Turkey by shallow shelves. Two island bridges, a southern one in the central Aegean and a northern one in the north Aegean Sea, connect mainland Greece with the landmass of Turkey.

The North Aegean shelf, north of the North Aegean Trough, is the widest one in the Aegean region. The largest rivers of the southern Balkan Peninsula flow from the mountains of northern Greece and outflow here, feeding the shelf with huge quantities of new sediments.

Further bathymetric and morphological data for the Mediterranean Sea can be obtained from the following websites:

- geodata.gov.gr (in Greek): contains open geospatial data and interactive maps of Greece including coastal zone maps and ortho-photographs.
- www.ciesm.org/marine/morphomap.htm: CIESM (Commission Internationale pour l'Exploration Scientifique de la mer Méditerranée) morphobathymetry for the Mediterranean.
- srtm.csi.cgiar.org/index.asp: SRTM 90-m Digital Elevation Database v4.1.
- portal.emodnet-hydrography.eu: Pilot Portal for digital bathymetry (not for navigation).
- portal.emodnet-hydrography.eu/#LayerPlace:300958: 0.25 arcminute data products e.g. Digital Terrain Models (DTM).
- www.eea.europa.eu/data-and-maps/data: Corine land cover (raster and vector) and others.
- ec.europa.eu/maritimeaffairs/atlas_en: European Atlas of the Seas. It includes GEBCO bathymetry (under "Bathymetry" — left frame) and more bathymetric maps/links (under "Hydrography" link- left frame).

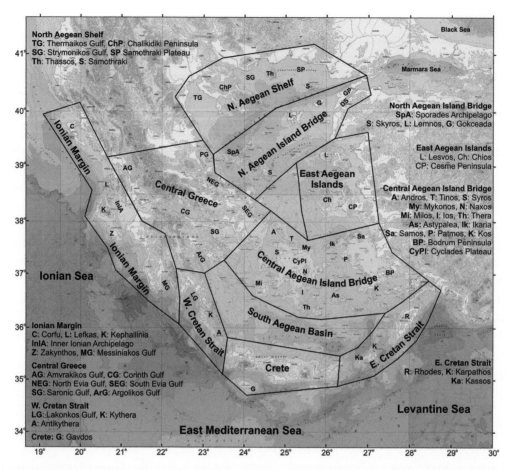

Fig. 15.4 Morpho-bathymetry map of the Aegean region (Aegean Sea & Hellenic Arc) extracted from the IOC-IBCM bathymetry of the Mediterranean (IBCM Project, Intergovernmental Oceanographic Commission, International Hydrographic Organization, and Head Department of Navigation and Oceanography, Russian Federation with permission), and divided into blocks according to the nature of the active morpho-tectonic processes, evolution and configuration of the coastal and submerged landscape. *Source*: www.ngdc.noaa.gov/mgg/ibcm/bathy/IBCM-Bathy-5M-254m-Shaded.jpg. Reproduced with permission from IOC/UNESCO.

– www.gebco.net: General Bathymetric Chart of the Oceans.

– maps.ngdc.noaa.gov/viewers/bathymetry/: National Oceanic and Atmospheric Administration (NOAA)/National Centers for Environmental Information (NCEI) bathymetry — ETOPO bathymetry.

Geological and Tectonic Background

The Aegean region belongs to the Alpine orogenic belt, a mountain chain which starts from the Pyrenees and the Alps in the west and continues over the Balkan Peninsula, Anatolia and the Caucasus Mountains to the Himalayas and further east. The geological structure of the Aegean region (Fig. 15.5) resulted from the successive closure of different parts of the Mesozoic Tethys Ocean and evolved in the course of four orogenic cycles from the Dogger (Middle Jurassic) to the Miocene. The orogenic activity and the associated mountain-building processes migrated from the internal (north/northeast) to the external (south/southwest) regions of the Hellenic Arc. The present arc-shaped configuration and rupturing derives from the overprinting of the present geodynamic regime on the Alpine structure.

The bulk geology of northern Greece, the north and east Aegean islands and northwestern Turkey comprises

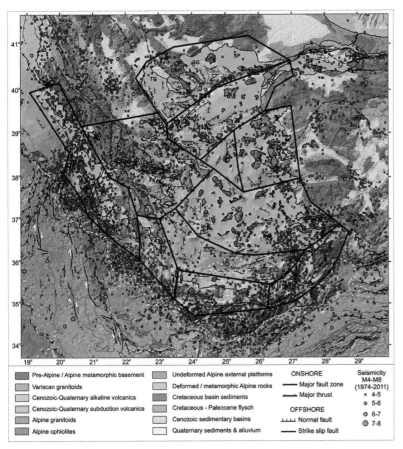

Fig. 15.5 Geological and morpho-tectonic map of the Aegean region, extracted and modified from the *Geological and Morpho-Tectonic Map of the Mediterranean Domain*. Mascle and Mascle (2012). Available online at ccgm.org/en/maps/131-carte-geomorphologique-et-tectonique-du-domaine-mediterraneen-9782917310137.html. Reproduced with permission from CGMW and UNESCO.

Alpine and Pre-Alpine metamorphic rocks of the internal metamorphic belt and sedimentary and volcanic rocks of Mesozoic and Cenozoic age which form the mountainous regions. Extended Quaternary and alluvial plains predominate close to the north Aegean coasts and are associated with major river-delta formations.

The eastern part of the Greek mainland, the central Aegean archipelago, and the Greek-facing coasts of Turkey are mostly built of Alpine sedimentary rocks and metamorphic rocks of the median metamorphic belt. Volcanic rocks of Quaternary age and active volcanoes occur in the central Aegean Island Bridge. The mountain chain of western Greece, from Epirus in the north to the southern Peloponnese to the south and through the islands of the Hellenic Arc to southwest Turkey, consists predominantly of folded and thrusted sedimentary rocks of Mesozoic to Cenozoic age.

Late and Post-Alpine, Oligocene to Miocene and Pliocene sedimentary deposits occur in various regions

between the Mesozoic mountain chains, filling back-arc and molassic basins arranged parallel to the Alpine trend or neotectonic, Plio-Quaternary grabens cutting across the Alpine trend.

The North Aegean Trough (NAT) is the most impressive and actively evolving structure. It developed as a series of isolated, deep basins along the westward prolongation of the North Anatolian Fault. The NAT separates the relatively non-deforming North Aegean shelf from the deforming Aegean region to the south. Similar structures parallel to the NAT occur further south, between and within the north and central Aegean island bridges.

Extensional tectonics in central Greece has created a series of neotectonic grabens, which at present form marine gulfs connected to the open sea through narrow and shallow straits. The Gulf of Corinth, the north and south Evia Gulf, the west Saronikos Gulf, the Amvrakikos Gulf and the Pagasitikos Gulf have developed in the

Quaternary, so that during the low sea-level period of the LGM (and probably during older glacial periods?) they were isolated from the open sea and became lakes. Volcanic explosive activity is one of the major issues in the geological evolution of the Aegean region during the Quaternary. The main volcanic centers are located along the volcanic arc in the south Aegean Sea. Large and minor volcanic eruptions have repeatedly modified the landscape of the areas adjacent to the volcanoes in the Late Pleistocene and Holocene.

The Hellenic Arc, extending from the Ionian Islands to the north-west through the south Peloponnese and Crete to Karpathos and Rhodes to the south-east, has developed above the north-northeastward subduction of the east Mediterranean crust and is one of the most active geological structures worldwide. Frequent large earthquakes nucleated on major active faults of the crust, as well as long-term tectonic movements, create a puzzle of uplifting or subsiding tectonic blocks. Vertical tectonic movements along the Hellenic Arc and in other areas of the Aegean region significantly modify the relative coastal effect of Quaternary sea-level fluctuations. Deposition of fine-grained sediments, mostly silt and clay, prevails in the deep-marine areas and basins of the north and south Aegean Sea during the Holocene.

Late Pleistocene Morphological Configuration

Recently, on the basis of systematic sub-bottom seismic profiling data (AirGun, Sparker and 3.5 kHz), Lykousis (2009) has been able to correlate up to five (vertically) successive Low-System-Tracts (LST) — prograded sequences below the Aegean marginal slopes. The LST sequences suggest shelf-break delta progradation at low sea-level stands (glacial periods) during Marine Isotopic Stages (MIS) 2, 6, 8, 10 and probably 12 (the last 400 kyr). Their upward conformable succession at increasing depths and the assumption that they have been deposited in comparable sea-level stillstands, indicates continuous subsidence of the Aegean margins during the last 400 kyr. Subsidence rates of the Aegean margins were

calculated from the vertical displacement of successive topset-to-foreset transitions (palaeoshelf break) of the LST-prograding sediment sequences.

A major aspect to be taken into consideration for the reconstruction of submerged landscapes and the elaboration of local sea-level curves in the Aegean region is the role of active tectonics (Pirazzoli 1988; Lambeck & Purcell 2005; Lykousis 2009). Firstly, because active tectonics are an important mechanism in landscape evolution, and secondly because differential vertical tectonic movements accommodated by faults have created a complex pattern of tectonically stable, uplifted and subsiding areas, which modifies the general effects of eustatic sea-level rise.

On the basis of available marine geological, sedimentological and tectonic data and the modeled or calculated sea-level fluctuations during the Middle and Late Pleistocene, Lykousis (2009) has reconstructed the paleomorphological configuration of the Aegean region for the most pronounced low sea-level periods of MIS 2, MIS 6, MIS 8 and MIS 10/12. In Fig. 15.6 we use paleomorphological reconstructions from Lykousis (2009) and draw the boundaries of the nine geographic blocks proposed in Fig. 15.4. In the following section, we show that the proposed blocks represent areas with distinct histories of landscape evolution since the Late Pleistocene.

LGM and Post-LGM Climate

The paleoclimatic records of the LGM in the eastern Mediterranean suggest that the region was generally cooler and more arid than present. The aridity most probably caused a retreat of the paleoglaciers, which appear smaller in comparison to the previous Pleistocene period (Pindus Mountains; Hughes & Woodward 2008). Annual temperature and precipitation in Greece were lower than today by ca. 10°C and ca. 550 mm respectively (Peyron et al. 1998; Tzedakis et al. 2002). The climate influenced the vegetation, and almost all pollen records from marine and terrestrial sediments suggest the dominance of steppic and semi-desert elements (Artemisia, Chenopodiaceae, Ephedra; Bottema 1974; Aksu et al. 1995; Digerfeldt et al. 2000; Lawson et al. 2004; Kotthoff

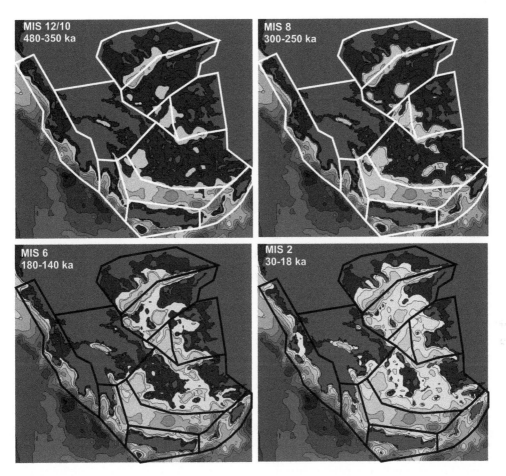

Fig. 15.6 Modified fig. 5 of Lykousis (2009): Paleogeographic reconstruction of the Aegean region during the major glacial stages of the last 400 kyr (MIS 10/12, 8, 6 and 2), and outline of the nine morphological blocks proposed in Fig. 15.4 superimposed on the four maps. Color code of exposed land: light gray for present landmass configuration; dark gray for exposed landmasses during low sea-level stages.

et al. 2008a; Geraga *et al.* 2010). However, the abundance of Chenopodiaceae and *Ephedra* during this period may also partly be connected to sea-level changes (van Andel & Lianos 1984; Geraga *et al.* 2008; Kotthoff *et al.* 2008a).

Despite the dominance of herbs, there was always a moderate abundance of tree taxa, especially oak and *Pinus* (Tzedakis *et al.* 2002). This vegetation pattern is not uniform and depends on local moisture availability, and local topography that provided shelter from incursions of polar air (Tzedakis *et al.* 2002). Therefore, in western Greece the relatively high precipitation caused by orographic uplift of moist air from the nearby Ionian and Adriatic seas resulted in the persistence of thermophilous trees at mid-altitude sites in contrast to eastern Greece (eastern Pindus mountains) where moisture was lower

and the presence of temperate taxa was minimal except probably along coastal plains.

The low temperature is also demonstrated by the low sea-surface temperatures (SST) that prevailed in this period, indicated by the heavy $\delta^{18}O$ values of surface plankton, ranging between 2‰ and 4.7‰ in the Aegean, Ionian and adjacent seas (Vergnaud-Grazzini *et al.* 1986; Aksu *et al.* 1995; Emeis *et al.* 2000; Geraga *et al.* 2000; Geraga *et al.* 2005), and the dominance of cold planktonic foraminifera and dinoflagellate cyst associations (Thunell 1979; Aksu *et al.* 1995; Casford *et al.* 2002; Geraga *et al.* 2008; 2010). Models based on the distribution of the planktonic associations and the stable isotopic signal indicate that the SST was lower than today by 4°C to 6°C (Thunell 1979; Aksu *et al.*

1995; Hayes et al. 2005; Kuhlemann et al. 2008). The water column was characterized by homogeneity as is seen in the similar isotopic signal obtained from planktonic specimens dwelling in different water masses (Casford et al. 2002), and high eutrophication based on microfaunal and microfloral compositions and the increase of marine biomarkers (Gogou et al. 2007). In addition, benthic foraminifera associations indicate moderate-to-high organic matter fluxes and well-ventilated conditions on the sea floor (Kuhnt et al. 2007). A general climatic improvement is indicated in the proxies of terrestrial and marine sediments during the Late Glacial Period. The climatic improvement is not gradual and monotonic, but presents brief variations between warm and/or humid interstadials and cold and/or arid stadials. These oscillations are associated with the prevalence of the interstadials GI 1a, GI 1c and GI 1e and the relative colder GI 1b and GI 1d and the stadial GS 1, according to the INTIMATE-stratigraphy (Lowe et al. 2008) (INTIMATE — INTegration of Ice-core, MArine and TErrestrial records group of the INQUA Palaeoclimatic Commission). The detection of this climatic variation during the Late Glacial Period suggests a direct response of the southern Balkans to centennial- and millennial-scale climatic variability recorded in northern latitudes. GI events correspond to the Meiendorf/Bølling/Allerød complex, and there is no clear discrimination between them in the climatic data. During the warm intervals (GI 1a, 1c, 1e), despite the persistent dominance of steppe elements in the pollen records, the increase of trees (mostly broadleaved taxa) suggests an increase in moisture availability (Digerfeldt et al. 2000; Lawson et al. 2004; Kotthoff et al. 2008a).

On the other hand, depletion in $\delta^{18}O$ values in combination with a reduction in the abundance of cold marine microfauna and microflora indicates an increase in SST (Geraga et al. 2008; 2010; Triantaphyllou et al. 2009). Alkenone SST estimation shows that the temperature of the surface waters ranged between 21°C and 24°C in the Ionian and Aegean seas (Emeis et al. 2000; Gogou et al. 2007). Higher abundance of terrestrial biomarkers (Gogou et al. 2007) indicates this relatively warm and humid interval in the Aegean Sea. Furthermore, the increase in freshwater budget (due to Termination

1a) would have increased surface buoyancy leading to the establishment of seasonal (summer) stratification (Casford et al. 2002) and to a change in benthic faunas from oxic to dysoxic indicator species (Kuhnt et al. 2007). GS 1 is the most pronounced in the climatic records of the Late Glacial Period and corresponds to the Younger Dryas event. GS 1 is characterized by a paleoglacier re-advance in high altitude areas (Hughes & Woodward 2008), the expansion of steppic elements (Digerfeldt et al. 2000; Lawson et al. 2004; Kotthoff et al. 2008a; Geraga et al. 2010), enrichment in $\delta^{18}O$ values accompanied with increases in the abundance of cold marine microfauna and microflora (Geraga et al. 2008; 2010; Triantaphyllou et al. 2009) and a decrease in alkenone SST to 14-16°C, in the Aegean and Ionian seas (Emeis et al. 2000; Gogou et al. 2007). Planktonic and benthic foraminifera data, together with the isotopic signal, suggest a strengthening of winter convection, at least in the Aegean Sea (Casford et al. 2002; Kuhnt et al. 2007).

Climatic instability also characterizes the warm Holocene epoch. Among the Holocene climatic events, the most pronounced is detected between 9.5 ka BP and 6.5 ka BP and is characterized by an increase in precipitation (>270 mm, during winter in northern Greece; Kotthoff et al. 2008b) and an increase in temperature (>4–6°C, during winter in northern Greece; Kotthoff et al. 2008b). The increased precipitation of that time has been related to the strengthening of the southern monsoonal system (Rossignol-Strick et al. 1982; Rossignol-Strick 1983; Aksu et al. 1995) and is reflected in pollen records by the increase of trees and the dominance of deciduous forests. In marine sediments it is evidenced by the deposition of sapropel S1 (Anastasakis & Stanley 1984). Increased temperature, and enhanced freshwater inputs from the outflow of rivers into the Aegean Sea which were later supplemented by the overflowing of the Black Sea (6.5 ka BP; Sperling et al. 2003), lowered the sea surface density and reduced the vertical water mass circulation resulting in reduced bottom-water ventilation.

The reduction of O_2 supply to the seabed in combination with increased fluxes of terrigenous and/or marine organic matter were responsible for the creation of dysoxic to anoxic conditions on the seabed, which led to the formation of sapropel S1 (Rohling & Gieskes 1989;

De Rijk *et al.* 1999; Geraga *et al.* 2000; Casford *et al.* 2002; Triantaphyllou *et al.* 2009). The establishment of warm temperatures and low salinity is documented by the large depletions in the isotopic signal (Vergnaud-Grazzini *et al.* 1986; Aksu *et al.* 1995; Emeis *et al.* 2000; Geraga *et al.* 2008) and the increase in eutrophication is documented by the increase of marine and terrestrial biomarkers (Gogou *et al.* 2007). Sapropel S1 appears in two layers (S1a and S1b). The interruption of S1 centered at 8.2 ka BP and the end of sapropel deposition coincides with the prevalence of stadials (Rohling *et al.* 1997; Geraga *et al.* 2000). The establishment of these two stadials is suggested in alkenone SST records, which show a 2.5°C reduction (Gogou *et al.* 2007), and in climatic models (based on pollen data), which show a 2°C reduction of winter temperatures (during both intervals), and weaker winter precipitation at the end of sapropel S1 (Kotthoff *et al.* 2008b). Almost all records suggest that climatic variations also prevailed during the last 6 kyr, although their exact dating is not yet clearly defined. However, it seems that the climate was warmer and more humid between 5.8 ka BP and 4 ka BP (Pavlopoulos *et al.* 2006; Triantaphyllou *et al.* 2009; Geraga *et al.* 2010) than in the previous and following intervals. Most of the Holocene stadials that have been detected in marine and terrestrial sediments coincide with the prevalence of stadials occurring at northern latitudes (Bond *et al.* 1997) suggesting a direct atmospheric link between the regions. Rohling *et al.* (2002), examining Holocene data from marine records, suggested that strengthening of the Siberian High results in stronger outbreaks of northerly air flows and thus in lower SSTs.

Overview of the Present Wind and Wave Climate

The overall description of the wind and wave climate of the seas in the Aegean region is based on a 10-year hindcast time series. The wind and wave data for the examined period (1995–2004) have been generated by a non-hydrostatic weather model (an improved version of the SKIRON/Eta model) and a 3rd generation wave model (WAM Cycle-4 model), using a spatial resolution of 0.1° × 0.1° and a temporal resolution of three hours. The results of the models (wind speed U_w, significant wave height H_s and spectral peak period T_p) were calibrated (corrected) with collocated *in situ* measurements referring to the joint time window 1999–2004. The length of the time series and the spatial and temporal resolution of the hindcasts ensure a statistically reliable assessment of the corresponding wind and wave climate.

The Aegean Sea is a semi-closed basin with many groups of islands. In combination with relatively short fetch durations and lengths and relatively low swells, the overall wind and wave climate of the Hellenic seas, on an annual basis, appears quite mild. Nevertheless, extreme weather and wave phenomena appear in specific areas, characterized by peak values and short durations. These areas are usually straits, where the wind speed and wave heights are significantly intensified due to the channeling effect. Relatively long fetch lengths are observed in the Libyan and Ionian seas, located on the periphery of the Hellenic Arc, as shown by the dashed line in Fig. 15.7a. The significant effects of the fetch lengths are evident specifically in the straits between Kythera and Crete (Area B, west Cretan Strait), Kasos and Crete, as well as Karpathos and Rhodes (Area C, east Cretan Strait). Swell waves are also observed in these specific areas and are characterized by relatively small values for wave height.

The most intense wind and wave conditions on a mean annual basis, as obtained from numerical models, are observed in three explicitly defined areas of the Hellenic seas (Fig. 15.7a):

– The area north of the Cyclades archipelago (central Aegean Island Bridge), and particularly the straits between Mykonos and Ikaria (Area A, Fig. 15.7a);
– The area outward (west and southwest of) the west Cretan Strait (Area B, Fig. 15.7a) and;
– The area inside (north of) and outward (south of) the east Cretan Strait (Area C, Fig. 15.7a).

A general feature is that wind and waves propagate from the edge of the Dardanelles Strait to the south Aegean Sea, initially with north–northeast directions at the

Fig. 15.7 (a) Areas of intense wind and wave regime for the Hellenic Seas; (b) Spatial distribution of mean annual wind speed; (c) Spatial distribution of mean annual significant wave height. Note that occurrence of maximum annual wind speed and significant wave height coincides largely with the spatial distribution of the mean values.

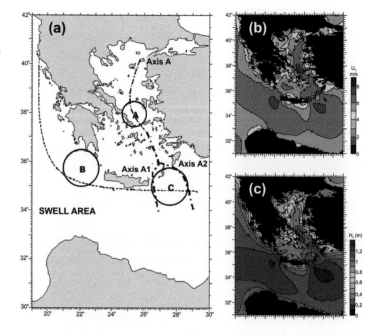

Dardanelles Strait, ending with northwest directions in the south Aegean (see Axis A, Fig. 15.7a). The directions of propagation of wind and waves are generally in good agreement. The weather and wave systems relax partly in the Cyclades archipelago, resulting in milder — on a mean annual basis — wind and wave conditions. On the other hand, while the northeast Cyclades act as a breaker to the propagation of waves and wind flow from the northern part of the Aegean Sea, they simultaneously drive and amplify, by means of a channeling effect, the corresponding wind and wave fields, mainly to the southeast areas. Thus, in specific straits of the Cyclades archipelago, wind and wave intensities higher than those in the neighboring areas are quite frequently observed. Along the south part of the propagation axis, which extends in an arc form from about the straits between Mykonos and Ikaria to the north-east, there is a specific area north of the Cyclades complex (in the central Aegean), where the intensity reaches its local maximum. In this area, the overall maximum of wind speed and significant wave height for all Hellenic seas during the 10-year hindcasts was detected.

The situation is simpler for the Ionian Sea, since it is an open sea area. It is characterized by wave propagation and wind-flow patterns of various directions. However the prevailing wind and wave systems are those originating from the east Italian coasts, Taranto Bay and the north Adriatic coasts heading towards the western coasts of the Peloponnese and Ionian islands. The fetches corresponding to those directions are significantly longer than in the Aegean (400–500 km from the east Italian coast, 800 km from the north Adriatic coast) as well as the ones from the south–southwest directions (700–800 km from the African coast). Consequently, the offshore areas of the Ionian Sea exhibit — on a mean annual basis — the highest wave potential. The local wind and wave maxima are detected along the Hellenic Arc, particularly in areas B and C. The annual spatial distribution of wind speed and significant wave height is presented in Fig. 15.7b and 15.7c.

For a detailed wind and wave climate analysis of the Hellenic seas see Soukissian *et al.* (2008a,b). More information on oceanographic and atmospheric data for the Mediterranean Sea can be retrieved from several websites including the following:

– www.poseidon.hcmr.gr/onlinedata.php: the online data page of POSEIDON System offers access to the most recent oceanographic data recorded by the buoy network in the Aegean and Ionian seas. The data are available either as time series graphs or as text-based format for the latest transmission;

– www.ncdc.noaa.gov/oa/rsad/air-sea/seawinds.html# data: the NOAA Blended Sea Winds contain globally gridded, high-resolution ocean surface vector winds and wind stresses on a global 0.25° grid and time resolution of 6-hourly. The period of record is 1987–present. The wind speeds were generated by blending observations from multiple satellites;

– www.mareografico.it/: this link leads to the National Tidegauge Network which offers access to the most recent, as well as historic, atmospheric and oceano-graphic data recorded by the coastal stations network of the Italian coasts;

– podaac.jpl.nasa.gov: This National Aeronautics and Space Administration (NASA) site offers access to a variety of atmospheric and oceanographic data obtained by satellites on a global basis;

– apps.ecmwf.int/datasets/: This European Centre for Medium-Range Weather Forecasts (ECMWF) site offers access to a variety of atmospheric and oceano-graphic data obtained by numerical simulation models. The period of simulations is 1979–present.

Preliminary Analysis of Submerged Landscapes and Survey Potential

Very few attempts have been undertaken so far in the Mediterranean and particularly in the Aegean region to explore and reconstruct submerged prehistoric landscapes with the use of modern seafloor mapping techniques (van Andel & Lianos 1984; Perissoratis & van Andel 1988; Perissoratis & Conispoliatis 2003). Most of these pioneering works resulted in the creation of maps and reconstructions generalized over large regions which smooth out topographic details, irregularities and local variabilities. Here we attempt to shed light on the environmental factors which characterize the shelf of the nine geographic blocks (Fig. 15.4) of the Aegean region and which have controlled the survival or destruction of submerged landscapes and prehistoric remains. A more detailed analysis of the role of active tectonics on the vertical movements and the development and preservation of submerged landscapes can be found in Sakellariou and Galanidou (2015).

North Aegean shelf

This is the area with the most extended shelf in the Aegean region. The maximum shelf width of 15 km to 20 km is associated with the gulfs of Thermaikos and Strymonikos and the Samothraki Plateau, between the islands of Thassos and Samothraki. Three large rivers, which drain the eastern part of the Pindus mountain chain, outflow into the Thermaikos Gulf. The Strymon and Nestos rivers flow through the central part of the metamorphic basement of the Rhodope Mountains and outflow into the Strymonikos Gulf and north of Thassos Island respectively. The Evros River drains the eastern part of the Rhodope Massif and outflows in the eastern part of the Samothraki Plateau. All of these rivers transport important quantities of fertile and erosional material from their mountainous catchment areas to the North Aegean shelf. Extensive delta areas with coastal lakes and lagoons have developed near to the river outflows. Narrow and very narrow shelves occur mostly along the three 'legs' of the southern Chalkidiki Peninsula, particularly along the two coasts of the Athos Peninsula, the third (eastern) leg.

The North Aegean shelf is located to the north of the northern boundary of the deforming Aegean region and belongs to the 'non-deforming', more rigid, Eurasian continent. Moderate seismic activity and earthquakes do occur in the area, associated either with extensional structures mainly on land or with the strike-slip faulting along the North Aegean Trough. Consequently, vertical tectonic movements do not significantly affect the mor-phological evolution and configuration of the coastal and shelf area.

Systematic sub-bottom profiling performed in the 1980s (Perissoratis & Mitropoulos 1989) indicates that about 5300 km^2 of the North Aegean shelf between the Chalkidiki Peninsula and the Samothraki Plateau was exposed above sea level during the Last Glacial Maximum. At least two permanent and a number of ephemeral lakes existed in this area, while Thassos and Samothraki were connected to the main landmass to the north. The presently submerged landscape of the North Aegean shelf is covered by Holocene sedimentary deposits with a thickness of up to 20 m to 25 m.

Following the spatial distribution of the mean annual wind speed and significant wave height of Fig. 15.7, it is evident that the inner parts of the North Aegean shelf, particularly the inner gulfs of Thermaikos and Strymonikos are exposed to weaker winds and lower waves than the outer parts. The southern coasts of the Chalkidiki Peninsula, and the islands of Thassos and Samothraki are exposed to significantly higher wind speed and waves. The preservation or erosion of possible shallow submerged sites is subject to many factors including local oceanographic and coastal dynamic conditions. The use of the mean annual wind speed and significant wave height offers a first approach to assess the potential for preservation or erosion of prehistoric remains on the shallow seafloor.

North Aegean Island Bridge

The Sporades archipelago, the islands of Skyros, Aghios Efstratios, Lemnos, Gokceada and other smaller islets, which are located between the North Aegean Trough to the north and the central Aegean to the south, constitute stepping stones between central Greece and northwestern Anatolia. Most of the islands are surrounded by shallow shelves and many of them were connected to each other during the low sea-level periods of the Late Pleistocene (Fig. 15.6).

During the low sea levels of MIS 12, MIS 10 and MIS 8, the North Aegean Trough was isolated from the open sea, and the north Aegean Island Bridge connected the mainland of Greece with that of northwestern Anatolia. The North Aegean Trough/Lake separated the North Aegean shelf from the north Aegean Island Bridge. During MIS 6, the Sporades archipelago remained connected to the Greek mainland to the west. A shallow sea separated the Sporades Ridge from the exposed land which is attached to northwestern Anatolia and includes the northeast Aegean islands (Lemnos, Gokceada, Aghios Efstratios). The latter remained connected to the Anatolian landmass during the LGM because of the extended exposed shelf.

Active tectonics in this area are more or less associated with the adjacent prolongation of the North Anatolian Fault into the North Aegean Trough and the parallel secondary structures, mostly strike-slip and transtensional. Evidence of vertical tectonic activity is known from Skyros Island. Skyros underwent subsidence during the Holocene either caused by gradual tectonic events or of co-seismic origin as testified by submerged notches (Evelpidou et al. 2012).

The eastern, extended shelf of the north Aegean Island Bridge, together with the North Aegean shelf, are the marine areas most exploited by the fishing industry in the Aegean Sea. Intensive trawling activity constitutes a potential threat to prehistoric remains when exposed on the sea floor. The north Aegean Island Bridge belongs to the area of highest mean annual wind velocity and significant wave heights in the Aegean Sea (Fig. 15.7). It is therefore reasonable to expect a highly dynamic regime along the coastal zones, especially those facing the open sea.

East Aegean islands

The two largest islands, Lesvos and Chios, remained connected to the Anatolia landmass during all the major low sea-level periods of the last 500 kyr. During the low sea levels of MIS 12, MIS 10 and MIS 8, a large lake occupied the central Aegean area, between the north Aegean Island Bridge and the eastern Aegean islands to the north and the central Aegean Island Bridge to the south. The lake consisted of several deep basins developed along major tectonic lineaments and separated the exposed eastern Aegean islands from the Greek landmass to the west. After MIS 8, the lakes of the central Aegean and the North Aegean Trough never formed again. The basins remained connected to the open sea through shallow sea areas.

Several harbor installations of the Classical and Hellenistic periods have been discovered along the coast of Lesvos Island, submerged at 1 m to 2 m below present sea level (Williams 2007; Theodoulou 2008). Relative subsidence of the island is mostly attributed to glacio-isostatic adjustment (GIA) associated with eustatic sea-level rise, although tectonically-driven vertical movements cannot be ruled out. The exposed shelf of the east Aegean Islands receives the fertile material discharged by two major rivers, the Bakir Çayi and Gediz, which drain the mountainous central part of western Anatolia.

Central Greece

The block of central Greece, as defined here, coincides roughly with the Central Greece Extensional Zone, an area of Quaternary and active extensional tectonics. A series of WNW–ESE trending neotectonic grabens have developed within this regime, and at present have the form of marine elongate gulfs cutting across the Alpine structure of the Hellenides mountain chain. The 900-m-deep Gulf of Corinth and the 450-m-deep north Evia Gulf are the largest and most active grabens in central Greece. The Amvrakikos, south Evia, Pagasitikos and west Saronikos gulfs are smaller, shallower and less active. These water bodies have one characteristic in common: at present they are connected to the open sea through narrow and shallow straits with sills which were exposed above sea level during the LGM. All of these gulfs were disconnected from the open sea during the LGM as demonstrated by the presence of lacustrine, aragonite-bearing deposits below the sea floor (Lykousis & Anagnostou 1993; Perissoratis et al. 1993; Lykousis et al. 2007; Sakellariou et al. 2007a,b).

These water bodies were all isolated lakes surrounded by fluvial plains and high mountains. With the exception of the Gulf of Corinth and the north Evia Gulf, the rest are characterized by relatively extended shallow areas which were exposed above sea level during the LGM. The first two gulfs are subject to fast, active, vertical tectonics, with subsiding and uplifting movements which modify significantly the effect of sea- or lake-level fluctuations (Armijo et al. 1996; Leeder et al. 2005; Lykousis et al. 2007; Sakellariou et al. 2007b; Evelpidou et al. 2011). These gulfs are effectively protected from strong winds and high wave activity. Although erosional processes do occur along parts of their coasts, they are mostly controlled by very local conditions and/or anthropogenic interventions.

The Argolikos Gulf is the only major gulf in the central Greece block which has always been connected

to the south Aegean Sea. It has developed as a deep neotectonic graben (Papanikolaou et al. 1994) with an extensive shelf along its eastern margin. Van Andel and Lianos (1984) and Perissoratis and van Andel (1988) have already surveyed this shelf, particularly off the Franchthi Cave, where traces of Mesolithic and Neolithic habitation have been found, and provided generalized reconstructions.

Central Aegean Island Bridge

The central Aegean Island Bridge is a major element in the Quaternary paleomorphological evolution of the Aegean region. It comprises the Cyclades archipelago and plateau, the northern part of the Dodecanese archipelago and the islands of Ikaria and Samos. The Cyclades archipelago is an extended shallow plateau at the southeastward prolongation of central Greece. The northern Dodecanese archipelago comprises many small and larger islands, including Ikaria and Samos, which are connected to each other by an extensive shelf attached to the coasts of the southwestern Anatolia landmass.

During the major low sea-level periods of MIS 12, MIS 10 and MIS 8, a continuous, elongate landmass was exposed in this area and connected central Greece with Anatolia (Fig. 15.6). In the Late Pleistocene, the central Aegean Island Bridge separated the central Aegean Lake to the north from the south Aegean enclosed sea to the south. The Hellenic Volcanic Arc built the southern edge of the exposed landmass. Continuous volcanic activity and violent eruptions at the main volcanic centers in Methana, Milos, Thera and Nisyros episodically modified the landscape during the Late Pleistocene. Since MIS 6, the central Aegean landmass has broken into two main parts. The western one includes the Cyclades Plateau, which initially remained connected to central Greece and was transformed into a large island during the LGM (Kapsimalis et al. 2009). The eastern part includes most of the north and east Dodecanese islands, Ikaria and Samos, and remained attached to the Anatolia landmass during all low-sea-level stages. Broodbank (2000: 112–13) discusses the potential Paleolithic occupation of the Cyclades, and the significance of the reduction in area from a single landmass to scattered islands.

Most of the islands of the Cyclades archipelago, as well as Ikaria and Samos, are built of metamorphic rocks, predominantly schists and marbles of the median metamorphic belt. Metamorphosed granitoids occur as well, especially in the northern Cyclades and Ikaria. The Dodecanese Islands are mostly built of sedimentary rocks, Mesozoic limestone, marls, radiolarites and Tertiary flysch. Post-Alpine clastic deposits, such as marls, sandstones and conglomerates, outcrop on the lower part of some of the islands.

The central Aegean is characterized by relatively low seismicity, an absence of large earthquakes, and minor faulting. Vertical tectonic movements are apparently of minor significance for the paleogeographic evolution of the central Aegean Island Bridge. Therefore, the evolution of the coastline during the Late Pleistocene and Holocene has been mostly controlled by eustatic sea-level fluctuations, and to a lesser extent by isostatic movements. Still, evidence of tectonic subsidence during the last 6000 years in the center of the Cyclades has been observed from evidence of submerged beachrocks (Desruelles *et al.* 2009).

The eastern part of the central Aegean Island Bridge receives the fertile material of two major rivers which drain the metamorphic Menderes Massif: the first of these, the Küçük Menderes or Cayster River, rises from the Bozdağ Mountains, flows westward and outflows into the Aegean Sea south of Izmir. The Holocene fluvial and deltaic deposits of the Küçük Menderes have silted up the area of ancient Ephesus, an important port in Antiquity, which is now several kilometers inland of the present coastline. The second of these, the Büyük (Great) Menderes or Maeander, is about 600 km long and the largest river in the Aegean region. It outflows south of Samos, feeding the shelf of the east Dodecanese archipelago with clastic sediments derived from the erosion of mountainous southwest Turkey.

The strait between the Cyclades Plateau and the east Dodecanese islands is an area of high annual mean wind velocity and significant wave height (Fig. 15.7). This is also the case for the area north of the central Aegean Island Bridge. Thus, the coasts which are facing these areas are subject to highly dynamic coastal processes related to the wave activity (Fig. 15.8).

Ionian Margin

The western part of the active Hellenic Arc, the Ionian Margin, belongs to the seismically most active areas of the Aegean region. Seismic activity and tectonic movements

Fig. 15.8 Morphological map of the Aegean region extracted from the IOC-IBCM bathymetry of the Mediterranean (IBCM Project, Intergovernmental Oceanographic Commission, International Hydrographic Organization, and Head Department of Navigation and Oceanography, Russian Federation, with permission). Areas of observed or suspected erosion and sediment accumulation in the shallow coastal zone are marked with rectangles. Circles and ellipses mark areas of coastal sediment accumulation. Data from the Hellenic Centre for Marine Research. Note: more information on coastal morphology and erosion risk at scale 1:100,000 can be retrieved from the CORINE coastal erosion database (Version 1990) at www.eea.europa.eu/data-and-maps/data/coastal-erosion. Reproduced with permission from IOC/UNESCO.

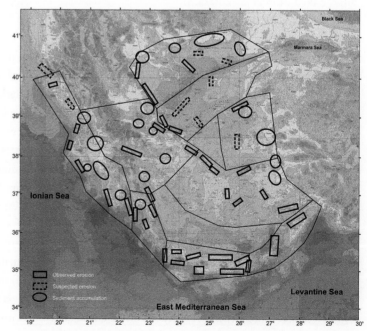

are associated with the thrusting of the overriding Aegean crustal block above the subducting Ionian lithosphere, and with the activity of the Kephallinia strike-slip fault (Fig. 15.3). The Ionian Islands and the landscape of the western part of the Greek mainland and of the Peloponnese have resulted from the active tectonic movements along these two major tectonic elements. Mesozoic evaporites, limestones, radiolarites, Tertiary flysch and Late Alpine clastic deposits and Quaternary Post-Alpine sediments form the bulk geology of the Ionian Margin area (Fig. 15.5).

A relatively extended shelf width characterizes the northern half of the Ionian Margin. Most of the Ionian islands were connected to the Greek mainland during the major low sea-level periods of the Late Pleistocene. Several major and minor rivers, which drain the western flanks of the Pindus mountain chain, outflow into the Ionian Sea. Holocene sediment thickness on the shelf close to the river mouths may reach some tens of meters. This is the case, for example, for the shelf area off the Acheloos Delta, in the eastern half of the inner Ionian archipelago (Fig. 15.4). The deltaic plain of the Acheloos River, developed during the Holocene, includes fairly extended lagoons and marshes, and has incorporated several basement hills which used to be islands before the seaward progradation of the delta.

The inner Ionian archipelago comprises many tens of islands of all sizes. It is a shallow, semi-enclosed marine area between the western coast of central Greece and the islands of Lefkas, Kephallinia and Zakynthos (Zante). Most of the islands of the archipelago were connected to each other and the mainland during the LGM and previous low sea-level stages. Opposite the western coasts of the Ionian Islands, the area of the inner Ionian archipelago is protected from strong winds and high waves. Numerous caves, partly or totally submerged, are known to exist along the coasts of the Ionian Islands. The calcareous composition of the prevailing sedimentary rocks favors the formation of karstic caves due to dissolution of limestones, especially in fractured rocks.

Further south, the west coasts of the Peloponnese are exposed to the open Ionian Sea and particularly high waves. The shelf is fairly narrow and becomes negligible along the coasts of the southwestern Peloponnese. Short rivers flow through the mountains of the western Peloponnese and outflow on the west and southwest coasts. Holocene sediment deposition on the narrow shelf is generally low. Mass gravity processes at the edge of the shelf transport significant parts of the sedimentary load to the deep basins of the Ionian Sea.

West Cretan Strait

The west Cretan Strait is one of the two sea straits which separate the island of Crete from the two adjacent landmasses of Greece and Turkey. It includes the southeastern part of the Peloponnese and the islands of Kythera and Antikythera, two stepping stones between the Peloponnese and Crete (Fig. 15.4). The area is known for very high seismic potential and long-term vertical tectonic movements. Uplifted Late Pleistocene marine terraces and Holocene shorelines (Pirazzoli *et al.* 1982) are evident along the coastline of the southeastern Peloponnese indicating continuous tectonic uplift during the Quaternary as a response to the ongoing subduction of the Ionian lithosphere underneath the overriding Aegean microplate and the deformation of the latter.

The mostly mountainous area of the southeastern Peloponnese is built of Mesozoic and Tertiary sedimentary and metamorphic rocks (limestones, flysch, schists and marbles) and Plio-Pleistocene sediments deposited in Post-Alpine basins. Late Pleistocene and Holocene deposits occur in the restricted lowlands of the river plains and deltas. The shelf off the coasts is fairly narrow and progresses very rapidly to steep submarine slopes leading to the deep basins of the Hellenic Trench to the southwest, or to the deep west Cretan Sea to the east. A shallow, northwest–southeast running ridge connects the islands of Kythera and Antikythera. The shallowest parts of this ridge were exposed during the major low sea-level periods of the Late Pleistocene and were probably connected to the southeastern Peloponnese. During these periods, the width of the strait separating Crete from the Peloponnese was reduced to only a few nautical miles (Fig. 15.6).

In parallel to the prevailing long- and short-term tectonic uplift of the landmasses in this region, local tectonics and faulting is responsible for the formation of the deep basins and trenches and the rough seafloor morphology of the west Cretan Strait. The submerged prehistoric city at Pavlopetri in Vatika Bay close to Elaphonissos (Flemming 1968a,b; 1978; Harding *et al.* 1969; Harding 1970) is clear evidence of local tectonic subsidence within a long-term uplifting region. The ruins of the submerged city survived on the shallow sea floor because they were covered by coastal sand deposits in a sheltered location protected from the very dynamic wind and wave climate of the west Cretan Strait (Fig. 15.7).

Crete

The largest island in the Aegean region, Crete is located on the leading edge of the south-southwestward-moving Aegean microplate and occupies the southern segment of the Hellenic Arc, which is believed to be in incipient collision with the Libyan promontory of the African continental plate (Mascle *et al.* 1999).

The geotectonic position of Crete during the active geodynamic regime of the Quaternary is responsible for the very prominent brittle deformation of the island through faulting and large magnitude earthquakes. The largest earthquake in the Mediterranean during the Historic Era occurred on a northeast-dipping fault-plane a few kilometers off southwest Crete (Shaw *et al.* 2008). That earthquake uplifted western Crete abruptly by up to 8-9 m in the fourth century AD (Pirazzoli *et al.* 1982).

Initial north–south extension followed by extension in an east–west direction formed the Cretan structural high and divided the island into several tectonic blocks (van Hinsbergen & Meulenkamp 2006) which display slightly different histories of tectonic movements. The high mountains of the island represent the faster uplifting blocks, while the lower parts between them exhibit slower uplift or even relative subsidence episodes (Pirazzoli 1988). This is the reason why, despite the overall trend of uplift of the island as a whole or of the individual tectonic blocks, whether faster or slower, longer or shorter term, there are still areas on the island undergoing subsidence where Bronze Age sites are submerged below present sea level.

Active tectonics is a major factor in the evolution of the landscape on Crete (Fassoulas & Nikolakakis 2005). Thus the steep southern flanks of the island have developed along major, east–west running faults which have created an enormous morphological discontinuity between the >2000-m-high Cretan mountains and the >2000-m-deep basins south of the island. Gavdos Island constitutes the summit of a tectonic block south of Crete, separated from it by a trench 1000 m to 3000 m deep (Fig. 15.3). The island is surrounded by a fairly extensive shelf which used to be exposed above sea level during previous low sea-level periods (Fig. 15.6) but has never been connected to Crete during the Quaternary. The shallow area off the west coast of the Messara basin in central southern Crete is the only place along the southern coastline of Crete with considerable shelf development. The linear morphology and rocky character of the steep west and east coasts of Crete are a good indication of their fault-related development. The sea floor of the straits west and east of Crete display tremendous irregularity with very deep trenches alternating with steep, elongate ridges developed perpendicular to the trend of the Hellenic Arc. Shelf development is more pronounced off the northern coasts of Crete, especially off the wide gulfs which are associated with the slower uplifting tectonic blocks of the island.

All sides of Crete are exposed to very strong winds and wave heights (Fig. 15.7). The rocky nature and steepness of the west, south and east coasts demonstrate that erosion is the dominant process along the coastal and shallow offshore zones. Erosion prevails along the northern coasts of the island too (Fig. 15.8), where it results from exposure to wave action and the numerous artificial interventions and constructions along the shoreline.

East Cretan Strait

The larger islands of Kasos, Karpathos, Rhodes and other smaller ones form an island chain between east Crete and the southwestern edge of the Turkish landmass. The islands of the east Cretan Strait represent the uplifted parts of the southeastern segment of the Hellenic Arc. Due to the progressive curvature of the Hellenic Arc (Fig. 15.3), the southeastern segment trends at low angle or parallel to the motion-path of the down-going plate and is characterized by active strike-slip tectonics (Jongsma 1977; Mascle *et al.* 1982) and oblique thrusting (Fig. 15.2) (Stiros *et al.* 2010). Subsequent deformation and uplift of the arc creates extensional tectonics of the upper plate which is manifested by numerous normal and oblique faults.

Like Crete, the east Cretan Strait is segmented into tectonic blocks, each exhibiting its own history of vertical movements. According to Pirazzoli *et al.* (1989), Rhodes can be characterized by considering a small number of crustal blocks, each one displaying a specific tectonic history. Signs of up to eight stepped Late Holocene shorelines have been recognized and studied along the east coast of Rhodes (Flemming 1978; Pirazzoli *et al.* 1982; 1989; Kontogianni *et al.* 2002).

The three main islands are surrounded by narrow shelves which, especially off the eastern coast of Karpathos, are negligible in width. The strait between Rhodes and Turkey is shallow and it is possible that the island was connected to Anatolia during earlier low sea-level periods (Fig. 15.6). Karpathos and Kasos were connected to each other during Late Pleistocene low sea-level periods but were always separated from Crete and Rhodes by deep-sea straits.

The east Cretan Strait is one of the three explicitly defined areas of the Aegean region with the most intense wind and wave conditions on a mean annual basis, as obtained from numerical models (Fig. 15.7). Wind speed values of >11 m/sec and significant wave heights of >4 m are statistically expected fairly frequently every year (Soukissian *et al.* 2008a,b).

Potential Areas for Future Work

Prehistoric human occupation (Paleolithic to Neolithic) has been documented in many places on the Greek mainland, and the Aegean and Ionian islands, even on remote islands like Crete, which has been separated from any landmass throughout human history (Fig. 15.9). Paleolithic findings include 700,000 year old stone tools from Corfu and Palaiokastro Kozani, the 300,000 year old Neanderthal skull from Petralona Cave in Chalkidiki, a *Homo sapiens* skull from Apidima Cave in Mani, stone tools from Kokkinopilos in Epirus, and animal traces, teeth, traces of fireplaces and other artifacts from various areas, including the Theopetra Cave close to Kalambaka in Thessaly. The Mesolithic period is mostly known from Franchthi Cave in Argolis and Theopetra in Thessaly.

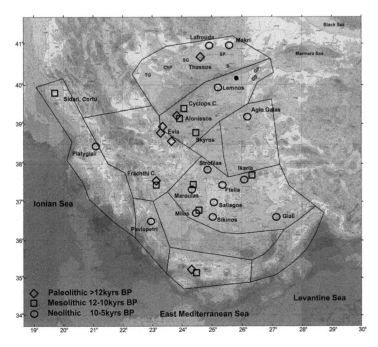

Fig. 15.9 Indicative (not exhaustive) map of prehistoric settlements in the Aegean region. Modified from Sampson (2006) with permission from Atrapos, Athens.

Further Mesolithic evidence has been found at Sidari on the island of Corfu, in the Cyclops Cave on the island of Gioura (Sporades archipelago) and in Maroulas on Kythnos with surviving parts of houses and graves. Neolithic settlements have been found in many places in Greece. Some of the most important excavated sites are in Thrace (Paradimi, Makri), Macedonia (Servia Kozani, Dispilio Kastoria, Drama, Olynthus, Makrygialos Pieria), Thessaly (Sesklo, Dimini, Theopetra, etc.), Attica (Nea Makri), the Peloponnese (Franchthi and Lerna in Argolis, Alepotripa in Mani) and on the Cycladic Islands (Strofilas in Andros, Ftelia in Mykonos, Saliagos in Antiparos, etc.).

On Crete (and the island of Gavdos) particularly, recent and older systematic, archaeological surveys have yielded artifacts proving Mesolithic occupation of the island, and evidence of possible Lower Paleolithic occupation dated to over 100,000 years ago (Strasser *et al.* 2010). The fact that Crete has always been an island during the last million years implies that the early inhabitants reached the island by using open-sea-going vessels and that the history of seafaring in the Mediterranean started in early prehistoric times.

In contrast to the many sites on land, few submerged prehistoric sites are known so far from the Aegean and Ionian seas. The most spectacular site is the Late Neolithic–Bronze Age (sixth to fourth millennium BP) city of Pavlopetri in Vatika Bay in the southeastern Peloponnese, which has survived on the sea floor at a depth of ca. 3.5 m. Further evidence of submerged prehistoric settlements has been found at shallow depths off Plytra in the Peloponnese, in Agios Petros islet (next to Kyra Panagia Island, northern Sporades), in Agios Georgios off the western coast of Corfu, in Platigiali close to Astakos, in Salanti next to the Franchthi Cave (Argolis) and a few other places.

Systematic survey and reconstruction of the submerged landscape of the Aegean region is expected to reveal significant new information on the drowned prehistoric archaeology of the region and will presumably bring to light many unknown sites under the sea. The coastal landscapes exposed at lowered sea level provided relatively fertile and productive refugia for plants, land mammals and humans during the cold, low sea-level periods when increased aridity would have reduced or deterred hinterland occupation (Bailey & Flemming 2008). Therefore, underwater investigation of the shelf is essential for the understanding of early human adaptation and dispersal.

Nevertheless, systematic survey of the shelf areas by means of modern technology is costly and time consuming. Certain criteria and consideration of the environmental factors which may have attracted human occupation can be used to define the most promising areas to be surveyed. The availability of freshwater resources is one of the main factors which make an area attractive for habitation. Water access points include rivers, lakes and freshwater springs. Reconstruction of the course of the rivers and identification of possible lakes can reasonably be a first task for the survey of extended shelves like the North Aegean shelf, the eastern part of the north Aegean Island Bridge, the central Aegean Island Bridge and the inner Ionian archipelago in the Ionian Margin.

Fresh and brackish water karstic springs, submerged at shallow depths below present sea level, are very common in the Aegean region, especially off rocky coasts composed of calcareous rocks (limestone, marble). Karstic springs occur commonly at the interface of the Mesozoic limestone or marble and impermeable formations like Eocene–Oligocene flysch and clastic formations of the Mio-Pliocene. During the Late Pleistocene low sea-level periods, these springs may have constituted attractive freshwater points on the exposed coastal landscape. Most of the known underwater karstic springs in the Aegean region are located on narrow shelves off moderate or steeply sloping coasts. Large underwater karstic springs have been discovered on the eastern shelf of the Messiniakos Gulf (Ionian Margin, southeast Peloponnese), the Ionian Islands (Kephallinia) and off the Ionian coastline of Greece (Ionian Margin), the western shelf and coast of the Argolikos Gulf (central Greece), off the southern and northern steep coasts of the Gulf of Corinth, the north and south gulfs of Evia (central Greece), the northern shelf of Crete and many other areas.

Submerged karstic caves are sites of potential prehistoric occupation. Like karstic springs, the caves are associated with calcareous rocks and their creation is due to the solution of limestone and marble by sea water.

Submerged caves have been discovered off the rocky coasts of the Ionian Islands (e.g. Meganisi — Ionian Margin), the south Peloponnese (Mani Peninsula — Ionian Margin and west Cretan Strait), Crete, the Sporades Islands (north Aegean Island Bridge) and many other places.

Opposing coasts on either side of narrow sea-straits may also be of major importance for prehistoric occupation and migration, and thus may host traces of prehistoric human presence. Narrow sea-straits can be found in many places in the Aegean region as major elements of the changing paleogeographic configuration during the Late Pleistocene and Holocene. The island bridges in the north and central Aegean, the east Aegean islands, the west and east Cretan straits and the Ionian Margin, particularly the Ionian Islands and the inner Ionian archipelago, display variable configurations during the successive low sea-level periods and were characterized by narrow sea-straits separating isolated islands from each other and from adjacent landmasses.

References

Aksu, A. E., Yaşar, D., Mudie, P. J. & Gillespie, H. 1995. Late glacial–Holocene paleoclimatic and paleoceanographic evolution of the Aegean Sea: micropaleontological and stable isotopic evidence. *Marine Micropaleontology* 25:1-28.

Anastasakis, G. C. & Stanley, D. J. 1984. Sapropels and organic-rich variants in the Mediterranean: sequence development and classification. In Stow, D. A. V., & Piper, D. J. W. (eds.) *Fine Grained Sediments: Deep-Water Processes and Facies. Geological Society Special Publication* 15:497-510.

Angelier, J., Lybéris, N., Le Pichon, X., Barrier, E. & Huchon P. 1982. The tectonic development of the Hellenic Arc and the Sea of Crete: A synthesis. *Tectonophysics* 86:159-196.

Armijo, R., Meyer, B., King, G. C. P., Rigo, A. & Papanastassiou, D. 1996. Quaternary evolution of the Corinth rift and its implications for the Late Cenozoic evolution of the Aegean. *Geophysical Journal International* 126:11-53.

Armijo, R., Flerit, F., King, G., & Meyer, B. 2004. Linear elastic fracture mechanics explains the past and present evolution of the Aegean. *Earth and Planetary Science Letters* 217:85-95.

Bailey, G. N. & Flemming, N. C. 2008. Archaeology of the continental shelf: Marine resources, submerged landscapes and underwater archaeology. *Quaternary Science Reviews* 27:2153-2165.

Bond, G., Showers, W., Cheseby, M. *et al.* 1997. A pervasive millennial-scale cycle in North Atlantic Holocene and glacial climates. *Science* 278:1257-1266.

Bottema, S. 1974. *Late Quaternary Vegetation History of Northwestern Greece.* Ph.D thesis. University of Groningen, Netherlands.

Broodbank, C. 2000. *An Island Archaeology of the Early Cyclades.* Cambridge University Press: Cambridge.

Brosolo, L., Mascle, J. & Loubtrieu, B. 2012. *Morpho-Bathymetric map of the Mediterranean Sea (scale 1:4,000,000)*, 1st Edition. CGMW/UNESCO.

Casford, J. S. L., Rohling, E. J., Abu-Zied, R. *et al.* 2002. Circulation changes and nutrient concentrations in the late Quaternary Aegean Sea: a nonsteady state concept for sapropel formation. *Paleoceanography* 17:1024-1034.

De Rijk, S., Hayes, A. & Rohling, E. J. 1999. Eastern Mediterranean sapropel S1 interruption: An expression of the onset of climatic deterioration around 7 ka BP. *Marine Geology* 153:337-343.

Desruelles, S., Fouache, E., Ciner, A. *et al.* 2009. Beachrocks and sea level changes since Middle Holocene: Comparison between the insular group of Mykonos–Delos–Rhenia (Cyclades, Greece) and the southern coast of Turkey. *Global and Planetary Change* 66:19-33.

Dewey, J. F. & Şengör, A. M. C. 1979. Aegean and surrounding regions: Complex multiplate and continuum tectonics in a convergent zone. *Geological Society of America Bulletin* 90:84-92.

Digerfeldt, G., Olsson, S. & Sandgren, P. 2000. Reconstruction of lake-level changes in lake Xinias, central Greece, during the last 40,000 years. *Palaeogeography, Palaeoclimatology, Palaeoecology* 158:65-82.

Emeis, K.-C., Struck, U., Schulz, H.-M. *et al.* 2000. Temperature and salinity variations of Mediterranean Sea surface waters over the last 16,000 years from records of planktonic stable oxygen isotopes and alkenone unsaturation ratios. *Palaeogeography, Palaeoclimatology, Palaeoecology* 158:259-280.

Evelpidou, N., Pirazzoli, P. A., Saliège, J.-F. & Vassilopoulos, A. 2011. Submerged notches and doline sediments as evidence for Holocene subsidence. *Continental Shelf Research* 31:1273-1281.

Evelpidou, N., Vassilopoulos, A. & Pirazzoli, P. A. 2012. Submerged notches on the coast of Skyros Island (Greece) as evidence for Holocene subsidence. *Geomorphology* 141–142:81-87.

Fassoulas, C. & Nikolakakis, M. 2005. Landscape response to the tectonic uplift of Crete, Greece. *Bulletin of the Geological Society of Greece* 37:201-207.

Flemming, N. C. 1968a. Holocene earth movements and eustatic sea level change in the Peloponnese. *Nature* 217:1031-1032.

Flemming, N. C. 1968b. Mediterranean sea level changes. *Science Journal* 4:51-55.

Flemming, N. C. 1978. Holocene eustatic changes and coastal tectonics in the Northeast Mediterranean: implications for models of crustal consumption. *Philosophical Transactions of the Royal Society, London, Series A* 289:405-458 (plus Appendix 1).

Geraga, M., Tsaila-Monopoli, S., Ioakim, C., Papatheodorou G. & Ferentinos G. 2000. An evaluation of palaeoenvironmental changes during the last 18,000 years in the Myrtoon Basin, SW Aegean Sea. *Palaeogeography, Palaeoclimatology, Palaeoecology* 156:1-17.

Geraga, M., Tsaila-Monopolis, S., Ioakim, C., Papatheodorou, G. & Ferentinos, G. 2005. Short-term climate changes in the southern Aegean Sea over the last 48,000 years. *Palaeogeography, Palaeoclimatology, Palaeoecology* 220:311-332.

Geraga, M., Mylona, G., Tsaila-Monopoli S., Papatheodorou, G. & Ferentinos, G. 2008. Northeastern Ionian Sea: Palaeoceanographic variability over the last 22 ka. *Journal of Marine Systems* 74:623-638.

Geraga, M., Ioakim, C., Lykousis, V., Tsaila-Monopolis, S. & Mylona, G. 2010. The high-resolution palaeoclimatic and palaeoceanographic history of the last 24,000 years in the central Aegean Sea, Greece. *Palaeogeography, Palaeoclimatology, Palaeoecology* 287:101-115.

Gogou, A., Bouloubassi, I., Lykousis, V., Arnaboldi, M., Gaitani, P & Meyers, P. A. 2007. Organic geochemical evidence of Late Glacial–Holocene climate instability in the North Aegean Sea. *Palaeogeography, Palaeoclimatology, Palaeoecology* 256:1-20.

Harding, A. F. 1970. Pavlopetri. A Mycenaean town underwater. *Archaeology* 23:242-250.

Harding, A. F., Cadogan, G. & Howell, R. 1969. Pavlopetri: an underwater Bronze Age town in Laconia. *The Annual of the British School at Athens* 64:113-142.

Hayes, A., Kucera, M., Kallel, N., Sbaffi, L. & Rohling, E. J. 2005. Glacial Mediterranean sea surface temperatures based on planktonic foraminiferal assemblages. *Quaternary Science Reviews* 24:999-1016.

Hughes, P. D. & Woodward, J. C. 2008. Timing of glaciation in the Mediterranean mountains during the last cold stage. *Journal of Quaternary Science* 23:575-588.

Jolivet, L. 2001. A comparison of geodetic and finite strain pattern in the Aegean, geodynamic implications. *Earth and Planetary Science Letters* 187:95-104.

Jongsma, D. 1977. Bathymetry and shallow structure of the Pliny and Strabo Trenches, south of the Hellenic Arc. *Geological Society of America Bulletin* 88:797-805.

Kapsimalis, V., Pavlopoulos, K., Panagiotopoulos, I. *et al.* 2009. Geoarchaeological challenges in the Cyclades continental shelf (Aegean Sea). *Zeitschrift für Geomorphologie* 53:169-190.

Kontogianni, V. A., Tsoulos, N. & Stiros S. C. 2002. Coastal uplift, earthquakes and active faulting of Rhodes Island (Aegean Arc): Modeling based on geodetic inversion. *Marine Geology* 186:299-317.

Kotthoff, U., Müller, U. C., Pross, J. *et al.* 2008a. Lateglacial and Holocene vegetation dynamics in the Aegean region: an integrated view based on pollen data from marine and terrestrial archives. *The Holocene* 18:1019-1032.

Kotthoff, U., Pross, J., Müller, U. C. *et al.* 2008b. Climate dynamics in the borderlands of the Aegean Sea during formation of sapropel S1 deduced from a marine pollen record. *Quaternary Science Reviews* 27:832-845.

Kreemer, C. & Chamot-Rooke, N. 2004, Contemporary kinematics of the southern Aegean and the

Mediterranean Ridge. *Geophysical Journal International* 157:1377-1392.

Kuhlemann, J., Rohling, E. J., Krumrei, I., Kubik, P., Ivy-Ochs, S. & Kucera, M. 2008. Regional synthesis of Mediterranean atmospheric circulation during the Last Glacial Maximum. *Science* 321:1338-1340.

Kuhnt, T., Schmiedl, G., Ehrmann, W., Hamann, Y. & Hemleben, C. 2007. Deep-sea ecosystem variability of the Aegean Sea during the past 22 kyr as revealed by Benthic Foraminifera. *Marine Micropaleontology* 64:141-162.

Lambeck, K. & Purcell, A. 2005. Sea-level change in the Mediterranean Sea since the LGM: model predictions for tectonically stable areas. *Quaternary Science Reviews* 24:1969-1988.

Lawson, I., Frogley, M., Bryant, C., Preece, R. & Tzedakis, P. 2004. The Lateglacial and Holocene environmental history of the Ioannina basin, northwest Greece. *Quaternary Science Reviews* 23:1599-1625.

Le Pichon, X. 1982. Land-locked oceanic basins and continental collision: The Eastern Mediterranean as a case example. In Hsu, K. J. (ed.) *Mountain Building Processes*. pp. 201-211. Academic Press: San Diego.

Le Pichon, X. & Angelier, J. 1979. The Hellenic arc and trench system: a key to the neotectonic evolution of the eastern Mediterranean area. *Tectonophysics* 60:1-42.

Le Pichon, X., Angelier, J. & Sibuet, J. -C. 1982. Plate boundaries and extensional tectonics. *Tectonophysics* 81:239-256.

Le Pichon, X., Chamon-Rooke, N., Lallemant, S., Noomen, R. & Veis, G. 1995. Geodetic determination of the kinematics of central Greece with respect to Europe: Implications for eastern Mediterranean tectonics. *Journal of Geophysical Research* 100:12675-12690.

Leeder, M. R., Portman, C., Andrews, J. E. *et al.* 2005. Normal faulting and crustal deformation, Alkyonides Gulf and Perachora peninsula, eastern Gulf of Corinth rift, Greece. *Journal of the Geological Society* 162:549-561.

Lowe, J. J., Rasmussen S. O., Björck, S. *et al.* 2008. Synchronisation of palaeoenvironmental events in the North Atlantic region during the Last Termination: a revised protocol recommended by the INTIMATE group. *Quaternary Science Reviews* 27:6-17.

Lykousis, V. 2009. Sea-level changes and shelf break prograding sequences during the last 400 ka in the Aegean margins: Subsidence rates and palaeogeographic implications. *Continental Shelf Research* 29:2037-2044.

Lykousis, V. & Anagnostou, C. 1993. Sedimentological and paleogeographic evolution of the Saronic Gulf during the Late Quaternary. *Bulletin of the Geological Society of Greece* 28(1):501-510 (in Greek).

Lykousis, V., Sakellariou, D., Moretti, I. & Kaberi, H. 2007. Late Quaternary basin evolution of the Gulf of Corinth: Sequence stratigraphy, sedimentation, fault-slip and subsidence rates. *Tectonophysics* 440:29-51.

Mascle, J. & Martin L. 1990. Shallow structure and recent evolution of the Aegean Sea: A synthesis based on continuous reflection profiles. *Marine Geology* 94:271-299.

Mascle, J. & Mascle, G. 2012. Geological and Morpho-Tectonic Map of the Mediterranean Domain. Published by the Commission For The Geological Map Of The World (CGMW) and UNESCO. Available online at ccgm.org/en/maps/131-carte-geomorphologi que-et-tectonique-du-domaine-mediterraneen-978291 7310137.html

Mascle, J., Jongsma, D., Campredon, R. *et al.* 1982. The Hellenic margin from eastern Crete to Rhodes: Preliminary Results. *Tectonophysics* 86:133-147.

Mascle, J., Huguen, C., Benkhelil, J. *et al.* 1999. Images may show start of European–African plate collision. *Eos, Transactions American Geophysical Union* 80:421-428.

McKenzie, D. P. 1970. Plate tectonics of the Mediterranean region. *Nature* 226:239-243.

McKenzie, D. 1978. Active tectonics of the Alpine-Himalayan belt: the Aegean Sea and surrounding regions. *Geophysical Journal International* 55:217-254.

Meijer, P. T. & Wortel, M. J. R. 1997. Present-day dynamics of the Aegean region: A model analysis of the horizontal pattern of stress and deformation. *Tectonics* 16:879-895.

Meulenkamp, J. E., Wortel, M. J. R., van Wamel, W. A., Spakman, W. & Hoogerduyn Strating, E. 1988. On the Hellenic subduction zone and the geodynamical evolution of Crete since the late Middle Miocene. *Tectonophysics* 146:203-215.

Papanikolaou, D., Chronis, G. & Metaxas, C. 1994. Neotectonic structure of the Argolic Gulf. *Bulletin of the Geological Society of Greece* 30(2):305-316.

Pavlopoulos, K., Karkanas, P., Triantaphyllou, M., Karymbalis, E., Tsourou, T. & Palyvos, N. 2006. Paleoenvironmental evolution of the coastal plain of Marathon, Greece, during the Late Holocene: Depositional environment, climate, and sea level changes. *Journal of Coastal Research* 22:424-438.

Perissoratis, C. & Conispoliatis, N. 2003. The impacts of sea-level changes during latest Pleistocene and Holocene times on the morphology of the Ionian and Aegean seas (SE Alpine Europe). *Marine Geology* 196:145-156.

Perissoratis, C. & Mitropoulos, D. 1989. Late Quaternary Evolution of the northern Aegean Shelf. *Quaternary Research* 32:36-50.

Perissoratis, C. & van Andel, T. H. 1988. Late Pleistocene uncomformity in the Gulf of Kavala, northern Aegean, Greece. *Marine Geology* 81:53-61.

Perissoratis, C., Piper, D. J. W. & Lykousis, V. 1993. Late Quaternary sedimentation in the Gulf of Corinth: the effects of marine-lake fluctuations driven by eustatic sea level changes. pp. 693-744. *Special Publications of the National Technical University of Athens (dedicated to Prof. A. Panagos).*

Peyron, O., Guiot, J., Cheddadi, R. *et al.* 1998. Climatic reconstruction in Europe for 18,000 years BP from pollen data. *Quaternary Research* 49:183-196.

Pirazzoli, P. A. 1988. Sea-level changes and crustal movements in the Hellenic Arc (Greece). The contribution of archaeological and historical data. In Raban, A. (ed.) *Archaeology of Coastal Changes. BAR International Series* 404:157-184.

Pirazzoli, P. A., Thommeret, J., Thommeret, Y., Laborel, J. & Montaggioni, L. F. 1982. Crustal block movements from Holocene shorelines: Crete and Antikythira (Greece). *Tectonophysics* 86:27-43.

Pirazzoli, P. A., Montaggioni, L. F., Saliege, J. F., Segonzac, G., Thommeret, Y. & Vergnaud-Grazzini, C. 1989. Crustal block movements from Holocene shorelines: Rhodes Island (Greece). *Tectonophysics* 170:89-114.

Rohling, E. J. & Gieskes, W. W. C. 1989. Late Quaternary changes in Mediterranean intermediate water density and formation rate. *Paleoceanography* 4:531-545.

Rohling, E. J., Jorissen, F. J. & De Stigter, H. C. 1997. 200 year interruption of Holocene sapropel formation in the Adriatic Sea. *Journal of Micropalaeontology* 16:97-108.

Rohling, E. J., Mayewski, P. A., Abu-Zied, R. H., Casford, J. S. L. & Hayes, A. 2002. Holocene atmosphere–ocean interactions: records from Greenland and the Aegean Sea. *Climate Dynamics* 18:587-593.

Rossignol-Strick, M. 1983. African monsoons, an immediate climate response to orbital insolation. *Nature* 304:46-49.

Rossignol-Strick, M., Nesteroff, W., Olive, P. & Vergnaud-Grazzini, C. 1982. After the deluge: Mediterranean stagnation and sapropel formation. *Nature* 295:105-110.

Sakellariou, D. & Galanidou, N. 2015. Pleistocene submerged landscapes and Palaeolithic archaeology in the tectonically active Aegean region. In Harff, J., Bailey, G. & Lüth, F. (eds.) *Geology and Archaeology: Submerged Landscapes of the Continental Shelf.* Geological Society, London, Special Publications, 411.

Sakellariou, D., Lykousis, V., Alexandri, S. *et al.* 2007a. Faulting, seismic-stratigraphic architecture and Late Quaternary evolution of the Gulf of Alkyonides Basin — East Gulf of Corinth, Central Greece. *Basin Research* 19:273-295.

Sakellariou D., Rousakis G., Kaberi H. *et al.* 2007b. Tectono-sedimentary structure and Late Quaternary evolution of the north Evia gulf basin, central Greece: Preliminary results. *Bulletin of the Geological Society of Greece* 40(1):451-462.

Sampson, A. (ed.) 2006. *The Prehistory of the Aegean Basin.* Atrapos: Athens.

Shaw, B., Ambraseys, N. N., England, P. C. *et al.* 2008. Eastern Mediterranean tectonics and tsunami hazard inferred from the AD 365 earthquake. *Nature Geoscience* 1:268-276.

Soukissian T., Prospathopoulos, A., Hatzinaki, M. & Kabouridou, M. 2008a. Assessment of the wind and wave climate of the Hellenic Seas using 10-year hindcast results. *The Open Ocean Engineering Journal* 1:1-12.

Soukissian T., Prospathopoulos A., Korres G., Papadopoulos A., Hatzinaki M. & Kambouridou M. 2008b. A new wind and wave atlas of the Hellenic Seas. In *Proceedings of the ASME 27th International Conference on Offshore Mechanics and Arctic Engineering* (vol. 4). 15th – 20th June 2008, Estoril, Portugal. pp. 791-799.

Sperling, M., Schmiedl, G., Hemleben, C., Emeis, K. C., Erlenkeuser, H. & Grootes, P. M. 2003. Black Sea impact on the formation of eastern Mediterranean sapropel S1? Evidence from the Marmara Sea. *Palaeogeography, Palaeoclimatology, Palaeoecology* 190:9-21.

Stiros, S., Blackman, D. & Pirazzoli, P. 2010. Late Holocene seismic coastal uplift and subsidence in Rhodes Island, SE Aegean Arc, Greece: evidence from ancient shipsheds. In *IGCP 521 - INQUA 501 Sixth Plenary Meeting and Field Trip (Abstracts Volume)*. 27th September – 5th October 2010, Rhodes. p. 211.

Strasser, T. F., Panagopoulou, E., Runnels, C. N. *et al.* 2010. Stone Age seafaring in the Mediterranean: Evidence from the Plakias region for Lower Palaeolithic and Mesolithic habitation of Crete. *Hesperia* 79:145-190.

Taymaz, T., Jackson, J. & McKenzie, D. 1991. Active tectonics of the north and central Aegean Sea. *Geophysical Journal International* 106:433-490.

ten Veen, J. H. & Kleinspehn, K. L. 2002. Geodynamics along an increasingly curved convergent plate margin: Late Miocene-Pleistocene Rhodes, Greece. *Tectonics* 21(3).

Theodoulou T., 2008. The harbour network of ancient Lesvos. First step of an underwater approach. In Tzalas, H. (ed.) *Tropis X. Proceedings of the 10th International Symposium on Ship Construction in Antiquity, Hydra 2002*. 28th August – 2nd September 2008, Hydra, Greece (in press).

Thunell, R. C. 1979. Eastern Mediterranean Sea during the Last Glacial Maximum, an 18,000 years BP reconstruction. *Quaternary Research* 11:353-372.

Triantaphyllou, M. V., Antonarakou, A., Kouli, K. *et al.* 2009. Late Glacial-Holocene ecostratigraphy of the south-eastern Aegean Sea, based on plankton and pollen assemblages. *Geo-Marine Letters* 29:249-267.

Tzedakis, P. C., Lawson, I. T., Frogley, M. R., Hewitt, G. M. & Preece, R. C. 2002. Buffered tree population changes in a Quaternary refugium: Evolutionary implications. *Science* 297:2044-2047.

van Andel, T. H. & Lianos, N. 1984. High-resolution seismic reflection profiles for the reconstruction of post-glacial transgressive shorelines. An example from Greece. *Quaternary Research* 22:31-45.

van Hinsbergen, D. J. J. & Meulenkamp, J. E. 2006. Neogene supradetachment basin development on Crete (Greece) during exhumation of the South Aegean core complex. *Basin Research* 18:103-124.

Vergnaud-Grazzini, C., Devaux, M. & Znaidi, J. 1986. Stable isotope "anomalies" in Mediterranean Pleistocene records. *Marine Micropaleontology* 10:35-69.

Williams H. 2007. The harbors of ancient Lesbos. In Betancourt, P. P. Nelson, M. C. & Williams, H. (eds.) *Krinoi kai Limenes: Studies in honor of Joseph and Maria Shaw*. pp. 107-116. INSTAP Press: Pennsylvania.

Chapter 16

Geological and Geomorphological Factors and Marine Conditions of the Azov-Black Sea Basin and Coastal Characteristics as They Determine Prospecting for Seabed Prehistoric Sites on the Continental Shelf

Valentina Yanko-Hombach,[1,2,3] Evgeny Schnyukov,[4] Anatoly Pasynkov,[5] Valentin Sorokin,[6] Pavel Kuprin,[6] Nikolay Maslakov,[4] Irena Motnenko[3] and Olena Smyntyna[7]

[1] *Interdisciplinary Scientific and Educational Center of Geoarchaeology, Marine and Environmental Geology, Paleonthological Museum, Odessa I.I. Mechnikov National University, Odessa, Ukraine*
[2] *Department of Physical and Marine Geology, Odessa I.I. Mechnikov National University, Odessa, Ukraine*
[3] *Avalon Institute of Applied Science, Winnipeg, Canada*
[4] *Department of Marine Geology and Mineral Resources of the National Academy of Sciences of Ukraine, Kiev, Ukraine*
[5] *Department of Physical Geography and Geomorphology, Geographical faculty, Federal Crimean University named after V.I. Vernadsky, Simferopol, Crimea*
[6] *Department of Lithology and Marine Geology, M.V. Lomonosov Moscow State University, Moscow, Russia*
[7] *Department of Archaeology and Ethnology, Odessa I.I. Mechnikov National University, Odessa, Ukraine*

Submerged Landscapes of the European Continental Shelf: Quaternary Paleoenvironments, First Edition.
Edited by Nicholas C. Flemming, Jan Harff, Delminda Moura, Anthony Burgess and Geoffrey N. Bailey.
© 2017 John Wiley & Sons Ltd. Published 2017 by John Wiley & Sons Ltd.

Introduction

The Black Sea lies at the junction of three major cultural areas: Europe, Central Asia, and the Near East. The history of primary occupation and cultural exploitation of the Black Sea basin goes back to 1.89 million years ago (Dmanisi, Georgia), as is documented by numerous open-air archaeological sites, the frequency of which indicates a high concentration of human activity from the Lower Paleolithic to the Early Iron Age (Özdoğan 2007). Comprehensive study of these sites contributes to some of the most interesting debates in European prehistory, among which are the spread of anatomically modern humans, the transition to an agricultural economy, the repercussions of early urbanization across Eurasia, and others, which play a crucial role in enduring discussions about the impact of complex Near Eastern societies on European societies. Fluctuations in sea level and the commensurate shrinking and expansion of littoral areas had considerable impact on the settlement pattern of prehistoric societies of the Black Sea region, and submerged archaeological landscapes are highly possible (Stanko 2007).

The details and taphonomic conditions of the Black Sea are unusual. It is the world's largest anoxic (oxygen-free) marine basin. Its strongly stratified water column possesses (1) a thin, well-oxygenated surface layer (20–30 m) with low salinity and warm temperatures, (2) a low-oxygen (suboxic) transition layer (30–150 m), and (3) a thick bottom layer of colder, denser, and more saline water lacking oxygen but high in sulfides. Few organisms feed on organic material in its oxygen-starved depths; this highly favorable underwater environment preserves archaeological material, such as shipwrecks, creating the world's largest underwater museum (Ballard 2008).

The Black Sea is the easternmost of the seas of the Atlantic Ocean basin. If we take the ratio of the sea volume to the summary area of the cross-sections of all its straits (which is 0.04 km^2 for the Bosporus and 0.02 km^2 for the Kerch Strait) as a measure of isolation of a sea basin, then the Black Sea can be considered the most isolated sea of the Global Ocean (Zubov 1956). Its maximum length (along 42° 29' N lat) and width

are 1148 km and 611 km, respectively. Its surface area (excluding estuaries, such as the Dnieper-Bug *liman* — liman is a local term for ancient estuaries in the Black Sea and Sea of Azov) and its volume are about 416,790 km^2 and 535,430 km^3, respectively, and the maximum depth is 2212 m (Ivanov & Belokopytov 2013). The Sea of Azov is connected to the Black Sea via the Kerch Strait and has an area of 39,000 km^2 and a volume of 290 km^3. The maximum length, width, and depth of the Sea of Azov are 360 km, 180 km, and 14 m, respectively. On average, the level of the Black Sea is 7–11 cm lower than that of the Sea of Azov and 30 cm higher than that of the Sea of Marmara.

Climatic conditions of the Black Sea basin are determined by its geographical location and general atmospheric circulation. The northern part of the basin is located in the temperate climate zone while the southern part is in the subtropical climate belt. In January, mean air temperature above the central basin is 8°C; it decreases to 0–3°C with an absolute minimum of −30°C in the northwest. Mean air temperature in July is 22°C to 24°C, sometimes reaching 35°C. In the Sea of Azov, cold winters (down to −33°C) can be followed by dry and sultry summers (up to 40°C). Annual precipitation over the Sea of Azov and the western and northwestern Black Sea is 300 mm to 500 mm; in the southern Black Sea, it is 700 mm to 800 mm; and to the east, it can be 1800 mm to 2500 mm. Over most of the Black Sea, northerly and northeasterly winds prevail, whereas southerly and southeasterly ones are typical in the south-east. Average monthly wind speed is maximal in January-February (7–8 m/sec) and minimal in June-July (4–6 m/sec). Constant maximum wind speed is observed in the western Black Sea, and the minimum wind speed is in the south-east.

During the Late Quaternary, the Black Sea was repeatedly isolated from the ocean due to varying climatic conditions (Yanko-Hombach 2007; Yanko-Hombach *et al.* 2007; 2011a,b; 2012a,b; 2013a,b). Geographical location and periodic connection of the Black Sea either with the Mediterranean or Caspian seas (Fig. 16.1) predetermined specific hydrogeological regimes in the basin, making it an excellent paleoenvironmental amplifier and a sensitive recorder of climatic events.

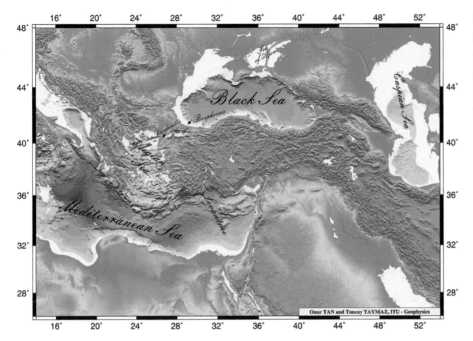

Fig. 16.1 Black Sea and connecting straits. Red = Manych Spillway which presumably connected the Caspian and Black seas at ~18 ka BP to 10 ka BP. Modified after O. Tan and T. Taymaz, Istanbul Technical University.

This chapter is focused on the description of the Late Pleistocene–Holocene environmental factors defining the Azov-Black Sea basin, and the identification of potential sample areas for seabed prehistoric site prospecting and landscape exploration on the Black Sea's northwestern continental shelf.

Earth Sciences Data

Main sources of data

Within the International Bathymetric Chart of the Mediterranean (IBCM) (www.ngdc.noaa.gov/mgg/ibcm/ibcm.html) there is a set of geological maps for the Black Sea (Sheet 5) with a scale of 1:1,000,000, published by the Charts Division of the Head Department of Navigation and Oceanography in Russia under the authority of the Intergovernmental Oceanographic Commission (IOC) of UNESCO. The set of maps with explanatory notes (Emelyanov *et al.* 2005) includes the basemap for a Geological/Geophysical series with Bouguer gravity anomalies [IBCM-G], seismicity [IBCM-S], unconsolidated bottom-surface sediments [IBCM-SED] (www.ngdc.noaa.gov/mgg/ibcm/ibcmsed.html), thickness of Plio-Quaternary sediments [IBCM-PQ] (www.ngdc.noaa.gov/mgg/ibcm/ibcmsedt.html), and magnetic anomalies [IBCM-M].

The IBCM-SED map of the Black Sea is satisfactory for both the deep-sea area and for the western, northwestern, and northern areas of the shelf. The Anatolian shelf is characterized by numerous gaps, as no detailed data were available to the compilers. For the Anatolian slope, the results of thorough lithological investigations were conducted by IORAS (Shirshov Institute of Oceanology Russian Academy of Sciences), its Siberian Branch (SBIORAS), the Institute of Geology, NASU (National Academy of Science of Ukraine) (Shnyukov 1981; 1982; 1983; 1984a,b; 1985; 1987), and M.V. Lomonosov Moscow State University (Kuprin & Sorokin 2007).

The Black Sea is surrounded by six countries: Ukraine, Russia, Georgia, Turkey, Romania and Bulgaria. Each country keeps the most recent earth science database for its own economic zone. Much of these data are not a public record.

The best source of most recent earth science data for the Ukrainian economic zone (Shchiptsov 1998) of the Black Sea and Sea of Azov is in a map series (each 594 × 841 mm) with a scale of 1:500,000. An overview of map locations (marked by "S") is shown in Fig. 16.2.

Fig. 16.2 Overview geological map (1:1,000,000) of pre-Quaternary sediments with identification of mineral resources of the Ukrainian economic zone of the Black Sea, Shnyukov (2011). The map includes seven-sheet set (each 594 × 841 mm) 1:500,000 scale. For the overview of their location see 'S' on the map. The map, geological profile A–A3 and tectonic chart Б1–Б8, utilize color schemes based on standards that are related to the timescale. FGDC (2006).

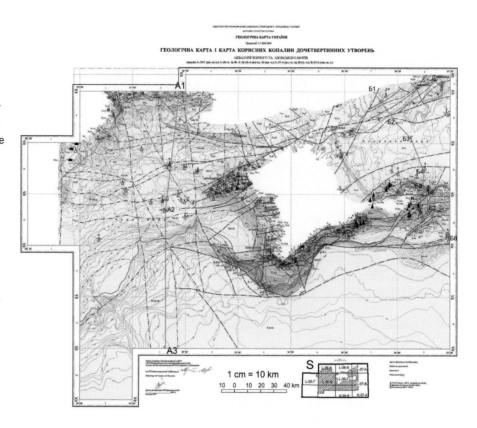

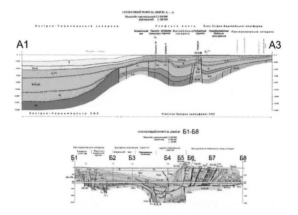

The map series includes: pre-Quaternary and Quaternary geology of the coast, shelf, continental slope, and abyssal plain, supplemented by information on the distribution of mineral resources, including lists of mineral fields and their projected areas; lithological and geomorphological maps of recent, Quaternary, and pre-Quaternary sediments; and charted neotectonic movements. The maps are compiled by the state-owned enterprise *Krymgeologiya* [Industrial Geological Association Crimean Geology] (crimeageology.ru) and Prichernomor SRGE [Black Sea Area State Regional Geologic Enterprise] (www.pgrgp.com.ua). The explanatory notes and legend were complied by Kakaranza *et al.* (2007).

The maps are based on geological survey results at different scales (from 1:1,000,000 to 1:50,000). The survey's primary goal was to identify promising areas for mineral exploration in the Azov-Black Sea basin. In total, 41 reports were written between 1972 and 2011 (e.g. Podoplelov *et al.* 1973-1975; Sibirchenko *et al.* 1983; Ivanov 1987; Avrametz *et al.* 2007). For the full list of reports see Pasynkov (2013). The survey area was divided into quadrats, within each of which a substantial number of gravity cores and/or vibrocores up to 5 m in length were recovered on grids of 2 × 2 km and 0.5 × 0.5 km. In addition, each quadrat also contained about 25 mapping boreholes (Fig. 16.3).

The survey was supplemented by geophysical data and studied by a multidisciplinary team (Shnyukov 1981; 1982; 1983; 1984a,b; 1985; 1987; Shnyukov *et al.* 2011 among many others), providing a vast amount of lithological, paleontological, micropaleontological and palynological data from the sediment column, and

radiocarbon dated in many cases (for dates see Appendix 1 and 2 in Balabanov 2007; Yanko-Hombach 2007a,b; Yanko-Hombach *et al.* 2013a,b).

Today, Prichernomor SRGE and Krymgeologiya are primary archives of geological and geophysical data. Other institutions possessing great amounts of earth science data include the Institute of Geological Sciences and Department of Marine Geology and Mineral Resources of the National Academy of Sciences of Ukraine; the Departments of Physical and Marine Geology, Engineering Geology and Hydrogeology, and Physical Geography, all at Odessa I.I. Mechnikov National University, Ukraine (onu.edu.ua/en/) (Konikov 2007; Shmuratko 2007; Shuisky 2007; Yanko-Hombach 2007a); the

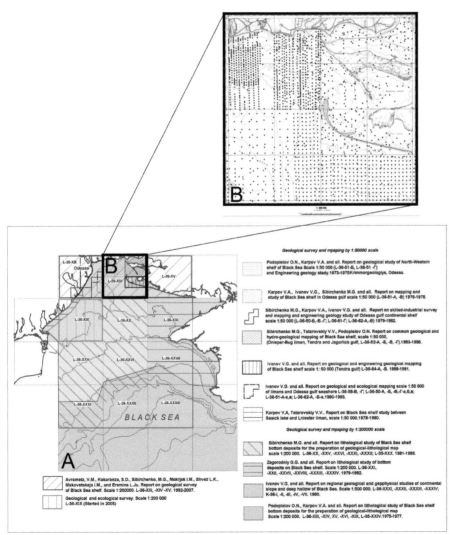

Fig. 16.3 Chart of medium- (1:200,000) to large- (1:50,000) scale geological survey of the Ukrainian part of the northwestern shelf: (A) shelf is divided into quadrats; (B) an example of the quadrat L-36-XIV where cores are shown by black dots. Citations provided in text.

Department of Lithology and Marine Geology, Moscow M.V. Lomonosov State University (Shcherbakov *et al.* 1978; Shcherbakov 1983; Kuprin & Sorokin 2007); the Southern Branch of the Shirshov Institute of Oceanology of the Russian Academy of Sciences (Shimkus 2005); GeoEcoMar, Romania (Panin & Popescu 2007); the Institute of Oceanology, Bulgaria (Malovitsky 1979), IFREMER (Lericolais *et al.* 2007a,b; 2011), and some others.

Evidence from around the world shows that wave energy and breaking surf are the principal forces that can destroy prehistoric sites, both during inundation and afterwards. Given the limited dimensions of the Black Sea basin, the wind fetch is limited, and the waves cannot be as large as in the open Mediterranean, and those affecting the Atlantic coast are larger still. The regions with the highest wave energy and the greatest threat of destruction of submerged archaeological sites (if any) include the southwestern part of the Black Sea and the Thracian shores of Turkey, especially the west side of Istanbul. The eastern part of the Black Sea is the least energetic in terms of wave power. In general, wave energy decreases along the coast from west to east. More information about annual wave energies (kWh/m) and breaking surf for different regions of the Black Sea can be found in Aydogan *et al.* (2013).

Geodynamic settings of the Black Sea

The Black Sea lies within the Anatolian sector of the Alpine-Himalayan orogenic system, located between the Eurasian plate to the north and the African-Arabian plates to the south (Fig. 16.4).

Global plate models (DeMets *et al.* 1990; 1994) and recent space geodetic measurements (Smith *et al.* 1994; Reilinger *et al.* 1997) indicate that in the surrounding region, the northward-moving African and Arabian plates are colliding with the Eurasian plate. From this collision, the Anatolian block is moving westward with a rotation pole located approximately to the north of the Sinai Peninsula (Tari *et al.* 2000). The northward movement of the Arabian plate and westward escape of the Anatolian block along the North and East Anatolian faults have been accompanied by several episodes of extension and

shortening since the Permian (Yilmaz 1997; Robertson *et al.* 2004) until now, as can be seen in seismic reflection data (McKenzie 1972; McClusky *et al.* 2000).

The crustal structure of the Black Sea reveals western and eastern deep basins (Fig. 16.4). The former is older than the latter; it was rifted with the dissection of a Late Jurassic to Early Cretaceous carbonate platform that had been established on the southern margin (Moesian Platform) of the northern supercontinent, Laurasia. The strong western Black Sea lithosphere apparently rifted for about 30 Myr, while rifting and spreading in the eastern Black Sea, characterized by a much weaker lithosphere, were completed within 8 Myr (Spadini *et al.* 1996).

The western Black Sea basin is underlain by oceanic to suboceanic crust and overlain by sediment cover with a thickness of about 19 km. The eastern Black Sea basin is underlain in the center by oceanic to suboceanic crust, surrounded by thinned continental crust approximately 10 km in thickness and overlain by sediment cover about 12 km thick (Nikishin *et al.* 2011). The basins are separated from each other by the Andrusov Ridge, formed from continental crust and overlain by sedimentary cover 5 km to 6 km thick (Tugolesov *et al.* 1985; Finetti *et al.* 1988; Beloussov & Volvovsky 1989; Robinson 1997; Nikishin *et al.* 2003). The compressional tectonic regime is still active in the eastern Black Sea region as can be seen from geological and geophysical evidence, including offshore seismic reflection profiles (Finetti *et al.* 1988), offshore morphology (Meisner *et al.* 1995), onshore geology and morphology (Okay & Sahinturk 1997 and references therein), and recent seismic activity (Neprochnov & Ross 1978; Barka & Reilinger 1997). The north–south motions of several mm/year are seen mostly in the eastern Black Sea, while in the Anatolian region, they are approximately 10 mm/year to 20 mm/year (Tari *et al.* 2000).

The isostatic anomaly and the isostatic residual anomaly along two transects in the western and eastern Black Sea basins, based on high-quality regional gravity data, show a clear discrepancy indicating a non-local isostatic compensation of the Black Sea lithosphere. The western basin appears to be in an overall upward state of flexure (undercompensated basin center and overcompensated basin flanks), and the eastern basin is

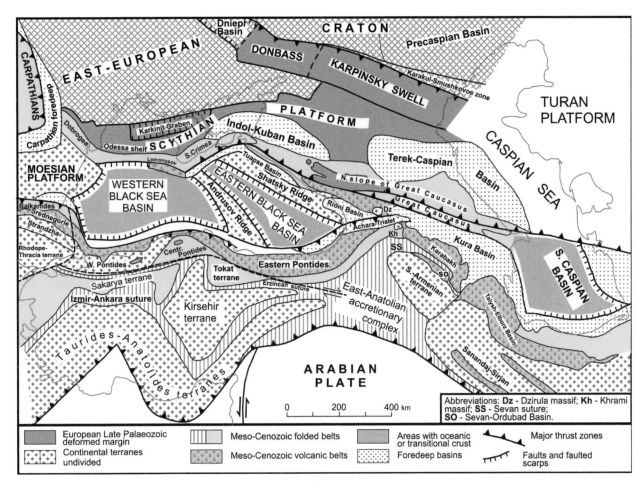

Fig. 16.4 Tectonic chart of the Azov–Black Sea region. Nikishin *et al.* (2011). Reproduced with permission from Turkish Journal from Earth Sciences.

in a downward state of flexure (overcompensated basin center and undercompensated basin flanks). This suggests that the level of necking involved in the deformation is deep in the west but shallow in the east (Spadini *et al.* 1996).

Vertical earth movements and rates of vertical coastal displacement

Geodetic and tide-gauge observations are probably the best sources of measured vertical movements. They provide information describing secular movements of the Earth's crust from north of the Scandinavian Peninsula to the Black Sea basin. The rate of recent crustal movements in this area is on average 2 mm/year to 4 mm/year. The

maximum rate of uplift (~10 mm/year) occurs in the central regions of Fennoscandia (the Baltic Shield), in the region of Krivoy Rog (the Ukrainian Shield), and in the near-Carpathian region (Artyushkov & Mescherikov 2013).

The map by Blagovolin and Pobedonostsev (1973) shows the crustal movements of the northern Black Sea region. Vertical earth movements are clearly visible in parts of the Black Sea bottom. However, studies of local tectonics in the Black Sea region encounter serious problems due to: (1) a limited number of reliably dated marine terrace deposits and insufficient use of oxygen isotope ($d^{18}O$) oceanic records in regional stratigraphic schemes; and (2) complicated tectonic pulses that, in addition to climatically-induced sea-level changes, contributed

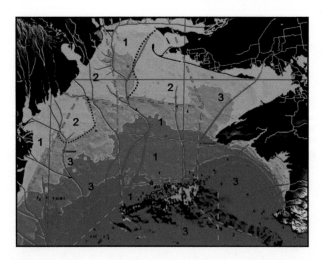

Fig. 16.5 Chart of neotectonic movements of the northwestern shelf. Compiled by Pasynkov (2013), based on data of Morgunov (1981). (1) Area of intensive subsidence; (2) Area of weak subsidence; (3) Area of relative uplift in areas (1) and (2). Dotted red line = faults.

significantly to repeated erosion of older marine terraces by younger transgressions resulting in numerous gaps in the geological record. Shelf morphology is intricately linked to major climatic fluctuations over the past million years (Shmuratko 2001; 2003). The number of steps in the staircase of marine terraces corresponds to the number of Ice Age cycles, each with an average duration of about 100,000 years. The lower marine terraces are useful for identifying shorelines over the course of the postglacial transgression (Shmuratko 2007).

The areas of intensive and weak subsidence, as well as relative uplifting of the continental crust in the northwestern Black Sea area calculated from thicknesses of the Pliocene–Quaternary sediments and instrumental observations, is shown in Fig. 16.5.

The rates of subsidence on the northwestern shelf vary between 1 and 5 mm/year to the north-west of Odessa; 0.8 and 0.9 mm/year in areas of Stanislav and Ochakov; and 0.5 mm/year in the area of Belgorod-Dniestrovsky. The shelf area adjacent to the western coast of Crimea also experienced an intense Pliocene-Quaternary subsidence that led to the development of accumulative landforms due to erosion of the beaches. The variety of elevations of Chaudian deposits (Fig. 16.6) between 35 m and 49 m below sea level (BSL) supports

an intense subsidence. Especially vulnerable are the tectonically-fractured sedimentary coasts of the Taman Peninsula and the estuaries in southern Ukraine. The Taman Peninsula shows subsidence rates of 0.4 m/kyr to 1.6 m/kyr (Brückner *et al.* 2010).

The greater part of the Caucasian shelf edge is marked by a scarp of presumably tectonic origin (Glebov & Shel'ting 2007, fig. 9). Depth at the shelf edge varies between 85 m and 115 m BSL due to different rates of tectonic movement.

Some areas of the continental slope experienced catastrophic subsidence due to seismotectonics from the movement of the Arabian plate, which generates strong earthquakes along the Caucasus and Anatolian coast. Supporting catastrophic subsidence is the wide range of elevations of the surfaces of Maikopian (Oligocene to Early Miocene) sediments on the upper continental slope. They vary between 1000 m and 1300 m BSL on the slope and 4000 m and 6000 m BSL in the deep basin.

The presence of lacustrine sediments around Karaburun on the southern shelf indicates an uplift of about 75 m along the northwestern coast of the Istanbul Peninsula during the Late Quaternary (Oktay *et al.* 2002).

Among other institutes, primary datasets on Black Sea geotectonics are held by the S.I. Subbotin Institute of Geophysics, National Academy of Sciences, Kiev, Ukraine (Starostenko *et al.* 2004), the Geological Faculty of Moscow State University (Nikishin *et al.* 2001; 2003; 2011), the Department of Marine Geology and Mineral Resources of the National Ukrainian Academy of Sciences (Shnyukov 1987), the Crimean Department of Ukrgeofizika (www.ukrgeofizika.kiev.ua/en/company.html), Kiev, Ukraine, *Chernomorneftegaz* [the State Joint Stock Company] (gas.crimea.ru), Simpheropol, Crimea, and the Institute of Marine Geology and Geoecology, GeoEcoMar, Romania (www.geoecomar.ro).

Solid geology of the Black Sea

Projects involving scientists from countries around the Black Sea as well as Canada, France, Germany, the USA and others, have been developed to study the solid geology of the Black Sea. It is difficult to indicate the main institutes holding extensive data on this subject.

Division	Section	Age, ka BP	MIS	Alps	Fedorov 1978 (based on mollusks)	Yanko 1990; Yanko-Hombach et al. 2006; Yanko-Hombach & Motnenko 2011; 2012 (based on foraminifera)	Time ka BP	BS isolation	MS-BS connection	CS overflow
Holocene	Upper	1.2, 1.6, 2.4, 2.8, 6.0, 6.4, 7.0	1	Holocene	Recent / Korsunian / Nymphaean / Phanagorian / Novochernomorian (New Black Sea)	Recent / Korsunian Regression / Nymphaean / Phanagorian Regression / Dzhemetinian / Egrissian Regression / Kalamitian / Pontian Regression	Present–9.4		+	
	Lower	7.4, 7.9, 8.1, 9.4			Drevnechernomorian (Old Black Sea)	Vityazevian / Kolkhidian Regression / Bugazian				
Neopleistocene	Upper	10	2?		Regression	Younger Dryas Regression	9.4–14.6	+		
		17	2	Würm	Upper Neoeuxinian	Upper Neoeuxinian	14.6–16.6			+
		27	3		Lower Neoeuxinian	Lower Neoeuxinian Regression				
		40			Tarkhankutian					
		56	4		Post-Karangatian Regression	Post-Karangatian Regression	16.6–73.7	+		
					Upper Karangatian	Upper Karangatian				
						Middle Karangatian ?				
			5	Riss-Würm	Lower Karangatian	Lower Karangatian	73.7–81.6		+	
							81.6–88.2*	+		+
							88.5–105.5		+	
							105.5–109.7*	+		+
		120					109.7–122.9		+	
			6	R₃	Regression		122.9–129.8	+		
							129.8–133.2			+
							188.9–198.7		+	
	Middle		7	R₂ / R₁	Uzunlarian	Upper Uzunlarian	198.7–202.1	+		
							202.1–215.0		+	
							215.0–216.2	+		
						Middle Uzunlarian	216.2–224.9			+
							224.9–228.4	+		
							228.4–235.6		+	
					Regression	Lower Uzunlarian	235.6–?	+		
					Paleouzunlarian	Post-Drevneeuxinian (Old Euxinian) Regression	?–287.0		+	
	Lower		8–11	Mindel-Riss	Drevneeuxinian (Old Euxinian)	Upper Old Euxinian	287.0–292.6	+		
						Lower Old Euxinian	292.6–307.0		+	
			12–14	Mindel	Regression	Post-Chaudian Regression	497.0–505.0		+	
				Günz-Mindel	Chaudian	Karadenizian	505.0–510.0	+		
							510.0–514.0			+
							514.0–516.0			
						Upper Chaudian	573.0–578.0		+	
							578.0–580.0			
			15–21			?	580.0–586.0			+
							586.0–592.0	+		
							592.0–616.0		+	
							616.0–619.0	+		
		780				Lower Chaudian	619.0–641.0			+
							641.0–651.0	+		
Eopleistocene	Upper		22–23	Günz	Gurian Chauda	Gurian				

Fig. 16.6 Stratigraphy of the Black Sea region (without proper scale) and its connections with the Mediterranean (MS) and the Caspian (CS). Light gray = MS–BS connection; dark gray = CS–BS connection; black = BS isolation; ** = suggested connection.

So far, there is no single concept of foundation time, origin, and sediment infill for the Black Sea, although discussions have been ongoing since the end of the nineteenth century.

The geological history of the Black Sea basin can be divided into three major stages: (1) a pre-Middle Cretaceous time, prior to the commencement of the Black Sea opening; (2) a Middle–Late Cretaceous and/or Paleocene–Eocene phase of complex basin opening; and (3) a post-rift development in Oligocene–Quaternary times, accompanied by orogenic activity in the surrounding areas, such as the Greater Caucasus, Balkans, and Pontides. Any reconstruction of the regional kinematics of the Black Sea area must also take into account the mechanism responsible for the profound subsidence of the western Black Sea basin (16 km) since the Early Tertiary (Robb et al. 1998).

Information on the age and lithology of stratigraphic units in the Black Sea region comes from drilling and onshore geologic mapping. Sediments as old as the Late Miocene have been sampled in the center of the Black Sea at three Deep Sea Drilling Project (DSDP) sites, and sediments as old as the Late Jurassic have been recovered from industrial wells at the margins of the Black Sea (Zonenshain & Le Pichon 1986; Banks et al. 1997).

In the eastern Black Sea (Fig. 16.7), seismic stratigraphic horizons have been tied to well control at the edges of the basin using 2D and 3D industry seismic

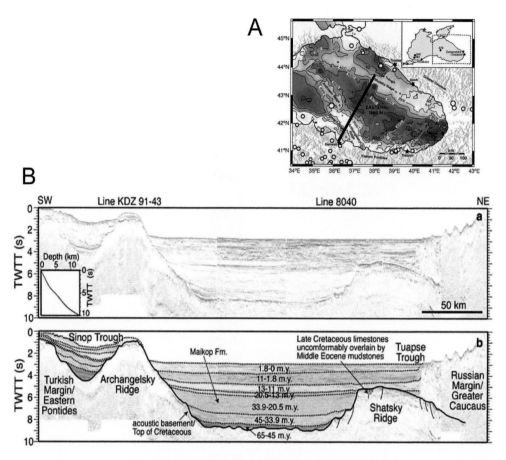

Fig. 16.7 (A) Map of the eastern Black Sea showing Cenozoic sediment thickness in the center of the basin and illuminated elevation from GEBCO (IOC IHO BODC (2003)) outside the basin. Sediment thickness is estimated from seismic reflection profiles. White circles indicate earthquakes with magnitudes >3 which occurred between August 2005 and August 2006, and are scaled by magnitude. The inset in the upper right-hand corner gives the location of the study area with respect to the entire Black Sea, and shows the locations of scientific and industrial bore holes around the Black Sea. From Shillington *et al.* (2008); (B) Seismic reflection profiles KDZ 91-43 and 8040 (Robinson 1997), which correspond with the subsidence analysis along the profile whose location is shown in (A). Inset (a) shows the time-depth relationship derived from stacking velocities, which was used to convert seismic stratigraphic horizons to depth. The black line indicates the time-depth function in the center of the basin averaged over 150 km; (b) Interpreted section showing the horizons and ages used for subsidence analysis and other major features. From Shillington *et al.* (2008). Reproduced with permission from Elsevier.

data sets in order to estimate the ages and lithologies of sedimentary units (Shillington *et al.* 2008).

The oldest geological formations are represented by Late Jurassic through Late Cretaceous sedimentary rocks recovered by drilling at the margins of the Black Sea, as well as time-correlative onshore units. They comprise a variety of lithologies, notably including shallow-water carbonate rocks with significant volcanic material dated to the following intervals: Early Paleocene to Middle

Eocene (65–45 Ma); Middle Eocene to the top of the Eocene (45–33.9 Ma); top of Eocene to Early Miocene (33.9–20.5 Ma); Early Miocene to Middle Miocene (base of Sarmatian) (20.5–13 Ma); Middle Miocene (base of Sarmatian) to Late Miocene (top of Sarmatian) (13–11 Ma); and Late Miocene (top of Sarmatian) to Pliocene (11–1.8 Ma). The latter are represented by sands and conglomerates of Pliocene age recovered by drilling onshore in Georgia and mapping in northeastern Turkey.

These units are typically non-marine and unlikely to be representative of lithologies in the basin center where chalk, siderite, clay and limestone were recovered by DSDP drilling. This interval also contains a thin unit comprising algal mats and peletal limestones, indicative of very shallow water depths. Although interpretations regarding the age and causes of these deposits are disputed, it appears that they correspond to a drop in sea level of over 2000 m, possibly related to the Messinian Salinity Crisis that affected the entire Mediterranean region (for references, see Shillington *et al.* 2008).

Pleistocene deposits have been recovered in many gravity cores and drilling; the former are also well studied in coastal outcrops. Lithologically, they are variable: clays, marls and occasional turbidites (for references, see Yanko-Hombach 2007a). High-resolution seismic and sonar images show primarily flat-lying, undisturbed sediments in the basin center, although the shallowest sediments often show disruption by gas (Ergün *et al.* 2002; Shnyukov & Yanko-Hombach 2009).

Bathymetry of the Black Sea

The most comprehensive source is the International Bathymetric Chart of the Mediterranean (IBCM), including the Black Sea (Sheet 5), which is also the basemap for a Geological/Geophysical series of maps (Fig. 16.8). Each of these chart series now has a published explanatory brochure available as a pdf file. All published maps, some 70 sheets at 1:1 million, and 7 sheets at 1:5 million, have now been scanned and are available digitally at different resolutions (Hall & Morelli 2007). The digitized marine contours are available on CD-ROM as part of the British Oceanographic Data Centre's

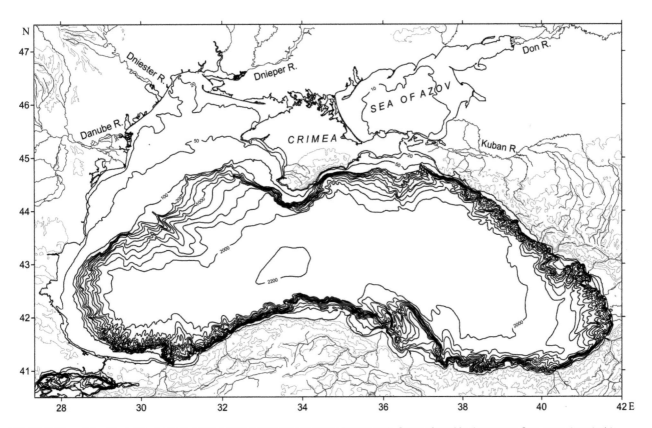

Fig. 16.8 Map of the Black Sea bottom relief, based on the International Bathymetric Chart of the Mediterranean Sea, reproduced with permission from IBCM (IBCM Project, Intergovernmental Oceanographic Commission, International Hydrographic Organization, and Head Department of Navigation and Oceanography, Russian Federation – Sheet 5).

Fig. 16.9 Digital high-resolution bathymetric map of the Black Sea and connecting straits. Modified after Bagrov *et al.* 2012. ∗ = names of underwater archaeological sites mentioned in the text.

GEBCO Digital Atlas (www.ngdc.noaa.gov/mgg/ibcm/ibcmbath.html).

Modern high-resolution bathymetric maps of the Black Sea are based on the IBCM map (e.g. Figs. 16.8 & 16.9).

Another source of bathymetric maps is the nautilus navigational charts of the Black Sea, divided into numerous sheets for each part of the basin, and published by the *Ukrmorcartographia* [State Hydrographic Service of Ukraine Branch] (www.charts.gov.ua/about_en.htm). Each sheet is a separate chart. Depending on their scale and purpose, these charts can provide a detailed representation of the coastline and natural features of the seabed (soundings and isobaths), which makes them very useful in searching for submerged archaeological sites. Ukrmorcartographia is a modern scientific-production complex, and the only specialized authority in Ukraine. Its main mission is to produce and issue official paper and electronic nautical charts, inland waterway charts, and guidance for navigation. Maps at a scale of 1:10,000 to 1:25,000 and larger cover mainly waters near ports, at a scale of 1:50,000 and 1:100,000 for coastal zones, and 1:200,000 and 1:500,000 for most of the economic zone of the Black Sea. Overview maps: 1:1,250,000 and smaller serve as navigational charts.

Modern coastline

The Black Sea coastline is slightly indented. Together with nearshore bars and spits, it is 4725 km in length, based on measurements from the 1:100,000 scale topographic map. The total length of the Sea of Azov shoreline is 1860 km. The best source for determination of the coastline length and area of the Black Sea using Geographic Information System (GIS) methods and Landsat 7 satellite images can be found in Stanchev *et al.* (2011).

The northwestern coastline is characterized by lowland shores indented by bays, gulfs, inlets, and estuaries, most of which do not penetrate far inland. The northeastern coast is mainly high mountains except for the lowland area between the Panagiia and Anaps'kyi capes on the coast of the Taman Peninsula. In the east, the vast Kolkhida lowland meets the sea. The southern coast is high and precipitous almost everywhere along the northern Tavr mountain system. It is interrupted by flat lowlands in the vicinity of the estuaries of the large rivers, such as the Kızılırmak, Yeşilırmak, Yenice, and Sakarya. The coast west of the Bosporus Strait is rather low and descends into lowlands. The Strandzhyts'ke and Medenrudnits'ke high coasts begin at the Rezovs'ka Estuary and reach Burgas Bay. Then, there is a low coast as far as the Varnensky Gulf. The capes in this area drop steeply toward the sea. From Cape Kaliakra to the great plain of the Danube Delta, the coast gradually descends. The coast noticeably rises to the east of Sevastopol Bay. The shores of the Kerch Peninsula are precipitous for almost the entire length, except for localities with barrier spits enclosing limans and lagoons. The shores of the Sea of Azov in the west, north, and east are mainly low, whereas in the south, they are steep. As in the Black Sea, the existence of limans and lagoons is a characteristic feature, especially in the northwestern part. The formation of large sand and shell spits, as well as barrier spits of different kinds is typical. They separate a number of shallow bays and limans from the sea. The largest of them, the Arabats'kaya Strelka, separates the shallow Syvash Gulf.

Among the largest bays and gulfs are: Odessa Bay and Karkinitsky Gulf on the northwestern coast; Novorossiysk Bay on the Caucasian coast; Sinop Bay, Bay of Samsun, Vaughn Bay, and the Gulf of İgneada on the Turkish coast; and Burgas and Varna bays on the Bulgarian coast. Except for the Crimean Peninsula, which projects well out into the sea, there are no other peninsulas and large capes. The largest island is Zmeiniy (Snake) Island (also known as Fidonisi) with a total area of 0.17 km^2 located not far from the mouth of the Danube River. The main geographical locations are shown in Fig. 16.10.

The main source of information on the Black Sea wetlands is the Directory of Azov-Black Sea Coastal Wetlands (Marushevsky 2003). The largest Black Sea wetlands are found in the coastal lowlands of Romania, Russia, and Ukraine, where the massive catchments of the rivers Danube, Dniester, Dnieper, Don, and Kuban support river deltas. In contrast, the Black Sea wetlands of Bulgaria, Turkey, and Georgia tend to be much smaller and have much more limited catchments, reflecting the mountainous hinterlands of these countries. There are about 94 wetlands with a total area of 2,486,372 ha. Thirty-five Black Sea coastal wetlands, totalling 1,953,576 ha, are of international importance and are designated as Ramsar sites. The best source on the Black Sea wetlands (Fig. 16.11)

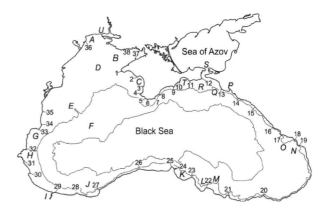

Fig. 16.10 A sketch of the Black Sea showing geographical locations: (A) Odessa Bay; (B) Karkinitsky Bay; (C) Kalamitsky Bay; (D) northwestern shelf; (E) Danube canyon; (F) paleo-Danube Fan; (G) Varna Bay; (H) Burgas Bay; (I) Bosporus; (J) Sakarya canyon; (K) Sinop Bay; (L) Samsun Bay; (M) Arkhangel'sky Ridge; (N) Kodori canyon; (O) Gudauta Bank; (P) Novorossiysk Bay; (Q) Kuban canyon; (R) Kerch–Taman shelf; (S) Kerch Peninsula; (T) Feodosia Bay; (U) Kinburn Strait; (1) Cape Tarkhankut; (2) Cape Yevpatoriysky; (3) Cape Lukull; (4) Cape Chersonesus; (5) Cape Aiya; (6) Cape Sarych; (7) Cape Ai-Todor; (8) Cape Ayu-Dag; (9) Cape Meganom; (10) Cape Kiik-Atloma; (11) Cape Chauda; (12) Cape Takil; (13) Cape Utrish; (14) Cape Idokopas; (15) Cape Kodosh; (16) Cape Konstantinovsky; (17) Cape Pitsunda; (18) Cape Sukhumi; (19) Cape Iskuria; (20) Cape Ishikli; (21) Cape Cham; (22) Cape Dzhyva; (23) Cape Bafra; (24) Cape Boztepe; (25) Cape Indzheburun; (26) Cape Kerempe; (27) Cape Oludzhe; (28) Cape Pazarbashi; (29) Cape Anadolu; (30) Cape Koru; (31) Cape Maslen Nos; (32) Cape Emine; (33) Cape Kaliakra; (34) Cape Shabla; (35) Cape Tuzla; (36) Cape Bolshoi Fontan; (37) Cape Peschaniy (isobaths of 200 m and 2000 m are plotted in thin lines); (38) Dzharylgachsky Gulf. Modified after Ivanov and Belokopytov (2013) with permission from Hydrophysical Institute NAS of Ukraine.

can be found at blacksearegion.wetlands.org/Portals/9/ BlackSea%20map.jpg.

The vast majority of limans and lagoons are present along the northwestern coastal zone (Fig. 16.12).

The Black Sea limans are unique geographical and geological features. Study of their depositional structure can aid in establishing the most recent stages of the basin's geological history. The limans are unique because:

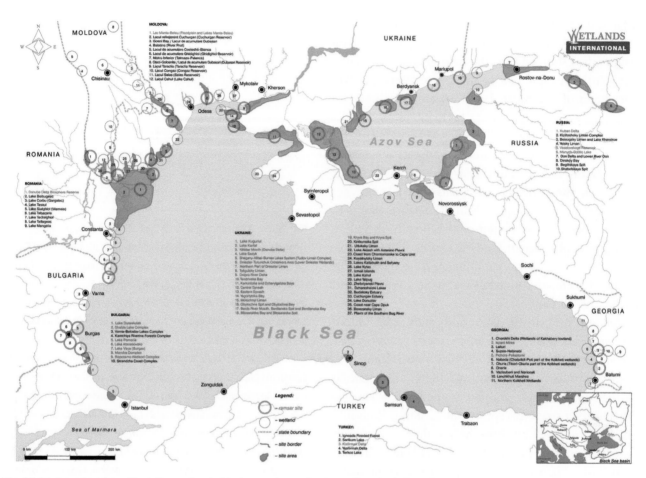

Fig. 16.11 Important Azov–Black Sea wetlands (blacksearegion.wetlands.org/Portals/9/BlackSea map.jpg).

1) the thickness of Neoeuxinian and Holocene deposits greatly exceeds their thickness on the shelf, continental slope, and the deep sea; 2) the lithological structure of deposits is diverse, and based on lithological and faunistic composition the deposits are clearly stratified; 3) the major faunal complexes range from freshwater to polyhaline; and 4) limans belong to the first belt of avalanche sedimentation and represent a zone of geochemical barrier (Konikov & Pedan 2006). There is an extensive bibliography on aspects of Black Sea liman geology (Shnyukov 1984b; Zaitsev *et al.* 2006).

There are two major deltas in the northwestern Black Sea: the Danube and Dnieper deltas (Fig. 16.13). The former is larger and divided into Romanian and Ukrainian parts, 4400 km² and 1240 km², respectively. A large data set on the Romanian part of the Danube Delta is held by GeoEcoMar, Bucharest, Romania (www.geoecomar.ro).

For the Ukrainian part of the delta, a large database is archived by the Ukrainian Scientific Center of Ecology of the Sea, Odessa, Ukraine (UkrSCES) (www.sea.gov.ua) (Berlinsky *et al.* 2006).

The Danube Delta is located between 44°24' and 45°40' N, and 28°14' and 29°46' E. It is marshy, contains a dense network of branches and lakes, and has three main arms: the St. George (I and II), the Sulina, and the Chilia or Kilia (Fig. 16.13a).

The delta is divided into two parts: ancient riverine and young marine. It includes a series of lakes, e.g. the Yalpug, Kugurluy, Katlabukh, Kitay, as well as a large area between the St. George I and Chilia branches, the Dranov Peninsula, and the Razelm-Sinoe lake-lagoon complex. All these areas are closely connected hydrologically, and in this sense they represent a single territory with similar climate, lithology, soil, and vegetation characteristics.

Fig. 16.12 Modified from Landsat satellite photo of the Black Sea's northwestern coast and its limans: (1) Sasyk; (2) Dzhantsheisky; (3) Maliy Sasyk; (4) Shagany; (5) Karachaus; (6) Alibey; (7) Khadzhider; (8) Kurudiol; (9) Burnas; (10) Budaksky; (11) Dniestrovsky; (12) Sukhoy; (13) Khadzhibeysky; (14) Kuyalnitsky; (15) Dofinovsky; (16) Grigor'evsky; (17) Tiligulsky; (18) Tuzly; (19) Berezansky; (20) Bugsky; (21) Dnieprovsky.

Source: zulu.ssc.nasa.gov/mrsid/mrsid.pl.

Sedimentological features of the delta are determined by its geographic position. Bounded on the north and south by the Budzhak and Dobrudza plateaux, it has a limited amount of sediment coming from those areas. Runoff from the plateau slopes into the lakes comes mainly during snowmelt or heavy rains via small rivers. Total annual runoff from these small rivers does not exceed 2000 million m^3, which is less than 0.1% of the average annual flow in the upper reaches of the Danube Delta. The evolution of the Danube Delta during the Holocene and corresponding coastline changes are shown in Fig. 16.13c.

The Dniester Delta is much smaller than the Danube Delta. It formed at the outlet of the river into the Dnieper-Bugsky liman (Fig. 16.13b). Sediment discharged by the river accumulates in the inner part of the delta forming spits, bars, and alluvial islands. The width of the floodplain terrace in the lower part between western Cairo and Kherson ranges from 2.5–3 km to 5 km. It widens below Kherson, at the outlet into the Dnieper Estuary, where the river splits into numerous arms forming a modern delta. Filling with sediment to varying degrees, the delta extends seaward. Deltaic sediments are represented by an alternation of sub-aerial and subaqueous facies. In the mouth of delta arms the crescent-shaped sand bars were formed. They collapse during floods with development of new bars seaward.

Morphometry and bottom topography of the Black Sea are important oceanographic features because they determine major characteristics of thermohaline structure and water circulation. Configuration of the shoreline, width of the shelf and continental slope, as well as the shape of the bottom profile, the presence of valleys, canyons, ridges, and depressions, influence the distribution of water masses, the direction and speed of currents, and the position and intensity of the topogenic eddies and coastal upwelling.

Morphometric characteristics and topography of the Black Sea bottom are described in detail in the geological literature and regional oceanographic research works (Arkhangel'sky & Strakhov 1938; Goncharov *et al.* 1978; Shnyukov 1987; Simonov & Altman 1991; Ignatov 2008; Eremeev & Symonenko 2009).

The Black Sea exhibits the standard oceanic provinces of continental shelf, slope, and abyssal plain (Fig. 16.14).

The continental shelf is part of the submerged coastal land and covers 25% of the sea area; isobath 200 m is commonly taken as the shelf boundary of the Global Ocean. The northwestern shelf extends outward 220 km, occupies 16% of the sea area (68,390 km^2) and 0.7% of the water volume (3555 km^3) between capes Chersonesus and Kaliakra (nos. 4 and 33 in Fig. 16.10). In the flattened and gently sloping part of the shelf adjacent to the shore, depths are 30 m to 40 m, and the bottom slope is 1° to 2°. Its steepness increases toward the shelf break to 10° to 12°. Against the flat plain of the shelf, several large, shallow paleoriver valleys are visible in Fig. 16.14, separated by low underwater hills (Ivanov & Belokopytov 2013). The bottom relief is largely smooth due to sediment discharge and distribution provided by

Fig. 16.13 (A) Landsat satellite photo of the Danube Delta. Photo generated using data from zulu.ssc.nasa.gov/mrsid/ mrsid.pl; (B) A Landsat satellite photo of the Dnieper Delta in the area of its confluence into the Dnieper–Bugsky liman; (C) Evolution of the Danube Delta during the Holocene and corresponding coastline changes. After Panin & Jipa (1997). (1) Initial formation of the Letea–Caraorman spit at 11.7 ka BP to 7.5 ka BP; (2) St. George I Delta, 9 ka BP to 7.2 ka BP; (3) Sulina Delta, 7.2 ka BP to 2 ka BP; (4) coastline at 100 AD; (5) St. George II and Chilia deltas, ~2.8 ka BP to the present; and (6) Cosna–Sinoie Delta, 3.5 ka BP to 1.5 ka BP.

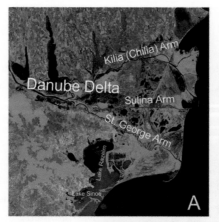

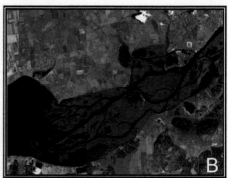

major lowland European rivers, such as the Danube, Dnieper, Dniester, and the Southern Bug, that together discharge 56.8 million tonnes of sediments annually (Panin & Jipa 1997). There are no known expressions of active tectonic movements that would influence the ancient shoreline positions and deposition of sediments in any appreciable way.

The other, less extensive shelf areas of the Black Sea include: the coastal zone of Bulgaria and western Turkey from Cape Kaliakra to the city of Ereğli (shelf width up to

50 km); the Kerch-Taman shelf (shelf width up to 50 km); the central Anatolian coast from Cape Kerempe to the city of Giresun (shelf width up to 35 km); the southern Crimean coast between capes Chersonesus and Ai-Todor (shelf width up to 30 km); and the Gudauta Bank in the vicinity of Ochamchira town (shelf width up 20 km) (Ivanov & Belokopytov 2013).

Narrow shelves with widths of several kilometers are located along the Caucasian and Anatolian coasts, as well as along the southern Crimean coast from Yalta to Cape

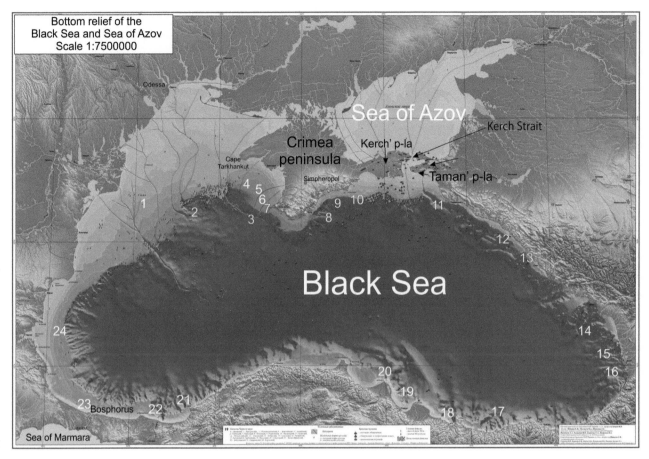

Fig. 16.14 Digital high-resolution map of relief. Modified after Bagrov *et al.* 2012: 1–24 canyons: (1) Dunaisky; (2) Dniestrovsky; (3) Peleokalanchaksky; (4) Donuslavsky; (5) Al'minsky; (6) Kachinsy; (7) Forossky; (8) Yaltinsky; (9) Khapkhal'sky; (10) Meganomsky; (11) Kubansky; (12) Novorossiysky; (13) Tuapsinsky; (14) Sochinsky; (15) Sukhumsky; (16) Rioniisky; (17) Batumsky; (18) Trabzonsky; (19) Ordossky; (20) Samsunsky; (21) Kysyl-Irmansky; (22) Sakar'iaksky; (23) Bosphorsky; (24) Burgazsky. Red, blue, and yellow triangles = mud volcanoes reliably detected, discovered based on geophysical data, and expected, respectively. Blue lines = river paleovalleys. Yellow cylinders = gas seeps. The map was compiled based on results of digitizing bathymetric maps of different scales (1:50,000, 1:100,000, 1:200,000, 1:500,000, 1:1,125,000) produced by Gosgidrografiya USSR. The Crimean shelf and slope, as well as the southern part of the northwestern shelf and slope, were adjusted based on the results of our own sonar and hydroacoustic soundings obtained during marine expeditions on R/Vs *Kiev*, *Professor Vodyanitsky*, *Mikhail Lomonosov*, *Ichthyander*, and *Vernadsky*. Reproduced with permission from Research Center of Sustainable Development, Ukraine.

Meganom. Their bottom slope is considerably steeper compared to broader shelves, ranging from 5–6° to 30°. The shelf break lies at depths from 100 m to 200 m, the slope is 1° to 2°. The depth of shelf break is close to 100 m while in areas with a broader shelf it can deepen to over 200 m. Most gas seepages occur along the shelf breaks in areas of fault scarps on the sea floor, shallow sub-surface faults, small grabens, and conical depressions or domes of

gently sloping anticlinal rises (Fig. 16.14) (Shnyukov & Yanko-Hombach 2009).

The predominantly flat bottom of the Sea of Azov descends gradually to the depression at its center. At the bottom, there are a few positive relief forms, the largest of them being the Pischana Bank.

The continental slope descends down to 1600 m to 1900 m of water depth with a considerable gradient from

11° to 13°, sometimes reaching 38° in the regions along the southern Crimean and Turkish coasts. The surface of the continental slope is complicated with blocks of the Earth's crust that often give it a graduated profile, and most of it reveals underwater canyons of different origin. They can begin in the coastal zone at a depth of 10 m to 15 m and reach a lower depth of 1600 m. The canyons are the most important route for the transfer of sedimentary material from the coast to the abyssal depression of the Black Sea (Fig. 16.14). In the deepest part of the canyons, at depths of 1600 m to 1900 m, the transferred sedimentary material forms big cones. Individual cones can coalesce to form the continental sub-slope. Thus, the morphogenesis of the slope is directly linked to selective erosion and denudation of rocks with different physical and mechanical properties. Erosive and denudation activities in the canyons caused the emergence of huge underwater amphitheaters forming deepwater fans and plumes of terrigenous sediments on the footslope.

The abyssal plain is bounded by isobath ~2000 m and occupies about 35% of the total sea area. It is a flat accumulative plain with a slight slope to the south. The bottom of the abyssal basin is characterized by hilly relief; slope angles vary from 0° to 1°. According to echo-sounding surveys, significantly large features of submarine relief are absent (Fig. 16.14). Deposits covering the abyssal plain form eleven material-genetic types. Six types are shallow, and five are deep water. Between all types of deposits, there is a continuous transfer conditioned by gradual change in their grain-size and material composition. The mean rate of accumulation at the bottom of the central abyssal depression is 30 mm/kyr to 40 mm/kyr.

There are about seventy mud volcanoes located mainly in the northern part of the western Black Sea, Sorokin trough, Tuapsinskaya trough, Shatskiy arch, and Kerch downfold (the area south of the Kerch Peninsula) (Fig. 16.14; Shnyukov *et al.* 2010, 2013).

Today, the Bosporus Strait (Fig. 16.14) is the only passage for exchange of water and organisms between the Black Sea and Sea of Marmara. This zigzagging strait is about 35 km in length, 0.7 km to 3.5 km in width, and 35.8 m deep on average, with a few elongate potholes

(about 110 m in depth each) on the bottom. It possesses two sills, one in the north at a water depth of 59 m and one in the south at a water depth of 34 m, each located about 3 km from the corresponding entrance to the strait. The two directions of waterflow within the strait overlap each other: the northward underflow (inflow) from the Sea of Marmara has an average salinity of 38 psu and a velocity of 5 cm/sec to 15 cm/sec, and the southward overflow (outflow) from the Black Sea has an average salinity of 18 psu and a velocity of 10 cm/sec to 30 cm/sec. Due to the sills, the interface between the two flow directions rises from −50 m at the northern end to −20 m at the southern end. The underflow is initiated by the difference in water density between the Black Sea and the Sea of Marmara; the pressure gradient pushes against the Black Sea and acts to power the underflow. The outflow is initiated by two main factors: (1) the 30 cm elevation of the Black Sea surface above that of the Sea of Marmara, which, in turn, is 5 cm to 27 cm above the level of the northern Aegean Sea, and (2) the positive balance of the Black Sea, where precipitation (575 km^3/year) exceeds evaporation (350 km^3/year), producing a discharge of about 600 km^3 of brackish water annually (Yanko-Hombach *et al.* 2007).

The Kerch Strait connects the Black Sea with the Sea of Azov (Fig. 16.14) and is 45 km long, 4.5 km wide, and up to 6 m deep. The shallowness of the strait results in reduced water exchange between the two basins, which is five to ten times smaller than that of the Bosporus.

Coastal geomorpho-dynamics, erosion, accumulation

Coastal zones of the Black Sea and Sea of Azov can be considered in two parts: surface and underwater. The former is referred to as the coastal zone, the latter as the submarine sea slope. Both parts are closely connected in their origin and develop simultaneously under the influence of common types and sources of energy. The main processes of coastal zone development are mor-phodynamic (topographic features) and lithodynamic (deposits).

Coastal geomorpho-dynamics are defined largely by the wave and current regimes that give the coastal zone a variety of borders and widths. The width of the coastal zone depends upon the gradient of the coast and continental slope. The form of coastal cross-section can be highly diversified: first, the shape of the profile depends upon the physical and mechanical properties of the rock and deposits, as well as their resistance to abrasion. In general, the coastline is significantly wider in plains areas and narrower in mountainous ones.

The coastal zone in the Black Sea and Sea of Azov basins is prone to constant change along its entire length. In some places, rough seas destroy the coast, while in other areas, they smooth the coast, creating new land areas and changing continental terrace topography. The mechanical energy of sea waves and wave currents encourages a variety of morphodynamic and lithodynamic processes leading to various topographic features, as well as the amount, composition, and mobility of the deposits and diversity of coastal zone components. In the Azov-Black Sea coastal zone, abrasion generally prevails over accumulation. Capes and coastal promontories that project into the sea undergo significant destruction; rough seas and landswells constantly seek to level the coastline. General coast erosion and degradation is slower in estuary areas with intense solid suspended sediment inflow; in the estuaries of big rivers (Danube, Dnipro, Psou, Bzyb, Inguri, Rioni, Chorokh), drifts and sand spits can grow up to 10 m during the river flood season. The formation of coastal landforms depends on the strength of the rocks. Igneous and some particularly strong metamorphic rocks are only slightly susceptible to the damaging effects of waves. Coastal slopes composed of such rocks are practically unaffected by waves, and thicknesses of generated sediment are negligible. In contrast, weakly cemented sedimentary rocks clearly show traces of wave and chemical weathering by sea water. Here, on the coast-sea edge, niches, caves, cracks, and 'stone boilers' are formed. Upon further exposure to waves, the overhanging niches collapse leading to cliff formation or abrasion ledges. Cliffs exposing non-cohesive sediments of loam, clay and sand are destroyed very quickly, and retreat at a rate of several meters per year. Different morpho-dynamic types of coastline are shown in Fig. 16.15.

The best source and digital archive for morphostructural zoning of the shelf is presently at the Department of Marine Geology and Mineral Resources of the National Academy of Sciences of Ukraine, as well as the Department of Physical Geography and Department of Engineering Geology and Hydrogeology at Odessa I.I. Mechnikov National University. Detailed characteristics of leading geological processes in the Black Sea coastal zone of Ukraine can be found in Pasynkov (2013).

Landscape regions of the northwestern Black Sea shelf

Based on distribution of statistical parameters describing water depth, Holocene sediment thickness, and percentage of silt and clay within the sediments, the following recent landscape areas have been delineated (Larchenkov & Kadurin 2007a,b; 2011): erosional coastal offshore slope; Danube pro-delta and paleovalley; depressions of river paleovalleys; terraces of the inner shelf; central shelf plain; and the outer shelf plain (Fig. 16.16).

Similar methods were used to reconstruct paleolandscapes for five crucial time slices (Fig. 16.17, modified after Yanko-Hombach et al. 2011a). These landscape regions, helpful in the search for submerged archaeological sites, were reconstructed in detail for the northwestern shelf (see Fig. 16.22). For the method used, refer to Larchenkov and Kadurin (2011).

Pleistocene and Holocene stratigraphy, sediment thickness on the continental shelf, modern sedimentation rates, data sources

Pleistocene sediments have a rare and patchy distribution (except for uppermost Neoeuxinian sediments described below) on the continental shelf, apparently due to erosion incurred over numerous transgression-regression cycles in which sediment accumulation prevails during the transgressive phase and erosion during the regressive, resulting in gaps in the sedimentary record. This patchy distribution prevents a reliable calculation of Pleistocene sediment thickness and rates of sedimentation. Pleistocene stratigraphy, sea levels, salinities, coastline migrations, climate characteristics, as well as the time and direction of water intrusions into the Black Sea from

Fig. 16.15 (A) abrasional coast of Cape Tarkhankut; (B) abrasional-avalanched coast of the western Crimea; (C) landslide coast of the Kerch Peninsula; (D) abrasional-creek coast of Cape Fiolent, Crimea; (E) abrasional-landslide southern coast of Crimea with shore protection structures; (F) the 'Eltigen' neostratotype exposed on abrasional-avalanched coast of tectonically elevated terrace, Kerch Peninsula, Crimea. From Yanko-Hombach *et al.* (2012b). This neostratotype contains the most complete and well-preserved marine sequence of the Karangatian transgression.

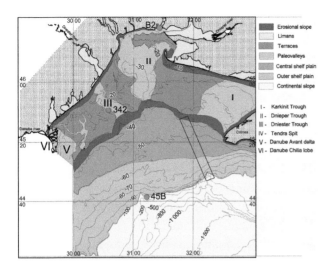

Fig. 16.16 Northwestern Black Sea area showing present bathymetry (in m BSL), paleovalleys I–III, locations of cores in Yanko-Hombach *et al.* (2013b), shelf section (in square) studied by Ryan *et al.* (1997), and other important geomorphological features marked by I–VI. Red circles mark location of key reference cores with ^{14}C ages studied by multidisciplinary methods described in Yanko-Hombach *et al.* (2013b). Reconstruction made on results of 1:200,000 marine geological survey outlined in Fig. 16.3. Reproduced with permission from Elsevier.

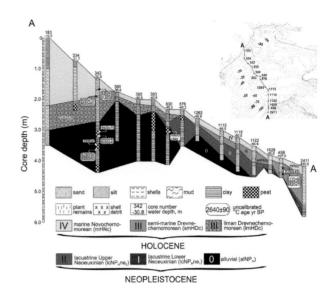

Fig. 16.17 Paleo-Dniester valley cross-shelf profile, showing sediment lithology and genetic type, depositional setting and stratigraphic ages. Inset shows shelf bathymetry, locations of transect cores and present-day limans at the seaward end of the Dnieper, Bug and Dniester rivers. Shell detrit = detrital shell fragments. Profile is made on results of 1:200,000 marine geological survey outlined in Fig. 16.3. Yanko-Hombach *et al.* (2013b). Reproduced with permission from Elsevier.

the Mediterranean and Caspian seas, are largely based on the study of outcrops exposed on tectonically elevated terraces of the Kerch and Taman peninsulas, Crimea, and Caucasus (Yanko 1990; Yanko-Hombach & Motnenko 2011; 2012).

The Russian subdivision of the Quaternary system includes the Eopleistocene (1.8–0.78 Ma), the Neopleistocene (0.78–0.01 Ma), and the Holocene (0.01–0.0 Ma) (Zhamoida 2004). The boundary between the Eopleistocene and Neopleistocene coincides with the Brunhes-Matuyama reversal (i.e. 780 kyr). This boundary is readily traced in the Black Sea at the bottom of the Chaudian horizon (Fig. 16.6).

The classical Quaternary stratification in the Black Sea region includes (from oldest to youngest) the Eopleistocene (Gurian), Neopleistocene (Chaudian, Drevneeuxinian [Old Euxinian], Uzunlarian, Karangatian, Neoeuxinian), and the Holocene (Drevnechernomorian and Novochernomorian) beds (Fig. 16.6). The Chaudian, Old Euxinian, and Karangatian beds were first proposed

by Andrusov (1918). They are exposed in stratotypes (except for the Old Euxinian which does not have a stratotype) in tectonically elevated terraces. These classic names were incorporated into the stratigraphic scale of Arkhangel'sky and Strakhov (1938), who also described the Uzunlarian horizon and identified its stratotype. Previously designated Euxinian sediments (Andrusov 1918) bearing brackish, Caspian-type fauna were divided by Arkhangel'sky and Strakhov (1938) into Drevneevksinian [Old Euxinian] and Novoevksinian [Neoeuxinian] beds. The former presently lies above sea level on tectonically elevated terraces, occasionally on the sea bottom, and it contains *Didacna pontocaspia*. The latter is distributed below sea level only and contains the mollusks *Dreissena* and *Monodacna*. The initial stratigraphic framework was later improved based on mollusks (Nevesskaya 1965), foraminifera and ostracoda (Yanko 1990; Yanko & Gramova 1990).

The absolute age of the sediments in the stratotypes varies depending on the method used, e.g., 230 U/Th, thermoluminescence, etc. For example, based on a few dozen thermoluminescence dates and 12 magnetic-polarity datum planes, Zubakov (1988) has assigned a numerical age to the horizons as follows: Chaudian = 1.1 Ma–600 ka; Uzunlarian = 580–300 ka; Karangatian = 300–50 ka. Tchepalyga (1984), Yanko (1990), Yanko-Hombach and Motnenko (2011; 2012), Yanko-Hombach *et al.* (2013b) suggest slightly different numerical ages for the Eo- and Neopleistocene geological sequences in the Black Sea.

Ecostratigraphy of the Pleistocene and Holocene sediments is based on taxonomy and spatial distribution of bivalve molluscs (Nevesskaya 1965), gastropods (Il'ina 1966), benthic foraminifera and ostracoda (Yanko 1990; Yanko & Gramova 1990; Yanko-Hombach 2007a,b) as well as palynological data supplemented by ^{14}C datings (e.g. Appendix 1 and 2 in Yanko-Hombach *et al.* 2007 and Balabanov 2007, respectively).

The paleosalinity in this chapter is described as follows: freshwater (<0.5 psu), semi-fresh (0.5–5 psu), brackish (>5–12 psu), semi-marine (>12–18 psu), and marine (>18–26 psu). The use of 'psu' (Practical Salinity Units) instead of former ‰ is explained in Yanko-Hombach *et al.* (2013a).

The Gurian (or Gurian Chauda) beds correspond to Marine Isotope Stage (MIS) 23–22 (Fig. 16.6). Its stratotype has a thickness exceeding 1000 m located in Tsvermaghala Mountain, Georgia. Except for the Bulgarian outer shelf (Malovitsky 1979), no Gurian deposits were discovered in other places of the Black Sea bottom. The Gurian basin was semi-freshwater with sea level somewhere around 100 m to 150 m BSL.

The Chaudian beds (780–500 ka) correspond to MIS 21–15 (Fig. 16.6). This stratotype was described by Andrusov (1889) on Cape Chauda, Kerch Peninsula. In the sea, the most complete geological sections were recovered in the southwestern part of the basin by DSDP drillhole 389, Bulgarian (Kuprin *et al.* 1980), Ukrainian shelf and Caucasian continental slope (Yanko-Hombach 2007a). On the western Georgian coast the thickness of Chaudian deposits is about 1000 m (Kitovani 1971; Kitovani *et al.* 1982). On the sea bottom, the thickness varies between 7.7 m (Galithin uplift, Karkinitsky Bay, northwestern shelf) and 380 m (DSDP drillhole 389).

On the Bulgarian outer shelf, the thickness is 13.7 m (Kuprin *et al.* 1984), while on the Caucasian shelf, it reaches 17 m (Yanko 1989). The level of the Chaudian basin varies according to different authors: either below the present one (Fedorov 1978; Yanina 2012); similar to the present one (Inozemtsev 2013); or above the present one at 15 m to 18 m above sea level (ASL) (Fedorov 1997) or 20 m ASL (Chepalyga 1997; Svitoch *et al.* 1998). The hydrological regime was as follows: at the beginning, it was a basin connected to the Caspian Sea with salinity about 5 psu to 7 psu; it increased to 7 psu to 8 psu in the Late Chaudian; and reached 17 psu to 18 psu during the Karadenizian with the first Pleistocene intrusion of Mediterranean water into the Chaudian basin (Yanko 1989; 1990; Yanko-Hombach & Motnenko 2011; 2012).

Chaudian beds are overlain by Drevneeuxinian [Old Euxinian] beds corresponding to MIS 14–8 (Fig. 16.6). The latter are widely distributed in tectonically elevated terraces on the Kerch and Taman peninsulas. However, their most complete geological section with a thickness of 12.8 m was recovered on the northwestern Black Sea shelf. The roof of the Drevneeuxinian beds (here and elsewhere calculated as a sum of water depth and sediment thickness) lies between isobaths 32.4 m and 55.7 m BSL on the northwestern shelf and between 18 m and 23 m BSL on the northeastern shelf, where most likely it is tectonically elevated. Sea level was likely around 30 m BSL (Yanko 1989), while Svitoch *et al.* (1998) consider it was slightly above the present one. The hydrological regime of the basin was as follows. In the Early Drevneeuxinian, the basin was connected to the Caspian Sea, which discharged its water into the Drevneeuxinian basin. The salinity did not exceed 7 psu. In the Late Drevneeuxinian, it increased to ~18 psu due to connection with the Mediterranean Sea (Fig. 16.6) (Yanko 1989; 1990).

The Drevneeuxinian beds with an erosional unconformity corresponding to the Paleouzunlarian regression are overlain by Uzunlarian beds (Fig. 16.6) corresponding to MIS 7. Uzunlarian beds are rarely found on the shelf but usually in coastal outcrops. The hydrological regime of the Uzunlarian basins was as follows: at the beginning, the basin was connected to the Caspian Sea, and its salinity did not exceed 8 psu to 10 psu; then it increased, probably to 18 psu, and dropped again to 12 psu to 13 psu by the end of the Uzunlarian. According to Fedorov (1997), the level of the Late Uzunlarian basin was about 10 m ASL. Svitoch *et al.* (1998) consider it was close to recent sea level, while Yanina (2012) believes it exceeded slightly the present one.

The Late Pleistocene is represented by the Karangatian (MIS 5), Tarkhankutian (MIS 4), and Neoeuxinian (MIS 3–2) beds (Fig. 16.6). Detailed descriptions of the Karangatian neostratotype 'Eltigen' (shown in Fig. 16.15e) are provided in Motnenko (1990) and Yanko *et al.* (1990). New ^{14}C dates for the Eltigen can be found in Nicholas *et al.* (2008). The Karangatian beds, which accumulated during the Riss-Würm (Mikulino Interstadial interglacial — a term proposed by A. I. Moskvitin in 1947), are widely distributed in both coastal areas and the sea bottom. Their stratigraphic position is clearly identified by the highest numbers of Mediterranean molluscan and foraminiferan species that do not live in the Black Sea today and are the warmest dwelling species among all others in Pleistocene and Holocene. The basin was connected to the Mediterranean Sea, and its salinity reached 30 psu. Sea level might have been around 7 m BSL (Fedorov 1978; 1997; Svitoch *et al.* 1998; Yanina 2012) or between –5 m and –10 m (Chepalyga 1997).

Tarkhankutian beds (Fig. 16.6) were first reported by Nevesskaya and Nevessky (1961) from Karkinitsky Bay, northwestern shelf, at water depths of 30 m to 35 m as sediments containing a mixture of Caspian and Mediterranean mollusks. Later, they were discovered in other places on the Black Sea coast, e.g. the Colchis Plain (Georgia) where they are overlain by sub-aerial peats dated ca. 31 ka BP at a sampling depth of 60 m (Dzhanelidze & Mikadze 1975). Popov and Zubakov (1975) and Popov (1983) recognized similar sediments as Surozhian. Svitoch *et al.* (1998) considered Tarkhankutian and Surozhian sediments as coeval, with an age of 40 ka to 25 ka BP. The Tarkhankutian transgression at 31,330±719 ka BP (Chepalyga 2002a,b) brought Mediterranean water and organisms into the Black Sea and increased salinity to about 8 psu to 11 psu (Nevesskaya 1965; Yanko 1989; 1990; Yanko-Hombach 2007a). Submerged accumulative coastal bars of synchronous age are located at water

depths of 22 m to 30 m on the Ukrainian (Chepalyga *et al.* 1989; Chepalyga 2002a,b) and Romanian (Caraivan *et al.* 1986) shelf, indicating that Tarkhankutian sea level was about 25 m BSL (Chepalyga 1997; Yanina 2012), 30 m BSL (Yanko-Hombach 2007a), or 45 m BSL (Fedorov 1997). Temporally, the Tarkhankutian sediments correspond to Unit 3 (Çağatay 2003) in the Sea of Marmara. This unit contains some marine mollusks and benthic foraminifera indicating a weak Mediterranean marine incursion during the early part of MIS 3. Interestingly, no similar sediments have yet been found in the Bosporus. Instead, they have been recovered in Izmit Bay and the Sakarya valley (Meriç *et al.* 1995; Yanko-Hombach *et al.* 2004).

The Tarkhankutian is overlain by Neoeuxinian beds (Yanko-Hombach 2007a). The latter correspond to MIS 2 and were formed during and soon after the Last Glacial Maximum (LGM). They can be divided into Lower and Upper Neoeuxinian beds. The Lower Neoeuxinian beds (Fig. 16.17) were formed between 27 ka BP and 17 ka BP and can be found below isobath ca. 100 m (Yanko-Hombach 2007a; Mudie *et al.* 2014; Yanko-Hombach *et al.* 2013a,b).

Lower Neoeuxinian sediments are distributed everywhere in the Black Sea below isobath 100 m (Kvasov 1975; Fedorov 1977; 1978; 1988; Shcherbakov *et al.* 1978, Abashin *et al.* 1982; Shcherbakov 1983; Shnyukov 1985; Svitoch *et al.* 1998; Kuprin 2002; Kuprin & Sorokin 2007; Yanko-Hombach 2007a; Yanko-Hombach *et al.* 2013b, fig. 3). Lithologically, they are represented by alternations of gray silt and gray striped clays enriched with hydrotriolite, sand (minor), and shells of *D. rostriformis distincta*. The level of the Early Neoeuxinian lake was ca. 100 m BSL (Kuprin & Sorokin 2007; Yanko-Hombach 2007a; Yanko-Hombach *et al.* 2011a,b; 2013a). Their recovered thickness ranges between 5 cm and 80 cm (Fig. 16.17).

The Lower Neoeuxinian beds are often overlapped by sub-aerial loams and further on by aquatic sediments with ostracoda *Candona, Candoniella,* and foraminifera *A. novoeuxinica.* This change indicates the transformation of the bottom from an erosional to subaquatic accumulative phase at the beginning of the Late Neoeuxinian transgression (Gozhik 1984b; Shnyukov 1985).

The Upper Neoeuxinian beds overlie the Lower Neoeuxinian beds, covering the Black Sea floor almost everywhere below isobath 39 m BSL on the northwestern shelf (Fig. 16.17) (Larchenkov & Kadurin 2011; Mudie *et al.* 2014; Yanko-Hombach *et al.* 2013a,b), 30 m BSL on the Bulgarian (Filipova-Marinova 2007), Crimean (Shnyukov 1985), and Caucasian shelves (Balabanov *et al.* 1981; Yanko & Gramova 1990), and 18 m BSL on the Turkish shelf. In some places (e.g. the western part of the Golitsin Uplift located at the mouth of Karkinitsky Bay), they are exposed on the sea floor (Tkachenko *et al.* 1970; Ishchenko 1974; Tkachenko 1974; Yanko 1974; 1975; 1989).

Their thickness can reach 25 m (Put' 1981). Lithologically, Upper Neoeuxinian beds on the shelf are rather monotonous and represented by light gray sandy coquina and/or bluish gray stiff clays that fill pre-Neoeuxinian depressions and paleoriver valleys (e.g. Arkhangel'sky & Strakhov 1938; Nevesskaya 1965; Semenenko & Kovalyukh 1973; Ostrovsky *et al.* 1977; Malovitsky 1979; Balabanov *et al.* 1981; Yanko 1982; Gozhik 1984a,b,d; Shnyukov 1985; Fedorov 1988; Gozhik *et al.* 1987; Yanko 1989; 1990; Yanko & Gramova 1990; Glebov *et al.* 1996). The stiff clay has a massive structure, high density (about 2.7 g/cm^3), and low water content. The interstitial water salinity is 7 psu (Konikov 2007). Mollusks are dominated by *D. polymorpha* and *D. rostriformis* on the inner and outer shelf, respectively. A paleosalinity for the Late Neoeuxinian lake was about 5 psu in the shallow areas; it could have reached 7 psu, which is typical of interstitial salinity (Manheim & Chan 1974), and even 11 psu to 12 psu (Nevesskaya 1965; Mudie *et al.* 2001; 2011; Marret *et al.* 2009; Yanko-Hombach *et al.* 2013a,b) in deeper parts of the basin. Despite a relatively high salinity, Mediterranean species are absent, and Caspian immigrants are abundant.

Holocene sediments cover the Black Sea shelf almost everywhere (Fig. 16.18). Their thickness and lithological characteristics vary significantly in different parts of the basin due to morphodynamics of the shelf (e.g. width, bottom relief), as well as tectonic activities, amount of sediment discharged into the Black Sea by rivers, as well as subsidence, compaction, and erosion of sediments during regressive stages — all greater on the narrow shelves

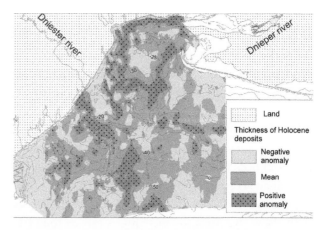

Fig. 16.18 The spatial distribution of Holocene sediment thickness on the northwestern Black Sea shelf. Larchenkov and Kadurin (2011). Reproduced with permission from GSA: Colorado.

compared to the wider northwestern shelf (Malovitsky 1979; Kuprin *et al.* 1980; Shcherbakov 1983; Shnyukov 1984b; Kuprin *et al.* 1985; Krystev *et al.* 1990; Yanko & Gramova 1990; Devdariani *et al.* 1992; Sorokin *et al.* 1998; Aksu *et al.* 2002; Shimkus 2005; Glebov & Shel'ting 2007; Konikov 2007; Kuprin & Sorokin 2007; Yanko-Hombach 2007a,b).

The spatial distribution of Holocene deposit thicknesses is non-uniform, with a distribution skewed toward smaller values; they can be subdivided into three general groupings: bathymetric rises, paleovalleys, and mid- to outer shelf deposits (Fig. 16.18). The normal thickness for each region corresponds to a mean value with a standard deviation, with deviations in thickness as positive or negative anomalies (Table 16.1).

The rate of sedimentation on the shelf varies significantly from place to place. On the northern shelf, it is 50 cm/kyr to 200 cm/kyr in the Danube pro-delta and Dneprovsky gutter; on the western Crimean shelf, it is 30 cm/kyr to 50 cm/kyr; and in the central part of the shelf at the 50 m isobath it is 10 cm/kyr to 20 cm/kyr. On the outer northwestern shelf at water depths of 70 m to 100 m, the rate of sedimentation decreases to 5–15 cm/kyr, remaining higher in its eastern part. In the upper part of the continental slope at water depths of 100 m to 400 m, the rate of sedimentation does not exceed 10 cm/kyr, falling to 5 cm/kyr and lower in the areas of intensive sediment erosion (Fig. 16.19).

TABLE 16.1 AVERAGE AND ANOMALOUS THICKNESSES OF UPPER PLEISTOCENE–HOLOCENE SEDIMENTS IN DIFFERENT GEOMORPHOLOGICAL AREAS OF THE SHELF. FROM LARCHENKOV & KADURIN (2011).

	Sediment thickness (m)		
Region	Mean	Negative anomaly	Positive anomaly
Bathymetric rise	0.4–3.1	<0.4	>3.1
Paleovalley	3.2–7.9	<3.2	>7.9
Middle to outer shelf	0.3–1.8	<0.3	>1.8

Sedimentologically, the Romanian shelf (between the Danube Delta and Cape Kaliakra) is a continuation of the Ukrainian shelf. Since solid sediments of the Danube River are carried out to the south by currents, this part of the shelf is sediment starved. Therefore, the rate of sedimentation on most of the Romanian shelf (excluding submerged parts of the Danube valley and narrow coastal zone) is 5 cm/kyr to 10 cm/kyr. At depths of 60 m to 70 m along the east–west boundary between Romania and Bulgaria, the 5-m-thick Holocene sediments indicate a sedimentation rate of 60 cm/kyr to 70 cm/kyr.

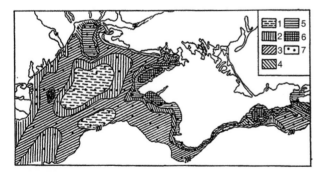

Fig. 16.19 Chart of distribution of sedimentation rate of Neoeuxinian and Holocene sediments on the northwestern shelf calculated on the basis of the ^{14}C data, cm/kyr: (1) <10; (2) 10–20; (3) 20–30, (4) 30–40; (5) 40–50; (6) >50; (7) black dots = sampling stations where sedimentation rate was calculated. Shnyukov (1985).

On the Bulgarian shelf, the highest sedimentation rates occur between capes Kaliakra and Emine. Between 30 m and 60 m depth, the rates of sedimentation vary between 200 cm/kyr to 350 cm/kyr. They decrease to 10–15 cm/kyr and less toward the shelf break and coastal slope. A high rate of sedimentation is typical for the southern shelf between Cape Emine and the Turkish–Bulgarian border. At Burgas Bay, it varies between 10 cm/kyr to 20 cm/kyr and 150 cm/kyr to 200 cm/kyr, reaching its maximum in buried river valleys. At depths of 40 m to 60 m, the sedimentation rate is about 100 cm/kyr. Toward the outer shelf, the rates decrease to 10–25 cm/kyr. In the upper continental slope, similar to the Romanian shelf, the distribution of sedimentation rates has a mosaic character due to a large number of underwater river valleys where erosion takes place. If no erosion of sediments occurred, sedimentation would be 15 cm/kyr to 20 cm/kyr.

The Turkish western shelf (west of the Bosporus) is covered by low thickness Holocene sediments. Judging from the limited number of studied cores and seismoacoustic data, the rate of sedimentation does not exceed 30 cm/kyr to 40 cm/kyr. At the shelf break and upper continental slope, it decreases to 10–20 cm/kyr.

On the Crimean shelf (between the Khersones Peninsula and Kerch Strait), the sedimentation rates vary significantly. The lowest values are typical of underwater coastal slopes, where a bench thinly covered by sandy-pebbly sediments is present almost everywhere. At depths between 15 m and 30 m, sedimentation reaches 250–400 cm/kyr. It decreases toward the middle part of the shelf to 40–60 cm/kyr and to 10–15 cm/kyr toward the shelf break at depths of 80 m to 100 m. The lowest sedimentation rates are in the eastern Crimean shelf.

The Caucasian shelf between Anapa and Batumi is characterized by reduced thickness of Quaternary sediments in general, and Holocene sediments in particular. In most of the shallow coastal zone, at depths of 5 m to 10 m, the bedrock is exposed. In the area of Novorossiysk-Gelendzhik, Adler-Pitsunda, and Poti-Kobuleti, sedimentation rate increases to 100–400 cm/kyr. It decreases to 50–100 cm/kyr and 10–40 cm/kyr in the central part of the shelf and at the shelf break, respectively. On the continental slope, the rates of sedimentation change from

0 cm/kyr to 20–30 cm/kyr due to its strong dissection and erosion by turbidity currents.

The best data sources on the spatial distribution of sedimentation rate and thicknesses of Late Pleistocene/Holocene sediments, including maps, can be found at the Department of Marine Geology and Mineral Resources of the National Academy of Sciences of Ukraine, the Department of Lithology and Marine Geology at Moscow State University, Prichernomor SRGE, and the Department of Engineering Geology and Hydrogeology, Odessa I.I. Mechnikov National University, Ukraine.

Post-LGM Climate, Sea Level, and Paleoshorelines

During the LGM between 27 ka BP and 17 ka BP, ice sheets spread over much of North America and northern Asia. Over Hudson Bay in Canada, the ice was 4 km thick. In Europe, ice sheets extended over much of the UK and as far south as Germany and Poland (Yanko-Hombach et al. 2012b). The spread of ice profoundly impacted the Earth's climate, causing drought, desertification, and a dramatic drop in sea levels in basins connected to the ocean. About 21 ka BP, the world's oceans were 120 m lower than today. So too was the Mediterranean Sea, which was connected to the Atlantic Ocean through the Gibraltar Strait. The north Adriatic Sea and the north Aegean Sea were completely exposed. On the edges of glacial areas, the entire East European Platform was covered with tundra-steppe vegetation (Yanko-Hombach et al. 2012b).

Early Neoeuxinian palynological diagrams are dominated by *Artemisia*, Chenopodiaceae, *Adonis*, and *Thalictrum* and are similar to those of the dry pine forests of Romania (Komarov et al. 1979; Pop 1957), the pine/birch forests and xerophytic steppe in southern Ukraine and Moldova (Artyushchenko et al. 1972; Kyrvel et al. 1976), and the steppe and forest-steppe on the Balkan Peninsula (Bottema 1974). All indicate a cold and dry climate (Nikonov & Pakhomov 1993). By implication, such a climate should lead to a dramatic decrease in river

discharge into the Early Neoeuxinian lake, causing in turn a dramatic drawdown of the lake level.

The Early Neoeuxinian lake was isolated from both the Mediterranean and Caspian seas. It was semifresh or brackish, aerobic (Degens & Ross 1974), and heavily populated by organisms, in particular those with calcareous shells ($CaCO^3$ in sediments: ~50%), e.g. mollusks, ostracoda, and on a much smaller scale, by foraminifera. During the Early Neoeuxinian, a large portion of the present shelf above present isobath 100 m was exposed and eroded. The northwestern shelf was downcut some 40 m to the basement by the Pre-Danube, Pre-Dnieper, and Pre-Dniester rivers, and covered by subaerial loams (e.g. Shcherbakov et al. 1978; Shcherbakov 1983; Inozemtsev et al. 1984; Fedorov 1988). River mouths were relocated 80 km to 100 km seaward (Gozhik 1984b; Shnyukov 1985), where they possessed poorly developed deltas and opened directly into the canyons on the continental slope (Fig. 16.20).

The river valleys and canyons were filled with thick (22–40 m) alluvial sediments (Kuprin & Sorokin 2007) of the stratigraphic unit Ant age (22,800–16,900 BP) containing 27 freshwater and 14 brackish water shallow ostracods dominated by *Cyprideis littoralis* and *Ilyocypris bradyi* (Gozhik 1984c). Direct palynology evidence from high-resolution marine cores (Mudie et al. 2002; 2007) suggests a marshy and mosquito-infested shoreline subject to periodic river flooding. This might have provided good hunting and fishing but poor conditions for settled farming because of brackish water and soils prone to salinization and waterlogging; these problems still limit coastal farming today.

Late Neoeuxinian palynological diagrams are dominated by *Quercus*, *Carpinus*, *Ulmus*, *Salix*, and *Betula*, with reduced concentration of *Pinus* and grass (Komarov et al. 1979; Kvavadze & Dzeiranshvili 1989). They are similar to the Late Glacial diagrams of the Balkan Peninsula (Bozilova 1973; 1975) and the Prichernomorian soil horizon (Veklich & Sirenko 1976) formed before 10.5 ka BP (Ivanova 1966). The climate warmed during Late Neoeuxinian times, as indicated by the replacement of pine by broad-leaved forests. In the warming climate of about 16 ka BP, a massive water discharge, most likely from the Caspian Sea via the Manych Spillway

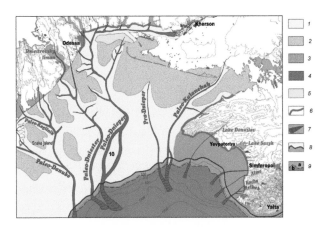

Fig. 16.20 Schematic geomorphological map of the northwestern shelf during the Early Neoeuxinian: (1) paleovalleys and large alluvial plains; (2) gently sloping hills formed by Late Pleistocene marine and continental landforms; (3) gently sloping piedmont uplands; (4) steep banks and seabed folded bedrock; (5) large relict accumulative landforms formed by sand; (6) paleoriver bed; (7) underwater debris cone; (8) coastline of 18 ka BP; (9) shelf edge: (a) outer shelf, (b) continental slope. Modified after Shnyukov (2002). Reconstructions were based on results of the 1:200,000 marine geological survey outlined in Fig. 16.3, as well as numerous onshore and offshore drillings, and geophysical profiles which enabled the discovery that deltas and valleys of the Pliocene Don, Dnieper, and other rivers migrated across the northwestern coast and adjacent shelf. Two buried paleovalleys of the Paleo-Kanchak and Pre-Dnieper were discovered as fragments of the migration of the ancient valley of the Dnieper River.

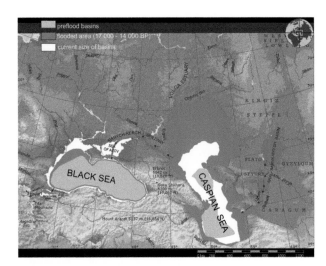

Fig. 16.21 Model showing the sequence of Ponto-Caspian Great Flood basins. Modified after Chepalyga (2007); from Yanko-Hombach *et al.* (2012b).

(Chepalyga 2007) increased the level of the Late Neoeuxinian lake to ca. −20 m. The latter must have overflowed, pouring its excess semi-fresh-to-brackish water into the Sea of Marmara and from there into the Mediterranean (Fig. 16.21).

The Late Neoeuxinian lake was aerobic and heavily populated by organisms with carbonate shells. Late Neoeuxinian sediments seem to be partially synchronous with the upper part of Unit 2 (Çağatay 2003) of the Sea of Marmara sediment column. The level of the Late Neoeuxinian lake was much higher (about −20 m) than that of the Sea of Marmara (about −85 m), and by implication, the Late Neoeuxinian lake discharged its waters into the Sea of Marmara. The Bosporus

continued to be a semi-fresh lake and might have served as a channel for southward water discharge from the Neoeuxinian lake. However, this discharge could have occurred through the Izmit Gulf–Sakarya valley, as indicated by the presence of fresh/brackish facies with an age of 14.6 ka BP in borehole KS2 (Kerey *et al.* 2004; Yanko-Hombach *et al.* 2004; 2007).

Upper Neoeuxinian beds, often with erosional unconformity, are overlapped by Bugazian beds containing the first Mediterranean immigrants. Bugazian sediments are widely distributed below the 17 m isobath. Their thickness increases from 0.03 m to 0.20 m on the slopes of submerged river valleys to 2.5 m on their bottom. Bugazian palynological diagrams are characterized by a sharp decrease in grassy elements (e.g. wormwood, goosefoot) and conifers (*Pinus, Picea, Juniperus*). Instead, broadleaf *Quercus, Corylus, Ulmus, Betula*, and even beech become dominant, indicating moderate climate conditions typical of the Boreal Ecozone (Komarov *et al.* 1979). A similar palynological diagram of the deep sediments of the Black Sea and peats of the Ril Massif in Bulgaria have ^{14}C ages of 10,737±315 BP (Shimkus *et al.* 1977) and 10,035±65 BP (Bozilova 1973). A summary of the palynological data from lakes over a wide area west and south of the Black Sea shows that oak-pistacio (*Quercus–Pistacia*) forests were present over most of the region by 10 ka BP, although local desert-steppe vegetation persisted until about 7 ka BP in the south-east, from Lake Van (Eastern Turkey) to the Caspian Sea (Mudie *et al.* 2002). These forests indicate the early establishment of mesic climatic conditions characterized by >600 mm/year of precipitation in excess of evapotranspiration, as is presently found in most of central and western Europe (Hiscott *et al.* 2007a).

Since the Bugazian, a series of low amplitude transgressive and regressive phases is clearly manifested on the inner shelf of the Black Sea (Fig. 16.6). Due to the low amplitude of the regressive phases, they cannot be traced in cores recovered from a depth of more than 50 m, thus giving the impression of a gradual increase in sea level and salinity. In general, the dynamic nature of the recent Black Sea level changes is a result of the joint influence of isostatic, eustatic, and tectonic processes, and to a lesser degree, anthropogenic factors.

The paleolandscapes of the northwestern shelf since the LGM are shown in Fig. 16.22.

There are three main scenarios for the post-LGM development of the Black Sea: gradual, oscillating, and catastrophic (also called rapid or prominent); they are discussed elsewhere (e.g. Yanko-Hombach 2007b; Yanko-Hombach *et al.* 2011a,b; see also Chapter 17, pages 485–487).

According to the gradual scenario, the Late Neoeuxinian lake was gradually transformed into a marine basin and re-colonized by Mediterranean organisms in the course of the postglacial transgression (Arkhangel'skiy & Strakhov 1938; Nevesskaya 1965; Il'ina 1966; Aksu *et al.* 2002; Hiscott *et al.* 2007a,b). The latter started at ca. 9.5 ka BP when a two-way interchange was established via the Bosporus (Stanley & Blanpied 1980).

According to the oscillating scenario (e.g. Ostrovsky *et al.* 1977; Fedorov 1978; 1982; Balabanov *et al.* 1981; Tchepalyga 1984; Shnyukov 1985; Balabanov & Izmailov 1988; Yanko 1990; Yanko & Gramova 1990; Shilik 1997; Svitoch *et al.* 1998; Chepalyga 2002a,b; Balabanov 2007; Konikov 2007; Yanko-Hombach 2007a,b; Mudie *et al.* 2014; Yanko-Hombach *et al.* 2011a,b; 2013a), the transformation of the Late Neoeuxinian lake into a marine basin was gradual but fluctuating.

According to the catastrophic (Ryan *et al.* 1997; Ryan 2007) / rapid (Lericolais *et al.* 2007a,b; 2011) / prominent (Nicholas *et al.* 2011) scenario, the transformation of the Neoeuxinian lake into a marine basin was not gradual but catastrophically rapid. At a rate in excess of 50 km^3 per day, Mediterranean salt water funneled through the narrow Bosporus and hit the Black Sea at 200 times the force of Niagara Falls, thereby sharply increasing salinity, refilling the Neoeuxinian lake with a rate of rise of level of 15 cm/day over two years, and replacing freshwater biota with marine organisms (Ryan *et al.* 1997). Later, instead of a single inundation, two lowstands (–120 m at 13.4–11 ka BP; and –95 m at 10–8.4 ka BP) and two catastrophic floods (sea-level rise from –120 m to –30 m at 11–10 ka BP; and from –95 m to –30 m at 8.4 ka BP) were proposed (Ryan *et al.* 2003). The second of these two major transgressions was labeled the Great Flood as described in the Bible. The initial Flood Hypothesis was based on evidence from seven short (about 1.25 m each), low-resolution sediment cores and 350 km of seismic profiles collected within a fairly restricted area of the Black Sea's northwestern shelf at water depths between 49 m and 140 m during a single mission in 1993 (for their location see Fig. 16.16).

Our latest multidisciplinary study of geological material recovered in areas of the northwestern, northeastern, and southwestern Black Sea shelf (Yanko-Hombach *et al.* 2013b; Mudie *et al.* 2014) confirm our previous data.

1 The level of the Late Neoeuxinan lake prior to the Early Holocene Mediterranean transgression stood around 40 m BSL, but not 100 m BSL or more, as suggested by advocates of the catastrophic/rapid/prominent flooding scenario.

2 Microfossil data examined from multiple shelf sites show that at all times, the Neoeuxinan lake was brackish with a salinity of about 7 psu prior to the Initial Marine Inflow (IMI) and Mediterranean transgression.

3 By 8.9 ka BP, the outer Black Sea shelf was already submerged by the Mediterranean transgression. An increase in salinity took place over 3600 years, with the rate of marine incursion estimated on the order of 0.05 cm/year to 1.7 cm/year.

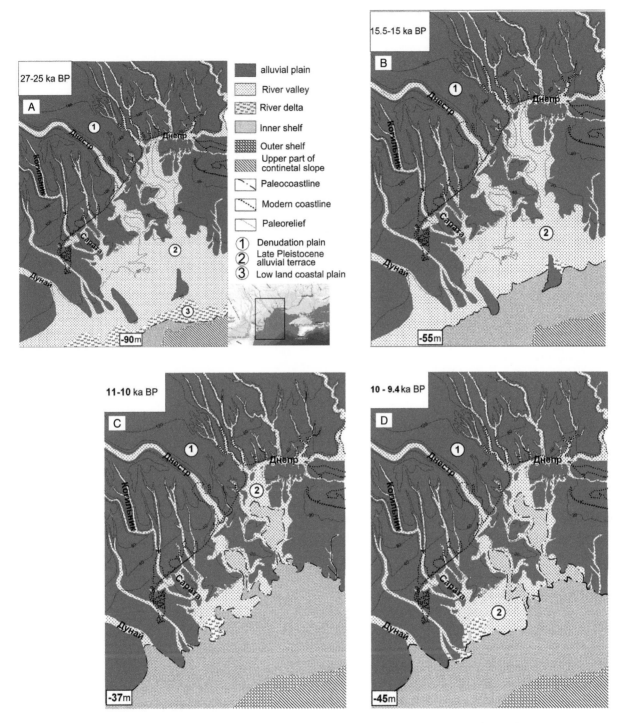

Fig. 16.22 Paleogeographic scheme of the Pontic Lowland and northwestern Black Sea shelf: (A) LGM (27–25 ka BP); (B) in the Late Neoeuxinian (15.5–15 ka BP); (C) at the Younger Dryas (11–10 ka BP); (D) just before the Holocene transgression (10–9.4 ka BP); (E) at the beginning of the Holocene transgression (9.4 ka BP); (F) during the Kalamitian (4 ka BP). Legend in Fig. 16.22(a) is also for Figs. 16.22(b–e). Yanko-Hombach et al. (2011a).

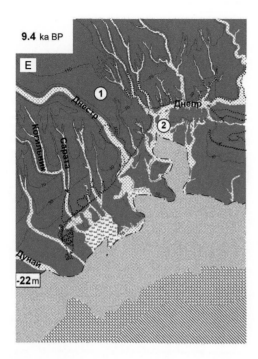
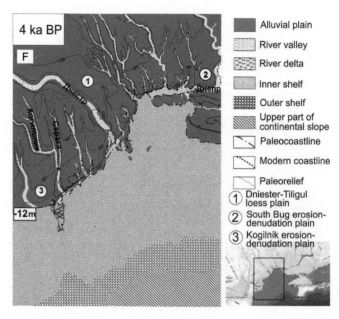

Fig. 16.22 *(Continued)*

4 The combined data from sedimentological character-istics and microfossil salinity evidence establish that the Holocene marine transgression was of a gradual, progressive but oscillating nature.

Lately, Badertscher *et al.* (2011) attempted to reconstruct the time and number of water intrusions into the Black Sea from the Mediterranean and Caspian seas using speleothems from Sofular Cave. According to their data, there were twelve and seven intrusions of Mediterranean and Caspian water respectively into the Black Sea. If their data are correct, we must then see an alternation of molluscan and foraminiferal assemblages in the coastal outcrops. Instead, we see that connections between the Mediterranean and Black seas occurred six times, while connections between the Caspian and Black seas occurred four times since the Brunhes-Matuyama reversal (i.e. the last 780 kyr), and in most cases, these connections did not occur synchronously with those of Badertscher *et al.* (2011) (Fig. 16.6). Namely, the Early Chaudian, Early Old Euxinian, Early Uzunlarian, and Late Neoeuxinian basins were connected to the Caspian Sea. The Karad-enizian, Late Old Euxinian, Middle–Late Uzunlarian,

Karangatian, Tarkhankutian, and Chernomorian (Old and New Black Sea) basins were connected to the Mediterranean Sea. In all stages, these connections had an oscillating character. As such, the speleothem-based conclusions for reconstructing water intrusions from the Mediterranean and Caspian into the Black Sea must be used with great caution (Yanko-Hombach & Motnenko 2011; 2012).

Evidence for Submerged Terrestrial Landforms and Ecology

Submerged terrestrial features are represented in the Black Sea by a variety of landforms. From the Danube Delta to Odessa Bay and further to Karkinitsky Bay, they are erosive-accumulative and accumulative sandy bars, spits, and barriers (Fig. 16.23a), with the exception of Snake Island, which was formed tectonically (Fig. 16.23b).

Underwater ridges, caves, and tunnels are widely distributed within the South Crimea littoral zone

Fig. 16.23 (A) Dzharylgachsky Gulf (for location see no. 38 in Fig. 16.10): (1) emerged barrier spit; (2) barrier beach; (3) lagoon; (4) barrier island; (5) dunes; (6) barrier shallows; (7) emerging spit; (B) Snake Island; (C) Cape Aya with underwater caves and freshwater springs; (D) Adalary Rocks mentioned in Homer's *The Odyssey*; (E) Protrusions formed by Sarmatian limestones in the form of islands called 'Stones-Ships', off the Kerch-Taman Peninsula; (F) Sonar profile across the southern underwater extension of Parpachsky Peak formed by folded Maikopian deposits with uplifted 'Stones-Ships'; (G) Seismic profile across a barkhan dune inside a trough. Lericolas *et al.* (2007b). Reproduced with permission.

(Fig. 16.23c). Erratic stacks, separated by erosion from the slopes of the Crimean Mountains, form underwater rock massifs sometimes rising above sea level in the form of islands, such as the Adalary Rocks mentioned in Homer's *The Odyssey* (Fig. 16.23d). A strong illustration of tectonic activity can be seen on the southeastern coast of Crimea, expressed as intense mudflows and landslide-crumbling landforms onshore, and canyons and landslides offshore. Protrusions in the form of islands can be seen off the

coast of the Kerch-Taman Peninsula near the crest of the submarine continuation of Parpachsky Full (Figs. 16.23e,f).

A number of underwater shorelines were digitally reconstructed based on relief trend analysis, statistical analysis of recent sea bottom relief, and GIS modeling using Kriging methods of interpolation within grid areas of 100 × 100 (Larchenkov & Kadurin 2006). One shoreline corresponding to the LGM is especially well

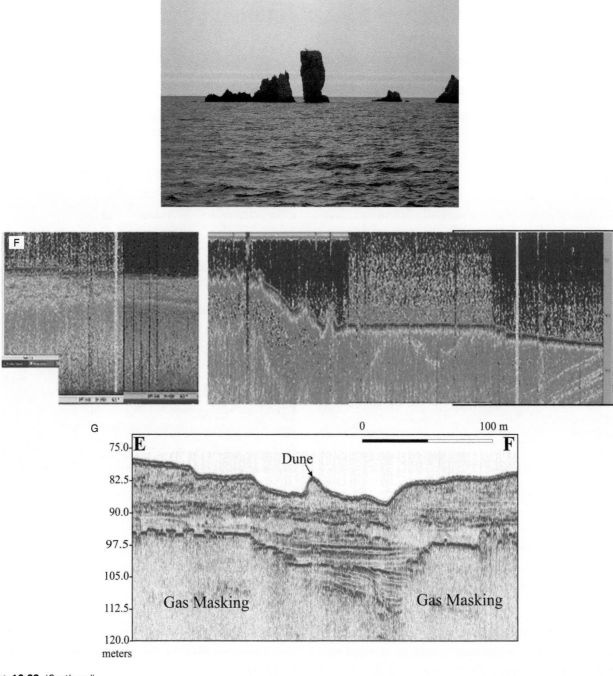

Fig. 16.23 (Continued)

pronounced. Ballard *et al.* (2000), Algan (2003), Algan *et al.* (2007) and Lericolais *et al.* (2007a,b; 2010; 2011) described a submerged coastline with wave-cut terraces and coastal paleodunes (Fig. 16.23g) at various depths ranging from 90 m on the Romanian shelf to 155 m

on the Turkish shelf near Sinop, in support of rapid transgression at the beginning of the Holocene. However, paleodune studies of Badyukova (2010) in the Caspian– Black Sea corridor established the following principles: (1) there are no opportunities for dunes to persist on the sea bottom in any transgressive scenario for the Black Sea corridor; (2) transgressive parasequences involve the accumulation of lagoonal and marine deposits during rising relative sea level and landward migration of a coastline over coastal plain deposits. Our core data are in full agreement with Badyukova's conclusion, and we find no lithological signs of drowned windblown dunes described by the above-mentioned authors (Yanko-Hombach et al. 2013b).

Potential for Archaeological Site Survival

The Pontic steppes are extremely rich in archaeological finds, which have been systematically studied by several generations of Soviet, Russian, and Ukrainian archaeologists (Dolukhanov 1979; Kremenetski 1991; Shilik 1997; Smyntyna 1999; Dergachev & Dolukhanov 2007; Dolukhanov & Shilik 2007; Stanko 2007). During the Upper Paleolithic, between 32,000 and 10,000 years ago, a diverse array of hunter-gatherer societies occupied the Pontic–Caspian steppes (Soffer & Praslov 1993; Sinitsyn 2003; Smyntyna 2004; Stanko 2007). The southern fringe of the Eurasian forest zone was, at that time, a periglacial steppe known as the famous 'mammoth steppe', where Ice-Age hunters built hide-covered huts over frameworks of mammoth tusks at places such as Mezhirich and Mezin on the middle Dnieper drainage and at Kostenki II on the middle Don. Reindeer hunters lived in the Carpathian foothills at Molodova V (Grigor'eva 1980). In what is now the steppe zone north of the Black and Caspian seas, bison and horse hunters lived at sites such as Amvrosievka and Anetovka II (Stanko 2007). The hunting patterns and tool kits of these hunters in the south were quite different from those of the colder mammoth steppe to the north. The area of the southern steppes was also much larger than it is now (Anthony 2007).

Over recent decades, geoarchaeological studies in the Black Sea region carried out within the framework of multidisciplinary international networking projects, such as IGCP 521 "Black Sea–Mediterranean corridor during the last 30 ky: Sea-level change and human

adaptation" (2005-2010), INQUA 0501 "Caspian-Black Sea-Mediterranean Corridor during last 30 ka: Sea level change and human adaptive strategies" (2005-2011), and IGCP 610 "From the Caspian to Mediterranean: Environmental Change and Human Response during the Quaternary" (Yanko-Hombach et al. 2012a,b). These geoarchaeological projects collected substantial material and focused on the possible influence of environmental change on human adaptive strategies. These investigations leave no doubt that, since early prehistoric times, subsistence and social dynamics of human groups in that area were directly affected by changes in climate, vegetation, and sea-level fluctuations, with commensurate shrinking and expansion of littoral areas, and had considerable impact on the settlement patterns of the prehistoric societies (Stanko 2007). Most researchers agree that during the LGM, the level of the Black Sea was at least 100 m BSL, and the area of the shelf above isobath 100 m was dry land. Geological and palynological data show it was an exposed steppe region (Yanko-Hombach et al. 2011a,b; 2013a) that could be exploited extensively by hunter-gatherers migrating in from west to east, and from northwest to south (Stanko 2007). This hypothesis is supported by (1) the similarity in artifacts from several Upper Paleolithic archaeological sites located today across the northwestern Black Sea in Romania, Ukraine, and Crimea (Fig. 16.24) — namely Liubymovka in the Lower Dnieper area, Ukraine, and Movileni and Lespezi in Romania, Mitoc-Malul Galben and Dobrudja in Romania, and Siuren 1 in Crimea, Molodova in the Middle Dniester region and Vishenne 2 in Crimea (Otte et al. 1997; Demidenko 2000-2001; Chabai 2007) and (2) the finding of several flint tools within boreholes taken in various places on the northwestern shelf (Stanko 2007).

However, no submerged prehistoric sites have been discovered so far, except one ceramic plate from −90 m off Varna, and photographs of boulders at −90 m depth off Sinop that, according to Ballard (2001), have a Neolithic age of over 8000 years. However, underwater artifacts and shipwrecks recovered to date from this region are of historical date (Ward & Ballard 2004; Ward & Horlings 2008).

There are some submerged archaeological sites of much younger age at water depths shallower than 10 m (Fig. 16.9), such as the Greco-Roman ruins near the Sea of

Fig. 16.24 Possible migration routes of Upper Paleolithic populations ca. 25 ka BP based on the similarity in artifacts from several Upper Paleolithic archaeological sites located today across the northwestern Black Sea in Romania, Ukraine, and Crimea, and described in Otte *et al.* (1997), Demidenko (2000-2001) and Chabai (2007).

Azov entrance in Taman Bay, near Phanagoria (Blavatsky 1972); in the Bug Estuary at Olbia Pontica (Kryzhitsky *et al.* 1999); at ancient Tyras (Samoilova 1988) in the Dniester Estuary; at Phanagoria off the southeastern Crimean Peninsula (Bolikhovskaya *et al.* 2004); at Cape Shabla north of Varna, in Lake Varna, or in the coastal Varna-Beloslav Lake and indirectly dated to the Late Eneolithic and Early Bronze Age (Peev 2009); and some others in Crimea (Zelenko 2008).

Potential Areas for Future Work

At the LGM (27–17 ka BP), when the level of the Early Neoeuxinian lake was at least 100 m BSL (Figs. 16.20; 16.22a), on the northwestern coast, we would expect Upper Paleolithic sites to be located within the deep valleys of small rivers. These valleys were flooded in the course of the Late Neoeuxinian transgression (17–10 ka BP); however, they are well expressed geomorphologically and can be easily traced on the present Black Sea shelf. This topographic information can be used to search for submerged Upper Paleolithic sites on the shelf, thereby helping to locate evidence for the transition among ancient human groups from hunting large herd

animals to small non-gregarious species. The beginning of the Mediterranean transgression occurred around 9.5 ka BP. Both the transgression and faunal migration occurred over the course of six transgressive-regressive stages. Mesolithic sites continued to be located along river valleys; they bear some evidence of the transition from hunting to gathering of edible plants. No signs of catastrophic flooding of the Black Sea in the Early Holocene have been found (Yanko-Hombach *et al.* 2011a).

One should bear in mind that searching for prehistoric archaeological objects, especially of Paleolithic age, will be difficult because the artifacts are small flints. Some approaches should be implemented: (1) to define the main principles of distribution of already known terrestrial archaeological sites; (2) to show their distribution on the coast with a focus on geomorphological features of the study area; (3) to define conformities in their terrestrial distribution.

Geoarchaeological modeling of the northwestern Black Sea region over the last 25 kyr with respect to paleoenvironment and settlement pattern mapping as an instrument for submerged prehistoric site prospecting was presented by Yanko-Hombach *et al.* at SPLASHCOS meetings in Berlin (2011) and by Yanko at Szczecin (2013); it was further developed by Kadurin and Kiosak (2013). GIS-aided mathematical modeling has indicated the most favorable areas on the northwestern shelf to search for submerged archaeological sites of Upper Paleolithic age based on the distribution of terrestrial sites and their locations along submerged valleys of small rivers as shown on geomorphological maps (such as Fig. 16.20, 16.22a for the Upper Paleolithic and Figs. 16.22b-f for younger ages).

Conclusion

The northwestern shelf of the Black Sea seems to be the most suitable area to search for submerged archaeological sites. It is the widest part of the shelf and the best studied geologically and geomorphologically. The finding of several flint tools retrieved from boreholes in various places on the northwestern shelf by Stanko (2007) is encouraging, and

> "…archaeological surveys targeted at final Paleolithic and Mesolithic sites on the northwest Black Sea shelf and along the submerged river valleys might be deemed promising…[they] might solve a number of major problems related to the character and chronology of the submergence, migrations, and the interrelationship between prehistoric groups of the Balkans, Central Europe, and Crimea (Stanko 2007:374)."

Recent archaeological studies in Turkey, Georgia, and Bulgaria, together with investigations of karstic caves on the Crimean Peninsula in Ukraine allow us to assume that steep, as well as abrupt and rocky Black Sea coasts, also deserve attention with respect to searches for submerged prehistoric sites, especially of Lower and Middle Paleolithic times, and of the Mesolithic period.

Acknowledgments

We are grateful to Prof. Allan Gilbert from Fordham University, USA for editing the English text and valuable comments. We appreciate the help of Prof. Dr. Tamara Yanina from Moscow State University for providing some data on Pleistocene sea-level change in the Black Sea.

This paper is a contribution to IGCP 521 "Black Sea-Mediterranean Corridor during the last 30 ky: sea level change and human adaptation"; INQUA 0501 "Caspian-Black Sea-Mediterranean Corridor during the last 30 ka: Sea-level change and human adaptive strategies"; IGCP 610 "From the Caspian to Mediterranean: Environmental Change and Human Response during the Quaternary"; the Russian–Ukrainian project No. Φ28/428-2009 "The Northwestern Black Sea Region and Global Climate Change: Environmental Evolution during the last 20 ka and Forecast for the 21st Century" sponsored by the State Fund for Fundamental Research, Ukraine; "Theories on interaction between the environment and human society in the northwestern Black Sea during the Late Pleistocene and Holocene" sponsored by the Ministry of Education and Science of Ukraine; and COST Action TD0902 SPLASHCOS project "Submerged Prehistoric Archaeology and Landscapes of the Continental Shelf".

In this paper, we have transliterated Cyrillic letters into the Latin alphabet according to the BGN/PCGN Romanization system for Russian used by Oxford University Press. Exceptions are the names of authors, which we have left in their own preferred transliterations, as well as geographical names as presented most commonly in the majority of English papers.

References

Abashin, A. A., Mel'nik, V. I. & Sidenko, O. G. 1982. Bottom morphology. Characteristics of relief. In Shnyukov, E. F. (ed.) *Geology of the Ukrainian Shelf. Environment, History and Methodology of the Study.* pp. 82-88. Naukova Dumka: Kie v (in Russian).

Aksu, A. E., Hiscott, R. N., Kaminski, M. A. *et al.* 2002. Last glacial–Holocene paleoceanography of the Black Sea and Marmara Sea: stable isotopic, foraminiferal and coccolith evidence. *Marine Geology* 190:119-149.

Algan, O. 2003. The connections between the Black Sea and Mediterranean during the last 30 ky. *Abstracts of the Geological Society of America Annual Meeting & Exposition.* 2nd – 5th November 2003, Seattle, p. 461.

Algan, O., Ergin, M., Keskin, S. *et al.* 2007. Sea-level changes during the Late Pleistocene-Holocene on the southern shelves of the Black Sea. In Yanko-Hombach, V., Gilbert, A. S., Panin, N. & Dolukhanov, P. M. (eds.) *The Black Sea Flood Question: Changes in Coastline, Climate, and Human Settlement.* pp. 603-632. Springer: Dordrecht.

Andrusov, N. I. 1889. About Pliocene deposits of Cape Chauda on the Kerch Peninsula. *Trudy Obschestva Ispytateley Prirody* [*Proceedings of the Society of Naturalists*] 20: XI. St. Petersburg (in Russian).

Andrusov, N. I. 1918. Geological structure of the bottom of the Kerch Strait]. *Novosti Akademii Nauk SSSR* [*News of Academy of Sciences of USSR*], Series 6, 12(1):23-28 (in Russian).

Anthony, D.W. 2007. Pontic-Caspian Mesolithic and Early Neolithic societies at the time of the Black Sea flood: a small audience and small effects. In Yanko-Hombach, V., Gilbert, A. S., Panin, N. & Dolukhanov, P. M. (eds) *The Black Sea Flood Question: Changes in Coastline, Climate and Human Settlement.* pp. 345-370. Springer: Dordrecht.

Arkhangel'sky, A. D. & Strakhov, N. M. 1938. *Geological Structure and Evolution of the Black Sea.* Geological Institute of the Russian Academy of Sciences of the USSR: Moscow-Leningrad (in Russian).

Artyushchenko, A.T., Pashkevich, G.A. & Kareva, E.V. 1972. Development of vegetation of the south of Ukraine in Anthropogenic on data of spore and pollen analysis]. *Byulleten' Komissii po izucheniyu Chetvertichnogo Perioda [Bulletin of Commission on the Study of Quaternary Period]* 39:82-89 (in Russian).

Artyushkov, E. V. & Mescherikov Yu, A. 2013. Recent movements of the Earth's crust and isostatic compensation. *Geophysical Monograph Series* 13: 379-390.

Avrametz, V. M., Kakaranza, S. D., Sibirchenko, M. G. *et al.* 2007. *Zvit z provedennja geologichnoi ziomky masshtabu 1:200,000 pivnichno-zakhidnoi chastyny shelfu Chornogo morja v mezhakh arkushiv L-36-XIII. –XIV, XV [Report on geological survey 1:200,000 within card segments L-36-XIII. L-36-XIV, L-36-XV].* Prichonomorske SRGE [Black Sea Area State Regional Geologic Enterprise]: Odessa (in Ukrainian).

Aydogan, B., Ayat, B. & Yüksel, Y. 2013. Black Sea wave energy atlas from 13 years hindcasted wave data. *Renewable Energy* 57: 436-447.

Badertscher, S., Fleitmann, D., Cheng, H. *et al.* 2011. Pleistocene water intrusions from the Mediterranean and Caspian seas into the Black Sea. *Nature Geoscience* 4:236-239.

Badyukova, E. N. 2010. Evolution of the Northern Caspian Sea Region and the Volga delta in the Late Pleistocene-Holocene. *Oceanology* 50:953-960. Original Russian Text © E.N. Badyukova, 2010, published in *Okeanologiya [Oceanology]* 50(6): 1002–1009 (in Russian).

Bagrov, N. V., Shnyukov, E. F., Maslakov, N. A. *et al.* 2012. *Map of Relief of the Black and Azov Seas, scale 1:500,000).* Department of Marine Geology and Mineral Resources of the National Academy of Sciences of Ukraine and Scientific - Research Center

of Sustainable Development, Taurida V.I. National University: Ukraine, Simpheropol (in Russian).

Balabanov, I. P. 2007. Holocene sea-level changes of the Black Sea. In Yanko-Hombach, V., Gilbert, A. S., Panin, N. & Dolukhanov, P. M (eds.) *The Black Sea Flood Question: Changes in Coastline, Climate and Human Settlement.* pp. 711-730. Springer: Dordrecht.

Balabanov, I. P. & Izmailov, Ya. A. 1988. Sea-level and hydrochemical changes of the Black Sea and Azov Sea during the last 20,000 years. *Vodnye Resursy [Water Resources]* 6:54-62 (in Russian).

Balabanov, I. P., Kvirkveliya, B. D. & Ostrovsky, A. B. 1981. *Recent History of the Development of Engineering-Geological Conditions and Long-Time Forecast for the Coastal Zone of the Pitsunda Peninsula.* Tbilisi: Metsnierba (in Russian).

Ballard, R. D. 2001. Black Sea mysteries. *National Geographic* 199(5):52-69.

Ballard, R. D. 2008. Introduction. In Ballard, R. D. (ed.) *Archaeological Oceanography.* pp. ix–x. Princeton University Press: Princeton.

Ballard, R. D., Coleman, D. F. & Rosenberg, G. 2000. Further evidence of abrupt Holocene drowning of the Black Sea shelf. *Marine Geology* 170:253-261.

Banks, C. J., Robinson, A. G. & Williams, M. P. 1997. Structure and regional tectonics of the Achara-Trialet fold belt and the adjacent Rioni and Kartli foreland basins. *American Association of Petroleum Geologists Memoir* 68:331-346.

Barka, A. & Reilinger, R. 1997. Active Tectonics of Eastern Mediterranean region: deduced from GPS, neotectonic and seismicity data. *Annali DiGeofisica* X2(3):587-610.

Beloussov, V. V. & Volvovsky, B. S. (eds). 1989. *Structure and Evolution of the Earth's Crust and Upper Mantle of the Black Sea.* Nauka: Moscow (in Russian).

Berlinsky, N., Bogatova, Yu. & Garkavaya, G. 2006. Estuary of the Danube. In Wangersky, P. J. (ed.) *Estuaries. The Handbook of Environmental Chemistry* (Volume 5 Part H). pp. 233-264. Springer-Verlag: Berlin-Heidelberg.

Blagovolin, N. S. & Pobedonostsev, S. V. 1973. Recent vertical movements of the shores of the Black and Azov Seas. *Geomorphology* 3:46-55 (in Russian).

Blavatsky, V. D. 1972. Submerged sectors of towns on the Black Sea coast. In *Underwater Archaeology, a Nascent Discipline.* pp. 117-119. UNESCO, Paris.

Bolikhovskaya, N., Kaitamba, M., Porotov, A. & Fouache, E. 2004. Environmental changes of the northeastern Black Sea's coastal region during the middle and late Holocene. In Scott, E. M. A., Alekseev, Yu. & Zaitseva, G. (eds.) *Impact of the Environment on Human Migration in Eurasia.* NATO Science Series IV: Earth and Environmental Sciences 42:209-223.

Bottema, S. 1974. *Late Quaternary Vegetation History of Northwestern Greece.* Ph.D. Thesis. University of Groningen, Netherlands.

Bozilova, E. 1973. Pollen analysis of a peatbog from N1-W1 Rila mountain of Bulgaria. In Khotinsky, N. A. & Koreneva, E. V. (eds) *Holocene Palynology and Marine Palynology.* pp. 44–46. Nauka: Moscow (in Russian).

Bozilova, E. 1975. Correlation of the vegetational development and climatic changes in the Rila and Pirinmountains in the Late Glacial and Post Glacial time. In Jordanov, D., Bondev, I., Kozuharov, S., Kuzmanov, B., Palamarev, E. & Velcev, V. (eds.) *Problems of Balkan Flora and Vegetation. Proceedings of the First International Symposium on Balkan Flora and Vegetation.* 7th – 14th June 1973, Varna, pp. 64-71. Pensoft: Sofia.

Brückner, H., Kelterbaum, D., Marunchak, O., Porotov, A. & Vött, A. 2010. The Holocene sea level story since 7500 BP — Lessons from the Eastern Mediterranean, the Black and the Azov Seas. *Quaternary International* 225:160-179.

Çağatay, M.N. 2003. Chronostratigraphy and sedimentology of the Marmara Sea over the last 40 kyrs. *Geophysical Research Abstracts* 5, 01883.

Caraivan, G., Herz, N. & Noakes, J. 1986. New proofs of the Black Sea rise during the Middle Würm Interstadial (Proceedings of the Institute of Geology and Physics no. 5). *Tectonica și Geologie Regionala [Regional Tectonics and Geology]* 70-71:57-62.

Chabai, V. P. 2007. The Middle Paleolithic and Early Upper Paleolithic in the Northern Black Sea Region. In Yanko-Hombach, V., Gilbert, A. S., Panin, N. & Dolukhanov, P. M. (eds.) *The Black Sea Flood Question:*

Changes in Coastline, Climate and Human Settlement. pp. 279-796. Springer: Dordrecht.

Chepalyga, A. L. 1997. Detailed events stratigraphy of the Pleistocene of the Black Sea. In Alekseev, M. N. & Khoreva, I. M. *Quaternary geology and paleogeography of Russia.* pp. 196-201. GEOS: Moscow (in Russian).

Chepalyga, A. L. 2002a. The Black Sea. In Velichko, A.A. (ed.) *Dinamika nazemnykh landshaftov i vnutrennikh morskikh besseynov v severnoy Evraazii v techenie 130,000 let [Dynamics of Terrestrial Landscape Components and Inner Marine Basins of Northern Eurasia during the Last 130,000 Years].* pp. 170-182. GEOS: Moscow (in Russian).

Chepalyga, A. L. 2002b. Specifics of the development of inner seas in the Pleistocene and Holocene. In Velichko, A. A. (ed.) *Dinamika nazemnykh landshaftov i vnutrennikh morskikh besseynov v severnoy Evraazii v techenie 130,000 let [Dynamics of Terrestrial Landscape Components and Inner Marine Basins of Northern Eurasia during the Last 130,000 Years].* pp. 208-213. GEOS: Moscow (in Russian).

Chepalyga, A. L. 2007. The late glacial great flood in the Ponto-Caspian basin. In Yanko-Hombach, V., Gilbert, A. S., Panin, N. & Dolukhanov, P. M. (eds.) *The Black Sea Flood Question: Changes in Coastline, Climate and Human Settlement.* pp. 119-148. Springer: Dordrecht.

Chepalyga, A. L., Mikhailesku, K., Izmailov, Ya. A., Markova, A. K., Kats, Yu. I. & Yanko, V. V. 1989. Problems of stratigraphy and paleogeography of the Black Sea. In Alekseev, M. N. & Nikiforova, K. V. (eds.) *Chetvertichniy period. Stratigrafiya. [Quaternary Period. Stratigraphy].* pp. 113-120. Nauka: Moscow (in Russian).

Degens, E. T. & Ross, D. A. (eds.). 1974. *The Black Sea Geology, Chemistry, and Biology. American Association of Petroleum Geologists Memoir* 20. Tulsa: Oklahoma.

DeMets, C., Gordon, R. G., Argus, D. F. & Stein, S. 1990. Current plate motions. *Geophysical Journal International* 101:425-478.

DeMets, C., Gordon, R. G., Argus, D. F. & Stein, S. 1994. Effects of recent revisions to the geomagnetic reversal time scale on estimates of current plate motions. *Geophysical Research Letters* 21:2191-2914.

Demidenko, Yu. E. 2000-2001. The European Early Aurignacian of Krems-Dufour type industries: a view from Eastern Europe. *Préhistoire européenne* 16-17:147-162.

Dergachev, V. & Dolukhanov, P. 2007. The Neolithization of the north Pontic area and the Balkans in the context of the Black Sea floods. In Yanko-Hombach, V., Gilbert, A. S., Panin, N. & Dolukhanov, P. M. (eds.) *The Black Sea Flood Question: Changes in Coastline, Climate and Human Settlement.* pp. 489-514. Springer: Dordrecht.

Devdariani, N. A., Sorokin, V. M., Starovoytov, A. V. & Kalinin, V. V. 1992. Structure of the upper part of the sedimentary cover Gudaut banks (Black Sea). *Vestnik of Moscow University [Moscow University Herald] Series 4, Geology* 6:74-80 (in Russian).

Dolukhanov, O. & Shilik, K. 2007. Environment, sea-level changes, and human migrations in the northern Pontic area during late Pleistocene and Holocene times. In Yanko-Hombach, V., Gilbert, A. S., Panin, N. & Dolukhanov, P. M. (eds.) *The Black Sea Flood Question: Changes in Coastline, Climate and Human Settlement.* pp. 297-318. Springer: Dordrecht.

Dolukhanov, P. M. 1979. *Ecology and Economy in Neolithic Eastern Europe.* Duckworth: London.

Dzhanelidze, Ch.P. & Mikadze, I. S. 1975. Evidence of Middle Würm transgression in Upper Pleistocene sediments of Colhis Plain. *Soobshcheniya Akademii Nauk Gruzinskoi SSR = Sak'art'velos SSR mec'nierebat'a akademia [Bulletin of the Academy of Sciences of the Georgian SSR]* 77:377-379 (in Russian).

Emelyanov, E. M., Shimkus, K. M. & Kuprin, P. N. 2005. Explanatory notes to the IBCM-SED map series Unconsolidated Sediments of the Mediterranean Sea and Black Sea. In Hall J. K., Krashenninkov V. A., Hirsch, F., Benjamini, C. & Flexer, A. (eds.) *Geological Framework of the Levant Vol.* II: *The Levantine Basin and Israel.* pp. 183-214. Historical Productions-Hall: Jerusalem.

Eremeev, V. M. and Symonenko, S. V. (eds.) 2009. *Oceanographic Atlas of the Black Sea and Sea of Azov.* Kyïv, Derzhhidrohrafiya [The State Hydrographic Service of Ukraine] (in Ukranian).

Ergün, M., Dondurur, D. & Çifçi, G. 2002. Acoustic evidence for shallow gas accumulations in the sediments of the Eastern Black Sea. *Terra Nova* 14:313-320.

FGDC (Federal Geographic Data Committee), Geologic Data Subcommittee 2006. FGDC digital cartographic standard for geologic map symbolization: Federal Geographic Data Committee Document Number FGDC–STD–013–2006, 290 p., 2 pls., available online at ngmdb.usgs.gov/fgdc_gds/.

Fedorov, P. V. 1977. Pozdnechetvertichnaia istoriia Chernogo moria i razvitie iuzhnykh morei Evropy [Late Quaternary history of the Black Sea and southern seas of Europe]. In Kaplin, P. A. & Shcherbakov, F. A. (eds.) *Paleogeografiia i otlozheniia pleistotsena iuzhnykh morei SSSR [Paleogeography and Deposits of the Pleistocene of the Southern Seas of the USSR].* pp. 25-32. Nauka: Moscow (in Russian).

Fedorov, P. V. 1978. *Pleistotsen Ponto-Kaspiya [The Pleistocene of the Ponto-Caspian].* Nauka: Moscow (in Russian).

Fedorov, P. V. 1982. Some debatable questions of the Pleistocene history of the Black Sea. *Byulleten Moscovskogo obshestva ispytateley prirody. Geologiya [Bulletin of Moscow Society of Naturalists, Geological Branch]* 57:108-117 (in Russian).

Fedorov, P. V. 1988. The problem of changes in the level of the Black Sea during the Pleistocene. *International Geology Review* 30:635-641.

Fedorov, P. V. 1997. Water exchange between the Caspian, Black and Mediterranean seas in the Pleistocene In Alekseev, M. N. & Khoreva, I. M. *Chetvertichnaya geologiya i paleogeografiya Rossii [Quaternary geology and paleogeography of Russia].* pp. 181-186. GEOS: Moscow (in Russian).

Filipova-Marinova, M. 2007. Archaeological and paleontological evidence of climate dynamics, sea-level change, and coastline migration in the Bulgarian sector of the Circum-Pontic Region In Yanko-Hombach, V., Gilbert, A. S., Panin, N. & Dolukhanov, P. M. (eds.) *The Black Sea Flood Question: Changes in Coastline, Climate and Human Settlement.* pp. 453-482. Springer: Dordrecht.

Finetti, I., Bricchi, G., Del Ben, A., Pipan, M. & Xuan, Z. 1988. Geophysical study of the Black Sea Area. *Bollettino di Geofisica Teorica e Applicata* 30:197-324.

Glebov, A. Yu. & Shel'ting, S. K. 2007. Sea-level changes and coastline migrations in the Russian sector of the Black Sea: application to the Noah's Flood Hypothesis. In Yanko-Hombach, V., Gilbert, A. S., Panin, N. & Dolukhanov, P. M. (eds.) *The Black Sea Flood Question: Changes in Coastline, Climate and Human Settlement.* pp. 731-774. Springer: Dordrecht.

Glebov, A. Yu., Shimkus, K. M., Komarov, A. V. & Chalenko, V. A. 1996. History and tendency of evolution of Caucasus region of the Black Sea. In Glumov, I. F. & Kochetkov, M. V. (eds.) *Tekhnogennoe zagriaznenie i protsessy estestvennogo samoochishcheniia Prikavkazskoi zony Chernogo moria [Man-made Pollution and Process of Natural Self-Cleaning of the Caucasus Zone of the Black Sea].* pp. 28-56. Nedra: Moscow (in Russian).

Goncharov, V. P., Neprochnov, Yu. P. & Neprochnova, I. A. 1978. *Donniy rel'ef i glubinnaya strauctura Chernogo moray [The Bottom Relief and Deep Structure of the Black Sea Basin].* Nauka: Moscow (in Russian).

Gozhik, P. F. 1984a. Radiouglerodnoe datirovanie [Radiocarbon dating]. In Shnyukov, E. F. (ed.) *Geologiya shel'fa USSR. Limany. [Geology of the Ukrainian Shelf. Limans].* pp. 38-40. Naukova Dumka: Kiev (in Russian).

Gozhik, P. F. 1984b. Istoria razvitia limanov [History of the development of the limans]. In Shnyukov, E. F. (ed.) *Geologiya shel'fa USSR. Limany. [Geology of the Ukrainian Shelf. Limans].* pp. 76-80. Naukova Dumka: Kiev (in Russian).

Gozhik, P. F. 1984c. Biostratigraficheskie kriterii. Fauna ostracod [Biostratigraphic Criteria. Ostracod Fauna]. In Shnyukov, E. F. (ed.) *Geologiya shel'fa USSR. Limany. [Geology of the Ukrainian Shelf. Limans].* pp. 35-37. Naukova Dumka: Kiev (in Russian).

Gozhik, P. F. 1984d. Stratigraficheskaia schema donnykh otlozeniy [Stratigraphic scheme of bottom sediments]. In Shnyukov, E. F. (ed.) *Geologiya shel'fa USSR. Limany. [Geology of the Ukrainian Shelf. Limans].* pp. 41-43. Naukova Dumka: Kiev (in Russian).

Gozhik, P. F., Karpov, V. A., Ivanov, V. G., & Sibirchenko, M. G. 1987. *Golotsen severo-zapadnoy chasti Chernogo moria [Holocene of the Northwestern Part of the Black Sea].* Geologicheskii Institut Ukrainskoi Akademii Nauk [Geological Institute of the Ukrainian Academy of Sciences]. Pre-Print 87-41 (in Russian).

Grigor'eva, G. V. 1980. Some sources for the study of the Upper Paleolithic in Moldavia. In Artemenko, I. I. (ed.) *Pervobytnaia Arkheologiia–poiski i Nakhodki [Primitive Archaeology–Research and Discoveries].* pp. 71–82. Naukova Dumka: Kiev (in Russian).

Hall, J. K. & Morelli, C. 2007. Status of the IHO-IOC IBCM-II 0.1' bathymetric grid for the Mediterranean. *Rapp. Comm. int. Mer Médit.* 38:24.

Hiscott, R. N., Aksu, A. E., Mudie, P. J. *et al.* 2007a. The Marmara Sea Gateway since ~16 Ka: non-catastrophic causes of paleoceanographic events in the Black Sea at 8.4 and 7.15 ka. In Yanko-Hombach, V., Gilbert, A. S., Panin, N. & Dolukhanov, P. M. (eds.) *The Black Sea Flood Question: Changes in Coastline, Climate and Human Settlement.* pp. 89-117. Springer: Dordrecht.

Hiscott, R. N., Aksu, A. E., Mudie, P. J. *et al.* 2007b. A gradual drowning of the southwestern Black Sea shelf: evidence for a progressive rather than abrupt Holocene reconnection with the eastern Mediterranean Sea through the Marmara Sea Gateway. *Quaternary International* 167–168:19-34.

Ignatov, V. S. 2008. Coastal and Bottom Topography. In Kostianoy, A & Kosarev, A (eds.) *The Black Sea Environment. The Handbook of Environmental Chemistry* (Vol. 5, Part Q). Springer-Verlag: Berlin Heidelberg.

Il'ina, L. B. 1966. Istoriya gastropod Chernogo morya [History of the Gastropods of the Black Sea]. *Trudy Paleontologicheskogo Instituta Akademii Nauk SSSR [Proceedings of the paleontological Institute of Academy of Sciences USSR]* 110: Moscow (in Russian).

Inozemtsev, Yu. I. 2013. *Quaternary deposits of Black Sea and Sea of Azov.* Doctoral Thesis. Institute of Geological Sciences of Ukrainian National Academy of Science: Kiev (in Ukrainian).

Inozemtsev, Yu. I., Lutsiv, Ya. K., Sobotovich, E. V., Kovalyukh, N. N. & Petrenko, L. V. 1984. Holocene geochronology and facies complexes of the Pontic area. In Shnyukov, E. F. (ed.) *Izuchenie geologicheskoi istorii i protsessov sovremennogo osadkoobrazovaniya Chernogo i Baltiyskogo morey [Study of the Geological History and Processes of Recent Sedimentation in the Black and Baltic*

Seas]. Part 1. pp. 103-113. Naukova Dumka: Kiev (in Russian).

IOC IHO BODC 2003. Centenary Edition of the GEBCO Digital Atlas, published on CD-ROM 693 on behalf of the Intergovernmental Oceanographic Commission and the International 694 Hydrographic Organization as part of the General Bathymetric Chart of the Oceans, 695 Liverpool, British Oceanographic Data Centre.

Ishchenko, L. V. 1974. Regularities of distribution of bottom sediments in near shore parts of the Black Sea's northwestern shelf. In *Mezhvedomstvennii respublikanskii naukovii sbornik "Geologiya uzberezhzhia i dna Chernogo ta Azovs'kogo moriv u mezhakh Ukrainskoii RSR" [Interdepartmental Republican Scientific Miscellanea "Geology of the Coast and Bottom of the Black Sea and Sea of Azov within the Ukrainian SSR"]* 4:123-130. Vishcha Shkola: Kiev (in Ukrainian).

Ivanov, V. G. (ed.) 1987. Report on the development of stratigraphic scheme and legend of Quaternary sediments of the Black Sea. *Reports of the Prichernomorskaya Expedition: Odessa* (in Russian).

Ivanov, V. A. & Belokopytov, V. N. 2013. *Oceanography of the Black Sea*. Hydrophysical Institute NAS of Ukraine: Sevastopol.

Ivanova, I. K. 1966. Stratigraphy of the Upper Pleistocene of the Middle and Eastern Europe on loesses. In Grichuk, V. P. (ed.) *Verkhniy Pleistotsen; stratigrafiya i absoliutnaya geokhronologiya [Upper Pleistocene Stratigraphy and Absolute Geochronology]*. pp. 32-66. Nauka: Moscow (in Russian).

Kadurin, V. & Kiosak, D. 2013. Perspectives of search of Palaeolithic sites on the Black Sea shelf. In Yanko-Hombach, V. (ed.) *Proceedings of the International Scientific-Practical conference "Environment of the Circumpontic region during the last 30 millennia: from past to forecast"*. 30th January – 1st February 2013, Odessa I.I.Mechnikov National University, pp. 48-51. ONU: Odessa (mostly in Russian but also includes papers written in Ukrainian & English).

Kakaranza, S. D., Avrametz, V. M., Volkov, V. A. 2007. *Explanatory notes to geological map of the northern and western part of the Black Sea and Sea of Azov bottom, scale 1:500,000, Sheets L-35-Г; L-36-A, -Б, -В, -Г;*

K-36-A, -Б; L-37-A, -B; K- 37-A;. Derzavna geologicna sluzhba Ukrainy [State Survey of Ukraine]: Kiev (in Ukrainian).

Kerey, I. E., Meriç, E., Tunoğlu, C., Kelling, G., Brenner, R. L. & Doğan, A. U. 2004. Black Sea–Marmara Sea Quaternary connections: new data from the Bosphorus, Istanbul, Turkey. *Palaeogeography, Palaeoclimatology, Palaeoecology* 204:277-295.

Kitovani, T. G. 1971. About Chaudian and Drevneuxinian deposits of the Black Sea. *Trudy instituta VNIGRI [Proccedings of the All-Russia Petroleum Research Exploration Institute VNIGRI]* 115: 87-99 (in Russian).

Kitovani, T. G., Kitovani, Sh. K., Imnadze, Z. A. & Torozov, R. I. 1982. New data about stratigraphy of Chaudian and yonger deposits of Georgia. In *Chetvertichnaya sistema Gruzii [Quaternary System of Georgia]*. pp. 26-39. Metzniereba: Tbilisi (in Russian).

Komarov, A., Bozilova, E., Filipova, M. & Oudintzeva, O. 1979. Palynological spectra and their stratigraphic interpretation. In Malovitsky, Ya. P. (ed.) *Geologiya i gidrologiya zapadnoy chasti Chernogo morya [Geology and Hydrology of the Western Part of the Black Sea]*. pp. 85-91. Bulgarian Academy of Sciences: Sofia (in Russian).

Konikov, E. G. 2007. Sea-level fluctuations and coastline migration in the northwestern Black Sea area over the last 18 ky based on high-resolution lithological-genetic analysis of sediment architecture. In Yanko-Hombach, V., Gilbert, A. S., Panin, N. & Dolukhanov, P. M. (eds.) *The Black Sea Flood Question: Changes in Coastline, Climate and Human Settlement*. pp. 435-436. Springer: Dordrecht.

Konikov, E. & Pedan, G. 2006. Geological-lithological structure of limans as a key to decoding Late Neoeuxinian and Holocene history of the Black Sea. In Yanko-Hombach, V., Gilbert, A., & Martin, R. (eds.) *Extended Abstracts of the Second Plenary Meeting and Field Trip of IGCP-521 Project "Black Sea - Mediterranean Corridor During Last 30 ky: Sea-level Change and Human Adaptation"*. 20th – 28th August 2006, Odessa National University, Odessa, Ukraine, pp. 89-91. Astroprint: Odessa.

Kremenetski, K. V. 1991. *Peleoecologya drevneyshkh zem-ledel'tsev I scotovodov Russkoy ravniny* [*Paleoecology of Ancient Farmers and Pastoralists of the Russian Plain*]. USSR Academy of Science: Moscow (in Russian).

Krystev, T. I., Limonov, A. V., Sorokin, V. M. & Starovoytov, A. V. 1990. Problem of the Chaudian on the Bulgarian shelf of the Black Sea. In Krystev, T. I. (ed.) *Geologicheskaia evoliutsiia zapadnoi chasti Chernomorskoi kotlovini v neogen-chetvertichnoe vremia* [*Geological Evolution of the Western Part of the Black Sea in Neogene-Quaternary Time*]. pp. 349–361. Institute of Oceanology Bulgarian Academy of Sciences: Sofia (in Russian).

Kryzhitsky, S. D., Rusyaiva, A. S., Krapivina, V. V., Leuipunskaya, N. A., Skrzhinskaya, M. V. & Anokhin, V. A. 1999. *Olvia. Drevnee poselenie na severnom poberez'e Cernogo morya* [*Olbia. Ancient State in the Northern Black Sea Region*]. Naukova dumka: Kiev (in Russian).

Kuprin, P. N. 2002. Lithology and paleogeography of the Neoeuxinian (Late Pleistocene) stage of the Black Sea]. *Biulleten' Moskovskogo Obshchestva ispytatelei prirody. Otdel geologicheskiy* [*Bulletin of Moscow Society of Naturalists, Geological branch*] 77:59-69 (in Russian).

Kuprin, P. N. & Sorokin, V. M. 2007. On the post-glacial changes in the level of the Black Sea. In Yanko-Hombach, V., Gilbert, A.S., Panin, N. & Dolukhanov, P. M. (eds.) *The Black Sea Flood Question: Changes in Coastline, Climate and Human Settlement.* pp. 205-220. Springer: Dordrecht.

Kuprin, P. N., Belberov, Z. K., Kalinin, A. V., Kanev, D. D. & Khrystev, T. I. (eds.) 1980. *Geologo-geofizicheskie issledovaniya bolgarskogo sektora Chernogo morya* [*Geological-Geophysical Investigations of the Bulgarian Shelf of the Black Sea*]. Bulgarian Academy of Sciences: Sofia (in Russian).

Kuprin, P. N., Sorokin, V. M., Babak, Y. V., Chernyshova, M. B. & Pirumova, L. G. 1984. Correlation of Quaternary geological sequences of the western part of the Black Sea. In Shnyukov, E. F. (ed.) *Trudy Mezh-dunarodnogo simposiuma "Izuchenie geologicheskoi istorii i protsessov sovremennogo osadkoobrazovaniia Chernogo i Baltiiskogo morey."* [*Study of Geological History and Processes of Recent Sedimentation in the Black and Baltic Seas*]. Part 1. pp. 116-122. Naukova Dumka: Kiev (in Russian).

Kuprin, P. N., Georgiev, V. M., Chochov, S. D., Limonov, A. F. & Polyakov, A. S. 1985. Characteristics and stratification of the Upper Quaternary sediments of the continental slope of the western and eastern part of the Black Sea. *Okeanologiia* 13:37-51 (in Russian).

Kvasov, D. D. 1975. *Pozdnechetvertichnaia istoriia krup-nykh ozer i vnutrennikh morei Vostochnoi Evropy* [*The Late Quaternary History of the Large Lakes and Inland Seas of Eastern Europe*]. Nauka: Moscow (in Russian).

Kvavadze, E. V. & Dzeiranshvili, V. G. 1989. Palyno-logical characteristics of the Upper Pleistocene and Holocene deposits of Kobuleti. *Soobshcheniia Akademii Nauk Gruzinskoi SSR* [*News of the Georgian Academy of Sciences*] 127:189-192 (in Russian).

Kyrvel, N. S., Pokatilov, V. P., Panchenko, G. A. & Ustinovskya, M. I. 1976. K palinologicheskoy kharak-teristike pleistotsenovykh otlozheniy Severnoy Mol-davii [On the palynological character of the Pleistocene deposits of Northern Moldova]. In Einor, O. L. (ed.) *Materialy po geologii, gidrogeologii i geokhimii Ukrainy, RSFSR i Moldovii* [*Materials on the Geology, Hydrology, and Geochemistry of Ukraine, RSFSR and Moldova*]. pp. 12-19. Naukova Dumka: Kiev (in Russian).

Larchenkov, E., & Kadurin, S. 2006. Reconstruction of north-western Black Sea coastline positions for the past 25 ky. In Yanko-Hombach, V., Buynevich, I., Chivas, A., Gilbert, A., Martin, R. & Mudie, P. (eds.) *Extended Abstracts of the Second Plenary Meeting and Field Trip of Project IGCP 521 "Black Sea–Mediterranean Corridor During Last 30 ky: Sea-level Change and Human Adaptation".* 20th – 28th August 2006, Odessa, Ukraine, p. 105. Astroprint: Odessa.

Larchenkov, E. & Kadurin, S. 2007a. Northwest-ern Black Sea shelf bottom landscapes. In Yanko-Hombach, V., Buinevich, I., Dolukhanov, P. *et al.* (eds.) *Extended Abstracts of the Joint Plenary Meeting and Field Trip of IGCP 521 and IGCP 481 projects.* 8th – 17th September 2007, Gelendzhik-Kerch, pp. 102–104. Rosselkhoz: Moscow.

Larchenkov, E. & Kadurin, S. 2007b. Paleogeographic reconstructions of Pontic Lowland and northwestern

Black Sea shelf for the past 25 k.y. *Geological Society of America Abstracts with Programs* 39:429.

Larchenkov, E. & Kadurin, S. 2011. Paleogeography of the Pontic Lowland and northwestern Black Sea shelf for the past 25 k.y. In Buynevich, I., Yanko-Hombach, V., Gilbert A. & Martin, R. (eds.) *Geology and Geoarchaeology of the Black Sea Region: Beyond the Flood Hypothesis. GSA Special Paper* 473:71-88. GSA: Colorado, USA.

Lericolais, G., Popescu, I., Guichard, F., Popescu, S.-M. & Manolakakis, L. 2007a. Water-level fluctuations in the Black Sea since the Last Glacial Maximum. In Yanko-Hombach, V., Gilbert, A. S., Panin, N. & Dolukhanov, P. M. (eds.) *The Black Sea Flood Question: Changes in Coastline, Climate and Human Settlement.* pp. 437-452. Springer: Dordrecht.

Lericolais, G., Popescu, I., Guichard, F. & Popescu, S. M. 2007b. A Black Sea lowstand at 8500 yr B.P. indicated by a relict coastal dune system at a depth of 90 m below sea level. In Harff, J. Hay, W. W. & Tetzlaff, D. M. (eds.) *Coastline Changes: Interrelation of Climate and Geological Processes. Special Paper* 426:171-188.

Lericolais, G., Guichard, F., Morigi, C., Minereau, A., Popescu, I. & Radan, S. 2010. A post Younger Dryas Black Sea regression identified from sequence stratigraphy correlated to core analysis and dating. *Quaternary International* 225:199-209.

Lericolais, G., Guichard, F., Morigi, C. *et al.* 2011. Assessment of Black Sea water-level fluctuations since the Last Glacial Maximum. In Buynevich, I., Yanko-Hombach, V., Gilbert, A. & Martin. R. (eds.) *Geology and Geoarchaeology of the Black Sea Region: Beyond the Flood Hypothesis. GSA Special Paper* 473:33-50. GSA: Colorado, USA.

Malovitsky, Ya. P. (ed.) 1979. *Geologiya i gidrologiya zapadnoy chasti Chernogo morya [Geology and Hydrology of the Western Part of the Black Sea].* pp. 85-91. Bulgarian Academy of Sciences: Sofia (in Russian).

Manheim, F. T. & Chan, K. M. 1974. Interstitial waters of Black Sea sediments: new data and review. In Degens, E. T. & Ross, D. A. (eds) *The Black Sea - Geology, Chemistry, and Biology American Association of Petroleum Geologists, Memoir* 20. pp. 155-180. Tulsa: Oklahoma.

Marret, F., Mudie, P., Aksu, A. & Hiscott, R. N. 2009. A Holocene dinocyst record of a two-step transformation of the Neoeuxinian brackish water lake into the Black Sea. *Quaternary International* 193:72-86.

Marushevsky, G. (ed.). 2003. *Directory of Azov-Black Sea Coastal Wetlands.* Wetlands International: Kiev.

McClusky, S., Balassanian, S., Barka, A. *et al.* 2000. Global Positioning System constraints on plate kinematics and dynamics in the eastern Mediterranean and Caucasus. *Journal of Geophysical Research* 105:5695-5719.

McKenzie, D. 1972. Active tectonics of the Mediterranean Region. *Geophysical Journal International.* 30:109-185.

Meisner, L. B., Gorshkoz, A. S. & Tugolesov, D. A. 1995. Neogene-Quaternary sedimentation in the Black Sea Basin. In Erler, A., Ercan, T., Bingül, E. & Örçen, S. (eds.) *Geology of the Black Sea Basin.* pp. 131-136. General Directorate of Mineral Research & Exploration: Ankara.

Meriç, E., Yanko, V. & Avşar, N. 1995. Izmit Körfenzi (Hersek Burnu – Kaba Burun) Kuvaternerstifinin foraminifer faunas [Foraminiferal fauna of the Quaternary sequence in the Gulf of Izmit (Herzek Burnu-Kaba Burun)]. In Meriç, E. (ed.) *Izmit Körfezi'nin Kuvaterner Istifi [Quaternary Sequence in the Gulf of Izmit].* pp. 105-152. Deniz Harp Okulu Komutanligi Basimevi: Istanbul (in Turkish).

Morgunov, Yu. G., Kalinin, A. V., Kuprin, P. N., Limonov, A. F., Pivovarov, B. L. & Scherbakov, F. A. 1981. *Tectonics and History of Development of the Northwestern Shelf of the Black Sea.* Nedra: Moscow (in Russian).

Motnenko, I. 1990. *Sedimentology of the Upper Pleistocene Sediments in Village Geroevskoye (Kerchenian Peninsula).* MSc Thesis. Department of Marine Geology and Sedimentology, Moscow State University, USSR.

Mudie, P. J., Aksu, A. E. & Yaşar, D. 2001. Late Quaternary dinoflagellate cysts from the Black, Marmara and Aegean seas: variations in assemblages, morphology and paleosalinity. *Marine Micropaleontology* 43:155-178.

Mudie, P. J., Rochon, A. & Aksu, A. E. 2002. Pollen stratigraphy of Late Quaternary cores from Marmara

Sea: land-sea correlation and paleoclimatic history. *Marine Geology* 190:233-260.

Mudie, P. J., Marret, F., Aksu, A. E., Hiscott, R. N. & Gillespie, H. 2007. Palynological evidence for climatic change, anthropogenic activity and outflow of Black Sea water during the Late Pleistocene and Holocene: centennial- to decadal-scale records from the Black and Marmara Seas. *Quaternary International* 167-168:73-90.

Mudie, P. J., Leroy, S. A. G., Marret, F. *et al.* 2011. Nonpollen palynomorphs: indicators of salinity and environmental change in the Caspian–Black Sea–Mediterranean corridor. In Buynevich, I., Yanko-Hombach, V., Gilbert, A. & Martin, R. (eds.) *Geology and Geoarchaeology of the Black Sea Region: Beyond the Flood Hypothesis. GSA Special Paper.* pp. 245-262. GSA: Colorado, USA.

Mudie, P. J., Yanko-Hombach, V. & Kadurin, S. 2014. The Black Sea dating game and Holocene marine transgression. *Open Journal of Marine Science* 4: 1-7.

Neprochnov, Y. P. & Ross, D. A. 1978. Black Sea geophysical framework. In Usher, J. L. & Supko, P. (eds.) *Initial Reports of the Deep Sea Drilling Project 42, Part 2* pp. 1043-1055.

Nevesskaya, L. A. 1965. Late Quaternary Bivalve Molluscs of the Black Sea, their Systematics and Ecology. *Trudy Paleontologicheskogo Instituta Akademii Nauk SSSR [Proceedings of Paleontologcal Institute of the USSR Academy of Sciences]* 105:1-391. Nauka: Moscow (in Russian).

Nevesskaya, L. A. & Nevessky, E. N. 1961. Correlation between the Karangatian and Neoeuxinian layers in littoral regions of the Black Sea. *Doklady Akademii Nauk SSSR [Proceedings of the USSR Academy of Sciences]* 137:934-937 (in Russian).

Nicholas, W. A., Chivas, A. R., Murray-Wallace, C. V. & Yanko-Hombach, V. 2008. Amino acid racemisation and AMS radiocarbon dating of Holocene Black Sea core sediments. In Gilbert A. & Yanko-Hombach V. (eds.) *Extended Abstracts of the Fourth Plenary Meeting and Field Trip of IGCP-521 - INQUA 0501.* 4th – 16th October 2008, Bucharest-Varna, pp. 125-127. GEO-ECO-MARINA: Bucharest.

Nicholas, W. A., Chivas, A. R., Murray-Wallace, C. V. & Fink, D. 2011. Prompt transgression and gradual salinisation of the Black Sea during the early Holocene constrained by amino acid racemization and radiocarbon dating. *Quaternary Science Review* 30:3769-3790.

Nikishin, A. M., Korotaev. M. V., Bolotov, S. H. & Yershov, A. V. 2001. Tectonic history of the Black Sea basin. *Byulleten Moscovskogo obshestva ispytateley prirody. Geologiya [Bulletin of Moscow Society of Naturalists, Geological Branch]* 76:3-18.

Nikishin, A. M., Korotaev. M. V., Yershov, A. V. & Brunet, M. F. 2003. The Black Sea basin: tectonic history and Neogene–Quaternary rapid subsidence modelling. *Sedimentary Geology* 156:149-168.

Nikishin, A. M., Ziegler, P. A., Bolotov, S. H. & Fokin, P. A. 2011. Late Palaeozoic to Cenozoic evolution of the Black Sea-Southern Eastern Europe Region: A view from the Russian Platform. *Turkish Journal of Earth Sciences* 20:571-634.

Nikonov, A. A. & Pakhomov, M. M. 1993. K paleogeografii poslekarangatskogo vremeni v basseine Azovskogo morya [To the post-Karangatian paleogeography of the Sea of Azov]. *Doklady Akademii Nauk SSSR* 133:753-756 (in Russian).

Okay, A. I. & Sahinturk, O. 1997. Geology of the Eastern Pontides. *American Association of Petroleum Geologists Memoir* 68:291:311.

Oktay, F. Y., Gökaşan, E., Sakinç, M., Yaltirak, C., Imren, C. & Demirbağ, E. 2002. The effects of the North Anatolian Fault Zone on the latest connection between Black Sea and Sea of Marmara. *Marine Geology* 190:367-382.

Ostrovsky, A. B., Izmailov, Ya. A., Balabanov, I. P. 1977. New data on the paleohydrological regime of the Black Sea in the Upper Pleistocene and Holocene. In Kaplin, P. A. & Shcherbakov, F. A. (eds.) *Paleogeografiia i otlozheniia pleistotsena iuzhnykh morei SSSR [Pleistocene Paleogeography and Sediments of the Southern Seas of the USSR].* pp. 131-140. Nauka: Moscow (in Russian).

Otte, M., Noiret, P. & López Bayón, I. 1997. Aspects of the Upper Palaeolithic in Central Europe. *Préhistoire européenne* 11:277-301.

Özdoğan, M. 2007. Coastal changes of the Black Sea and Sea of Marmara in archaeological perspective.

In Yanko-Hombach, V., Gilbert, A. S., Panin, N. & Dolukhanov, P. M. (eds.) *The Black Sea Flood Question: Changes in Coastline, Climate and Human Settlement.* pp. 651-670. Springer: Dordrecht.

Panin, N. & Jipa, D. 1997. Danube river sediment input and its interaction with the northwestern Black Sea: Results of EROS 2000 and EROS 2001 projects. *GEO-ECO-MARINA* 3:23-35.

Panin, N. & Popescu, I. 2007. The northwestern Black Sea: climatic and sea-level changes in the late Quaternary. In Yanko-Hombach, V., Gilbert, A. S., Panin, N. & Dolukhanov, P. M (eds.) *The Black Sea Flood Question: Changes in Coastline, Climate and Human Settlement.* pp. 387-404. Springer: Dordrecht.

Pasynkov, A. A. 2013. *Morphostructural Zoning of the Ukranian Part of the Azov-Black Sea Basin and Future Development of the Region.* Doctoral Thesis. Geological Institute of the National Academy of Ukraine: Kiev (in Ukrainian).

Peev, P. I. 2009. The Neolithisation of the Eastern Balkan Peninsula and fluctuations of the Black Sea level. *Quaternary International* 197:87-92.

Podoplelov, O. N., Karpov, V. A., Ivanov, V. G., Mokrjak, I. M. & Zagorodnij, G. G. 1973-1975. *Otchet o geologicheskoi s'emki shelfa severo-zapadnoj chasti Chernogo morja masshtaba 1:50,000 [Report on geological survey 1:50,000 within northwestern part of the Black Sea].* Krymgeologya: Odessa (in Russian).

Pop, E. 1957. Palynological investigations in Romania and their main results. *Botanicheskii zhurnal [Botanical Journal]* 42:72-79 (in Russian).

Popov, G. I. 1983. *Pleistotsen Chernomorsko-Kaspiiskikh prolivov [Pleistocene of the Black Sea-Caspian Straits].* Nauka: Moscow (in Russian).

Popov, G. I. & Zubakov, V. A. 1975. O vozraste surozhskoi transgressii Prichernomor'ia [On the age of the Surozhian transgression of the Pontic region]. In Zubakov, V. A. (ed.) *Kolebaniia urovnia mirovogo okeana v pleistotsene [Sea-level Changes in the World Ocean in the Pleistocene].* pp. 113-116. Geograficheskoe obshchestvo SSSR: Moscow (in Russian).

Put', L. L. 1981. Stratigrafiya chetvertichnykh otlozheniy [Stratigraphy of Quaternary sediments]. In Shnyukov, E. F. (ed.) *Geologiya shel'fa USSR: Kerchensky proliv [Geology of Ukrainian shelf: Kerch Strait].* pp. 33-42, Naukova Dumka: Kiev (in Russian).

Reilinger, R. E., McClusky, S. C. & Souter, B. J. 1997. Preliminary estimates of plate convergence in the Caucasus collision zone from GPS measurements. *Geophysical Reearch Letters* 24:1815-1818.

Robb, A., Nicolas, E. S. & Williams, G. 1998. Integrated structural, sequence stratigraphic and geodynamic modeling of the Western Black Sea: early Mesozoic to recent. *3rd International Conference on the Petroleum Geology and Hydrocarbon Potential of the Black and Caspian Seas Area (abstracts).* 13th – 15th September 1998, Neptun, Constanta, Bucharest, Romania, pp. 59-63.

Robertson, A. H. F., Ustaömer, T., Pickett, E. A., Collins, A. S., Andrew, T., & Dixon, J. E. 2004. Testing models of Late-Palaeozoic-Early Mesozoic orogeny in Western Turkey: support for an evolving open-Tethys model. *Journal of Geological Society* 161:501-511.

Robinson, A. G. (ed.) 1997. *Regional and Petroleum Geology of the Black Sea and Surrounding Region. American Association of Petroleum Geologists Memoir 68.* AAPG: Tulsa.

Ryan, W. B. F. 2007. Status of the Black Sea flood hypothesis. In Yanko-Hombach, V., Gilbert, A. S., Panin, N. & Dolukhanov, P. M. (eds.) *The Black Sea Flood Question: Changes in Coastline, Climate, and Human Settlement.* pp. 63-88. Springer: Dordrecht.

Ryan, W. B. F., Pitman, W. C. III, Major, C. O. *et al.* 1997. An abrupt drowning of the Black Sea shelf. *Marine Geology* 138:119-126.

Ryan, W. B. F., Major, C. O., Lericolais, G. & Goldstein, S. L. 2003. Catastrophic flooding of the Black Sea. *Annual Review of Earth and Planetary Sciences* 31:525-554.

Samoilova, T. L. 1988. *Tyras in 6th-1st Centuries BCE.* Naukova Dumka: Kiev (in Russian).

Semenenko, V. N. & Kovalyukh, N. N. 1973. Absolute age of Upper Quaternary deposits of the Azov-Black Sea basins based on radiocarbon analyses. *Geologicheskiy zhurnal [Geological Journal]* 33:91-97 (in Russian).

Shcherbakov, F. A. 1983. *Materikovye okrainy v pozdnem pleistotsene i golotsene [Continental Margins in the Late Pleistocene and Holocene].* Nauka: Moscow (in Russian).

Shcherbakov, F. A., Kuprin, P. N., Potapova, L. I., Polyakov, A. S., Zabelina, E. K. & Sorokin, V. M. 1978. *Osadkonakoplenie na kontinental'noi okraine Chernogo moria [Sedimentation on the Continental Shelf of the Black Sea]*. Nauka: Moscow (in Russian).

Shchiptsov, O. A. 1998. *Ukraina-mosrskaya derzhava [Ukraine — Marine State]*. Naukova Dymka: Kiev (in Ukranian).

Shilik, K. K. 1997. Oscillations of the Black Sea and ancient landscapes. In Chapman, J. C. & Dolukhanov, P. M. (eds.) *Landscape in Flux. Central and Eastern Europe in Antiquity*. pp. 115-130. Oxbow Books: Oxford.

Shillington, D. J., White, N., Minshull, T. A. 2008. Cenozoic evolution of the eastern Black Sea: A test of depth-dependent stretching models. *Earth and Planetary Science Letters* 265:360-378.

Shimkus, K. M. 2005. Processes of Sedimentation in the Mediterranean and Black Seas in the Late Cenozoic., Lisitzin, A.P. & Emel'yanov, E.M (eds). Scientific World Publications: Moscow (in Russian).

Shimkus, K., Komarov, A. & Grakova, I. 1977. On the stratigraphy of deep-water Upper Quaternary sediments of the Black Sea. *Okeanologiya [Oceanology]* 7:675-678 (in Russian).

Shmuratko, V. I. 2001. *Gravitatsionno-rezonansnaiia ekzotektonika [Gravity-resonance exotectonic]*. Astroprint: Odessa (in Russian).

Shmuratko, V. I. 2003. The question of the neotectonic and paleogeographic evolution of the northwestern Black Sea shelf in the Pleistocene. *Vestnik Odesskogo natsional'nogo universiteta [Odessa National University Herald]* 11:151-164 (in Russian).

Shmuratko, V. I. 2007. The post-glacial transgression of the Black Sea. In Yanko-Hombach, V., Gilbert, A.S., Panin, N. & Dolukhanov, P. M. (eds) *The Black Sea Flood Question: Changes in Coastline, Climate and Human Settlement*. pp. 221-250. Springer: Dordrecht.

Shnyukov, E. F. (ed.) 1981 *Geologiya shel'fa USSR: Kerchensky proliv [Geology of Ukrainian Shelf: Kerch Strait]*. Naukova Dumka: Kiev (in Russian).

Shnyukov, E. F. (ed.) 1982. *Geologiya shel'fa USSR: Sreda, istoria i metodika izucheniia [Geology of Ukrainian shelf: Environment, History and Methods of Study]*. Naukova Dumka: Kiev (in Russian).

Shnyukov, E. F. (ed.) 1983. *Geologiya shel'fa USSR: Tverdye poleznye iskopaemye [Geology of Ukrainian Shelf: Hard Minerals]*. Naukova Dumka: Kiev (in Russian).

Shnyukov, E. F. (ed.) 1984a. *Geologiya shel'fa USSR: Stratigrafiia (Shel'f i poberezh'ia Chernogo moria) [Geology of the Ukrainian shelf: Stratigraphy (Shelf and coast of the Black Sea]*. Naukova: Dumka: Kiev (in Russian).

Shnyukov, E. F. (ed.) 1984b. *Geologiya shel'fa USSR: Limany [Geology of Ukrainian Shelf: Limans]*. Naukova Dumka: Kiev (in Russian).

Shnyukov, E. F. (ed.) 1985. *Geologiya shel'fa USSR: Litologiya [Geology of Ukrainian shelf: Lithology]*. Naukova Dumka: Kiev (in Russian).

Shnyukov, E. F. (ed.) 1987. *Geologiya shel'fa USSR: Tektonika [Geology of Ukrainian shelf: Tectonics]*. Naukova Dumka: Kiev (in Russian).

Shnyukov, E. F. (ed.) 2002. *Geologicheskaya otsenka trassy podvodnogo kabelya Sevastopol'-Zatoka [Geological evaluation of the route of submarine cable connection Sevastopol-Zatoka]*. Logos: Kiev (in Russian).

Shnyukov, E. F. (ed.) 2011. *Geological map of the northern and western part of the Black Sea and Sea of Azov bottom, scale 1:500,000, Sheets L-35-Г; L-36-A, -Б, -В, -Г; K-36-A, -Б; L-37-A, -В; K- 37-A;*. Derzavna geologicna sluzhba Ukrainy [State Survey of Ukraine]: Kiev.

Shnyukov, E. F. & Yanko-Hombach, V. 2009. Degassing of the Black Sea: A review. In Gilbert A. and Yanko-Hombach V. (eds.) *Extended Abstracts of the Fifth Plenary Meeting and Field Trip of IGCP-521-INQUA 0501*. 22nd – 31st August 2009, Izmir-Çanakkale, Turkey, pp. 161-162. DEU Publishing House: Izmir.

Shnyukov, E. F., Maslakov, M. & Yanko-Hombach, V. 2010. Mud volcanoes of the Azov-Black Sea basin, onshore and offshore. In Yanko-Hombach V. & Gilbert A. (eds.) *Abstract Volume of the Sixth Plenary Meeting and Field Trip of IGCP-521-INQUA 0501*. 27th September – 5th October 2010, Rhodes, Greece, pp. 190-194.

Shnyukov, E. F., Pasynkov, A. A., Bashkirtseva, E. V. & Menasova, A. Sh. 2011. Manifestations of large scale degassing in the Black Sea and Sea of Azov. *Mineral'nye resursy Ukrainy [Mineral Resourses of Ukraine]* 4:35-80 (in Russian).

Shnyukov, E. F., Kobolev, V. P. & Pasynkov, A. A. 2013. *Gasoviy vulkanizm Chernogo morya [Gas Volcanism of the Black Sea]*. Logos: Kiev (In Russian).

Shuisky, Yu. 2007. Climate dynamics, sea-level change, and shoreline migration in the Ukrainian sector of the Circum-Pontic Region. In Yanko-Hombach, V., Gilbert, A. S., Panin, N. & Dolukhanov, P. M. (eds.) *The Black Sea Flood Question: Changes in Coastline, Climate and Human Settlement*. pp. 251-278. Springer: Dordrecht.

Sibirchenko, M. G., Karpov, V. A., Ivanov, V. G., Mokrjak, I. M., Zjultsle, V. G. & Tatarovsky, V. V. 1983, *Otchet po izucheniiu litilogicheskogo sostava donnykh otlozheniy shelfa Chernogo moria s zeliju sostavleniia geologolitologicheskoy karty masshtaba 1:200,000 [Report on Lithological Study of the Black Sea Shelf for the Preparation of Geological-Lithological Map 1:200,000]*. Krymgeologia: Odessa (in Russian).

Simonov, A. I. & Altman, E. N. (eds.) 1991. *Hydrometeorology and hydrochemistry of seas in the USSR (Vol IV) Black Sea, Issue 1. Hydrometeorological conditions*. Gidrometeoizdat: Leningrad (in Russian).

Sinitsyn, A. A. 2003. A Palaeolithic 'Pompeii' at Kostenki, Russia. *Antiquity* 77:9-14.

Smith, D. E., Kolenkiewics, R., Robbins, P. J. W., Dunn, J. & Torrence, M. H. 1994. Horizontal crustal motion in the central and eastern Mediterranean inferred from satellite laser ranging measurements. *Geophysical Research Letters* 21: 1979-1982.

Smyntyna, O. V. 1999. Human migrations and means of cultural-historic adaptations: problems of correlation (with the materials from Ukrainian Mesolithic sites). *Vita Antiqua* 2:13-37 (in Russian).

Smyntyna, O. V. 2004. Ecological explanation of Hunter-Gatherers behavior: an attempt of historical overview. *Social Evolution and History* 3:3-24.

Soffer, O. & Praslov, N. D. 1993. *From Kostenki to Clovis: Upper Paleolithic–Paleo-Indian Adaptations*. Plenum Press: New York.

Sorokin, V. M., Starovoytov, A. V. & Devdariani, N. A. 1998. Structure of the upper part of the sedimentary cover of Adjara Black Sea shelf. *Vestnik of Moscow University [Moscow University Herald] Series 4, Geology* 4:55-59 (in Russian).

Spadini, G., Robinson, A. & Cloetingh, S. 1996. Western versus Eastern Black Sea tectonic evolution: pre-rift lithospheric controls on basin formation. *Tectonophysics* 266:139-154.

Stanchev, H., Palazov, A., Stancheva, M. & Aposto Lov, A. 2011. Determination of the Black Sea area and coastline length using GIS methods and Landsat 7 satellite images. *GEO-ECO-MARINA* 17:27-31.

Stanko, V. N. 2007. Fluctuations in the level of the Black Sea and Mesolithic settlement of the northern Pontic area. In Yanko-Hombach, V., Gilbert, A. S., Panin, N. & Dolukhanov, P. M. (eds.) *The Black Sea Flood Question: Changes in Coastline, Climate and Human Settlement*. pp. 371-386. Springer: Dordrecht.

Stanley, D. J. & Blanpied, c. 1980. Late Quaternary water exchange between the eastern Mediterranean and the Black Sea. *Nature* 285:537-541.

Starostenko, V., Buryanov, V., Makarenko, I. *et al.* 2004. Topography of the crust–mantle boundary beneath the Black Sea Basin. *Tectonophysics* 381:211-233.

Svitoch, A. A., Selivanov, A. O. & Yanina, T. A. 1998. *Paleogeographic Events of the Ponto-Caspian and Mediterranean in Pleistocene: Data on Reconstruction and Correlation]*. Moscow State University: Moscow (in Russian).

Tari, E., Sahin, M., Barka, A. *et al.* 2000. Active tectonics of the Black Sea with GPS. *Earth Planets Space* 52:747-751.

Tchepalyga, A. L. (also Chepalyga) 1984. Inland sea basins. In Velichko, A. A. (ed.) *Late Quaternary Environments of the Soviet Union*. English edition Wright, H. E. Jr. & Barnowsky, C. W. (eds.). pp. 229-247. University of Minnesota Press: Minneapolis.

Tkachenko, G. G. 1974. Neotectonic properties of the bottom geological structure in dependence with deep structure reflection of the oil-gas bearing areas illustrated by the Golitsin Uplift, Black Sea. *Mezhvedomstvennii respublikanskii naukovii sbornik "Geologiya uzberezhzhia i dna Chernogo ta Azovs'kogo moriv u mezhakh Ukrainskoi RSR" [Interdepartmental Republican Scientific Miscellanea "Geology of the Coast and Bottom of the Black Sea and Sea of Azov within the Ukrainian SSR"]* 7:19-26 (in Ukrainian).

Tkachenko, G. G., Krasnoshchok, A. Ya., Pazyuk, L. I., Samsonov, A. I. & Tkachenko, V. F. 1970. On the role of the newest disjunctive tectonics on the formation of shoreline and morphology of the major areas of the Black and Azov Seas. *Mezhvedomstvennii respublikanskii naukovii sbornik "Geologiya uzberezhzhia i dna Chernogo ta Azovs'kogo moriv u mezhakh Ukrainskoi RSR"* [Interdepartmental Republican Scientific Miscellanea "Geology of the Coast and Bottom of the Black Sea and Sea of Azov within the Ukrainian SSR"] 4:24-33 (in Ukrainian).

Tugolesov, D. A., Gorshkov, A. S., Meysner, L. B. *et al.* 1985. *Tectonics of the Mesozoic Sediments of the Black Sea Basin*. Nedra: Moscow (in Russian).

Veklich, M. F. & Sirenko, N. A. 1976. Pliocene and Pleistocene of the Left Side of the Lower Dnieper and Crimean Plain. Naukova Dumka: Kiev (in Russian).

Ward, C. & Ballard, R. 2004. Black Sea shipwreck survey 2000. *International Journal of Nautical Archaeology* 33:2-13.

Ward, C. & Horlings, R. 2008. The remote exploration and archaeological survey of four Byzantine ships in the Black Sea. In Ballard, R. (ed.) *Archaeological Oceanography*. pp. 148-173. Princeton University Press: Princeton.

Yanina, T. A. 2012. *Neopleistocene Ponto-Caspian: biostratigraphy, paleogeography, correlation*. Moscow State University: Moscow.

Yanko, V. V. 1974. Some data about foraminifera of bottom sediments from some plots of the Northwestern and Caucasian shelf of the Black Sea]. *In Mezhvedomstvennii respublikanskii naukovii sbornik "Geologia uzberezh'ia I dna Chernogo ta Azovskogo moriv u mezhakh Ukrainskoi RSR"* [Interdepartmental Republican Scientific Miscellanea "Geology of Coastal Zone and Bottom of the Black Sea and Sea of Azov within the Ukrainian SSR"] 7:39-43 (in Ukrainian),

Yanko, V. V. 1975. *Late Quaternary Benthic Foraminifera of the Northwestern Shelf of the Black Sea*. PhD thesis. Odessa State University, Odessa (in Russian).

Yanko, V. V. 1982. Stratigraphy of the upper Quaternary sediments of the Black Sea northwestern shelf based on benthic foraminifera. In Zhuze, A. P. (ed.) *Morskaya mikropaleontologiya [Marine Micropaleontology]*. pp. 126-131. Nauka: Moscow (in Russian).

Yanko, V. V. 1989. *Quaternary Foraminifera of the Pontian-Caspian Region (the Black Sea, the Sea of Azov, the Caspian Sea and the Aral Sea): Taxonomy, Biostratigraphy, History, Ecology*. Doctoral Thesis. Moscow State University: Moscow (in Russian).

Yanko, V. V. 1990. Stratigraphy and paleogeography of marine Pleistocene and Holocene deposits of the southern seas of the USSR. *Memorie della Società Geologica Italiana* 44:167-187.

Yanko, V.V. 2013. Late Pleistocene environmental factors defining the Black Sea and the identification of potential example areas for seabed prehistoric sites and landscapes on the Black Sea continental shelf. In *Book of Abstracts of the International Conference "Under the Sea: Archaeology and Palaeolandscapes"*. 23rd – 27th September 2013, Szczecin, Poland, p. 22.

Yanko, V. V. & Gramova, L. V. 1990. Stratigraphy of the Quaternary sediments of the Caucasian shelf and continental slope of the Black Sea based on microfauna–foraminifera and ostracoda. *Sovetskaya Geologiya [Soviet Geology]* 2:60-72 (in Russian).

Yanko, V., Frolov, V. & Motnenko, I. 1990. Foraminifera and lithology of the Karangat stratotype (Quaternary, Kerchenian Peninsula). *Byulleten Moscovskogo obshestva ispytateley prirody. Geologiya [Bulletin of Moscow Society of Naturalists, Geological Branch]* 65:83-97 (in Russian).

Yanko-Hombach, V. V. 2007a. Controversy over Noah's Flood in the Black Sea: geological and foraminiferal evidence from the shelf. In Yanko-Hombach, V., Gilbert, A. S., Panin, N. & Dolukhanov, P. M. (eds.) *The Black Sea Flood Question: Changes in Coastline, Climate and Human Settlement*. pp. 149-204. Springer: Dordrecht.

Yanko-Hombach, V. 2007b. Late Quaternary history of the Black Sea: an overview with respect to the Noah's Flood hypothesis. In Erkut, G. & Mitchell, S. (eds.) *The Black Sea: Past, Present and Future. British Institute at Ankara Monograph* 42:5-20.

Yanko-Hombach, V. & Motnenko, I. 2011. Pleistocene water intrusions from the Mediterranean and Caspian Seas into the Black Sea: Reconstructions based on foraminifera. In Yanko-Hombach V. & Gilbert A. (eds). *Abstract Volume of the Seventh Plenary Meeting and Field Trip of INQUA 501.* 21st – 28th August 2011, Odessa, Ukraine, pp. 187-194.

Yanko-Hombach, V. & Motnenko, I. 2012. Reconstructions of Pleistocene water intrusions from the Mediterranean and Caspian seas into the Black sea based on Foraminifera. In *Proceedings of XV All-Russia Micropaleontological Meeting "Recent Micropaleontology".* 12th – 16th September 2012, Gelendgik, Russia, pp. 186-188.

Yanko-Hombach, V., Meriç, E., Avşar, N., Kerey, E. & Görmüs, M. 2004. Micropaleontological evidence of the Black Sea–Marmara Sea connection for the last 800 ka BP. In Yanko-Hombach, V., Görmüs, M., Ertunç, A. *et al.* (eds.). *Extended Abstracts of the 4th EMMM'2004.* 13th – 18th September 2004, Isparta, Turkey, pp. 228-230.

Yanko-Hombach, V., Gilbert, A., & Martin, R. (eds.) 2006. *Extended Abstracts of the Second Plenary Meeting and Field Trip of IGCP-521 Project "Black Sea — Mediterranean Corridor During the Last 30 ky: Sea Level Change and Human Adaptation".* 20th – 28th August 2006, Odessa National University, Odessa, Ukraine. Astroprint: Odessa.

Yanko-Hombach, V., Gilbert, A. S. & Dolukhanov, P. M. 2007. Critical overview of the Flood Hypotheses in the Black Sea in light of geological, paleontological, and archaeological evidence. *Quaternary International* 167-168:91-113.

Yanko-Hombach, V., Smyntyna, O., Kadurin, S. V. *et al.* 2011a. Black Sea level changes and adaptation strategies of prehistoric populations during the last 30 ka. *Geologiya I mineral'nye resursy Mirovogo Okeana [Geology and Mineral Resources of the World Ocean]* 2:61-94 (in Russian).

Yanko-Hombach, V., Mudie, P. & Gilbert, A. S. 2011b. Was the Black Sea catastrophically flooded during the post-glacial? Geological evidence and archaeological impacts. In Benjamin, J., Bonsall, C., Pickard, C. & Fischer, A. (eds.) *Submerged Prehistory.* pp. 245-262. Oxbow Books: Oxford.

Yanko-Hombach, V., Yilmaz, Y. & Dolukhanov, P. 2012a. *IGCP 521: The Black Sea-Mediterranean Corridor during the last 30 ka: Sea level change and human adaptation. Tales set in stone 40 years International Geoscience Programme, UNESCO Global Earth Observation Section.* pp. 34-40. France: Paris.

Yanko-Hombach, V., Leroy, S., Sintubin, M. & Schneegans, S. 2012b. Tales set in stone. *A World of Science* 10:2-11.

Yanko-Hombach, V., Yanina T. A., Motnenko I. 2013a. Neopleistocene stratigraphy of the Ponto-Caspian Corridors, In Gilbert, A. & Yanko-Hombach, V. *Proceedings of IGCP 610 First Plenary Meeting and Field Trip.* 12th – 19th October 2013, Tbilisi, Georgia, pp. 170-176.

Yanko-Hombach, V., Mudie, P. J., Kadurin S. & Larchenkov E. 2013b. Holocene marine transgression in the Black Sea: New evidence from the northwestern Black Sea shelf. *Quaternary International* 345:100-118.

Yilmaz, Y. 1997 Geology of Western Anatolia. Active tectonics of northwestern Anatolia. In Schindler, C., Pfister, M. & Aksoy, A. (eds.) *Active Tectonics of Northwestern Anatolia: the MARMARA Poly-Project: a Multidisciplinary Approach by Space-Geodesy, Geology, Hydrogeology, Geothermics and Seismology.* pp. 210. Vdf, Hochschulverlag AG an der ETH: Zurich.

Zaitsev, Yu. P., Aleksandrov, B. G. & Minicheva, G.G. (eds.) 2006. *North-Western Part of the Black Sea: Biology and Evolution.* Naukova Dumka: Kiev.

Zelenko, S. M. 2008. *Underwater Archaeology of Crimea.* Stilos: Kiev (in Russian).

Zhamoida, A. I. 2004. Problems related to the international (standard) stratigraphic scale and its perfection. *Stratigraphy and Geological Correlation* 12:321-330. (Translated from *Stratigrafiya. Geologicheskaya Korrelyatsiya* 12:3-13).

Zonenshain, L. P. & Le Pichon, X. 1986. Deep basins of the Black Sea and Caspian Sea as remnants of Mesozoic back-arc basins. *Tectonophysics* 123:181-211.

Zubakov, V. A. 1988. Climatostratigraphic scheme of the Black Sea Pleistocene and its correlation with the oxygen isotope scale and glacial events. *Quaternary Research* 29:1-24.

Zubov, N. N. 1956. *Fundamentals of the studies of World Ocean Straits.* Geographgiz: Moscow (in Russian).

Chapter 17

Late Pleistocene Environmental Factors defining the Black Sea, and Submerged Landscapes on the Western Continental Shelf

Gilles Lericolais

IFREMER, DAEI, Issy-les-Moulineaux, France

Introduction

The Black Sea semi-enclosed basin (Fig. 17.1) is bounded by Europe, Asia Minor and the Caucasus and is ultimately connected to the Atlantic Ocean via the Mediterranean and Aegean seas and various straits. The Bosporus Strait connects it to the Sea of Marmara, and the Dardanelles Strait connects it to the Aegean Sea region of the Mediterranean. These waters separate eastern Europe and western Asia. The Black Sea also connects to the Sea of Azov by the Kerch Strait (see Chapter 16, this volume for additional detail). The Black Sea water-level fluctuations are directly linked to changes in climate, river input and the balance between evaporation and precipitation, without any hysteresis effect compared to the Global Ocean (see discussion of sea level in Chapter 2, pages 34 and 42).

As a consequence of such fundamental climatic changes, there were modifications of landscapes and ecosystems extending into the Carpathian–Danubian–Pontian space, and corresponding adaptations by the human inhabitants. These changes were especially marked during the Epipaleolithic and Neolithic periods, associated with the final hunter-gatherer populations of the region and the spread of agriculture. Domestication of crops and animals is thought to have started in the

Submerged Landscapes of the European Continental Shelf: Quaternary Paleoenvironments, First Edition.
Edited by Nicholas C. Flemming, Jan Harff, Delminda Moura, Anthony Burgess and Geoffrey N. Bailey.
© 2017 John Wiley & Sons Ltd. Published 2017 by John Wiley & Sons Ltd.

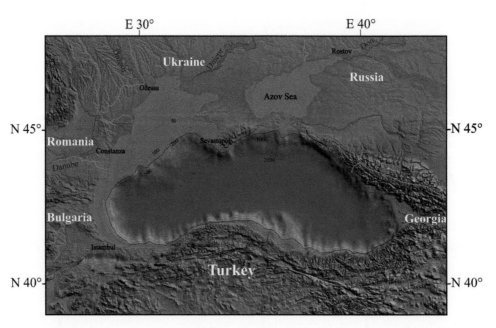

Fig. 17.1 Regional situation of the Black Sea showing simplified bathymetry, neighboring countries, and main river inputs. DTM obtained from Ryan *et al.* (2009).

Near East around 11 ka, with a spread to Anatolia and the southern part of the Black Sea by about 10 ka (Ivanov & Avramova 2000).

The northwestern Black Sea shelf (Fig. 17.1) has been exposed as much as the European shelves were during sea-level lowstands, but with a regionally distinct timing due to the isolation of the basin. This now-submerged Black Sea shelf probably provided a crucial arena for the survival and dispersal of some of Europe's earliest inhabitants during the Stone Age, the early development of prehistoric societies, the initial spread of agriculture from the Near East, and the foundations for the earliest civilizations. The oldest hominin archaeological site outside Africa is located in Georgia, 200 km inland from the Black Sea coast at Dmanisi (see for example Lordkipanidze *et al.* (2013) for descriptions of skull remains about 1.89 million years old at this site). Much of the key evidence relating to these Pleistocene developments during multiple glaciations now lies buried on the sea floor. However, the water-level fluctuations of the Black Sea behaved independently from the global sea level during its period of disconnection from the Global Ocean in successive glaciations. In the aftermath of the last ice age, water levels in the Black Sea and the Aegean Sea rose independently until they were high

enough to exchange water over the Bosporus sill. The exact timeline of this development has been subject to debate until recent dating of the re-connection obtained by Soulet *et al.* (2011b) based upon modeling experiments and micropaleontological reconstructions. For these authors, the Black Sea 'Lake' reconnection occurred in two steps, as follows: 1) Initial Marine Inflow (IMI) dated at 9000 cal BP followed by 2) a period of increasing basin salinity that led to the Disappearance of Lacustrine Species (DLS), a process lasting between 900 and 1000 years after the first reconnection, as also confirmed by Nicholas *et al.* (2011).

Regional Geology of the Black Sea Basin

The Black Sea is a land-locked basin, located between Europe and Asia Minor. It is generally considered to be the result of a back-arc extension associated with Mesozoic northward subduction of the Tethys Ocean beneath the Eurasian continent (Letouzey *et al.* 1978; Zonenshain & Pichon 1986; Finetti *et al.* 1988; Okay *et al.* 1994). At the end of the Eocene, the paleogeographic reorganization stemming from the closure of the Tethys, and the associated collision of continental blocks, resulted in the individualization of two new sedimentary realms on both sides of the Alpine orogenic belts: the Mediterranean Sea to the south, and the Paratethys to the north. The wide intracontinental Paratethys Sea extended through central Europe from the western Alpine foredeeps, toward the Aral Sea in Asia (Steininger & Papp 1979). Its present remnants are the French Rhône and Swiss Molasse basins (western Paratethys), the Pannonian basin (central Paratethys), the Dacian, the Euxinic (i.e. Black Sea) and the Aralo-Caspian basins (eastern Paratethys). The Late Eocene to Middle Miocene paleoenvironmental and paleogeographical evolution of the Paratethys was characterized by a long-term trend of decreasing marine influence and a correlated reduction in size of the sedimentation domains, both resulting from the Alpine orogenic activity (Rögl 1999; Meulenkamp & Sissingh 2003). After the Middle Miocene, Paratethyan conditions evolved into drastically restricted marine environments (Fig. 17.2).

This paleogeographical reshaping culminated in overall uplift around Paratethys, leading to the progressive isolation, dislocation and, during the Pliocene, to the final infilling of most of the western Paratethyan basins (Meulenkamp & Sissingh 2003). This Neogene evolution was characterized by several successive closure episodes of the Paratethys marine connections towards the Mediterranean Sea and Indian Ocean (Rögl 1999; Meulenkamp & Sissingh 2003). On the one hand, severance of these connections resulted in the development of largely endemic faunas and floras which led to the establishment of specific Neogene stratigraphic scales for the Paratethyan sub-basins (Papp *et al.* 1974; Papaianopol & Marinescu 1995; Rögl 1998; Chumakov 2000). On the other hand, the episodic closures of the open seaways led to potential eustatic responses within the isolated basins. Depending on the hydraulic budget of the basin, its base level would evolve toward two main tendencies during isolation phases: either a positive hydraulic budget with a rapid rise in water, or a negative hydraulic budget with a

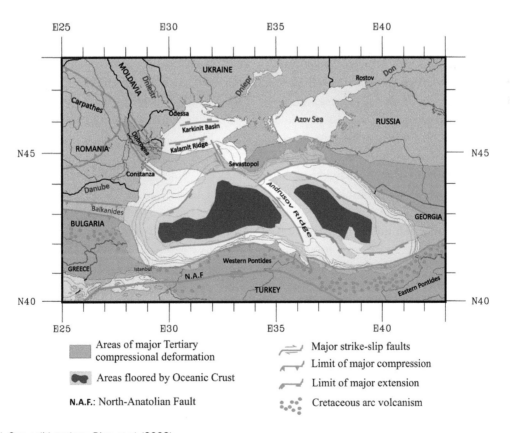

Fig. 17.2 Black Sea solid geology. Dinu *et al.* (2002).

drastic fall in level. Because the basins have relatively small superficial areas, these eustatic responses could reach large amplitudes in a very short time. In the Late Miocene, just before the Messinian Salinity Crisis in the Mediterranean Sea (Hsü *et al.* 1973), the eastern Paratethys, including the Black Sea and Dacic basin, was connected to the Mediterranean realm by a shallow sill north of the Aegean Sea (Rögl 1999; Meulenkamp & Sissingh 2003). Presence of such a connection is supported by the influx of Mediterranean fauna (NN11) recorded in the Dacic basin (Mărunţeanu 1992; Clauzon *et al.* 2005). With regard to this paleogeographical situation, it has been proposed that the Mediterranean Messinian Salinity Crisis resulted in complete isolation of the eastern Paratethys.

Bathymetry and High Resolution Data

Global bathymetry of the Black Sea (Fig. 17.1) can be obtained through the International Bathymetric Chart of the Mediterranean (IBCM), which is an intergovernmental project created to produce regional-scale bathymetric maps and data sets, together with geological/geophysical overlays, of the Mediterranean region including the Black Sea. Sponsorship of the IBCM project comes from the Intergovernmental Oceanographic Commission (IOC), a branch of UNESCO. As seafloor bathymetric data acquired with modern swath echo sounders provide coverage for only a small fraction of the global seabed, new global composite bathymetry has been built up. In 2009, a method for compilation of global seafloor bathymetry that preserves the inherent resolution of swath sonar raw data was published by Ryan *et al.* (2009). This Global Multi-Resolution Topography synthesis consists of a hierarchy of tiles with digital elevations and shaded relief imagery spanning nine magnification doublings from pole-to-pole (www.marine-geo.org/portals/gmrt). The compilation is updated and accessible as surveys are contributed, edited, and added to the tiles. Access to the bathymetry tiles is via web services and with WMS-enabled client applications such as GeoMapApp, Virtual Ocean, NASA World Wind, and Google Earth (Ryan

et al. 2009). More recently the FES2012 project from the Laboratoire d'Etudes en Géophysique et Océanographie Spatiales (LEGOS, Toulouse France) built a composite bathymetry based on ETOPO1 (Amante & Eakins 2009) and on a 1-km grid of a morpho-bathymetric map of the Mediterranean Sea derived from multibeam swath sonar surveys provided to the National Geophysical Data Center (NGDC, now the National Centers for Environmental Information (NCEI)) by Benoit Loubrieu (Institut Français de Recherche pour l'Exploitation de la Mer — IFREMER) and published and promoted by the Commission Internationale pour l'Exploration Scientifique de la mer Méditerranée (CIESM) (Loubrieu *et al.* 2008). Patches are then performed to update the original bathymetry with the most accurate local depths (or those believed to be so). Details on global and regional databases can be found on the LEGOS bathymetry database web page. All global databases have been updated before patching (www.legos.obs-mip.fr/recherches/equipes/ecola/projets/fes2012/bathymetry).

Here, part of the data used in the CIESM map was acquired in the Black Sea by the ASSEMBLAGE 5th European Project. Acquisition and reduction of multibeam bathymetry and imagery data sets were carried out to determine the sediment deposited since the last sea-level rise. Construction of Digital Terrain Models (DTM) of the bathymetry were generated to produce major maps of the western part of the Black Sea (Fig. 17.3).

Marine Sedimentology Research

For about 15 years the sedimentary systems of the northwestern part of the Black Sea extending from the continental shelf and slope down to the deep-sea zone have been studied using geophysical and coring techniques. These results (Popescu *et al.* 2001; 2004; Lericolais *et al.* 2007a,b; 2009; 2011; Popescu & Lericolais 2009; Soulet *et al.* 2010; 2011a,b; 2013) provide a robust record of water-level fluctuations in the Black Sea since the Last Glacial Maximum (LGM) and thereby shed new light on its disputed aspects. Recently, a wide range of work carried out in the Black Sea attempted

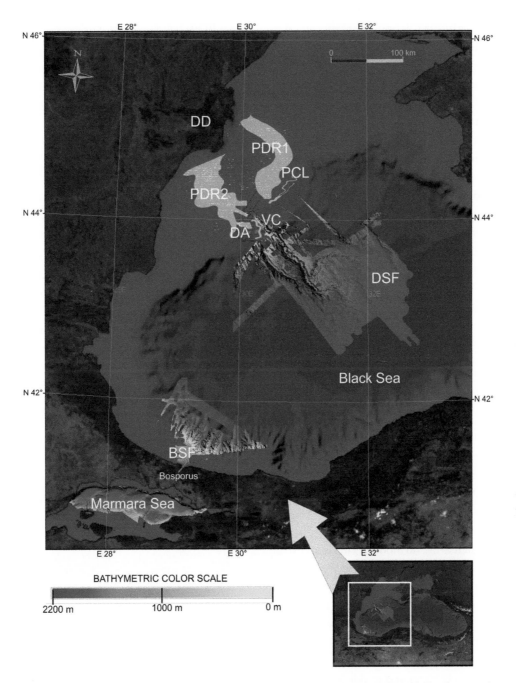

Fig. 17.3 Western Black Sea shelf presenting the location map of the geomorphological interpretation resulting from previous work. Popescu *et al.* (2001; 2004); Lericolais *et al.* (2007a). DD = Danube Delta; PCL = Paleocoastline; PDR1 = Paleo-Danube River 1; PDR2 = Paleo-Danube River 2; DA = Dunes area; VC = Viteaz canyon; DSF = Deep-sea fan; BSF = Bosporus-shelf fan.

to assess the last cycle of sea-level rise, and provide scenarios to assist in quantifying the processes governing the transition from a low-salinity to a marine state while also addressing the variability expressed by this system (see also Chapter 16, pages 458–460). Six major critical factors were established prior to reconstructing the Black Sea water-level fluctuations since the LGM (Lericolais *et al.* 2011).

1 The existence of a LGM lowstand wedge at the shelf edge off the coasts of Romania, Bulgaria and Turkey (Lericolais *et al.* 2011). This observation includes evidence of a second small lowstand wedge dated between 12,000 cal BP and 9000 cal BP at water depths from 100 m to 120 m, identified during ASSEMBLAGE cruises on the outer shelf of Romania and Bulgaria (Lericolais *et al.* 2007a,b; 2009; 2011), and described on the Turkish shelf by Algan *et al.* (2002). This wedge is associated with the recovery of strata immediately below an observed unconformity consisting of dense mud with low water content, and containing desiccation cracks, plant roots, and sand lenses rich in freshwater mollusks (*Dreissena rostriformis*) with both valves still joined together.

2 Information on the building of the Danube Delta/prodelta, showing that a former pro-delta built up at –40 m after the post-LGM meltwater pulses.

3 Mapping of meandering river channels capped by a regional unconformity, and extending seaward across the Romanian shelf to the vicinity of the –100 m isobath.

4 The presence of submerged shorelines with wave-cut terraces and coastal dunes, or delta mouth bars at depths between 80 m to 100 m below the Holocene Bosporus and Dardanelles Strait outlet sill to the Global Ocean.

5 Evidence on the western part of the Black Sea continental shelf of a shelf-wide ravinement surface, visible in high- to very-high-resolution seismic reflection profiles.

6 The presence of a uniform drape of sediment beginning at the same time above the unconformity, with practically the same thickness over nearby elevations and depressions and with no visible indication of coastal-directed onlap across the outer and middle shelf, except in the vicinity of the Danube Delta where this mud drape is overlapped by recent Danube sediments.

These critical factors enable quantification of the processes governing the transition from a semi-freshwater lake to a marine state and a better understanding of the last sea-level rise in the Black Sea. The scenario starts at the LGM about 21,000 years ago, when the Black

Sea was probably a giant freshwater lake (Soulet *et al.* 2010).

Post-LGM Climate, Sea Level, and Paleoshorelines

Water-level fluctuation scenario

The deep-sea fan studies (Popescu *et al.* 2001; Lericolais *et al.* 2012; 2013) demonstrate that the last channel–levée system on the Danube fan developed during the LGM with a water level about 120 m lower than today. The proximity of the Scandinavian-Russian ice cap supplied glacial melt water into the Black Sea through the major drainage system of the larger European rivers (Danube, Dnieper, Dniester, and Bug), and this is registered as brownish layers identifiable in cores (Fig. 17.4) (Major *et al.* 2002a; Bahr *et al.* 2005; Soulet *et al.* 2013). The volume of water brought to the Black Sea after Meltwater Pulse 1A (MWP-1A) at approximately 14,500 cal BP (Soulet *et al.* 2011b) was sufficient to raise the water level from –40 m to –20 m. The –40 m limit has been

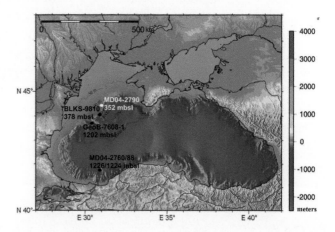

Fig. 17.4 Location of the cores used by Soulet *et al.* (2011a). The white dot indicates the location of the MD04 2790 coring site (44 12.8 N, 30 59.6 E) studied by Soulet *et al.* (2011a). Black dots represent the coring locations for previously published cores as follows: BLKS-9810 (Major *et al.* 2002b); GeoB 7608-1 (Bahr *et al.* 2005); and MD04-2760 and MD04-2788 (Kwiecien *et al.* 2008). The coring depths, in meters below sea level (MBSL), are indicated below the coring sites.

inferred from the Danube prodelta building (Lericolais *et al.* 2009), but this is not definitive and Yanko (1990) gives evidence for the −20 m limit. This last value for the transgression's upper limit would have brought the level of the Black Sea even higher in relation to the Bosporus sill, with a possible influx of marine species, such as Mediterranean dinoflagellates (Popescu 2004). Nevertheless, the rise in the water level of the Black Sea, which maintained fresh to brackish conditions, stopped deep-sea fan sedimentation.

Palynological studies conducted on BlaSON cores (Fig. 17.4) (Popescu 2004) show that, from the Bølling/Allerød to the Younger Dryas (i.e. from 14.7 ka to 12.7 ka), a cooler and drier climate prevailed. The flow from northeastern European rivers converged into the North Sea and the Baltic Ice Lake (Jensen *et al.* 1999), resulting in reduced river input to the Black Sea and a receding shoreline. These observations are consistent with an evaporative drawdown of the Black Sea and correlate with evidence of an authigenic aragonite layer present in all the cores studied (Strechie *et al.* 2002; Giunta *et al.* 2007). This drawdown is also confirmed by the forced regression-like reflectors recognized on the dune field mosaics (Lericolais *et al.* 2009). The presence of coastal sand dunes and wave-cut terraces confirms this lowstand. This had already been observed by several Russian authors, who considered a sea-level lowstand at about 90 m depth. Their observations were based on the location of offshore sand ridges described at the shelf edge south of Crimea. Around the Viteaz canyon (Danube canyon), the paleocoastline was forming a wide gulf in which two rivers (PDR1 and 2) were flowing (Fig. 17.3). Earlier studies had already proposed a depth of 105 m for this lowstand according to a regional erosional truncation recognized on the southern coast of the Black Sea (Demirbağ *et al.* 1999; Görür *et al.* 2001), but also based on the presence of a terrace on the northern shelf edge (Major *et al.* 2002b; Lericolais *et al.* 2007a).

On the Romanian shelf, preservation of sand dunes and small, buried incised valleys can be linked to a rapid transgression during which the ravinement processes related to water-level rise had insufficient time to erode the sea bottom to any substantial extent (Ahmed Benan & Kocurek 2000; Lericolais *et al.* 2004). Around

9000 cal BP, the surface waters of the Black Sea suddenly attained present-day conditions owing to an abrupt flooding of the Black Sea by Mediterranean waters, as shown by dinoflagellate cyst records (Popescu 2004) and recently demonstrated by Soulet *et al.* (2011b) and Nicholas *et al.* (2011). The inflow of marine water is confirmed by the abrupt replacement of fresh-to-brackish species by marine species. Furthermore, Soulet *et al.* (2011b) and Nicholas *et al.* (2011) demonstrate that the Black Sea 'Lake' reconnection occurred in two steps, with an Initial Marine Inflow (IMI) dated at 9000 cal BP, followed by a ca. 1000-year period of increasing basin salinity (as noted above in the Introduction, page 480). This last event can also be related to the beginning of the sapropel deposit which is widespread and synchronous across the basin slope and floor. The Black Sea basin would have been flooded in ~1000 years, equalizing water levels in the Black Sea and Sea of Marmara. Such a sudden flood at a rate of about 10 m per century would have preserved lowstand coastal marks on the Black Sea's northwestern shelf.

From these syntheses, the water-level fluctuation diagram proposed (Fig. 17.5) is judged to be the one that best fits the recently published observations.

Discussion of sea-level curve and its archaeological implications

Many decades of work carried out by scientists from Russia and other eastern European countries (Andrusov 1893; Yaranov 1938; Nevesskiy 1961; Fedorov 1963; Kvasov 1968; Muratov *et al.* 1974; Ostrovskiy *et al.* 1977; Arslanov *et al.* 1978; Shimkus *et al.* 1980; Balabanov 1984) led to the publication of different hypotheses and curves for Holocene sea-level changes in the Black Sea. Using these different works a first synthesis effort was produced by Pirazzoli (1991). However, Ryan *et al.* (1997) published a hypothesis according to which a massive flood through the Bosporus occurred in ancient times. They claim that the Black Sea was a vast freshwater lake, but then about 7500 BP, later corrected to 8400 BP (Ryan *et al.* 2003; Ryan 2007), the Mediterranean spilled over a sill at the Bosporus, creating the current communication between the Black and Mediterranean seas. Subsequent work has

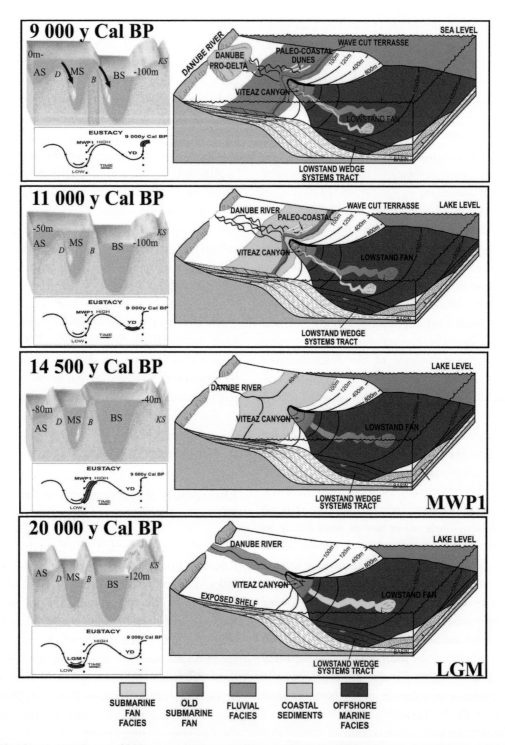

Fig. 17.5 Modified from Lericolais *et al.* (2009), where dates were calibrated using IntCAL09 with Soulet *et al.* (2011b) work and reservoir ages. The schematic scenario is inspired by Posamentier and Vail (1988) and shows water-level fluctuation in the Black Sea since the LGM, deduced from geomorphological results, supported by the Danube deep-sea fan functioning (Popescu *et al.* 2001), the results from palynology and dinoflagellates (Popescu *et al.* 2004) and the paleocoastline position. Reproduced with permission from Elsevier.

both supported and discredited this hypothesis, and it is still a matter of active debate (Aksu *et al.* 1999; 2002; Ballard *et al.* 2000; Kerr 2000; Uchupi & Ross 2000; Görür *et al.* 2001; Major *et al.* 2002b; 2006; Ryan *et al.* 2003; Algan *et al.* 2007; Balabanov 2007; Hiscott *et al.* 2007; Giosan *et al.* 2009; Lericolais *et al.* 2009; 2010). Such a late reconnection would lead to a longer exposure of the Black Sea shelf allowing human settlements to become established near the coast. This has led some to associate the supposedly catastrophic flooding of the shelf with prehistoric flood myths, and this has become one of the most visible scientific debates of recent years, and one that has fascinated the public imagination.

This controversy was one of the triggers for the installation of IGCP (International Geoscience Programme) Project 521 'Black Sea-Mediterranean Corridor during the last 30 ky: sea level change and human adaptation'. The resulting book edited by Yanko-Hombach *et al.* (2007) is a good record of the state of research from geology to archaeology carried out in the Black Sea. However, many conclusions of studies presented in the volume should be considered cautiously as it is evident that a vigorous debate is ongoing and much research remains to be done (see also Chapter 16, pages 458–460).

Recent studies supported by the European Commission (described above) led to a synthesis on the assessment of the last sea-level rise in the Black Sea obtained from observations collected by a series of expeditions carried out between 1998 to 2005, in particular, the evidence of recent submerged landscapes, the erosion of which was due to wave action around 9000 cal BP (Soulet *et al.* 2011b). Based on these results, a sea-level curve was first published in 2009 by Lericolais *et al.* (2009). This curve (Fig. 17.6) is here adapted using recent calibration of radiocarbon dates (Soulet *et al.* 2011b). Before these important results, numerous Russian authors indicated a sea-level lowstand at about 90 m depth, based on the location of offshore sand ridges described at the shelf edge south of Crimea. The unique wave-cut terrace on the outer Romanian shelf presenting an upper surface varying between –95 m and –100 m is therefore consistent with a major lowstand level situated somewhere around 100 m depth and evidence that the Black Sea shelf had emerged at the beginning of the Holocene.

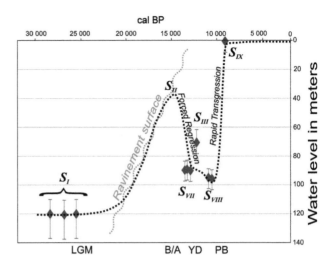

Fig. 17.6 Water-level fluctuation in the Black Sea since the Last Glacial Maximum. LGM = Last Glacial Maximum; B/A = Bølling/Allerød; YD = Younger Dryas; PB = Pre-Boreal. S_I to S_{IX} are the sequences interpreted and dated from the Romanian Black Sea shelf. Lericolais *et al.* (2009), revised version in Lericolais *et al.* (2011). Reproduced with permission from Elsevier.

Modern Coastline and Coastal Processes: the Danube Delta

Before entering the Black Sea, the Danube forms a wide, branching delta, the second largest river delta in Europe. It is located between longitude E28°45′ and E29°46 and latitude N44°25′ and N 45°37′ covering a surface of almost 5800 km² (Fig. 17.3 'DD'), confined by the Bugeac Plateau to the north and by the Dobrogea region to the south (Panin 1996). The development of the delta starts at the point where the Danube River's main channel divides into three main branches: the northern Chilia, the central Sulina and the southern St. George (Sfântu Gheorghe). A mosaic of shallow lakes and channels, fringed by reeds, lies between these branches. The Danube Delta is classified as a fluvial-dominated delta starting in a Black Sea embayment sheltered by a barrier (Panin 1983). After reaching the coast, the three main branches of the Danube Delta constitute four laterally offset lobes. Open-coast delta lobes are wave-dominated (Fig. 17.7), with the exception of Chilia III, the youngest lobe, which has a primarily fluvial-dominated morphology (Giosan

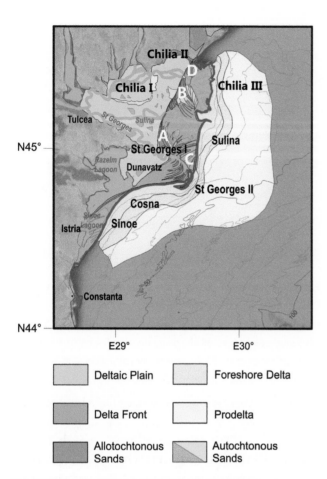

Fig. 17.7 Danube Delta morphology and lobe development sequence. After Panin *et al.* (1983); Giosan *et al.* (2006). Postulated locations for Cosna and Sinoe lobes are from Panin *et al.* (1983). Major beach ridge plains are (A) Caraorman; (B) Letea; (C) Saraturile; and (D) Jebrieni. Barrier systems of Zmeica, Lupilor, Istria, and Chituc-segment Razelm-Sinoe lagoons. Location of the ancient city of Istria is indicated.

et al. 2006). The wave-dominated lobes exhibit an asymmetric morphology (Bhattacharya & Giosan 2003). Longshore drift obstruction at distributary mouths led to development of an extensive system of beach ridges and plains on the updrift side of these lobes, whereas on the downdrift side, barrier plains developed as a succession of sandy ridges separated by elongated marshes and/or lakes (Giosan *et al.* 2006). Early interpretations of the pattern of beach ridges established a relative chronology for the open-coast lobes (de Martonne 1931; Zenkovich 1956). The St. George I was the first lobe formed at the open coast, followed by Sulina; subsequently, the St.

George arm was reactivated, developing a second lobe (Fig. 17.7). Panin (1983) explained the rapid growth of the Sulina lobe as a forced regression during a postulated Black Sea Phanagorian regression around 3 ka to 2 ka (Chepalyga 1984). The subsequent reactivation of the St. George branch was attributed to a channel slope increase at the Phanagorian lowstand (Panin 1983). Within the remnant shallow basins of the Danube embayment, the northernmost distributary, the Chilia, developed two successive lacustrine deltas before building into the Black Sea (Fig. 17.7). The Danube Delta Plain extends southward into several generations of bay-mouth barriers (Zmeica, Lupilor, Chituc, Istria), delineating the Razelm-Sinoe lagoon system (Fig. 17.7).

Coastal and Shelf Geomorpho-Dynamics, Erosion, and Accumulation

The physiographic provinces represented in Figure 17.8 are divided into four main areas: the continental shelf, the continental slope, the glacis, and the bathyal plain.

The continental shelf is well developed in the north-western part of the basin. The shelf is well marked 140 km seaward off the Danube mouth, reaching 190 km westward off Crimea with its width decreasing southerly to almost 40 km in front of the Bulgarian coast. The continental shelf is nearly absent in the southern part of the Black Sea near the Sakarya canyon. The other parts of the continental shelf are very narrow and reach no more than 20 km in width, except for the area south of the Sea of Azov where the continental shelf can reach 40 km in width. The shelf break is located at around a depth of 100 m (Ross & Degens 1974); in front of the mouth of the Danube it reaches −120 m to −140 m south of the Viteaz canyon (or Danube canyon), but can be deeper than 170 m in the northern part of the canyon, probably because of recent tectonic activity.

The continental slope related to the narrow shelf of the south, east and north-east of the Black Sea basin and south of Crimea is relatively steep (2.5% according to Ross & Degens 1974). In contrast to the continental shelf

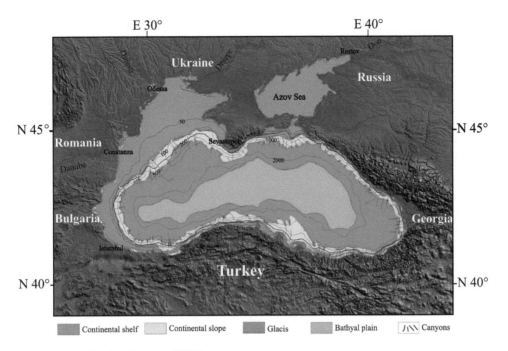

Fig. 17.8 Black Sea physiography. Panin & Popescu (2007).

the slope is incised by numerous canyons. Slopes with a lower gradient are situated in the northwestern part of the Black Sea (Danube and Dnieper deep-sea fans) and to the south of the Sea of Azov (Don and Kuban deep-sea fans). This parameter is linked to the high rate of sedimentation due to fluvial sediment inflows. In these areas wide canyons are present. The Viteaz canyon and the Dnieper canyons have incised the continental slope, and also the continental shelf landward over a distance of 20 km.

The glacis forms the area along the margin where the terrestrial inflows are deposited. With slopes ranging from 0.1% to 2.5%, its width is a function of the volume of the inflows. The width reaches its maximum at the Danube and Dnieper fan location.

The bathyal plain is located in the center of the basin showing a slope less than 0.1% and a maximal depth of 2212 m.

The present-day Black Sea catchment area is dominated by the Danube River and by the rivers from the north of the Black Sea (Dnieper, Dniester, Southern Bug). At the easternmost part of the Black Sea catchment area, the River Don approaches the Volga River. The Kuban is the biggest river bringing water from the Caucasus into the Black Sea via the Sea of Azov. The annual discharge of these six biggest rivers (Danube, Dnieper, Dniester, Southern Bug, Don and Kuban) is 270.3 km³/year (Fekete *et al.* 2000). The annual mean contribution of rivers from the Anatolian mountains is only 36 km³/year and they are not included in the report of Fekete *et al.* (2000). Most of the river input running into the Caspian Sea is due to the Volga, Ural and Kura (Stolberg *et al.* 2006). The Aral Sea receives fresh water from the Syr Darya and Amu Darya; their discharge is 60.5 km³/year (Fekete *et al.* 2000). The Ob and Yenisei (at present-day) discharge 965.1 km³/year of water into the Arctic Ocean.

Conclusion

The Black Sea is surrounded by high-folded mountain chains, i.e. the Balkanides-Pontides belts to the south/south-west, the Great and Little Caucasus to the east and by the Crimea Mountains to the north, and the Danube Delta lowland is one of its main features. This sea is one of the largest almost-enclosed seas in the world, having roughly an oval shape. The Black Sea is locked in between the southernmost tip of the Crimea and Cape Kerempe on the Turkish coast, and is connected to the Mediterranean Sea through the Istanbul-Canakkale (Bosporus-Dardanelles) straits to the west and the Sea of Azov through the Taman–Kerch Strait to the north. Such a physiography is inherited from a peculiar pattern of previous specific sea-level fluctuations. The updated results obtained from oceanographic surveys carried out in the Black Sea have led to the proposal of a scenario for the transition of the Black Sea system from a lacustrine to a marine environment.

Aside from the controversy about the conditions of the last reconnection of the Black Sea, these recent syntheses have improved the chronology of the last reconnection, indicating that it occurred around 9000 years ago. This finding shows that the reconnection was not related to catastrophic drainage of the ice-dammed Lake Agassiz. Moreover, now it is possible to confirm that the replacement of lacustrine by marine biota needed almost 1000 years, the time required for the onset of the two-way flow circulation currently observed in the Sea of Marmara gateway. These results also suggest that the level of the isolated Black Sea was below the former Bosporus sill depth. The recent results obtained from pore water analyses (Soulet *et al.* 2010) suggest that the Black Sea was a freshwater lake prior to its reconnection with the Sea of Marmara and show that microfossils are often tolerant to a wide range of salinities. For example, the taxon that Marret *et al.* (2004) used as evidence of Black Sea salinity (*S. cruciformis*), has also been found in modern sediment from the brackish Caspian Sea. Moreover, this taxon shows extreme morphological variability (Wall & Dale 1974; Mudie *et al.* 2001). Such variability may be linked to fluctuations in salinity (Dale 1996). However, no clear relationship between the different morphotypes and surface salinity has been established (Kouli *et al.* 2001; Mudie *et al.* 2002).

The debate about the effect of possible salinity variations on the probability of human occupation of the Black Sea shelf during the Neolithic does not necessarily determine whether people lived there or not. Farmers have always valued good, flat, alluvial land, and modern artisanal agriculture in the Mediterranean region shows that crops will grow very close to salt water on sheltered coastlines. Thus the proposal that the fresh Black Sea 'Lake' (Ballard *et al.* 2000; Major *et al.* 2002b; Ryan *et al.* 2003) would have allowed coastal farming on exposed

shelves is perfectly true, but a brackish or salty Black Sea (e.g. Mudie *et al.* 2001; Yanko-Hombach *et al.* 2007) would not have prevented settlement along the Black Sea coast. In either case people could have lived on the shelf, but, up till now, we do not have *in situ* evidence for such occupation.

If Neolithic farmers did live on the Black Sea shelf, a rapid flooding of the shelf would have accelerated their dispersal onto higher ground over a period of about 30 years. In spite of speculation that such a retreat from rising sea level would have influenced the rate of the spread of agriculture from the Middle East to northwest Europe, the genetic data which have been used to derive the spread of the so-called 'Neolithic Revolution' do not show a perturbation in the region of the Black Sea. The contours showing the 'frontier' or dates of Neolithization were first plotted by Ammerman and Cavalli-Sforza (1984), and have been refined and analyzed by numerous researchers more recently (e.g. Fort 2012), and there is no sign of a significant increase in rate or distortion in the process around the Black Sea. The spread of agriculture is now considered to have been caused proportionately 60% to 70% by demographic movement, and 30% to 40% by cultural transfer.

The specific sea-level fluctuations encountered in the Black Sea after the LGM allow us to think that early Neolithic populations could have lived near the Black Sea 'Lake' and experienced the rise of water level when this water body reconnected with the Global Ocean. If the level of the Black Sea drew down to a depth of 100 m below its present level between 14 ka and 10 ka, it is conceivable that one might find some buried Neolithic or earlier artifacts on the shelf of the Black Sea. Until archaeological remains are found *in situ on* the sea floor, the question remains unresolved.

References

Ahmed Benan, C. A. & Kocurek, G. 2000. Catastrophic flooding of an aeolian dune field: Jurassic Entrada and Todilto Formations, Ghost Ranch, New Mexico, USA. *Sedimentology* 47:1069-1080.

Aksu, A. E., Hiscott, R. N. & Yasar, D. 1999. Oscillating Quaternary water levels of the Marmara Sea and vigorous outflow into the Aegean Sea from the Marmara Sea-Black Sea drainage corridor. *Marine Geology* 153:275-302.

Aksu, A. E., Hiscott, R. N., Mudie, P. J. *et al.* 2002. Persistent Holocene outflow from the Black Sea to the eastern Mediterranean contradicts Noah's Flood hypothesis. *GSA Today* 12(5):4-9.

Algan, O., Gokasan, E., Gazioglu, C. *et al.* 2002. A high-resolution seismic study in Sakarya Delta and Submarine Canyon, southern Black Sea shelf. *Continental Shelf Research* 22:1511-1527.

Algan, O., Ergin, M., Keskin, S. *et al.* 2007. Sea-level changes during the Late Pleistocene-Holocene on the southern shelves of the Black Sea. In Yanko-Hombach, V., Gilbert, A. S., Panin, N. & Dolukhanov, P. M. (eds.) *The Black Sea Flood Question: Changes in Coastline, Climate, and Human Settlement.* pp. 603-632. Springer: Dordrecht.

Amante, C. & Eakins, B. W. 2009. *ETOPO1 1 Arc-minute global relief model: procedures, data sources and analysis.* National Geophysical Data Center: Boulder, Colorado.

Ammerman, A. J. & Cavalli-Sforza, L. L. 1984. *The Neolithic Transition and the Genetics of Populations in Europe.* Princeton University Press: Princeton.

Andrusov, N. I., 1893. Sur l'état du bassin de la Mer Noire pendant l'époque pliocène. *Akademiia Nauk Saint Petersburg Bulletin* 3:437-448.

Arslanov, K. A., Izmaylov, Y. A., Ostrovskiy, A. B., Tertychnyy, N. I. & Shcheglov, A. P. 1978. Radiometric age of "Karangatian" terraces of the western Caucasus and the Kerch' strait. *Doklady Akademii Nauk SSSR* [Reports of the Academy of Sciences of the USSR] 226:25-27 (in Russian).

Bahr, A., Lamy, F., Arz, H. & Wefer, G. 2005. Rapid hydrological and paleoenvironmental changes in the Black Sea during the Late Glacial and Holocene, recorded with trace element and stable isotope data. Paper presented at the *2nd ASSEMBLAGE Workshop.* 17th – 19th November 2005, Hamburg.

Balabanov, I. P. 1984. Change of wave conditions of Black Sea in Holocene. *Proceedings of the Academy of USSR Sciences* 5:70-81 (in Russian).

Balabanov, I. P. 2007. Holocene sea-level changes of the Black Sea. In Yanko-Hombach, V., Gilbert, A. S., Panin, N. & Dolukhanov, P. M. (eds.) *The Black Sea Flood Question: Changes in Coastline, Climate, and Human Settlement.* pp. 711-730. Springer: Dordrecht.

Ballard, R. D., Coleman, D. F. & Rosenberg, G. 2000. Further evidence of abrupt Holocene drowning of the Black Sea shelf. *Marine Geology* 170:253-261.

Bhattacharya, J. P. & Giosan, L. 2003. Wave-influenced deltas: geomorphological implications for facies reconstruction. *Sedimentology* 50:187-210.

Chepalyga, A. L. 1984. Inland sea basins. In Velichko, A. A., Wright, H. E. J. & Barnosky-Cathy C. W. (eds.) *Late Quaternary environments of the Soviet Union.* pp. 229-247. University of Minnesota Press: Minneapolis.

Chumakov, I. S., 2000. The problem of the Miocene-Pliocene boundary in the Euxinian region. *Stratigraphy and Geological Correlation* 8:396-404.

Clauzon, G., Suc, J.-P., Popescu, S.-M. *et al.* 2005. Influence of Mediterranean sea-level changes on the Dacic Basin (Eastern Paratethys) during the late Neogene: the Mediterranean Lago Mare facies deciphered. *Basin Research* 17:437-462.

Dale, B. 1996. Dinoflagellate cyst ecology: modelling and geological applications. In Jansonius, J. & McGregor, D. C. (eds.) *Palynology: principles and applications.* pp. 1249-1275. American Association of Stratigraphic Palynologists Foundation: Dallas.

De Martonne, E. 1931. *Europe Centrale* (2nd Ed.). Armand Colin: Paris.

Demirbağ, E., Gökasan, E., Oktay, F. Y., Simsek, M. & Yüce, H. 1999. The last sea level changes in the Black Sea: evidence from the seismic data. *Marine Geology* 157:249-265.

Dinu, C., Wong, H. K. & Tambrea, D. 2002. Stratigraphic and tectonic syntheses of the Romanian Black Sea shelf and correlation with major land structures. In Dinu, C. & Mocanu, V. (eds.) *Geology and Tectonics of the Romanian Black Sea Shelf and its Hydrocarbon Potential* (BGF vol. 2). pp. 101-117. Bucharest Geoscience Forum: Bucharest.

Fedorov, P. V. 1963. Stratigraphy of Quaternary sediments on the coast of the Crimea and Caucasus and some problems connected with the geological history of the Black Sea. *Akademiia Nauk SSSR Geologicheskii Institutii Trudy* 88:159 (in Russian).

Fekete, B. M., Voeroesmarty, C. J. & Grabs, W. 2000. *Global composite runoff fields on observed river discharge and simulated water balances (Report No. 22)*. Global Runoff Data Centre: Koblenz, Germany.

Finetti, I. R., Bricchi, G., Del Ben, A., Pipan, M. & Xuan, Z. 1988. Geophysical study of the Black Sea area. In Finetti, I. R. (ed.) Monograph on the Black Sea. *Bollettino di Geofisica Teorica e Applicata* 30:197-324.

Fort, J. 2012. Synthesis between demic and cultural diffusion in the Neolithic transition in Europe. *Proceedings of the National Academy of Sciences* 109:18669-18673.

Giosan, L., Donnelly, J. P., Constantinescu, S. *et al.* 2006. Young Danube delta documents stable Black Sea level since the middle Holocene: Morphodynamic, paleogeographic, and archaeological implications. *Geology* 34:757-760.

Giosan, L., Filip, F. & Constatinescu, S. 2009. Was the Black Sea catastrophically flooded in the early Holocene? *Quaternary Science Reviews* 28:1-6.

Giunta, S., Morigi, C., Negri, A., Guichard, F. & Lericolais, G. 2007. Holocene biostratigraphy and paleoenvironmental changes in the Black Sea based on calcareous nannoplankton. *Marine Micropaleontology* 63:91-110.

Görür, N., Çağatay, N. M., Emre, Ö. *et al.* 2001. Is the abrupt drowning of the Black Sea shelf at 7150 yr BP a myth? *Marine Geology* 176:65-73.

Hiscott, R. N., Aksu, A. E., Mudie, P. J. *et al.* 2007. A gradual drowning of the southwestern Black Sea shelf: Evidence for a progressive rather than abrupt Holocene reconnection with the eastern Mediterranean Sea through the Marmara Sea Gateway. *Quaternary International* 167-168:19-34.

Hsü, K. J., Ryan, W. B. F. & Cita, M. B. 1973. Late Miocene Desiccation of the Mediterranean. *Nature* 242:240-244.

Ivanov, I. S. & Avramova, M. 2000. *Varna necropolis: the Dawn of European Civilization*. Agat'o: Sofia.

Jensen, J. B., Bennike, O., Witkowski, A., Lemke, W. & Kuijpers, A. 1999. Early Holocene history of the southwestern Baltic Sea: The Ancylus Lake stage. *Boreas* 28:437-453.

Kerr, R. A. 2000. A victim of the Black Sea Flood found. *Science* 289:2021.

Kouli, K., Brinkhuis, H. & Dale, B. 2001. Spiniferites cruciformis: a fresh water dinoflagellate cyst? *Review of Palaeobotany and Palynology* 113:273-286.

Kvasov, D. D. 1968. Paleohydrology of eastern Europe in late Quaternary time. In Yezhegodnkkh, V., Chetniyakh, D. & Pamyati, L. S. (eds.) *Berga Doklady*. pp. 65-81. Izd Nauka: Moscow (in Russian).

Kwiecien, O., Arz, H. W., Lamy, F. *et al.* 2008. Estimated reservoir ages of the Black Sea since the last glacial. *Radiocarbon* 50:99-118.

Lericolais, G., Chivas, A. R., Chiocci, F. L. *et al.* 2004. Rapid transgressions into semi-enclosed basins since the Last Glacial Maximum. In *32nd International Geological Congress Proceedings (Abstract vol. 2)*. 20th – 27th August 2004, Florence, p. 1124. Electronic version posted on-line on July 20, 2004 ©32nd International Geological Congress.

Lericolais, G., Popescu, I., Guichard, F. & Popescu, S. M. 2007a. A Black Sea lowstand at 8500 yr B.P. indicated by a relict coastal dune system at a depth of 90 m below sea level. In Harff, J., Hay, W. W., & Tetzlaff, D. M. (eds.) *Coastline Changes: Interrelation of Climate and Geological Processes (Special Paper 426)*. pp. 171-188. Geological Society of America: Boulder.

Lericolais, G., Popescu, I., Guichard, F., Popescu, S-M. & Manolakakis, L. 2007b. Water-level fluctuations in the Black Sea since the Last Glacial Maximum. In Yanko-Hombach, V., Gilbert, A. S., Panin, N. & Dolukhanov, P. M. (eds.) *The Black Sea Flood Question: Changes in Coastline, Climate, and Human Settlement*. pp. 437-452. Springer: Dordrecht.

Lericolais, G., Bulois, C., Gillet, H. & Guichard, F. 2009. High frequency sea level fluctuations recorded in the Black Sea since the LGM. *Global and Planetary Change* 66:65-75.

Lericolais, G., Guichard, F., Morigi, C., Minereau, A., Popescu, I. & Radan, S. 2010. A post Younger Dryas Black Sea regression identified from sequence stratigraphy correlated to core analysis and dating. *Quaternary International* 225:199-209.

Lericolais, G., Guichard, F., Morigi, C. *et al.* 2011. Assessment of the Black Sea water-level fluctuations since the Last Glacial Maximum. In Buynevich, I. V., Yanko-Hombach, V., Gilbert, A. S. & Martin, R. E.

(eds) *Geology and Geoarchaeology of the Black Sea Region: Beyond the Flood Hypothesis (Special Paper 473).* pp. 33-50. Geological Society of America: Boulder.

Lericolais, G., Popescu, I, Bourget, J. *et al.* 2012. The "Sink" of the Danube River Basin: The Distal Danube Deep-Sea Fan. In Rosen, N. C., Weimer, P., Coutes dos Anjos, S. M. *et al.* (eds.) *New Understanding of the Petroleum Systems of Continental Margins of the World (32nd Annual GCSSEPM Foundation Annual Bob F. Perkins Research Conference Proceedings.* (vol. 32). 2nd – 5th December 2012, Houston (Texas), pp. 701-735.

Lericolais, G., Bourget, J., Popescu, I. *et al.* 2013. Late Quaternary deep-sea sedimentation in the western Black Sea: New insights from recent coring and seismic data in the deep basin. *Global and Planetary Change* 103:232-247.

Letouzey, J., Gonnard, R., Khristchev, K., Montadert, L. & Dorkel, A. 1978. Black Sea: Geological setting and recent deposits distribution from seismic reflection data. In Ross, D. A. *et al. Initial Reports of the Deep Sea Drilling Project* (vol. 42). pp. 1077-1084. USGPO: Washington.

Lordkipanidze, D., de Leon, M. S. P., Margvelashvili, A. *et al.* 2013. A complete skull from Dmanisi, Georgia, and the evolutionary biology of Early Homo. *Science* 342:326-331.

Loubrieu, B., Mascle, J., Benkhelil, J. *et al.* 2008. Morpho-bathymetry of the Mediterranean Sea. In CIESM (ed.) *CIESM Maps.* Ifremer Medimap Group: Monaco.

Major, C., Goldstein, S. L., Ryan, W., Piotrowski, A. & Lericolais, G. 2002a. Climate change in the black sea region through termination I from Sr and O isotopes. *Geochimica Et Cosmochimica Acta,* 66(15A):A476-A476.

Major, C., Ryan, W., Lericolais, G. & Hajdas, I. 2002b. Constraints on Black Sea outflow to the Sea of Marmara during the last glacial-interglacial transition. *Marine Geology* 190:19-34.

Major, C. O., Goldstein, S. L., Ryan, W. B. F., Lericolais, G., Piotrowski, A. M. & Hajdas, I. 2006. The co-evolution of Black Sea level and composition through the last deglaciation and its paleoclimatic significance. *Quaternary Science Reviews* 25:2031-2047.

Marret, F., Leroy, S., Chalié, F. & Françoise F. 2004. New organic-walled dinoflagellate cysts from recent sediments of Central Asian seas. *Review of Palaeobotany and Palynology* 129:1-20.

Mărunţeanu, M. 1992. Distribution of the calcareous nannofossils in the intra- and extra- Carpathian areas of Romania. *Knihovnicka Zemniho Plynu Nafty* 14b:247-261.

Meulenkamp, J. E. & Sissingh, W. 2003. Tertiary palaeogeography and tectonostratigraphic evolution of the Northern and Southern Peri-Tethys platforms and the intermediate domains of the African–Eurasian convergent plate boundary zone. *Palaeogeography, Palaeoclimatology, Palaeoecology* 196:209-228.

Mudie, P. J., Aksu, A. E. & Yasar, D. 2001. Late Quaternary dinoflagellate cysts from the Black, Marmara and Aegean seas: variations in assemblages, morphology and paleosalinity. *Marine Micropaleontology* 43:155-178.

Mudie, P. J., Rochon, A., Aksu, A. E. & Gillespie, H. 2002. Dinoflagellate cysts, freshwater algae and fungal spores as salinity indicators in Late Quaternary cores from Marmara and Black seas. *Marine Geology* 190:203-231.

Muratov, V. M., Ostrovskiy, A. B. & Fridenberg, E. O. 1974. Quaternary stratigraphy and palaeogeography on the Black Sea coast of Western Caucasus. *Boreas* 3:49-60.

Nevesskiy, E. N. 1961. Postglacial transgressions of the Black Sea. *Doklady Akademii Nauk SSSR* 137:667-670 (in Russian).

Nicholas, W. A., Chivas, A. R., Murray-Wallace, C. V. & Fink, D. 2011. Prompt transgression and gradual salinisation of the Black Sea during the early Holocene constrained by amino acid racemization and radiocarbon dating. *Quaternary Science Reviews* 30:3769-3790.

Okay, A. I., Celal Sengör, A. M. & Görür, N. 1994. Kinematic history of the opening of the Black Sea and its effect on the surrounding regions. *Geology* 22:267-270.

Ostrovskiy, A. B., Izmaylov, Y. A., Sccheglo, A. P., Arslanov, S. A. & Shchelinskiy, V. Y. 1977. New data on the stratigraphy and geochronology of Pleistocene marine terraces of the Black Sea coast, Caucasus, and Kerch-Taman region. In Kaplin, P. A. & Shcherbakov, F. A. (eds.) *Paleogeography and Deposits of the Pleistocene of the Southern seas of the USSR*. pp. 61-99. Nauka-Press: Moscow (in Russian).

Panin, N. 1983. Black Sea coastline changes in the last 10,000 years: a new attempt at identifying the Danube mouths as described by the ancients. *Dacia* 27:175-184.

Panin, N. 1996. Danube Delta: genesis, evolution, geological setting and sedimentology. *Geo-Eco-Marina* 1:7-23.

Panin, N. & Popescu, I. 2007. The northwestern Black Sea: climatic and sea level changes in the Upper Quaternary. In Yanko-Hombach, V., Gilbert, A. S., Panin, N. & Dolukhanov, P. M. (eds.) *The Black Sea Flood Question: Changes in Coastline, Climate, and Human Settlement*. pp. 387-404. Springer: Dordrecht.

Panin, N., Panin, S., Herz, N. & Noakes, J. E. 1983. Radiocarbon dating of Danube Delta deposits. *Quaternary Research* 19:249-255.

Papaianopol, I. & Marinescu, F. 1995. Lithostratigraphy and age of Neogene deposits on the Moesian Platform, between Olt and Danube Rivers. *Romanian Journal of Stratigraphy* 76:67-70.

Papp, A., Cicha, I., Rögl, F., Senes, J., Steininger, F. & Baldi, T. 1974. Principes de la subdivision stratigraphique de la Paratéthys Centrale. *Mémoire du BRGM* 78:767-774.

Pirazzoli, P. A. (ed.) 1991. *World Atlas of Holocene Sea-Level Changes. Elsevier Oceanography Series*, (vol. 58). Elsevier: Amsterdam.

Popescu, I. & Lericolais, G. 2009. An atypical sediment failure event in the Danube deep-sea fan (Black Sea). In Chiocci, F. L. (ed.) *International Conference on Seafloor Mapping for Geohazard Assessment*. 11th – 13th May 2009, Forio d'Ischia (Italy), pp. 182-185.

Popescu, I., Lericolais, G., Panin, N., Wong, H. K. & Droz, L. 2001. Late Quaternary channel avulsions on the Danube deep-sea fan, Black Sea. *Marine Geology* 179:25-37.

Popescu, I., Lericolais, G., Panin, N., Normand, A., Dinu, C. & Le Drezen, E. 2004. The Danube submarine canyon (Black Sea): morphology and sedimentary processes. *Marine Geology* 206:249-265.

Popescu, S. M. 2004. Sea-level changes in the Black Sea region since 14 ka BP. In *32nd International Geological Congress Proceedings* (Abstract vol. 2). 20th – 27th August 2004, Florence, p. 1426. Electronic version posted on-line on July 20, 2004 ©32nd International Geological Congress.

Posamentier, H. W. & Vail, P. R. 1988. Eustatic controls on clastic deposition II - sequence and systems tracts models. In Wilgus, W. K., Hastings, B. S., Posamentier, H. *et al.* (eds.) *Sea-level changes — An integrated approach*. pp. 125-154. SEPM: Tulsa.

Rögl, F. 1998. Paleogeographic considerations for Mediterranean and Paratethys seaways (Oligocene to Miocene). *Annalen des Naturhistorischen Museums in Wein* 99(A):279-310.

Rögl, F. 1999. Mediterranean and Paratethys. Facts and hypotheses of an Oligocene to Miocene paleogeography (short overview). *Geologica Carpathica-Bratislava* 50:339-349.

Ross, D. A. & Degens, E. T. 1974. Recent sediments of the Black Sea. In Degens, E. T. & Ross, D. A. (eds.) *The Black Sea - Geology, Chemistry and Biology*. pp. 183-199. American Association of Petroleum Geologists: Tulsa.

Ryan, W. B. F. 2007. Status of the Black Sea flood hypothesis. In Yanko-Hombach, V., Gilbert, A. S., Panin, N. & Dolukhanov, P. M. (eds.) *The Black Sea Flood Question: Changes in Coastline, Climate, and Human Settlement*. pp. 63-88. Springer: Dordrecht.

Ryan, W. B. F., Pitman, W. C., Major, C. O. *et al.* 1997. An abrupt drowning of the Black Sea shelf. *Marine Geology* 138:119-126.

Ryan, W. B. F., Major, C. O., Lericolais, G. & Goldstein, S. L. 2003. Catastrophic flooding of the Black Sea. *Annual Review of Earth and Planetary Sciences* 31:525-554.

Ryan, W. B. F., Carbotte, S. M., Coplan, J. O. *et al.* 2009. Global Multi-Resolution Topography synthesis. *Geochemistry, Geophysics, Geosystems* 10:Q03014.

Shimkus, K. M., Evsyukov, Y. D. & Solovjeva, R. N. 1980. Submarine terraces of the lower shelf zone and their nature. In Malovitsky, Y. P. & Shimkus, K. M. (eds.) *Geological and Geophysical Studies of the Pre-Oceanic Zone*. pp. 81-92. P.P. Shirshov Institute of Oceanology: Moscow (in Russian).

Soulet, G., Delaygue, G., Vallet-Coulomb, C. *et al.* 2010. Glacial hydrologic conditions in the Black Sea reconstructed using geochemical pore water profiles. *Earth and Planetary Science Letters* 296:57-66.

Soulet, G., Ménot, G., Garreta, V. *et al.* 2011a. Black Sea "Lake" reservoir age evolution since the Last Glacial — Hydrologic and climatic implications. *Earth and Planetary Science Letters* 308:245-258.

Soulet, G., Ménot, G., Lericolais, G. & Bard, E. 2011b. A revised calendar age for the last reconnection of the Black Sea to the global ocean. *Quaternary Science Reviews* 30:1019-1026.

Soulet, G., Ménot, G., Bayon, G. *et al.* 2013. Abrupt drainage cycles of the Fennoscandian Ice Sheet. *Proceedings of the National Academy of Sciences (U.S.A.)* 110:6682-6687.

Steininger, F. F. & Papp, A. 1979. Current biostratigraphic and radiometric correlations of Late Miocene Central Paratethys stages (Sarmatian s.str., Pannonian s.str., and Pontian) and Mediterranean stages (Tortonian and Messinian) and the Messinian Event in the Paratethys. *Newsletters on Stratigraphy* 8:100-110.

Stolberg, F., Borysova, O., Mitrofanov, I., Barannik, V. & Eghtesadi, P. 2006. *GIWA Regional Assessment 23 — Caspian Sea.* University of Kalmar: Sweden.

Strechie, C., André, F., Jelinowska, A. *et al.* 2002. Magnetic minerals as indicators of major environmental change in holocene black sea sediments: preliminary results. *Physics and Chemistry of the Earth, Parts A/B/C* 27:1363-1370.

Uchupi, E. & Ross, D. A. 2000. Early Holocene Marine Flooding of the Black Sea. *Quaternary Research* 54:68-71.

Wall, D. & Dale, B., 1974. Dinoflagellates in the late Quaternary deep-water sediments of the Black Sea. In Degens, E. T. & Ross, D. A. (eds.) *The Black Sea — Geology, Chemistry and Biology.* pp. 364-380. American Association of Petroleum Geologists: Tulsa.

Yanko, V. V. 1990. Stratigraphy and paleogeography of marine Pleistocene and Holocene deposits of the southern seas of the USSR. *Memorie Societa Geologica Italiana* 44:167-187.

Yanko-Hombach, V., Gilbert, A. S., Panin, N. & Dolukhanov, P. M. (eds.) 2007. *The Black Sea Flood Question: Changes in Coastline, Climate, and Human Settlement.* Springer: Dordrecht.

Yaranov, D. 1938. Essai sur le climat de la Bulgarie pendant le pliocène et le quaternaire (contribution à l'étude paléoclimatologique de la région méditerranéenne). *Bulletin de l'académie bulgare des sciences* 53:1-29.

Zenkovich, V. P. 1956. Zagadka Dunaiskoi Delty. *Priroda* 45:86-90 (in Russian).

Zonenshain, L. P. & Pichon, X. 1986. Deep basins of the Black Sea and Caspian Sea as remnants of Mesozoic back-arc basins. *Tectonophysics* 123:181-211.

Chapter 18

Submerged Prehistoric Heritage Potential of the Romanian Black Sea Shelf

Glicherie Caraivan,[1] Valentina Voinea,[2] Daniela Popescu[3] and Corneliu Cerchia[1]

[1] National Research and Development Institute for Marine Geology and Geoecology, Constanta Branch, Romania
[2] National Museum of History and Archaeology of Constanta, Romania
[3] Basin Water Administration – Dobrogea Littoral, Constanta, Romania

Introduction

The Black Sea is one of the largest enclosed seas in the world, covering an area of about 4.2×10^5 km^2 and possessing a maximum depth of 2212 m. Its total volume is 534,000 km^3, but most of the water (the 423,000 km^3 that lies below a depth of 150–200 m) is anoxic and contaminated with H$_2$S. The Bosporus and Dardanelles straits provide the sole connection between the Black and Mediterranean seas. The Bosporus is narrow (0.76–3.6 km) and shallow (at present 32–34 m at the sill). It restricts the two-way water exchange between the very saline eastern Mediterranean Sea (with a salinity of 38–39‰) and the more brackish Black Sea (about 17‰ at the surface and 22‰ at the bottom).

The northwestern part of the Black Sea is especially suitable for a study of sedimentation and coastline migration during the Late Quaternary. Here, the continental shelf widens dramatically and encompasses about 25% of the total area of the sea. It is also here where two of the largest rivers in Europe deliver their water and sediment load: the Danube, with a water discharge of about 200 km^3/year, and the Dnieper, contributing 54 km^3/year. The discharges of smaller rivers, such as the Dniester (9.77 km^3/year) and southern

Submerged Landscapes of the European Continental Shelf: Quaternary Paleoenvironments, First Edition.
Edited by Nicholas C. Flemming, Jan Harff, Delminda Moura, Anthony Burgess and Geoffrey N. Bailey.
© 2017 John Wiley & Sons Ltd. Published 2017 by John Wiley & Sons Ltd.

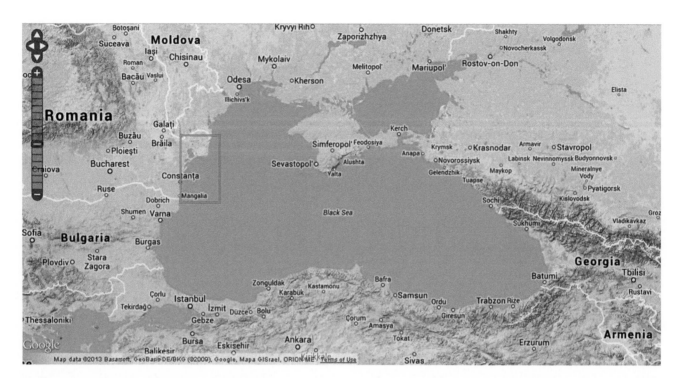

Fig. 18.1 Romanian Black Sea shelf. Source Google Maps.

Bug (2.58 km³/year), add a little more, bringing the total inflow into the northwestern Black Sea to about 266 km³/year (Fig. 18.1).

Romanian Black Sea Coast

Hydro-meteorological regime

Air temperature

The coastal climate is characterized by the existence of four seasons. The stabilizing effect of the Black Sea means that the variations in air temperature are lower than in adjacent regions. The average annual temperature is 11.3°C; the average temperature for the summer (June–August) period is about 21°C, and for winter (December–February) is 1°C.

Precipitation

The average rainfall measures about 380 mm/year. Monthly averages fall within a relatively narrow range (27.1–38.4 mm/month). However, in recent years, the values were significantly different from the multi-annual record, and varied significantly from year to year. For example, in August 2003, there was only 0.2 mm, while in

August 2004, the values increased to 259 mm, which led to landslides and erosion processes of several cliff sectors, and especially in Eforie Nord. In September 2005, an exceptionally heavy rainfall (over 1,000 mm in a few hours) was recorded, causing major floods and damage in Tuzla and Costinesti beach, as well as in other parts of the coastal area.

Wind regime

The dominating and strongest air currents are from the north, north-east and north-west. In the fall, over 42% of winds are northerly and northeastern, while in the winter over 36% blow from the north and west and in spring 32% of the winds blow from the north-east and southeast. Annual frequency of days with calm conditions as measured at Sulina is 2.8%. Depending on wind speed, on about 45% of the days winds blow with speeds of 1 m/sec to 5 m/sec, 42% between 5 m/sec to 10 m/sec, 9% between 10 m/sec and 15 m/sec, and 2% faster than 15 m/sec (Bondar *et al.* 1967). At winds above 15 m/sec, the coastal waters carry around 80% to 90% of the sediments brought by the Danube to the south.

An important feature of the strong winds on the Romanian Black Sea coast is the marine storms, with wind speeds exceeding 10 m/sec. Northeastern storms last for an average of 107 hours, of which about 47 hours is at peak speeds of over 28 m/sec (Panin 1997).

Sea level

In Constanta, where sea-level measurements have been made since 1933, the record reveals an increase in the annual average of approximately 2.2 mm/year. The Black Sea level is dependent on global climate change and Global Ocean level, the water intake from the Danube, and the astronomical sea and wind regime. The interseasonal variations of sea level in response to changes in the Danube water input are of amplitudes of about 24 cm. Sea-level semidiurnal astronomical variations occur with periods of about 12 hours and 25 minutes and the maximum amplitude is approximately 11 cm. Atmospheric pressure variations combined with the flux or reflux produce amplitudes in sea-level variations of up to 120 cm. In addition, there are some resonance oscillations (seiche) with significant values. Seiches (sudden oscillatory changes in sea level due to variations in atmospheric pressure) have an average vertical change of 50 cm. The highest value of a seiche is cited by Bondar and Emanoil (1963) and measured in December 1960 at Sulina, indicating a sea-level rise of 2 m.

Waves

In front of the deltaic coastal area, down to depths of 15 m (5–15 km from shore), waves can reach heights of 2.5 m and lengths of up to 35 m, with 59% of the waves propagating from the north, north-east and east, and 41% from the south-east and south. Bondar (1972) considers that once every 10 years, at Sulina, waves can reach typical hurricane features (produced by winds with speeds of 22 m/sec).

Measurements made at the Gloria drilling platform show that the overall value of the open sea 100-year return wave height (with statistical repeatability once every 100 years), depending on the direction, is 14.2 m from the north and 5.7 m from the south-west.

Currents and sediment transport

On the Romanian Black Sea coast, due to prevailing wind direction and sea basin configuration, currents are predominantly north–south, parallel to the coast. In periods of calm air, the measured value of the longitudinal north–south current is 3 cm/sec to 50 cm/sec (Bondar & Roventa 1967). During winds from the north and north-east (14–15 m/sec), current speeds can reach 1 m/sec at the surface and 0.2 m/sec to 0.3 m/sec at the bottom. Currents in the opposite direction occur during winds from the south and south-east. At the deltaic mouths, fresh water flows out perpendicular to the shoreline and fans out into the sea in plumes reaching up to 3 km from land (Gâştescu 1986).

In the Sulina–St. George section, net sediment transport along the coast takes place in two directions:

1 Oriented mainly north–south, just south of the Sulina piers (Fig. 18.2) (induced by the tombolo effect), at an annual rate of 190,000 m³/year transport (ranging from 130,000 m³/year at 6 m depth and 250,000 m³/year at 12 m depth (Giosan *et al.* 1997).

2 Clockwise current diverted by the Sulina piers, oriented northward and progressively closer to the shore. Net transport of sediments increases from zero at the point that separates the two tracks up to 800,000 m³/year in Casla Vadanei–St. George, remaining constant up to St. George (Fig. 18.2).

Danube water and sediment input

When it reaches the Danube Delta, the Danube water input has an annual average of about 198.3 km³, corresponding to a water flow of about 6283 m³/sec. Most of the Danube water input (approximately 90.1%) flows directly into the Black Sea through the three mouths of the Chilia, Sulina and St. George branches. The remaining water (about 9.9%) first passes through the inner delta and Razelm-Sinoe lagoon complex, then flows into the Black Sea through channels, streams and the rest of the Danube. In 2000, the distribution of the Danube water input in the Black Sea through the mouths of the three main branches was about 51.6% (Chilia), 19.9% (Sulina) and 24.4% (St. George). The Danube water input in the Black Sea is dominant, accounting for about 60% of all tributary rivers (Bondar & Roventa 1967).

When it enters the Danube Delta, the Danube's alluvial annual input averages 53.27 million tons,

Fig. 18.2 Longshore sediment transport model for the Danube Delta coast (the transport along Chilia lobe is from Shuisky and Vykhovanets (1984). (1) to (11) represent coastal sectors. High numbers represent transport rates in thousands of cubic meters per year. Circled + and − represent advancing and retreating sectors respectively. Giosan *et al.* (1997). Reproduced with permission.

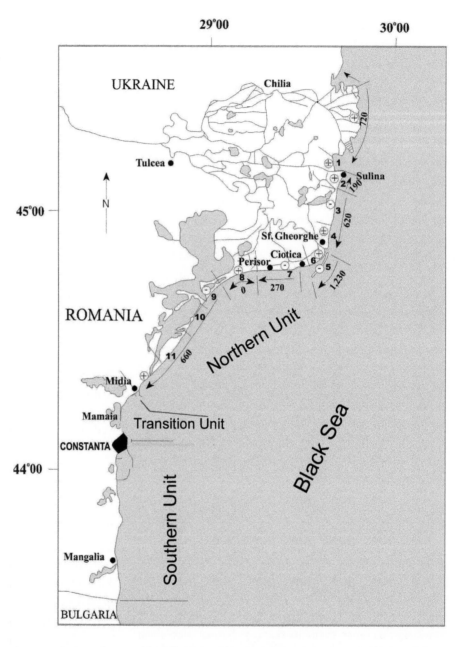

corresponding to a flow of about 1688 kg/sec of silt and a concentration of suspended sediment of about 269 g/m³. Most of the Danube's contribution of sediments (about 91%) flows into the Black Sea through the mouths of the three rivers: Chilia, Sulina and St. George. The remaining sediment (about 9%) accumulates inside the Danube Delta. In the last three decades, the decreasing trend of the silt quantities transmitted from the Danube to the Black Sea was estimated at a rate of about 7.35 kg/sec each year. In 2000, the distribution of silt input in the Black Sea via

the Danube's main branches was about 53.3% (Chilia), 5.8% (Sulina) and 21.9% (St. George). Recent regulation of the St. George branch caused a significant increase in its alluvial transport capacity.

Geomorphological data

The Romanian coast is situated between the southern-most part of the Chilia secondary delta, in the north, and the border with Bulgaria (Vama Veche), in the

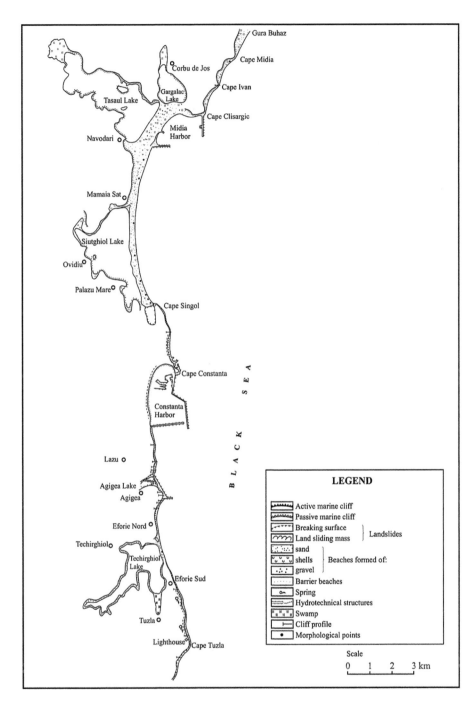

Fig. 18.3 Geomorphological map of the Romanian coast between Cape Midia and Cape Tuzla. Caraivan (2010).

LEGEND

Active marine cliff	
Passive marine cliff	
Breaking surface	} Landslides
Land sliding mass	
sand	
shells	} Beaches formed of:
gravel	
Barrier beaches	
Spring	
Hydrotechnical structures	
Swamp	
Cliff profile	
Morphological points	

Scale

0   1   2   3 km

south. Its length totals about 243 km (Fig. 18.3). From a geological, sedimentological and geomorphological point of view, the coast can be divided in two sections. The limit between them is conventionally located at Cape Midia. Nevertheless, Cape Midia represents the limit between the two sections only from the geological and geomorphological point of view (northernmost limit of the cliffs), whilst when taking into account the source of the littoral sediments, the original limit is situated more southwards. Thus, until the building of the harbor at Midia, the littoral section under the influence of the Danube had also included Mamaia Bay.

The existing relationship between the sediment quantity available for transport, processing, and accumulation, and sea energy, in terms of waves and marine currents, determines coastal evolution in both sections. The coastal current that redistributes the littoral sediments is oriented north–south, parallel to the shoreline. Each section has its own specific sources of sediments and energetic influence from the sea.

The northern unit

With a length of about 160 km, the northern unit is located between the border with Ukraine and Cape Midia. This unit represents the beaches in front of the Danube Delta, consisting of sandy littoral bars that set the limit between the inner part of the delta and the sea, or cut off the former lagoons and sand bars. Danubian-borne sediments, redistributed by the littoral currents, represent the main source of sediments for this unit. The essential feature of the sediments is an arenite mineral fraction mainly made of quartz, with some local heavy minerals in addition. The carbonate ratio, represented by shells and mollusk shells fragments from the beach sediments, increases from north to south, from a few percentages (Sulina) to over 90% (Periboina). This variation is due to the increasing distance from the Danube mouth. The northern part of the littoral is included in the Danube Delta Biosphere Reserve (Fig. 18.2).

The southern unit

With a total length of about 80 km, the southern unit is located between Cape Midia and Vama Veche. This unit consists of cliffs separated by low sandy shores (Mamaia, Eforie, Costinesti, Olimp–Mangalia). According to Shepard's generic classification (1954), the southern part of the Romanian littoral belongs to the secondary shores category with two main sub-types: erosional type (with cliffs) and depositional type (barrier type shores).

The geology of the coastal zone affects the amount of beach sediments, and influences the entire morphology of the southern coastal zone. The shore presents distinct configurations, reflecting local differences between the evolution of coastal processes and the coastal zone geology. Thus, the main feature of the southern littoral

area is morpodynamic processes associated with the evolution of cliffs. Another important characteristic of this area is the gradual transition from the Danubian terrigenous facies to the organogenous one. This feature, together with other specific geomorphological features, allows the separation of two distinct regions:

Cape Midia–Cape Singol, with some transitional features between the northern and the southern units; the main element is the presence of a big sandy complex barrier beach developed between the active cliffs (Fig. 18.3).

Cape Singol–Vama Veche, with dominant active cliff shores as a distinctive feature, separated by sandy littoral bars.

Erosional coasts

The most important geomorphological element for this type of coast is cliffs. The presence and evolution of the cliffs in the southern zone is the result of the interaction between geological and tectonic-structural factors (specific for central and southern Dobrogea) with other environmental factors controlling coastal processes. Relatively high cliffs shaped by waves and currents during a slow sea-level rise are a characteristic feature of the coast between Cape Singol and Vama Veche. In some areas cliffs are actively retreating, in other areas they are protected by accumulations of sand at the base of the cliff.

The lithological and structural irregularities of the cliff line determine the general curved configuration of the coast, with distinct changes in shoreline orientation. Active cliff shores are extensive, occurring in the northern Eforie–Belona, Cape Turcului–Cape Tuzla and Cape Tuzla–Vama Veche areas. In this latter section (Vama Veche), the cliff is discontinuous due to the presence of several littoral sand accumulations in front of Costinesti, Tatlageacul Mare and Mangalia Lake (as well as the Mangalia and Comorova swamps). The headlands are most prominent in hard-rock areas and are most exposed to marine abrasion. From north to south, the main actively eroding capes are Midia, Ivan, Turcului, Tuzla and Aurora. Other capes, such as Clisargic, Singol and Constanta, have a different configuration due to the construction of large harbor breakwaters for the port of

Midia at the town of Navodari, for the port of Pescarie at Cape Singol, and at Constanta harbor.

Accumulative coasts

There are two types of accumulative shores that can be found between Cape Midia and the Vama Veche area: beach-barriers (Techirghiol, Costinesti, Tatlageac, Mangalia littoral bars) and complex beach-barriers (the accumulative shores between Cape Midia and Cape Ivan, Cape Ivan and Cape Clisargic, and the littoral bar extending between Cape Clisargic and Cape Singol). All these shores are depositional in origin, being built up during a substantial and continuous input of sediments. Nowadays, erosional processes with different intensities from one coastal sector to another affect all of them. A steady decrease in the Danubian sediment input has generated an important sedimentary deficit in the southern part of the Romanian littoral during recent decades (especially after the beginning of the 1980s, and the building of the Midia harbor jetties).

Coastal zone geology

Stratigraphy

The littoral zone is a continuously changing geomorphological and geological unit, located at the boundary between the marine and terrestrial domains. For the Romanian littoral, the terrestrial domain consists of Dobrogea and the Danube Delta. More specifically, the littoral components are south Dobrogea (the South Dobrogea Platform), central Dobrogea (Central Dobrogea Massif), north Dobrogea (North Dobrogea orogeny) and the Predobrogean Depression.

South Dobrogea is limited to the north by the Capidava-Ovidiu Fault. The crystalline basis of this unit varies from Precambrian to Quaternary, interrupted by a few sedimentary lacunae. The evolution and characteristics of the littoral zone are equally determined by current dynamic factors and by geodynamic factors related to the behavior of the geological units mentioned above. Most of the Romanian littoral is not in direct contact with the geological formation outcrops in Dobrogea. Therefore, in its northern unit, the Romanian littoral represents

the front of the Danube Delta (with the Razim-Sinoe lagoon complex). The deposits in the littoral zone, as well as the deltaic and lacustrine ones near the shore, belong to the Quaternary. Outcrops of pre-Quaternary deposits can be encountered in the littoral zone from Cape Midia to the south, as far as Vama Veche and further on, into Bulgaria. Therefore, the Late Precambrian green schist outcrops at Histria Fortress, near Vadu village, on the banks of Tasaul Lake and at Cape Midia. The Late Jurassic (Oxfordian and Kimmeridgian) limestone surfaces at Cape Ivan and Cape Clisargic and on the banks of Tasaul Lake (Chiriac 1960; Drăgănescu 1976). Around Siutghiol Lake there are Cretaceous deposits belonging to the Barremian, Aptian and Senonian (Chiriac 1960).

The Paleogene in the South Dobrogea Platform follows uplift at the end of the Cretaceous. The Eocene has been paleontologically dated and the Oligocene identified on the basis of lithological criteria. The study of lithology in cores revealed a series of clays and bituminous schist, found south of Mangalia and around Vama Veche. It is believed that the sulfurous water springs around Mangalia are related in some way to this formation (Mutihac & Ionesi 1974). The Sarmatian represents the most widespread formation of the south Dobrogea shore aside from the Quaternary, and forms the base of the marine cliffs. The studies carried out by Chiriac (1960) revealed the presence of the Middle Sarmatian (Bessarabian) and the Late Sarmatian (Kersonian).

The Early Pleistocene from the southern part of the Romanian littoral consists of green, white and red clays with gypsum concretions and gravel intercalations (Liteanu & Ghenea 1966; Conea 1970). Thick loess deposits represent the Middle and Late Pleistocene on the Black Sea shore. The loess deposits in Dobrogea show that the wind was the main transport agent for the detrital material. The older loess layers have similar grain size throughout Dobrogea. The last and the second-to-last layer of loess indicates the sorting of the detritus material from very fine sand on the Danubian margin of Dobrogea, to silt close to the sea. Following the same direction, a decrease in the thickness of the layers can be observed. Another source of detrital material for the loess deposits was the littoral zone and its sediments. In this way, we may explain the larger grain size of the last loess layer right next to the shore. In the Quaternary, south Dobrogea evolved as a region of uplift. A layer of loess and red-brown clays covers the pre-Quaternary formations from central and southern Dobrogea. At the base of the loess layer, on top of a weathered area from older geological formations, there are red or yellow-red clays with several calcareous and iron-manganese concretions, gypsum crystals and fragments of pre-existing Dobrogean rocks (Conea 1970).

The total lack of fossils makes the task of dating the clays very difficult. Macarovici (1968) compared them to the *Scythian* clays from Russia, which, based on fossil mammal remnants, were attributed to the Apşeronian (Conea 1970). Conea also attributes the same Early Pleistocene age to the red clays (Conea 1970). Caraivan (1982) considers that during the Early Pleistocene, climatic oscillations occurred under severe terrestrial weather conditions. Thus, soil strata (to become afterwards the red clays) formed in warmer periods, while in the colder periods, limestone gravels accumulated in a red clay matrix.

Loess deposits are widespread, covering most of the region (with only rare outcrops of pre-Quaternary deposits). In the loess deposits, Conea (1970) identified seven groups of soils, starting with the latest level positioned right under the youngest loess. Some of the soil groups are not well developed (probably interstadials), while others are well developed, representing interglacial stages. Regarding the grain size, the loess is predominantly silty. Nevertheless, the loess deposits surfacing near the Danube river bank have a more sandy character. The mineralogical characteristics of the loess deposits have a net predominance of light minerals, with quartz, feldspars, and muscovite present in approximately equal proportions. Based on mineralogical differences, the Dobrogean loess formations can be divided into three provinces: (1) a Danubian province (the role of Danubian alluvium is predominant); (2) a central province (area of green schists — central Dobrogea), which has compositional similarities with southern Dobrogea; and (3) a series represented by zones of limited extent in the northern part of Dobrogea, characterized by loess deposits with an important local input (Babadag basin, the area neighboring the Razelm-Sinoe lagoon complex).

The sources of loess material are multiple. Northeastern winds brought material from glacio-fluvial deposits in the southern part of Russia. The Danube alluvium was a significant source for only the two youngest loess layers. Marine sediments represent another source, which explains the coarser structure of the most recent loess layer, along a narrow band near the coast. The weathering crust represents another source. In south Dobrogea, the loess generally covers the surface with more or less horizontal layers. In central Dobrogea, the loess covers only the watershed area between the Danube and the Black Sea basins, as well as along its western part. Secondary, partially stratified loess can also be found in depressions, where it is formed from remnant material transported by rain and wind. The wind factor has played a secondary role in the present-day aspect of the loess cover in central Dobrogea. The material was initially wind-transported, then eroded from the watershed and re-deposited in depressions, where it accumulated in thick layers. Here, torrents cut deep gullies revealing the loess stratification.

Tectonics

Due to the Baikalian orogenesis, the bedrock of the South Dobrogea Platform is divided into several blocks parallel with the Palazu Fault. The highest block is the northern compartment — Palazu Mare — where the crystalline bedrock was found at depths of 1 km. Southwards, the bedrock lowers, while further south, at the border with Bulgaria, a new uplift was noted, called the Mangalia Uplift. Overall, the South Dobrogea Platform is uplifted along the Danube Fault.

The oldest deposits dated from the base belong to the Late Silurian. The first sedimentation cycle, during which the central part of the platform functioned as a depression area with very active subsidence, ends in the Devonian. The movements from the Ardenian phase, when a reactivation of the older faults took place, influenced the fractured bedrock, causing a weak deformation of the Silurian cover. Towards the end of the Paleozoic and during a significant part of the Mesozoic, the South Dobrogea Platform functioned as an emerged region liable to denudation because of the Hercynian orogenesis. A new cycle of sedimentation began in the

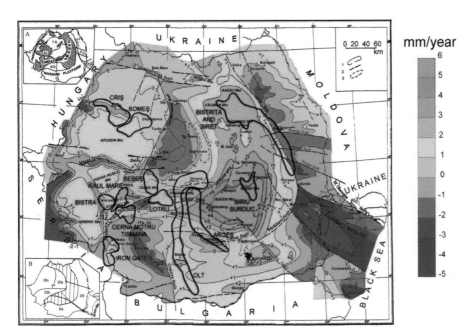

Fig. 18.4 Present vertical crust movement (mm) map from Dobrogea. Polonic *et al.* (1999) in Dimitriu and Sava (2007). Reproduced with permission.

Middle Jurassic and continued in the Cretaceous. During this entire time, the South Dobrogea Platform functioned as a typical epicontinental region, forming carbonated and reef facies, with vertical oscillations, which led to frequent discontinuities in sedimentation and to the development of evaporites and continental deposits. The South Dobrogea Platform was deeply influenced by the Meso-Cretaceous orogenesis, because, while the Late Cretaceous deposits have a horizontal or quasi-horizontal position, those of the Jurassic and Early Cretaceous are slightly folded.

Vertical earth movements

The present vertical crust movement (mm/year) map made by Polonic *et al.* (1999) illustrates the Dobrogea sector (Fig. 18.4). The geotectonic blocks of the pre-Dobrogean Depression, in north and central Dobrogea, marked by the St. George, Peceneaga-Camena and Capidava-Ovidiu faults, suffer subsidence movements to varying extent. In contrast, the southern Dobrogea tectonic block, marked by the Capidava-Ovidiu Fault and the Intra-Moesic Fault is relatively stable geodynamically, undergoing slight uplift near the Bulgarian border. In all these tectonic blocks, the rate of subsidence in the littoral areas increases up to 4 mm/year.

Romanian Black Sea Shelf

Shelf geomorphology

Three major geological units can be identified, aligned from north to south: the Predobrogean Depression, the North Dobrogea orogeny and the Moesia Platform. Two major fault lines mark these units: the St. George Fault and the Peceneaga-Camena Fault. The Capidava-Ovidiu Fault is also important.

Bathymetric, seismic-acoustic and sedimentological studies made by GeoEcoMar on the Romanian Black Sea shelf permit the identification of three distinct units: the littoral zone, the inner shelf, and the outer shelf. In addition to these sections, a very distinctive unit can be defined: the Danube Delta.

Inner shelf

The Romanian Black Sea inner shelf is very well defined, having a width of 10 km to 15 km in the northern area and about 1 km to 5 km south of Constanta (Fig. 18.5). Modern sediments locally mask the relict geomorphologic structures. Northwards of Cape Midia, the bottom slope varies between 1.1° and 4.0°, while southwards of the Constanta section, the relict structures

Fig. 18.5 Physiographical map of the Romanian Black Sea shelf. Vadu–Costineşti zone, Caraivan (2010).

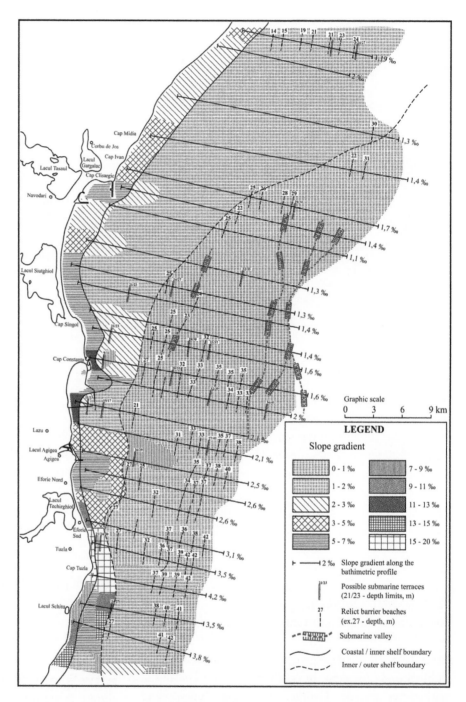

are better preserved, especially the submarine terraces, where the slope is steeper (1.6–6.0°).

Eastwards, the inner shelf boundary is marked by the 27-m to 30-m isobaths. On the inner shelf, bedload movement dominates sedimentary processes, with deposition of fine sediments in calmer conditions, and sheets of sand in stormy conditions. Considering its hydrodynamic and sedimentological structures, the

Danube Delta front is the equivalent of the inner shelf unit.

Outer shelf

From its western edge, along the 27-m to 30-m isobaths, the outer shelf develops a very gentle slope (below 1.0°) extending eastwards to its limit located

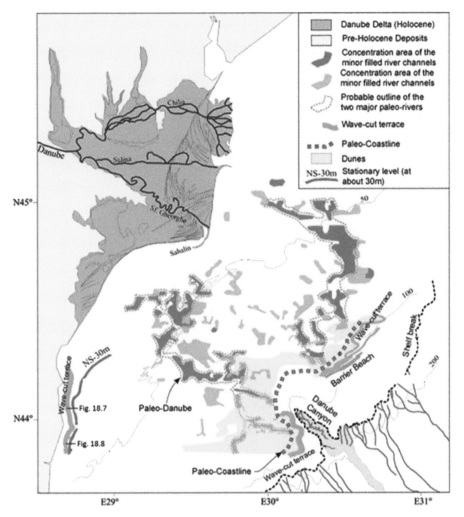

Fig. 18.6 Paleogeographical map of the Romanian continental shelf during the last active period of the Danube Channel. GeoEcoMar data reproduced with permission from Popescu (2008). Positions of sub-bottom profiles (Figs. 18.7 & 18.8) are marked.

at about −120 m. Modern sedimentation rates are significantly reduced on the outer shelf surface. The most spectacular structure is the Viteaz canyon (Danube canyon), connected in its origin with the Danubian St. George branch (Fig. 18.6). The Danube pro-delta of the outer shelf displays all the known types of deformation processes for non-consolidated sediments (Panin 1997).

Relict geomorphologic structures

The Romanian Black Sea shelf reveals the presence of geomorphologic structures of both positive and negative relief, witnesses of older coastal and terrestrial environments, such as submarine terraces, barrier beaches, river valleys, etc.

Submarine relict terraces

It is well known that during the Quaternary, sea level underwent several vertical oscillations. Consequently, the shelf surface was marked, at certain critical depths, by successive lines of wave-cut terraces:

Terrace 1: mean depth: 10 m; slope 10–20°;
Terrace 2: mean depth: 13 m; slope 7–20°;
Terrace 3: mean depth: 17 m; slope 8–20°;
Terrace 4: mean depth: 23 m; slope 13–20°;
Terrace 5: mean depth: 27 m; slope 15–30°
Terrace 6: mean depth: 32 m; slope 12–25°;
Terrace 7: mean depth: 33 m; slope 12–15°;
Terrace 8: mean depth: 35 m; slope ca. 20°.

The relict terraces placed at 10 m and 13 m could correspond to the Surozhian (Fig. 18.7).

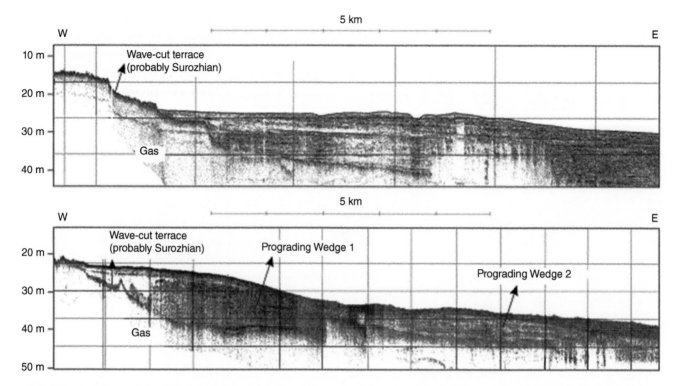

Fig. 18.7 Sub-bottom profiles (3.5 kHz) on the proximal shelf showing the Surozhian wave-cut terrace. GeoEcoMar data reproduced with permission from Popescu (2008). Location in Fig. 18.6.

There are also submarine relict terraces at depths greater than 40 m across the surface of the outer shelf (Fig. 18.6). We estimate that the modern shelf configuration is mainly the result of the last Holocene transgression. Older geomorphological features could be preserved as well. The most obvious wave-cut terrace can be identified on the seismic records along the shelf break at 100 km (Popescu 2008). This terrace borders the Viteaz canyon at depths ranging from 78 m to 112 m, marking the lowest Neoeuxinian basin of the Black Sea during the LGM (Fig. 18.8).

Relict barrier beaches

Panin (1983) considered the convex-upward geomorphologic structures found along the shelf surface as traces of the older barrier beaches. On the inner shelf surface, where sedimentation rates are greater, the relict structures were partly blurred or smoothed out. During the Holocene transgression, the sea level remained stationary at times, inducing the formation of accumulative coastlines of the 'barrier beach' type. Meanwhile, erosional cliffs formed in front of the promontories. Usually, the vertical

Fig. 18.8 Sub-bottom profiles (3.5 kHz) on the distal shelf showing the Neoeuxinian wave-cut terrace. GeoEcoMar data. Location in Fig. 18.6.

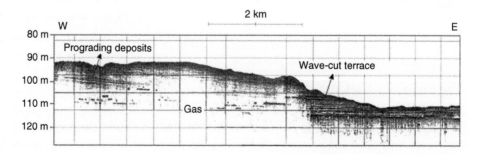

amplitude of the relict barrier beaches is 1 m to 6 m and their width is 150 m to 200 m, but sometimes they extend to 1500 m; the seaward side is longer, with a gentler slope than the landward side, suggesting that their formation occurred during periods of rising sea level.

The submerged barrier beaches consist of sandy deposits set by wave action along the shore during periods of relatively stable sea level. Several generations of relict barrier beaches can be identified at the critical depths of 23 m, 25 m, 27 m, 28 m, 32 m, 35 m, 37 m, and 42 m, forming successive shorelines. Seawards and parallel to the Neoeuxinian terrace from the shelf break, a 20-km-long accumulative structure develops.

Other geomorphologic structures

Seismic-acoustic prospecting of the shelf surface reveals the presence of many other geomorphological structures, such as:

- Depressions with sub-bottom sediments, suggesting old lacustrine or lagoon environments;
- Several river channels can be identified at 30 m to 80 m depth on the shelf. The river valleys are completely covered with sediments, so they cannot be identified in the bathymetry. They correspond to two of the drainage systems: the paleo-Danube and the paleo-Dniester. The old terraces of the Casimcea, Techirghiol and Mangalia rivers are hardly recognizable (Figs. 18.7 & 18.8);
- Undulating surfaces suggesting former littoral dune plains (Fig. 18.8).

From depths of 35 m to 40 m seawards, the bottom slope is very gentle. Its surface is very irregular, owing to the relict geomorphologic structures created in terrestrial environments.

Late Quaternary Shelf Evolution

The lithological studies of cores from the Mamaia barrier beach (Caraivan 1982; Caraivan *et al.* 1986; 2012), corroborated with geomorphological data from the Romanian Black Sea shelf (Panin & Popescu 2007; Strechie-Sliwinski 2007), allow the sequential reconstruction of

paleogeographic and sedimentary environments during the Late Quaternary period. The first marine sequence is at the 38 m to 22 m level in the F6 core sequence (Mamaia North), with marine and brackish fauna (Fig. 18.9). The ^{14}C absolute age of the shell material found at 21 m to 23 m (Zone E 'beach rock') places it in MIS 3 (Würm Paudorf), corresponding to the Tarhankutian Beds of Shcherbakov *et al.* (1978). In this case, the subjacent packed silty clays (Zones D1–D4) can be considered to date back to the transgressive Surozhian stage (according to Popov 1955; Shcherbakov 1978; Caraivan 2010). Strechie-Sliwinski (2007) identified the Surozhian level in two cores taken on the outer shelf at depths of 135 m and 100 m. The Surozhian sea level was –10 m, compared to the modern one, and was marked by wave-cut terraces in front of the headlands (Fig. 18.10).

The LGM induced the retreat of the sea level down to about 100 m below present (27–17 ka BP in Yanko-Hombach 2007), followed by an advancement of the shoreline to the present position. During this period, Zone E beach deposits, identified in the Mamaia North core (Tarhankutian Beds), were cemented under continental conditions. The lowest positions of the Neoeuxinian basin are demonstrated by the wave-cut terrace developed along the Romanian Black Sea shelf break.

The Holocene (MIS 1) is marked in the Black Sea by the re-connection with the Mediterranean, reached at the same level, close to the present-day one, at about 9 ka BP to 7.5 ka BP. The sea then returned to the Mamaia area, marking the beginning of the 'Black Sea' stage (Figs. 18.9 & 18.10). Consequently, Zone F continental deposits with scarce brackish fauna could match the Bugazian strata. Subsequently, the Vityazevian Beds accumulated in coastal inshore conditions. The next stage (Zone H — Lower Kalamitian Beds) marks a freshening process of the coastal waters. The accumulation of the Upper Kalamitian strata (Zone I) is synchronous with a rapid rise of the sea level, which eventually surpassed the current one by 2 m to 4 m. Meanwhile, the Mamaia Gulf western cliff was exposed to marine abrasion. In the conditions of a slight decrease of sea level (the Phanagorian regression), the closing of the Mamaia Gulf by a sandy barrier beach began. In the newly created freshwater environment,

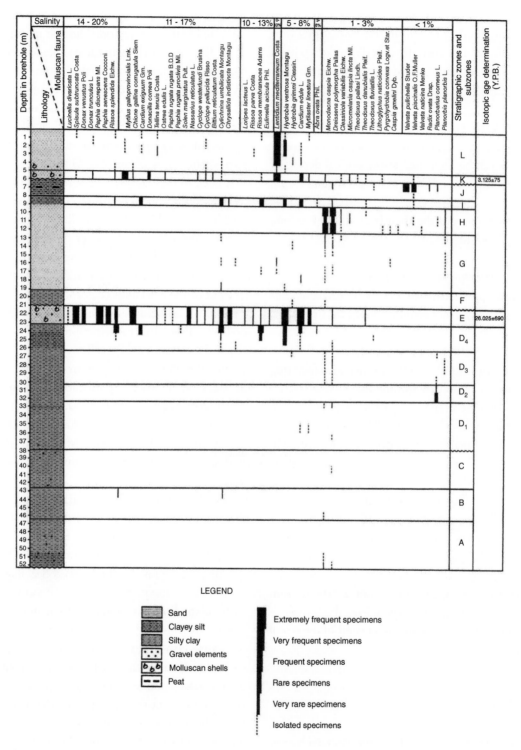

Fig. 18.9 Stratigraphic column of Quaternary deposits (Mamaia North drill F6). GeoEcoMar data.

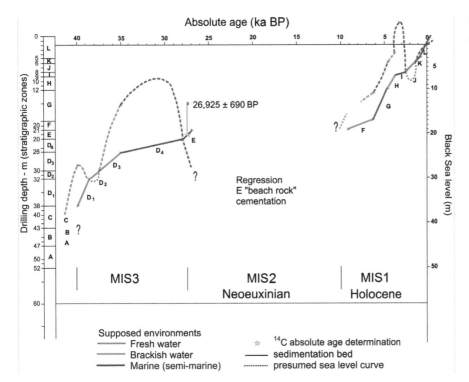

Fig. 18.10 Upper Quaternary changes of sea level in the Mamaia region of the Romanian coast. Caraivan *et al.* (2012).

peat-type deposits accumulated. The last lithological sequence evolved during the Nymphaean transgression in nearshore marine conditions, similar to the current ones.

A special note concerning the Danube Delta

The main stages of the Danube Delta evolution during the Holocene were highlighted and dated using the corroboration of geomorphologic, structural, textural, geochemical, mineralogical, and faunal analyses, and radiocarbon dating (Panin *et al.* 1983; Fig. 18.11). These phases can be summarized as follows: (1) the formation of the Letea-Caraorman initial spit, 11.7 ka BP to 7.5 ka BP; (2) the St. George I Delta, 9 ka BP to 7.2 ka BP; (3) the Sulina Delta, 7.2 ka BP to 2 ka BP; (4) the St. George II and Chilia deltas, ~2.8 ka BP to present; (5) the Cosna-Sinoe Delta, 3.5 ka BP to 1.5 ka BP. This model, if the radiocarbon dating is correct, leads to a scenario of a rather elevated highstand by 11.7 ka BP (very close to the present-day level) as the delta coastline was represented by the Letea-Caraorman initial spit, located at about 25 km to 30 km westwards of the present coastline. Since this

time, no catastrophic event (such as the sea level dropping to –156 m) can be recognized in the delta territory. The subsequent successive phases of delta development are perfectly continuous and the different progression phases of the delta lobes can be observed without gaps.

Dynamics of Neo-Eneolithic Settlements in the Western Part of the Black Sea Region

Over the millennia, coastal areas have laid down a record of climate and geomorphological changes. Morphological and geoarchaeological research undertaken in recent decades in the Black Sea basin has shown that prehistoric habitations from the fertile valleys of the western Black Sea coast were more sensitive to climate changes (temperature, precipitation) than further inland. During the fifth millennium BC, Neo-Eneolithic (the uppermost part of the Neolithic) communities preferred the extensive and productive landscape associated with marine lagoons, lakes and islands, and the Danube Delta. The intensive use of maritime routes (Aegean–Mediterranean

Fig. 18.11 The Danube Delta evolution during the Holocene and correspondent coastline position changes. (1) Initial spit 11.7 ka BP to 7.5 ka BP; (2) St. George I Delta 9 ka BP to 7.2 ka BP; (3) Sulina Delta 7.2 ka BP to 2 ka BP; (4) Coastline position at 100 AD; (5) St. George II Delta and Chilia Delta 2.8 ka BP–present; (6) Cosna-Sinoe Delta 3.5 ka BP to 1.5 ka BP. Panin (1997). Adapted with permission from GEO-ECO.

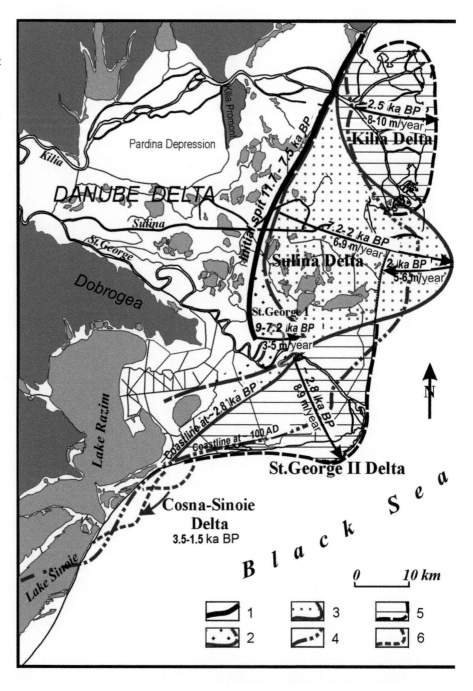

and Circumpontic) and rivers communicating with the Black Sea (the Casimcea River for instance), favored an extraordinary development of these communities.

The absence of Mesolithic and Early Neolithic traces in Dobrogea is explained by environmental factors such as the flooding of paleorivers by waters from the western part of the Black Sea. In the absence of Early Neolithic indicators, many researchers have attempted to fill this hiatus by extending the previous culture period (i.e. the Late Mesolithic), to overlap with the Hamangia culture. Starting from 5200 BC to 5000 BC, the Hamangia communities intensively settled the mouths of the rivers along the western Black Sea coast. Their settlements were placed around coastal lakes like Mangalia, Tatlageac,

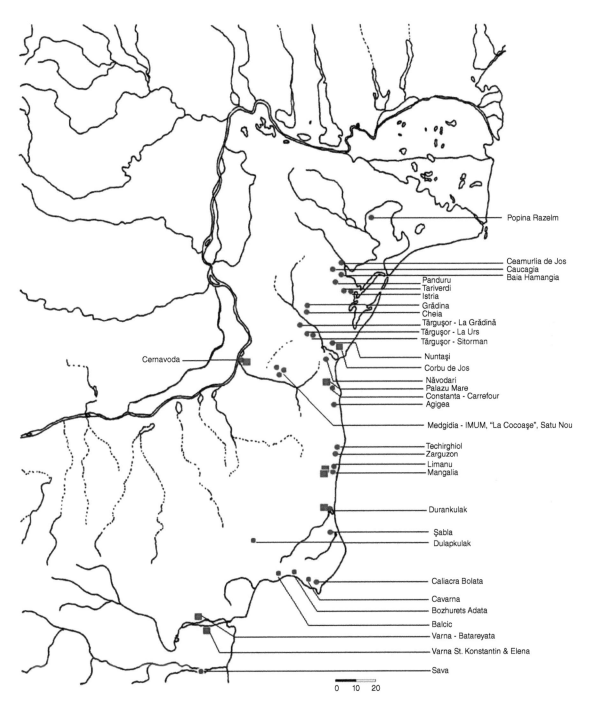

Popina Razelm
Ceamurlia de Jos
Caucagia
Baia Hamangia
Panduru
Tariverdi
Istria
Grădina
Cheia
Tărgușor - La Grădină
Tărgușor - La Urs
Tărgușor - Sitorman
Nuntași
Corbu de Jos
Năvodari
Palazu Mare
Constanta - Carrefour
Agigea
Medgidia - IMUM, "La Cocoașe", Satu Nou
Techirghiol
Zarguzon
Limanu
Mangalia
Durankulak
Șabla
Dulapkulak
Caliacra Bolata
Cavarna
Bozhurets Adata
Balcic
Varna - Batareyata
Varna St. Konstantin & Elena
Sava
Cernavoda

0 10 20

Fig. 18.12 Distribution of sites of the Hamangia culture (5000–4500 BC).

Techirghiol-Zarguzon, Agigea, Siutghiol, Tașaul, Gargalâc, Sinoe, Istria, Golovita, along the Casimcea valley and Carasu valley (Fig. 18.12).

Circumpontic–Aegean–Anatolian cultural exchanges are clearly demonstrated by the presence of both exotic raw materials, and by artifacts from Aegean–Anatolian settlements in the Balkans: *Spondylus* ornaments, Dentalium, marble, obsidian, common types of figurines, and pots. The marine currents and winds from the western part of the Black Sea were favorable for cabotage shipping,

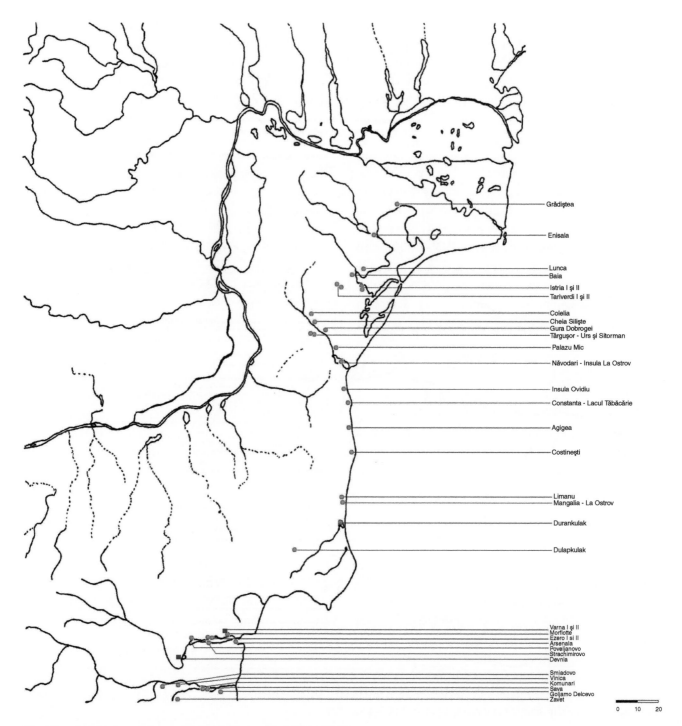

Grădiştea

Enisala

Lunca
Baia
Istria I şi II
Tariverdi I şi II

Colelia
Cheia Silişte
Gura Dobrogei
Târguşor - Urs şi Sitorman
Palazu Mic

Năvodari - Insula La Ostrov

Insula Ovidiu

Constanta - Lacul Tăbăcărie

Agigea

Costineşti

Limanu
Mangalia - La Ostrov

Durankulak

Dulapkulak

Varna I şi II
Morflotté
Ezero I şi II
Arsenala
Poveljanovo
Strachimirovo
Devnia

Smiadovo
Vinica
Komunari
Sava
Goljamo Delcevo
Zavet

0   10   20

Fig. 18.13 Distribution of sites of the Gumelniţa culture (4500–4000 BC).

facilitating links along the sea coast and up river valleys to the hinterland.

The Gumelniţa settlement discovered on 'La Ostrov' Island in Taşaul Lake provides new data on changes at the end of the Eneolithic (Marinescu-Bîlcu *et al.* 2000–2001; Voinea 2001; 2004–2005; 2005). Stratigraphical features on the northwest side of the island indicate the flooding of a Gumelniţa settlement. The causes for the end of the

Gumelniţa civilization (4500–4000 BC) are complex, the hypothesis of violent attack by the north Pontic tribes being very plausible. The western Black Sea coast area is marked by the discovery of cemeteries at Varna (Fol & Lichardus 1988), and the rapid disappearance of Neolithic settlements without the cultural developments so common in other regions (Fig. 18.13).

According to the chronology established by Bulgarian researchers, the abandonment of settlements took place at the end of the Eneolithic. This coincides with climate changes at the end of the Atlantic Period between 6050–5600 BP, characterized by a sharp warming climate with warm and long summers (Tomescu 1998–2000), associated with a marine transgression in coastal areas. The Neolithic Black Sea transgression was most likely not an isolated phenomenon, because similar and synchronous events are reported elsewhere.

Conclusion

In the Late Quaternary, the Black Sea level underwent important variations due to global climate changes. In MIS 3, 40 ka to 25 ka, the sea level was close to the present one, probably about 10 m lower, in conditions of temperate climate. Environmental conditions were probably more variable for Paleolithic communities during this period compared to the previous glacial. The coastline probably followed the contours of the Surozhian terraces and barrier beaches. The subsequent regression started about 25 ka BP.

During the LGM (MIS 2), living conditions became harsher, with Paleolithic communities sheltering in natural caves in Dobrogea. The sea level dropped dramatically, reaching the modern 100-m to 120-m isobath, and exposing the entire continental shelf. The Neoeuxinian lowstand is marked on the distal edge of the Romanian shelf by a wave-cut terrace, found today at 98 m to 115 m depth (Panin & Popescu 2007). The Dobrogean rivers and the Danube were continuing their flow as far as the shelf break, while incising deep valleys and canyons (e.g. the Viteaz canyon). About 16 ka BP to 15 ka BP, postglacial and ice melting began. The direct and continuous supply of melting water from the glaciers to the Pontic basin caused the Neoeuxinian water level to rise very rapidly up to a depth of approximately 30 m ('stationary level'), reaching close to the Bosporus sill level by 12 ka BP. Therefore, the prehistoric communities from Asia Minor were able to migrate to the western part of the Black Sea shelf through the Bosporus and Dardanelles straits, which were exposed terrestrial areas at that time. These communities established settlements along the borders of inland-penetrating gulfs, created by the flooded Dobrogean river valleys, or on the Danube Delta sand bars.

The Holocene (MIS 1) is marked in the Black Sea by the re-connection with the Mediterranean Sea, reached at the same level, close to the present one, about 9 ka to 7.5 ka. The sea returned to the Mamaia area, marking the beginning of the 'Black Sea' stage (Figs. 18.9 & 18.10). The Hamangia culture started to develop along the western part of the Black Sea. The accumulation of the Upper Kalamitian strata (Zone I) is synchronous with a rapid rise of sea level, which eventually surpassed the current level by 2 m to 4 m. Meanwhile, the Mamaia Gulf's western cliff was exposed to marine abrasion. It was at this point that the Gumelniţa culture reached its apogee.

In the conditions of a slight decrease of sea level, the closing of the Mamaia Gulf by a sandy barrier beach began. In the newly created freshwater environment, peat-type deposits accumulated. This Phanagorian regression is coeval with the first Greek colonization of the Black Sea coast. The last lithological sequence evolves during the Nymphaean transgression in nearshore marine conditions, similar to the current ones. By about the tenth century AD, the Black Sea level experienced a decline of 1 m to 2 m, after which a slow rise commenced which continues to the present day.

References

Bondar, C. 1972. Contribution to hydraulic studies of Danube river mouths. *Studii de hidrologie* [Hydrological Studies] 32:1-467 (in Romanian).

Bondar, C. & Emanoil, G. 1963. Contributions to Romanian Black Sea coastal waters. *Studii de hidrologie* [Hydrological Studies] 4:89-160 (in Romanian).

Bondar, C. & Roventa, V. 1967. Longshore currents along the Black Sea coastal waters and their influence on water body's stratification. *Studii de hidrologie* [Hydrological Studies] 19:5-21 (in Romanian).

Bondar, C., Besnea, P. & Dumitrascu, A. 1967. Wave height assessment measured by visual methods along the Romanian coastal waters. *Studii de hidrologie* [Hydrological Studies] 19:35-46 (in Romanian).

Caraivan, G. 1982. The Upper Quaternary evolution of the Mamaia zone. *Pontica* 15:15-32 (in Romanian).

Caraivan, G. 2010. *Sedimentological Study of Beach and Black Sea Inner Shelf Sediments between Portita and Tuzla*. Editura Ex Ponto: Constanta (in Romanian).

Caraivan. G., Herz. N. & Noakes J. 1986. New proofs of the Black Sea rise during the Middle Würm Interstadial (Proceedings of the Institute of Geology and Physics no. 5). *Tectonică și Geologie Regională* [Regional Tectonics and Geology] 70–71:57-62.

Caraivan, G., Fulga, C. & Opreanu, P. 2012. Upper Quaternary evolution of the Mamaia Lake area (Romanian Black Sea shore). *Quaternary International* 261:14-20.

Chiriac, M. 1960. Preliminary information on Sarmatian deposits from Dobrogea. *Com. Acad. R. P. R.* [Proceedings of the Romanian Academy] 10:613-623 (in Romanian).

Conea, A. 1970. *Quaternary formations in Dobrogea (Loess and paleo soils)*. Editura Academiei: Bucharest (in Romanian).

Drăgănescu, A. 1976. Constructional to corpuscular spongalgal, algal and coralgal facies in the Upper Jurassic carbonate Formation of Central Dobrogea (the Casimcea Formation). In Patrulius D., Drăgănescu, A., Baltres, A. & Popescu B. (eds.) *Carbonate Rocks and Evaporites — Guidebook*. pp. 13-42. Institute of Geology & Geophysics: Bucharest.

Dimitriu, R. G. & Sava, C. S. 2007. Considerations on current geodynamic processes in Dobrogea — an appeal for achieving "Geodynamics Polygon Dobrogea". In Oaie, Gh. (ed.) *Hazard Natural: Evenimente Tsunami in Marea Neagră*[Natural Hazard: Tsunami Events in the Black Sea]. pp. 62-71. Geo-Eco-Marina (Special Publication): Paris(in Romanian).

Fol, A. & Lichardus, J. (eds.) 1988. *Macht, Herrschaft und Gold: Das Gräberfeld von Varna (Bulgarien) und die Anfänge einer neuen europäischen Zivilisation*. Saarland Museum: Saarbrücken.

Gâştescu, P. (ed.) 1986. *Morpho-hydrographic Changes of the Romanian Black Sea Accumulation Coasts*. Institutul de Geografie: Bucharest (in Romanian).

Giosan, L., Bokuniewicz, H., Panin, N. & Postolache I. 1997. Longshore sediment transport pattern along Romanian Danube Delta Coast. In *GEO-ECO-MARINA Proceedings of the International Workshop on 'Fluvial-Marine Interactions'* (vol. 2). 1st – 7th October 1996, Malnas (Romania), pp. 11-23.

Liteanu, E. & Ghenea, C. 1966. *Quaternary from Romania, Technical and Economic Studies (H Series)*. Comitetul Geologic: Bucharest (in Romanian).

Macarovici, N. 1968. *Quaternary Geology*. Editura Didactică și Pedagogică: Bucharest (in Romanian).

Marinescu-Bîlcu, S., Voinea, V., Dumitrescu, S., Haita, C, Moise, D. & Radu, V. 2000–2001. "La Ostrov" Island Eneolithic settlement, Tasaul Lake (Navodari, Constantza County). Preliminary report — 1999-2000 campaigns. *Pontica* 33-34:123-170 (in Romanian).

Mutihac, V. & Ionesi, L. 1974. *Geology of Romania*. Editura Tehnică: Bucharest (in Romanian).

Panin, N. 1997. On the geomorphologic and the geologic evolution of the River Danube — Black Sea interaction zone. In *GEO-ECO-MARINA Proceedings of the International Workshop on 'Fluvial-Marine Interactions'* (vol. 2). 1st – 7th October 1996, Malnas (Romania), pp. 31-40.

Panin, N. & Popescu, I. 2007. The northwestern Black Sea: climatic and sea level changes in the Late Quaternary. In Yanko-Hombach, V., Gilbert, A. S., Panin, N. & Dolukhanov, P. M. (eds.) *The Black Sea Flood Question: Changes in Coastline, Climate, and Human Settlement*. pp. 387-404. Springer: Dordrecht.

Panin, N., Panin, S., Herz, N. & Noakes, E. 1983. Radiocarbon dating of Danube Delta deposits. *Quaternary Research* 19:249-255.

Polonic, G., Zugravescu, D., Horomnea, M. & Dragomir, V. 1999. Crustal vertical recent movements and the geodynamic compartments of Romanian territory. In *Second Balkan Geophysical Congress & Exhibition (Abstracts)*. 5th – 9th July 1999, Istanbul, pp. 300-301.

Popescu, I. 2008. *Processus sédimentaires récents dans l'éventail profond du Danube (Mer Noire) (Geo-Eco-Marina, Special Publication no.2.)*. National Institute of Marine Geology & Geo-ecology: Bucharest.

Popov, G. I. 1955, History of the Manych Strait as related to the stratigraphy of the Black Sea and Caspian Deposits. *Biulletin MOIP [Bulletin of the Moscow Society of Nature Investigators, Geology]* 20:31-49 (in Russian).

Shcherbakov, F. A., Kuprin, P. N., Potapova, I. I., Polyakov, A. S., Zabelina, E. K. & Sorokin, V. M. 1978. *Sedimentation on the Continental Shelf of the Black Sea.* Nauka Press: Moscow (in Russian).

Shepard, F. P. 1954. Nomenclature based on sand-silt-clay ratios. *Journal of Sedimentary Research* 24:151-158.

Shuisky, Y. D. & Vykhovanets, G. V. 1984. Studies of beaches on abrasional shores of the Black and Azov seas. *Inghenernaia gheologhia* [Engineering Geology] 2:73-80 (in Russian).

Strechie-Sliwinski, C. 2007. *Changements environnementaux récents dans la zone de Nord-Ouest de la Mer Noire (Geo-Eco-Marina, no. 13/2007, Special Issue).*

National Institute of Marine Geology & Geo-ecology: Bucharest.

Tomescu, M. 1998–2000. The Holocene — Chronology and Climate. *Cercetări Arheologice* [Archaeological Research] 11:235-270 (in Romanian).

Voinea, V. 2001. Gumelnita culture in central and southern Dobrogea. In *O civilizaţie "necunoscută": Gumelnita* [A 'Lost 'Civilzation: Gumelnita] CD-ROM. CIMEC: Bucharest (in Romanian). Available online at www.cimec.ro/arheologie/gumelnita/cd/default.htm

Voinea, V. 2004–2005. Causes concerning the end of the Eneolithic in the west-Pontic littoral. "La Ostrov" island settlement, Tasaul Lake (Navodari, Constanta County). *Pontica* 37-38:21-46 (in Romanian).

Voinea V. 2005. *Gumelnita Ceramics Cultural Complex - Karanovo VI. Phases A1 and A2.* Ex Ponto: Constanţa (in Romanian).

Yanko-Hombach, V. V. 2007. Controversy over Noah's Flood in the Black Sea: geological and foraminiferal evidence from the shelf. In Yanko-Hombach, V., Gilbert, A. S., Panin, N. & Dolukhanov, P. (eds.) *The Black Sea Flood Question: Changes in Coastline, Climate and Human Settlement.* pp. 149-204. Springer: Dordrecht.

Glossary of Acronyms

ADCP	Acoustic Doppler Current Profiler
ADRAMMAR	Association pour le Développement de la Recherche en Archéologie MARitime (France)
AEA	Association for Environmental Archaeology
AFBI	Agri-Food & Biosciences Institute (UK – Northern Ireland)
ALSF	Aggregates Levy Sustainability Fund
AMCG	Applied Modelling Computation Group
AMH	Anatomically Modern Humans
AOCGCM	Atmosphere-Ocean Coupled General Circulation Models
AOGCM	Atmosphere-Ocean General Circulation Model
ARQUA	Museo Nacional de Arqueología Subacuática (Spain)
AWI	Alfred Wegener Institute
BGR	Bundesanstalt für Geowissenschaften und Rohstoffe (Germany)
BGS	British Geological Survey
BIFROST	Baseline Inferences from Fennoscandian Rebound Observations, Sea-level and Tectonics
BIL	Baltic Ice Lake
BMAPA	British Marine Aggregate Producers Association
BODC	British Oceanographic Data Centre
BP	Before Present
BRGM	Bureau de Recherches Géologiques et Minières (France)
BSBD	Baltic Sea Bathymetry Database

BSH	Bundesamt für Seeschiffahrt und Hydrographie (Germany)
BSHC	Baltic Sea Hydrographic Commission
BSL	Below Sea Level
BUGCEP	Bugs Coleopteran Ecology Package
CANDHIS	Centre d'Archivage National de Données de Houle In Situ (France)
CARG	Cartografia Geologica project (Italy)
CAS	Centre d'Arqueologia Subaquàtica (and various branches thereof) (Spain)
CCO	Channel Coastal Observatory
CERSAT	Centre ERS d'Archivage et de Traitement (France)
CGMW	Commission for the Geological Map of the World
CHP	Civil Hydrography Programme (United Kingdom)
CIESM	Commission Internationale pour l'Exploration Scientifique de la mer Méditerranée
CMIPS	Climate Model Intercomparison Project
CNR	Consiglio Nazionale delle Ricerche (Italy)
CNRS	Centre National de la Recherche Scientifique (France)
CReAAH	Centre de Recherche en Archéologie, Archéosciences, Histoire (France)
D-O	Dansgaard-Oeschger (events)
DAERA	Department of Agriculture, Environment and Rural Affairs (UK – Northern Ireland)
DBEIS	Department for Business, Energy and Industrial Strategy (UK)

Submerged Landscapes of the European Continental Shelf: Quaternary Paleoenvironments, First Edition.
Edited by Nicholas C. Flemming, Jan Harff, Delminda Moura, Anthony Burgess and Geoffrey N. Bailey.
© 2017 John Wiley & Sons Ltd. Published 2017 by John Wiley & Sons Ltd.

DEM	Digital Elevation Model	HSC	Historic Seascape Characterization
DESM	Dynamic Equilibrium Shore Model	IBCM	International Bathymetric Chart of the Mediterranean
DG MARE	Directorate-General for Maritime Affairs and Fisheries	ICGC	Institut Cartogràfic i Geològic de Catalunya (Spain)
DISPERSE	Dynamic Landscapes, Coastal Environments and Human Dispersals	IFREMER	Institut Français de Recherche pour l'Exploitation de la Mer (France)
DLS	Disappearance of Lacustrine Species	IGCP	International Geoscience Programme
DOENI	Department of Environment, Northern Ireland (now the DAERA)	IGM	Istituto Geografico Militare (Italy)
DRASSM	Département des Recherches Archéologiques Subaquatiques et Sous-Marines (France)	IGME	Instituto Geológico Y Minero de España (Spain)
DSDP	Deep Sea Drilling Project	IGN	(a) Institut national de l'information géographique et forestière (France) (b) Instituto Geográfico Nacional (Spain)
DSM	Digital Surface Model		
DSWC	Dense Shelf Water Cascading		
DTM	Digital Terrain Model	IGP	Instituto Geográfico Português (Portugal)
ECDIS	Electronic Chart and Display System		
ECMWF	European Centre for Medium-Range Weather Forecasts	IHO	International Hydrographic Organization
EMODnet	European Marine Observation and Data Network	IKAW	Indicative Map of Archaeological Values
ERDF	European Regional Development Fund	IKUWA	International Congress on Underwater Archaeology
ESEAS	European Sea-Level Service	IMI	Initial Marine Inflow
ESL	Equivalent Sea Level	INFOMAR	The INtegrated Mapping FOr the Sustainable Development of Ireland's MArine Resource
ESM	Earth System Model		
ESRI	Environmental Systems Research Institute		
EUROSION	European initiative for sustainable coastal erosion management	INIS-Hydro	Ireland-Northern Ireland-Scotland Hydrographic survey
FAGS	Federation of Astronomical and Geophysical Data Analysis Services	INQUA	The International Union for Quaternary Science
FOMAR	Fondos Marinos	INTIMATE	INTegration of Ice-core, MArine and TErrestrial records group
GEBCO	General Bathymetric Chart of the Oceans	IOC	Intergovernmental Oceanographic Commission
GEUS	Geological Survey of Denmark and Greenland	IOLR	Israel Oceanographic and Limnological Research Institute
GIA	Glacio-Isostatic Adjustment	IORAS	Institute of Oceanology Russian Academy of Sciences
GIS	Geographic Information System		
GMP	Galicia Mud Patch	IPEAN	Irish Palaeoecology and Environmental Archaeology Network
GPS	Global Positioning System		
GSHHG	Global Self-consistent, Hierarchical, High-resolution Geography Database	IPMA	Instituto Português do Mar e da Atmosfera (Portugal)
GSI	Geological Survey of Ireland	IPOL	Irish Pollen Database
GSNI	Geological Survey of Northern Ireland	IQUA	Irish Quaternary Association

IRD	Ice Rafted Debris	MONGOOS	Mediterranean Oceanography Network for the Global Ocean Observing System
ISPRA	Istituto Superiore per la Protezione e la Ricerca Ambientale (Italy)	MoU	Memorandum of Understanding
JIBS	Joint Irish Bathymetric Survey (Northern Ireland, Republic of Ireland)	MWP	Meltwater Pulse
JNCC	Joint Nature Conservation Committee (UK)	NADW	North Atlantic Deep Water
		NAF	North Anatolian Fault
KF	Kephallinia Transform Fault	NAO	North Atlantic Oscillation
LAT	Lowest Astronomical Tide	NASA	National Aeronautics and Space Administration (USA)
LEGOS	Laboratoire d'Etudes en Géophysique et Océanographie Spatiales (France)	NASU	National Academy of Science of Ukraine
LGC	Last Glacial Cycle	NAT	North Aegean Trough
LGM	Last Glacial Maximum	NCEI	National Centers for Environmental Information (formerly the NGDC) (USA)
LGP	Last Glacial Period		
LIA	Little Ice Age		
LiDAR	Light Detection and Ranging	NERC	Natural Environment Research Council (UK)
LM	Local Magnitude		
LNEG	Laboratório Nacional de Energia e Geologia (Portugal)	NGDC	National Geophysical Data Center (now the NCEI) (USA)
LPS	Land and Property Services (UK – Northern Ireland)	NGU	Norwegian Geological Survey
		NIEA	Northern Ireland Environment Agency
LST	(a) Low System Tracts (b) Littoral Sand Transport (c) Longshore Transport	NOAA	National Oceanic and Atmospheric Administration (USA)
MAD	(a) Marine Aggregate Deposits (b) Mid Atlantic Depression	NOC	National Oceanography Centre (UK)
		OIS	Oxygen Isotope Stage
MAGIC	Marine Geohazards along the Italian Coasts (Italy)	OPW	Office of Public Works (Republic of Ireland)
MAI	Marine Aggregate Industry (Protocol)	ORPAD	Offshore Renewables Protocol for Archaeological Discoveries
MAREMAP	Marine Environmental Mapping Programme		
		OS	Ordnance Survey (UK)
MBES	Multibeam Echosounder	OSI	Ordnance Survey of Ireland (Republic of Ireland)
MCA	(a) Maritime and Coastguard Agency (UK) (b) Medieval Climate Anomaly		
MEDFLOOD	MEDiterranean sea level change and projection for future FLOODing	PALSEA2	PALeo-constraints on SEA-level rise 2
		PCN	Portale Cartografico Nazionale (Italy)
MEDIN	Marine Environmental Data & Information Network (UK)	PISCES	Partnerships Involving Stakeholders in the Celtic Sea Eco-System
MFS	Mediterranean ocean Forecasting System	POL	Proudman Oceanographic Laboratories (now NOC – Liverpool)
MGCZ	Mid-German Crystalline Zone	PPNC	Pre-Pottery Neolithic
MHWS	Mean High Water Spring	PSL	Present Sea Level
MI	Marine Institute (Republic of Ireland)	PSMSL	Permanent Service for Mean Sea Level
MIS	Marine Isotope Stage	PSU	Practical Salinity Unit

PWMCF	Portuguese Western Meso-Cenozoic Fringe	SOMLIT	Service d'Observation du Milieu Littoral (France)
QRA	Quaternary Research Association	SPLASHCOS	Submerged Prehistoric Archaeology and Landscapes of the Continental Shelf
RCDS	Raster Chart Display System		
RCZAS	Rapid Coastal Zone Assessment Surveys	SRGE	State Regional Geologic Enterprise (Ukraine)
REC	Regional Environmental Characterization	SST	Sea Surface Temperature
		TEN-T	(EU) Trans-European Transport Network
RSL	Relative sea-level		
SBES	Single-Beam Echosounder	TEOS-10	Thermodynamic Equation of Seawater 2010
SBR	Station biologique de Roscoff (France)		
SCAPE	Scottish Coastal Archaeology and the Problem of Erosion	UKHO	United Kingdom Hydrographic Office
		UKMMAS	UK Marine Monitoring and Assessment Strategy
SCOPAC	Standing Conference on Problems Associated with the Coastline (UK)		
SEA	Strategic Environment Assessment	UkrSCES	Ukrainian Scientific Center of Ecology of the Sea
SELRC	Severn Estuary Levels Research Committee	UNESCO	United Nations Educational, Scientific and Cultural Organization
SGF	Société Géologique de France	VLIZ	Vlaams Instituut voor de Zee (Netherlands)
SHOM	Service Hydrographique et Océanographique de la Marine (France)		
		WEAO	Western European Armaments Organisation
SISMER	Systèmes d'Informations Scientifiques pour la Mer (France)		
		WMS	Web Map Services
SLE	Sea-Level Equation	XIOM	Xarxa d'Instruments Oceanogràfics i Meteorològics (Spain)
SLIPs	Sea-Level Index Points		
SNIRH	Sistema Nacional de Informação de Recursos Hídricos (Portugal)		

Index

Submerged Landscapes of the European Continental Shelf: Quaternary Paleoenvironments, First Edition.
Edited by Nicholas C. Flemming, Jan Harff, Delminda Moura, Anthony Burgess and Geoffrey N. Bailey.
© 2017 John Wiley & Sons Ltd. Published 2017 by John Wiley & Sons Ltd.